Android 传感器开发与智能设备案例实战

朱元波◆编著

人民邮电出版社
北京

图书在版编目（CIP）数据

Android传感器开发与智能设备案例实战 / 朱元波编著. -- 北京 : 人民邮电出版社, 2016.5（2024.1重印）
ISBN 978-7-115-41474-8

Ⅰ. ①A… Ⅱ. ①朱… Ⅲ. ①移动终端－应用程序－程序设计 Ⅳ. ①TN929.53

中国版本图书馆CIP数据核字(2016)第080065号

内 容 提 要

本书主要介绍 Android 传感器和外设的开发，共 29 章，主要包括 Android 开发技术基础、获取并编译源码、Android 技术核心框架分析、Android 传感器系统分析、使用地图定位、光线传感器详解、接近警报传感器详解、磁场传感器详解、加速度传感器详解、方向传感器详解、陀螺仪传感器详解、旋转向量传感器详解、距离传感器详解、气压传感器详解、温度传感器详解、湿度传感器详解、Android 蓝牙系统概述、低功耗蓝牙技术详解、语音识别技术详解、手势识别技术详解、NFC 近场通信技术详解、拍照解析条形码技术详解、基于图像处理的人脸识别技术详解、行走轨迹记录器、手势音乐播放器和智能家居系统等。本书几乎涵盖了 Android 传感器和外设开发所需的所有主要内容，讲解方法通俗易懂。

本书适合 Android 初学者、Android 爱好者以及 Android 底层开发人员、传感器开发人员、智能设备开发人员、Android 外部设备开发工程师学习，也可以作为相关培训学校和大专院校相关专业的教学用书。

◆ 编　　著　朱元波
责任编辑　张　涛
责任印制　焦志炜

◆ 人民邮电出版社出版发行　　北京市丰台区成寿寺路 11 号
邮编　100164　　电子邮件　315@ptpress.com.cn
网址　https://www.ptpress.com.cn
北京天宇星印刷厂印刷

◆ 开本：787×1092　1/16
印张：47　　　　2016 年 5 月第 1 版
字数：1203 千字　　　　2024 年 1 月北京第 14 次印刷

定价：108.00 元

读者服务热线：(010)81055410　印装质量热线：(010)81055316
反盗版热线：(010)81055315
广告经营许可证：京东市监广登字 20170147 号

前　言

2007年11月5日谷歌公司正式向外界展示了基于Linux平台的开源手机操作系统Android。该平台由操作系统、中间件、用户界面和应用软件组成，号称首个为移动终端打造的真正开放和完整的移动软件平台。随着Android系统应用的不断扩大，Android系统在移动设备、可穿戴设备上的开发越来越多，本书就是为了适应新智能设备的开发特意策划编写的。

本书的内容

本书共计29章，全面讲解了基于Android系统开发传感器程序和智能设备应用的技术，主要内容为Android技术核心框架分析、Android传感器系统分析、光线传感器详解、接近警报传感器详解、磁场传感器详解、加速度传感器详解、方向传感器详解、陀螺仪传感器详解和基于图像处理的人脸识别技术详解，以及智能家居系统等知识。

本书特色

本书内容丰富，分析细致。我们的目标是通过一本图书，提供多本图书的价值，在内容的编写上，本书具有以下特色。

（1）遵循“基础讲解—技术剖析—实战演练”这一主线，全面剖析了与Android传感器和外设应用开发相关的核心技术，并通过具体实例进行了实践演练和举例说明。

（2）实用性强。本书彻底摒弃了枯燥的理论和简单的操作，注重实用性和可操作性，用通俗的语言详细讲解了各个知识点的基本原理。

（3）内容全面。本书堪称市面上“内容较全面的一本Android传感器和外设应用开发书”，无论是获取源码，还是各个相关技术的运作机制和具体使用方法，在本书中您都能找到解决问题的答案。

读者对象

初学Android编程的自学者
Linux开发人员
大中专院校的老师和学生
毕业设计的学生
Android编程爱好者
相关培训机构的老师和学员
从事Android开发的程序员
Android智能穿戴设备开发工程师
Android外部设备开发工程师

本书在编写过程中，得到了人民邮电出版社工作人员的大力支持，正是各位编辑的求实、耐心和效率，才使得本书在这么短的时间内出版。另外，也十分感谢我的家人在我写作的时候给予的巨大支持。由于本人水平有限，书中的纰漏和不尽如人意之处在所难免，诚请读者提出意见或建议，以便修订并使之更臻完善。另外，我们也提供了售后支持和源程序下载网站：http://www.toppr.net/，读者如有疑问可以在此提出，一定会得到满意的答复。编辑联系邮箱：zhangtao@ptpress.com.cn。

编　者

目 录

第1章　Android开发技术基础

Android是科技界巨头谷歌公司推出的一款智能设备操作系统，是以Linux开源系统架构为基础的。Android功能十分强大。从2008年推出到现在，Android一直在全球智能手机操作系统市场中占居第一的宝座。本章简单介绍Android的发展历程和背景，并介绍智能设备和Android的密切关系，为读者步入本书后面知识的学习打下基础。

1.1 智能手机操作系统介绍

在Android系统诞生之前，智能手机这个新鲜事物大大丰富了人们的生活，得到了广大手机用户的青睐。各大手机厂商在利益的驱动之下，纷纷建立了各种智能手机操作系统，并全力抢夺市场份额。Android系统就是在这个风起云涌的时代背景下诞生的。

何谓智能手机

智能手机是指具有像个人计算机那样强大的功能，拥有独立的操作系统，用户可以自行安装应用软件、游戏等第三方服务商提供的程序，并且可以通过移动通信网络接入到无线网络中。在Android系统诞生之前已经有很多优秀的智能手机操作系统产品，例如，家喻户晓的Symbian系列和微软的Windows Mobile系列等。

对于初学者来说，可能还不知道怎样来区分什么是智能手机。某大型专业统计公司曾经为智能手机的问题做过一项市场调查，经过大众讨论并投票之后，总结出了智能手机所必须具备的功能标准，其中下面是当时投票后得票率最高的前5个选项：

（1）操作系统必须支持新应用的安装；

（2）高速度处理芯片；

（3）支持播放式的手机电视；

（4）具有大容量存储芯片和存储扩展能力；

（5）支持GPS导航。

根据大众投票结果，手机联盟制定了一个标准。并根据这个标准为基础，总结出了如下智能手机的主要特点：

（1）具备普通手机的全部功能，例如可以进行正常的通话和发短信等手机应用；

（2）是一个开放性的操作系统，在系统平台上可以安装更多的应用程序，从而实现功能的无限扩充；

（3）具备上网功能；

（4）具备PDA的功能，实现个人信息管理、日程记事、任务安排、多媒体应用和浏览网页；

（5）可以根据个人需要扩展机器的功能；

（6）扩展性能强，并且可以支持很多第三方软件。

1.2 Android 的巨大优势

为什么 Android 能在这么多的智能系统中脱颖而出，成为市场占有率第一的手机系统呢？要想分析其原因，需要先了解它的巨大优势，分析究竟是哪些优点吸引了厂商和消费者的青睐。本节将对上述问题一一解答。

1.2.1　系出名门

Android 是出身于 Linux 世家，是一款开源的手机操作系统。Android 功成名就之后，各大手机联盟纷纷加入，这个联盟由中国移动、摩托罗拉、高通、HTC 和 T-Mobile 在内的 30 多家技术和无线应用的领军企业组成。通过与运营商、设备制造商、开发商和其他有关各方结成深层次的合作伙伴关系，希望借助建立标准化、开放式的移动电话软件平台，在移动产业内形成一个开放式的生态系统。

1.2.2　强大的开发团队

Android 的研发队伍阵容强大，包括摩托罗拉、Google、HTC（宏达电子）、PHILIPS、T-Mobile、高通、魅族、三星、LG 以及中国移动在内的 34 家企业，这都是在手机开发生产领域中享誉盛名的企业。它们都将基于该操作平台开发手机的新型业务，应用之间的通用性和互联性将在最大程度上得到保持。此外，相关生产厂家还成立了手机开放联盟，联盟中的成员名单如下所示。

1. 手机制造商

中国台湾宏达国际电子（HTC）（Palm 等多款智能手机的代工厂）、摩托罗拉（美国最大的手机制造商）、韩国三星电子（仅次于诺基亚的全球第二大手机制造商）、韩国 LG 电子、中国移动（全球最大的移动运营商）、日本 KDDI（2900 万用户）、日本 NTT DoCoMo（5200 万用户）、美国 Sprint Nextel（美国第三大移动运营商，5400 万用户）、意大利电信（Telecom Italia）（意大利主要的移动运营商，3400 万用户）、西班牙 Telefónica（在欧洲和拉美有 1.5 亿用户）、T-Mobile（德意志电信旗下公司，在美国和欧洲有 1.1 亿用户）。

2. 半导体公司

Audience Corp（声音处理器公司）、Broadcom Corp（无线半导体主要提供商）、英特尔（Intel）、Marvell Technology Group、Nvidia（图形处理器公司）、SiRF（GPS 技术提供商）、Synaptics（手机用户界面技术）、德州仪器（Texas Instruments）、高通（Qualcomm）、惠普 HP（Hewlett-Packard Development Company,L.P）。

3. 软件公司

Aplix、Ascender、eBay 的 Skype、Esmertec、Living Image、NMS Communications、Noser Engineering AG、Nuance Communications、PacketVideo、SkyPop、Sonix Network、TAT-The Astonishing Tribe、Wind River Systems。

1.2.3　Android 系统开源

开源意味着对开发人员和手机厂商来说，Android 是完全无偿免费使用的。因为源代码公开的原因，所以吸引了全世界各地无数程序员的热情。于是很多手机厂商都纷纷采用 Android 作为自己产品的系统。因为免费，所以降低了成本，提高了利润。而对于开发人员来说，众多厂商的采用就意味着人才需求大，所以纷纷加入到 Android 开发大军中来。于是有一些干的还可以的程序员禁不住高薪的诱惑，都纷纷改行做 Android 开发。至于“混”的不尽如人意的程序员，就更加坚定了“改行做 Android 手机开发”，目的是想寻找自己程序员生涯的转机。而像笔者这样遇到发展瓶颈的程序员，也决定做 Androd 开发，因为这样可以学习一门新技术，使自己的未来更加有保障。

第 2 章　搭建 Android 应用开发环境

“工欲善其事，必先利其器”出自《论语》，意思是要想高效地完成一件事，需要有一个合适的工具。对于安卓开发人员来说，开发工具同样至关重要。作为一项新兴技术，在进行开发前首先要搭建一个对应的开发环境。而在搭建开发环境前，需要了解安装开发工具所需要的硬件和软件配置条件。Android 开发包括底层开发和应用开发，底层开发大多数是指和硬件相关的开发，并且是基于 Linux 环境的，例如开发驱动程序。应用开发是指开发能在 Android 系统上运行的程序，例如游戏、地图等程序。本书的重点是讲解多媒体应用开发，即使讲一些底层的知识，也是为上层的应用服务的。因为开发 Android 智能设备程序既需要底层开发知识，也需要上层应用开发的知识。所以在本书需要讲解底层和应用开发环境的搭建知识。本章将首先介绍在 Windows 下搭建 Android 应用开发环境的过程。

2.1　安装 Android SDK 的系统要求

在搭建之前，一定先确定基于 Android 应用软件所需要开发环境的要求，具体如表 2-1 所示。

表 2-1　开发系统所需求参数

项　目	版本要求	说　明	备　注
操作系统	Windows XP 或 Vista Mac OS X 10.4.8+Linux Ubuntu Drapper	根据自己的计算机自行选择	选择自己最熟悉的操作系统
软件开发包	Android SDK	选择最新版本的 SDK	截止到目前，最新手机版本是 2.3
IDE	Eclipse IDE+ADT	Eclipse3.3（Europa），3.4（Ganymede）ADT（Android Development Tools）开发插件	选择“for Java Developer”
其他	JDK Apache Ant	Java SE Development Kit 5 或 6 Linux 和 Mac 上使用 Apache Ant 1.6.5+，Windows 上使用 1.7+版本	（单独的 JRE 是不可以的，必须要有 JDK），不兼容 Gnu Java 编译器（gcj）

Android 工具是由多个开发包组成的，具体说明如下。

- JDK：可以到网址 http://java.sun.com/javase/downloads/index.jsp 处下载。
- Eclipse（Europa）：可以到网址 http://www.eclipse.org/downloads/下载 Eclipse IDE for Java Developers。
- Android SDK：可以到网址 http://developer.android.com 下载。
- 还有对应的开发插件。

2.2　安装 JDK

JDK（Java Development Kit）是整个 Java 的核心，包括了 Java 运行环境、Java 工具和 Java

基础的类库。JDK 是学好 Java 的第一步，是开发和运行 Java 环境的基础，当用户要对 Java 程序进行编译的时候，必须先获得对应操作系统的 JDK，否则将无法编译 Java 程序。在安装 JDK 之前需要先获得 JDK，获得 JDK 的操作流程如下所示。

（1）登录 Oracle 官方网站，网址为 http://developers.sun.com/downloads/，如图 2-1 所示。

（2）在图 2-1 中可以看到有很多版本，在此选择当前最新的版本 Java 7，下载页面如图 2-2 所示。

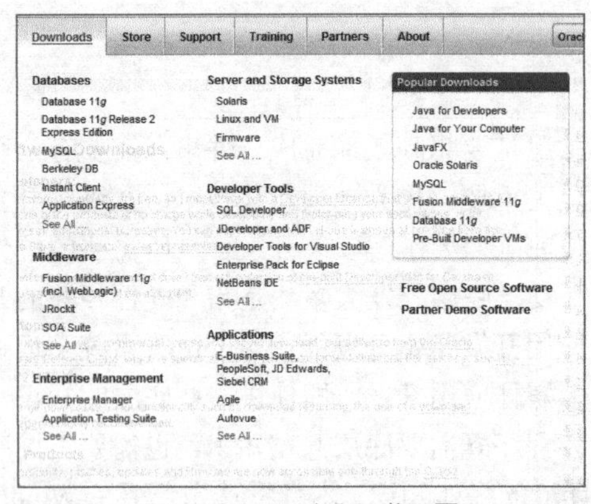

▲图 2-1　Oracle 官方下载页面

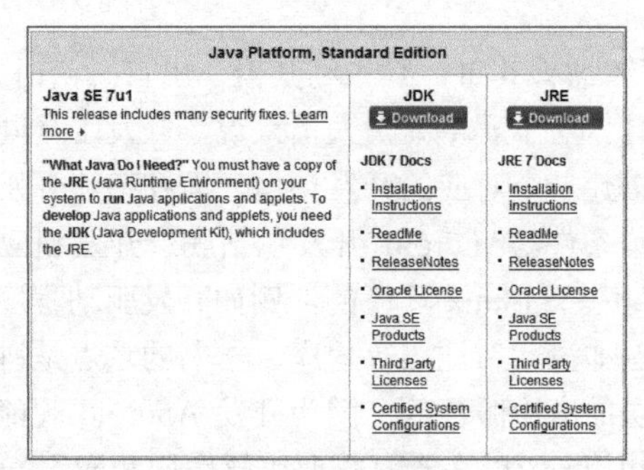

▲图 2-2　JDK 下载页面

（3）在图 2-2 中单击 JDK 下方的“Download”按钮，在弹出的新界面中选择将要下载的 JDK，本书在此选择的是 Windows X86 版本，如图 2-3 所示。

（4）下载完成后双击下载的“.exe”文件开始进行安装，将弹出“安装向导”对话框，在此单击“下一步”按钮，如图 2-4 所示。

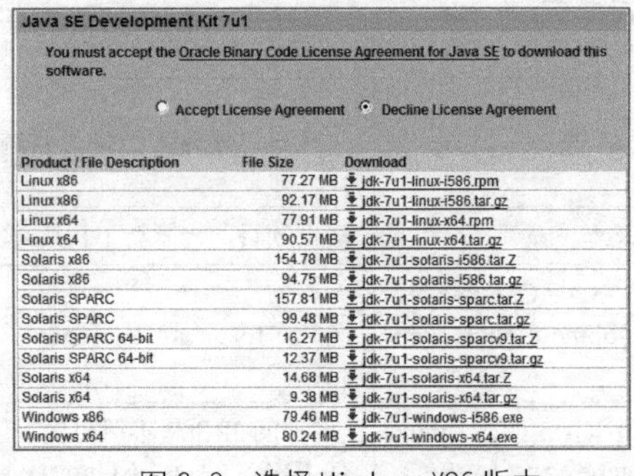

▲图 2-3　选择 Windows X86 版本

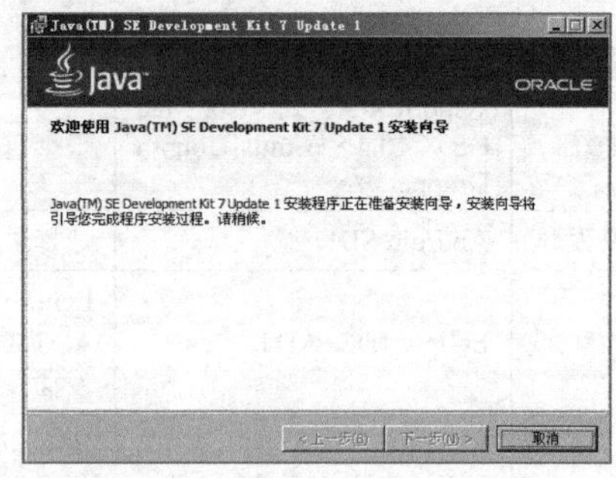

▲图 2-4　“许可证协议”对话框

（5）弹出“安装路径”对话框，在此选择文件的安装路径，如图 2-5 所示。

（6）在此设置安装路径是“E:\jdk1.7.0_01\”，然后单击“下一步”按钮开始在安装路径解压缩下载的文件，如图 2-6 所示。

（7）完成后弹出“目标文件夹”对话框，在此选择要安装的位置，如图 2-7 所示。

（8）单击“下一步”按钮后开始正式安装，如图 2-8 所示。

（9）完成后弹出“完成”对话框，单击“完成”按钮后完成整个安装过程，如图 2-9 所示。

完成安装后可以检测是否安装成功，检测方法是依次单击“开始”｜“运行”，在运行框中输入“cmd”并按下“Enter”键，在打开的 CMD 窗口中输入“java –version”，如果显示图 2-10 所示的提示信息，则说明安装成功。

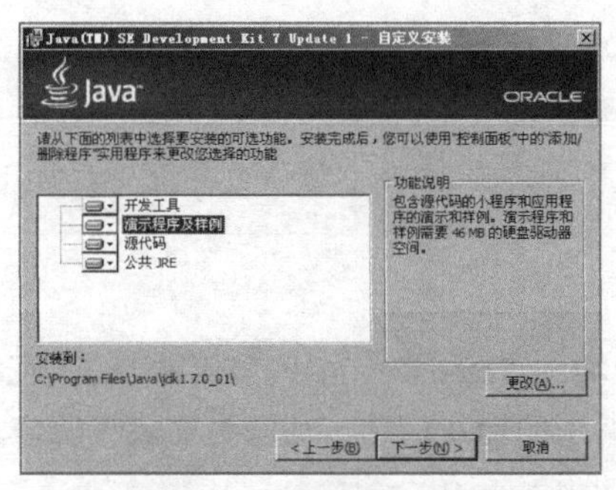

▲图 2-5 “安装路径”对话框

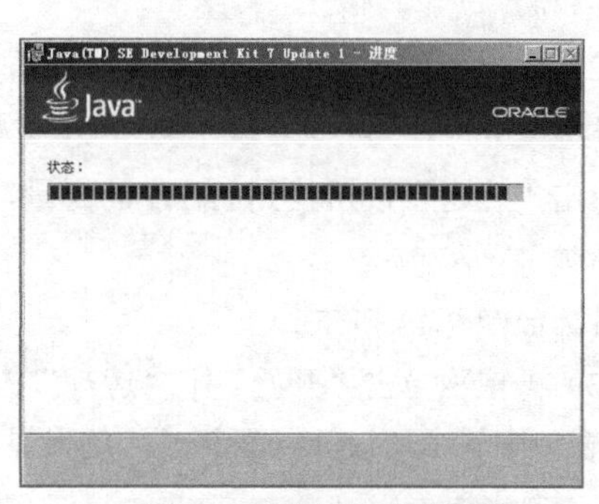

▲图 2-6 解压缩下载的文件

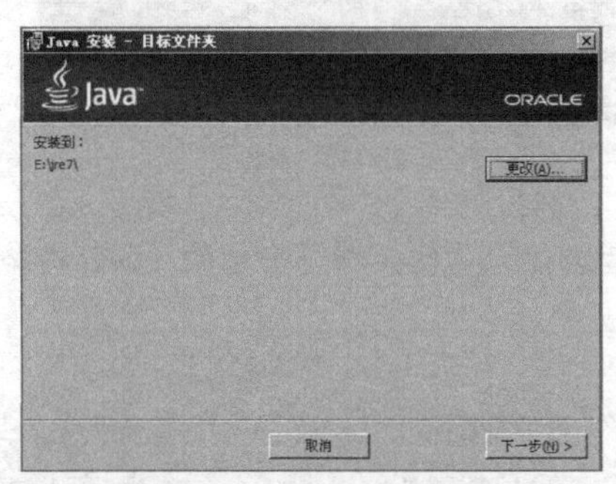

▲图 2-7 “目标文件夹”对话框

▲图 2-8 继续安装

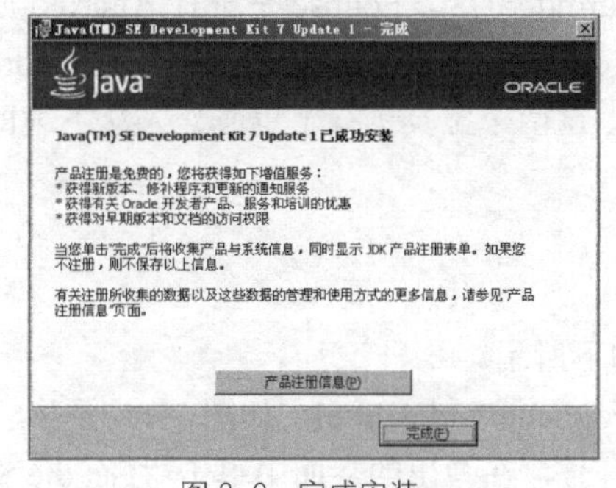

▲图 2-9 完成安装

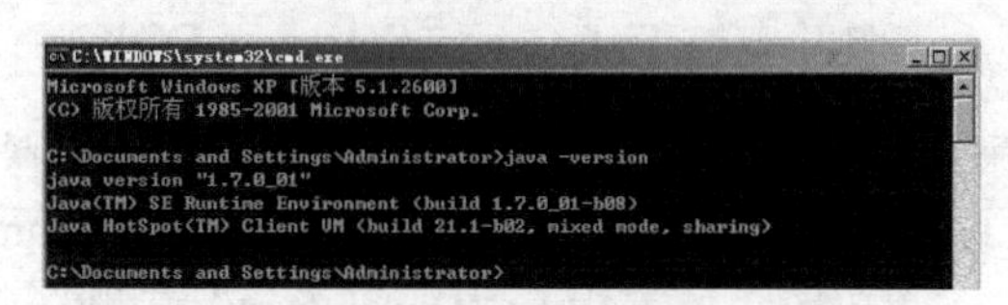

▲图 2-10 CMD 窗口

注意 完成安装后可以检测是否安装成功，方法是依次单击“开始” | “运行”，在运行框中输入“cmd”并按下“Enter”键，在打开的 CMD 窗口中输入“java –version”，如果显示图 2-11 所示的提示信息，则说明安装成功。

如果检测没有安装成功，需要将其目录的绝对路径添加到系统的 PATH 中。具体做法如下所示。

（1）右键单击“我的电脑”，依次选择“属性” | “高级”，单击下面的“环境变量”，在下面的“系统变量”处选择新建，在变量名处输入 JAVA_HOME，变量值中输入刚才的目录，例如设置为“F:\Java\jdk1.6.0_22”。如图 2-12 所示。

▲图 2-11 CMD 窗口

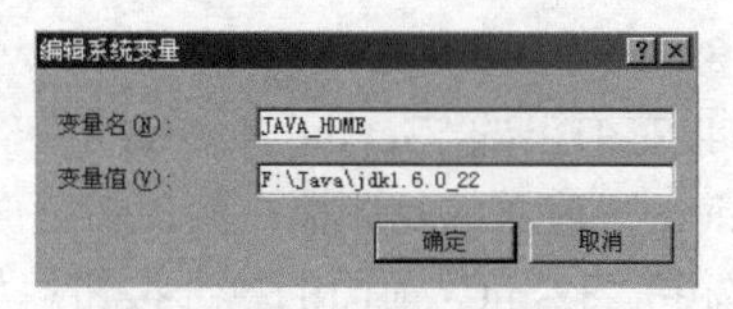

▲图 2-12 设置系统变量

（2）再次新建一个变量名为 classpath，其变量值如下所示。

```
.;%JAVA_HOME%/lib/rt.jar;%JAVA_HOME%/lib/tools.jar
```

单击“确定”按钮找到 PATH 的变量，双击或单击“编辑”，在变量值最前面添加如下值。

```
%JAVA_HOME%/bin;
```

具体如图 2-13 所示。

（3）再依次单击“开始”|“运行”，在运行框中输入“cmd”并按下“Enter”键，在打开的 CMD 窗口中输入“java –version”，如果显示图 2-14 所示的提示信息，则说明安装成功。

▲图 2-13　设置系统变量

▲图 2-14　CMD 界面

> **注意**　上述变量设置中，是按照个人的安装路径设置的，这里安装的 JDK 的路径是 C:\Program Files\Java\jdk1.7.0_02。

2.3　安装 Eclipse 和 Android SDK

在安装好 JDK 后，接下来需要安装 Eclipse 和 Android SDK。Eclipse 是进行 Android 应用开发的一个集成工具，而 Android SDK 是开发 Android 应用程序所必须具备的框架。在 Android 官方公布的最新版本中，已经将 Eclipse 和 Android SDK 这两个工具进行了集成，一次下载即可同时获得这两个工具。

2.3.1　获取并安装 Eclipse 和 Android SDK

获取并安装 Eclipse 和 Android SDK 的具体步骤如下所示。

（1）登录 Android 的官方网站 http://developer.android.com/index.html，如图 2-15 所示。

（2）单击图 2-15 左上方“Developers”右侧的 ⌄ 符号，在弹出的界面中单击“Get the SDK”链接。如图 2-16 所示。

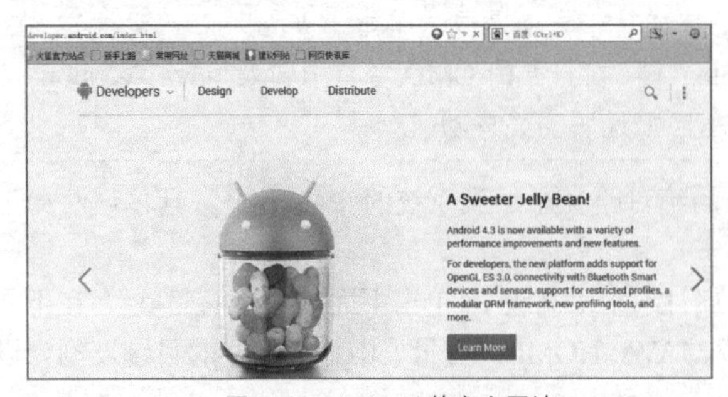

▲图 2-15　Android 的官方网站

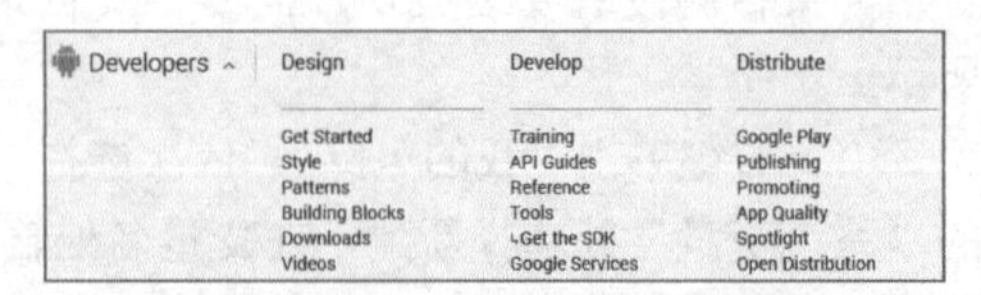

▲图 2-16　单击“Get the SDK”链接

（3）在弹出的新页面中单击“Download the SDK”按钮，如图 2-17 所示。

（4）在弹出的“Get the Android SDK”界面中选中“I have read and agree with the above terms and conditions”前面的复选框，然后在下面的单选按钮中选择系统的位数。如果机器是 32 位的，就可以选中“32-bit”前面的单选按钮。如图 2-18 所示。

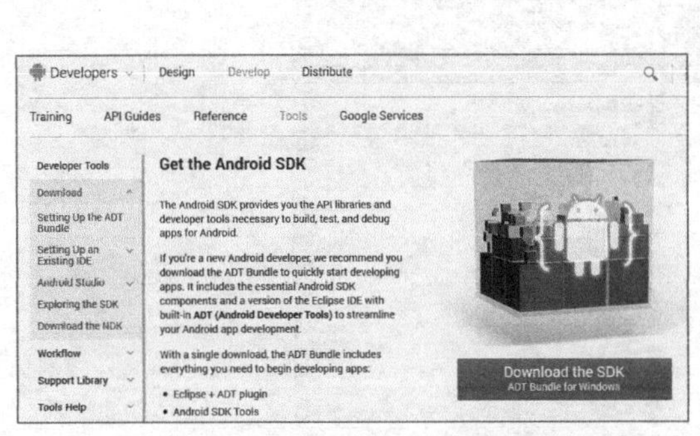

▲图 2-17 单击“Download the SDK”按钮

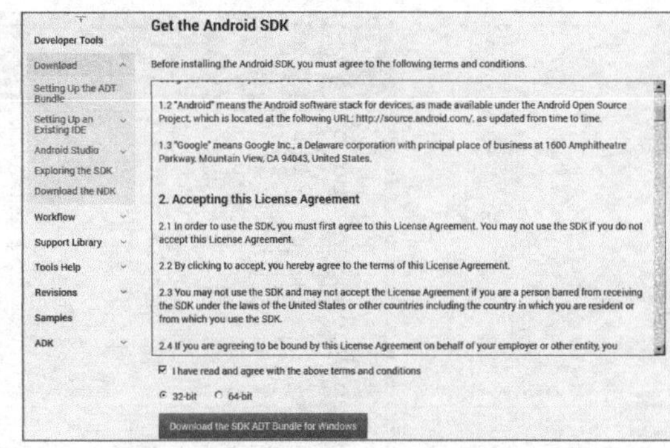

▲图 2-18 “Get the Android SDK”界面

（5）单击图 2-18 中的Download the SDK ADT Bundle for Windows按钮后开始下载工作，下载的目标文件是一个压缩包。如图 2-19 所示。

（6）将下载得到的压缩包进行解压，解压后的目录结构如图 2-20 所示。

▲图 2-19 开始下载目标文件压缩包

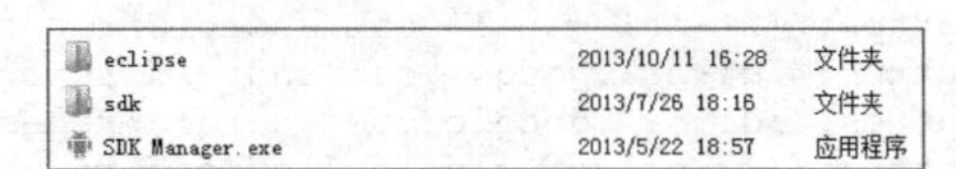

▲图 2-20 解压后的目录结构

由此可见，Android 官方已经将 Eclipse 和 Android SDK 实现了集成。双击“eclipse”目录中的“eclipse.exe”可以打开 Eclipse，界面效果如图 2-21 所示。

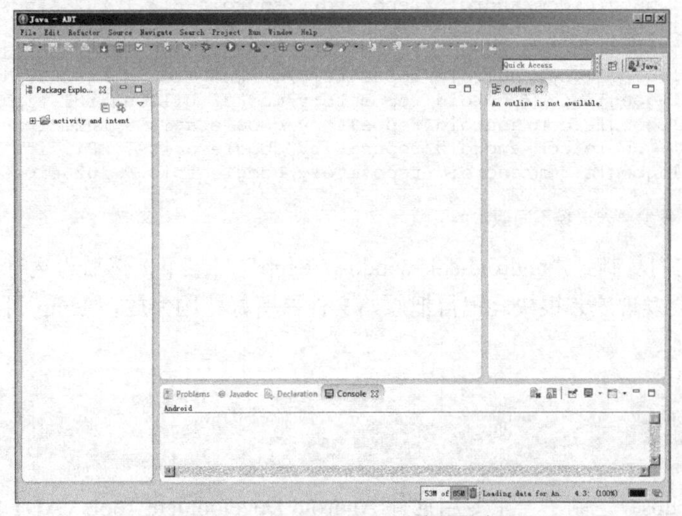

▲图 2-21 打开 Eclipse 后的界面效果

（7）打开 Android SDK 的方法有两种，第一种是双击下载目录中的“SDK Manager.exe”文件，第二种在是 Eclipse 工具栏中单击图标。打开后的效果如图 2-22 所示。

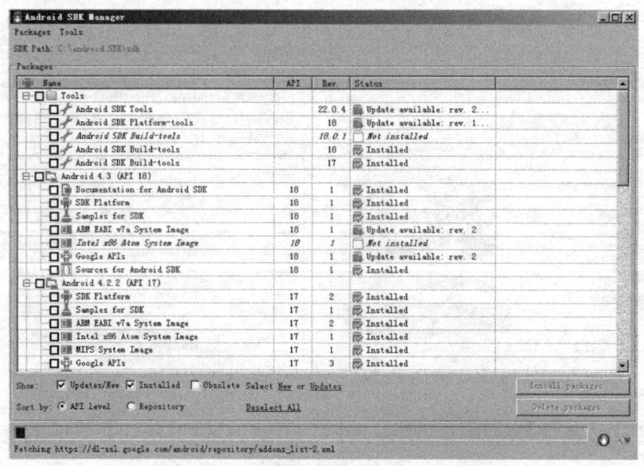

▲图 2-22　打开 Android SDK 后的界面效果

2.3.2　快速安装 SDK

通过 Android SDK Manager 在线安装的速度非常慢，而且有时容易断开。其实可以先从网络中寻找到 SDK 资源，用迅雷等下载工具下载后，将其存储到指定目录后就可以完成安装。具体方法是先下载可以更新的 android-sdk-windows 文件，然后在 android-sdk-windows 下双击 setup.exe，在更新的过程中会发现安装 Android SDK 的速度是 1 Kb/s，此时打开迅雷，分别输入下面的地址：

```
https://dl-ssl.google.com/android/repository/platform-tools_r05-windows.zip
https://dl-ssl.google.com/android/repository/docs-3.1_r02-linux.zip
https://dl-ssl.google.com/android/repository/android-2.2_r02-windows.zip
https://dl-ssl.google.com/android/repository/android-2.3.3_r02-linux.zip
https://dl-ssl.google.com/android/repository/android-2.1_r02-windows.zip
https://dl-ssl.google.com/android/repository/samples-2.3.3_r02-linux.zip
https://dl-ssl.google.com/android/repository/samples-2.2_r02-linux.zip
https://dl-ssl.google.com/android/repository/samples-2.1_r02-linux.zip
https://dl-ssl.google.com/android/repository/compatibility_r02.zip
https://dl-ssl.google.com/android/repository/tools_r12-windows.zip
https://dl-ssl.google.com/android/repository/google_apis-10_r02.zip
https://dl-ssl.google.com/android/repository/android-2.3.1_r02-linux.zip
https://dl-ssl.google.com/android/repository/usb_driver_r04-windows.zip
https://dl-ssl.google.com/android/repository/googleadmobadssdkandroid-4.1.0.zip
https://dl-ssl.google.com/android/repository/market_licensing-r01.zip
https://dl-ssl.google.com/android/repository/market_billing_r01.zip
https://dl-ssl.google.com/android/repository/google_apis-8_r02.zip
https://dl-ssl.google.com/android/repository/google_apis-7_r01.zip
https://dl-ssl.google.com/android/repository/google_apis-9_r02.zip
……
可以继续根据自己开发要求选择不同版本的API
```

下载完后将它们复制到“android-sdk-windows/temp”目录下，然后再运行 setup.exe，选中需要的 API 选项，会发现安装很快。记得把原始文件保留好，因为放在 temp 目录下的文件装好后会立刻消失。

2.4 安装 ADT

Android 为 Eclipse 定制了一个专用插件 Android Development Tools（ADT），此插件为用户提供了一个强大的开发 Android 应用程序的综合环境。ADT 扩展了 Eclipse 的功能，可以让用户快速地建立 Android 项目，创建应用程序界面。要安装 Android Development Tools plug-in，首先需要打开 Eclipse IDE，然后进行如下操作。

（1）打开 Eclipse 后，依次单击菜单栏中的“Help”|“Install New Software”选项，如图 2-23 所示。

（2）在弹出的对话框中单击“Add”按钮，如图 2-24 所示。

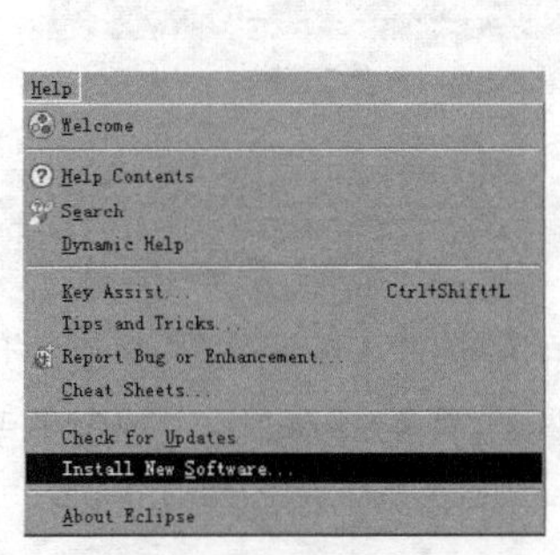

▲图 2-23 添加插件

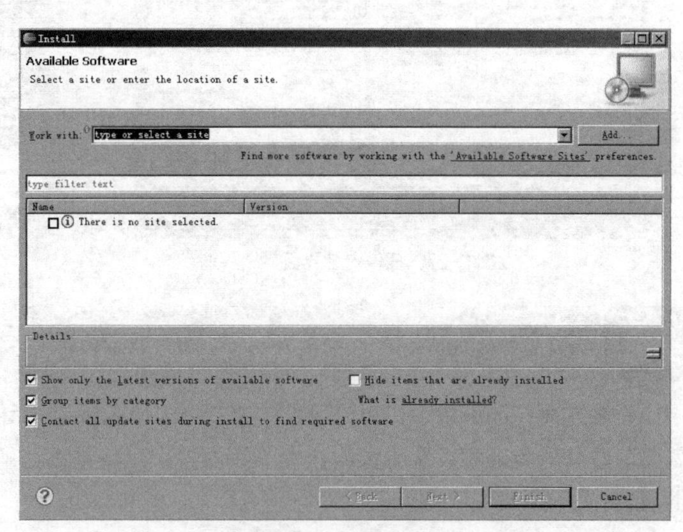

▲图 2-24 添加插件

（3）在弹出的“Add Site”对话框中分别输入名字和地址，名字可以自己命名，例如“123”，但是在 Location 中必须输入插件的网络地址 http://dl-ssl.google.com/Android/eclipse/。如图 2-25 所示。

（4）单击“OK”按钮，此时在“Install”界面将会显示系统中可用的插件。如图 2-26 所示。

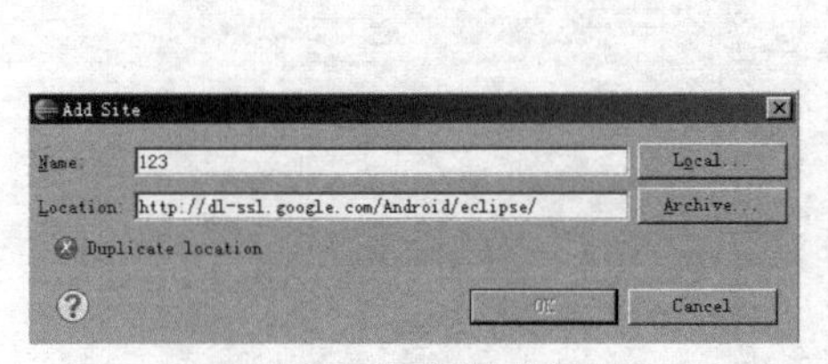

▲图 2-25 设置地址

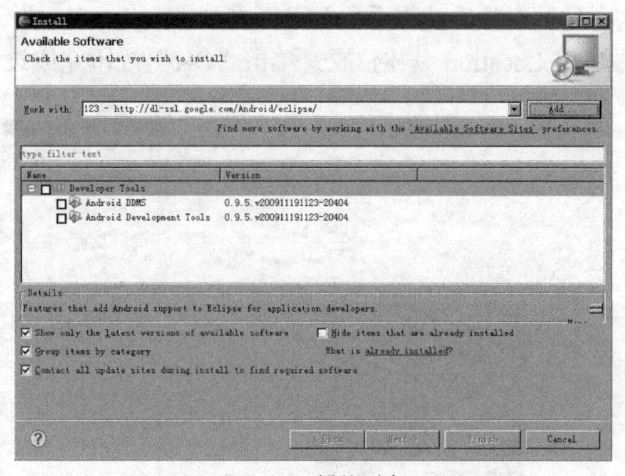

▲图 2-26 插件列表

（5）选中“Android DDMS”和“Android Development Tools”，然后单击“Next”按钮进入安装界面。如图 2-27 所示。

（6）选中“I accept the terms of the license agreement”选项，单击“Finish”按钮，开始进行安装。如图 2-28 所示。

注意　在上面步骤中，可能会发生计算插件占用资源情况，过程有点慢。完成后会提示重启 Eclipse 来加载插件，等重启后即可。并且不同版本的 Eclipse 安装插件的方法和步骤是不同的，但是都大同小异，读者可以根据操作提示能够自行解决。

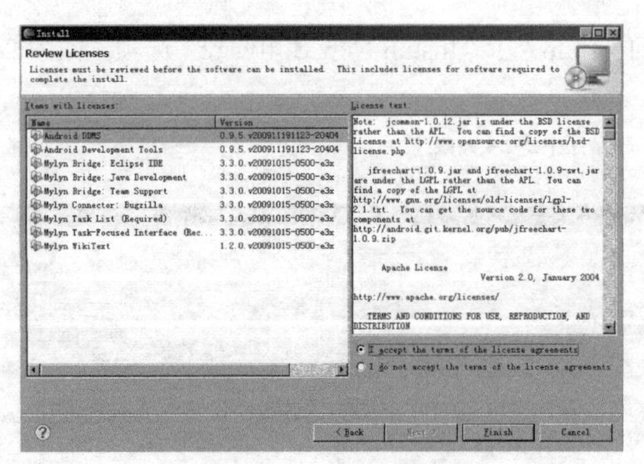

▲图 2-27　插件安装界面

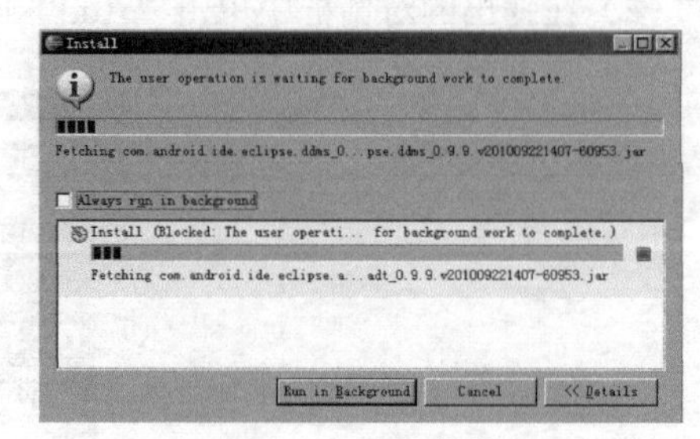

▲图 2-28　开始安装

2.5 验证设置

在本章前面的内容中，已经讲解了搭建安装 Android 基本环境的知识。在完成安装之后，还需要一些具体验证和设置工作。在本节的内容中，将详细讲解验证和设置 Android 开发环境的基本知识。

2.5.1　设定 Android SDK Home

当完成上述插件装备工作后，此时还不能使用 Eclipse 创建 Android 项目，还需要在 Eclipse 中设置 Android SDK 的主目录。

（1）打开 Eclipse，在菜单中依次单击“Windows”｜“Preferences”项，如图 2-29 所示。

（2）在弹出的界面左侧可以看到“Android”项，选中 Android 后，在右侧设定 Android SDK 的 SDK Location 具体位置，单击“OK”按钮完成设置。如图 2-30 所示。

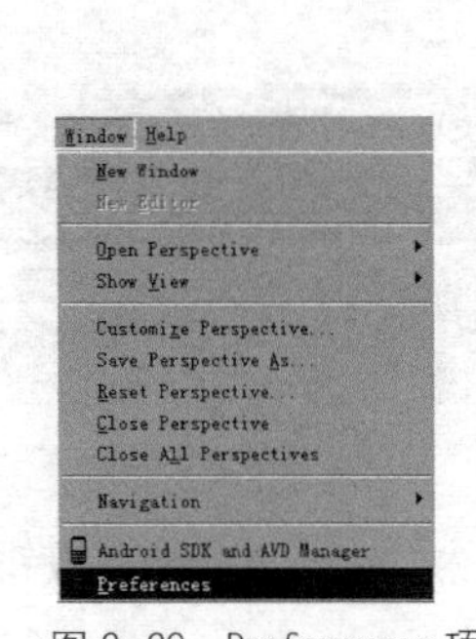

▲图 2-29　Preferences 项

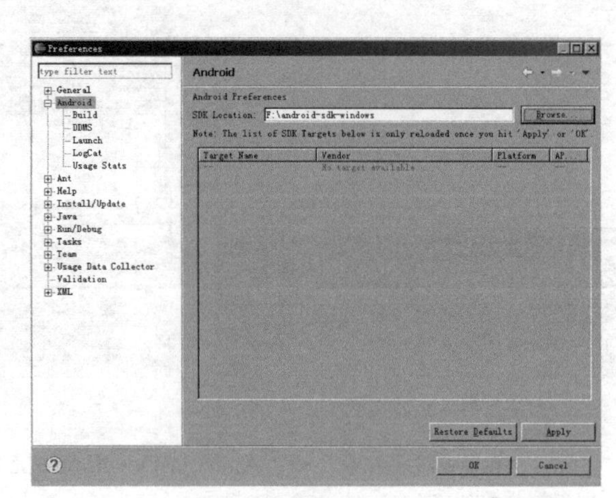

▲图 2-30　Preferences 项

2.5.2　验证开发环境

按照前面讲解的步骤操作后，一个基本的 Android 开发环境算是搭建完成了。实践是检验真理的唯一标准，下面通过新建一个项目来验证当前的环境是否可以正常工作。

（1）打开 Eclipse，在菜单中依次选择“File”｜“New”｜“Project”项，在弹出的对话框上可以看到 Android 类型的选项，如图 2-31 所示。

（2）在图 2-31 上选择“Android”，单击“Next”按钮后打开“New Android Application”对话框，在对应的文本框中输入必要的信息，如图 2-32 所示。

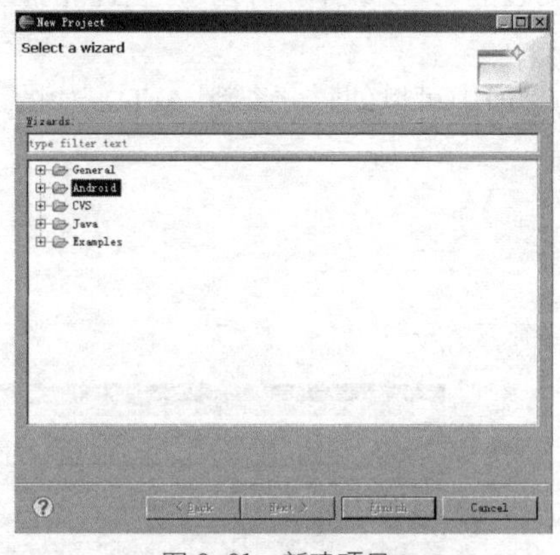

▲图 2-31 新建项目

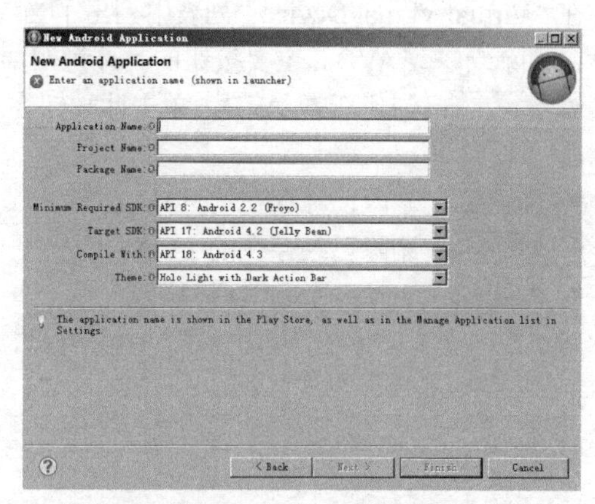

▲图 2-32 “New Android Application”对话框

（3）单击“Finish”按钮后 Eclipse 会自动完成项目的创建工作，最后会看到图 2-33 所示的项目结构。

▲图 2-33 项目结构

2.5.3 创建 Android 虚拟设备（AVD）

程序开发需要调试，只有经过调试之后才能知道程序是否正确运行。作为一款手机操作系统，如何能在计算机平台之上调试 Android 程序呢？不用担心，谷歌提供了模拟器来解决这个问题。所谓模拟器，就是指在计算机上模拟 Android 系统，可以用这个模拟器来调试并运行开发的 Android 程序。开发人员不需要一个真实的 Android 手机，只需要通过计算机即可模拟运行一个手机，即可开发出应用在手机上面的程序。

AVD 全称为 Android 虚拟设备（Android Virtual Device），每个 AVD 模拟了一套虚拟设备来运行 Android 平台，这个平台至少要有自己的内核、系统图像和数据分区，还可以有自己的的 SD 卡和用户数据以及外观显示等。创建 AVD 的基本步骤如下所示。

（1）单击 Eclips 菜单中的图标，如图 2-34 所示。

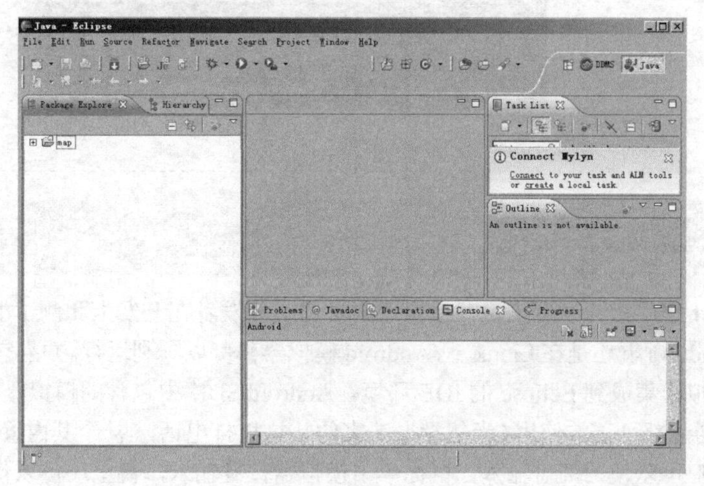

▲图 2-34 Eclipse

（2）在弹出的“Android AVD Manager”界面的左侧导航中选择“Android Virtual Device”选项，如图 2-35 所示。

在“Virtual Virtual Device”列表中列出了当前已经安装的 AVD 版本，可以通过右侧的按钮来创建、删除或修改 AVD。主要按钮的具体说明如下所示。

- New...：创建新的 AVD，单击此按钮在弹出的界面中可以创建一个新 AVD，如图 2-36 所示。
- Edit...：修改已经存在的 AVD。
- Delete...：删除已经存在的 AVD。
- Start...：启动一个 AVD 模拟器。

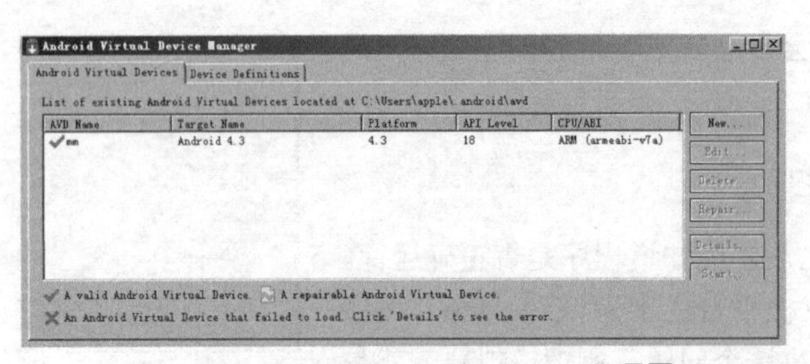

图 2-35　“Android SDK and AVD Manager”界面

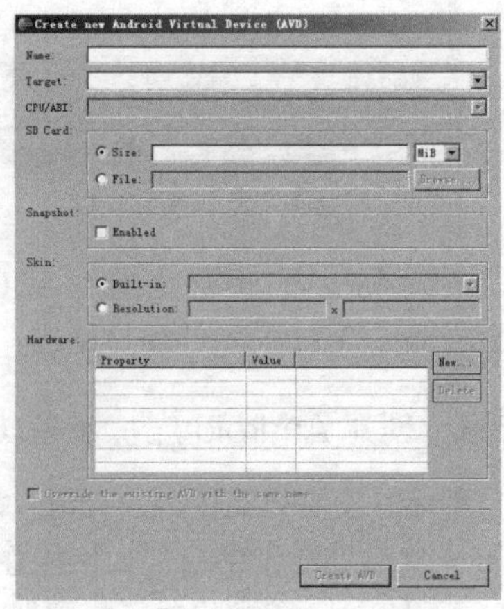

▲图 2-36　新建 AVD 界面

注意

人们可以在 CMD 中创建或删除 AVD，例如可以按照如下 CMD 命令创建一个 AVD。

```
android create avd --name <your_avd_name> --target <targetID>
```

其中“your_avd_name”是需要创建的 AVD 的名字，CMD 窗口界面如图 2-37 所示。

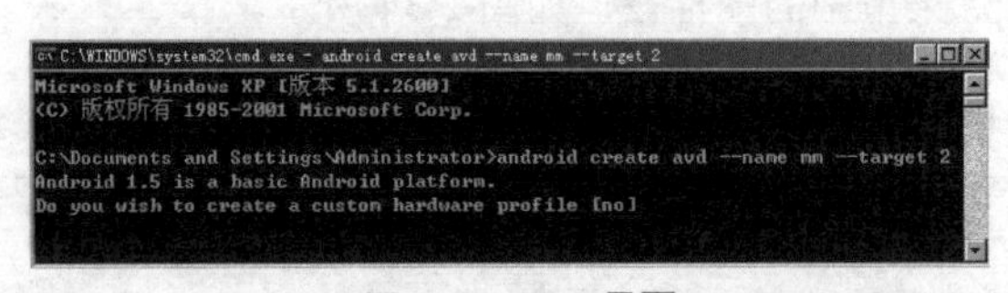

▲图 2-37　CMD 界面

2.6 启动 AVD 模拟器

对于 Android 程序的开发者来说，模拟器的推出给开发者在开发上和测试上带来了很大的便利。无论是在 Windows 下还是在 Linux 下，Android 模拟器都可以顺利运行。并且官方提供了 Eclipse 插件，可以将模拟器集成到 Eclipse 的 IDE 环境。Android SDK 中包含的模拟器的功能非常齐全，电话本、通话等功能都可正常使用（当然没办法真的从这里打电话）。甚至其内置的浏览器和 Maps 都可以联网。用户可以使用键盘输入，鼠标单击模拟器按键输入，甚至还可以使用鼠标单击、拖动屏幕进行操纵。模拟器在计算机上模拟运行的效果如图 2-38 所示。

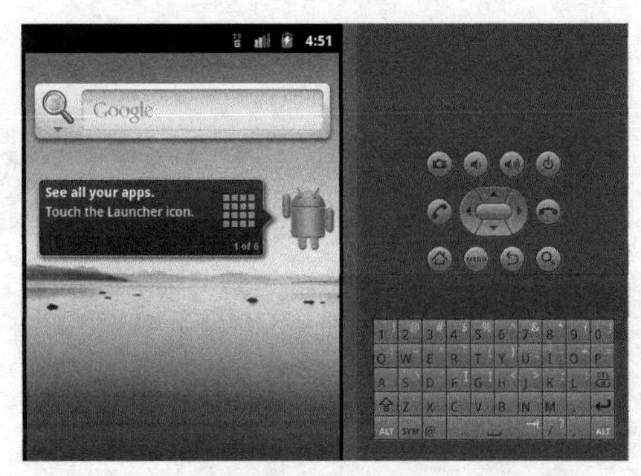

▲图 2-38 模拟器

2.6.1 模拟器和真机究竟有何区别

当然 Android 模拟器不能完全替代真机，具体来说有如下差异。

- 模拟器不支持呼叫和接听实际来电，但可以通过控制台模拟电话呼叫（呼入和呼出）；
- 模拟器不支持 USB 连接；
- 模拟器不支持相机/视频捕捉；
- 模拟器不支持音频输入（捕捉），但支持输出（重放）；
- 模拟器不支持扩展耳机；
- 模拟器不能确定连接状态；
- 模拟器不能确定电池电量水平和交流充电状态；
- 模拟器不能确定 SD 卡的插入/弹出；
- 模拟器不支持蓝牙。

2.6.2 启动 AVD 模拟器的基本流程

在调试的时候需要启动 AVD 模拟器，启动 AVD 模拟器的基本流程如下所示。

（1）选择图 2-35 列表中名为“mm”的 AVD，单击 Start... 按钮后弹出“Launch Options”界面。如图 2-39 所示。

（2）单击“Launch”按钮后将会运行名为“mm”的模拟器，运行界面效果如图 2-40 所示。

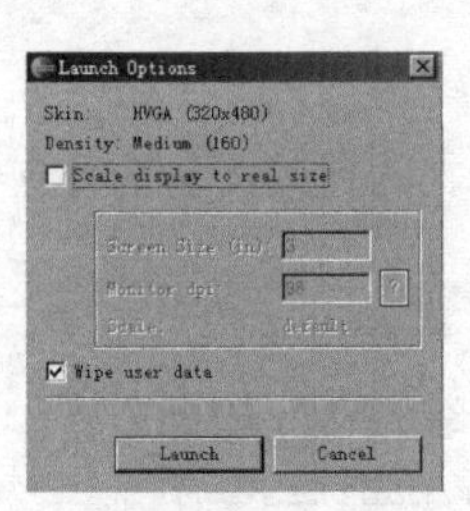

▲图 2-39 “Launch Options”对话框

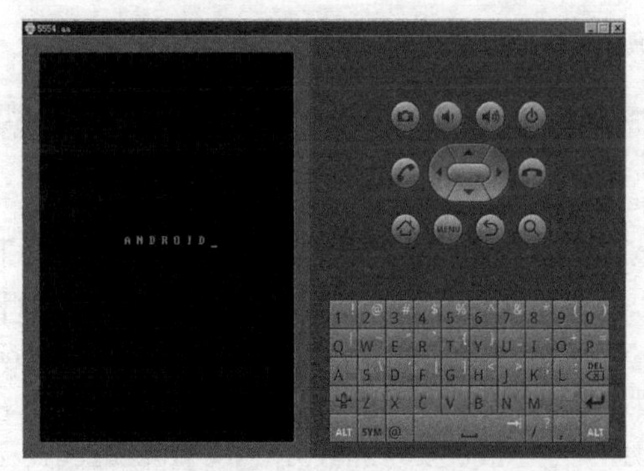

▲图 2-40 模拟运行成功

第3章　获取并编译源码

要想掌握 Android 智能设备底层开发相关技术，需要先了解 Android 系统源码的基本知识。在了解 Android 系统源码之前，需要先获取其具体源码。因为目前市面上主流的操作系统有 Windows、Linux 和 Mac OS 的操作系统，由于 Mac OS 源自于 Linux 系统，所以本章讲解分别在 Windows 系统和 Linux 系统中获取 Android 源码的知识，并讲解编译 Android 源码的具体过程，为读者步入本书后面知识的学习打下基础。

3.1　在 Linux 系统获取 Android 源码

在 Linux 系统中，通常使用 Ubuntu 来下载和编译 Android 源码。由于 Android 的源码内容很多，Google 采用了 git 的版本控制工具，并对不同的模块设置不同的 git 服务器，可以用 repo 自动化脚本来下载 Android 源码，下面介绍如何一步一步地获取 Android 源码的过程。

1. 下载 repo

在用户目录下，创建 bin 文件夹，用于存储 repo，并把该路径设置到环境变量中去，命令如下：

```
$ mkdir ~/bin
$ PATH=~/bin:$PATH
```

下载 repo 的脚本，用于执行 repo，命令如下：

```
$ curl https://dl-ssl.google.com/dl/googlesource/git-repo/repo > ~/bin/repo
```

设置可执行权限，命令如下：

```
$ chmod a+x ~/bin/repo
```

2. 初始化一个 repo 的客户端

在用户目录下，创建一个空目录，用于存储 Android 源码，命令如下：

```
$ mkdir AndroidCode
$ cd AndroidCode
```

进入 AndroidCode 目录，并运行 repo 下载源码，下载主线分支的代码，主线分支包括最新修改的 bug 以及并未正式发出版本的最新源码，命令如下：

```
$ repo init -u https://android.googlesource.com/platform/manifest
```

下载其他分支，正式发布的版本，可以通过添加-b 参数来下载，命令如下：

```
$ repo init -u https://android.googlesource.com/platform/manifest -b
android-4.4_r1
```

在下载过程中需要填写 Name 和 E-mail，填写完毕之后，选择 Y 进行确认，最后提示 repo 初始化完成，这时可以开始同步 Android 源码，同步过程很漫长，需要耐心的等待，执行下面命令开始同步代码：

```
$ repo sync
```

经过上述步骤后，便开始下载并同步 Android 源码，界面效果如图 3-1 所示。

```
Checking out files: 100% (2497/2497), done. out files:  31% (790/2497)
Checking out files: 100% (1654/1654), done. out files:  39% (649/1654)
Checking out files: 100% (3471/3471), done.
Checking out files: 100% (24607/24607), done.ut files:  21% (5269/24607)
Checking out files: 100% (2431/2431), done. out files:  32% (800/2431)
Checking out files: 100% (18696/18696), done.
Checking out files: 100% (4276/4276), done. out files:  48% (2058/4276)
Checking out files: 100% (2216/2216), done. out files:  49% (1093/2216)
Checking out files: 100% (857/857), done.ng out files:  13% (115/857)
Checking out files: 100% (1141/1141), done. out files:   2% (28/1141)
Checking out files: 100% (431/431), done.ng out files:  10% (46/431)
Checking out files: 100% (175/175), done.ng out files:  10% (19/175)
Checking out files: 100% (135/135), done.
Checking out files: 100% (378/378), done.
Checking out files: 100% (433/433), done.
Checking out files: 100% (2407/2407), done. out files:  30% (724/2407)
Checking out files: 100% (2489/2489), done.
Checking out files: 100% (2493/2493), done.
Checking out files: 100% (177/177), done.ng out files:  15% (27/177)
Checking out files: 100% (137/137), done.
Checking out files: 100% (40775/40775), done.ut files:   5% (2199/40775)
Checking out files: 100% (93/93), done.
Checking out files: 100% (450/450), done.
Checking out files: 100% (5265/5265), done. out files:  41% (2167/5265)
Syncing work tree: 100% (329/329), done.
```

▲图 3-1　下载同步

3.2　在 Windows 平台获取 Android 源码

Windows 平台上获取源码和 Linux 原理相同，但是需要预先在 Windows 平台上面搭建一个 Linux 环境，此处需要用到 cygwin 工具。cygwin 的作用是构建一套在 Windows 上的 Linux 模拟环境，下载 cygwin 工具的地址如下：

```
http://cygwin.com/install.html
```

下载成功后会得到一个名为“setup.exe”可执行文件，通过此文件可以更新和下载最新的工具版本，具体流程如下所示。

（1）启动 cygwin，如图 3-2 所示。

（2）单击“下一步”按钮，选择第一个选项：从网络下载安装，如图 3-3 所示。

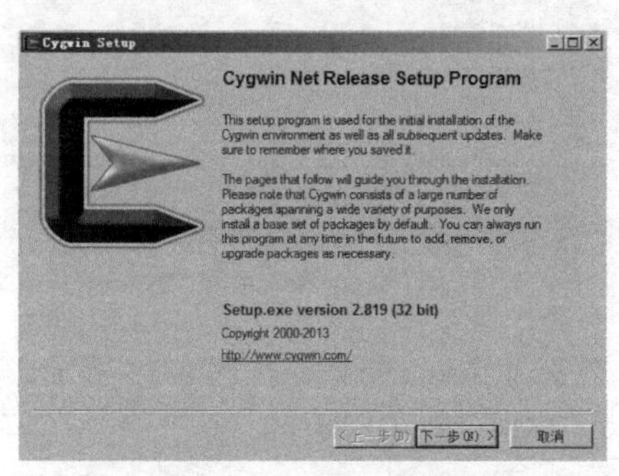

▲图 3-2　启动 cygwin

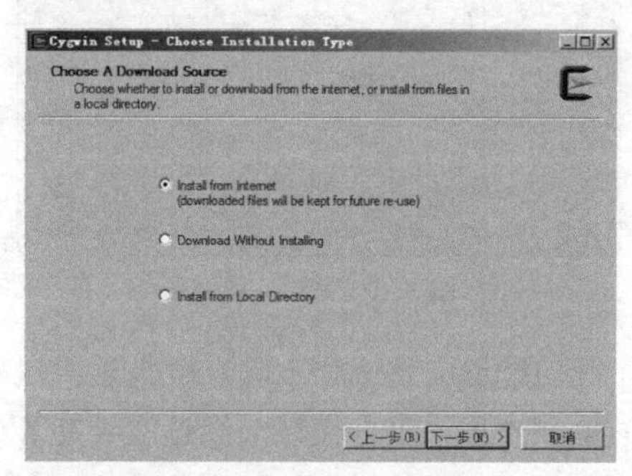

▲图 3-3　选择从网络下载安装

（3）单击“下一步”按钮，选择安装根目录，如图 3-4 所示。

（4）单击“下一步”按钮，选择临时文件目录，如图 3-5 所示。

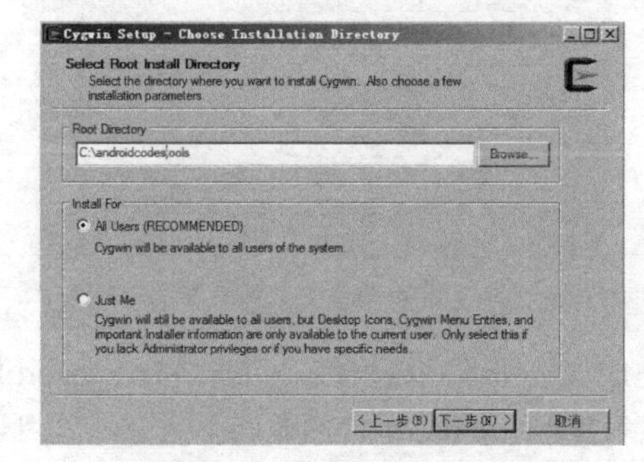

▲图 3-4　选择安装根目录

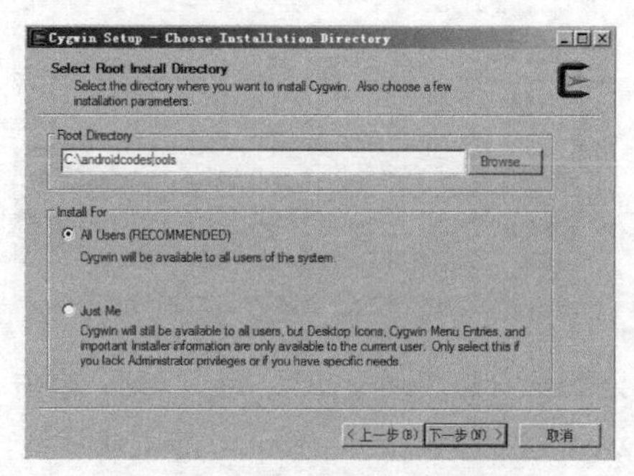

▲图 3-5　选择临时文件目录

（5）单击“下一步”按钮，设置网络代理。如果所在网络需要代理，则在这一步进行设置，如果不用代理，则选择直接下载，如图 3-6 所示。

（6）单击“下一步”按钮，选择下载站点。一般选择比较近的站点，速度会比较快，如图 3-7 所示。

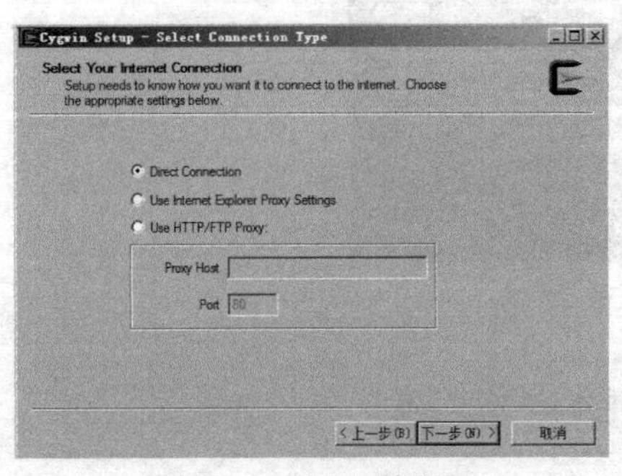

▲图 3-6　设置网络代理

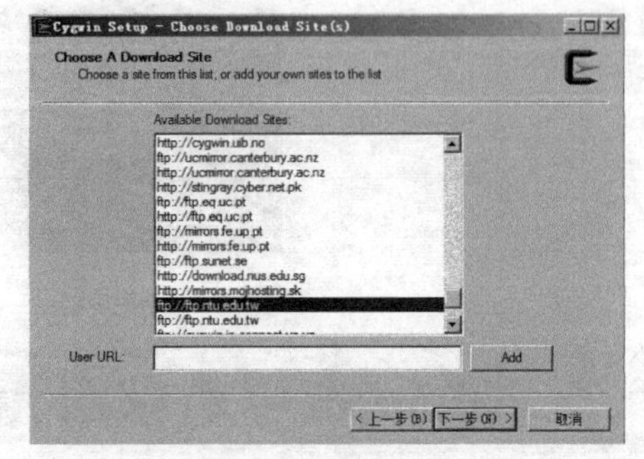

▲图 3-7　选择下载站点

（7）单击“下一步”按钮，开始更新工具列表，如图 3-8 所示。

（8）单击“下一步”按钮，选择需要下载的工具包。在此需要依次下载 curl、git 和 python 这些工具。如图 3-9 所示。

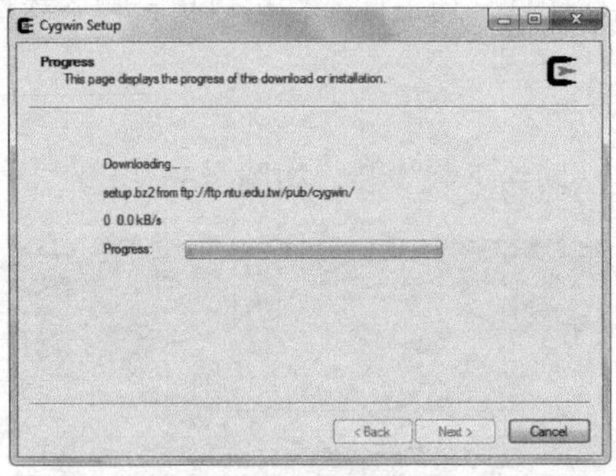

▲图 3-8　更新工具列表

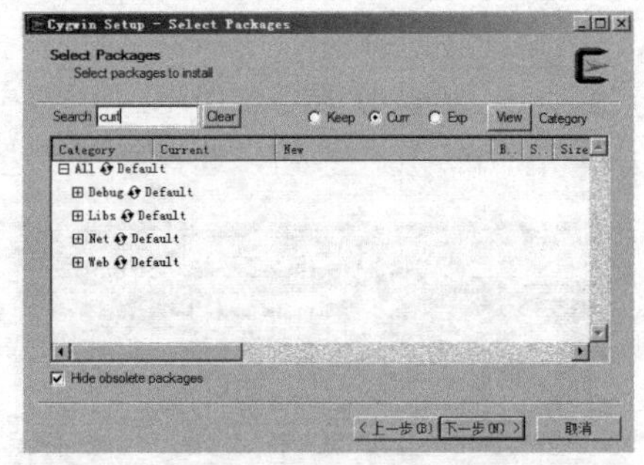

▲图 3-9　依次下载工具

为了确保能够安装上述工具，一定要用鼠标双击变为 Install 形式，如图 3-10 所示。

（9）单击“下一步”按钮，经过漫长的等待过程，如图 3-11 所示。

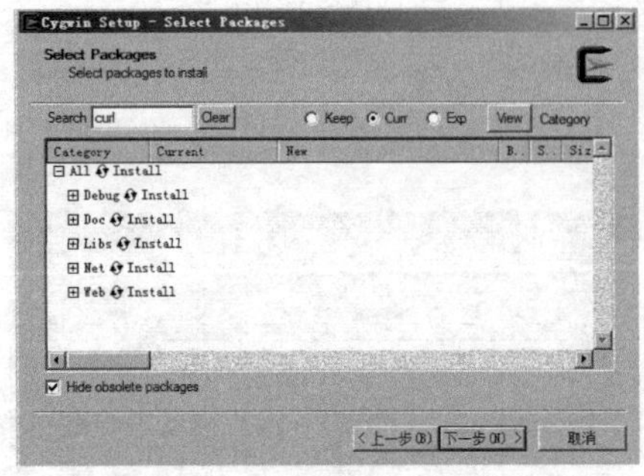

▲图 3-10　务必设置为 Install 形式

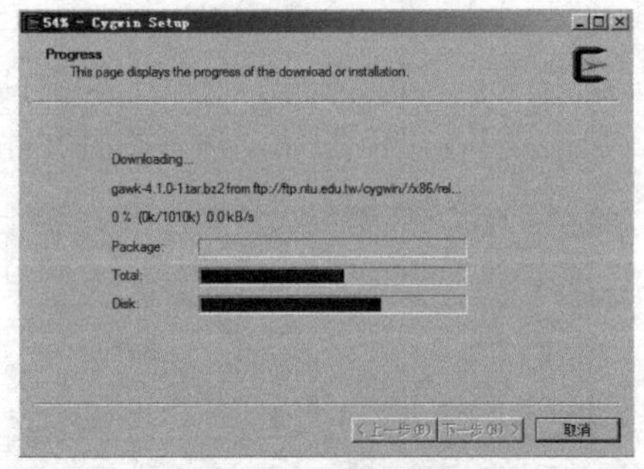

▲图 3-11　下载进度条

如果下载安装成功会出现提示信息，单击“完成”按钮即完成安装。当安装好 cygwin 后，打开 cygwin，会模拟出一个 Linux 的工作环境，然后按照 Linux 平台的源码下载方法就可以下载

Android 源码了。

建议读者在下载 Android 源码时，严格按照官方提供的步骤进行，地址是：http://source.android.com/source/downloading.html，这一点对初学者来说尤为重要。另外，整个下载过程比较漫长，需要大家耐心的等待。例如图 3-12 所示是下载 Android 4.4 时的机器命令截图。

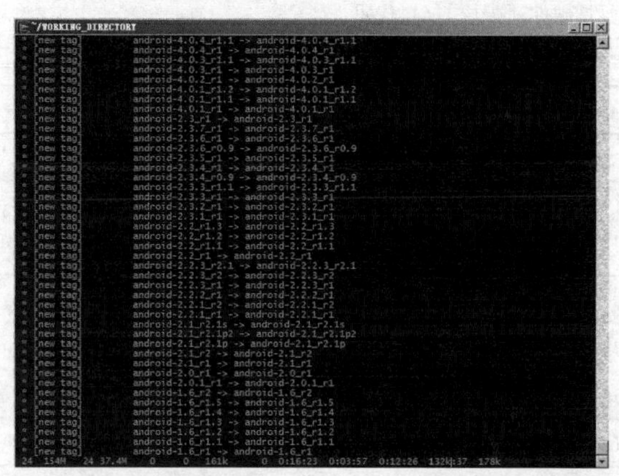

▲图 3-12　在 Windows 中用 cygwin 工具下载 Android 源码的截图

3.3 分析 Android 源码结构

获得 Android 源码后，可以将源码的全部工程分为如下 3 个部分。

- Core Project：核心工程部分，这是建立 Android 系统的基础，被保存在根目录的各个文件夹中。
- External Project：扩展工程部分，可以使其他开源项目具有扩展功能，被保存在“external”文件夹中。
- Package：包部分，提供了 Android 的应用程序、内容提供者、输入法和服务，被保存在“package”文件夹中。

无论是 Android 1.5 还是 Android 4.4，各个版本的源码目录基本类似。在里面包含了原始 Android 的目标机代码、主机编译工具和仿真环境。解压缩下载的 Android 4.4 源码包后，第一级别目录结构的具体说明如表 3-1 所示。

表 3-1　Android 4.4 源码的根目录

Android 源码根目录	描　述
abi	abi 相关代码，abi:application binary interface，应用程序二进制接口
bionic	bionic C 库
bootable	启动引导相关代码
build	存储系统编译规则及 generic 等基础开发配置包
cts	Android 兼容性测试套件标准
dalvik	dalvik Java 虚拟机
development	应用程序开发相关
device	设备相关代码
docs	介绍开源的相关文档
external	android 使用的一些开源的模组

续表

Android 源码根目录	描　　述
frameworks	核心框架——Java 及 C++语言，是 Android 应用程序的框架
gdk	即时通信模块
hardware	主要是硬件适配层 HAL 代码
kernel	Linux 的内核文件
libcore	核心库相关
libnativehelper	是 Support functions for Android's class libraries 的缩写，表示动态库，是实现 JNI 库的基础
ndk	ndk 相关代码。Android NDK（Android Native Development Kit）是一系列的开发工具，允许程序开发人员在 Android 应用程序中嵌入 C/C++语言编写的非托管代码
out	编译完成后的代码输出在此目录
packages	应用程序包
pdk	Plug Development Kit 的缩写，是本地开发套件
prebuilts	x86 和 arm 架构下预编译的一些资源
sdk	sdk 及模拟器
system	文件系统和应用及组件，是用 C 语言实现的
tools	工具文件夹
vendor	厂商定制代码
Makefile	全局的 Makefile

3.3.1　应用程序

应用程序主要是 UI 界面的实现，广大开发者基于 SDK 上开发的一个个独立的 APK 包，都是属于应用程序这一层的，应用程序在 Android 系统中处于最上层的位置。源码结构中的 packages 目录用于实现系统的应用程序，packages 的目录结构如下所示。

```
packages /
├── apps //应用程序库
│    ├── BasicSmsReceiver           //基础短信接收
│    ├── Bluetooth                  //蓝牙
│    ├── Browser                    //浏览器
│    ├── Calculator                 //计算器
│    ├── Calendar                   //日历
│    ├── Camera                     //照相机
│    ├── CellBroadcastReceiver      //单元广播接收
│    ├── CertInstaller              //被调用的包，在 Android 中安装数字签名
│    ├── Contacts                   //联系人
│    ├── DeskClock                  //桌面时钟
│    ├── Email                      //电子邮件
│    ├── Exchange                   //Exchange 服务
│    ├── Gallery                    //图库
│    ├── Gallery2                   //图库 2
│    ├── HTMLViewer                 //HTML 查看器
│    ├── KeyChain                   //密码管理
│    ├── Launcher2                  //启动器 2
│    ├── Mms                        //彩信
│    ├── Music                      //音乐
│    ├── MusicFX                    //音频增强
│    ├── Nfc                        //近场通信
│    ├── PackageInstaller           //包安装器
│    ├── Phone                      //电话
│    ├── Protips                    //主屏幕提示
│    ├── Provision                  //引导设置
```

```
│   ├── QuickSearchBox                 //快速搜索框
│   ├── Settings                       //设置
│   ├── SoundRecorder                  //录音机
│   ├── SpareParts                     //系统设置
│   ├── SpeechRecorder                 //录音程序
│   ├── Stk                            //sim 卡相关
│   ├── Tag                            //标签
│   ├── VideoEditor                    //视频编辑
│   └── VoiceDialer                    //语音编号
├── experimental                       //非官方的应用程序
│   ├── BugReportSender                //bug 的报告程序
│   ├── Bummer
│   ├── CameraPreviewTest              //照相机预览测试车程序
│   ├── DreamTheater
│   ├── ExampleImsFramework
│   ├── LoaderApp
│   ├── NotificationLog
│   ├── NotificationShowcase
│   ├── procstatlog
│   ├── RpcPerformance
│   └── StrictModeTest
├── inputmethods                       //输入法
│   ├── LatinIME                       //拉丁文输入法
│   ├── OpenWnn                        //OpenWnn 输入法
│   └── PinyinIME                      //拼音输入法
├── providers                          //提供器
│   ├── ApplicationsProvider           //应用程序提供器，提供应用程序所需要的界面
│   ├── CalendarProvider               //日历提供器
│   ├── ContactsProvider               //联系人提供器
│   ├── DownloadProvider               //下载管理提供器
│   ├── DrmProvider                    //数据库相关
│   ├── GoogleContactsProvider         //Google 联系人提供器
│   ├── MediaProvider                  //媒体提供器
│   ├── TelephonyProvider              //彩信提供器
│   └── UserDictionaryProvider         //用户字典提供器
├── screensavers                       //屏幕保护
│   ├── Basic                          //基本屏幕保护
│   ├── PhotoTable                     //照片方格
│   ├── WebView                        //网页
└── wallpapers                         //墙纸
├── Basic                              //系统内置墙纸
├── Galaxy4                            //S4 内置墙纸
├── HoloSpiral                         //手枪皮套墙纸
├── LivePicker
├── MagicSmoke
├── MusicVisualization
├── NoiseField
└── PhaseBeam
```

通过上面的目录结构可以看出，package 目录主要存储的是 Android 系统应用层相关的内容，包括应用程序相关的包或者资源文件，其中既包括系统自带的应用程序，又有第三方开发的应用程序，还有屏幕保护和墙纸等应用，所以源码中 package 目录对应着系统的应用层。

3.3.2 应用程序框架

应用程序框架是 Android 系统中的核心部分，也就是 SDK 部分，它会提供接口给应用程序使用，同时应用程序框架又会和系统服务、系统程序库、硬件抽象层有关联，所以其作用十分重要，应用程序框架的实现代码大部分都在/frameworks/base 和/frameworks/av 目录下，/frameworks/base 的目录结构如下所示。

```
frameworks/base
├── api            //全是 XML 文件，定义了 API
├── cmds           //Android 中的重要命令（am、app_proce 等）
├── core           //核心库
├── data           //声音字体等数据文件
```

```
├── docs            //文档
├── drm             //数字版权管理
├── graphics        //图形图像
├── icu4j           //用于解决国际化问题
├── include         //头文件
├── keystore        //数字签名证书相关
├── libs            //库
├── location        //地理位置
├── media           //多媒体
├── native          //本地库
├── nfc-extras      //NFC 相关
├── obex            //蓝牙传输
├── opengl          //opengl 相关
├── packages        //设置、TTS、VPN 程序
├── policy          //锁屏界面相关
├── sax             //XML 解析器
├── services        //Android 的服务
├── telephony       //电话相关
├── test-runner     //测试相关
├── tests           //测试相关
├── tools           //工具
├── voip            //可视通话
└── wifi            //无线网络
```

以上这些文件夹包含了应用程序框架层的大部分代码，正是这些目录下的文件构成了 Android 的应用程序框架层，提供接口给应用程序调用，同时衔接系统程序库和硬件抽象层，形成一个由上至下的调用过程。在/frameworks/base 目录下也涉及系统服务、程序库中的一些代码，将在后面的两个小节中会详细分析。

3.3.3　系统服务

在 3.3.2 中介绍了应用程序框架层的内容，了解到大部分的实现代码保存在"/frameworks/base"目录下。其实在这个目录中还有一个名为"service"的目录，里面的代码是用于实现 Android 系统服务的。接下来将详细介绍 service 目录下的内容，其目录结构如下所示。

```
frameworks/base/services
├── common_time          //时间日期相关的服务
├── input                //输入系统服务
├── java                 //其他重要服务的 Java 层
├── jni                  //其他重要服务的 JNI 层
└── tests                //测试相关
```

其中 java 和 jni 两个目录分别是一些其他的服务的 Java 层和 JNI 层实现，java 目录下更详细的目录结构以及其他 Android 系统服务的说明如下所示。

```
frameworks/base/services/java/com/android/server
├── accessibility
├── am
├── connectivity
├── display
├── dreams
├── drm
├── input
├── location
├── net
├── pm
├── power
├── updates
├── usb
├── wm
├── AlarmManagerService.java          //闹钟服务
├── AppWidgetService.java             //应用程序小工具服务
├── AppWidgetServiceImpl.java
├── AttributeCache.java
├── BackupManagerService.java         //备份服务
```

```
├── BatteryService.java                   //电池相关服务
├── BluetoothManagerService.java          //蓝牙
├── BootReceiver.java
├── BrickReceiver.java
├── CertBlacklister.java
├── ClipboardService.java
├── CommonTimeManagementService.java      //时间管理服务
├── ConnectivityService.java
├── CountryDetectorService.java
├── DevicePolicyManagerService.java
├── DeviceStorageMonitorService.java      //设备存储器监听服务
├── DiskStatsService.java                 //磁盘状态服务
├── DockObserver.java                     //底座监视服务
├── DropBoxManagerService.java
├── EntropyMixer.java
├── EventLogTags.logtags
├── INativeDaemonConnectorCallbacks.java
├── InputMethodManagerService.java        //输入法管理服务
├── IntentResolver.java
├── IntentResolverOld.java
├── LightsService.java
├── LocationManagerService.java           //地理位置服务
├── MasterClearReceiver.java
├── MountService.java                     //挂载服务
├── NativeDaemonConnector.java
├── NativeDaemonConnectorException.java
├── NativeDaemonEvent.java
├── NetworkManagementService.java         //网络管理服务
├── NetworkTimeUpdateService.java
├── NotificationManagerService.java       //通知服务
├── NsdService.java
├── PackageManagerBackupAgent.java
├── PreferredComponent.java
├── ProcessMap.java
├── RandomBlock.java
├── RecognitionManagerService.java
├── SamplingProfilerService.java
├── SerialService.java                    //NFC 相关
├── ServiceWatcher.java
├── ShutdownActivity.java
├── StatusBarManagerService.java          //状态栏管理服务
├── SystemBackupAgent.java
├── SystemServer.java
├── TelephonyRegistry.java
├── TextServicesManagerService.java
├── ThrottleService.java
├── TwilightCalculator.java
├── TwilightService.java
├── UiModeManagerService.java
├── UpdateLockService.java                //锁屏更新服务
├── VibratorService.java                  //震动服务
├── WallpaperManagerService.java          //壁纸服务
├── Watchdog.java                         //看门狗
├── WifiService.java                      //无线网络服务
└── WiredAccessoryManager.java            //无线设备管理服务
```

从上面的文件夹和文件可以看出，Android 中涉及的服务种类非常多，包括界面、网络、电话等核心模块基本上都有其专属的服务，这些是属于系统级别的服务，这些系统服务一般都会在 Android 系统启动的时候加载，在系统关闭的时候结束，受到系统的管理，应用程序并没有权力去打开或者关闭，它们会随着系统的运行一直在后台运行，供应用程序和其他的组件来使用。

另外在 frameworks/av/下面也有一个 services 目录，这个目录下存储的是音频和照相机的服务的实现代码，目录结构如下所示。

```
frameworks/av/services
├── audioflinger            //音频管理服务
└── camera                  //照相机的管理服务
```

这个 av/services 目录下的文件主要是用于支持 Android 系统中的音频和照相机服务的，这是两个非常重要的系统服务，在开发应用程序时会经常依赖这两个服务的。

3.3.4　系统程序库

Android 的系统程序库类型非常多，功能也非常强大，正是有了这些程序库，Android 系统才能运行多种多样的应用程序。在接下来的内容中，本书选择了一些很常用的也是很重要的系统程序库来分析它们在源码中所处的位置。

1. 系统 C 库

Android 系统采用的是一个从 BSD 继承而来的标准的系统函数库 bionic，在源码根目录下有这个文件夹，其目录结构如下所示。

```
bionic/
├── libc             //C 库
├── libdl            //动态链接库相关
├── libm             //数学库
├── libstdc++        //C++实现库
├── libthread_db     //线程库
├── linker           //连接器相关
└── test             //测试相关
```

2. 媒体库

Android 中的媒体库在 2.3 之前是由 OpenCore 实现的，2.3 之后 Stragefright 被替换了，OpenCore 成为了新的多媒体的实现库。同时 Android 也自带了一些音视频的管理库，用于管理多媒体的录制、播放、编码和解码等功能。Android 的多媒体程序库的实现代码主要在/frameworks/av/media 目录下，其目录结构如下所示。

```
frameworks/av/media/
├── common_time              //时间相关
├── libeffects               //多媒体效果
├── libmedia                 //多媒体录制、播放
├── libmedia_native          //里面只有一个 Android.mk，用来编译 native 文件
├── libmediaplayerservice    //多媒体播放服务的实现库
├── libstagefright           //stagefright 的实现库
├── mediaserver              //跨进程多媒体服务
└── mtp                      //mtp 协议的实现（媒体传输协议）
```

3. 图层显示库

Android 中的图层显示库主要负责对显示子系统的管理，负责图层的渲染、叠加、绘制等功能，提供了 2D 和 3D 图层的无缝融合，是整个 Android 系统显示的“大脑中枢”，其代码在/frameworks/native/services/surfaceflinger/目录下，其目录结构如下所示。

```
frameworks/native/services/surfaceflinger/
├── DisplayHardware          //显示底层相关
├── tests                    //测试
├── Android.mk               //MakeFile 文件
├── Barrier.h
├── Client.cpp               //显示的客户端实现文件
├── Client.h
├── clz.cpp
├── clz.h
├── DdmConnection.cpp
├── DdmConnection.h
├── DisplayDevice.cpp        //显示设备相关
├── DisplayDevice.h
├── EventThread.cpp          //消息线程
```

```
├── EventThread.h
├── GLExtensions.cpp          //opengl 扩展
├── GLExtensions.h
├── Layer.cpp                 //图层相关
├── Layer.h
├── LayerBase.cpp             //图层基类
├── LayerBase.h
├── LayerDim.cpp              //图层相关
├── LayerDim.h
├── LayerScreenshot.cpp       //图层相关
├── LayerScreenshot.h
├── MessageQueue.cpp          //消息队列
├── GLExtensions.h
├── MessageQueue.h
├── MODULE_LICENSE_APACHE2    //证书
├── SurfaceFlinger.cpp        //图层管理者，图层管理的核心类
├── SurfaceFlinger.h
├── SurfaceTextureLayer.cpp   //文字图层
├── SurfaceTextureLayer.h
├── Transform.cpp
└── Transform.h
```

4. 网络引擎库

网络引擎库主要是用来实现 Web 浏览器的引擎，支持 Android 的 Web 浏览器和一个可嵌入的 Web 视图，这个是采用第三方开发的浏览器引擎 Webkit 实现的，Webkit 的代码在/external/webkit/目录下，其目录结构如下所示。

```
external/webkit/
├── Examples                  //Webkit 例子
├── LayoutTests               //布局测试
├── PerformanceTests          //表现测试
├── Source                    //Webkit 源代码
├── Tools                     //工具
├── WebKitLibraries           //Webkit 用到的库
├── Android.mk                //Makefile
├── bison_check.mk
├── CleanSpec.mk
├── MODULE_LICENSE_LGPL       //证书
├── NOTICE
└── WEBKIT_MERGE_REVISION     //版本信息
```

5. 3D 图形库

Android 中的 3D 图形渲染是采用 Opengl 来实现的，Opengl 是开源的第三方图形渲染库，使用该库可以实现 Android 中的 3D 图形硬件加速或者 3D 图形软件加速功能，是一个非常重要的功能库。从 Android 4.3 开始，支持最新、最强大的 OpenGL ES 3.0。其实现代码在/frameworks/native/opengl 中，其目录结构如下所示。

```
frameworks/native/opengl/
├── include                   //Opengl 中的头文件
├── libagl                    //在 mac os 上的库
├── libs                      //Opengl 的接口和实现库
├── specs                     //Opengl 的文档
├── tests                     //测试相关
└── tools                     //工具库
```

6. SQLite

SQLite 是 Android 系统自带的一个轻量级关系数据库，其实现源代码已经在网上开源。SQLite 的具有操作简单方便、运行速度较快和占用资源较少等特点，比较适合在嵌入式设备上面使用。SQLite 是 Android 系统自带的实现数据库功能的核心库，其代码实现分为 Java 和 C 两个部分，Java 部分的代码在/frameworks/base/core/java/android/database，目录结构如下所示。

```
frameworks/base/core/java/android/database/
├── sqlite                                    //SQLite 的框架文件
├── AbstractCursor.java                       //游标的抽象类
├── AbstractWindowedCursor.java
├── BulkCursorDescriptor.java
├── BulkCursorNative.java
├── BulkCursorToCursorAdaptor.java            //游标适配器
├── CharArrayBuffer.java
├── ContentObservable.java
├── ContentObserver.java                      //内容观察者
├── CrossProcessCursor.java
├── CrossProcessCursorWrapper.java            //CrossProcessCursor 的封装类
├── Cursor.java                               //游标实现类
├── CursorIndexOutOfBoundsException.java      //游标出界异常
├── CursorJoiner.java
├── CursorToBulkCursorAdaptor.java            //适配器
├── CursorWindow.java                         //游标窗口
├── CursorWindowAllocationException.java      //游标窗口异常
├── CursorWrapper.java                        //游标封装类
├── DatabaseErrorHandler.java                 //数据库错误句柄
├── DatabaseUtils.java                        //数据库工具类
├── DataSetObservable.java
├── DataSetObserver.java
├── DefaultDatabaseErrorHandler.java          //默认数据库错误句柄
├── IBulkCursor.java
├── IContentObserver.aidl                     //aidl 用于跨进程通信
├── MatrixCursor.java
├── MergeCursor.java
├── Observable.java
├── package.html
├── SQLException.java                         //数据库异常
└── StaleDataException.java
```

Java 层的代码主要是实现 SQLite 的框架和接口，方便用户开发应用程序的时候能很简单地操作数据库，并且捕获数据库异常。

C++层的代码在/external/sqlite 路径下，其目录结构如下所示。

```
external/sqlite/
├── android          //Android 数据库的一些工具包
└── dist             //Android 数据库底层实现
```

从上面 Java 和 C 部分的代码目录结构可以看出，SQLite 在 Android 中还是有很重要的地位的，并且在 SDK 中会有开放的接口让应用程序可以很简单方便地操作数据库，例如对数据进行存储和删除。

3.3.5　系统运行库

众所周知，Android 系统的应用层是采用 Java 开发的，由于 Java 语言的跨平台特性，Java 代码必须运行在虚拟机中。正是因为这个特性，Android 系统自己也实现了一个类似 JVM 但是更适用于嵌入式平台的 Java 虚拟机，这被称为 dalvik。

dalvik 功能等同于 JVM，为 Android 平台上的 Java 代码提供了运行环境，dalvik 本身是由 C++语言实现的，在源码中根目录下有 dalvik 文件夹，里面存储的是 dalvik 虚拟机的实现代码，其目录结构如下所示。

```
./
├── dalvikvm                 //入口目录
├── dexdump                  //dex 反汇编
├── dexgen                   //dex 生成相关
├── dexlist                  //dex 列表
├── dexopt                   //与验证和优化
├── docs                     //文档
├── dvz                      //zygot 相关
├── dx                       //dx 工具，将多个 Java 转换为 dex
```

```
├── hit
├── libdex                          //dex 库的实现代码
├── opcode-gen
├── tests                           //测试相关
├── tools                           //工具
├── unit-tests                      //测试相关
├── vm                              //虚拟机的实现
├── Android.mk                      //Makefile
├── CleanSpec.mk
├── MODULE_LICENSE_APACHE2
├── NOTICE
└── README.txt
```

正是有上面这些代码实现的 Android 虚拟机，所以应用程序生成的二进制执行文件能够快速、稳定地运行在 Android 系统上。

3.3.6 硬件抽象层

Android 的硬件抽象是各种功能的底层实现，理论上不同的硬件平台会有不同的硬件抽象层实现，这一个层次也是与驱动层和硬件层有紧密的联系的，起着承上启下的作用，对上要实现应用程序框架层的接口，对下要实现一些硬件基本功能以及调用驱动层的接口。需要注意的是，这一层也是广大 OEM 厂商改动最大的一层，因为这一层的代码跟终端采用什么样硬件的硬件平台有很大关系。源码中存储的是硬件抽象层框架的实现代码和一些平台无关性的接口的实现。硬件抽象层代码存储在源码根目录下的 hardware 文件夹中，其目录结构如下所示。

```
hardware/
├── libhardware                 //新机制硬件库
├── libhardware_legacy          //旧机制硬件库
└── ril                         //ril 模块相关的底层实现
```

从上面的目录结构可以看出，硬件抽象层中主要是实现了一些底层的硬件库，用来实现应用层框架层中的功能，具体硬件库中有哪些内容，可以继续细分其目录结构，例如 libhardware 目录下的结构如下。

```
hardware/libhardware/
├── include                     //入口目录
├── modules                     //dex 反汇编
│   ├── audio                   //音频相关底层库
│   ├── audio_remote_submix     //音频混合相关
│   ├── gralloc                 //帧缓冲
│   ├── hwcomposer              //音频相关
│   ├── local_time              //本地时间
│   ├── nfc                     //nfc 功能
│   ├── nfc-nci                 //nfc 的接口
│   ├── power                   //电源
│   ├── usbaudio                //USB 音频设备
│   ├── Android.mk              //Makefile
│   └── README.android
├── tests                       //dex 生成相关
├── dexlist                     //dex 列表
├── dexopt                      //与验证和优化
└── docs                        //文档
```

从上面的目录结构可以分析出，libhardware 目录主要是 Android 系统的某些功能的底层实现，包括 audio、nfc 和 power。

libhardware_legacy 的目录与 libhardware 大同小异，只是针对旧的实现方式做的一套硬件库，其目录下还有 uevent、wifi 以及虚拟机的底层实现。这两个目录下的代码一般会由设备厂家根据自身的硬件平台来实现符合 Android 机制的硬件库。

ril 目录下存储的是无线硬件设备与电话的实现，其目录结构如下所示。

```
hardware/ril/
```

```
├── include                    //头文件
├── libril                     //libril 库
├── mock-ril
├── reference-ril              //reference ril 库
├── rild                       //ril 守护进程
└── CleanSpec.mk
```

3.4 编译源码

编译 Android 源码的方法非常简单，只需要使用 Android 源码根目录下的 Makefile，执行 make 命令即可轻松实现。当然在编译 Android 源码之前，首先要确定已经完成同步工作。进入 Android 源码目录使用 make 命令进行编译，使用此命令的格式如下所示。

```
$: cd ~/Android4.3（这里的“Android4.3”就是下载源码的保存目录）
$: make
```

编译 Android 源码可以得到“~/project/android/cupcake/out”目录。

整个编译过程是非常漫长的，需要读者耐心等待。

3.4.1 搭建编译环境

在编译 Android 源码之前，需要先进行环境搭建工作。在接下来的内容中，以 Ubuntu 系统为例讲解搭建编译环境以及编译 Android 源码的方法。具体流程如下所示。

（1）安装 JDK，编译 Android 4.3 的源码需要 JDK1.6，下载 jdk-6u23-linux-i586.bin 后进行安装，对应命令如下：

```
$ cd /usr
$ mkdir java
$ cd java
$ sudo cp jdk-6u23-linux-i586.bin 所在目录 ./
$ sudo chmod 755 jdk-6u23-linux-i586.bin
$ sudo sh jdk-6u23-linux-i586.bin
```

（2）设置 JDK 环境变量，将如下环境变量添加到主文件夹目录下的.bashrc 文件中，然后用 source 命令使其生效，加入的环境变量代码如下：

```
export JAVA_HOME=/usr/java/jdk1.6.0_23
export JRE_HOME=$JAVA_HOME/jre
export CLASSPATH=.:$JAVA_HOME/lib:$JRE_HOME/lib:$CLASSPATH
export PATH=$PATH:$JAVA_HOME/bin:$JAVA_HOME/bin/tools.jar:$JRE_HOME/bin
export ANDROID_JAVA_HOME=$JAVA_HOME
```

（3）安装需要的包，读者可以根据编译过程中的提示进行选择，可能需要的包的安装命令如下：

```
$ sudo apt-get install git-core bison zlib1g-dev flex libx13-dev gperf sudo aptitude
install git-core gnupg flex bison gperf libsdl-dev libesd0-dev libwxgtk2.6-dev
build-essential zip curl libncurses5-dev zlib1g-dev
```

3.4.2 开始编译

当所依赖的包安装完成之后，就可以开始编译 Android 源码了，具体步骤如下所示。

（1）首先进行编译初始化工作，在终端中执行下面的命令：

```
source build/envsetup.sh
```

或：

```
. build/envsetup.sh
```

执行后将会输出：

```
source build/envsetup.sh
```

```
including device/asus/grouper/vendorsetup.sh
including device/asus/tilapia/vendorsetup.sh
including device/generic/armv7-a-neon/vendorsetup.sh
including device/generic/armv7-a/vendorsetup.sh
including device/generic/mips/vendorsetup.sh
including device/generic/x86/vendorsetup.sh
including device/samsung/maguro/vendorsetup.sh
including device/samsung/manta/vendorsetup.sh
including device/samsung/toroplus/vendorsetup.sh
including device/samsung/toro/vendorsetup.sh
including device/ti/panda/vendorsetup.sh
including sdk/bash_completion/adb.bash
```

（2）然后选择编译目标，命令是：

```
lunch full-eng
```

执行后会输出如下所示的提示信息：

```
============================================
PLATFORM_VERSION_CODENAME=REL
PLATFORM_VERSION=4.3
TARGET_PRODUCT=full
TARGET_BUILD_VARIANT=eng
TARGET_BUILD_TYPE=release
TARGET_BUILD_APPS=
TARGET_ARCH=arm
TARGET_ARCH_VARIANT=armv7-a
HOST_ARCH=x86
HOST_OS=linux
HOST_OS_EXTRA=Linux-2.6.33-45-generic-x86_64-with-Ubuntu-10.04-lucid
HOST_BUILD_TYPE=release
BUILD_ID=JOP40C
OUT_DIR=out
============================================
```

（3）接下来开始编译代码，在终端中执行下面的命令：

```
make -j4
```

其中“-j4”表示用4个线程进行编译。整个编译进度根据不同机器的配置而需要不同的时间。例如这里计算机为intel i5-2300四核2.8 GHz、4 GB内存，经过近4小时才编译完成。当出现下面的信息时表示编译完成：

```
target Java: ContactsTests (out/target/common/obj/APPS/ContactsTests_intermediates/
classes)
target Dex: Contacts
Done!
Install: out/target/product/generic/system/app/Browser.odex
Install: out/target/product/generic/system/app/Browser.apk
Note: Some input files use or override a deprecated API.
Note: Recompile with -Xlint:deprecation for details.
Copying: out/target/common/obj/APPS/Contacts_intermediates/noproguard.classes.dex
target Package: Contacts (out/target/product/generic/obj/APPS/Contacts_intermediates/
package.apk)
 'out/target/common/obj/APPS/Contacts_intermediates/classes.dex' as 'classes.dex'...
Processing target/product/generic/obj/APPS/Contacts_intermediates/package.apk
Done!
Install: out/target/product/generic/system/app/Contacts.odex
Install: out/target/product/generic/system/app/Contacts.apk
build/tools/generate-notice-files.py out/target/product/generic/obj/NOTICE.txt out/
target/product/generic/obj/NOTICE.html "Notices for files contained in the filesystem
images in this directory:" out/target/product/generic/obj/NOTICE_FILES/src
Combining NOTICE files into HTML
Combining NOTICE files into text
Installed file list: out/target/product/generic/installed-files.txt
Target system fs image: out/target/product/generic/obj/PACKAGING/systemimage_
intermediates/system.img
Running: mkyaffs2image -f out/target/product/generic/system out/target/product/generic/
obj/PACKAGING/systemimage_intermediates/system.img
```

```
Install system fs image: out/target/product/generic/system.img
DroidDoc took 5331 sec. to write docs to out/target/common/docs/doc-comment-check
```

3.4.3　在模拟器中运行

之后在模拟器中运行的步骤就比较简单了，只需要在终端中执行下面的命令即可：

```
emulator
```

运行成功后的效果如图 3-13 所示。

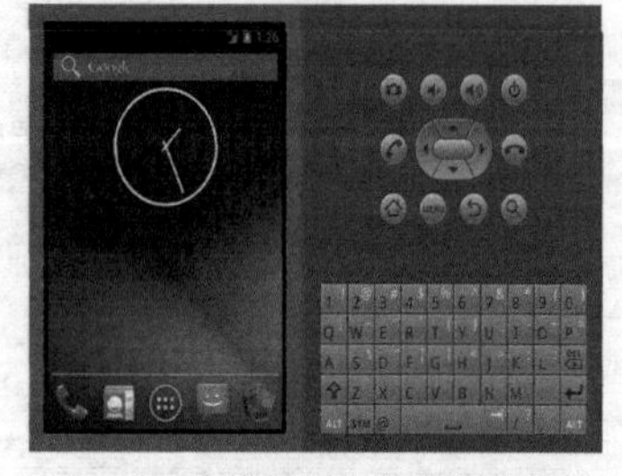

▲图 3-13　在模拟器中的编译执行效果

3.4.4　常见的错误分析

虽然编译方法非常简单，但是作为初学者来说很容易出错，在下面列出了其中常见的编译错误类型。

1. 缺少必要的软件

进入到 Android 目录下，使用 make 命令编译，可能会出现如下错误提示。

```
host C: libneo_cgi <= external/clearsilver/cgi/cgi.c
external/clearsilver/cgi/cgi.c:22:18: error: zlib.h: No such file or directory
```

上述错误是因为缺少 zlib1g-dev，需要使用 apt-get 命令从软件仓库中安装 zlib1g-dev，具体命令如下所示。

```
sudo apt-get install zlib1g-dev
```

同理需要安装下面的软件，否则也会出现上述类似的错误。

```
sudo apt-get install flex
sudo apt-get install bison
sudo apt-get install gperf
sudo apt-get install libsdl-dev
sudo apt-get install libesd0-dev
sudo apt-get install libncurses5-dev
sudo apt-get install libx13-dev
```

2. 没有安装 Java 环境 JDK

当安装所有上述软件后，运行 make 命令再次编译 Android 源码。如果在之前忘记安装 Java 环境 JDK，则此时会出现很多 Java 文件无法编译的错误，如果打开 Android 的源码，可以看到在下面目录中发现有很多 Java 源文件。

```
android/dalvik/libcore/dom/src/test/java/org/w3c/domts
```

这充分说明在编译 Android 之前必须先安装 Java 环境 JDK，安装流程如下所示。

- 从 Oracle 官方网站下载 jdk-6u16-linux-i586.bin 文件，然后安装。

在 Ubuntu 8.04 中，“/etc/profile”文件是全局的环境变量配置文件，它适用于所有的 shell。在登录 Linux 系统时应该先启动“/etc/profile”文件，然后再启动用户目录下的“~/.bash_profile”“~/.bash_login”或“~/.profile”文件中的其中一个，执行的顺序和上面的排序一样。如果“~/.bash_profile”文件存在话，则还会执行“~/.bashrc”文件。在此只需要把 JDK 的目录存储到“/etc/profile”目录下即可。

```
JAVA_HOME=/usr/local/src/jdk1.6.0_16
PATH=$PATH:$JAVA_HOME/bin:/usr/local/src/android-sdk-linux_x86-1.1_r1/tools:~/bin
```

- 重新启动机器，输入 java –version 命令，输出下面的信息则表示配置成功。

```
ava version "1.6.0_16"
Java(TM) SE Runtime Environment (build 1.6.0_16-b01)
Java HotSpot(TM) Client VM (build 13.3-b01, mixed mode, sharing)
```

当成功编译 Android 源码后，在终端会输出如下提示。

```
Target system fs image: out/target/product/generic/obj/PACKAGING/systemimage_unopt_
intermediates/system.img
Install system fs image: out/target/product/generic/system.img
Target ram disk: out/target/product/generic/ramdisk.img
Target userdata fs image: out/target/product/generic/userdata.img
Installed file list: out/target/product/generic/installed-files.txt
root@dfsun2009-desktop:/bin/android#
```

3.4.5 实践演练——演示两种编译 Android 程序的方法

Android 编译环境本身比较复杂，并且不像普通的编译环境那样只有顶层目录下才有 Makefile 文件，而其他的每个 component 都使用统一标准的 Android.mk 文件。不过这并不是人们熟悉的 Makefile，而是经过 Android 自身编译系统的很多处理。所以说要真正理清楚其中的联系还比较复杂，不过这种方式的好处在于，编写一个新的 Android.mk 给 Android 增加一个新的 component 会变得比较简单。为了使读者更加深入地理解在 Linux 环境下编译 Android 程序的方法，在接下来的内容中，将分别演示两种编译 Android 程序的方法。

1. 编译 Native C（本地 C 程序）的 helloworld 模块

编译 Java 程序可以直接采用 Eclipse 的集成环境来完成，实现方法非常简单，在这里就不再重复了。接下来将主要针对 C/C++进行说明，通过一个例子来讲解在 Android 中增加一个 C 程序的 Hello World 的方法。

（1）在“$(YOUR_ANDROID)/development”目录下创建一个名为“hello”的目录，并用“$(YOUR_ANDROID)”指向 Android 源代码所在的目录。

```
- # mkdir $(YOUR_ANDROID)/development/hello
```

（2）在目录“$(YOUR_ANDROID)/development/hello/”下编写一个名为“hello.c”的 C 语言文件，文件 hello.c 的实现代码如下所示。

```
#include <stdio.h>
int main()
{
    printf("Hello World!\n");//输出 Hello World
return 0;
}
```

（3）在目录“$(YOUR_ANDROID)/development/hello/”下编写 Android.mk 文件。这是 Android Makefile 的标准命名，不能更改。文件 Android.mk 的格式和内容可以参考其他已有的 Android.mk 文件的写法，针对 helloworld 程序的 Android.mk 文件内容如下所示。

```
LOCAL_PATH:= $(call my-dir)
include $(CLEAR_VARS)
LOCAL_SRC_FILES:= \
    hello.c
LOCAL_MODULE := helloworld
include $(BUILD_EXECUTABLE)
```

上述各个内容的具体说明如下所示。

- LOCAL_SRC_FILES：用来指定源文件用；
- LOCAL_MODULE：指定要编译的模块的名字，在下一步骤编译时将会用到；
- include $(BUILD_EXECUTABLE)：表示要编译成一个可执行文件，如果想编译成动态库则可用 BUILD_SHARED_LIBRARY，这些具体用法可以在“$(YOUR_ANDROID)/build/core/config.mk”查到。

（4）回到Android源代码顶层目录进行编译。

```
# cd $(YOUR_ANDROID) && make helloworld
```

在此需要注意，make helloworld中的目标名helloworld就是上面Android.mk文件中由LOCAL_MODULE指定的模块名。最终的编译结果如下所示。

```
target thumb C: helloworld <= development/hello/hello.c
target Executable: helloworld (out/target/product/generic/obj/EXECUTABLES/helloworld_
intermediates/LINKED/helloworld)
target  Non-prelinked:  helloworld  (out/target/product/generic/symbols/system/bin/
helloworld)
target  Strip:  helloworld  (out/target/product/generic/obj/EXECUTABLES/helloworld_
intermediates/helloworld)
Install: out/target/product/generic/system/bin/helloworld
```

（5）如果和上述编译结果相同，则编译后的可执行文件存储在如下目录：

```
out/target/product/generic/system/bin/helloworld
```

这样通过”adb push”将它传送到模拟器上，再通过”adb shell”登录到模拟器终端后就可以执行了。

2．手工编译C模块

在前面讲解了通过标准的Android.mk文件来编译C模块的具体流程，其实可以直接运用gcc命令行来编译C程序，这样可以更好地了解Android编译环境的细节。具体流程如下所示。

（1）在Android编译环境中，提供了”showcommands”选项来显示编译命令行，可以通过打开这个选项来查看一些编译时的细节。

（2）在具体操作之前需要使用如下命令把前面中的helloworld模块清除。

```
# make clean-helloworld
```

上面的“make clean-$(LOCAL_MODULE)”命令是Android编译环境提供的make clean的方式。

（3）使用showcommands选项重新编译helloworld，具体命令如下所示。

```
# make helloworld showcommands
build/core/product_config.mk:229: WARNING: adding test OTA key
target thumb C: helloworld <= development/hello/hello.c
prebuilt/linux-x86/toolchain/arm-eabi-4.3.1/bin/arm-eabi-gcc  -I system/core/include
-I hardware/libhardware/include  -I hardware/ril/include  -I dalvik/libnativehelper/
include  -I frameworks/base/include  -I external/skia/include  -I out/target/product/
generic/obj/include  -I bionic/libc/arch-arm/include  -I bionic/libc/include  -I
bionic/libstdc++/include  -I bionic/libc/kernel/common  -I bionic/libc/kernel/arch-
arm  -I bionic/libm/include  -I bionic/libm/include/arch/arm  -I bionic/libthread_
db/include  -I development/hello  -I out/target/product/generic/obj/EXECUTABLES/
helloworld_intermediates  -c  -fno-exceptions -Wno-multichar -march=armv5te -mtune=
xscale  -msoft-float  -fpic  -mthumb-interwork  -ffunction-sections  -funwind-tables
-fstack-protector -D__ARM_ARCH_5__  -D__ARM_ARCH_5T__  -D__ARM_ARCH_5E__  -D__ARM_ARCH_
5TE__ -include system/core/include/arch/linux-arm/AndroidConfig.h -DANDROID -fmessage-
length=0 -W -Wall -Wno-unused -DSK_RELEASE -DNDEBUG -O2 -g -Wstrict-aliasing=2 -finline-
functions -fno-inline-functions-called-once -fgcse-after-reload -frerun-cse-after-loop
-frename-registers  -DNDEBUG  -UDEBUG  -mthumb  -Os  -fomit-frame-pointer  -fno-strict-
aliasing  -finline-limit=64   -MD  -o  out/target/product/generic/obj/EXECUTABLES/
helloworld_intermediates/hello.o development/hello/hello.c

target Executable: helloworld (out/target/product/generic/obj/EXECUTABLES/helloworld_
intermediates/LINKED/helloworld)

prebuilt/linux-x86/toolchain/arm-eabi-4.3.1/bin/arm-eabi-g++ -nostdlib -Bdynamic -Wl,
-T,build/core/armelf.x -Wl,-dynamic-linker,/system/bin/linker -Wl,--gc-sections -Wl,-z,
nocopyreloc  -o  out/target/product/generic/obj/EXECUTABLES/helloworld_ intermediates/
LINKED/helloworld  -Lout/target/product/generic/obj/lib  -Wl,-rpath-link= out/target/
product/generic/obj/lib -lc -lstdc++ -lm  out/target/product/generic/obj/ lib/crtbegin_
dynamic.o out/target/product/generic/obj/EXECUTABLES/helloworld_ intermediates/hello.o
```

```
-Wl,--no-undefined prebuilt/linux-x86/toolchain/arm-eabi-4.3.1/ bin/../lib/gcc/arm-eabi/4.3.1/interwork/libgcc.a out/target/product/generic/obj/lib/crtend_android.o

target Non-prelinked: helloworld (out/target/product/generic/symbols/system/bin/helloworld)

out/host/linux-x86/bin/acp -fpt out/target/product/generic/obj/EXECUTABLES/helloworld_intermediates/LINKED/helloworld out/target/product/generic/symbols/system/bin/helloworld

target Strip: helloworld (out/target/product/generic/obj/EXECUTABLES/helloworld_intermediates/helloworld)

out/host/linux-x86/bin/soslim --strip --shady --quiet out/target/product/generic/symbols/system/bin/helloworld --outfile out/target/product/generic/obj/EXECUTABLES/helloworld_intermediates/helloworld

Install: out/target/product/generic/system/bin/helloworld

out/host/linux-x86/bin/acp -fpt out/target/product/generic/obj/EXECUTABLES/helloworld_intermediates/helloworld out/target/product/generic/system/bin/helloworld
```

从上述命令行可以看到，Android 编译环境所用的交叉编译工具链如下所示。

```
prebuilt/linux-x86/toolchain/arm-eabi-4.3.1/bin/arm-eabi-gcc
```

其中参数“-I”和“-L”分别指定了所用的C库头文件和动态库文件路径，分别是“bionic/libc/include”和“out/target/product/generic/obj/lib”，其他还包括很多编译选项以及-D所定义的预编译宏。

（4）此时可以利用上面的编译命令手工编译 helloworld 程序，首先手工删除上次编译得到的 helloworld 程序。

```
# rm out/target/product/generic/obj/EXECUTABLES/helloworld_intermediates/hello.o
# rm out/target/product/generic/system/bin/helloworld
```

然后再用 gcc 编译以生成目标文件。

```
# prebuilt/linux-x86/toolchain/arm-eabi-4.3.1/bin/arm-eabi-gcc -I bionic/libc/arch-arm/include -I bionic/libc/include -I bionic/libc/kernel/common -I bionic/libc/kernel/arch-arm -c -fno-exceptions -Wno-multichar -march=armv5te -mtune=xscale -msoft-float -fpic -mthumb-interwork -ffunction-sections -funwind-tables -fstack-protector -D__ARM_ARCH_5__ -D__ARM_ARCH_5T__ -D__ARM_ARCH_5E__ -D__ARM_ARCH_5TE__ -include system/core/include/arch/linux-arm/AndroidConfig.h -DANDROID -fmessage-length=0 -W -Wall -Wno-unused -DSK_RELEASE -DNDEBUG -O2 -g -Wstrict-aliasing=2 -finline-functions -fno-inline-functions-called-once -fgcse-after-reload -frerun-cse-after-loop -frename-registers -DNDEBUG -UDEBUG -mthumb -Os -fomit-frame-pointer -fno-strict-aliasing -finline-limit=64 -MD -o out/target/product/generic/obj/ EXECUTABLES/helloworld_intermediates/hello.o development/hello/hello.c
```

如果此时与 Android.mk 编译参数进行比较，会发现上面主要减少了不必要的参数“-I”。

（5）接下来开始生成可执行文件。

```
# prebuilt/linux-x86/toolchain/arm-eabi-4.3.1/bin/arm-eabi-gcc -nostdlib -Bdynamic -Wl,-T,build/core/armelf.x -Wl,-dynamic-linker,/system/bin/linker -Wl,--gc-sections -Wl,-z,nocopyreloc -o out/target/product/generic/obj/EXECUTABLES/helloworld_ intermediates/LINKED/helloworld -Lout/target/product/generic/obj/lib -Wl,-rpath-link=out/target/product/generic/obj/lib -lc -lm out/target/product/generic/obj/EXECUTABLES/helloworld_intermediates/hello.o out/target/product/generic/obj/lib/ crtbegin_dynamic.o -Wl,--no-undefined ./prebuilt/linux-x86/toolchain/arm-eabi-4.3.1/ bin/../lib/gcc/arm-eabi/4.3.1/interwork/libgcc.a out/target/product/generic/obj/lib/crtend_android.o
```

在此需要特别注意的是参数“-Wl,-dynamic-linker,/system/bin/linker”，它指定了 Android 专用的动态链接器是“/system/bin/linker”，而不是平常使用的“ld.so”。

（6）最后可以使用命令 file 和 readelf 来查看生成的可执行程序。

```
# file out/target/product/generic/obj/EXECUTABLES/helloworld_intermediates/LINKED/helloworld
out/target/product/generic/obj/EXECUTABLES/helloworld_intermediates/LINKED/helloworl
```

```
d: ELF 33-bit LSB executable, ARM, version 1 (SYSV), dynamically linked (uses shared libs),
not stripped
#   readelf  -d  out/target/product/generic/obj/EXECUTABLES/helloworld_intermediates/
LINKED/helloworld |grep NEEDED
 0x00000001 (NEEDED)                     Shared library: [libc.so]
 0x00000001 (NEEDED)                     Shared library: [libm.so]
```

这就是 ARM 格式的动态链接可执行文件，在运行时需要 libc.so 和 libm.so。当提示“not stripped”时表示它还没被 STRIP（剥离）。嵌入式系统中为节省空间通常将编译完成的可执行文件或动态库进行剥离，即去掉其中多余的符号表信息。在前面“make helloworld showcommands”命令的最后也可以看到，Android 编译环境中使用了“out/host/linux-x86/bin/soslim”工具进行 STRIP。

3.5 编译 Android Kernel

编译 Android Kernel 代码就是编译 Android 内核代码，在进行具体编译工作，之前需要先了解在 Android 开源系统中包含的如下 3 部分代码。

- 仿真器公共代码：对应的工程名是 kernel/common.get；
- MSM 平台的内核代码：对应的工程名是 kernel/msm.get；
- OMAP 平台的内核代码：对应的工程名是 kernel/omap.get。

在本节的内容中，将详细讲解编译上述 Android Kernel 的基本知识。

3.5.1 获取 Goldfish 内核代码

Goldfish 是一种虚拟的 ARM 处理器，通常在 Android 的仿真环境中使用。在 Linux 的内核中，Goldfish 作为 ARM 体系结构的一种“机器”。在 Android 的发展过程中，Goldfish 内核的版本也从 Linux 2.6.25 升级到了 Linux 3.4，此处理器的 Linux 内核和标准的 Linux 内核有以下 3 个方面的差别。

- Goldfish 机器的移植；
- Goldfish 一些虚拟设备的驱动程序；
- Android 中特有的驱动程序和组件。

Goldfish 处理器有两个版本，分别是 ARMv5 和 ARMv7，在一般情况下，只需要使用 ARMv5 版本即可。在 Android 开源工程的代码仓库中，使用 git 工具得到 Goldfish 内核代码的命令如下所示。

```
$ git clone git://android.git.kernel.org/kernel/common.git
```

在其 Linux 源代码的根目录中，配置和编译 Goldfish 内核的过程如下所示。

```
$make ARCH=arm goldfish_defconfig .config
$make ARCH=arm CROSS_COMPILE={path}/arm-none-linux-gnueabi-
```

其中，CROSS_COMPILE 的 path 值用于指定交叉编译工具的路径。

编译结果如下所示。

```
LD vmlinux
SYSMAP system.map
SYSMAP .tmp_system.map
OBJCOPY arch/arm/boot/Image
Kernel: arch/arm/boot/Image is ready
AS arch/arm/boot/compressed/head.o
GZIP arch/arm/boot/compressed/piggy.gz
AS arch/arm/boot/compressed/piggy.o
CC arch/arm/boot/compressed/misc.o
LD arch/arm/boot/compressed/vmlinux
   OBJCONPY arch/arm/boot/zImage
   Kernel: arch/arm/boot/zImage is ready
```

- vmlinux：是 Linux 进行编译和连接之后生成的 Elf 格式的文件；
- Image：是未经过压缩的二进制文件；
- piggy：是一个解压缩程序；
- zImage：是解压缩程序和压缩内核的组合。

在 Android 源代码的根目录中，vmlinux 和 zImage 对应 Android 代码 prebuilt 中的预编译的 arm 内核。使用 zImage 可以替换 prebuilt 中的"prebuilt/android-arm/"目录下的 goldfish_defconfig，此文件的主要片断如下所示。

```
CONFIG_ARM=y
#
# System Type
#
CONFIG_ARCH_GOLDFISH=y
#
# Goldfish options
#
CONFIG_MACH_GOLDFISH=y
# CONFIG_MACH_GOLDFISH_ARMV7 is not set
```

因为 GoldFish 是 ARM 处理器，所以 CONFIG_ARM 宏需要被使用，CONFIG_ARCH_GOLDFISH 和 CONFIG_MACH_GOLDFISH 宏是 GoldFish 处理器这类机器使用的配置宏。

在 gildfish_defconfig 中，与 Android 系统相关的宏如下所示。

```
#
# android
#
CONFIG_ANDROID=y
CONFIG_ANDROID_BUNDER_IPC=y #binder ipc 驱动程序
CONFIG_ANDROID_LOGGER=y #log 记录器驱动程序
# CONFIG_ANDROID_RAM_CONSOLE is not set
CONFIG_ANDROID_TIMED_OUTPUT=y #定时输出驱动程序框架
CONFIG_ANDROID_LOW_MEMORY_KILLER=y
CONFIG_ANDROID_PMEM=y #物理内存驱动程序
CONFIG_ASHMEM=y #匿名共享内存驱动程序
CONFIG_RTC_INTF_ALARM=y
CONFIG_HAS_WAKELOCK=y 电源管理相关的部分 wakelock 和 earlysuspend
CONFIG_HAS_EARLYSUSPEND=y
CONFIG_WAKELOCK=y
CONFIG_WAKELOCK_STAT=y
CONFIG_USER_WAKELOCK=y
CONFIG_EARLYSUSPEND=y
goldfish_defconfig 配置文件中，另外有一个宏是处理器虚拟设备的"驱动程序"，其内容如下所示。
CONFIG_MTD_GOLDFISH_NAND=y
CONFIG_KEYBOARD_GOLDFISH_EVENTS=y
CONFIG_GOLDFISH_TTY=y
CONFIG_BATTERY_GOLDFISH=y
CONFIG_FB_GOLDFISH=y
CONFIG_MMC_GOLDFISH=y
CONFIG_RTC_DRV_GOLDFISH=y
```

在 Goldfish 处理器的各个配置选项中，体系结构和 Goldfish 的虚拟驱动程序基于标准 Linux 的内容的驱动程序框架，但是这些设备在不同的硬件平台的移植方式不同；Android 专用的驱动程序是 Android 中特有的内容，非 Linux 标准，但是和硬件平台无关。

和原 Linux 内核相比，Android 内核增加了 Android 的相关 Driver，对应的目录如下所示。

```
kernel/drivers/android
```

主要分为以下几类 Driver。

- Android IPC 系统：Binder (binder.c)；
- Android 日志系统：Logger (logger.c)；
- Android 电源管理：Power (power.c)；

- Android 闹钟管理：Alarm (alarm.c)；
- Android 内存控制台：Ram_console (ram_console.c)；
- Android 时钟控制的 gpio：Timed_gpio (timed_gpio.c)。

对于本书讲解的驱动程序开发来说，人们比较关心的是 GoldFish 平台下相关的驱动文件，具体说明如下所示。

1. 字符输出设备

```
kernel/drivers/char/goldfish_tty.c
```

2. 图像显示设备（Frame Buffer）

```
kernel/drivers/video/goldfishfb.c
```

3. 键盘输入设备文件

```
kernel/drivers/input/keyboard/goldfish_events.c
```

4. RTC 设备（Real Time Clock）文件

```
kernel/drivers/rtc/rtc-goldfish.c
```

5. USB Device 设备文件

```
kernel/drivers/usb/gadget/android_adb.c
```

6. SD 卡设备文件

```
kernel/drivers/mmc/host/goldfish.c
```

7. FLASH 设备文件

```
kernel/drivers/mtd/devices/goldfish_nand.c
kernel/drivers/mtd/devices/goldfish_nand_reg.h
```

8. LED 设备文件

```
kernel/drivers/leds/ledtrig-sleep.c
```

9. 电源设备

```
kernel/drivers/power/goldfish_battery.c
```

10. 音频设备

```
kernel/arch/arm/mach-goldfish/audio.c
```

11. 电源管理

```
kernel/arch/arm/mach-goldfish/pm.c
```

12. 时钟管理

```
kernel/arch/arm/mach-goldfish/timer.c
```

3.5.2 获取 MSM 内核代码

在当前市面中，谷歌的手机产品 G1 是基于 MSM 内核的，MSM 是高通公司的应用处理器，在 Android 代码库中公开了对应的 MSM 的源代码。在 Android 开源工程的代码仓库中，使用 git 工具得到 MSM 内核代码的命令如下所示。

```
$ git clone git://android.git.kernel.org/kernel/msm.git
```

3.5.3 获取 OMAP 内核代码

OMAP 是德州仪器公司的应用处理器，为 Android 使用的是 OMAP3 系列的处理器。在 Android 代码库中公开了对应的 MSM 的源代码，使用 git 工具得到 MSM 内核代码的命令如下所示。

```
$ git clone git://android.git.kernel.org/kernel/omap.git
```

3.5.4 编译 Android 的 Linux 内核

了解了上述 3 类 Android 内核后，下面开始讲解编译 Android 内核的方法。在此假设以 Ubuntu 8.10 为例，完整编译 Android 内核的流程如下所示。

1. 构建交叉编译环境

Android 的默认硬件处理器是 ARM，因此需要在自己的机器上构建交叉编译环境。交叉编译器 GNU Toolchain for ARM Processors 下载地址如下所示。

```
http://www.codesourcery.com/gnu_toolchains/arm/download.html
```

单击 GNU/Linux 对应的链接，之后单击“Download Sourcery CodeBench Lite 5.1 2012.03-117”链接后直接下载，如图 3-14 所示。

把 arm-2008q3-73-arm-none-linux-gnueabi-i686-pc-linux-gnu.tar.bz2 解压到一目录下，例如“~/programes/”，并加入 PATH 环境变量：

```
vim ~/.bashrc
```

然后添加：

```
ARM_TOOLCHIAN=~/programes/arm-2008q3/bin/
export PATH=${PATH}:${ARM_TOOLCHIAN};
```

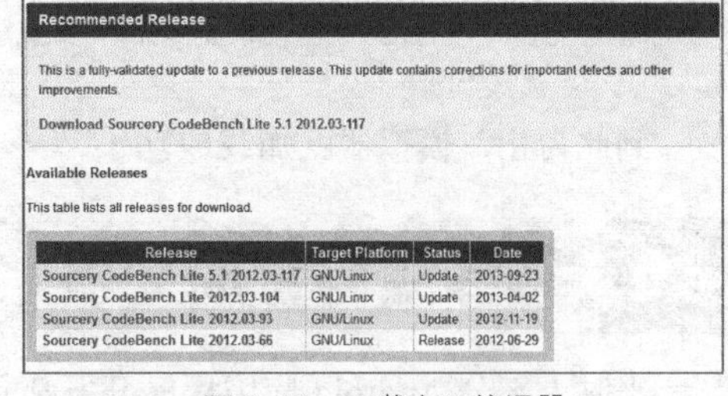

▲图 3-14 下载交叉编译器

保存后并入 source ~/.bashrc。

2. 获取内核源码

源码地址如下所示。

```
http://code.google.com/p/android/downloads/list
```

选择的内核版本要与选用的模拟器版本尽量一致。下载后并解压后得到 kernel.git 文件夹：

```
tar -xvf ~/download/linux-3.2.5-android-4.3_r1.tar.gz
```

3. 获取内核编译配置信息文件

编译内核时需要使用 configure，通常 configure 有很多选项，人们往往不知道需要哪些选项。在运行 Android 模拟器时，有一个文件“/proc/config.gz”，这是当前内核的配置信息文件，把 config.gz 获取并解压到“kernel.git/”下，然后改名为.config。命令如下所示。

```
cd kernel.git/
emulator &
adb pull /proc/config.gz
gunzip config.gz
mv config .config
```

4. 修改 Makefile

修改 195 行的代码：

```
CROSS_COMPILE    = arm-none-linux-gnueabi-
```

将 CROSS_COMPILE 值改为 arm-none-linux-gnueabi-，这是安装的交叉编译工具链的前缀，修改此处意在告诉 make 在编译的时候要使用该工具链。然后注释掉 562 和 563 行的如下代码：

```
#LDFLAGS_BUILD_ID = $(patsubst -Wl$(comma)%,%,/
#$(call ld-option, -Wl$(comma)--build-id,))
```

必须将上述代码中的 build id 值注释掉，因为目前版本的 Android 内核不支持该选项。

5. 编译

使用 make 进行编译，并同时生成 zImage。

```
LD      arch/arm/boot/compressed/vmlinux
OBJCOPY arch/arm/boot/zImage
Kernel: arch/arm/boot/zImage is ready
```

这样生成 zImage 大小为 1.23 MB，android- sdk-linux_x86-4.3_r1/tools/lib/images/kernel-qemu 是 1.24 MB。

6. 使用模拟器加载内核测试

命令如下所示。

```
cd android/out/cupcake/out/target/product/generic
emulator   -image   system.img   -data   userdata.img   -ramdisk   ramdisk.img   -kernel
~/project/android/kernel.git/arch/arm/boot/zImage &
```

到此为止，模拟器就加载成功了。

3.6 编译源码生成 SDK

平时大部分 Android 应用程序开发都是基于 SDK 实现的，其过程是使用 SDK 中的接口实现各种各样的功能。可以在 Android 的官方网站上面直接下载最新的 SDK 版本，不过也可以从源码中生成 SDK，因为源码里面也包含有 SDK 的代码。

在下载的 Android 4.3 的源码的根目录下有个 SDK 目录，所有的 SDK 相关的代码都存储在这个目录，包括镜像文件、模拟器和 ADB 等常用工具以及 SDK 中的开发包的文档，可以通过编译的方式来生成开发需要的 SDK，编译命令如下所示。

```
$ Make SDK
```

当编译完成后，会在/out/host/linux-x86/sdk/目录下生成 SDK，这个 SDK 是完全跟源码同步的，与官网上下载的 SDK 功能完全相同，会有开发用的 JAR 包、模拟器管理工具和 ADB 调试工具，可以使用这个编译生成的 SDK 来开发应用程序。

对于 Android 系统的开发，基本可以分为如下两种开发方式。

- 基于 SDK 的开发；
- 基于源码的开发。

在一般的情况下，开发的应用程序都是基于 SDK 的开发，比较方便而且兼容性比较好。基于源码的开发相对于基于 SDK 的开发要求对源码的架构认识更深刻，一般用于需要修改系统层面的场合。两种方式应用场景不同，各有优缺点，在本节将主要介绍基于 SDK 的开发。

如果想基于 SDK 开发 Android 的应用程序，需要 JDK、SDK 和一个开发环境，JDK 和 SDK 在不同的平台下有不同的版本，本章主要讨论 Windows 7 平台下的开发环境搭建。

1. 安装 JDK

由于 Android 的应用程序是使用 Java 语言开发的，所以首先需要安装 Java 的 JDK，下载链接为 http://java.sun.com/javase/downloads/index.jsp，进入后选择合适的平台以及下载最新版本的 JDK，安装成功后在命令行下可以查看 JDK 版本，如图 3-15 所示。

2. 安装 Eclipse

Eclipse 是开发 Android 应用程序的 IDE 环境，有非常丰富的插件可以使用，单击 http://www.eclipse.org/downloads/可以下载合适平台的最新版本 Eclipse。

3. 安装 Android SDK

Android SDK 是 Google 对外发布的专门用于 Android 开发的工具包，里面有各种版本的开发框架和工具以及丰富的文档，打开 http://developer.android.com/sdk/index.html 可以下载最新版本的针对 Window 7 平台的 SDK。

当下载完成上述 3 个工具之后，需要对开发环境进行如下所示的配置。

1. 配置 Eclipse

第 1 步：打开 Eclipse，在菜单栏上选择“help”｜“Install New SoftWare”，出现图 3-16 所示的界面。

```
C:\Windows\system32>java -version
java version "1.6.0_25"
Java(TM) SE Runtime Environment (build 1.6.0_25-b06)
Java HotSpot(TM) Client VM (build 20.0-b11, mixed mode, sharing)
```

▲图 3-15　成功安装的 JDK

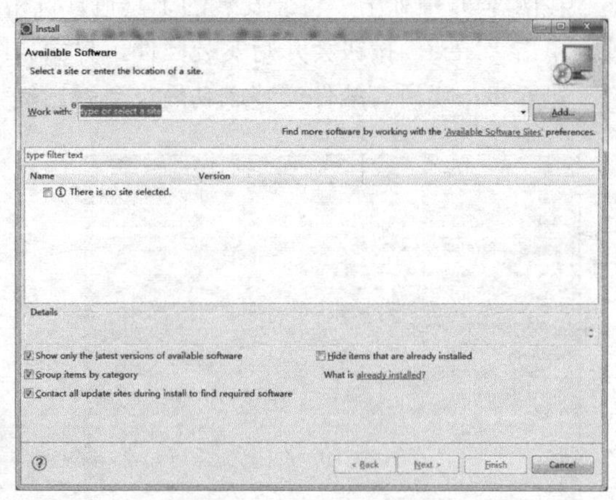

▲图 3-16　“Install”界面

第 2 步：单击“Add”按钮，会出现图 3-17 所示的界面。

第 3 步：在 Name 栏里面输入“Android”或者自定义任何名字，在 Location 里面输入“https://dl-ssl.google.com/android/eclipse/”，输入后的效果如图 3-18 所示。

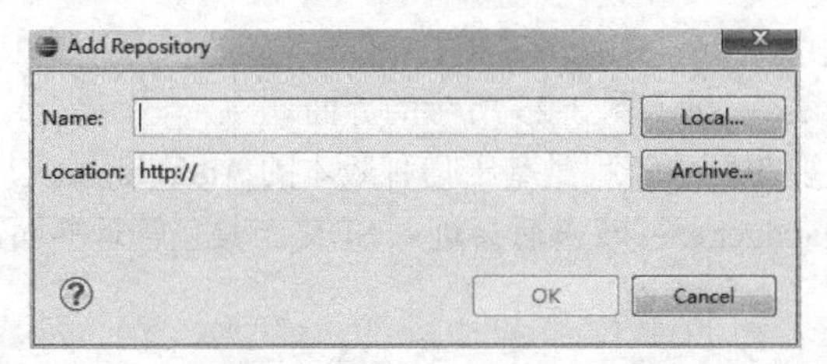

▲图 3-17　“Add Repository”界面

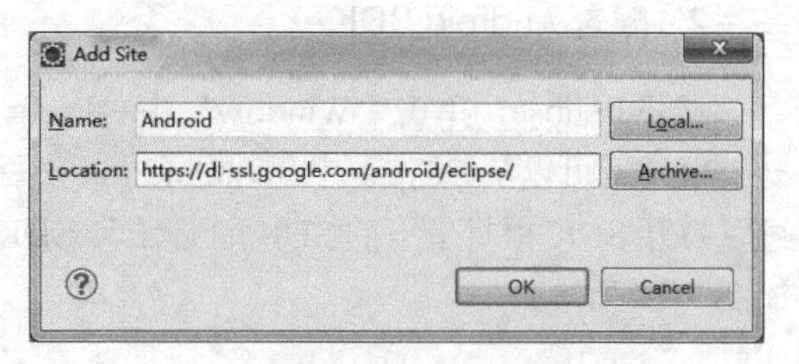

▲图 3-18　“Add Site”界面

第 4 步：如果发现 https://无法使用，可以改成 http://尝试下，当输入完名字和地址之后，单击“OK”按钮，会出现图 3-19 所示的界面。

图 3-19 中的两个插件都是开发 Android 必不可少的工具包，Android DDMS 是可以用来调试、管理 Android 进程和存储器、查看日志的工具，Android Development Tool 简称 ADT，是开发 Android 的插件，只有装了 ADT 才能创建 Android 工程。

第 5 步：单击“Next”按钮，出现图 3-20 所示的界面。

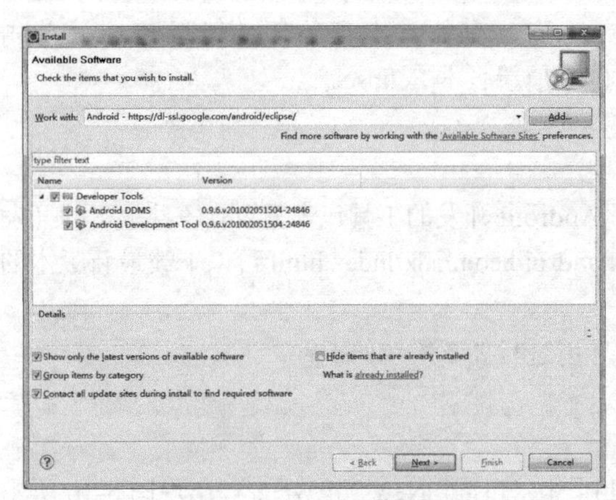

▲图 3-19　“Install”界面

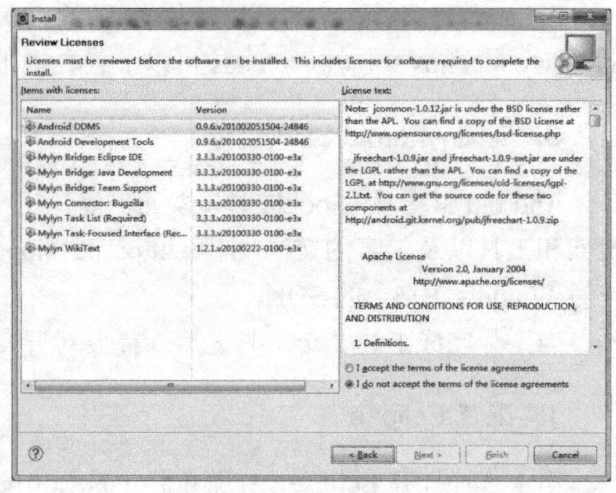

▲图 3-20　选择安装

在图 3-20 中列出了将会安装的工具包，选中“I accept the terms of the license agreements”选项，单击“Next”按钮会开始安装插件，界面如图 3-21 所示。

第 6 步：当所有插件安装成功后，会弹出提示界面，如图 3-22 所示。

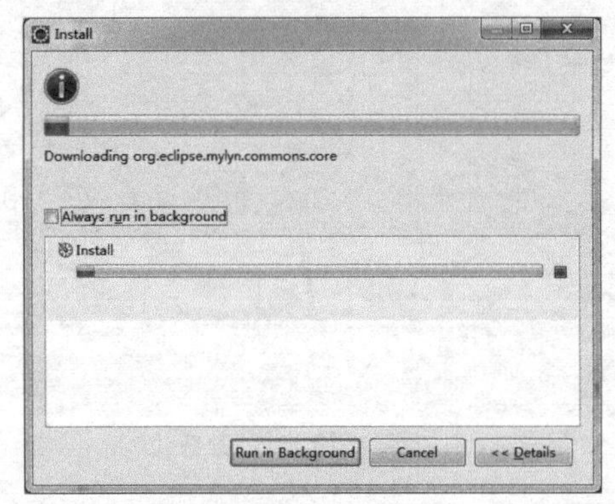

▲图 3-21　开始安装

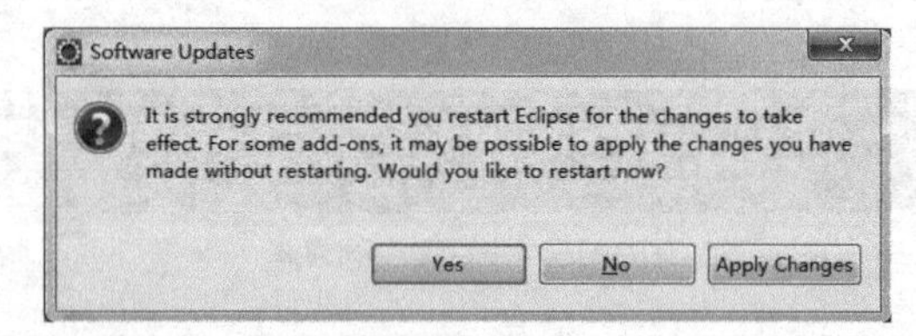

▲图 3-22　成功

这时需要单击“Yes”按钮重启 Eclipse 让所有插件生效。

2. 配置 Android SDK

打开 Eclipse，单击“Window”|“Preferences”，进入图 3-23 所示的界面。

这样就可以从 Eclipse 中新建 Android 工程，要想新建工程是基于何种版本的 Android 系统，可以打开 SDK 根目录下的 SDK 管理工具 SDK Manager.exe，双击后会进入 SDK 工具包管理界面，如图 3-24 所示。

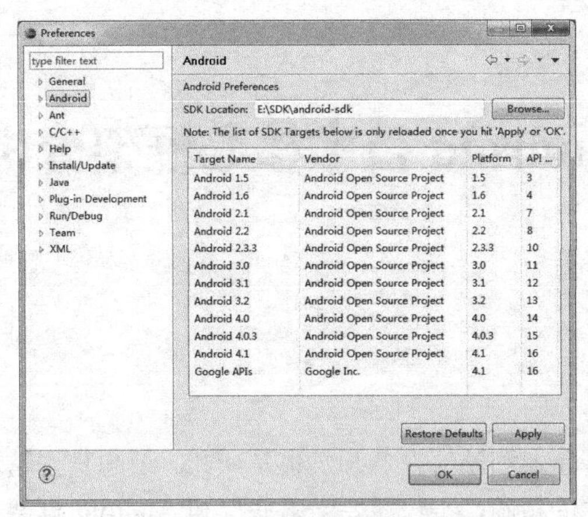

▲图 3-23 配置界面

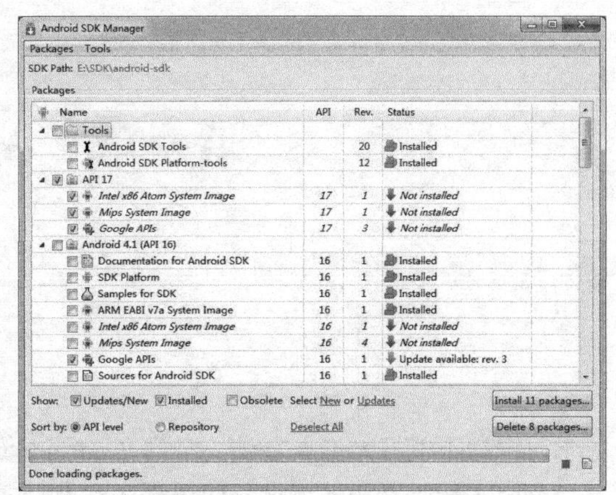

▲图 3-24 Android SDK 管理

在图 3-24 中可以看到，很清晰地列出了当前版本 SDK 中包含的工具包以及已经安装的和没有安装的版本。可以继续单击“Install 11 packages”或者“Delete 8 packages”按钮安装和删除 SDK 中的工具包。如果是安装，则过程会比较慢，与网速的关系比较密切。当 SDK 中的工具包安装完毕，同时也就完成了 Eclipse 和 SDK 的配置工作，至此在 Windows 7 平台下基于 SDK 的 Android 的开发环境搭建全部完成。

第4章　Android技术核心框架分析

学习编程不能打无把握之仗，学习Android传感器和外部设备开发也是如此。要想真正精通开发Android传感器和外部设备应用程序开发技术的精髓，不仅需要学习底层和Android框架方面的知识，还需要掌握Android顶层应用程序开发的基本知识。本章将详细讲解Android系统的体系结构，为读者步入本书后面高级知识的学习打下基础。

4.1 分析Android的系统架构

为了更加深入地理解Android系统的精髓，初学者们很有必要了解Android系统的整体架构，了解它的具体组成。只有这样才能知道Android究竟能干什么，人们所要学的是什么。

4.1.1 Android体系结构介绍

Android是一个移动设备的开发平台，其软件层次结构包括操作系统（OS）、中间件（MiddleWare）和应用程序（Application）。根据Android的软件框图，其软件层次结构自下而上分为以下4层：

（1）操作系统层（OS）；

（2）各种库（Libraries）和Android运行环境（RunTime）；

（3）应用程序框架（Application Framework）；

（4）应用程序（Application）。

上述各个层的具体结构如图4-1所示。

1. 操作系统层（OS）——最底层

因为Android源于Linux，使用了Linux内核，所以Android使用Linux 2.6作为操作系统。Linux 2.6是一种标准的技术，Linux也是一个开放的操作系统。Android对操作系统的使用包括核心和驱动程序两部分，Android的Linux核心为标准的Linux 2.6内核，Android更多的是需要一些与移动设备相关的驱动程序。主要的驱动如下所示。

- 显示驱动（Display Driver）：常用基于Linux的帧缓冲（Frame Buffer）驱动；
- Flash内存驱动（Flash Memory Driver）：是基于MTD的Flash驱动程序；
- 照相机驱动（Camera Driver）：常用基于Linux的v4l（Video for）驱动；
- 音频驱动（Audio Driver）：常用基于ALSA（Advanced Linux Sound Architecture，高级Linux声音体系）驱动；
- Wi-Fi驱动（Camera Driver）：基于IEEE 802.11标准的驱动程序；
- 键盘驱动（KeyBoard Driver）：作为输入设备的键盘驱动；
- 蓝牙驱动（Bluetooth Driver）：基于IEEE 802.15.1标准的无线传输技术；

● Binder IPC 驱动：Android 一个特殊的驱动程序，具有单独的设备节点，提供进程间通信的功能；

● Power Management（能源管理）：管理电池电量等信息。

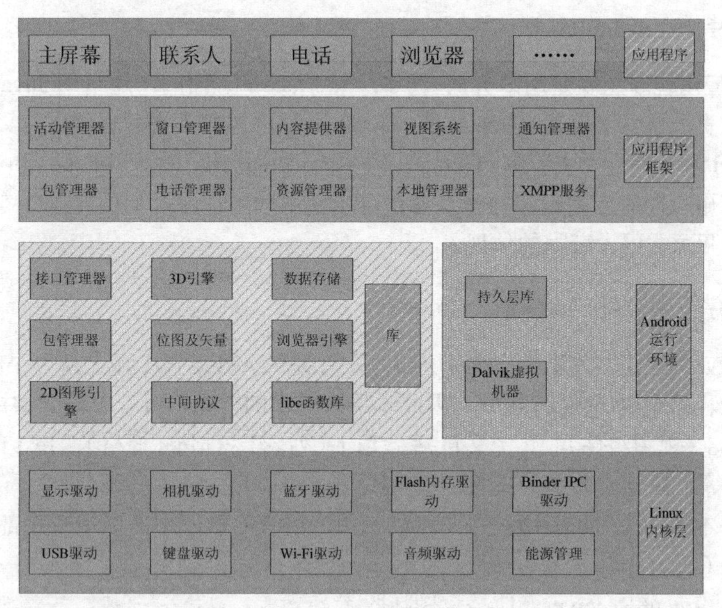

▲图 4-1　Android 操作系统的组件结构图

2. 各种库（Libraries）和 Android 运行环境（RunTime）——中间层

本层次对应一般嵌入式系统，相当于中间件层次。Android 的本层次分成两个部分，一个是各种库，另一个是 Android 运行环境。本层的内容大多是使用 C 实现的。其中包含的各种库如下所示。

● C 库：C 语言的标准库，也是系统中一个最为底层的库，C 库是通过 Linux 的系统调用来实现；

● 多媒体框架（MediaFrameword）：这部分内容是 Android 多媒体的核心部分，基于 PacketVideo（即 PV）的 OpenCORE，从功能上本库一共分为两大部分，一个部分是音频、视频的回放（PlayBack），另一部分是则是音视频的记录（Recorder）；

● SGL：2D 图像引擎；

● SSL：即 Secure Socket Layer 位于 TCP/IP 协议与各种应用层协议之间，为数据通信提供安全支持；

● OpenGL ES 1.0：提供了对 3D 的支持；

● 界面管理工具（Surface Management）：提供了对管理显示子系统等功能；

● SQLite：一个通用的嵌入式数据库；

● WebKit：网络浏览器的核心；

● FreeType：位图和矢量字体的功能；

Android 的各种库一般是以系统中间件的形式提供的，它们均有的一个显著特点就是与移动设备的平台的应用密切相关。

Android 运行环境主要是指的虚拟机技术——Dalvik。Dalvik 虚拟机和一般 Java 虚拟机（Java VM）不同，它执行的不是 Java 标准的字节码（Bytecode），而是 Dalvik 可执行格式（.dex）中执行文件。在执行的过程中，每一个应用程序即一个进程（Linux 的一个 Process）。二者最大的区别在于 Java VM 是基于栈（Stack-based）的虚拟机，而 Dalvik 是基于寄存器（Register-based）的虚拟机。显然，后者最大的好处在于可以根据硬件实现更大的优化，这更适合移动设备的特点。

3. 应用程序（Application）

Android 的应用程序主要是用户界面（User Interface）方面的，通常用 Java 语言编写，其中还可以包含各种资源文件（放置在 res 目录中）。Java 程序和相关资源在经过编译后，会生成一个 APK 包。Android 本身提供了主屏幕（Home）、联系人（Contact）、电话（Phone）和浏览器（Browers）等众多的核心应用。同时应用程序的开发者还可以使用应用程序框架层的 API 实现自己的程序。这也是 Android 开源的巨大潜力的体现。

4. 应用程序框架（Application Framework）

Android 的应用程序框架为应用程序层的开发者提供 APIs，它实际上是一个应用程序的框架。由于上层的应用程序是以 Java 构建的，因此本层次提供的首先包含了 UI 程序中所需要的各种控件，例如：Views（视图组件），其中又包括了 List（列表）、Grid（栅格）、Text Box（文本框）和 Button（按钮）等。甚至一个嵌入式的 Web 浏览器。

作为一个基本的 Android 应用程序，可以利用应用程序框架中的以下 5 个部分来构建：

- Activity（活动）；
- Broadcast Intent Receiver（广播意图接收者）；
- Service（服务）；
- Content Provider（内容提供者）；
- Intent and Intent Filter（意图和意图过滤器）。

4.1.2　Android 应用工程文件组成

讲解完 Android 的整体结构之后，接下来开始讲解 Android 工程文件的组成。在 Eclipse 中，一个基本的 Android 项目的目录结构如图 4-2 所示。

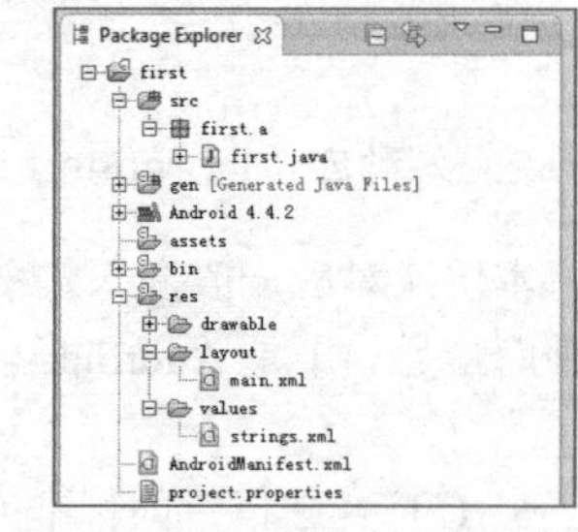

▲图 4-2　Android 应用工程文件组成

1. src 目录

在里面保存了开发人员编写的程序文件。和一般的 Java 项目一样，“src”目录下保存的是项目的所有包及源文件（.java），“res”目录下包含了项目中的所有资源。例如，程序图标（drawable）、布局文件（layout）和常量（values）等。不同的是，在 Java 项目中没有“gen”目录，也没有每个 Android 项目都必须有的 AndroidManfest.xml 文件。

“.java”格式文件是在建立项目时自动生成的，这个文件是只读模式，不能更改。R.java 文件是定义该项目所有资源的索引文件。例如下面是某项目中 R.java 文件的代码。

```
package com.yarin.Android.HelloAndroid;
public final class R {
    public static final class attr {
    }
    public static final class drawable {
        public static final int icon=0x7f020000;
    }
    public static final class layout {
        public static final int main=0x7f030000;
    }
    public static final class string {
        public static final int app_name=0x7f040001;
```

```
        public static final int hello=0x7f040000;
    }
}
```

在上述代码中定义了很多常量，并且这些常量的名字都与 res 文件夹中的文件名相同，这再次证明.java 文件中所存储的是该项目所有资源的索引。有了这个文件，在程序中使用资源将变得更加方便，可以很快地找到要使用的资源，由于这个文件不能被手动编辑，所以在项目中加入了新的资源时，只需要刷新一下该项目，.java 文件便自动生成了所有资源的索引。

2. 设置文件 AndroidManfest.xml

文件 AndroidManfest.xml 是一个控制文件，在里面包含了该项目中所使用的 Activity、Service 和 Receiver。例如下面是某项目中文件 AndroidManfest.xml 的代码。

```
<?xml version="1.0" encoding="utf-8"?>
<manifest xmlns:android="http://schemas.android.com/apk/res/android"
     package="com.yarin.Android.HelloAndroid"
     android:versionCode="1"
     android:versionName="1.0">
   <application android:icon="@drawable/icon"
android:label="@string/app_name">
       <activity android:name=".HelloAndroid"
                 android:label="@string/app_name">
           <intent-filter>
               <action android:name="android.intent.action.MAIN" />
               <category android:name="android.intent.category.LAUNCHER" />
           </intent-filter>
       </activity>
   </application>
   <uses-sdk android:minSdkVersion="9" />
</manifest>
```

在上述代码中，intent-filters 描述了 Activity 启动的位置和时间。每当一个 Activity（或者操作系统）要执行一个操作时，它将创建出一个 Intent 的对象，这个 Intent 对象可以描述想做什么、想处理什么数据、数据的类型以及一些其他信息。Android 会和每个 Application 所出现的 intent-filter 的数据进行比较，找到最合适 Activity 来处理调用者所指定的数据和操作。下面来仔细分析 AndroidManfest.xml 文件，如表 4-1 所示。

表 4-1 AndroidManfest.xml 分析

参　数	说　明
manifest	根节点，描述了 package 中所有的内容
xmlns:android	包含命名空间的声明。xmlns:android=http://schemas.android.com/apk/res/android，使得 Android 中各种标准属性能在文件中使用，提供了大部分元素中的数据
Package	声明应用程序包
application	包含 package 中 application 级别组件声明的根节点。此元素也可包含 application 的一些全局和默认的属性，如标签、icon、主题和必要的权限等。一个 manifest 能包含零个或一个此元素（不能大余一个）
android:icon	应用程序图标
android:label	应用程序名字
activity	activity 是与用户交互的主要工具，是用户打开一个应用程序的初始页面，大部分被使用到的其他页面也由不同的 activity 所实现，并声明在另外的 activity 标记中。注意，每一个 activity 必须有一个<activity>标记对应，无论它给外部使用或是只用于自己的 package 中。如果一个 activity 没有对应的标记，将不能运行它。另外，为了支持运行时查找 activity，可包含一个或多个<intent-filter>元素来描述 activity 所支持的操作
android:name	应用程序默认启动的 activity

续表

参　　数	说　　明
intent-filter	声明了指定的一组组件支持的 Intent 值，从而形成了 Intent Filter。除了能在此元素下指定不同类型的值，属性也能存储在这里来描述一个操作所需要的唯一的标签、icon 和其他信息
action	组件支持的 Intent action
category	组件支持的 Intent Category。这里指定了应用程序默认启动的 activity
uses-sdk	与该应用程序所使用的 sdk 版本相关

3. 常量定义文件

下面介绍一在资源文件中对常量的定义，例如文件 String.xml 的代码如下所示。

```
<?xml version="1.0" encoding="utf-8"?>
<resources>
 <string name="hello">Hello World, HelloAndroid!</string>
   <string name="app_name">HelloAndroid</string>
</resources>
```

上述常量定义文件的代码非常简单，只定义了两个字符串资源，请不要小看上面的几行代码。它们的内容很“露脸”，里面的字符直接显示在手机屏幕中，就像动态网站中的 HTML 一样。

4. 布局文件

布局（layout）文件一般位于“res\layout\main.xml”目录，通过其代码能够生成一个显示界面。例如下面的代码。

```
<?xml version="1.0" encoding="utf-8"?>
<LinearLayout xmlns:android="http://schemas.android.com/apk/res/android"
    android:orientation="vertical"
    android:layout_width="fill_parent"
    android:layout_height="fill_parent"
    >
<TextView
    android:layout_width="fill_parent"
    android:layout_height="wrap_content"
    android:text="@string/hello"
    />
</LinearLayout>
```

在上述代码中，有以下几个布局和参数。

- <LinearLayout></LinearLayout>：在这个标签中，所有元件都是按由上到下的队列排成的。
- android:orientation：表示这个介质的版面配置方式是从上到下垂直地排列其内部的视图。
- android:layout_width：定义当前视图在屏幕上所占的宽度，fill_parent 即填充整个屏幕。
- android:layout_height：定义当前视图在屏幕上所占的高度，fill_parent 即填充整个屏幕。
- wrap_content：随着文字栏位的不同而改变这个视图的宽度或高度。

在上述布局代码中，使用了一个 TextView 来配置文本标签 Widget（构件），其中设置的属性 android:layout_width 为整个屏幕的宽度，android:layout_height 可以根据文字来改变高度，而 android:text 则设置了这个 TextView 要显示的文字内容，这里引用了@string 中的 hello 字符串，即 String.xml 文件中的 hello 所代表的字符串资源。hello 字符串的内容 Hello World, HelloAndroid!这就是人们在 HelloAndroid 项目运行时看到的字符串。

注意　上面介绍的文件只是主要文件，在项目中需要自行编写。在项目中还有很多其他的文件，那些文件很少需要自己编写，所以在此就不进行讲解了。

4.2 Android 的五大组件

一个典型的 Android 应用程序通常由 5 个组件组成，这 5 个组件构成了 Android 的核心功能。本节将一一讲解这五大组件的基本知识，为读者步入本书后面知识的学习打下基础。

4.2.1　Activity 界面组件

Activitiy 是这 5 个组件中最常用的一个组件。程序中 Activity 通常的表现形式是一个单独的界面（screen）。每个 Activitiy 都是一个单独的类，它扩展实现了 Activity 基础类。这个类显示为一个由 Views 组成的用户界面，并响应事件。大多数程序有多个 Activity。例如，一个文本信息程序有以下几个界面：显示联系人列表界面、写信息界面、查看信息界面或者设置界面等。每个界面都是一个 Activity。切换到另一个界面就是载入一个新的 Activity。在某些情况下，一个 Activity 可能会给前一个 Activity 返回值，例如，一个让用户选择相片的 Activity 会把选择到的相片返回给其调用者。

打开一个新界面后，前一个界面就被暂停，并放入历史栈中（界面切换历史栈）。使用者可以回溯前面已经打开的存储在历史栈中的界面。也可以从历史栈中删除没有界面价值的界面。Android 在历史栈中保留程序运行产生的所有界面：从第一个界面，到最后一个。

4.2.2　Intent 切换组件

Android 通过一个专门的 Intent 类来进行界面的切换。Intent 描述了程序想做什么（Intent 翻译为意图、目的、意向）。Intent 类还有一个相关类 IntentFilter。Intent 是一个请求来做什么事情，IntentFilter 则描述了一个 Activity（或下文的 IntentReceiver）能处理什么意图。显示某人联系信息的 Activity 使用了一个 IntentFilter，就是说它知道如何处理应用到此人数据的 VIEW 操作。Activities 在文件 AndroidManifest.xml 中使用 IntentFilters。

通过解析 Intents 可以实现 Activity 的切换，人们可以使用 startActivity（myIntent）启用新的 Activity。系统会考察所有安装程序的 IntentFilters，然后找到与 myIntent 匹配最好的 IntentFilters 所对应的 Activity。这个新 Activity 能够接收 Intent 传来的消息，并因此被启用。解析 Intents 的过程发生在 startActivity 被实时调用的时候，这样做有如下两个好处。

（1）Activity 仅发出一个 Intent 请求，便能重用其他组件的功能。

（2）Activity 可以随时被替换为有等价 IntentFilter 的新 Activity。

4.2.3　Service 服务组件

Service 是一个没有 UI 且长驻系统的代码，最常见的例子是媒体播放器从播放列表中播放歌曲。在媒体播放器程序中，可能有一个或多个 Activities 让用户选择播放的歌曲。然而在后台播放歌曲时无须 Activity 干涉，因为用户希望在音乐播放的同时能够切换到其他界面。既然这样，媒体播放器 Activity 需要通过 Context.startService()启动一个 Service，这个 Service 在后台运行以保持继续播放音乐。在媒体播放器被关闭之前，系统会保持音乐后台播放 Service 的正常运行。可以用 Context.bindService()方法连接到一个 Service 上（如果 Service 未运行的话，连接后还会启动它），连接后就可以通过一个 Service 提供的接口与 Service 进行通话。对音乐 Service 来说，提供了暂停和重放等功能。

1．如何使用服务

在 Android 系统中有如下两种使用服务的方法。

（1）通过调用 Context.startServece()启动服务，调用 Context.stopService()结束服务，startService()可以传递参数给 Service。

（2）通过调用 Context.bindService()启动，调用 Context.unbindService()结束，还可以通过 ServiceConnection 访问 Service。二者可以混合使用，例如可以先调用 startService()再调用 unbindService()。

2. Service 的生命周期

在 startService()后，即使调用 startService()的进程结束了，Service 还仍然存在，一直到有进程调用 stoptService()或者 Service 自己结束（stopSelf()）为止。

在 bindService()后，Service 就和调用 bindService()的进程同生共死了，也就是说当调用 bindService()的进程结束了，那么它绑定的 Service 也要跟着被结束，当然期间也可以调用 unbind Service()让 Service 结束。

当混合使用上述两种方式时，例如你启动 startService()，我启动 bindService()，那么只有你调用 stoptService()而且我也调用 unbindService()，这个 Service 才会被结束。

3. 进程生命周期

Android 系统将会尝试保留那些启动的或者绑定的服务进程，具体说明如下所示。

（1）如果该服务正在进程的 onCreate()、onStart()或者 onDestroy()这些方法中执行时，那么主进程将会成为一个前台进程，以确保此代码不会被停止。

（2）如果服务已经开始，那么它的主进程的重要性会低于所有的可见进程，但是会高于不可见进程。由于只有少数几个进程是用户可见的，所以只要不是内存特别低，该服务就不会停止。

（3）如果有多个客户端绑定了服务，只要客户端中的一个对于用户是可见的，就可以认为该服务是可见的。

4.2.4 用 Broadcast/Receiver 广播机制组件

当要执行一些与外部事件相关的代码时，例如来电响铃时或者半夜时就可能用到 IntentReceiver。尽管 IntentReceiver 使用 NotificationManager 来通知用户一些好玩的事情发生，如果没有 UI.IntentReceivers 可以在文件 AndroidManifest.xml 中声明，也可以使用 Context.register Receiver()来声明。当一个 IntentReceiver 被触发时，如果需要系统自然会自动启动程序。程序也可以通过 Context.broadcastIntent()来发送自己的 Intent 广播给其他程序。

4.2.5 ContentProvider 存储组件

应用程序把数据存储一个 SQLite 数据库格式文件里，或者存储在其他有效设备里。如果想让其他程序能够使用程序中的数据，此时 Content Provider 就很有用了。Content Provider 是一个实现了一系列标准方法的类，这个类使得其他程序能存储、读取某种 Content Provider 可处理的数据。

4.3 进程和线程

进程和线程很容易理解，计算机中有一个进程管理器，当打开后，会显示当前运行的所有程序。同样在 Android 中也有进程，当某个组件第一次运行的时候，Android 会启动一个进程。在默认情况下，所有的组件和程序运行在这个进程和线程中，也可以安排组件在其他的进程或者线程中运行。

4.3.1 什么是进程

组件运行的进程是由 manifest file 控制的。组件的节点一般都包含一个 process 属性，例如<activity>、<service>、<receiver>和<provider>节点。属性 process 可以设置组件运行的进程，可以配置组件在一个独立进程中运行，或者多个组件在同一个进程中运行，甚至可以多个程序在一个进程中运行，当然前提是这些程序共享一个 User ID 并给定同样的权限。另外<application>节点也包含了 process 属性，用来设置程序中所有组件的默认进程。

当更加常用的进程无法获取足够内存时，Android 会智能地关闭不常用的进程。当下次启动程序的时候会重新启动这些进程。当决定哪个进程需要被关闭的时候，Android 会考虑哪个对用户更加有用。例如 Android 会倾向于关闭一个长期不显示在界面的进程来支持一个经常显示在界面的进程。是否关闭一个进程决定于组件在进程中的状态。

4.3.2 什么是线程

当用户界面需要很快对用户进行响应，就需要将一些费时的操作，如网络连接、下载或者非常占用服务器时间的操作等放到其他线程。也就是说，即使为组件分配了不同的进程，有时候也需要再分配线程。

线程是通过 Java 的标准对象 Thread 来创建的，在 Android 中提供了如下方便地管理线程的方法：

（1）Looper 在线程中运行一个消息循环；

（2）Handler 传递一个消息；

（3）HandlerThread 创建一个带有消息循环的线程；

（4）Android 让一个应用程序在单独的线程中，指导它创建自己的线程；

（5）应用程序组件（Activity、Service、Broadcast receiver）所有都在理想的主线程中实例化；

（6）当被系统调用时，没有一个组件应该执行长时间或是阻塞操作（例如网络呼叫或是计算循环），这将中断所有在该进程的其他组件；

（7）可以创建一个新的线程来执行长期操作。

4.3.3 应用程序的生命周期

自然界的事物都有自己的生命周期，例如人的生、老、病、死。作为一个 Android 应用程序也如同自然界的生物一样，也有自己的生命周期。人们开发一个程序的目的是为了完成一个功能，例如银行计算加息的软件，每当一个用户去柜台办理取款业务时，银行工作人员便启动了这个程序的生命，当用这个软件完成利息计算时，这个软件当前的任务就完成了，此时就需要结束自己的使命。肯定有人提出疑问：生生死死多么麻烦，就让这个程序一直是“活着”的状态，一个用户办理完取款业务后，继续等着下一个用户办理取款业务，这样这个程序就“长生不老”了。其实谁都想自己的程序“长生不老”，但是很不幸，不能这样做。原因是计算机的处理性能是一定的，一个人、两个人、三个人计算机可以处理这个任务。但是一个安装这个软件的机器一天会处理成千上万个取款业务，如果它们都一直“活着”，一台有限配置的计算机能承受得了吗？

由此可见，应用程序的生命周期就是一个程序的存活时间，即在什么时间内有效。Android 是一个构建在 Linux 之上的开源移动开发平台，在 Android 中，多数情况下每个程序都是在各自独立的 Linux 进程中运行的。当一个程序或其某些部分被请求时，它的进程就“出生”了；当这个程序没有必要再运行下去且系统需要回收这个进程的内存用于其他程序时，这个进程就“死亡”了。可以看出，Android 程序的生命周期是由系统控制的而非程序自身直接控制的。这和编写桌面应用程序时的思维有一些不同，一个桌面应用程序的进程也是在其他进程或用户请求时被创建的，但是往往是在程序自身收到关闭请求后执行一个特定的动作（例如从 main 函数中返回）而导致进程结束的。要想做好某种类型的程序或者某种平台下的程序的开发，最关键的就是要弄清楚这种类型的程序或整个平台下的程序的一般工作模式并熟记在心。在 Android 中，程序的生命周期控制就是属于这个范畴。

开发者必须理解不同的应用程序组件，尤其是 Activity、Service 和 Intent Receiver，需要了解这些组件是如何影响应用程序的生命周期的。如果不能正确地使用这些组件，可能会导致系统终止正在执行重要任务的应用程序进程。

一个常见的进程生命周期漏洞的例子是 Intent Receiver（意图接收器），当 Intent Receiver 在 onReceive 方法中接收到一个 Intent（意图）时，它会启动一个线程，然后返回。一旦返回，系统将认为 Intent Receiver 不再处于活动状态，因而 Intent Receiver 所在的进程也就不再有用了（除非该进程中还有其他的组件处于活动状态）。因此，系统可能会在任意时刻终止该进程以回收占有的内存。这样进程中创建出的那个线程也将被终止。解决这个问题的方法是从 Intent Receiver 中启动一个服务，让系统知道进程中还有处于活动状态的工作。为了使系统能够正确决定在内存不足时应该终止哪个进程，Android 根据每个进程中运行的组件及组件的状态把进程放入一个“Importance Hierarchy（重要性分级）”中。

进程的类型多种多样，按照重要的程度主要包括如下几类进程。

1. 前台进程（Foreground）

前台进程是看得见的，与用户当前正在做的事情密切相关，不同的应用程序组件能够通过不同的方法将它的宿主进程移到前台。在下面的任何一个条件下系统都会把进程移动到前台。

- 进程正在屏幕的最前端运行一个与用户交互的活动（Activity），它的 onResume 方法被调用；
- 进程有一正在运行的 Intent Receiver（它的 IntentReceiver.onReceive 方法正在执行）；
- 进程有一个服务（Service），并且在服务的某个回调函数（Service.onCreate、Service.onStart 或 Service.onDestroy）内有正在执行的代码。

2. 可见进程（Visible）

可见进程也是可见的，它有一个可以被用户从屏幕上看到的活动，但不在前台（它的 onPause 方法被调用）。假如前台的活动是一个对话框，以前的活动隐藏在对话框之后就会现这种进程了。可见进程非常重要，一般不允许被终止，除非是了保证前台进程的运行而不得不终止它。

3. 服务进程（Service）

服务进程是无法看见的，拥有一个已经用 startService()方法启动的服务。虽然用户无法直接看到这些进程，但它们做的事情却是用户所关心的（如后台 MP3 回放或后台网络数据的上传、下载）。所以系统将一直运行这些进程，除非内存不足以维持所有的前台进程和可见进程。

4. 后台进程（Background）

后台进程也是看不见的，只有打开之后才能看见。例如迅雷下载，可以将其最小化，虽然在桌面上看不见了，但是它一直在进行下载的工作。拥有一个当前用户看不到的活动（它的 onStop() 方法被调用）。这些进程对用户体验没有直接的影响。如果它们正确执行了活动生命周期，系统可以在任意时刻终止该进程以回收内存，并提供给前面 3 种类型的进程使用。系统中通常有很多这样的进程在运行，因此要将这些进程保存在 LRU 列表中，以确保当内存不足时用户最近看到的进程最后一个被终止。

5. 空进程（Empty）

空进程是指不拥有任何活动的应用程序组件的进程。保留这种进程的唯一原因是在下次应用程序的某个组件需要运行时，不需要重新创建进程，这样可以提高启动速度。系统将以进程中当前处于活动状态组件的重要程度为基础对进程进行分类。进程的优先级可能也会根据该进程与其他进程的依赖关系而增长。假如进程 A 通过在进程 B 中设置 Context.BIND_AUTO_CREATE 标记或使用 ContentProvider 绑定到一个服务（Service），那么进程 B 在分类时至少要被看成与进程 A 同等重要。

例如 Activity 的状态转换图如图 4-3 所示。

图 4-3 所示的状态的变化是由 Android 内存管理器决定的，Android 会首先关闭那些包含 Inactive Activity 的应用程序，然后关闭 Stopped 状态的程序。只有在极端情况下才会移除 Paused 状态的程序。

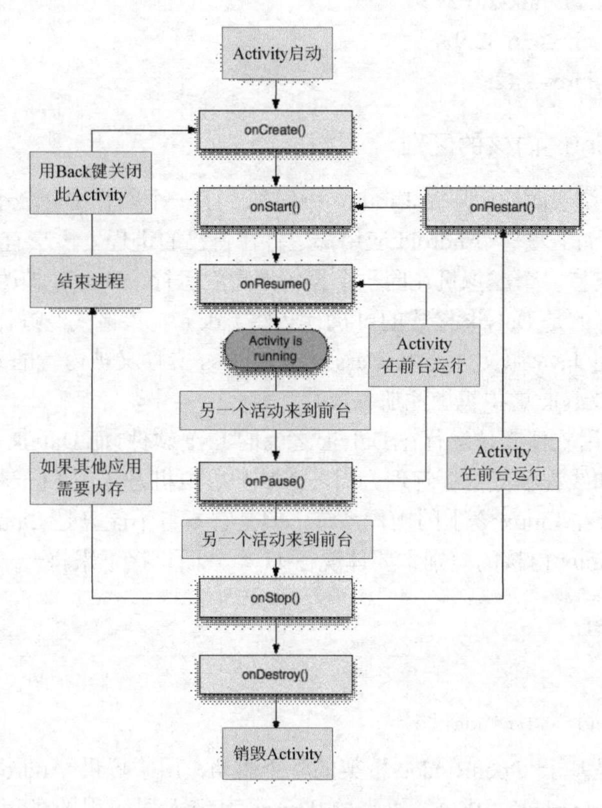

▲图 4-3 Activity 状态转换图

4.4 Android 和 Linux 的关系

在了解 Linux 和 Android 的关系之前，首先需要明确如下 3 点。

（1）Android 采用 Linux 作为内核。

（2）Android 对 Linux 内核做了修改，以适应其在移动设备上的应用。

（3）Andorid 开始是作为 Linux 的一个分支，后来由于无法并入 Linux 的主开发树，曾经被

Linux 内核组从开发树中删除。2012 年 5 月 18 日，Linux kernel 3.3 发布后来又被加入。

4.4.1　Android 继承于 Linux

Android 是在 Linux 的内核基础之上运行的，提供的核心系统服务包括安全、内存管理、进程管理、网络组和驱动模型等内容。内核部分还相当于一个介于硬件层和系统中其他软件组之间的一个抽象层次。但是严格来说它不算是 Linux 操作系统。

因为 Android 内核是由标准的 Linux 内核修改而来的，所以继承了 Linux 内核的诸多优点，保留了 Linux 内核的主题架构。同时 Android 按照移动设备的需求，在文件系统、内存管理、进程间通信机制和电源管理方面进行了修改，添加了相关的驱动程序和必要的新功能。但是和其他精简的 Linux 系统相比（例如 uClinux），Android 很大程度地保留了 Linux 的基本架构，因此 Android 的应用性和扩展性更强。当前 Android 版本对应的 Linux 内核版本如下所示。

- Android 1.5：Linux-2.6.27。
- Android 1.6：Linux-2.6.29。
- Android 2.0,2.1：Linux-2.6.29。
- Android 2.2：Linux-2.6.32.9。
- Android 4.3：Linux-3.4。

4.4.2　Android 和 Linux 内核的区别

Android 系统的系统层面的底层是 Linux，中间加上了一个叫作 Dalvik 的 Java 虚拟机，表面层上面是 Android 运行库。每个 Android 应用都运行在自己的进程上，享有 Dalvik 虚拟机为它分配的专有实例。为了支持多个虚拟机在同一个设备上高效运行，Dalvik 被改写过。

Dalvik 虚拟机执行的是 Dalvik 格式的可执行文件（.dex），该格式经过优化，以降低内存耗用到最低。Java 编译器将 Java 源文件转为 class 文件，class 文件又被内置的 dx 工具转化为 dex 格式文件，这种文件在 Dalvik 虚拟机上注册并运行。

Android 系统的应用软件都是运行在 Dalvik 之上的 Java 软件，而 Dalvik 是运行在 Linux 中的，在一些底层功能，例如线程和低内存管理方面，Dalvik 虚拟机是依赖 Linux 内核的。由此可见，可以说 Android 是运行在 Linux 之上的操作系统，但是它本身不能算是 Linux 的某个版本。

Android 内核和 Linux 内核的差别主要体现在 11 个方面，接下来将一一简要介绍。

1. Android Binder

其源代码位于：

```
drivers/staging/android/binder.c
```

Android Binder 是基于 OpenBinder 框架的一个驱动，用于提供 Android 平台的进程间通信（Inter-Process Communication，IPC）。原来的 Linux 系统上层应用的进程间通信主要是 D-bus（Desktop bus），采用消息总线的方式来进行 IPC。

2. Android 电源管理（PM）

Android 电源管理是一个基于标准 Linux 电源管理系统的轻量级的 Android 电源管理驱动，针对嵌入式设备做了很多优化。利用锁和定时器来切换系统状态，控制设备在不同状态下的功耗，以达到节能的目的。

Android 电源管理的源代码分别位于：

```
kernel/power/earlysuspend.c
kernel/power/consoleearlysuspend.c
kernel/power/fbearlysuspend.c
kernel/power/wakelock.c
kernel/power/userwakelock.c
```

3. 低内存管理器（Low Memory Killer）

Android 中的低内存管理器和 Linux 标准的 OOM（Out Of Memory）相比，其机制更加灵活，它可以根据需要杀死进程来释放需要的内存。Low Memory Killer 的代码很简单，关键的一个函数是 Lowmem_shrinker。作为一个模块在初始化时调用 register_shrinke 注册了个 lowmem_shrinker，它会被 VM 在内存紧张的情况下调用。Lowmem_shrinker 完成具体操作。简单说就是寻找一个最合适的进程将其杀死，从而释放它占用的内存。

低内存管理器的源代码位于 drivers/staging/android/lowmemorykiller.c。

4. 匿名共享内存（Ashmem）

匿名共享内存为进程间提供大块共享内存，同时为内核提供回收和管理这个内存的机制。如果一个程序尝试访问 Kernel 释放的一个共享内存块，它将会收到一个错误提示，然后重新分配内存并重载数据。

匿名共享内存的源代码位于 mm/ashmem.c。

5. Android PMEM（Physical）

PMEM 用于向用户空间提供连续的物理内存区域，DSP 和某些设备只能工作在连续的物理内存上。驱动中提供了 mmap、open、release 和 ioctl 等接口。

Android PMEM 的源代码位于 drivers/misc/pmem.c。

6. Android Logger

Android Logger 是一个轻量级的日志设备，用于抓取 Android 系统的各种日志，是 Linux 所没有的。

Android Logger 的源代码位于 drivers/staging/android/logger.c。

7. Android Alarm

Android Alarm 提供了一个定时器用于把设备从睡眠状态唤醒，同时它也提供了一个即使在设备睡眠时也会运行的时钟基准。

Android Alarm 的源代码位于：

- drivers/rtc/alarm.c；
- drivers/rtc/alarm-dev.c。

8. USB Gadget 驱动

USB Gadget 驱动是一个基于标准 Linux USB gadget 驱动框架的设备驱动，Android 的 USB 驱动是基于 gadget 框架的。

USB Gadget 驱动的源代码位于：

- drivers/usb/gadget/android.c；
- drivers/usb/gadget/f_adb.c；
- drivers/usb/gadget/f_mass_storage.c。

9. Android Ram Console

为了提供调试功能，Android 允许将调试日志信息写入一个被称为 RAM Console 的设备里，它是一个基于 RAM 的 Buffer。

Android Ram Console 的源代码位于 drivers/staging/android/ram_console.c。

10. Android timed device

Android timed device 提供了对设备进行定时控制功能，目前仅仅支持 vibrator 和 LED 设备。

Android timed device 的源代码位于 drivers/staging/android/timed_output.c(timed_gpio.c)。

11. YAFFS2 文件系统

在 Android 系统中，采用 YAFFS2 作为 MTD nand flash 文件系统。YAFFS2 是一个快速稳定的应用于 NAND 和 NOR Flash 的跨平台的嵌入式设备文件系统，同其他 Flash 文件系统相比，YAFFS2 使用更小的内存来保存它的运行状态，因此它占用内存小；YAFFS2 的垃圾回收非常简单而且快速，因此能达到更好的性能；YAFFS2 在大容量的 NAND Flash 上性能表现尤为明显，非常适合大容量的 Flash 存储。

YAFFS2 文件系统源代码位于 fs/yaffs2/目录下。

Android 是在 Linux 的内核基础之上运行的，提供的核心系统服务包括安全、内存管理、进程管理、网络组和驱动模型等内容。内核部分还相当于一个介于硬件层和系统中其他软件组之间的一个抽象层次。但是严格来说它不算是 Linux 操作系统。

4.5 第一段 Android 程序

本实例的功能是在手机屏幕中显示问候语“你好我的朋友!”，在具体开始之前先做一个简单的流程规划，如图 4-4 所示。

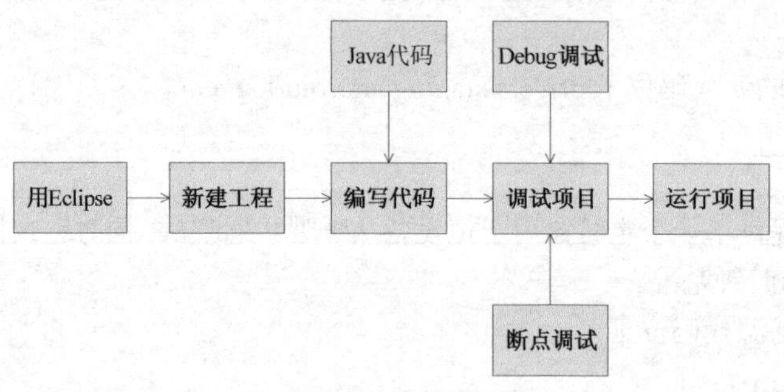

▲图 4-4　规划流程图

题目	目的	源码路径
实例 4-1	演示第一个 Android 应用程序	\daima\4\first

在本节的内容中，将详细讲解本实例的具体实现流程。

4.5.1　新建 Android 工程

（1）在 Eclipse 中依次单击“File”｜“New”｜“Project”新建一个工程文件。如图 4-5 所示。

（2）选择“Android Project”选项，单击“Next”按钮。

（3）在弹出的“New Android Project”对话框中，设置工程信息。如图 4-6 所示。

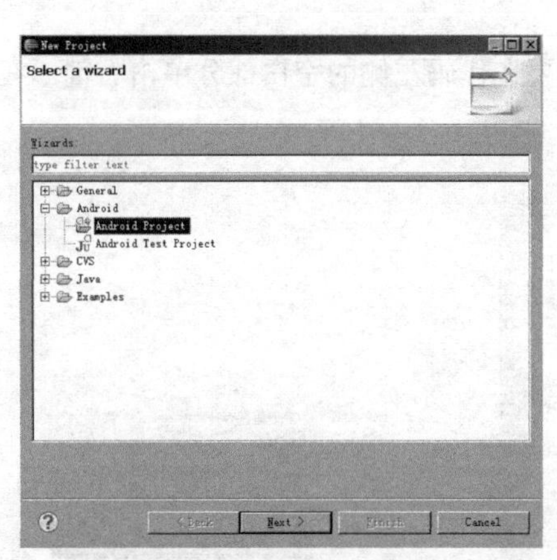

▲图 4-5　新建工程文件

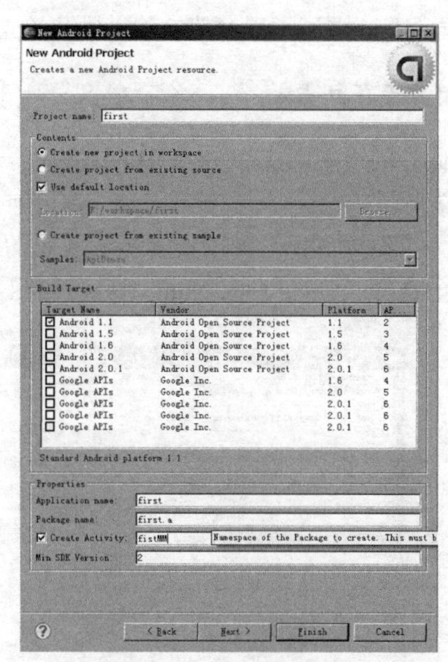

▲图 4-6　设置工程

在图 4-6 所示的界面中依次设置工程名称、包名称、Activity 名称和应用名称。

现在已经创建了一个名为“first”的工程文件，现在打开文件 first.java，会显示自动生成的如下代码。

```
package first.a;
import android.app.Activity;
import android.os.Bundle;
public class fistMM extends Activity {
    /** Called when the activity is first created. */
    @Override
    public void onCreate(Bundle savedInstanceState) {
        super.onCreate(savedInstanceState);
        setContentView(R.layout.main);
    }
}
```

如果此时运行程序，将不会显示任何内容。此时可以对上述代码进行适当的修改，让程序输出“HelloWorld”。具体代码如下所示。

```
package first.a;
import android.app.Activity;
import android.os.Bundle;
import android.widget.TextView;

public class fistMM extends Activity {
    /** Called when the activity is first created. */
    @Override
    public void onCreate(Bundle savedInstanceState) {
        super.onCreate(savedInstanceState);
        setContentView(R.layout.main);
        TextView tv = new TextView(this);
        tv.setText("你好我的朋友！");
        setContentView(tv);
    }
}
```

经过上述代码改写后，应该可以在屏幕中输出“你好我的朋友！”，完全符合预期的要求。

4.5.2　调试程序

Android 调试一般分为 3 个步骤，分别是设置断点、Debug 调试和断点调试。

1．设置断点

此处的设置断点和 Java 中的方法是一样的，可以通过双击代码左侧的区域进行断点设置。如图 4-7 所示。

为了调试方便，可以设置显示代码的行数。只需要在代码左侧的空白部分单击右键，在弹出命令中选择“Show Line Numbers”。如图 4-8 所示。

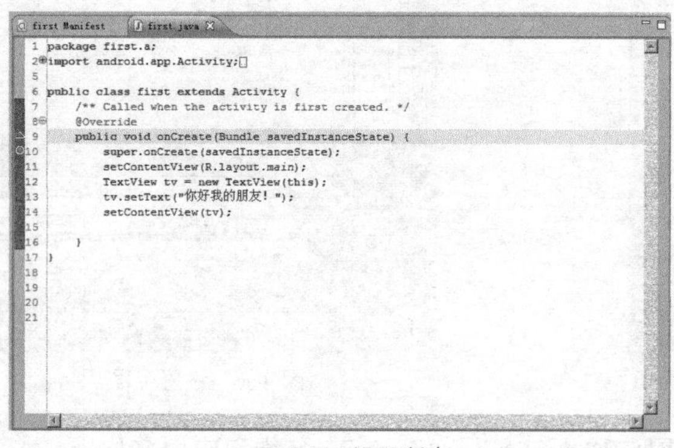

▲图 4-7　设置断点

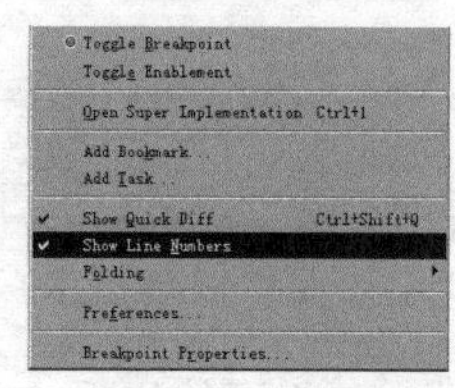

▲图 4-8　显示行数

2．Debug 调试

Debug Android 调试项目的方法和普通 Debug Java 调试项目的方法类似，唯一的不同是在选择调试项目时选择“Android Application”命令。具体方法是右键单击项目名，在弹出命令中依次选择“Debug As”｜“Android Application”命令。如图 4-9 所示。

3．断点调试

可以进行单步调试，具体调试方法和调试普通 Java 程序的方法类似，调试界面如图 4-10 所示。

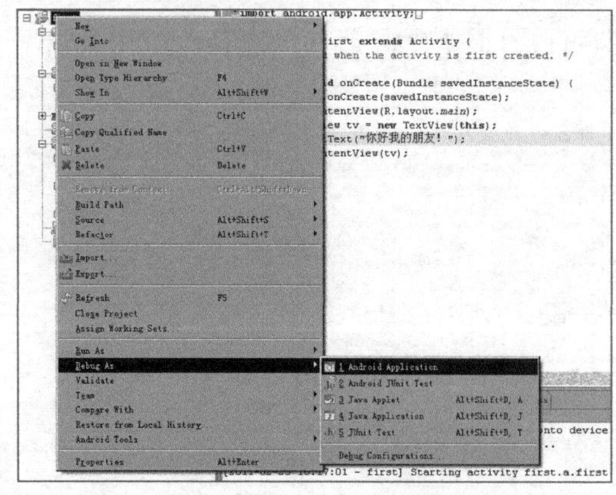

▲图 4-9　Debug 项目

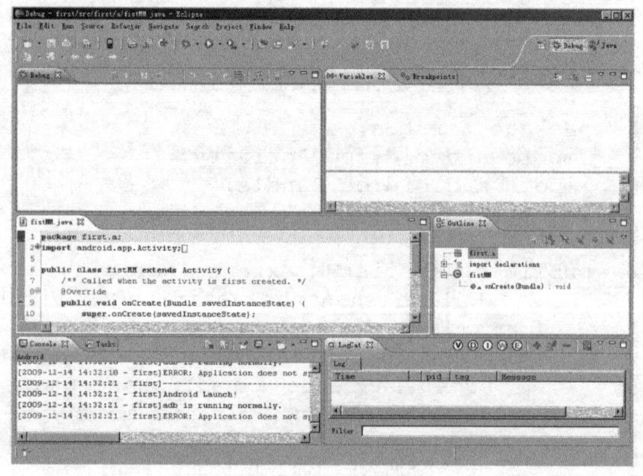

▲图 4-10　调试界面

4.5.3　运行程序

将上述代码保存后就可运行这段程序了，具体过程如下所示。

（1）右键单击项目名，在弹出命令中依次选择“Run As”|“Android Application”。如图 4-11 所示。

（2）此时工程开始运行，运行完成后在屏幕中输出“你好我的朋友！”这段文字。如图 4-12 所示。

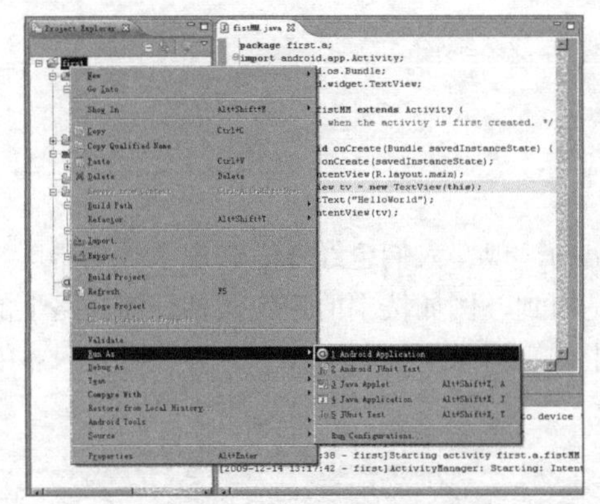

▲图 4-11　开始调试

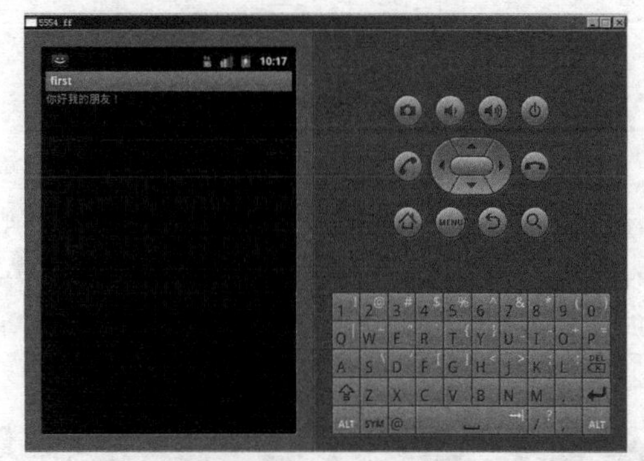

▲图 4-12　运行结果

第 5 章　Android 传感器系统分析

传感器是近年来随着物联网这一概念的流行而推出的，现在人们已经逐渐认识了传感器这一概念。其实传感器在大家日常的生活中经常见到甚至是用到，例如楼宇的声控楼梯灯和马路上的路灯等。在本章的内容中，将详细讲解 Android 系统中传感器系统的基本知识，为读者步入本书后面知识的学习打下基础。

5.1 Android 传感器系统概述

在 Android 系统中提供的主要传感器有加速度、磁场、方向、陀螺仪、光线、压力、温度和接近传感器等。传感器系统会主动对上层报告传感器精度和数据的变化，并且提供了设置传感器精度的接口，这些接口可以在 Java 应用和 Java 框架中使用。

Android 传感器系统的基本层次结构如图 5-1 所示。

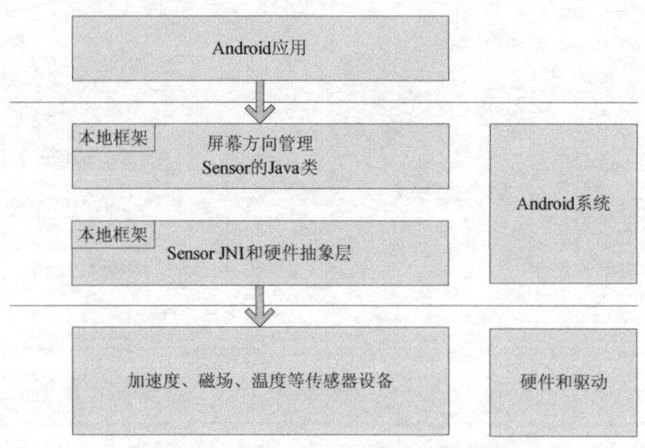

▲图 5-1　传感器系统的层次结构

根据图 5-1 所示的结构，Android 传感器系统从上到下分别是 Java 应用层、Java 框架对传感器的应用、传感器类、传感器硬件抽象层和传感器驱动。各个层的具体说明如下所示。

（1）传感器系统的 Java 部分

代码路径是：

```
frameworks/base/include/core/java/android/hardware
```

此部分对应的实现文件是 Sensor*.java。

（2）传感器系统的 JNI 部分

代码路径是：

```
frameworks/base/core/jni/android_hardware_SensorManager.cpp
```

在此部分中提供了对类 android.hardware.Sensor.Manage 的本地支持。

（3）传感器系统 HAL 层

头文件路径是：

```
hardware/libhardware/include/hardware/sensors.h
```

在 Android 系统中，传感器系统的硬件抽象层需要特定编码实现。

（4）驱动层

驱动层的代码路径是：

```
kernel/driver/hwmon/$(PROJECT)/sensor
```

在库 sensor.so 中提供了如下所示的 8 个 API 函数。

- 控制方面：在结构体 sensors_control_device_t 中定义，包括下面所示的函数。
  - ➢ int (*open_data_source)(struct sensors_control_device_t *dev);
  - ➢ int (*activate)(struct sensors_control_device_t *dev, int handle, int enabled);
  - ➢ int (*set_delay)(struct sensors_control_device_t *dev, int32_t ms);
  - ➢ int (*wake)(struct sensors_control_device_t *dev)。
- 数据方面：在结构体 sensors_data_device_t 中定义，包括下面所示的函数。
  - ➢ int (*data_open)(struct sensors_data_device_t *dev, int fd);
  - ➢ int (*data_close)(struct sensors_data_device_t *dev);
  - ➢ int (*poll)(struct sensors_data_device_t *dev, sensors_data_t* data)。
- 模块方面：在结构体 sensors_module_t 中定义，包括下面一个函数。

```
int (*get_sensors_list)(struct sensors_module_t* module, struct sensor_t const** list)
```

在 Android 系统的 Java 层中，Sensor 的状态是由 SensorService 来负责控制的，其 Java 代码和 JNI 代码分别位于如下文件中。

```
frameworks/base/services/java/com/android/server/SensorService.java
frameworks/base/services/jni/com_android_server_SensorService.cpp
```

SensorManager 负责在 Java 层 Sensor 的数据控制，它的 Java 代码和 JNI 代码分别位于如下文件中。

```
frameworks/base/core/java/android/hardware/SensorManager.java
frameworks/base/core/jni/android_hardware_SensorManager.cpp
```

在 Android 的 Framework 中，是通过文件 sensorService.java 和 sensorManager.java 实现与 Sensor 传感器通信的。文件 sensorService.java 的通信功能是通过 JNI 调用 sensorService.cpp 中的方法实现的。

文件 sensorManager.java 的具体通信功能是通过 JNI 调用 sensorManager.cpp 中的方法实现的。文件 sensorService.cpp 和 sensorManager.cpp 通过文件 hardware.c 与 sensor.so 通信。其中文件 sensorService.cpp 实现对 sensor 的状态控制，文件 sensorManger.cpp 实现对 sensor 的数据控制。

库 sensor.so 通过 ioctl 控制 sensor driver 的状态，通过打开 sensor driver 对应的设备文件读取 G-sensor 采集的数据。

5.2 分析 Java 层

在 Android 系统中，传感器系统的 Java 部分的实现文件是：

```
\sdk\apps\SdkController\src\com\android\tools\sdkcontroller\activities\SensorActivit
y.java
```

通过阅读文件 SensorActivity.java 的源码可知，在应用程序中使用传感器需要用到 hardware

包中的 SensorManager、SensorListener 等相关的类，具体实现代码如下所示。

```
public class SensorActivity extends BaseBindingActivity
        implements android.os.Handler.Callback {

    @SuppressWarnings("hiding")
    public static String TAG = SensorActivity.class.getSimpleName();
    private static boolean DEBUG = true;

    private static final int MSG_UPDATE_ACTUAL_HZ = 0x31415;

    private TableLayout mTableLayout;
    private TextView mTextError;
    private TextView mTextStatus;
    private TextView mTextTargetHz;
    private TextView mTextActualHz;
    private SensorChannel mSensorHandler;

    private final Map<MonitoredSensor, DisplayInfo> mDisplayedSensors =
        new HashMap<SensorChannel.MonitoredSensor, SensorActivity.DisplayInfo>();
    private final android.os.Handler mUiHandler = new android.os.Handler(this);
    private int mTargetSampleRate;
    private long mLastActualUpdateMs;

    /** 第一次创建 activity 是调用. */
    @Override
    public void onCreate(Bundle savedInstanceState) {
        super.onCreate(savedInstanceState);
        setContentView(R.layout.sensors);
        mTableLayout = (TableLayout) findViewById(R.id.tableLayout);
        mTextError  = (TextView) findViewById(R.id.textError);
        mTextStatus = (TextView) findViewById(R.id.textStatus);
        mTextTargetHz = (TextView) findViewById(R.id.textSampleRate);
        mTextActualHz = (TextView) findViewById(R.id.textActualRate);
        updateStatus("Waiting for connection");

        mTextTargetHz.setOnKeyListener(new OnKeyListener() {
            @Override
            public boolean onKey(View v, int keyCode, KeyEvent event) {
                updateSampleRate();
                return false;
            }
        });
        mTextTargetHz.setOnFocusChangeListener(new OnFocusChangeListener() {
            @Override
            public void onFocusChange(View v, boolean hasFocus) {
                updateSampleRate();
            }
        });
    }

    @Override
    protected void onResume() {
        if (DEBUG) Log.d(TAG, "onResume");
        // BaseBindingActivity 绑定后套服务
        super.onResume();
        updateError();
    }

    @Override
    protected void onPause() {
        if (DEBUG) Log.d(TAG, "onPause");
        // BaseBindingActivity.onResume will unbind from (but not stop) the service
        super.onPause();
    }

    @Override
    protected void onDestroy() {
        if (DEBUG) Log.d(TAG, "onDestroy");
```

```
    super.onDestroy();
    removeSensorUi();
}

// ----------

@Override
protected void onServiceConnected() {
    if (DEBUG) Log.d(TAG, "onServiceConnected");
    createSensorUi();
}

@Override
protected void onServiceDisconnected() {
    if (DEBUG) Log.d(TAG, "onServiceDisconnected");
    removeSensorUi();
}

@Override
protected ControllerListener createControllerListener() {
    return new SensorsControllerListener();
}

// ----------

private class SensorsControllerListener implements ControllerListener {
    @Override
    public void onErrorChanged() {
        runOnUiThread(new Runnable() {
            @Override
            public void run() {
                updateError();
            }
        });
    }

    @Override
    public void onStatusChanged() {
        runOnUiThread(new Runnable() {
            @Override
            public void run() {
                ControllerBinder binder = getServiceBinder();
                if (binder != null) {
                    boolean connected = binder.isEmuConnected();
                    mTableLayout.setEnabled(connected);
                    updateStatus(connected ? "Emulated connected" : "Emulator disconnected");
                }
            }
        });
    }
}

private void createSensorUi() {
    final LayoutInflater inflater = getLayoutInflater();

    if (!mDisplayedSensors.isEmpty()) {
        removeSensorUi();
    }

    mSensorHandler = (SensorChannel) getServiceBinder().getChannel(Channel.SENSOR_
    CHANNEL);
    if (mSensorHandler != null) {
        mSensorHandler.addUiHandler(mUiHandler);
        mUiHandler.sendEmptyMessage(MSG_UPDATE_ACTUAL_HZ);

        assert mDisplayedSensors.isEmpty();
        List<MonitoredSensor> sensors = mSensorHandler.getSensors();
        for (MonitoredSensor sensor : sensors) {
```

```
            final TableRow row = (TableRow) inflater.inflate(R.layout.sensor_row,
                                                    mTableLayout,
                                                    false);
            mTableLayout.addView(row);
            mDisplayedSensors.put(sensor, new DisplayInfo(sensor, row));
        }
    }
}

private void removeSensorUi() {
    if (mSensorHandler != null) {
        mSensorHandler.removeUiHandler(mUiHandler);
        mSensorHandler = null;
    }
    mTableLayout.removeAllViews();
    for (DisplayInfo info : mDisplayedSensors.values()) {
        info.release();
    }
    mDisplayedSensors.clear();
}

private class DisplayInfo implements CompoundButton.OnCheckedChangeListener {
    private MonitoredSensor mSensor;
    private CheckBox mChk;
    private TextView mVal;

    public DisplayInfo(MonitoredSensor sensor, TableRow row) {
        mSensor = sensor;

        // Initialize displayed checkbox for this sensor, and register
        // checked state listener for it
        mChk = (CheckBox) row.findViewById(R.id.row_checkbox);
        mChk.setText(sensor.getUiName());
        mChk.setEnabled(sensor.isEnabledByEmulator());
        mChk.setChecked(sensor.isEnabledByUser());
        mChk.setOnCheckedChangeListener(this);

        //初始化显示该传感器的文本框
        mVal = (TextView) row.findViewById(R.id.row_textview);
        mVal.setText(sensor.getValue());
    }

    /**
     *相关的复选框选中状态变化处理。当复选框被选中时会注册传感器变化。
     *如果不加以控制会取消传感器的变化
     */
    @Override
    public void onCheckedChanged(CompoundButton buttonView, boolean isChecked) {
        if (mSensor != null) {
            mSensor.onCheckedChanged(isChecked);
        }
    }

    public void release() {
        mChk = null;
        mVal = null;
        mSensor = null;

    }

    public void updateState() {
        if (mChk != null && mSensor != null) {
            mChk.setEnabled(mSensor.isEnabledByEmulator());
            mChk.setChecked(mSensor.isEnabledByUser());
        }
    }

    public void updateValue() {
        if (mVal != null && mSensor != null) {
```

```
            mVal.setText(mSensor.getValue());
        }
    }
}

/**实现回调处理程序*/
@Override
public boolean handleMessage(Message msg) {
    DisplayInfo info = null;
    switch (msg.what) {
    case SensorChannel.SENSOR_STATE_CHANGED:
        info = mDisplayedSensors.get(msg.obj);
        if (info != null) {
            info.updateState();
        }
        break;
    case SensorChannel.SENSOR_DISPLAY_MODIFIED:
        info = mDisplayedSensors.get(msg.obj);
        if (info != null) {
            info.updateValue();
        }
        if (mSensorHandler != null) {
            updateStatus(Integer.toString(mSensorHandler.getMsgSentCount())+"events sent");

            //如果值已经修改则更新 "actual rate"
            long ms = mSensorHandler.getActualUpdateMs();
            if (ms != mLastActualUpdateMs) {
                mLastActualUpdateMs = ms;
                String hz = mLastActualUpdateMs <= 0 ? "--" :
                        Integer.toString((int) Math.ceil(1000. / ms));
                mTextActualHz.setText(hz);
            }
        }
        break;
    case MSG_UPDATE_ACTUAL_HZ:
        if (mSensorHandler != null) {
            //如果值已经修改则更新 "actual rate"
            long ms = mSensorHandler.getActualUpdateMs();
            if (ms != mLastActualUpdateMs) {
                mLastActualUpdateMs = ms;
                String hz = mLastActualUpdateMs <= 0 ? "--" :
                        Integer.toString((int) Math.ceil(1000. / ms));
                mTextActualHz.setText(hz);
            }
            mUiHandler.sendEmptyMessageDelayed(MSG_UPDATE_ACTUAL_HZ, 1000 /*1s*/);
        }
    }
    return true; // we consumed this message
}

private void updateStatus(String status) {
    mTextStatus.setVisibility(status == null ? View.GONE : View.VISIBLE);
    if (status != null) mTextStatus.setText(status);
}

private void updateError() {
    ControllerBinder binder = getServiceBinder();
    String error = binder == null ? "" : binder.getServiceError();
    if (error == null) {
        error = "";
    }

    mTextError.setVisibility(error.length() == 0 ? View.GONE : View.VISIBLE);
    mTextError.setText(error);
}

private void updateSampleRate() {
    String str = mTextTargetHz.getText().toString();
    try {
```

```
            int hz = Integer.parseInt(str.trim());

            // Cap the value. 50 Hz is a reasonable max value for the emulator
            if (hz <= 0 || hz > 50) {
                hz = 50;
            }

            if (hz != mTargetSampleRate) {
                mTargetSampleRate = hz;
                if (mSensorHandler != null) {
                    mSensorHandler.setUpdateTargetMs(hz <= 0 ? 0 : (int)(1000.0f / hz));
                }
            }
        } catch (Exception ignore) {}
    }
}
```

通过上述代码可知，整个 Java 层利用了大家熟悉的观察者模式对传感器的数据进行了监听处理。

5.3 分析 Frameworks 层

在 Android 系统中，传感器系统的 Frameworks 层的代码路径是：

```
frameworks/base/include/core/java/android/hardware
```

Frameworks 层是 Android 系统提供的应用程序开发接口和应用程序框架。应用程序的调用是通过类实例化或类继承进行的。对应用程序来说，最重要的就是把 SensorListener 注册到 SensorManager 上，从而才能以观察者身份接收到数据的变化，因此，要把目光落在 SensorManager 的构造函数、RegisterListener 函数和通知机制相关的代码上。

本节将详细讲解传感器系统的 Frameworks 层的具体实现流程。

5.3.1　监听传感器的变化

在 Android 传感器系统的 Frameworks 层中，文件 SensorListener.java 用于监听从 Java 应用层中传递过来的变化。文件 SensorListener.java 比较简单，具体代码如下所示。

```
package android.hardware;
@Deprecated
public interface SensorListener {
    public void onSensorChanged(int sensor, float[] values);
    public void onAccuracyChanged(int sensor, int accuracy);
}
```

5.3.2　注册监听

当文件 SensorListener.java 监听到变化之后，会通过文件 SensorManager.java 来向服务注册监听变化，并调度 Sensor 的具体任务。例如在开发 Android 传感器应用程序时，在上层的通用开发流程如下所示。

（1）通过“getSystemService(SENSOR_SERVICE);”语句得到传感器服务。这样得到一个用来管理分配调度处理 Sensor 工作的 SensorManager。SensorManager 并不服务而是运行于后台，真正属于 Sensor 的系统服务是 SensorService，在终端下的“#service list”中可以看到 sensorservice: [android.gui.SensorServer]。

（2）通过“getDefaultSensor(Sensor.TYPE_GRAVITY);”得到传感器类型，当然还有各种功能不同的传感器，具体可以查阅 Android 官网 API 或者源码 Sensor.java。

（3）注册监听器 SensorEventListener。在应用程序中打开一个监听接口，专门用于处理传感器的数据。

（4）通过回调函数 onSensorChanged 和 onAccuracyChanged 实现实时监听。例如对重力感应器的 *xyz* 值经过算法变换得到左右上下前后方向等，就由这个回调函数实现。

综上所述，传感器顶层的处理流程如图 5-2 所示。

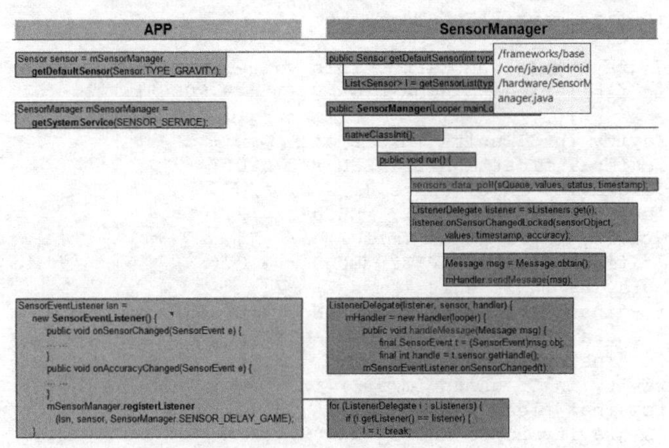

▲图 5-2　传感器顶层的处理流程

文件 SensorManager.java 的具体实现流程如下所示。

（1）定义类 SensorManager，然后设置各种传感器的初始变量值，具体代码如下所示。

```
public abstract class SensorManager {
    protected static final String TAG = "SensorManager";
    private static final float[] mTempMatrix = new float[16];

    // Cached lists of sensors by type.  Guarded by mSensorListByType
    private final SparseArray<List<Sensor>> mSensorListByType =
            new SparseArray<List<Sensor>>();

    // Legacy sensor manager implementation. Guarded by mSensorListByType during initialization
    private LegacySensorManager mLegacySensorManager;
    @Deprecated
    public static final int SENSOR_ORIENTATION = 1 << 0;
    @Deprecated
    public static final int SENSOR_ACCELEROMETER = 1 << 1;
    @Deprecated
    public static final int SENSOR_TEMPERATURE = 1 << 2;
    @Deprecated
    public static final int SENSOR_MAGNETIC_FIELD = 1 << 3;
    @Deprecated
    public static final int SENSOR_LIGHT = 1 << 4;
    @Deprecated
    public static final int SENSOR_PROXIMITY = 1 << 5;
    @Deprecated
    public static final int SENSOR_TRICORDER = 1 << 6;
    @Deprecated
    public static final int SENSOR_ORIENTATION_RAW = 1 << 7;
    @Deprecated
    public static final int SENSOR_ALL = 0x7F;
    @Deprecated
    public static final int SENSOR_MIN = SENSOR_ORIENTATION;
    @Deprecated
    public static final int SENSOR_MAX = ((SENSOR_ALL + 1)>>1);

    @Deprecated
    public static final int DATA_X = 0;
    @Deprecated
    public static final int DATA_Y = 1;
    @Deprecated
    public static final int DATA_Z = 2;
    @Deprecated
```

```
public static final int RAW_DATA_INDEX = 3;
@Deprecated
public static final int RAW_DATA_X = 3;
@Deprecated
public static final int RAW_DATA_Y = 4;
@Deprecated
public static final int RAW_DATA_Z = 5;

/** Standard gravity (g) on Earth. This value is equivalent to 1G */
public static final float STANDARD_GRAVITY = 9.80665f;

/** Sun's gravity in SI units (m/s^2) */
public static final float GRAVITY_SUN = 275.0f;
/** Mercury's gravity in SI units (m/s^2) */
public static final float GRAVITY_MERCURY = 3.70f;
/** Venus' gravity in SI units (m/s^2) */
public static final float GRAVITY_VENUS = 8.87f;
/** Earth's gravity in SI units (m/s^2) */
public static final float GRAVITY_EARTH = 9.80665f;
/** The Moon's gravity in SI units (m/s^2) */
public static final float GRAVITY_MOON = 1.6f;
/** Mars' gravity in SI units (m/s^2) */
public static final float GRAVITY_MARS = 3.71f;
/** Jupiter's gravity in SI units (m/s^2) */
public static final float GRAVITY_JUPITER = 23.12f;
/** Saturn's gravity in SI units (m/s^2) */
public static final float GRAVITY_SATURN = 8.96f;
/** Uranus' gravity in SI units (m/s^2) */
public static final float GRAVITY_URANUS = 8.69f;
/** Neptune's gravity in SI units (m/s^2) */
public static final float GRAVITY_NEPTUNE = 11.0f;
/** Pluto's gravity in SI units (m/s^2) */
public static final float GRAVITY_PLUTO = 0.6f;
/** Gravity (estimate) on the first Death Star in Empire units (m/s^2) */
public static final float GRAVITY_DEATH_STAR_I = 0.000000353036145f;
/** Gravity on the island */
public static final float GRAVITY_THE_ISLAND = 4.815162342f;


/** Maximum magnetic field on Earth's surface */
public static final float MAGNETIC_FIELD_EARTH_MAX = 60.0f;
/** Minimum magnetic field on Earth's surface */
public static final float MAGNETIC_FIELD_EARTH_MIN = 30.0f;


/** Standard atmosphere, or average sea-level pressure in hPa (millibar) */
public static final float PRESSURE_STANDARD_ATMOSPHERE = 1013.25f;


/** Maximum luminance of sunlight in lux */
public static final float LIGHT_SUNLIGHT_MAX = 120000.0f;
/** luminance of sunlight in lux */
public static final float LIGHT_SUNLIGHT = 110000.0f;
/** luminance in shade in lux */
public static final float LIGHT_SHADE = 20000.0f;
/** luminance under an overcast sky in lux */
public static final float LIGHT_OVERCAST = 10000.0f;
/** luminance at sunrise in lux */
public static final float LIGHT_SUNRISE = 400.0f;
/** luminance under a cloudy sky in lux */
public static final float LIGHT_CLOUDY = 100.0f;
/** luminance at night with full moon in lux */
public static final float LIGHT_FULLMOON = 0.25f;
/** luminance at night with no moon in lux*/
public static final float LIGHT_NO_MOON = 0.001f;


/** get sensor data as fast as possible */
public static final int SENSOR_DELAY_FASTEST = 0;
```

```
/** rate suitable for games */
public static final int SENSOR_DELAY_GAME = 1;
/** rate suitable for the user interface */
public static final int SENSOR_DELAY_UI = 2;
/**（默认值）适合屏幕方向的变化*/
public static final int SENSOR_DELAY_NORMAL = 3;


/**
 *返回的值，该传感器是不可信的，需要进行校准或环境不允许读数
 */
public static final int SENSOR_STATUS_UNRELIABLE = 0;

/**
 *该传感器是报告的低精度的数据，与环境的校准是必要的
 */
public static final int SENSOR_STATUS_ACCURACY_LOW = 1;

/**
 * This sensor is reporting data with an average level of accuracy,
 * calibration with the environment may improve the readings
 */
public static final int SENSOR_STATUS_ACCURACY_MEDIUM = 2;

/** This sensor is reporting data with maximum accuracy */
public static final int SENSOR_STATUS_ACCURACY_HIGH = 3;

/** see {@link #remapCoordinateSystem} */
public static final int AXIS_X = 1;
/** see {@link #remapCoordinateSystem} */
public static final int AXIS_Y = 2;
/** see {@link #remapCoordinateSystem} */
public static final int AXIS_Z = 3;
/** see {@link #remapCoordinateSystem} */
public static final int AXIS_MINUS_X = AXIS_X | 0x80;
/** see {@link #remapCoordinateSystem} */
public static final int AXIS_MINUS_Y = AXIS_Y | 0x80;
/** see {@link #remapCoordinateSystem} */
public static final int AXIS_MINUS_Z = AXIS_Z | 0x80;
```

（2）定义各种设备类型方法和设备数据的方法，这些方法非常重要，在编写的应用程序中，可以通过 AIDL 接口远程调用（RPC）的方式得到 SensorManager。这样通过在类 SensorManager 中的方法，可以得到底层的各种传感器数据。上述方法的具体实现代码如下所示。

```
public int getSensors() {
    return getLegacySensorManager().getSensors();
}
public List<Sensor> getSensorList(int type) {
    // cache the returned lists the first time
    List<Sensor> list;
    final List<Sensor> fullList = getFullSensorList();
    synchronized (mSensorListByType) {
        list = mSensorListByType.get(type);
        if (list == null) {
            if (type == Sensor.TYPE_ALL) {
                list = fullList;
            } else {
                list = new ArrayList<Sensor>();
                for (Sensor i : fullList) {
                    if (i.getType() == type)
                        list.add(i);
                }
            }
            list= Collections.unmodifiableList(list);
            mSensorListByType.append(type, list);
        }
    }
    return list;
```

```
    }
    public Sensor getDefaultSensor(int type) {
        // TODO: need to be smarter, for now, just return the 1st sensor
        List<Sensor> l = getSensorList(type);
        return l.isEmpty() ? null : l.get(0);
    }
    @Deprecated
    public boolean registerListener(SensorListener listener, int sensors) {
        return registerListener(listener, sensors, SENSOR_DELAY_NORMAL);
    }
    @Deprecated
    public boolean registerListener(SensorListener listener, int sensors, int rate) {
        return getLegacySensorManager().registerListener(listener, sensors, rate);
    }
    @Deprecated
    public void unregisterListener(SensorListener listener) {
        unregisterListener(listener, SENSOR_ALL | SENSOR_ORIENTATION_RAW);
    }
    @Deprecated
    public void unregisterListener(SensorListener listener, int sensors) {
        getLegacySensorManager().unregisterListener(listener, sensors);
    }
    public void unregisterListener(SensorEventListener listener, Sensor sensor) {
        if (listener == null || sensor == null) {
            return;
        }
        unregisterListenerImpl(listener, sensor);
    }
    public void unregisterListener(SensorEventListener listener) {
        if (listener == null) {
            return;
        }
        unregisterListenerImpl(listener, null);
    }
    protected abstract void unregisterListenerImpl(SensorEventListener listener, Sensor
    sensor);
    public boolean registerListener(SensorEventListener listener, Sensor sensor, int rate)
{
        return registerListener(listener, sensor, rate, null);
    }
    public boolean registerListener(SensorEventListener listener, Sensor sensor, int rate,
            Handler handler) {
        if (listener == null || sensor == null) {
            return false;
        }

        int delay = -1;
        switch (rate) {
            case SENSOR_DELAY_FASTEST:
                delay = 0;
                break;
            case SENSOR_DELAY_GAME:
                delay = 20000;
                break;
            case SENSOR_DELAY_UI:
                delay = 66667;
                break;
            case SENSOR_DELAY_NORMAL:
                delay = 200000;
                break;
            default:
                delay = rate;
                break;
        }
        return registerListenerImpl(listener, sensor, delay, handler);
    }
    protected abstract boolean registerListenerImpl(SensorEventListener listener, Sensor
    sensor,int delay, Handler handler);
```

```
public static boolean getRotationMatrix(float[] R, float[] I,
        float[] gravity, float[] geomagnetic) {
    // TODO: move this to native code for efficiency
    float Ax = gravity[0];
    float Ay = gravity[1];
    float Az = gravity[2];
    final float Ex = geomagnetic[0];
    final float Ey = geomagnetic[1];
    final float Ez = geomagnetic[2];
    float Hx = Ey*Az - Ez*Ay;
    float Hy = Ez*Ax - Ex*Az;
    float Hz = Ex*Ay - Ey*Ax;
    final float normH = (float)Math.sqrt(Hx*Hx + Hy*Hy + Hz*Hz);
    if (normH < 0.1f) {
        // device is close to free fall (or in space?), or close to
        // magnetic north pole. Typical values are  > 100
        return false;
    }
    final float invH = 1.0f / normH;
    Hx *= invH;
    Hy *= invH;
    Hz *= invH;
    final float invA = 1.0f / (float)Math.sqrt(Ax*Ax + Ay*Ay + Az*Az);
    Ax *= invA;
    Ay *= invA;
    Az *= invA;
    final float Mx = Ay*Hz - Az*Hy;
    final float My = Az*Hx - Ax*Hz;
    final float Mz = Ax*Hy - Ay*Hx;
    if (R != null) {
        if (R.length == 9) {
            R[0] = Hx;     R[1] = Hy;     R[2] = Hz;
            R[3] = Mx;     R[4] = My;     R[5] = Mz;
            R[6] = Ax;     R[7] = Ay;     R[8] = Az;
        } else if (R.length == 16) {
            R[0]  = Hx;    R[1]  = Hy;    R[2]  = Hz;    R[3]  = 0;
            R[4]  = Mx;    R[5]  = My;    R[6]  = Mz;    R[7]  = 0;
            R[8]  = Ax;    R[9]  = Ay;    R[10] = Az;    R[11] = 0;
            R[12] = 0;     R[13] = 0;     R[14] = 0;     R[15] = 1;
        }
    }
    if (I != null) {
        // compute the inclination matrix by projecting the geomagnetic
        // vector onto the Z (gravity) and X (horizontal component
        // of geomagnetic vector) axes
        final float invE = 1.0f / (float)Math.sqrt(Ex*Ex + Ey*Ey + Ez*Ez);
        final float c = (Ex*Mx + Ey*My + Ez*Mz) * invE;
        final float s = (Ex*Ax + Ey*Ay + Ez*Az) * invE;
        if (I.length == 9) {
            I[0] = 1;     I[1] = 0;     I[2] = 0;
            I[3] = 0;     I[4] = c;     I[5] = s;
            I[6] = 0;     I[7] =-s;     I[8] = c;
        } else if (I.length == 16) {
            I[0] = 1;     I[1] = 0;     I[2] = 0;
            I[4] = 0;     I[5] = c;     I[6] = s;
            I[8] = 0;     I[9] =-s;     I[10]= c;
            I[3] = I[7] = I[11] = I[12] = I[13] = I[14] = 0;
            I[15] = 1;
        }
    }
    return true;
}
public static float getInclination(float[] I) {
    if (I.length == 9) {
        return (float)Math.atan2(I[5], I[4]);
    } else {
        return (float)Math.atan2(I[6], I[5]);
    }
}
```

```
public static boolean remapCoordinateSystem(float[] inR, int X, int Y,
        float[] outR)
{
    if (inR == outR) {
        final float[] temp = mTempMatrix;
        synchronized(temp) {
            // we don't expect to have a lot of contention
            if (remapCoordinateSystemImpl(inR, X, Y, temp)) {
                final int size = outR.length;
                for (int i=0 ; i<size ; i++)
                    outR[i] = temp[i];
                return true;
            }
        }
    }
    return remapCoordinateSystemImpl(inR, X, Y, outR);
}

private static boolean remapCoordinateSystemImpl(float[] inR, int X, int Y,
        float[] outR)
{
    /*
     * X and Y define a rotation matrix 'r':
     *
     * (X==1)?((X&0x80)?-1:1):0 (X==2)?((X&0x80)?-1:1):0 (X==3)?((X&0x80)?-1:1):0
     * (Y==1)?((Y&0x80)?-1:1):0 (Y==2)?((Y&0x80)?-1:1):0 (Y==3)?((X&0x80)?-1:1):0
     *                          r[0] ^ r[1]
     *
     * where the 3rd line is the vector product of the first 2 lines
     *
     */

    final int length = outR.length;
    if (inR.length != length)
        return false;  // invalid parameter
    if ((X & 0x7C)!=0 || (Y & 0x7C)!=0)
        return false;  // invalid parameter
    if (((X & 0x3)==0) || ((Y & 0x3)==0))
        return false;  // no axis specified
    if ((X & 0x3) == (Y & 0x3))
        return false;  // same axis specified

    // Z is "the other" axis, its sign is either +/- sign(X)*sign(Y)
    // this can be calculated by exclusive-or'ing X and Y; except for
    // the sign inversion (+/-) which is calculated below
    int Z = X ^ Y;

    // extract the axis (remove the sign), offset in the range 0 to 2
    final int x = (X & 0x3)-1;
    final int y = (Y & 0x3)-1;
    final int z = (Z & 0x3)-1;

    // compute the sign of Z (whether it needs to be inverted)
    final int axis_y = (z+1)%3;
    final int axis_z = (z+2)%3;
    if (((x^axis_y)|(y^axis_z)) != 0)
        Z ^= 0x80;

    final boolean sx = (X>=0x80);
    final boolean sy = (Y>=0x80);
    final boolean sz = (Z>=0x80);

    // Perform R * r, in avoiding actual muls and adds
    final int rowLength = ((length==16)?4:3);
    for (int j=0 ; j<3 ; j++) {
        final int offset = j*rowLength;
        for (int i=0 ; i<3 ; i++) {
            if (x==i)   outR[offset+i] = sx ? -inR[offset+0] : inR[offset+0];
```

```
            if (y==i)  outR[offset+i] = sy ? -inR[offset+1] : inR[offset+1];
            if (z==i)  outR[offset+i] = sz ? -inR[offset+2] : inR[offset+2];
        }
    }
    if (length == 16) {
        outR[3] = outR[7] = outR[11] = outR[12] = outR[13] = outR[14] = 0;
        outR[15] = 1;
    }
    return true;
}
public static float[] getOrientation(float[] R, float values[]) {

    if (R.length == 9) {
        values[0] = (float)Math.atan2(R[1], R[4]);
        values[1] = (float)Math.asin(-R[7]);
        values[2] = (float)Math.atan2(-R[6], R[8]);
    } else {
        values[0] = (float)Math.atan2(R[1], R[5]);
        values[1] = (float)Math.asin(-R[9]);
        values[2] = (float)Math.atan2(-R[8], R[10]);
    }
    return values;
}
public static float getAltitude(float p0, float p) {
    final float coef = 1.0f / 5.255f;
    return 44330.0f * (1.0f - (float)Math.pow(p/p0, coef));
}

public static void getAngleChange( float[] angleChange, float[] R, float[] prevR) {
    float rd1=0,rd4=0, rd6=0,rd7=0, rd8=0;
    float ri0=0,ri1=0,ri2=0,ri3=0,ri4=0,ri5=0,ri6=0,ri7=0,ri8=0;
    float pri0=0, pri1=0, pri2=0, pri3=0, pri4=0, pri5=0, pri6=0, pri7=0, pri8=0;

    if(R.length == 9) {
        ri0 = R[0];
        ri1 = R[1];
        ri2 = R[2];
        ri3 = R[3];
        ri4 = R[4];
        ri5 = R[5];
        ri6 = R[6];
        ri7 = R[7];
        ri8 = R[8];
    } else if(R.length == 16) {
        ri0 = R[0];
        ri1 = R[1];
        ri2 = R[2];
        ri3 = R[4];
        ri4 = R[5];
        ri5 = R[6];
        ri6 = R[8];
        ri7 = R[9];
        ri8 = R[10];
    }

    if(prevR.length == 9) {
        pri0 = prevR[0];
        pri1 = prevR[1];
        pri2 = prevR[2];
        pri3 = prevR[3];
        pri4 = prevR[4];
        pri5 = prevR[5];
        pri6 = prevR[6];
        pri7 = prevR[7];
        pri8 = prevR[8];
    } else if(prevR.length == 16) {
        pri0 = prevR[0];
        pri1 = prevR[1];
        pri2 = prevR[2];
```

```
            pri3 = prevR[4];
            pri4 = prevR[5];
            pri5 = prevR[6];
            pri6 = prevR[8];
            pri7 = prevR[9];
            pri8 = prevR[10];
        }

        // calculate the parts of the rotation difference matrix we need
        // rd[i][j] = pri[0][i] * ri[0][j] + pri[1][i] * ri[1][j] + pri[2][i] * ri[2][j];

        rd1 = pri0 * ri1 + pri3 * ri4 + pri6 * ri7; //rd[0][1]
        rd4 = pri1 * ri1 + pri4 * ri4 + pri7 * ri7; //rd[1][1]
        rd6 = pri2 * ri0 + pri5 * ri3 + pri8 * ri6; //rd[2][0]
        rd7 = pri2 * ri1 + pri5 * ri4 + pri8 * ri7; //rd[2][1]
        rd8 = pri2 * ri2 + pri5 * ri5 + pri8 * ri8; //rd[2][2]

        angleChange[0] = (float)Math.atan2(rd1, rd4);
        angleChange[1] = (float)Math.asin(-rd7);
        angleChange[2] = (float)Math.atan2(-rd6, rd8);
    }
    public static void getRotationMatrixFromVector(float[] R, float[] rotationVector) {

        float q0;
        float q1 = rotationVector[0];
        float q2 = rotationVector[1];
        float q3 = rotationVector[2];

        if (rotationVector.length == 4) {
            q0 = rotationVector[3];
        } else {
            q0 = 1 - q1*q1 - q2*q2 - q3*q3;
            q0 = (q0 > 0) ? (float)Math.sqrt(q0) : 0;
        }

        float sq_q1 = 2 * q1 * q1;
        float sq_q2 = 2 * q2 * q2;
        float sq_q3 = 2 * q3 * q3;
        float q1_q2 = 2 * q1 * q2;
        float q3_q0 = 2 * q3 * q0;
        float q1_q3 = 2 * q1 * q3;
        float q2_q0 = 2 * q2 * q0;
        float q2_q3 = 2 * q2 * q3;
        float q1_q0 = 2 * q1 * q0;

        if(R.length == 9) {
            R[0] = 1 - sq_q2 - sq_q3;
            R[1] = q1_q2 - q3_q0;
            R[2] = q1_q3 + q2_q0;

            R[3] = q1_q2 + q3_q0;
            R[4] = 1 - sq_q1 - sq_q3;
            R[5] = q2_q3 - q1_q0;

            R[6] = q1_q3 - q2_q0;
            R[7] = q2_q3 + q1_q0;
            R[8] = 1 - sq_q1 - sq_q2;
        } else if (R.length == 16) {
            R[0] = 1 - sq_q2 - sq_q3;
            R[1] = q1_q2 - q3_q0;
            R[2] = q1_q3 + q2_q0;
            R[3] = 0.0f;

            R[4] = q1_q2 + q3_q0;
            R[5] = 1 - sq_q1 - sq_q3;
            R[6] = q2_q3 - q1_q0;
            R[7] = 0.0f;

            R[8] = q1_q3 - q2_q0;
```

```
            R[9] = q2_q3 + q1_q0;
            R[10] = 1 - sq_q1 - sq_q2;
            R[11] = 0.0f;

            R[12] = R[13] = R[14] = 0.0f;
            R[15] = 1.0f;
        }
    }
    public static void getQuaternionFromVector(float[] Q, float[] rv) {
        if (rv.length == 4) {
            Q[0] = rv[3];
        } else {
            Q[0] = 1 - rv[0]*rv[0] - rv[1]*rv[1] - rv[2]*rv[2];
            Q[0] = (Q[0] > 0) ? (float)Math.sqrt(Q[0]) : 0;
        }
        Q[1] = rv[0];
        Q[2] = rv[1];
        Q[3] = rv[2];
    }
    public boolean requestTriggerSensor(TriggerEventListener listener, Sensor sensor) {
        return requestTriggerSensorImpl(listener, sensor);
    }
    protected abstract boolean requestTriggerSensorImpl(TriggerEventListener listener,
            Sensor sensor);
    public boolean cancelTriggerSensor(TriggerEventListener listener, Sensor sensor) {
        return cancelTriggerSensorImpl(listener, sensor, true);
    protected abstract boolean cancelTriggerSensorImpl(TriggerEventListener listener,
            Sensor sensor, boolean disable);
    private LegacySensorManager getLegacySensorManager() {
        synchronized (mSensorListByType) {
            if (mLegacySensorManager == null) {
                Log.i(TAG,"This application is using deprecated SensorManager API which will "
                        + "be removed someday.  Please consider switching to the new API.");
                mLegacySensorManager = new LegacySensorManager(this);
            }
            return mLegacySensorManager;
        }
    }
}
```

上述方法的功能非常重要，其实就是人们在开发传感器应用程序时用到的 API 接口。有关上述方法的具体说明，读者可以查阅官网 SDK API 中对于类 android.hardware.SensorManager 的具体说明。

5.4 分析 JNI 层

在 Android 系统中，传感器系统的 JNI 部分的代码路径是：

```
frameworks/base/core/jni/android_hardware_SensorManager.cpp
```

在此文件中提供了对类 android.hardware.Sensor.Manage 的本地支持。上层和 JNI 层的调用关系如图 5-3 所示。

在上图所示的调用关系中涉及了下面所示的 API 接口方法。

- nativeClassInit()：在 JNI 层得到 android.hardware.Sensor 的 JNI 环境指针；
- sensors_module_init()：通过 JNI 调用本地框架，得到 SensorService 和 SensorService 初始化控制流各功能；
- new Sensor()：建立一个 Sensor 对象，具体可查阅官网 API android.hardware.Sensor；
- sensors_module_get_next_sensor()：上层得到设备支持的所有 Sensor，并放入 SensorList 链表；
- new SensorThread()：创建 Sensor 线程，当应用程序 registerListener()注册监听器的时候开启线程 run()，注意当没有数据变化时线程会阻塞。

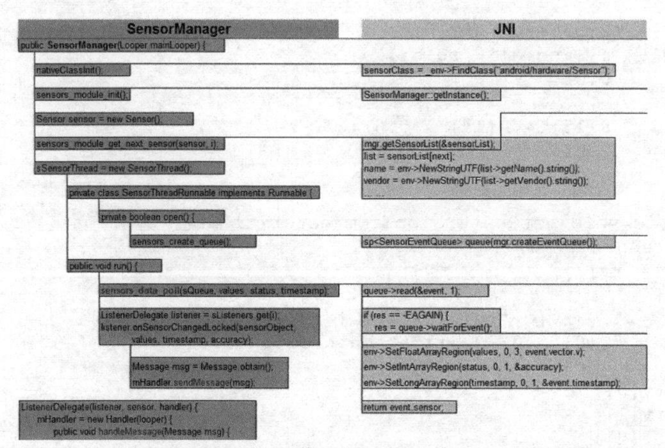

▲图 5-3　上层和 JNI 层的调用关系

5.4.1　分析 android_hardware_SensorManager.cpp

文件 android_hardware_SensorManager.cpp 的功能是实现文件 SensorManager.java 中的 native（本地）函数，主要是通过调用文件 SensorManager.cpp 和文件 SensorEventQueue.cpp 中的相关类来完成相关的工作的。文件 android_hardware_SensorManager.cpp 的具体实现代码如下所示。

```
static struct {
    jclass clazz;
    jmethodID dispatchSensorEvent;
} gBaseEventQueueClassInfo;

namespace android {

struct SensorOffsets
{
    jfieldID    name;
    jfieldID    vendor;
    jfieldID    version;
    jfieldID    handle;
    jfieldID    type;
    jfieldID    range;
    jfieldID    resolution;
    jfieldID    power;
    jfieldID    minDelay;
} gSensorOffsets;


/*
 * The method below are not thread-safe and not intended to be
 */

static void
nativeClassInit (JNIEnv *_env, jclass _this)
{
    jclass sensorClass = _env->FindClass("android/hardware/Sensor");
    SensorOffsets& sensorOffsets = gSensorOffsets;
    sensorOffsets.name = _env->GetFieldID(sensorClass, "mName", "Ljava/lang/String;");
    sensorOffsets.vendor = _env->GetFieldID(sensorClass, "mVendor", "Ljava/lang/String;");
    sensorOffsets.version = _env->GetFieldID(sensorClass, "mVersion", "I");
    sensorOffsets.handle = _env->GetFieldID(sensorClass, "mHandle", "I");
    sensorOffsets.type = _env->GetFieldID(sensorClass, "mType", "I");
    sensorOffsets.range = _env->GetFieldID(sensorClass, "mMaxRange", "F");
    sensorOffsets.resolution = _env->GetFieldID(sensorClass, "mResolution","F");
    sensorOffsets.power = _env->GetFieldID(sensorClass, "mPower", "F");
    sensorOffsets.minDelay = _env->GetFieldID(sensorClass, "mMinDelay", "I");
}
```

```
static jint
nativeGetNextSensor(JNIEnv *env, jclass clazz, jobject sensor, jint next)
{
    SensorManager& mgr(SensorManager::getInstance());

    Sensor const* const* sensorList;
    size_t count = mgr.getSensorList(&sensorList);
    if (size_t(next) >= count)
        return -1;

    Sensor const* const list = sensorList[next];
    const SensorOffsets& sensorOffsets(gSensorOffsets);
    jstring name = env->NewStringUTF(list->getName().string());
    jstring vendor = env->NewStringUTF(list->getVendor().string());
    env->SetObjectField(sensor, sensorOffsets.name, name);
    env->SetObjectField(sensor, sensorOffsets.vendor, vendor);
    env->SetIntField(sensor, sensorOffsets.version, list->getVersion());
    env->SetIntField(sensor, sensorOffsets.handle, list->getHandle());
    env->SetIntField(sensor, sensorOffsets.type, list->getType());
    env->SetFloatField(sensor, sensorOffsets.range, list->getMaxValue());
    env->SetFloatField(sensor, sensorOffsets.resolution, list->getResolution());
    env->SetFloatField(sensor, sensorOffsets.power, list->getPowerUsage());
    env->SetIntField(sensor, sensorOffsets.minDelay, list->getMinDelay());

    next++;
    return size_t(next) < count ? next : 0;
}
//----------------------------------------------------------------------------

class Receiver : public LooperCallback {
    sp<SensorEventQueue> mSensorQueue;
    sp<MessageQueue> mMessageQueue;
    jobject mReceiverObject;
    jfloatArray mScratch;
public:
    Receiver(const sp<SensorEventQueue>& sensorQueue,
            const sp<MessageQueue>& messageQueue,
            jobject receiverObject, jfloatArray scratch) {
        JNIEnv* env = AndroidRuntime::getJNIEnv();
        mSensorQueue = sensorQueue;
        mMessageQueue = messageQueue;
        mReceiverObject = env->NewGlobalRef(receiverObject);
        mScratch = (jfloatArray)env->NewGlobalRef(scratch);
    }
    ~Receiver() {
        JNIEnv* env = AndroidRuntime::getJNIEnv();
        env->DeleteGlobalRef(mReceiverObject);
        env->DeleteGlobalRef(mScratch);
    }
    sp<SensorEventQueue> getSensorEventQueue() const {
        return mSensorQueue;
    }

    void destroy() {
        mMessageQueue->getLooper()->removeFd( mSensorQueue->getFd() );
    }

private:
    virtual void onFirstRef() {
        LooperCallback::onFirstRef();
        mMessageQueue->getLooper()->addFd(mSensorQueue->getFd(), 0,
                ALOOPER_EVENT_INPUT, this, mSensorQueue.get());
    }

    virtual int handleEvent(int fd, int events, void* data) {
        JNIEnv* env = AndroidRuntime::getJNIEnv();
        sp<SensorEventQueue> q = reinterpret_cast<SensorEventQueue *>(data);
        ssize_t n;
```

```
        ASensorEvent buffer[16];
        while ((n = q->read(buffer, 16)) > 0) {
            for (int i=0 ; i<n ; i++) {

                env->SetFloatArrayRegion(mScratch, 0, 16, buffer[i].data);

                env->CallVoidMethod(mReceiverObject,
                        gBaseEventQueueClassInfo.dispatchSensorEvent,
                        buffer[i].sensor,
                        mScratch,
                        buffer[i].vector.status,
                        buffer[i].timestamp);

                if (env->ExceptionCheck()) {
                    ALOGE("Exception dispatching input event.");
                    return 1;
                }
            }
        }
        if (n<0 && n != -EAGAIN) {
            // FIXME: error receiving events, what to do in this case?
        }

        return 1;
    }
};

static jint nativeInitSensorEventQueue(JNIEnv *env, jclass clazz, jobject eventQ, jobject
msgQ, jfloatArray scratch) {
    SensorManager& mgr(SensorManager::getInstance());
    sp<SensorEventQueue> queue(mgr.createEventQueue());

    sp<MessageQueue> messageQueue = android_os_MessageQueue_getMessageQueue(env, msgQ);
    if (messageQueue == NULL) {
        jniThrowRuntimeException(env, "MessageQueue is not initialized.");
        return 0;
    }

    sp<Receiver> receiver = new Receiver(queue, messageQueue, eventQ, scratch);
    receiver->incStrong((void*)nativeInitSensorEventQueue);
    return jint(receiver.get());
}

static jint nativeEnableSensor(JNIEnv *env, jclass clazz, jint eventQ, jint handle, jint
us) {
    sp<Receiver> receiver(reinterpret_cast<Receiver *>(eventQ));
    return receiver->getSensorEventQueue()->enableSensor(handle, us);
}

static jint nativeDisableSensor(JNIEnv *env, jclass clazz, jint eventQ, jint handle) {
    sp<Receiver> receiver(reinterpret_cast<Receiver *>(eventQ));
    return receiver->getSensorEventQueue()->disableSensor(handle);
}

static void nativeDestroySensorEventQueue(JNIEnv *env, jclass clazz, jint eventQ, jint
handle) {
    sp<Receiver> receiver(reinterpret_cast<Receiver *>(eventQ));
    receiver->destroy();
    receiver->decStrong((void*)nativeInitSensorEventQueue);
}

//----------------------------------------------------------------------------

static JNINativeMethod gSystemSensorManagerMethods[] = {
    {"nativeClassInit",
            "()V",
            (void*)nativeClassInit },
```

```
    {"nativeGetNextSensor",
            "(Landroid/hardware/Sensor;I)I",
            (void*)nativeGetNextSensor },
};

static JNINativeMethod gBaseEventQueueMethods[] = {
    {"nativeInitBaseEventQueue",

"(Landroid/hardware/SystemSensorManager$BaseEventQueue;Landroid/os/MessageQueue;[F)I",
            (void*)nativeInitSensorEventQueue },

    {"nativeEnableSensor",
            "(III)I",
            (void*)nativeEnableSensor },

    {"nativeDisableSensor",
            "(II)I",
            (void*)nativeDisableSensor },

    {"nativeDestroySensorEventQueue",
            "(I)V",
            (void*)nativeDestroySensorEventQueue },
};

}; // namespace android

using namespace android;

#define FIND_CLASS(var, className) \
        var = env->FindClass(className); \
        LOG_FATAL_IF(! var, "Unable to find class " className); \
        var = jclass(env->NewGlobalRef(var));

#define GET_METHOD_ID(var, clazz, methodName, methodDescriptor) \
        var = env->GetMethodID(clazz, methodName, methodDescriptor); \
        LOG_FATAL_IF(! var, "Unable to find method " methodName);

int register_android_hardware_SensorManager(JNIEnv *env)
{
    jniRegisterNativeMethods(env, "android/hardware/SystemSensorManager",
            gSystemSensorManagerMethods, NELEM(gSystemSensorManagerMethods));

    jniRegisterNativeMethods(env,
"android/hardware/SystemSensorManager$BaseEventQueue",
            gBaseEventQueueMethods, NELEM(gBaseEventQueueMethods));

    FIND_CLASS(gBaseEventQueueClassInfo.clazz,
"android/hardware/SystemSensorManager$BaseEventQueue");

    GET_METHOD_ID(gBaseEventQueueClassInfo.dispatchSensorEvent,
            gBaseEventQueueClassInfo.clazz,
            "dispatchSensorEvent", "(I[FIJ)V");

    return 0;
}
```

5.4.2 处理客户端数据

文件 frameworks\native\libs\gui\SensorManager.cpp 功能是提供了对传感器数据的部分操作，实现了“sensor_data_XXX()”格式的函数。另外在 Native 层的客户端，文件 SensorManager.cpp 还负责与服务端 SensorService.cpp 之间的通信工作。文件 SensorManager.cpp 的具体实现代码如下所示。

```
// ----------------------------------------------------------------------------
namespace android {
// ----------------------------------------------------------------------------

ANDROID_SINGLETON_STATIC_INSTANCE(SensorManager)
```

```
SensorManager::SensorManager()
    : mSensorList(0)
{
    // okay we're not locked here, but it's not needed during construction
    assertStateLocked();
}

SensorManager::~SensorManager()
{
    free(mSensorList);
}

void SensorManager::sensorManagerDied()
{
    Mutex::Autolock _l(mLock);
    mSensorServer.clear();
    free(mSensorList);
    mSensorList = NULL;
    mSensors.clear();
}

status_t SensorManager::assertStateLocked() const {
    if (mSensorServer == NULL) {
        // try for one second
        const String16 name("sensorservice");
        for (int i=0 ; i<4 ; i++) {
            status_t err = getService(name, &mSensorServer);
            if (err == NAME_NOT_FOUND) {
                usleep(250000);
                continue;
            }
            if (err != NO_ERROR) {
                return err;
            }
            break;
        }

        class DeathObserver : public IBinder::DeathRecipient {
            SensorManager& mSensorManger;
            virtual void binderDied(const wp<IBinder>& who) {
                ALOGW("sensorservice died [%p]", who.unsafe_get());
                mSensorManger.sensorManagerDied();
            }
        public:
            DeathObserver(SensorManager& mgr) : mSensorManger(mgr) { }
        };

        mDeathObserver = new DeathObserver(*const_cast<SensorManager *>(this));
        mSensorServer->asBinder()->linkToDeath(mDeathObserver);

        mSensors = mSensorServer->getSensorList();
        size_t count = mSensors.size();
        mSensorList = (Sensor const**)malloc(count * sizeof(Sensor*));
        for (size_t i=0 ; i<count ; i++) {
            mSensorList[i] = mSensors.array() + i;
        }
    }

    return NO_ERROR;
}


ssize_t SensorManager::getSensorList(Sensor const* const** list) const
{
    Mutex::Autolock _l(mLock);
    status_t err = assertStateLocked();
    if (err < 0) {
        return ssize_t(err);
    }
```

```
    *list = mSensorList;
    return mSensors.size();
}

Sensor const* SensorManager::getDefaultSensor(int type)
{
    Mutex::Autolock _l(mLock);
    if (assertStateLocked() == NO_ERROR) {
        // For now we just return the first sensor of that type we find
        // in the future it will make sense to let the SensorService make
        // that decision
        for (size_t i=0 ; i<mSensors.size() ; i++) {
            if (mSensorList[i]->getType() == type)
                return mSensorList[i];
        }
    }
    return NULL;
}

sp<SensorEventQueue> SensorManager::createEventQueue()
{
    sp<SensorEventQueue> queue;

    Mutex::Autolock _l(mLock);
    while (assertStateLocked() == NO_ERROR) {
        sp<ISensorEventConnection> connection =
                mSensorServer->createSensorEventConnection();
        if (connection == NULL) {
            // SensorService just died
            ALOGE("createEventQueue: connection is NULL. SensorService died.");
            continue;
        }
        queue = new SensorEventQueue(connection);
        break;
    }
    return queue;
}

// ----------------------------------------------------------------------------
}; // namespace android
```

5.4.3 处理服务端数据

文件 frameworks\native\services\sensorservice\SensorService.cpp 功能是实现了 Sensor 真正的后台服务，是服务端的数据处理中心。在 Android 的传感器系统中，SensorService 作为一个轻量级的 System Service，运行于 SystemServer 内，即在 system_init<system_init.cpp>中调用了 SensorService::instantiate()。SensorService 主要功能如下所示。

（1）通过 SensorService::instantiate 创建实例对象，并增加到 ServiceManager 中，然后创建并启动线程，并执行 threadLoop；

（2）threadLoop 从 sensor 驱动获取原始数据，然后通过 SensorEventConnection 把事件发送给客户端；

（3）BnSensorServer 的成员函数负责让客户端获取 sensor 列表和创建 SensorEventConnection。

文件 SensorService.cpp 的具体实现代码如下所示。

```
namespace android {

const char* SensorService::WAKE_LOCK_NAME = "SensorService";

SensorService::SensorService()
    : mInitCheck(NO_INIT)
{
}

void SensorService::onFirstRef()
```

```
{
    ALOGD("nuSensorService starting...");

    SensorDevice& dev(SensorDevice::getInstance());

    if (dev.initCheck() == NO_ERROR) {
        sensor_t const* list;
        ssize_t count = dev.getSensorList(&list);
        if (count > 0) {
            ssize_t orientationIndex = -1;
            bool hasGyro = false;
            uint32_t virtualSensorsNeeds =
                    (1<<SENSOR_TYPE_GRAVITY) |
                    (1<<SENSOR_TYPE_LINEAR_ACCELERATION) |
                    (1<<SENSOR_TYPE_ROTATION_VECTOR);

            mLastEventSeen.setCapacity(count);
            for (ssize_t i=0 ; i<count ; i++) {
                registerSensor( new HardwareSensor(list[i]) );
                switch (list[i].type) {
                    case SENSOR_TYPE_ORIENTATION:
                        orientationIndex = i;
                        break;
                    case SENSOR_TYPE_GYROSCOPE:
                        hasGyro = true;
                        break;
                    case SENSOR_TYPE_GRAVITY:
                    case SENSOR_TYPE_LINEAR_ACCELERATION:
                    case SENSOR_TYPE_ROTATION_VECTOR:
                        virtualSensorsNeeds &= ~(1<<list[i].type);
                        break;
                }
            }

            // it's safe to instantiate the SensorFusion object here
            // (it wants to be instantiated after h/w sensors have been
            // registered)
            const SensorFusion& fusion(SensorFusion::getInstance());

            if (hasGyro) {
                // Always instantiate Android's virtual sensors. Since they are
                // instantiated behind sensors from the HAL, they won't
                // interfere with applications, unless they looks specifically
                // for them (by name)

                registerVirtualSensor( new RotationVectorSensor() );
                registerVirtualSensor( new GravitySensor(list, count) );
                registerVirtualSensor( new LinearAccelerationSensor(list, count) );

                // these are optional
                registerVirtualSensor( new OrientationSensor() );
                registerVirtualSensor( new CorrectedGyroSensor(list, count) );
            }

            // build the sensor list returned to users
            mUserSensorList = mSensorList;

            if (hasGyro) {
                // virtual debugging sensors are not added to mUserSensorList
                registerVirtualSensor( new GyroDriftSensor() );
            }

            if (hasGyro &&
                    (virtualSensorsNeeds & (1<<SENSOR_TYPE_ROTATION_VECTOR))) {
                // if we have the fancy sensor fusion, and it's not provided by the
                // HAL, use our own (fused) orientation sensor by removing the
                // HAL supplied one form the user list
                if (orientationIndex >= 0) {
                    mUserSensorList.removeItemsAt(orientationIndex);
```

```
            }
        }

        // debugging sensor list
        for (size_t i=0 ; i<mSensorList.size() ; i++) {
            switch (mSensorList[i].getType()) {
                case SENSOR_TYPE_GRAVITY:
                case SENSOR_TYPE_LINEAR_ACCELERATION:
                case SENSOR_TYPE_ROTATION_VECTOR:
                    if (strstr(mSensorList[i].getVendor().string(), "Google")) {
                        mUserSensorListDebug.add(mSensorList[i]);
                    }
                    break;
                default:
                    mUserSensorListDebug.add(mSensorList[i]);
                    break;
            }
        }

        run("SensorService", PRIORITY_URGENT_DISPLAY);
        mInitCheck = NO_ERROR;
    }
  }
}

void SensorService::registerSensor(SensorInterface* s)
{
    sensors_event_t event;
    memset(&event, 0, sizeof(event));

    const Sensor sensor(s->getSensor());
    // add to the sensor list (returned to clients)
    mSensorList.add(sensor);
    // add to our handle->SensorInterface mapping
    mSensorMap.add(sensor.getHandle(), s);
    // create an entry in the mLastEventSeen array
    mLastEventSeen.add(sensor.getHandle(), event);
}

void SensorService::registerVirtualSensor(SensorInterface* s)
{
    registerSensor(s);
    mVirtualSensorList.add( s );
}

SensorService::~SensorService()
{
    for (size_t i=0 ; i<mSensorMap.size() ; i++)
        delete mSensorMap.valueAt(i);
}

static const String16 sDump("android.permission.DUMP");

status_t SensorService::dump(int fd, const Vector<String16>& args)
{
    const size_t SIZE = 1024;
    char buffer[SIZE];
    String8 result;
    if (!PermissionCache::checkCallingPermission(sDump)) {
        snprintf(buffer, SIZE, "Permission Denial: "
                "can't dump SurfaceFlinger from pid=%d, uid=%d\n",
                IPCThreadState::self()->getCallingPid(),
                IPCThreadState::self()->getCallingUid());
        result.append(buffer);
    } else {
        Mutex::Autolock _l(mLock);
        snprintf(buffer, SIZE, "Sensor List:\n");
        result.append(buffer);
        for (size_t i=0 ; i<mSensorList.size() ; i++) {
```

```
            const Sensor& s(mSensorList[i]);
            const sensors_event_t& e(mLastEventSeen.valueFor(s.getHandle()));
            snprintf(buffer, SIZE,
                    "%-48s| %-32s | 0x%08x | maxRate=%7.2fHz | "
                    "last=<%5.1f,%5.1f,%5.1f>\n",
                    s.getName().string(),
                    s.getVendor().string(),
                    s.getHandle(),
                    s.getMinDelay() ? (1000000.0f / s.getMinDelay()) : 0.0f,
                    e.data[0], e.data[1], e.data[2]);
            result.append(buffer);
        }
        SensorFusion::getInstance().dump(result, buffer, SIZE);
        SensorDevice::getInstance().dump(result, buffer, SIZE);

        snprintf(buffer, SIZE, "%d active connections\n",
                mActiveConnections.size());
        result.append(buffer);
        snprintf(buffer, SIZE, "Active sensors:\n");
        result.append(buffer);
        for (size_t i=0 ; i<mActiveSensors.size() ; i++) {
            int handle = mActiveSensors.keyAt(i);
            snprintf(buffer, SIZE, "%s (handle=0x%08x, connections=%d)\n",
                    getSensorName(handle).string(),
                    handle,
                    mActiveSensors.valueAt(i)->getNumConnections());
            result.append(buffer);
        }
    }
    write(fd, result.string(), result.size());
    return NO_ERROR;
}

void SensorService::cleanupAutoDisabledSensor(const sp<SensorEventConnection>& connection,
        sensors_event_t const* buffer, const int count) {
    SensorInterface* sensor;
    status_t err = NO_ERROR;
    for (int i=0 ; i<count ; i++) {
        int handle = buffer[i].sensor;
        if (getSensorType(handle) == SENSOR_TYPE_SIGNIFICANT_MOTION) {
            if (connection->hasSensor(handle)) {
                sensor = mSensorMap.valueFor(handle);
                err = sensor ?sensor->resetStateWithoutActuatingHardware(connection.
                get(), handle): status_t(BAD_VALUE);
                if (err != NO_ERROR) {
                    ALOGE("Sensor Inteface: Resetting state failed with err: %d", err);
                }
                cleanupWithoutDisable(connection, handle);
            }
        }
    }
}

bool SensorService::threadLoop()
{
    ALOGD("nuSensorService thread starting...");

    const size_t numEventMax = 16;
    const size_t minBufferSize = numEventMax + numEventMax * mVirtualSensorList.size();
    sensors_event_t buffer[minBufferSize];
    sensors_event_t scratch[minBufferSize];
    SensorDevice& device(SensorDevice::getInstance());
    const size_t vcount = mVirtualSensorList.size();

    ssize_t count;
    bool wakeLockAcquired = false;
    const int halVersion = device.getHalDeviceVersion();
    do {
        count = device.poll(buffer, numEventMax);
```

```
if (count<0) {
    ALOGE("sensor poll failed (%s)", strerror(-count));
    break;
}

// Poll has returned. Hold a wakelock
// Todo(): add a flag to the sensors definitions to indicate
// the sensors which can wake up the AP
for (int i = 0; i < count; i++) {
    if (getSensorType(buffer[i].sensor) == SENSOR_TYPE_SIGNIFICANT_MOTION) {
        acquire_wake_lock(PARTIAL_WAKE_LOCK, WAKE_LOCK_NAME);
        wakeLockAcquired = true;
        break;
    }
}

recordLastValue(buffer, count);

// handle virtual sensors
if (count && vcount) {
    sensors_event_t const * const event = buffer;
    const DefaultKeyedVector<int, SensorInterface*> virtualSensors(
            getActiveVirtualSensors());
    const size_t activeVirtualSensorCount = virtualSensors.size();
    if (activeVirtualSensorCount) {
        size_t k = 0;
        SensorFusion& fusion(SensorFusion::getInstance());
        if (fusion.isEnabled()) {
            for (size_t i=0 ; i<size_t(count) ; i++) {
                fusion.process(event[i]);
            }
        }
        for (size_t i=0 ; i<size_t(count) && k<minBufferSize ; i++) {
            for (size_t j=0 ; j<activeVirtualSensorCount ; j++) {
                if (count + k >= minBufferSize) {
                    ALOGE("buffer too small to hold all events: "
                            "count=%u, k=%u, size=%u",
                            count, k, minBufferSize);
                    break;
                }
                sensors_event_t out;
                SensorInterface* si = virtualSensors.valueAt(j);
                if (si->process(&out, event[i])) {
                    buffer[count + k] = out;
                    k++;
                }
            }
        }
        if (k) {
            // record the last synthesized values
            recordLastValue(&buffer[count], k);
            count += k;
            // sort the buffer by time-stamps
            sortEventBuffer(buffer, count);
        }
    }
}

// handle backward compatibility for RotationVector sensor
if (halVersion < SENSORS_DEVICE_API_VERSION_1_0) {
    for (int i = 0; i < count; i++) {
        if (getSensorType(buffer[i].sensor) == SENSOR_TYPE_ROTATION_VECTOR) {
            // All the 4 components of the quaternion should be available
            // No heading accuracy. Set it to -1
            buffer[i].data[4] = -1;
        }
    }
}
```

```
        // send our events to clients
        const SortedVector< wp<SensorEventConnection> > activeConnections(
                getActiveConnections());
        size_t numConnections = activeConnections.size();
        for (size_t i=0 ; i<numConnections ; i++) {
            sp<SensorEventConnection> connection(
                    activeConnections[i].promote());
            if (connection != 0) {
                connection->sendEvents(buffer, count, scratch);
                // Some sensors need to be auto disabled after the trigger
                cleanupAutoDisabledSensor(connection, buffer, count);
            }
        }

        // We have read the data, upper layers should hold the wakelock
        if (wakeLockAcquired) release_wake_lock(WAKE_LOCK_NAME);

    } while (count >= 0 || Thread::exitPending());

    ALOGW("Exiting SensorService::threadLoop => aborting...");
    abort();
    return false;
}

void SensorService::recordLastValue(
        sensors_event_t const * buffer, size_t count)
{
    Mutex::Autolock _l(mLock);

    // record the last event for each sensor
    int32_t prev = buffer[0].sensor;
    for (size_t i=1 ; i<count ; i++) {
        // record the last event of each sensor type in this buffer
        int32_t curr = buffer[i].sensor;
        if (curr != prev) {
            mLastEventSeen.editValueFor(prev) = buffer[i-1];
            prev = curr;
        }
    }
    mLastEventSeen.editValueFor(prev) = buffer[count-1];
}

void SensorService::sortEventBuffer(sensors_event_t* buffer, size_t count)
{
    struct compar {
        static int cmp(void const* lhs, void const* rhs) {
            sensors_event_t const* l = static_cast<sensors_event_t const*>(lhs);
            sensors_event_t const* r = static_cast<sensors_event_t const*>(rhs);
            return l->timestamp - r->timestamp;
        }
    };
    qsort(buffer, count, sizeof(sensors_event_t), compar::cmp);
}

SortedVector< wp<SensorService::SensorEventConnection> >
SensorService::getActiveConnections() const
{
    Mutex::Autolock _l(mLock);
    return mActiveConnections;
}

DefaultKeyedVector<int, SensorInterface*>
SensorService::getActiveVirtualSensors() const
{
    Mutex::Autolock _l(mLock);
    return mActiveVirtualSensors;
}

String8 SensorService::getSensorName(int handle) const {
```

```
    size_t count = mUserSensorList.size();
    for (size_t i=0 ; i<count ; i++) {
        const Sensor& sensor(mUserSensorList[i]);
        if (sensor.getHandle() == handle) {
            return sensor.getName();
        }
    }
    String8 result("unknown");
    return result;
}

int SensorService::getSensorType(int handle) const {
    size_t count = mUserSensorList.size();
    for (size_t i=0 ; i<count ; i++) {
        const Sensor& sensor(mUserSensorList[i]);
        if (sensor.getHandle() == handle) {
            return sensor.getType();
        }
    }
    return -1;
}


Vector<Sensor> SensorService::getSensorList()
{
    char value[PROPERTY_VALUE_MAX];
    property_get("debug.sensors", value, "0");
    if (atoi(value)) {
        return mUserSensorListDebug;
    }
    return mUserSensorList;
}

sp<ISensorEventConnection> SensorService::createSensorEventConnection()
{
    uid_t uid = IPCThreadState::self()->getCallingUid();
    sp<SensorEventConnection> result(new SensorEventConnection(this, uid));
    return result;
}

void SensorService::cleanupConnection(SensorEventConnection* c)
{
    Mutex::Autolock _l(mLock);
    const wp<SensorEventConnection> connection(c);
    size_t size = mActiveSensors.size();
    ALOGD_IF(DEBUG_CONNECTIONS, "%d active sensors", size);
    for (size_t i=0 ; i<size ; ) {
        int handle = mActiveSensors.keyAt(i);
        if (c->hasSensor(handle)) {
            ALOGD_IF(DEBUG_CONNECTIONS, "%i: disabling handle=0x%08x", i, handle);
            SensorInterface* sensor = mSensorMap.valueFor( handle );
            ALOGE_IF(!sensor, "mSensorMap[handle=0x%08x] is null!", handle);
            if (sensor) {
                sensor->activate(c, false);
            }
        }
        SensorRecord* rec = mActiveSensors.valueAt(i);
        ALOGE_IF(!rec, "mActiveSensors[%d] is null (handle=0x%08x)!", i, handle);
        ALOGD_IF(DEBUG_CONNECTIONS,
                "removing connection %p for sensor[%d].handle=0x%08x",
                c, i, handle);

        if (rec && rec->removeConnection(connection)) {
            ALOGD_IF(DEBUG_CONNECTIONS, "... and it was the last connection");
            mActiveSensors.removeItemsAt(i, 1);
            mActiveVirtualSensors.removeItem(handle);
            delete rec;
            size--;
        } else {
```

```
            i++;
        }
    }
    mActiveConnections.remove(connection);
    BatteryService::cleanup(c->getUid());
}

status_t SensorService::enable(const sp<SensorEventConnection>& connection,
        int handle)
{
    if (mInitCheck != NO_ERROR)
        return mInitCheck;

    Mutex::Autolock _l(mLock);
    SensorInterface* sensor = mSensorMap.valueFor(handle);
    SensorRecord* rec = mActiveSensors.valueFor(handle);
    if (rec == 0) {
        rec = new SensorRecord(connection);
        mActiveSensors.add(handle, rec);
        if (sensor->isVirtual()) {
            mActiveVirtualSensors.add(handle, sensor);
        }
    } else {
        if (rec->addConnection(connection)) {
            // this sensor is already activated, but we are adding a
            // connection that uses it. Immediately send down the last
            // known value of the requested sensor if it's not a
            // "continuous" sensor
            if (sensor->getSensor().getMinDelay() == 0) {
                sensors_event_t scratch;
                sensors_event_t& event(mLastEventSeen.editValueFor(handle));
                if (event.version == sizeof(sensors_event_t)) {
                    connection->sendEvents(&event, 1);
                }
            }
        }
    }

    if (connection->addSensor(handle)) {
        BatteryService::enableSensor(connection->getUid(), handle);
        // the sensor was added (which means it wasn't already there)
        // so, see if this connection becomes active
        if (mActiveConnections.indexOf(connection) < 0) {
            mActiveConnections.add(connection);
        }
    } else {
        ALOGW("sensor %08x already enabled in connection %p (ignoring)",
            handle, connection.get());
    }

    // we are setup, now enable the sensor
    status_t err = sensor ? sensor->activate(connection.get(), true) : status_t(BAD_VALUE);

    if (err != NO_ERROR) {
        // enable has failed, reset state in SensorDevice
        status_t resetErr = sensor ? sensor->resetStateWithoutActuatingHardware
        (connection.get(),handle) : status_t(BAD_VALUE);
        // enable has failed, reset our state
        cleanupWithoutDisable(connection, handle);
    }
    return err;
}

status_t SensorService::disable(const sp<SensorEventConnection>& connection,
        int handle)
{
    if (mInitCheck != NO_ERROR)
        return mInitCheck;
```

```
    status_t err = cleanupWithoutDisable(connection, handle);
    if (err == NO_ERROR) {
        SensorInterface* sensor = mSensorMap.valueFor(handle);
        err = sensor ? sensor->activate(connection.get(), false) : status_t(BAD_VALUE);
    }
    return err;
}

status_t SensorService::cleanupWithoutDisable(const sp<SensorEventConnection>& connection,
        int handle) {
    Mutex::Autolock _l(mLock);
    SensorRecord* rec = mActiveSensors.valueFor(handle);
    if (rec) {
        // see if this connection becomes inactive
        if (connection->removeSensor(handle)) {
            BatteryService::disableSensor(connection->getUid(), handle);
        }
        if (connection->hasAnySensor() == false) {
            mActiveConnections.remove(connection);
        }
        // see if this sensor becomes inactive
        if (rec->removeConnection(connection)) {
            mActiveSensors.removeItem(handle);
            mActiveVirtualSensors.removeItem(handle);
            delete rec;
        }
        return NO_ERROR;
    }
    return BAD_VALUE;
}

status_t SensorService::setEventRate(const sp<SensorEventConnection>& connection,
        int handle, nsecs_t ns)
{
    if (mInitCheck != NO_ERROR)
        return mInitCheck;

    SensorInterface* sensor = mSensorMap.valueFor(handle);
    if (!sensor)
        return BAD_VALUE;

    if (ns < 0)
        return BAD_VALUE;

    nsecs_t minDelayNs = sensor->getSensor().getMinDelayNs();
    if (ns < minDelayNs) {
        ns = minDelayNs;
    }

    if (ns < MINIMUM_EVENTS_PERIOD)
        ns = MINIMUM_EVENTS_PERIOD;

    return sensor->setDelay(connection.get(), handle, ns);
}

// ---------------------------------------------------------------------------

SensorService::SensorRecord::SensorRecord(
        const sp<SensorEventConnection>& connection)
{
    mConnections.add(connection);
}

bool SensorService::SensorRecord::addConnection(
        const sp<SensorEventConnection>& connection)
{
    if (mConnections.indexOf(connection) < 0) {
        mConnections.add(connection);
```

```
        return true;
    }
    return false;
}
bool SensorService::SensorRecord::removeConnection(
        const wp<SensorEventConnection>& connection)
{
    ssize_t index = mConnections.indexOf(connection);
    if (index >= 0) {
        mConnections.removeItemsAt(index, 1);
    }
    return mConnections.size() ? false : true;
}

// ---------------------------------------------------------------------------

SensorService::SensorEventConnection::SensorEventConnection(
        const sp<SensorService>& service, uid_t uid)
    : mService(service), mChannel(new BitTube()), mUid(uid)
{
}

SensorService::SensorEventConnection::~SensorEventConnection()
{
    ALOGD_IF(DEBUG_CONNECTIONS, "~SensorEventConnection(%p)", this);
    mService->cleanupConnection(this);
}

void SensorService::SensorEventConnection::onFirstRef()
{
}

bool SensorService::SensorEventConnection::addSensor(int32_t handle) {
    Mutex::Autolock _l(mConnectionLock);
    if (mSensorInfo.indexOf(handle) < 0) {
        mSensorInfo.add(handle);
        return true;
    }
    return false;
}

bool SensorService::SensorEventConnection::removeSensor(int32_t handle) {
    Mutex::Autolock _l(mConnectionLock);
    if (mSensorInfo.remove(handle) >= 0) {
        return true;
    }
    return false;
}

bool SensorService::SensorEventConnection::hasSensor(int32_t handle) const {
    Mutex::Autolock _l(mConnectionLock);
    return mSensorInfo.indexOf(handle) >= 0;
}

bool SensorService::SensorEventConnection::hasAnySensor() const {
    Mutex::Autolock _l(mConnectionLock);
    return mSensorInfo.size() ? true : false;
}

status_t SensorService::SensorEventConnection::sendEvents(
        sensors_event_t const* buffer, size_t numEvents,
        sensors_event_t* scratch)
{
    // filter out events not for this connection
    size_t count = 0;
    if (scratch) {
        Mutex::Autolock _l(mConnectionLock);
        size_t i=0;
```

```
        while (i<numEvents) {
            const int32_t curr = buffer[i].sensor;
            if (mSensorInfo.indexOf(curr) >= 0) {
                do {
                    scratch[count++] = buffer[i++];
                } while ((i<numEvents) && (buffer[i].sensor == curr));
            } else {
                i++;
            }
        }
    } else {
        scratch = const_cast<sensors_event_t *>(buffer);
        count = numEvents;
    }

    // NOTE: ASensorEvent and sensors_event_t are the same type
    ssize_t size = SensorEventQueue::write(mChannel,
            reinterpret_cast<ASensorEvent const*>(scratch), count);
    if (size == -EAGAIN) {
        // the destination doesn't accept events anymore, it's probably
        // full. For now, we just drop the events on the floor
        //ALOGW("dropping %d events on the floor", count);
        return size;
    }

    return size < 0 ? status_t(size) : status_t(NO_ERROR);
}

sp<BitTube> SensorService::SensorEventConnection::getSensorChannel() const
{
    return mChannel;
}

status_t SensorService::SensorEventConnection::enableDisable(
        int handle, bool enabled)
{
    status_t err;
    if (enabled) {
        err = mService->enable(this, handle);
    } else {
        err = mService->disable(this, handle);
    }
    return err;
}

status_t SensorService::SensorEventConnection::setEventRate(
        int handle, nsecs_t ns)
{
    return mService->setEventRate(this, handle, ns);
}

// ---------------------------------------------------------------------------
}; // namespace android
```

通过上述实现代码，可以了解 SensorService 服务的创建、启动过程，整个过程的“C/S”通信架构如图 5-4 所示。

在此需要注意，BpSensorServer 并没有在系统中被用到，即使从 ISensorServer.cpp 中把它删除也不会对 Sensor 的工作有任何影响。这是因为它的工作已经被 SensorManager.cpp 所取代，ServiceManager 会直接获取上面 System_init 文件中添加的 SensorService 对象。

5.4.4 封装 HAL 层的代码

在 Android 系统中，通过文件 frameworks\native\services\sensorservice\SensorDevice.cpp 封装了 HAL 层的代码，主要包含的功能如下所示：

- 获取 sensor 列表（getSensorList）；
- 获取 sensor 事件（poll）；
- Enable 或 Disable sensor（activate）；
- 设置 delay 时间。

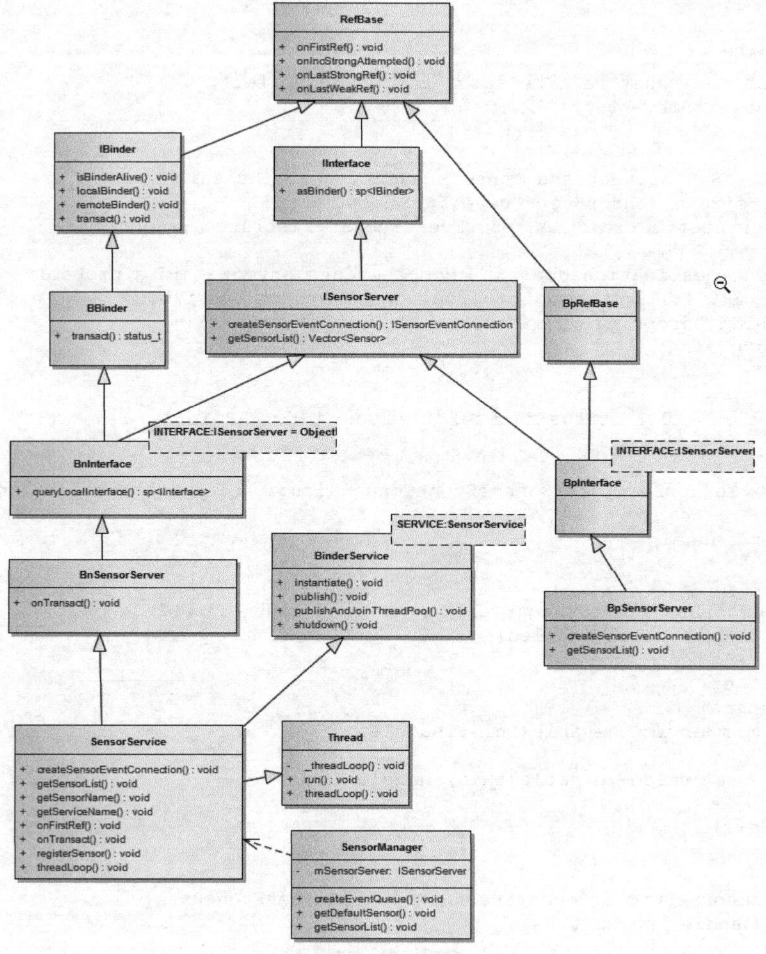

▲图 5-4 “C/S”通信架构图

文件 SensorDevice.cpp 的具体实现代码如下所示。

```
namespace android {
// ---------------------------------------------------------------------------

ANDROID_SINGLETON_STATIC_INSTANCE(SensorDevice)

SensorDevice::SensorDevice()
    :  mSensorDevice(0),
       mSensorModule(0)
{
    status_t err = hw_get_module(SENSORS_HARDWARE_MODULE_ID,
            (hw_module_t const**)&mSensorModule);

    ALOGE_IF(err, "couldn't load %s module (%s)",
            SENSORS_HARDWARE_MODULE_ID, strerror(-err));

    if (mSensorModule) {
        err = sensors_open(&mSensorModule->common, &mSensorDevice);
```

```
        ALOGE_IF(err, "couldn't open device for module %s (%s)",
                SENSORS_HARDWARE_MODULE_ID, strerror(-err));

        if (mSensorDevice) {
            sensor_t const* list;
            ssize_t count = mSensorModule->get_sensors_list(mSensorModule, &list);
            mActivationCount.setCapacity(count);
            Info model;
            for (size_t i=0 ; i<size_t(count) ; i++) {
                mActivationCount.add(list[i].handle, model);
                mSensorDevice->activate(mSensorDevice, list[i].handle, 0);
            }
        }
    }
}

void SensorDevice::dump(String8& result, char* buffer, size_t SIZE)
{
    if (!mSensorModule) return;
    sensor_t const* list;
    ssize_t count = mSensorModule->get_sensors_list(mSensorModule, &list);

    snprintf(buffer, SIZE, "%d h/w sensors:\n", int(count));
    result.append(buffer);

    Mutex::Autolock _l(mLock);
    for (size_t i=0 ; i<size_t(count) ; i++) {
        const Info& info = mActivationCount.valueFor(list[i].handle);
        snprintf(buffer, SIZE, "handle=0x%08x, active-count=%d, rates(ms)={ ",
                list[i].handle,
                info.rates.size());
        result.append(buffer);
        for (size_t j=0 ; j<info.rates.size() ; j++) {
            snprintf(buffer, SIZE, "%4.1f%s",
                    info.rates.valueAt(j) / 1e6f,
                    j<info.rates.size()-1 ? ", " : "");
            result.append(buffer);
        }
        snprintf(buffer, SIZE, " }, selected=%4.1f ms\n",  info.delay / 1e6f);
        result.append(buffer);
    }
}

ssize_t SensorDevice::getSensorList(sensor_t const** list) {
    if (!mSensorModule) return NO_INIT;
    ssize_t count = mSensorModule->get_sensors_list(mSensorModule, list);
    return count;
}

status_t SensorDevice::initCheck() const {
    return mSensorDevice && mSensorModule ? NO_ERROR : NO_INIT;
}

ssize_t SensorDevice::poll(sensors_event_t* buffer, size_t count) {
    if (!mSensorDevice) return NO_INIT;
    ssize_t c;
    do {
        c = mSensorDevice->poll(mSensorDevice, buffer, count);
    } while (c == -EINTR);
    return c;
}

status_t SensorDevice::resetStateWithoutActuatingHardware(void *ident, int handle)
{
    if (!mSensorDevice) return NO_INIT;
    Info& info( mActivationCount.editValueFor(handle));
    Mutex::Autolock _l(mLock);
    info.rates.removeItem(ident);
```

```
    return NO_ERROR;
}

status_t SensorDevice::activate(void* ident, int handle, int enabled)
{
    if (!mSensorDevice) return NO_INIT;
    status_t err(NO_ERROR);
    bool actuateHardware = false;

    Info& info( mActivationCount.editValueFor(handle) );


    ALOGD_IF(DEBUG_CONNECTIONS,
            "SensorDevice::activate: ident=%p, handle=0x%08x, enabled=%d, count=%d",
            ident, handle, enabled, info.rates.size());

    if (enabled) {
        Mutex::Autolock _l(mLock);
        ALOGD_IF(DEBUG_CONNECTIONS, "... index=%ld",
                info.rates.indexOfKey(ident));

        if (info.rates.indexOfKey(ident) < 0) {
            info.rates.add(ident, DEFAULT_EVENTS_PERIOD);
            if (info.rates.size() == 1) {
                actuateHardware = true;
            }
        } else {
            // sensor was already activated for this ident
        }
    } else {
        Mutex::Autolock _l(mLock);
        ALOGD_IF(DEBUG_CONNECTIONS, "... index=%ld",
                info.rates.indexOfKey(ident));

        ssize_t idx = info.rates.removeItem(ident);
        if (idx >= 0) {
            if (info.rates.size() == 0) {
                actuateHardware = true;
            }
        } else {
            // sensor wasn't enabled for this ident
        }
    }

    if (actuateHardware) {
        ALOGD_IF(DEBUG_CONNECTIONS, "\t>>> actuating h/w");

        err = mSensorDevice->activate(mSensorDevice, handle, enabled);
        ALOGE_IF(err, "Error %s sensor %d (%s)",
                enabled ? "activating" : "disabling",
                handle, strerror(-err));
    }

    { // scope for the lock
        Mutex::Autolock _l(mLock);
        nsecs_t ns = info.selectDelay();
        mSensorDevice->setDelay(mSensorDevice, handle, ns);
    }

    return err;
}

status_t SensorDevice::setDelay(void* ident, int handle, int64_t ns)
{
    if (!mSensorDevice) return NO_INIT;
    Mutex::Autolock _l(mLock);
    Info& info( mActivationCount.editValueFor(handle) );
    status_t err = info.setDelayForIdent(ident, ns);
    if (err < 0) return err;
```

```
    ns = info.selectDelay();
    return mSensorDevice->setDelay(mSensorDevice, handle, ns);
}

int SensorDevice::getHalDeviceVersion() const {
    if (!mSensorDevice) return -1;

    return mSensorDevice->common.version;
}

// ---------------------------------------------------------------------------

status_t SensorDevice::Info::setDelayForIdent(void* ident, int64_t ns)
{
    ssize_t index = rates.indexOfKey(ident);
    if (index < 0) {
        ALOGE("Info::setDelayForIdent(ident=%p, ns=%lld) failed (%s)",
                ident, ns, strerror(-index));
        return BAD_INDEX;
    }
    rates.editValueAt(index) = ns;
    return NO_ERROR;
}

nsecs_t SensorDevice::Info::selectDelay()
{
    nsecs_t ns = rates.valueAt(0);
    for (size_t i=1 ; i<rates.size() ; i++) {
        nsecs_t cur = rates.valueAt(i);
        if (cur < ns) {
            ns = cur;
        }
    }
    delay = ns;
    return ns;
}

// ---------------------------------------------------------------------------
}; // namespace android
```

这样 SensorSevice 会把任务交给 SensorDevice，而 SensorDevice 会调用标准的抽象层接口。由此可见，Sensor 架构的抽象层接口是最标准的一种，它很好地实现了抽象层与本地框架的分离。

5.4.5　消息队列处理

在 Android 的传感器系统中，文件 frameworks\native\libs\gui\SensorEventQueue.cpp 实现了处理消息队列的功能。此文件能够在创建其实例的时候传入 SensorEventConnection 的实例，SensorEventConnection 继承于 ISensorEventConnection。SensorEventConnection 其实是客户端调用 SensorService 的 createSensorEventConnection()方法创建的，是客户端与服务端沟通的桥梁，通过这个桥梁，可以完成如下所示的任务：

- 获取管道的句柄；
- 往管道读写数据；
- 通知服务端对 Sensor 可用。

文件 frameworks\native\libs\gui\SensorEventQueue.cpp 的具体实现代码如下所示。

```
// ---------------------------------------------------------------------------
namespace android {
// ---------------------------------------------------------------------------

SensorEventQueue::SensorEventQueue(const sp<ISensorEventConnection>& connection)
    : mSensorEventConnection(connection)
{
}
```

```
SensorEventQueue::~SensorEventQueue()
{
}

void SensorEventQueue::onFirstRef()
{
   mSensorChannel = mSensorEventConnection->getSensorChannel();
}

int SensorEventQueue::getFd() const
{
   return mSensorChannel->getFd();
}


ssize_t SensorEventQueue::write(const sp<BitTube>& tube,
      ASensorEvent const* events, size_t numEvents) {
   return BitTube::sendObjects(tube, events, numEvents);
}

ssize_t SensorEventQueue::read(ASensorEvent* events, size_t numEvents)
{
   return BitTube::recvObjects(mSensorChannel, events, numEvents);
}

sp<Looper> SensorEventQueue::getLooper() const
{
   Mutex::Autolock _l(mLock);
   if (mLooper == 0) {
      mLooper = new Looper(true);
      mLooper->addFd(getFd(), getFd(), ALOOPER_EVENT_INPUT, NULL, NULL);
   }
   return mLooper;
}

status_t SensorEventQueue::waitForEvent() const
{
   const int fd = getFd();
   sp<Looper> looper(getLooper());

   int events;
   int32_t result;
   do {
      result = looper->pollOnce(-1, NULL, &events, NULL);
      if (result == ALOOPER_POLL_ERROR) {
         ALOGE("SensorEventQueue::waitForEvent error (errno=%d)", errno);
         result = -EPIPE; // unknown error, so we make up one
         break;
      }
      if (events & ALOOPER_EVENT_HANGUP) {
         // the other-side has died
         ALOGE("SensorEventQueue::waitForEvent error HANGUP");
         result = -EPIPE; // unknown error, so we make up one
         break;
      }
   } while (result != fd);

   return  (result == fd) ? status_t(NO_ERROR) : result;
}

status_t SensorEventQueue::wake() const
{
   sp<Looper> looper(getLooper());
   looper->wake();
   return NO_ERROR;
}
```

```
status_t SensorEventQueue::enableSensor(Sensor const* sensor) const {
    return mSensorEventConnection->enableDisable(sensor->getHandle(), true);
}

status_t SensorEventQueue::disableSensor(Sensor const* sensor) const {
    return mSensorEventConnection->enableDisable(sensor->getHandle(), false);
}

status_t SensorEventQueue::enableSensor(int32_t handle, int32_t us) const {
    status_t err = mSensorEventConnection->enableDisable(handle, true);
    if (err == NO_ERROR) {
        mSensorEventConnection->setEventRate(handle, us2ns(us));
    }
    return err;
}

status_t SensorEventQueue::disableSensor(int32_t handle) const {
    return mSensorEventConnection->enableDisable(handle, false);
}

status_t SensorEventQueue::setEventRate(Sensor const* sensor, nsecs_t ns) const {
    return mSensorEventConnection->setEventRate(sensor->getHandle(), ns);
}

// ----------------------------------------------------------------------------
}; // namespace android
```

由此可见，SensorManager 负责控制流，通过 C/S 的 Binder 机制与 SensorService 实现通信。具体过程如图 5-5 所示。

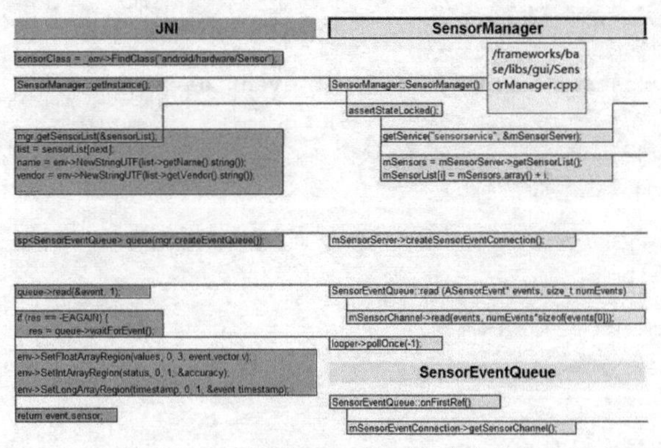

▲图 5-5　SensorManager 控制流的处理流程

而 SensorEventQueue 负责数据流，功能是通过管道机制来读写底层的数据。具体过程如图 5-6 所示。

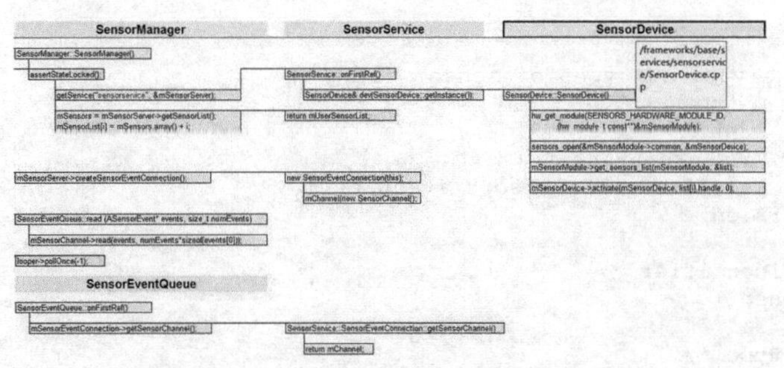

▲图 5-6　SensorEventQueue 数据流的处理流程

5.5　分析 HAL 层

在 Android 系统中，在 HAL 层中提供了 Android 独立于具体硬件的抽象接口。其中 HAL 层的头文件路径如下。

```
hardware/libhardware/include/hardware/sensors.h
```

而具体实现文件需要开发者个人编写，具体可以参考下面内容。

```
hardware\invensense\libsensors_iio\sensors_mpl.cpp
```

文件 sensors.h 的主要实现代码如下所示。

```
typedef struct {
    union {
        float v[3];
        struct {
            float x;
            float y;
            float z;
        };
        struct {
            float azimuth;
            float pitch;
            float roll;
        };
    };
    int8_t status;
    uint8_t reserved[3];
} sensors_vec_t;

/**
 * uncalibrated gyroscope and magnetometer event data
 */
typedef struct {
  union {
    float uncalib[3];
    struct {
      float x_uncalib;
      float y_uncalib;
      float z_uncalib;
    };
  };
  union {
    float bias[3];
    struct {
      float x_bias;
      float y_bias;
      float z_bias;
    };
  };
} uncalibrated_event_t;

/**
 * Union of the various types of sensor data
 * that can be returned.
 */
typedef struct sensors_event_t {
    /* must be sizeof(struct sensors_event_t) */
    int32_t version;

    /* sensor identifier */
    int32_t sensor;

    /* sensor type */
    int32_t type;
```

```
    /* reserved */
    int32_t reserved0;

    /* time is in nanosecond */
    int64_t timestamp;

    union {
        float data[16];

        /* acceleration values are in meter per second per second (m/s^2) */
        sensors_vec_t acceleration;

        /* magnetic vector values are in micro-Tesla (uT) */
        sensors_vec_t magnetic;

        /* orientation values are in degrees */
        sensors_vec_t orientation;

        /* gyroscope values are in rad/s */
        sensors_vec_t gyro;

        /* temperature is in degrees centigrade (Celsius) */
        float temperature;

        /* distance in centimeters */
        float distance;

        /* light in SI lux units */
        float light;

        /* pressure in hectopascal (hPa) */
        float pressure;

        /* relative humidity in percent */
        float relative_humidity;

        /* step-counter */
        uint64_t step_counter;

        /* uncalibrated gyroscope values are in rad/s */
        uncalibrated_event_t uncalibrated_gyro;

        /* uncalibrated magnetometer values are in micro-Teslas */
        uncalibrated_event_t uncalibrated_magnetic;
    };
    uint32_t reserved1[4];
} sensors_event_t;

struct sensor_t;
struct sensors_module_t {
    struct hw_module_t common;
    int (*get_sensors_list)(struct sensors_module_t* module,
            struct sensor_t const** list);
};

struct sensor_t {

    /* Name of this sensor.
     * All sensors of the same "type" must have a different "name".
     */
    const char* name;

    /* vendor of the hardware part */
    const char* vendor;
    int version;
    int handle;

    /* this sensor's type */
```

```
    int type;

    /* maximum range of this sensor's value in SI units */
    float maxRange;

    /* smallest difference between two values reported by this sensor */
    float resolution;

    /* rough estimate of this sensor's power consumption in mA */
    float power;
    int32_t minDelay;

    /* reserved fields, must be zero */
    void* reserved[8];
};

struct sensors_poll_device_t {
    struct hw_device_t common;
    int (*activate)(struct sensors_poll_device_t *dev,
            int handle, int enabled);
    int (*setDelay)(struct sensors_poll_device_t *dev,
            int handle, int64_t ns);
    int (*poll)(struct sensors_poll_device_t *dev,
            sensors_event_t* data, int count);
};
typedef struct sensors_poll_device_1 {
    union {
        struct sensors_poll_device_t v0;

        struct {
            struct hw_device_t common;
            int (*activate)(struct sensors_poll_device_t *dev,
                    int handle, int enabled);

            int (*setDelay)(struct sensors_poll_device_t *dev,
                    int handle, int64_t period_ns);

            int (*poll)(struct sensors_poll_device_t *dev,
                    sensors_event_t* data, int count);
        };
    };
    int (*batch)(struct sensors_poll_device_1* dev,
            int handle, int flags, int64_t period_ns, int64_t timeout);

    void (*reserved_procs[8])(void);

} sensors_poll_device_1_t;



/** convenience API for opening and closing a device */

static inline int sensors_open(const struct hw_module_t* module,
        struct sensors_poll_device_t** device) {
    return module->methods->open(module,
            SENSORS_HARDWARE_POLL, (struct hw_device_t**)device);
}

static inline int sensors_close(struct sensors_poll_device_t* device) {
    return device->common.close(&device->common);
}

static inline int sensors_open_1(const struct hw_module_t* module,
        sensors_poll_device_1_t** device) {
    return module->methods->open(module,
            SENSORS_HARDWARE_POLL, (struct hw_device_t**)device);
}

static inline int sensors_close_1(sensors_poll_device_1_t* device) {
```

```
    return device->common.close(&device->common);
}

__END_DECLS

#endif  // ANDROID_SENSORS_INTERFACE_H
```

而具体的实现文件是Linux Kernel层，也就是具体的硬件设备驱动程序，例如可以将其命名为“sensors.c”，然后编写如下定义struct sensors_poll_device_t的代码。

```
struct sensors_poll_device_t {
    struct hw_device_t common;

    // Activate/deactivate one sensor
    int (*activate)(struct sensors_poll_device_t *dev,
            int handle, int enabled);

    // Set the delay between sensor events in nanoseconds for a given sensor
    int (*setDelay)(struct sensors_poll_device_t *dev,
            int handle, int64_t ns);

    // Returns an array of sensor data
    int (*poll)(struct sensors_poll_device_t *dev,
            sensors_event_t* data, int count);
};
```

也可以编写如下定义struct sensors_module_t的代码。

```
struct sensors_module_t {
    struct hw_module_t common;

    /**
     * Enumerate all available sensors. The list is returned in "list".
     * @return number of sensors in the list
     */
    int (*get_sensors_list)(struct sensors_module_t* module,
            struct sensor_t const** list);
};
```

也可以编写如下定义struct sensor_t 的代码。

```
struct sensor_t {
    /* name of this sensors */
    const char*name;
    /* vendor of the hardware part */
    const char*vendor;
    /* version of the hardware part + driver. The value of this field
     * must increase when the driver is updated in a way that changes the
     * output of this sensor. This is important for fused sensors when the
     * fusion algorithm is updated.
     */
    int version;
    /* handle that identifies this sensors. This handle is used to activate
     * and deactivate this sensor. The value of the handle must be 8 bits
     * in this version of the API.
     */
    int handle;
    /* this sensor's type. */
    int type;
    /* maximaum range of this sensor's value in SI units */
    float maxRange;
    /* smallest difference between two values reported by this sensor */
    float resolution;
    /* rough estimate of this sensor's power consumption in mA */
    float power;
    /* minimum delay allowed between events in microseconds. A value of zero
     * means that this sensor doesn't report events at a constant rate, but
     * rather only when a new data is available */
    int32_t minDelay;
    /* reserved fields, must be zero */
```

```
    void*reserved[8];
};
```

也可以编写如下代码定义 struct sensors_event_t。

```
typedef struct {
    union {
        float v[3];
        struct {
            float x;
            float y;
            float z;
        };
        struct {
            float azimuth;
            float pitch;
            float roll;
        };
    };
    int8_t status;
    uint8_t reserved[3];
} sensors_vec_t;

/**
 * Union of the various types of sensor data
 * that can be returned.
 */
typedef struct sensors_event_t {
    /* must be sizeof(struct sensors_event_t) */
    int32_t version;

    /* sensor identifier */
    int32_t sensor;

    /* sensor type */
    int32_t type;

    /* reserved */
    int32_t reserved0;

    /* time is in nanosecond */
    int64_t timestamp;

    union {
        float data[16];

        /* acceleration values are in meter per second per second (m/s^2) */
        sensors_vec_t acceleration;

        /* magnetic vector values are in micro-Tesla (uT) */
        sensors_vec_t magnetic;

        /* orientation values are in degrees */
        sensors_vec_t orientation;

        /* gyroscope values are in rad/s */
        sensors_vec_t gyro;

        /* temperature is in degrees centigrade (Celsius) */
        float temperature;

        /* distance in centimeters */
        float distance;

        /* light in SI lux units */
        float light;

        /* pressure in hectopascal (hPa) */
        float pressure;
```

```
        /* relative humidity in percent */
        float relative_humidity;
    };
    uint32_t reserved1[4];
} sensors_event_t;
```

也可以编写如下代码定义 struct sensors_module_t。

```
static const struct sensor_t sSensorList[] = {
        { "MMA8452Q 3-axis Accelerometer",
                "Freescale Semiconductor",
                1, SENSORS_HANDLE_BASE+ID_A,
                SENSOR_TYPE_ACCELEROMETER, 4.0f*9.81f, (4.0f*9.81f)/256.0f, 0.2f, 0, { } },
        { "AK8975 3-axis Magnetic field sensor",
                "Asahi Kasei",
                1, SENSORS_HANDLE_BASE+ID_M,
                SENSOR_TYPE_MAGNETIC_FIELD, 2000.0f, 1.0f/16.0f, 6.8f, 0, { } },
        { "AK8975 Orientation sensor",
                "Asahi Kasei",
                1, SENSORS_HANDLE_BASE+ID_O,
                SENSOR_TYPE_ORIENTATION, 360.0f, 1.0f, 7.0f, 0, { } },

    { "ST 3-axis Gyroscope sensor",
          "STMicroelectronics",
          1, SENSORS_HANDLE_BASE+ID_GY,
          SENSOR_TYPE_GYROSCOPE, RANGE_GYRO, CONVERT_GYRO, 6.1f, 1190, { } },

    { "AL3006Proximity sensor",
        "Dyna Image Corporation",
        1, SENSORS_HANDLE_BASE+ID_P,
        SENSOR_TYPE_PROXIMITY,
        PROXIMITY_THRESHOLD_CM, PROXIMITY_THRESHOLD_CM,
        0.5f, 0, { } },

        { "AL3006 light sensor",
                "Dyna Image Corporation",
                1, SENSORS_HANDLE_BASE+ID_L,
                SENSOR_TYPE_LIGHT, 10240.0f, 1.0f, 0.5f, 0, { } },

};

static int open_sensors(const struct hw_module_t* module, const char* name,
        struct hw_device_t** device);

static int sensors__get_sensors_list(struct sensors_module_t* module,
        struct sensor_t const** list)
{
    *list = sSensorList;
    return ARRAY_SIZE(sSensorList);
}

static struct hw_module_methods_t sensors_module_methods = {
    .open = open_sensors
};

const struct sensors_module_t HAL_MODULE_INFO_SYM = {
    .common = {
        .tag = HARDWARE_MODULE_TAG,
        .version_major = 1,
        .version_minor = 0,
        .id = SENSORS_HARDWARE_MODULE_ID,
        .name = "MMA8451Q & AK8973A & gyro Sensors Module",
        .author = "The Android Project",
        .methods = &sensors_module_methods,
    },
    .get_sensors_list = sensors__get_sensors_list
};

static int open_sensors(const struct hw_module_t* module, const char* name,
        struct hw_device_t** device)
```

```
{
    return init_nusensors(module, device); //待后面讲解
}
```

到此为止，整个 Android 系统中传感器模块的源码分析完毕。由此可见，整个传感器系统的总体调用关系如图 5-7 所示。

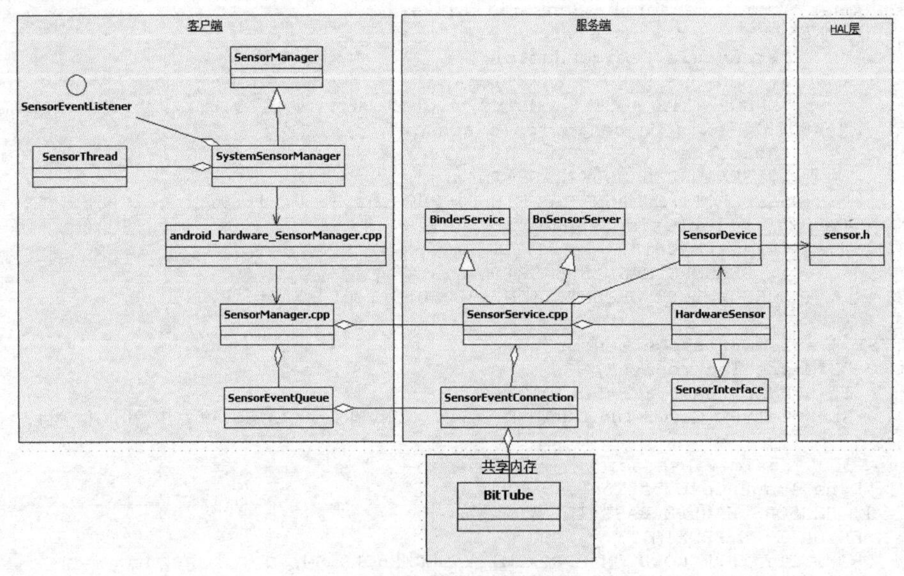

▲图 5-7　传感器系统的总体调用关系

客户端读取数据时的调用时序如图 5-8 所示。

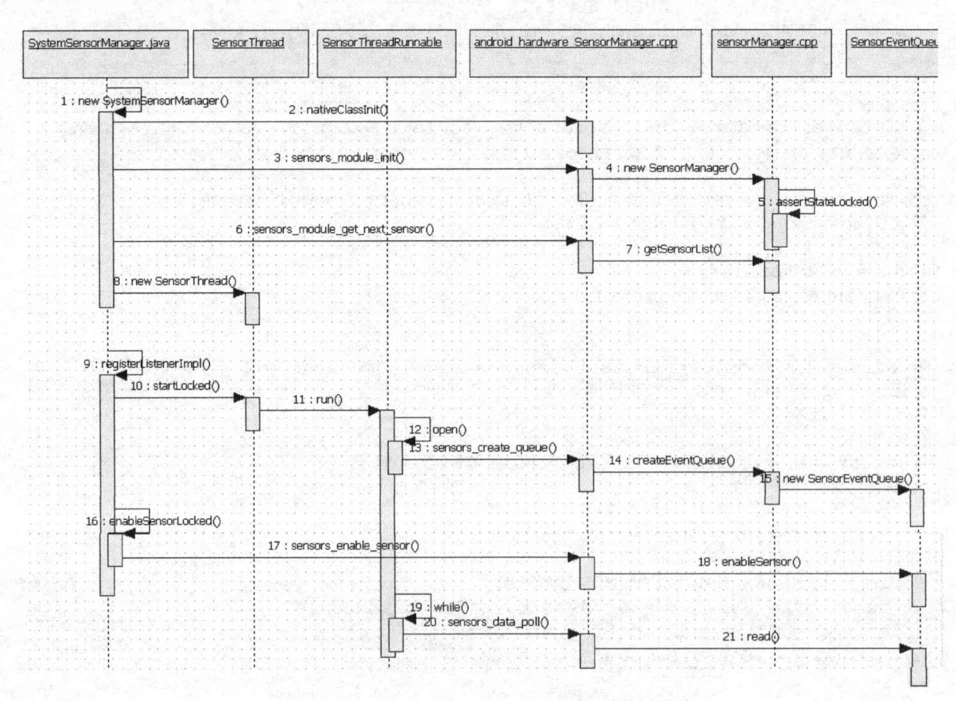

▲图 5-8　客户端读取数据时的调用时序图

服务器端的调用时序如图 5-9 所示。

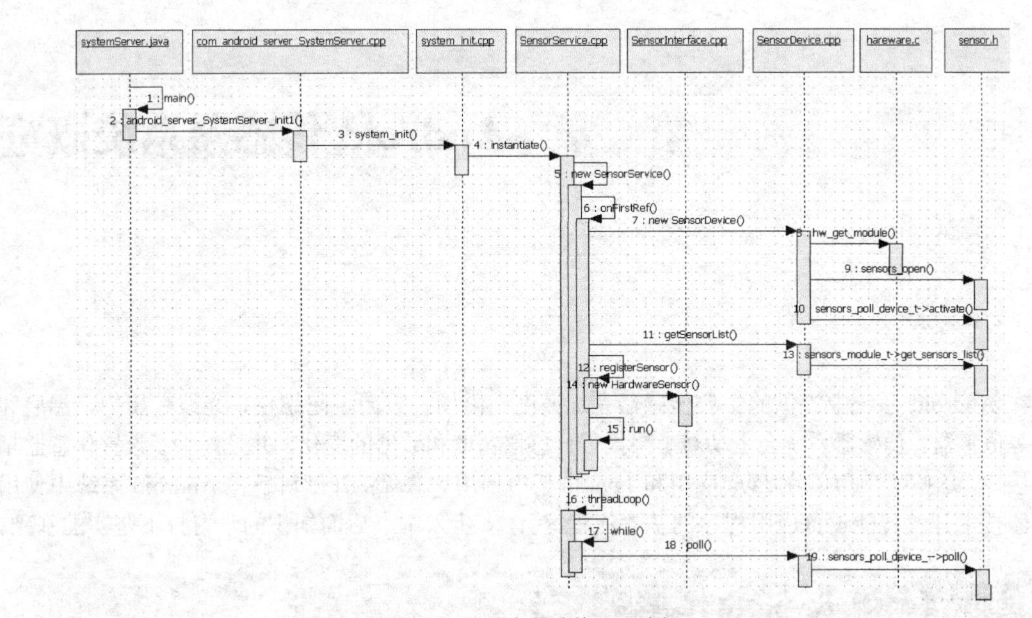

▲图 5-9 服务器端的调用时序图

第 6 章　Android 传感器系统概览

在 Android 系统中提供的主要传感器有加速度、磁场、方向、陀螺仪、光线、压力、温度和接近等传感器。传感器系统会主动对上层报告传感器精度和数据的变化，并且提供了设置传感器精度的接口，这些接口可以在 Java 应用和 Java 框架中使用。本章将详细讲解在 Android 系统中使用传感器技术开发智能设备应用程序的基本流程，为读者步入本书后面知识的学习打下坚实的基础。

6.1 Android 设备的传感器系统

本书在前面第 5 章中，已经详细讲解了 Android 传感器系统的基本架构。本节将详细讲解调用传感器 API 接口，在 Android 智能设备中开发应用程序的方法。

6.1.1 包含的传感器

在安装 Android SDK 后，依次打开安装目录中下面的帮助文件。

```
android SDK/sdk/docs/reference/android/hardware/Sensor.html
```

在此文件中列出了 Android 传感器系统所包含的所有传感器类型，如图 6-1 所示。

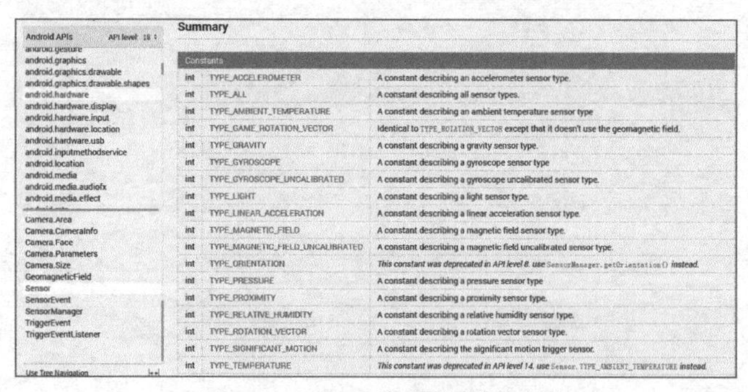

▲图 6-1　Android 传感器系统的类型

另外，也可以直接在线登录 http://developer.android.com/reference/android/hardware/Sensor.html 来查看。由此可见，在当前最新（作者写稿时最新）版本 Android 4.4 中一共提供了 18 种传感器 API 接口。各个类型的具体说明如下所示。

（1）TYPE_ACCELEROMETER：加速度传感器，单位是 m/s^2，测量应用于设备 x、y、z 轴上的加速度，又叫作 G-sensor。

（2）TYPE_AMBIENT_TEMPERATURE：温度传感器，单位是℃，能够测量并返回当前的温度。

（3）TYPE_GRAVITY：重力传感器，单位是 m/s^2，用于测量设备 x、y、z 轴上的重力，也叫作 GV-sensor，地球上的数值是 9.8 m/s^2。

（4）TYPE_GYROSCOPE：陀螺仪传感器，单位是 rad/s^2，能够测量设备 x、y、z 三轴的角加速度数据。

（5）TYPE_LIGHT：光线感应传感器，单位是 lx，能够检测周围的光线强度，在手机系统中主要用于调节 LCD 亮度。

（6）TYPE_LINEAR_ACCELERATION：线性加速度传感器，单位是 m/s^2，能够获取加速度传感器去除重力的影响得到的数据。

（7）TYPE_MAGNETIC_FIELD：磁场传感器，单位是μT（微特斯拉），能够测量设备周围 3 个物理轴（x，y，z）的磁场。

（8）TYPE_ORIENTATION：方向传感器，用于测量设备围绕 3 个物理轴（x，y，z）的旋转角度，在新版本中已经使用 SensorManager.getOrientation()替代。

（9）TYPE_PRESSURE：气压传感器，单位是 hPa（百帕斯卡），能够返回当前环境下的压强。

（10）TYPE_PROXIMITY：距离传感器，单位是 cm，能够测量某个对象到屏幕的距离。可以在打电话时判断人耳到电话屏幕距离，以关闭屏幕而达到省电功能。

（11）TYPE_RELATIVE_HUMIDITY：湿度传感器，能够测量周围环境的相对湿度。

（12）TYPE_ROTATION_VECTOR：旋转向量传感器，旋转矢量代表设备的方向，是一个将坐标轴和角度混合计算得到的数据。

（13）TYPE_TEMPERATURE：温度传感器，在新版本中被 TYPE_AMBIENT_TEMPERATURE 替换。

（14）TYPE_ALL：返回所有的传感器类型。

（15）TYPE_GAME_ROTATION_VECTOR：除了不能使用地磁场之外，和 TYPE_ROTATION_VECTOR 的功能完全相同。

（16）TYPE_GYROSCOPE_UNCALIBRATED：提供了能够让应用调整传感器的原始值，定义了一个描述未校准陀螺仪的传感器类型。

（17）TYPE_MAGNETIC_FIELD_UNCALIBRATED：和 TYPE_GYROSCOPE_UNCALIBRATED 相似，也提供了能够让应用调整传感器的原始值，定义了一个描述未校准陀螺仪的传感器类型。

（18）TYPE_SIGNIFICANT_MOTION：运动触发传感器，应用程序不需要为这种传感器触发任何唤醒锁。能够检测当前设备是否运动，并发送检测结果。

6.1.2 检测当前设备支持的传感器

在接下来的实例中，将演示在 Android 设备中检测当前设备支持传感器类型的方法。

实例	功能	源码路径
实例 6-1	检测当前设备支持的传感器	\daima\6\Sensor

本实例的功能是检测当前设备支持的传感器类型，具体实现流程如下所示。

布局文件 main.xml 的具体实现代码如下所示。

```
<linearlayout android:layout_height="fill_parent" android:layout_width="fill_parent"
android:orientation="vertical" xmlns:android="http://schemas.android.com/apk/res/android">
<textview android:layout_height="wrap_content"
 android:layout_width="fill_parent" android:text=""
 android:id="@+id/TextView01"
>
</textview>
</linearlayout>
```

主程序文件 MainActivity.java 的具体实现代码如下所示。

```
public class MainActivity extends Activity {
```

```
    /** Called when the activity is first created. */
    @SuppressWarnings("deprecation")
    @Override
    public void onCreate(Bundle savedInstanceState) {
        super.onCreate(savedInstanceState);
        setContentView(R.layout.main);

        //准备显示信息的 UI 组建
        final TextView tx1 = (TextView) findViewById(R.id.TextView01);

        //从系统服务中获得传感器管理器
        SensorManager sm = (SensorManager) getSystemService(Context.SENSOR_SERVICE);

        //从传感器管理器中获得全部的传感器列表
        List<Sensor> allSensors = sm.getSensorList(Sensor.TYPE_ALL);

        //显示有多少个传感器
        tx1.setText("经检测该手机有" + allSensors.size() + "个传感器，它们分别是：\n");

        //显示每个传感器的具体信息
        for (Sensor s : allSensors) {

            String tempString = "\n" + " 设备名称：" + s.getName() + "\n" + " 设
            备版本：" + s.getVersion() + "\n" + " 供应商：" + s.getVendor() + "\n";

            switch (s.getType()) {
            case Sensor.TYPE_ACCELEROMETER:
                tx1.setText(tx1.getText().toString() + s.getType() + " 加速
                度传感器 accelerometer" + tempString);
                break;
            case Sensor.TYPE_GYROSCOPE:
                tx1.setText(tx1.getText().toString() + s.getType() + " 陀螺
                仪传感器 gyroscope" + tempString);
                break;
            case Sensor.TYPE_LIGHT:
                tx1.setText(tx1.getText().toString() + s.getType() + " 光线
                感应传感器 light" + tempString);
                break;
            case Sensor.TYPE_MAGNETIC_FIELD:
                tx1.setText(tx1.getText().toString() + s.getType() + " 电磁
                场传感器 magnetic field" + tempString);
                break;
            case Sensor.TYPE_ORIENTATION:
                tx1.setText(tx1.getText().toString() + s.getType() + " 方向
                传感器 orientation" + tempString);
                break;
            case Sensor.TYPE_PRESSURE:
                tx1.setText(tx1.getText().toString() + s.getType() + " 压力
                传感器 pressure" + tempString);
                break;
            case Sensor.TYPE_PROXIMITY:
                tx1.setText(tx1.getText().toString() + s.getType() + " 距离
                传感器 proximity" + tempString);
                break;
            case Sensor.TYPE_AMBIENT_TEMPERATURE :
                tx1.setText(tx1.getText().toString() + s.getType() + " 温度
                传感器 temperature" + tempString);
                break;
            default:
                tx1.setText(tx1.getText().toString() + s.getType() + " 未知
                传感器" + tempString);
                break;
            }
        }

    }
}
```

上述实例代码需要在真机中运行，执行后将会列表显示当前设备所支持的传感器类型。如图 6-2 所示。

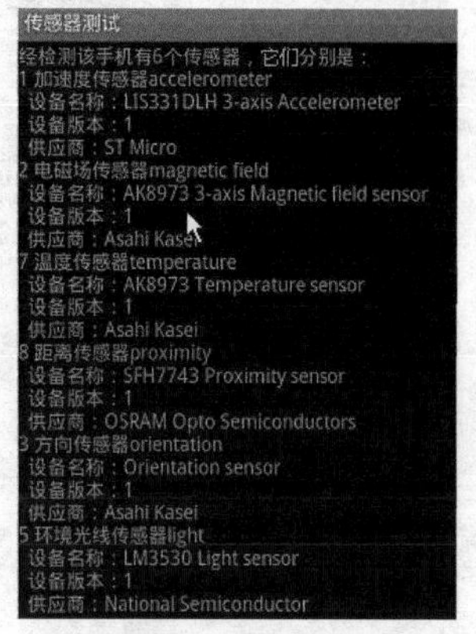

▲图 6-2　执行效果

6.2　使用 SensorSimulator

在进行和传感器相关的开发工作时，使用 SensorSimulator 可以提高开发效率。SensorSimulator 是一个开源免费的传感器小型工具，通过该工具可以在模拟器中调试传感器的应用。搭建 SensorSimulator 开发环境的基本流程如下所示。

（1）下载 SensorSimulator，读者可从 http://code.google.com/p/openintents/wiki/SensorSimulator 网站找到该工具的下载链接。这里下载的是 sensorsimulator-1.1.1.zip 版本，如图 6-3 所示。

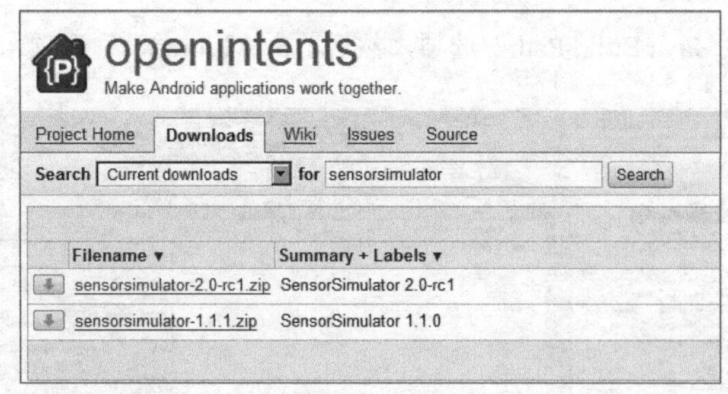

▲图 6-3　下载 sensorsimulator-1.1.1.zip

（2）将下载好的 SensorSimulator 解压到本地根目录，例如 C 盘的根目录。

（3）向模拟器安装 SensorSimulatorSettings-1.1.1.apk。首先在操作系统中依次选择“开始”|“运行”，进入“运行”对话框。

（4）在“运行”对话框输入“cmd”进入 cmd 命令行，之后通过 cd 命令将当前目录导航到 SensorSimulatorSettings-1.1.1.apk 目录下，然后输入下列命令向模拟器安装该 apk。

```
adb install SensorSimulatorSettings-1.1.1.apk
```

如图 6-4 所示。在此需要注意的是，安装 apk 时，一定要保证模拟器正在运行才可以，安装成功后会输出“Success”提示。

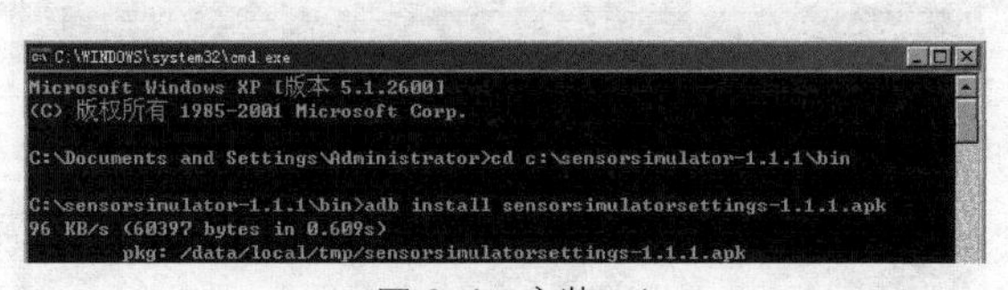

▲图 6-4　安装 apk

接下来开始配置应用程序，假设要在项目“jiaSCH”中使用 SensorSimulator，则配置流程如下所示。

（1）在 Eclipse 中打开项目“jiaSCH”，然后为该项目添加 JAR 包，使其能够使用 SensorSimulator 工具的类和方法。添加方法非常简单，在 Eclipse 的 Package Explorer 中找到该项目的文件夹“jiaSCH”，然后右键单击该文件夹并选择“Properties”选项，弹出图 6-5 所示的“Properties for jiaS”窗口。

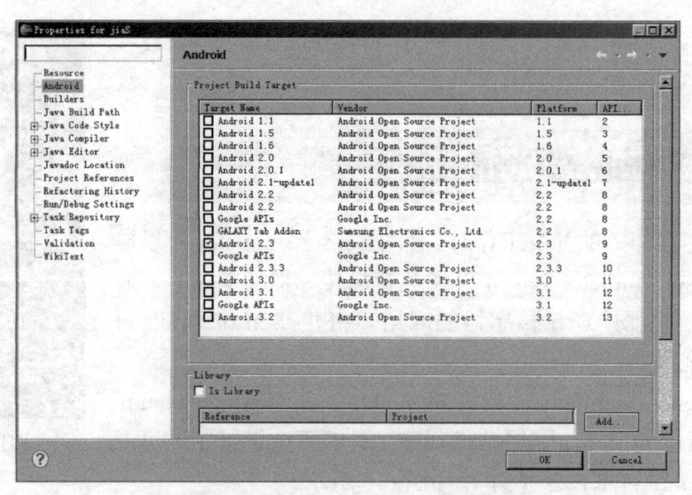

▲图 6-5　“Properties for jiaS”窗口

（2）选择左面的“Java Build Path”选项，然后单击“Libraries”选项卡，如图 6-6 所示。

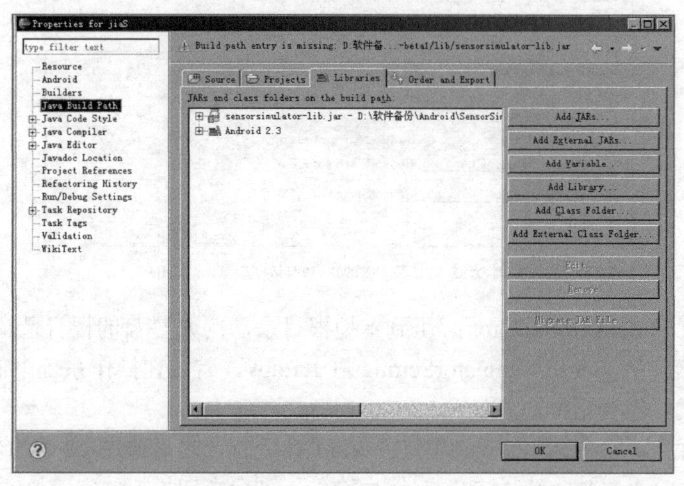

▲图 6-6　“Libraries”选项卡

（3）单击“Add External JARs”按钮，在弹出的“JAR Selection”对话框中找到 Sensorsimulator 安装目录下的 sensorsimulator-lib-1.1.1.jar，并将其添加到该项目中，如图 6-7 所示。

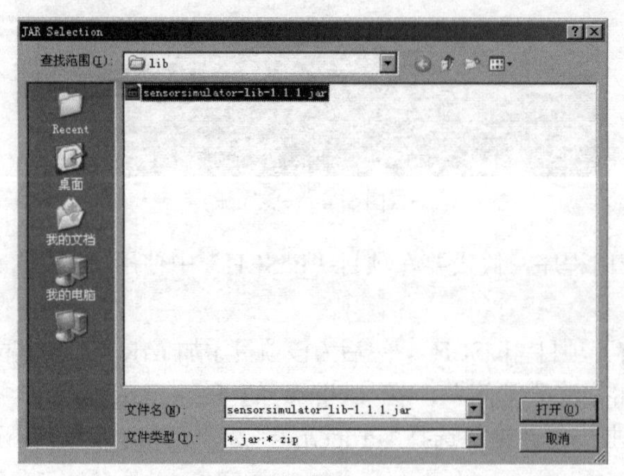

▲图 6-7　添加需要的 JAR 包

（4）开始启动 sensorsimulator.jar，并对手机模拟器上的 SensorSimulator 进行必要的配置。首先在“C:\sensorsimulator-1.1.1\bin”目录下找到 sensorsimulator.jar 并启动，运行后的界面如图 6-8 所示。

（5）接下来开始进行手机模拟器和 SensorSimulator 的连接配置工作，运行手机模拟器上安装好的 SensorSimulatorSettings.apk，如图 6-9 所示。

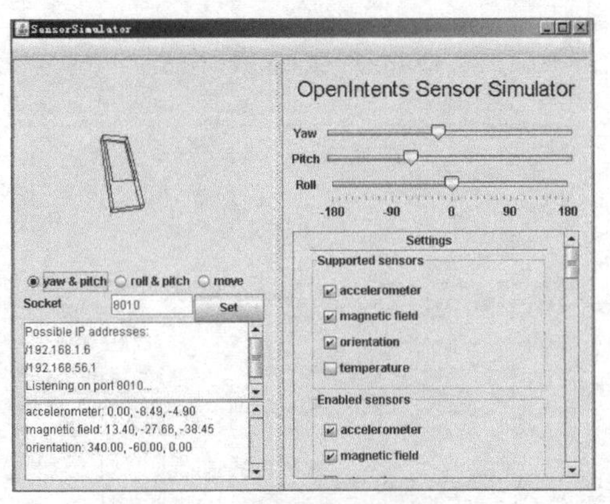

▲图 6-8　传感器的模拟器

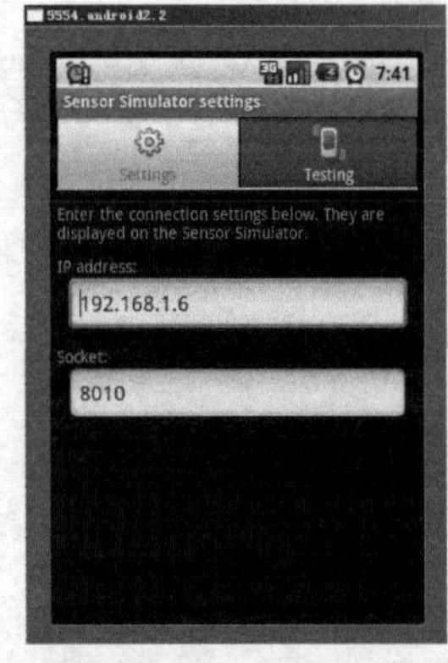

▲图 6-9　运行手机模拟器上的 SensorSimulatorSettings-1.1.1.apk

（6）在图 6-9 中输入 SensorSimulator 启动时显示的 IP 地址和端口号，单击屏幕右上角的“Testing”按钮后转到测试连接界面。如图 6-10 所示。

（7）单击屏幕上的“Connect”按钮进入下一界面，如图 6-11 所示。在此界面中可以选择需要监听的传感器，如果能够从传感器中读取到数据，说明 SensorSimulator 与手机模拟器连接成功，就可以测试自己开发的应用程序了。

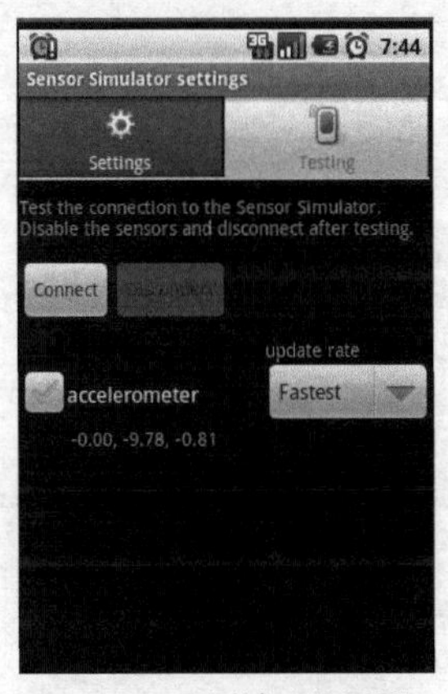

▲图 6-10　测试连接界面

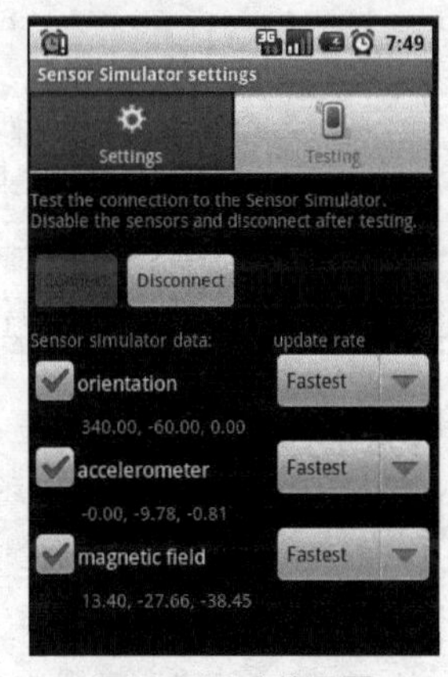

▲图 6-11　连接界面

到此为止，使用 Eclipse 结合 SensorSimulator 配置传感器应用程序的基本流程介绍完毕。

6.3　查看传感器的相关信息

在接下来的实例中，将演示在 Android 智能设备中查看传感器数据的方法。

实例	功能	源码路径
实例 6-2	查看传感器的相关信息	https://github.com/gast-lib/gast-lib

本实例的功能是查看设备中传感器的基本信息，本实例源码来源于：

```
https://github.com/gast-lib/gast-lib
```

读者可以自行登录下载。本实例的具体实现流程如下所示。

（1）编写用户交互界面 SensorListActivity，实现了两个用户界面交互。其中主界面 sensor_main.xml 用于列表显示系统中的传感器，功能和前面的实例 6-1 类似。文件 sensor_main.xml 的主要实现代码如下所示。

```
<LinearLayout xmlns:android="http://schemas.android.com/apk/res/android"
   android:layout_width="fill_parent"
   android:layout_height="fill_parent"
   >
    <fragment
        android:id="@+id/frag_sensor_select"
        android:layout_width="fill_parent"
        android:layout_height="fill_parent"
        class="root.gast.playground.sensor.SensorSelectorFragment" />

    <fragment
        android:id="@+id/frag_sensor_view"
        android:layout_width="fill_parent"
        android:layout_height="fill_parent"
        class="root.gast.playground.sensor.SensorDisplayFragment" />
</LinearLayout>
```

文件 SensorListActivity.java 的具体实现代码如下所示。

```
public class SensorListActivity extends FragmentActivity
{
    @Override
    protected void onCreate(Bundle savedInstanceState)
    {
        super.onCreate(savedInstanceState);
        setContentView(R.layout.sensor_main);

        // wire up the fragments so selector
        // can call display
        SensorDisplayFragment sensorDisplay =
                (SensorDisplayFragment) getSupportFragmentManager()
                        .findFragmentById(R.id.frag_sensor_view);
        SensorSelectorFragment sensorSelect =
                (SensorSelectorFragment) getSupportFragmentManager()
                        .findFragmentById(R.id.frag_sensor_select);
        sensorSelect.setSensorDisplay(sensorDisplay);
    }
}
```

通过上述代码可知，传感器类型列表显示功能是通过调用文件 SensorSelectorFragment.java 实现的，具体实现代码如下所示。

```
public class SensorSelectorFragment extends ListFragment
{
    private static final String TAG = "SensorSelectorFragment";

    private SensorDisplayFragment sensorDisplay;

    /**
     * connect with a display fragment to call later when user clicks a sensor
     * name, also setup the ListAdapter to show all the Sensors
     */
    public void setSensorDisplay(SensorDisplayFragment sensorDisplay)
    {
        this.sensorDisplay = sensorDisplay;

        SensorManager sensorManager =
                (SensorManager) getActivity().getSystemService(
                        Activity.SENSOR_SERVICE);
```

```
        List<Sensor> sensors = sensorManager.getSensorList(Sensor.TYPE_ALL);
        this.setListAdapter(new SensorListAdapter(getActivity()
                .getApplicationContext(), android.R.layout.simple_list_item_1,
                sensors));
    }

    /**
     * hide the list of sensors and show the sensor display fragment
     * add these changes to the backstack
     */
    private void showSensorFragment(Sensor sensor)
    {
        sensorDisplay.displaySensor(sensor);
        FragmentTransaction ft =
                getActivity().getSupportFragmentManager().beginTransaction();
        ft.hide(this);
        ft.show(sensorDisplay);
        ft.addToBackStack("Showing sensor: " + sensor.getName());
        ft.commit();
    }

    /**
     * list view adapter to show sensor names and respond to clicks.
     */
    private class SensorListAdapter extends ArrayAdapter<Sensor>
    {
        public SensorListAdapter(Context context, int textViewResourceId,
                List<Sensor> sensors)
        {
            super(context, textViewResourceId, sensors);
        }

        /**
         * create a text view containing the sensor name
         */
        @Override
        public View getView(final int position, View convertView,
                ViewGroup parent)
        {
            final Sensor selectedSensor = getItem(position);
            if (convertView == null)
            {
                convertView =
                        LayoutInflater.from(getContext()).inflate(
                                android.R.layout.simple_list_item_1, null);
            }

            ((TextView) convertView).setText(selectedSensor.getName());

            convertView.setOnClickListener(new View.OnClickListener()
            {
                @Override
                public void onClick(View v)
                {
                    if (BuildConfig.DEBUG)
                    {
                        Log.d(TAG,
                                "display sensor! " + selectedSensor.getName());
                    }

                    showSensorFragment(selectedSensor);
                }
            });
            return convertView;
        }
    }
}
```

（2）再看第 2 个界面，当用户在主界面选择列表中的某一个传感器时，会显示这个传感器的详细信息。界面文件 sensors_north_main.xml 的主要实现代码如下所示。

```
<ScrollView xmlns:android="http://schemas.android.com/apk/res/android"
    android:layout_width="match_parent"
    android:layout_height="match_parent"
    android:fillViewport="true">
    <RelativeLayout
        android:layout_width="wrap_content"
        android:layout_height="wrap_content"
        android:orientation="vertical" >

        <RadioGroup android:id="@+id/sensorRateSelector"
            android:layout_width="match_parent"
            android:layout_height="wrap_content"
            android:layout_alignParentTop="true" >

            <RadioButton android:id="@+id/delayFastest"
                android:layout_width="match_parent"
                android:layout_height="wrap_content"
                android:text="SENSOR_DELAY_FASTEST"
                android:checked="false"/>

            <RadioButton android:id="@+id/delayGame"
                android:layout_width="match_parent"
                android:layout_height="wrap_content"
                android:text="SENSOR_DELAY_GAME"
                android:checked="false"/>

            <RadioButton android:id="@+id/delayNormal"
                android:layout_width="match_parent"
                android:layout_height="wrap_content"
                android:text="SENSOR_DELAY_NORMAL"
                android:checked="true"/>

            <RadioButton android:id="@+id/delayUi"
                android:layout_width="match_parent"
                android:layout_height="wrap_content"
                android:text="SENSOR_DELAY_UI"
                android:checked="false"/>
        </RadioGroup>

        <View android:id="@+id/seperator"
            style="@style/line_separator"
            android:layout_below="@id/sensorRateSelector" />

        <TextView android:id="@+id/nameLabel"
            android:layout_width="wrap_content"
            android:layout_height="wrap_content"
            android:layout_below="@id/seperator"
            android:layout_alignParentLeft="true"
            android:text="Name:"
            android:layout_marginRight="5dip" />

        <TextView android:id="@+id/name"
            android:layout_width="wrap_content"
            android:layout_height="wrap_content"
            android:layout_toRightOf="@id/nameLabel"
            android:layout_alignTop="@id/nameLabel"
            android:layout_alignBottom="@id/nameLabel" />

        <TextView android:id="@+id/typeLabel"
            android:layout_width="wrap_content"
            android:layout_height="wrap_content"
            android:layout_alignLeft="@id/nameLabel"
            android:layout_below="@id/nameLabel"
            android:text="Type:"
            android:layout_marginRight="5dip" />

        <TextView android:id="@+id/type"
            android:layout_width="wrap_content"
            android:layout_height="wrap_content"
```

```
        android:layout_toRightOf="@id/typeLabel"
        android:layout_alignTop="@id/typeLabel"
        android:layout_alignBottom="@id/typeLabel"/>

    <TextView android:id="@+id/maxRangeLabel"
        android:layout_width="wrap_content"
        android:layout_height="wrap_content"
        android:layout_alignLeft="@id/nameLabel"
        android:layout_below="@id/typeLabel"
        android:text="Max Range:"
        android:layout_marginRight="5dip" />

    <TextView android:id="@+id/maxRange"
        android:layout_width="wrap_content"
        android:layout_height="wrap_content"
        android:layout_toRightOf="@id/maxRangeLabel"
        android:layout_alignTop="@id/maxRangeLabel"
        android:layout_alignBottom="@id/maxRangeLabel"/>

    <TextView android:id="@+id/minDelayLabel"
        android:layout_width="wrap_content"
        android:layout_height="wrap_content"
        android:layout_alignLeft="@id/maxRangeLabel"
        android:layout_below="@id/maxRangeLabel"
        android:text="Min Delay:"
        android:layout_marginRight="5dip" />

    <TextView android:id="@+id/minDelay"
        android:layout_width="wrap_content"
        android:layout_height="wrap_content"
        android:layout_toRightOf="@id/minDelayLabel"
        android:layout_alignTop="@id/minDelayLabel"
        android:layout_alignBottom="@id/minDelayLabel"/>

    <TextView android:id="@+id/powerLabel"
        android:layout_width="wrap_content"
        android:layout_height="wrap_content"
        android:layout_alignLeft="@id/minDelayLabel"
        android:layout_below="@id/minDelayLabel"
        android:text="Power:"
        android:layout_marginRight="5dip" />

    <TextView android:id="@+id/power"
        android:layout_width="wrap_content"
        android:layout_height="wrap_content"
        android:layout_toRightOf="@id/powerLabel"
        android:layout_alignTop="@id/powerLabel"
        android:layout_alignBottom="@id/powerLabel"
        android:layout_marginRight="5dip"/>

    <TextView
        android:layout_width="wrap_content"
        android:layout_height="wrap_content"
        android:layout_toRightOf="@id/power"
        android:layout_alignTop="@id/power"
        android:layout_alignBottom="@id/power"
        android:text="mA"/>

    <TextView android:id="@+id/resolutionLabel"
        android:layout_width="wrap_content"
        android:layout_height="wrap_content"
        android:layout_alignLeft="@id/powerLabel"
        android:layout_below="@id/powerLabel"
        android:text="Resolution:"
        android:layout_marginRight="5dip" />

    <TextView android:id="@+id/resolution"
        android:layout_width="wrap_content"
        android:layout_height="wrap_content"
```

```
    android:layout_toRightOf="@id/resolutionLabel"
    android:layout_alignTop="@id/resolutionLabel"
    android:layout_alignBottom="@id/resolutionLabel"/>

<TextView android:id="@+id/vendorLabel"
    android:layout_width="wrap_content"
    android:layout_height="wrap_content"
    android:layout_alignLeft="@id/resolutionLabel"
    android:layout_below="@id/resolutionLabel"
    android:text="Vendor:"
    android:layout_marginRight="5dip" />

<TextView android:id="@+id/vendor"
    android:layout_width="wrap_content"
    android:layout_height="wrap_content"
    android:layout_toRightOf="@id/vendorLabel"
    android:layout_alignTop="@id/vendorLabel"
    android:layout_alignBottom="@id/vendorLabel"/>

<TextView android:id="@+id/versionLabel"
    android:layout_width="wrap_content"
    android:layout_height="wrap_content"
    android:layout_toRightOf="@id/versionLabel"
    android:layout_alignLeft="@id/vendorLabel"
    android:layout_below="@id/vendorLabel"
    android:text="Version:"
    android:layout_marginRight="5dip" />

<TextView android:id="@+id/version"
    android:layout_width="wrap_content"
    android:layout_height="wrap_content"
    android:layout_toRightOf="@id/versionLabel"
    android:layout_alignTop="@id/versionLabel"
    android:layout_alignBottom="@id/versionLabel"/>

<TextView android:id="@+id/accuracyLabel"
    android:layout_width="wrap_content"
    android:layout_height="wrap_content"
    android:layout_alignLeft="@id/versionLabel"
    android:layout_below="@id/versionLabel"
    android:text="Accuracy:"
    android:layout_marginRight="5dip" />

<TextView android:id="@+id/accuracy"
    android:layout_width="wrap_content"
    android:layout_height="wrap_content"
    android:layout_toRightOf="@id/accuracyLabel"
    android:layout_alignTop="@id/accuracyLabel"
    android:layout_alignBottom="@id/accuracyLabel"/>

<!-- timestamp -->
<TextView android:id="@+id/timestampLabel"
    android:layout_width="wrap_content"
    android:layout_height="wrap_content"
    android:layout_alignLeft="@id/accuracyLabel"
    android:layout_below="@id/accuracyLabel"
    android:layout_marginRight="5dip"
    android:text="Timestamp:" />

<TextView android:id="@+id/timestamp"
    android:layout_width="wrap_content"
    android:layout_height="wrap_content"
    android:layout_toRightOf="@id/timestampLabel"
    android:layout_alignTop="@id/timestampLabel"
    android:layout_alignBottom="@id/timestampLabel"
    android:layout_marginRight="5dip"
    android:visibility="gone" />

<TextView android:id="@+id/timestampUnits"
```

```
        android:layout_width="wrap_content"
        android:layout_height="wrap_content"
        android:layout_toRightOf="@id/timestamp"
        android:layout_alignTop="@id/timestamp"
        android:layout_alignBottom="@id/timestamp"
        android:visibility="gone"
        android:text="(ns)" />

    <TextView android:id="@+id/dataLabel"
        android:layout_width="wrap_content"
        android:layout_height="wrap_content"
        android:layout_alignLeft="@id/accuracyLabel"
        android:layout_below="@id/timestampLabel"
        android:layout_marginRight="5dip"
        android:visibility="gone"/>

    <TextView android:id="@+id/dataUnits"
        android:layout_width="wrap_content"
        android:layout_height="wrap_content"
        android:layout_toRightOf="@id/dataLabel"
        android:layout_alignTop="@id/dataLabel"
        android:layout_alignBottom="@id/dataLabel"
        android:visibility="gone" />

    <TextView android:id="@+id/singleValue"
        android:layout_width="wrap_content"
        android:layout_height="wrap_content"
        android:layout_toRightOf="@id/dataUnits"
        android:layout_alignTop="@id/dataUnits"
        android:layout_alignBottom="@id/dataUnits"
        android:visibility="gone" />

    <!-- X axis -->
    <TextView android:id="@+id/xAxisLabel"
        android:layout_width="wrap_content"
        android:layout_height="wrap_content"
        android:layout_alignLeft="@id/dataLabel"
        android:layout_below="@id/dataLabel"
        android:layout_marginRight="5dip"
        android:visibility="gone"
        android:text="@string/xAxisLabel" />

    <TextView android:id="@+id/xAxis"
        android:layout_width="wrap_content"
        android:layout_height="wrap_content"
        android:layout_toRightOf="@id/xAxisLabel"
        android:layout_alignTop="@id/xAxisLabel"
        android:layout_alignBottom="@id/xAxisLabel"
        android:layout_marginRight="5dip"
        android:visibility="gone" />

    <!-- Y axis -->
    <TextView android:id="@+id/yAxisLabel"
        android:layout_width="wrap_content"
        android:layout_height="wrap_content"
        android:layout_alignLeft="@id/xAxisLabel"
        android:layout_below="@id/xAxisLabel"
        android:layout_marginRight="5dip"
        android:visibility="gone"
        android:text="@string/yAxisLabel" />

    <TextView android:id="@+id/yAxis"
        android:layout_width="wrap_content"
        android:layout_height="wrap_content"
        android:layout_toRightOf="@id/yAxisLabel"
        android:layout_alignTop="@id/yAxisLabel"
        android:layout_alignBottom="@id/yAxisLabel"
        android:layout_marginRight="5dip"
        android:visibility="gone" />
```

```
    <!-- Z axis -->
    <TextView android:id="@+id/zAxisLabel"
        android:layout_width="wrap_content"
        android:layout_height="wrap_content"
        android:layout_alignLeft="@id/yAxisLabel"
        android:layout_below="@id/yAxisLabel"
        android:layout_marginRight="5dip"
        android:visibility="gone"
        android:text="@string/zAxisLabel" />

    <TextView android:id="@+id/zAxis"
        android:layout_width="wrap_content"
        android:layout_height="wrap_content"
        android:layout_toRightOf="@id/zAxisLabel"
        android:layout_alignTop="@id/zAxisLabel"
        android:layout_alignBottom="@id/zAxisLabel"
        android:layout_marginRight="5dip"
        android:visibility="gone" />

    <!-- cos value (for rotation vector only) -->
    <TextView android:id="@+id/cosLabel"
        android:layout_width="wrap_content"
        android:layout_height="wrap_content"
        android:layout_alignLeft="@id/zAxisLabel"
        android:layout_below="@id/zAxisLabel"
        android:layout_marginRight="5dip"
        android:visibility="gone"
        android:text="cos(\u0398/2):" />

    <TextView android:id="@+id/cos"
        android:layout_width="wrap_content"
        android:layout_height="wrap_content"
        android:layout_toRightOf="@id/cosLabel"
        android:layout_alignTop="@id/cosLabel"
        android:layout_alignBottom="@id/cosLabel"
        android:layout_marginRight="5dip"
        android:visibility="gone" />
</RelativeLayout>
```

某个传感器的详细信息功能是通过文件SensorDisplayFragment.java实现的，主要代码如下所示。

```
public class SensorDisplayFragment extends Fragment implements SensorEventListener
{
    private static final String TAG = "SensorDisplayFragment";
    private static final String THETA = "\u0398";
    private static final String ACCELERATION_UNITS = "m/s\u00B2";

    private SensorManager sensorManager;
    private Sensor sensor;
    private TextView name;
    private TextView type;
    private TextView maxRange;
    private TextView minDelay;
    private TextView power;
    private TextView resolution;
    private TextView vendor;
    private TextView version;
    private TextView accuracy;
    private TextView timestampLabel;
    private TextView timestamp;
    private TextView timestampUnits;
    private TextView dataLabel;
    private TextView dataUnits;
    private TextView xAxis;
    private TextView xAxisLabel;
    private TextView yAxis;
    private TextView yAxisLabel;
    private TextView zAxis;
    private TextView zAxisLabel;
```

```
    private TextView singleValue;
    private TextView cosLabel;
    private TextView cos;

    @Override
    public View onCreateView(LayoutInflater inflater, ViewGroup container,
            Bundle savedInstanceState)
    {
        View layout = inflater.inflate(R.layout.sensor_view, null);

        sensorManager =
                (SensorManager) getActivity().getSystemService(Context.SENSOR_SERVICE);

        name = (TextView) layout.findViewById(R.id.name);
        type = (TextView) layout.findViewById(R.id.type);
        maxRange = (TextView) layout.findViewById(R.id.maxRange);
        minDelay = (TextView) layout.findViewById(R.id.minDelay);
        power = (TextView) layout.findViewById(R.id.power);
        resolution = (TextView) layout.findViewById(R.id.resolution);
        vendor = (TextView) layout.findViewById(R.id.vendor);
        version = (TextView) layout.findViewById(R.id.version);
        accuracy = (TextView) layout.findViewById(R.id.accuracy);
        timestampLabel = (TextView) layout.findViewById(R.id.timestampLabel);
        timestamp = (TextView) layout.findViewById(R.id.timestamp);
        timestampUnits = (TextView) layout.findViewById(R.id.timestampUnits);
        dataLabel = (TextView) layout.findViewById(R.id.dataLabel);
        dataUnits = (TextView) layout.findViewById(R.id.dataUnits);
        xAxis = (TextView) layout.findViewById(R.id.xAxis);
        xAxisLabel = (TextView) layout.findViewById(R.id.xAxisLabel);
        yAxis = (TextView) layout.findViewById(R.id.yAxis);
        yAxisLabel = (TextView) layout.findViewById(R.id.yAxisLabel);
        zAxis = (TextView) layout.findViewById(R.id.zAxis);
        zAxisLabel = (TextView) layout.findViewById(R.id.zAxisLabel);
        singleValue = (TextView) layout.findViewById(R.id.singleValue);
        cosLabel = (TextView) layout.findViewById(R.id.cosLabel);
        cos = (TextView) layout.findViewById(R.id.cos);

        layout.findViewById(R.id.delayFastest).setOnClickListener(new
OnClickListener()
        {
            @Override
            public void onClick(View v)
            {
                sensorManager.unregisterListener(SensorDisplayFragment.this);
                sensorManager.registerListener(SensorDisplayFragment.this,
                        sensor,
                        SensorManager.SENSOR_DELAY_FASTEST);
            }
        });

        layout.findViewById(R.id.delayGame).setOnClickListener(new OnClickListener()
        {
            @Override
            public void onClick(View v)
            {
                sensorManager.unregisterListener(SensorDisplayFragment.this);
                sensorManager.registerListener(SensorDisplayFragment.this,
                        sensor,
                        SensorManager.SENSOR_DELAY_GAME);
            }
        });

        layout.findViewById(R.id.delayNormal).setOnClickListener(new OnClickListener()
        {
            @Override
            public void onClick(View v)
            {
                sensorManager.unregisterListener(SensorDisplayFragment.this);
                sensorManager.registerListener(SensorDisplayFragment.this,
```

```
                        sensor,
                        SensorManager.SENSOR_DELAY_NORMAL);
            }
        });

        layout.findViewById(R.id.delayUi).setOnClickListener(new OnClickListener()
        {
            @Override
            public void onClick(View v)
            {
                sensorManager.unregisterListener(SensorDisplayFragment.this);
                sensorManager.registerListener(SensorDisplayFragment.this,
                        sensor,
                        SensorManager.SENSOR_DELAY_UI);
            }
        });

        return layout;
    }

    public void displaySensor(Sensor sensor)
    {
        if (BuildConfig.DEBUG)
        {
            Log.d(TAG, "display the sensor");
        }

        this.sensor = sensor;

        name.setText(sensor.getName());
        type.setText(String.valueOf(sensor.getType()));
        maxRange.setText(String.valueOf(sensor.getMaximumRange()));
        minDelay.setText(String.valueOf(sensor.getMinDelay()));
        power.setText(String.valueOf(sensor.getPower()));
        resolution.setText(String.valueOf(sensor.getResolution()));
        vendor.setText(String.valueOf(sensor.getVendor()));
        version.setText(String.valueOf(sensor.getVersion()));

        sensorManager.registerListener(this,
                sensor,
                SensorManager.SENSOR_DELAY_NORMAL);
    }

    /**
     * @see android.hardware.SensorEventListener#onAccuracyChanged(android.hardware.
Sensor, int)
     */
    @Override
    public void onAccuracyChanged(Sensor sensor, int accuracy)
    {
        switch(accuracy)
        {
            case SensorManager.SENSOR_STATUS_ACCURACY_HIGH:
                this.accuracy.setText("SENSOR_STATUS_ACCURACY_HIGH");
                break;
            case SensorManager.SENSOR_STATUS_ACCURACY_MEDIUM:
                this.accuracy.setText("SENSOR_STATUS_ACCURACY_MEDIUM");
                break;
            case SensorManager.SENSOR_STATUS_ACCURACY_LOW:
                this.accuracy.setText("SENSOR_STATUS_ACCURACY_LOW");
                break;
            case SensorManager.SENSOR_STATUS_UNRELIABLE:
                this.accuracy.setText("SENSOR_STATUS_UNRELIABLE");
                break;
        }
    }

    /**
     *  @see  android.hardware.SensorEventListener#onSensorChanged(android.hardware.
```

```
SensorEvent)
     */
    @Override
    public void onSensorChanged(SensorEvent event)
    {
        onAccuracyChanged(event.sensor, event.accuracy);

        timestampLabel.setVisibility(View.VISIBLE);
        timestamp.setVisibility(View.VISIBLE);
        timestamp.setText(String.valueOf(event.timestamp));
        timestampUnits.setVisibility(View.VISIBLE);

        switch (event.sensor.getType())
        {
            case Sensor.TYPE_ACCELEROMETER:
                showEventData("Acceleration - gravity on axis",
                        ACCELERATION_UNITS,
                        event.values[0],
                        event.values[1],
                        event.values[2]);
                break;

            case Sensor.TYPE_MAGNETIC_FIELD:
                showEventData("Abient Magnetic Field",
                        "uT",
                        event.values[0],
                        event.values[1],
                        event.values[2]);
                break;
            case Sensor.TYPE_GYROSCOPE:
                showEventData("Angular speed around axis",
                        "radians/sec",
                        event.values[0],
                        event.values[1],
                        event.values[2]);
                break;
            case Sensor.TYPE_LIGHT:
                showEventData("Ambient light",
                        "lux",
                        event.values[0]);
                break;
            case Sensor.TYPE_PRESSURE:
                showEventData("Atmospheric pressure",
                        "hPa",
                        event.values[0]);
                break;
            case Sensor.TYPE_PROXIMITY:
                showEventData("Distance",
                        "cm",
                        event.values[0]);
                break;
            case Sensor.TYPE_GRAVITY:
                showEventData("Gravity",
                        ACCELERATION_UNITS,
                        event.values[0],
                        event.values[1],
                        event.values[2]);
                break;
            case Sensor.TYPE_LINEAR_ACCELERATION:
                showEventData("Acceleration (not including gravity)",
                        ACCELERATION_UNITS,
                        event.values[0],
                        event.values[1],
                        event.values[2]);
                break;
            case Sensor.TYPE_ROTATION_VECTOR:

                showEventData("Rotation Vector",
                        null,
```

```
                    event.values[0],
                    event.values[1],
                    event.values[2]);

            xAxisLabel.setText("x*sin(" + THETA + "/2)");
            yAxisLabel.setText("y*sin(" + THETA + "/2)");
            zAxisLabel.setText("z*sin(" + THETA + "/2)");

            if (event.values.length == 4)
            {
                cosLabel.setVisibility(View.VISIBLE);
                cos.setVisibility(View.VISIBLE);
                cos.setText(String.valueOf(event.values[3]));
            }

            break;
        case Sensor.TYPE_ORIENTATION:
            showEventData("Angle",
                    "Degrees",
                    event.values[0],
                    event.values[1],
                    event.values[2]);

            xAxisLabel.setText(R.string.azimuthLabel);
            yAxisLabel.setText(R.string.pitchLabel);
            zAxisLabel.setText(R.string.rollLabel);

            break;
        case Sensor.TYPE_RELATIVE_HUMIDITY:
            showEventData("Relatice ambient air humidity",
                    "%",
                    event.values[0]);
            break;
        case Sensor.TYPE_AMBIENT_TEMPERATURE:
            showEventData("Ambien temperature",
                    "degree Celcius",
                    event.values[0]);
            break;
    }
}

private void showEventData(String label, String units, float x, float y, float z)
{
    dataLabel.setVisibility(View.VISIBLE);
    dataLabel.setText(label);

    if (units == null)
    {
        dataUnits.setVisibility(View.GONE);
    }
    else
    {
        dataUnits.setVisibility(View.VISIBLE);
        dataUnits.setText("(" + units + "):");
    }

    singleValue.setVisibility(View.GONE);

    xAxisLabel.setVisibility(View.VISIBLE);
    xAxisLabel.setText(R.string.xAxisLabel);
    xAxis.setVisibility(View.VISIBLE);
    xAxis.setText(String.valueOf(x));

    yAxisLabel.setVisibility(View.VISIBLE);
    yAxisLabel.setText(R.string.yAxisLabel);
    yAxis.setVisibility(View.VISIBLE);
    yAxis.setText(String.valueOf(y));

    zAxisLabel.setVisibility(View.VISIBLE);
```

```
        zAxisLabel.setText(R.string.zAxisLabel);
        zAxis.setVisibility(View.VISIBLE);
        zAxis.setText(String.valueOf(z));
    }

    private void showEventData(String label, String units, float value)
    {
        dataLabel.setVisibility(View.VISIBLE);
        dataLabel.setText(label);

        dataUnits.setVisibility(View.VISIBLE);
        dataUnits.setText("(" + units + "):");

        singleValue.setVisibility(View.VISIBLE);
        singleValue.setText(String.valueOf(value));

        xAxisLabel.setVisibility(View.GONE);
        xAxis.setVisibility(View.GONE);

        yAxisLabel.setVisibility(View.GONE);
        yAxis.setVisibility(View.GONE);

        zAxisLabel.setVisibility(View.GONE);
        zAxis.setVisibility(View.GONE);
    }

    /**
     * @see android.support.v4.app.Fragment#onHiddenChanged(boolean)
     */
    @Override
    public void onHiddenChanged(boolean hidden)
    {
        super.onHiddenChanged(hidden);

        if (hidden)
        {
            if (BuildConfig.DEBUG)
            {
                Log.d(TAG, "Unregistering listener");
            }

            sensorManager.unregisterListener(this);
        }
    }

    /**
     * @see android.support.v4.app.Fragment#onPause()
     */
    @Override
    public void onPause()
    {
        super.onPause();

        if (BuildConfig.DEBUG)
        {
            Log.d(TAG, "onPause");
            Log.d(TAG, "Unregistering listener");
        }

        sensorManager.unregisterListener(this);
    }
}
```

到此为止，整个实例介绍完毕，执行后的效果如图 6-12 所示。

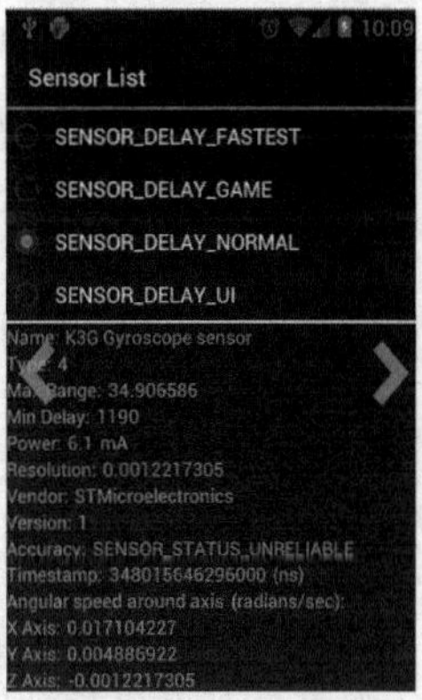

▲图 6-12 执行效果

第 7 章　地图定位

Map 地图对大家来说应该不算陌生，它让人们体会到了高科技的奥妙。作为谷歌官方旗下产品之一的 Android 系统，可以非常方便地使用 Google 地图实现位置定位功能。本章将详细讲解在 Android 设备中使用位置服务和地图 API 的基本流程，为读者步入本书后面知识的学习打下坚实的基础。

7.1 位置服务

在 Android 系统中可以非常容易地获取当前的位置信息，此功能是通过谷歌地图实现的。Android 系统可以无缝地支持 GPS 和谷歌网络地图，人们通常将各种不同的定位技术称之为 LBS。LBS 是基于位置的服务（Location Based Service）的英文缩写，它是通过电信移动运营商的无线电通信网络（如 GSM 网、CDMA 网）或外部定位方式（如 GPS）获取移动终端用户的位置信息（地理坐标或大地坐标），在地理信息系统（Geographic Information System，GIS）平台的支持下，为用户提供相应服务的一种增值业务。

7.1.1 android.location 功能类

在 Android 设备中，可以使用类 android.location 来实现定位功能。

1. Google Map API

Android 系统提供了一组访问 Google Map 的 API，借助 Google Map 及定位 API，就可以在地图上显示用户当前的地理位置。在 Android 中定义了一个名为 com.google.android.maps 的包，其中包含了一系列用于在 Google Map 上显示、控制和层叠信息的功能类，下面是该包中最重要的几个类。

- MapActivity：用于显示 Google Map 的 Activity 类，它需要连接底层网络。
- MapView：用于显示地图的 View 组件，它必须和 MapActivity 配合使用。
- MapController：用于控制地图的移动。
- Overlay：是一个可显示于地图之上的可绘制的对象。
- GeoPoint：是一个包含经纬度位置的对象。

2. Android Location API

在 Android 设备中，实现定位功能的相关类如下所示。

- LocationManager：本类提供访问定位服务的功能，也提供了获取最佳定位提供者的功能。另外，临近警报功能也可以借助该类来实现。
- LocationProvider：该类是定位提供者的抽象类。定位提供者具备周期性报告设备地理位置的功能。

- LocationListener：提供定位信息发生改变时的回调功能。必须事先在定位管理器中注册监听器对象。
- Criteria：该类使得应用能够通过在LocationProvider中设置的属性来选择合适的定位提供者。

7.1.2 实现定位服务功能

在Android设备中，实现定位处理的基本流程如下所示。

1. 先准备Activity类

此步骤的目的是使用Google Map API来显示地图，然后使用定位API来获取设备的当前定位信息以在Google Map上设置设备的当前位置，用户定位会随着用户的位置移动而发生改变。

首先需要一个继承MapActivity的Activity类，例如下面的代码。

```
class MyGPSActivity extends MapActivity {
        …
        }
```

要成功引用Google Map API，还必须先在AndroidManifest.xml中定义如下信息。

```
<uses-library android:name="com.google.android.maps" />
```

2. 然后使用MapView

要让地图显示，需要将MapView加入到应用中。例如在布局文件（main.xml）中加入如下代码。

```
<com.google.android.maps.MapView
            android:id="@+id/myGMap"
            android:layout_width="fill_parent"
            android:layout_height="fill_parent"
            android:enabled="true"
            android:clickable="true"
            android:apiKey="API_Key_String"
/>
```

另外，要使用Google Map服务，还需要一个API key。可以通过如下方式获取API key。

- 找到USER_HOME\Local Settings\Application Data\Android目录下的debug.keystore文件。
- 使用keytool工具来生成认证信息（MD5），使用如下命令行。

```
keytool  -list  -alias  androiddebugkey  -keystore  <path_to_debug_keystore>.keystore
-storepass
android -keypass android
```

- 打开“Sign Up for the Android Maps API”页面，输入之前生成的认证信息（MD5）后将获取到API key。
- 使用刚才获取的API key替换上面AndroidManifest.xml配置文件中“API_Key_String”。

> 注意　上面获取API key的介绍比较简单，后面将会通过一个具体实例来演示获取API key的方法。

接下来继续补全MyGPSActivity类的代码，在此使用MapView，例如下面的代码。

```
class MyGPSActivity extends MapActivity {
    @Override
    public void onCreate(Bundle savedInstanceState) {
```

```
    //创建并初始化地图
      gMapView = (MapView) findViewById(R.id.myGMap);
        GeoPoint p = new GeoPoint((int) (lat * 1000000), (int) (long * 1000000));
         gMapView.setSatellite(true);
         mc = gMapView.getController();
         mc.setCenter(p);
         mc.setZoom(14);
    }
}
```

另外，必须先设置一些权限后才能使用定位信息，在文件 AndroidManifest.xml 中的配置方式如下。

```
<uses-permission android:name="android.permission.INTERNET"></uses-permission>
<uses-permission
android:name="android.permission.ACCESS_COARSE_LOCATION"></uses-permission>
<uses-permission
android:name="android.permission.ACCESS_FINE_LOCATION"></uses-permission>
```

3. 实现定位管理器

可以使用 Context.getSystemService()方法实现定位管理器功能，并传入 Context.LOCATION_SERVICE 参数来获取定位管理器的实例，例如下面的代码。

```
LocationManager lm = (LocationManager) getSystemService(Context.LOCATION_SERVICE);
```

接下来需要将原先的 MyGPSActivity 做一些修改，让它实现一个 LocationListener 接口，使其能够监听定位信息的改变。

```
class MyGPSActivity extends MapActivity implements LocationListener {
…
public void onLocationChanged(Location location) {}
public void onProviderDisabled(String provider) {}
public void onProviderEnabled(String provider) {}
public void onStatusChanged(String provider, int status, Bundle extras) {}
protected boolean isRouteDisplayed() {
return false;
}
}
```

接下来需要初始化 LocationManager，并在它的 onCreate()方法中注册定位监听器，例如下面的代码。

```
@Override
public void onCreate(Bundle savedInstanceState) {
LocationManager lm = (LocationManager)getSystemService(Context.LOCATION_SERVICE);
lm.requestLocationUpdates(LocationManager.GPS_PROVIDER, 1000L, 500.0f, this);
        }
```

这样代码中的方法 onLocationChanged()会在用户的位置发生 500 m 距离的改变之后进行调用。这里默认使用的 LocationProvider 是“gps”（GSP_PROVIDER），但是可以根据需要，使用特定的 Criteria 对象调用 LocationManger 类的 getBestProvider 方法获取其他的 LocationProvider。以下代码是 onLocationChanged()方法的参考实现。

```
public void onLocationChanged(Location location) {
    if (location != null) {
    double lat = location.getLatitude();
    double lng = location.getLongitude();
     p = new GeoPoint((int) lat * 1000000, (int) lng * 1000000);
     mc.animateTo(p);
   }
}
```

通过上面的代码，获取了当前的新位置并在地图上更新位置显示。还可以为应用程序添加一些例如缩放效果、地图标注和文本等功能。

4. 添加缩放控件

```
//将缩放控件添加到地图上
ZoomControls zoomControls =  (ZoomControls) gMapView.getZoomControls();
zoomControls.setLayoutParams(new ViewGroup.LayoutParams(LayoutParams.WRAP_CONTENT,
LayoutParams.WRAP_CONTENT));
gMapView.addView(zoomControls);
gMapView.displayZoomControls(true);
```

5. 添加 Map Overlay

最后一步是添加 Map Overlay，例如通过下面的代码可以定义一个 Overlay。

```
 class MyLocationOverlay extends com.google.android.maps.Overlay {
  public boolean draw(Canvas canvas, MapView mapView, boolean shadow, long when) {
 super.draw(canvas, mapView, shadow);
 Paint paint = new Paint();
 // 将经纬度转换成实际屏幕坐标
 Point myScreenCoords = new Point();
 mapView.getProjection().toPixels(p, myScreenCoords);
 paint.setStrokeWidth(1);
 paint.setARGB(255, 255, 255, 255);
 paint.setStyle(Paint.Style.STROKE);
 Bitmap bmp = BitmapFactory.decodeResource(getResources(), R.drawable.marker);
 canvas.drawBitmap(bmp, myScreenCoords.x, myScreenCoords.y, paint);
 canvas.drawText("how are you…", myScreenCoords.x, myScreenCoords.y, paint);
 return true;
}
}
```

通过上面的 Overlay 会在地图上显示一段文本，接下来可以把这个 Overlay 添加到地图上。

```
MyLocationOverlay myLocationOverlay = new MyLocationOverlay();
List<Overlay> list = gMapView.getOverlays();
list.add(myLocationOverlay);
```

7.1.3 实战演练——在 Android 设备中实现 GPS 定位

在本节的内容中，将通过具体实例来演示在 Android 设备中实现 GPS 定位功能的基本流程。

题目	目的	源码路径
实例 7-1	用 GPS 定位技术获取当前的位置信息	\daima\7\GPSLocation

本实例的具体实现流程如下所示。

（1）在文件 AndroidManifest.xml 中添加 ACCESS_FINE_LOCATION 权限，具体代码如下所示。

```
<uses-permission android:name="android.permission.ACCESS_FINE_LOCATION"/>
```

（2）在 onCreate（Bundle savedInstanceState）中获取当前位置信息，具体代码如下所示。

```
public void onCreate(Bundle savedInstanceState) {
   super.onCreate(savedInstanceState);
   setContentView(R.layout.main);
   LocationManager locationManager;
   String serviceName = Context.LOCATION_SERVICE;
   locationManager = (LocationManager)getSystemService(serviceName);
   Criteria criteria = new Criteria();
   criteria.setAccuracy(Criteria.ACCURACY_FINE);
   criteria.setAltitudeRequired(false);
   criteria.setBearingRequired(false);
   criteria.setCostAllowed(true);
   criteria.setPowerRequirement(Criteria.POWER_LOW);
   String provider = locationManager.getBestProvider(criteria, true);

   Location location = locationManager.getLastKnownLocation(provider);
   updateWithNewLocation(location);
    /*每隔1000 ms更新一次*/
```

```
        locationManager.requestLocationUpdates(provider, 2000, 10,
                locationListener);
    }
```

在上述代码中，LocationManager 用于周期获得当前设备的一个类。要想获取 LocationManager 实例，必须调用 Context.getSystemService()方法并传入服务名 LOCATION_SERVICE("location")。创建 LocationManager 实例后，就可以通过调用 getLastKnownLocation()方法将上一次 LocationManager 获得有效位置信息以 Location 对象的形式返回。getLastKnownLocation()方法需要传入一个字符串参数来确定使用定位服务类型，本实例传入的是静态常量 LocationManager.GPS_PROVIDER，这表示使用 GPS 技术定位。最后还需要使用 Location 对象将位置信息以文本方式显示到用户界面。

（3）定义方法 updateWithNewLocation(Location location)来更新显示用户界面，具体代码如下所示。

```
        private void updateWithNewLocation(Location location) {
            String latLongString;
            TextView myLocationText;
            myLocationText = (TextView)findViewById(R.id.myLocationText);
            if (location != null) {
            double lat = location.getLatitude();
            double lng = location.getLongitude();
            latLongString = "纬度是:" + lat + "\n 经度是:" + lng;
            } else {
            latLongString = "失败";
            }
            myLocationText.setText("获取的当前位置是:\n" +
            latLongString);
        }
    }
```

（4）定义 LocationListener 对象 locationListener，当坐标改变时触发此函数。如果 Provider 传入相同的坐标，它就不会被触发。具体代码如下所示。

```
    private final LocationListener locationListener = new LocationListener() {
        public void onLocationChanged(Location location) {
        updateWithNewLocation(location);
        }
        public void onProviderDisabled(String provider){
        updateWithNewLocation(null);
        }
        public void onProviderEnabled(String provider){ }
        public void onStatusChanged(String provider, int status,
        Bundle extras){ }
    };
```

下面开始测试，因为模拟器上没有 GPS 设备，所以需要在 Eclipse 的 DDMS 工具中提供模拟的 GPS 数据。即依次单击“DDMS”|“Emulator Control”，在弹出对话框中找到“Location Control”选项，在此输入坐标，完成后单击“Send”按钮，如图 7-1 所示。

因为用到了 Google API，所以要在项目中引入 Google API，右键单击项目选择“Properties”，在弹出对话框中选择相应的 Google API 版本，如图 7-2 所示。

▲图 7-1　设置坐标

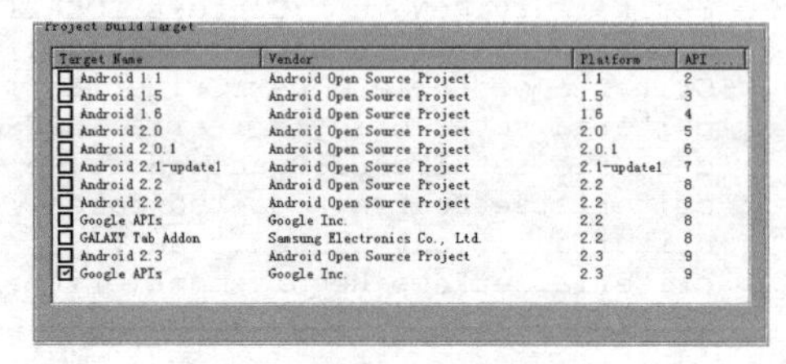

▲图 7-2　引用 Google API

这样模拟器运行后，会显示当前的坐标。

7.2 及时更新位置信息

到此为止，可能会有读者疑问：如果GPS的位置信息变化后，是不是也能够及时地显示新的位置信息到界面上呢？是不是只有退出系统，然后重新启动系统后才能得到最新的GPS信息呢？其实不然，可以通过编程的方式来及时获取并更新当前的位置信息，本节将讲解这部分内容。

7.2.1 Maps 库类

在Maps库类中提供了十几个类，通过这些类可以实现位置更新功能。在这些库类中，最为常用的类包括MapView、MapController和MapActivity等。

1. MapController

控制地图移动、伸缩、以某个GPS坐标为中心，控制MapView中的View组件，管理Overlay，提供View的基本功能。使用多种地图模式（地图模式（某些城市可实时对交通状况进行更新）、卫星模式、街景模式）来查看Google Map。

常用方法有animateTo(GeoPoint point)、setCenter(GeoPoint point)和setZoom(int zoomLevel)等。

2. MapView

Mapview是用来显示地图的View，它派生自android.view.ViewGroup。当MapView获得焦点时，可以控制地图的移动和缩放。Android中的地图可以以不同的形式显示出来，如街景模式、卫星模式等。

MapView只能被MapActivity来创建，这是因为MapView需要通过后台的线程来连接网络或者文件系统，而这些线程要由MapActivity来管理。常用方法有getController()、getOverlays()、setSatellite(boolean)、setTraffic(boolean)、setStreetView(boolean)和setBuiltInZoomControls(boolean)等。

3. MapActivity

MapActivity是一个抽象类，任何想要显示MapView的activity都需要派生自MapActivity。并且在其派生类的onCreate()中，都要创建一个MapView实例，可以通过MapViewconstructor（然后添加到View中ViewGroup.addView(View)）或者通过layout XML来创建。

4. Overlay

Overlay是覆盖到MapView的最上层，可以扩展其ondraw接口，自定义在MapView中显示一些自己的内容。MapView通过MapView.getOverlays()对Overlay进行管理。

除了Overlay这个基类，Google还扩展了如下两个比较有用的Overlay。

- MylocationOverlay：集成了Android.location中接收当前坐标的接口，集成SersorManager中CompassSensor的接口。只需要enableMyLocation()，enableCompass就可以让程序拥有实时的MyLocation以及Compass功能（Activity.onResume()中）。
- ItemlizedOverlay：管理一个OverlayItem链表，用图片等资源在地图上做风格相同的标记。

5. Projection

MapView中GPS坐标与设备坐标的转换（GeoPoint和Point）。

7.2.2　使用 LocationManager 及时监听

LocationManager 支持监听器模式，通过调用 requestLocationUpdates()方法能够为其设置一个位置监听器 LocationListener。同时方法 requestLocationUpdates()还需要指定要使用的位置服务类型、位置更新时间和最新位移，这样可以确保在满足用户需求的前提下最低的电量消耗。

例如在下面的代码中，设置了更新位置信息的最小间隔为 2 s，位移变化在 10 m 以上。如果 GPS 位置超过 10 m，且时间间隔超过 2 s 时，LocationListener 的回调方法 onLocationChanged（）就会被调用，应用程序可以通过 onLocationChanged 来反映位置信息的变化。

```
public class CurrentLocationWithMap extends MapActivity {
    public void onCreate(Bundle savedInstanceState) {
        super.onCreate(savedInstanceState);
        setContentView(R.layout.main);
        LocationManager locationManager;
        String context = Context.LOCATION_SERVICE;
        locationManager = (LocationManager)getSystemService(context);
        //String provider = LocationManager.GPS_PROVIDER;
        /*Location Provider 查询条件*/
        Criteria criteria = new Criteria();
        criteria.setAccuracy(Criteria.ACCURACY_FINE);
        criteria.setAltitudeRequired(false);
        criteria.setBearingRequired(false);
        criteria.setCostAllowed(true);
        criteria.setPowerRequirement(Criteria.POWER_LOW);
        String provider = locationManager.getBestProvider(criteria, true);
        Location location = locationManager.getLastKnownLocation(provider);
        updateWithNewLocation(location);
        /*设置更新位置信息的最小间隔为 2 秒，位移变化在 10 米以上。*/
        locationManager.requestLocationUpdates(provider, 2000, 10,
                locationListener);
    }
   /*Location 发生变化时被调用*/
   private final LocationListener locationListener = new LocationListener() {
        public void onLocationChanged(Location location) {
        updateWithNewLocation(location);
        }
        public void onProviderDisabled(String provider){
        updateWithNewLocation(null);
        }
        public void onProviderEnabled(String provider){ }
        public void onStatusChanged(String provider, int status,
        Bundle extras){ }
    };
    private void updateWithNewLocation(Location location) {
        String latLongString;
        TextView myLocationText;
        myLocationText = (TextView)findViewById(R.id.myLocationText);
        if (location != null) {
            double lat = location.getLatitude();
            double lng = location.getLongitude();
            latLongString = "纬度是:" + lat + "\n 经度是:" + lng;
            ctrlMap.animateTo(new GeoPoint((int)(lat*1E6),(int)(lng*1E6)));
        } else {
            latLongString = "失败";
        }
        myLocationText.setText("获取的当前位置是:\n" +latLongString);

    }
```

在使用类 LocationManager 时，通常需要用到如下所示的方法。

- getLatitude()：获取经度值。
- getLongitude()：获取纬度值。
- getAltitude()：获取海拔的值。

7.2.3　实战演练——在 Android 设备中显示当前位置的坐标和海拔

在本节的内容中，将通过具体实例来演示在 Android 设备中显示当前位置的坐标和海拔的基本方法。

题目	目的	源码路径
实例 7-2	显示当前位置的坐标和海拔	\daima\7\GPS

本实例的具体实现流程如下所示。

（1）在文件 AndroidManifest.xml 中添加 ACCESS_FINE_LOCATION 权限和 ACCESS_LOCATION_EXTRA_COMMANDS 权限，具体代码如下所示。

```
<uses-permission android:name="android.permission.ACCESS_FINE_LOCATION" />
<uses-permission android:name="android.permission.ACCESS_LOCATION_EXTRA_COMMANDS"/>
```

（2）编写布局文件 main.xml，设置在屏幕中分别显示当前位置的经度、纬度、速度和海拔等信息。文件 main.xml 的具体实现代码如下所示。

```
<LinearLayout xmlns:android="http://schemas.android.com/apk/res/android"
    android:layout_width="fill_parent"
    android:layout_height="fill_parent"
    android:background="#008080"
    android:id="@+id/mainlayout" android:orientation="vertical">

    <gps.mygps.paintview android:id="@+id/iddraw"
        android:layout_width="fill_parent"
        android:layout_height="300dip"
    />

    <TableLayout android:layout_width="fill_parent"
            android:layout_height="wrap_content">

        <TableRow>
            <TextView android:id="@+id/speed"
                    android:layout_width="wrap_content"
                    android:layout_height="wrap_content"
                    android:text="速度"
                    style="@style/smalltext"
                    android:gravity="center"
                    android:layout_weight="33"/>
            <TextView android:id="@+id/altitude"
                    android:layout_width="wrap_content"
                    android:layout_height="wrap_content"
                    android:text="海拔"
                    style="@style/smalltext"
                    android:gravity="center"
                    android:layout_weight="33"/>
            <TextView android:id="@+id/bearing"
                    android:layout_width="wrap_content"
                    android:layout_height="wrap_content"
                    android:text="航向"
                    style="@style/smalltext"
                    android:gravity="center"
                    android:layout_weight="34"/>
        </TableRow>
        <TableRow>
            <TextView android:id="@+id/speedvalue"
                    android:layout_width="wrap_content"
                    android:layout_height="wrap_content"
                    style="@style/normaltext"
                    android:gravity="center"
                    android:layout_weight="33"/>
            <TextView android:id="@+id/altitudevalue"
                    android:layout_width="wrap_content"
                    android:layout_height="wrap_content"
```

```
                style="@style/normaltext"
                android:layout_weight="33"
                android:gravity="center"/>
        <TextView android:id="@+id/bearvalue"
                android:layout_width="wrap_content"
                android:layout_height="wrap_content"
                style="@style/normaltext"
                android:gravity="center"
                android:layout_weight="34"/>
    </TableRow>
</TableLayout>

 <TableLayout android:layout_width="fill_parent"
        android:layout_height="wrap_content">
    <TableRow>
        <TextView android:layout_width="wrap_content"
                android:layout_height="wrap_content"
                android:text="维度"
                android:gravity="center"
                android:layout_weight="50"
                style="@style/smalltext"/>
        <TextView android:layout_width="wrap_content"
                android:layout_height="wrap_content"
                android:text="卫星"
                android:gravity="center"
                android:layout_weight="50"
                style="@style/smalltext"/>
        <TextView android:layout_width="wrap_content"
                android:layout_height="wrap_content"
                android:text="经度"
                style="@style/smalltext"
                android:gravity="center"
                android:layout_weight="50"/>
    </TableRow>
    <TableRow>
        <TextView android:id="@+id/latitudevalue"
                android:layout_width="wrap_content"
                android:layout_height="wrap_content"
                style="@style/normaltext"
                android:gravity="center"
                android:layout_weight="33"/>
        <TextView android:id="@+id/satellitevalue"
                android:layout_width="wrap_content"
                android:layout_height="wrap_content"
                style="@style/normaltext"
                android:gravity="center"
                android:layout_weight="33"/>
        <TextView android:id="@+id/longitudevalue"
                android:layout_width="wrap_content"
                android:layout_height="wrap_content"
                style="@style/normaltext"
                android:gravity="center"
                android:layout_weight="34"/>
    </TableRow>
</TableLayout>
<TableLayout android:layout_width="fill_parent"
        android:layout_height="wrap_content">
        <TableRow>
           <TextView android:id="@+id/time"
                android:layout_width="wrap_content"
                android:layout_height="wrap_content"
                android:text="时间:"
                style="@style/normaltext"
                />
           <TextView android:id="@+id/timevalue"
                android:layout_width="wrap_content"
                android:layout_height="wrap_content"
                style="@style/normaltext"
                />
```

```
        </TableRow>
    </TableLayout>

    <RelativeLayout android:layout_width="fill_parent"
            android:layout_height="wrap_content">
        <Button android:id="@+id/close"
            android:layout_width="wrap_content"
            android:layout_height="wrap_content"
            android:text="关闭"
            android:textSize="20sp"
            android:layout_alignParentRight="true"></Button>
        <Button android:id="@+id/open"
            android:layout_height="wrap_content"
            android:layout_width="wrap_content"
            android:text="打开"
            android:textSize="20sp"
               android:layout_toLeftOf="@id/close"></Button>
    </RelativeLayout>

    <TextView android:id="@+id/error"
            android:layout_width="fill_parent"
            android:layout_height="wrap_content"
            style="@style/smalltext"
            />
</LinearLayout>
```

（3）编写程序文件 Mygps.java，功能是监听用户单击屏幕按钮的事件，获取当前位置的定位信息。文件 Mygps.java 的具体实现代码如下所示。

```
public class Mygps extends Activity {

    protected static final String TAG = null;
    //位置类
    private Location location;
    // 定位管理类
    private LocationManager locationManager;
    private String provider;
    //监听卫星变量
    private GpsStatus gpsStatus;
    Iterable<GpsSatellite> allSatellites;
    float satellitedegree[][] = new float[24][3];

    float alimuth[] = new  float[24];
    float elevation[] = new float[24];
    float snr[] = new float[24];

    private boolean status=false;
    protected Iterator<GpsSatellite> Iteratorsate;
    private float bear;

    //获取手机屏幕分辨率的类
    private DisplayMetrics dm;


    paintview layout;
    Button openbutton;
    Button closebutton;
    TextView latitudeview;
    TextView longitudeview;
    TextView altitudeview;
    TextView speedview;
    TextView timeview;
    TextView errorview;
    TextView bearingview;
    TextView satcountview;

    /** Called when the activity is first created. */

    @Override
```

```
    public void onCreate(Bundle savedInstanceState) {
        super.onCreate(savedInstanceState);

       requestWindowFeature(Window.FEATURE_NO_TITLE);
       getWindow().setFlags(WindowManager.LayoutParams.FLAG_FULLSCREEN, WindowManager.
LayoutParams.FLAG_FULLSCREEN);

       setContentView(R.layout.main);

        findview();

        openbutton.setOnClickListener(new View.OnClickListener() {

            @Override
            public void onClick(View v) {
                // TODO Auto-generated method stub
                if(!status)
                 {
                    openGPSSettings();
                     getLocation();
                     status = true;
                 }
            }
        });

        closebutton.setOnClickListener(new View.OnClickListener() {

            @Override
            public void onClick(View v) {
                // TODO Auto-generated method stub
                closeGps();
            }
        });
    }

    private void findview() {
        // TODO Auto-generated method stub
        openbutton = (Button)findViewById(R.id.open);
        closebutton = (Button)findViewById(R.id.close);
        latitudeview = (TextView)findViewById(R.id.latitudevalue);
        longitudeview = (TextView)findViewById(R.id.longitudevalue);
        altitudeview = (TextView)findViewById(R.id.altitudevalue);
        speedview = (TextView)findViewById(R.id.speedvalue);
        timeview = (TextView)findViewById(R.id.timevalue);
        errorview = (TextView)findViewById(R.id.error);
        bearingview = (TextView)findViewById(R.id.bearvalue);
        layout=(gps.mygps.paintview)findViewById(R.id.iddraw);
        satcountview = (TextView)findViewById(R.id.satellitevalue);
    }


    protected void closeGps() {
        // TODO Auto-generated method stub

        // dm = new DisplayMetrics();
        // getWindowManager().getDefaultDisplay().getMetrics(dm);
        // heightp = dm.heightPixels;
        // widthp = dm.widthPixels;
        // alimuth[0] = 60;
        // elevation[0] = 20;
        // snr[0] = 60;
        // alimuth[1] = 260;
        // elevation[1] = 10;
        // snr[1] = 50;
        // layout.redraw(240,alimuth,elevation,snr, widthp,heightp, 2);

        if(status == true)
        {
            locationManager.removeUpdates(locationListener);
```

```
            locationManager.removeGpsStatusListener(statusListener);
            errorview.setText("");
            latitudeview.setText("");
            longitudeview.setText("");
            speedview.setText("");
            timeview.setText("");
            altitudeview.setText("");
            bearingview.setText("");
            satcountview.setText("");
            status = false;
        }
    }



    //定位监听类负责监听位置信息的变化情况
    private final LocationListener locationListener = new LocationListener()
    {

        @Override
        public void onLocationChanged(Location location)
        {
        // 获取GPS信息，获取位置提供者provider中的位置信息
        // location = locationManager.getLastKnownLocation(provider);
        // 通过GPS获取位置
            updateToNewLocation(location);
            //showInfo(getLastPosition(), 2);
        }
        @Override
        public void onProviderDisabled(String arg0)
        {

        }

        @Override
        public void onProviderEnabled(String arg0)
        {

        }

        @Override
        public void onStatusChanged(String arg0, int arg1, Bundle arg2)
        {
            updateToNewLocation(null);
        }
    };

    //添加监听卫星
    private final GpsStatus.Listener statusListener= new GpsStatus.Listener(){

        @Override
        public void onGpsStatusChanged(int event) {
            // TODO Auto-generated method stub
            // 获取GPS卫星信息
            gpsStatus = locationManager.getGpsStatus(null);

            switch(event)
            {
            case GpsStatus.GPS_EVENT_STARTED:

            break;
                //第一次定位时间
            case GpsStatus.GPS_EVENT_FIRST_FIX:

            break;
                //收到的卫星信息
```

```
            case GpsStatus.GPS_EVENT_SATELLITE_STATUS:
                DrawMap();

            break;

            case GpsStatus.GPS_EVENT_STOPPED:
            break;
            }
        }
    };
    private int heightp;
    private int widthp;

    private void openGPSSettings()
    {

        // dm = new DisplayMetrics();
        // getWindowManager().getDefaultDisplay().getMetrics(dm);
        // heightp = dm.heightPixels;
        // widthp = dm.widthPixels;
        // alimuth[0] = 60;
        // elevation[0] = 70;
        // snr[0] = 60;
        // alimuth[1] = 260;
        // elevation[1] = 10;
        // snr[1] = 50;
        // layout.redraw(50,alimuth,elevation,snr, widthp,heightp, 2);

        // 获取位置管理服务
        locationManager = (LocationManager)this.getSystemService(Context.LOCATION_SERVICE);
        if (locationManager.isProviderEnabled(android.location.LocationManager.GPS_PROVIDER))
        {
            Toast.makeText(this, "GPS 模块正常", Toast.LENGTH_SHORT).show();
            return;
        }
        status = false;
        Toast.makeText(this, "请开启 GPS! ", Toast.LENGTH_SHORT).show();
        Intent intent = new Intent(Settings.ACTION_SECURITY_SETTINGS);
        startActivityForResult(intent,0); //此为设置完成后返回到获取界面     }
    }


    protected void DrawMap() {
        // TODO Auto-generated method stub

        int i = 0;
        //获取屏幕信息
         dm = new DisplayMetrics();
         getWindowManager().getDefaultDisplay().getMetrics(dm);
         heightp = dm.heightPixels;
         widthp = dm.widthPixels;

        //获取卫星信息
        allSatellites = gpsStatus.getSatellites();
        Iteratorsate = allSatellites.iterator();

        while(Iteratorsate.hasNext())
        {
            GpsSatellite satellite = Iteratorsate.next();
            alimuth[i] = satellite.getAzimuth();
            elevation[i] = satellite.getElevation();
            snr[i] = satellite.getSnr();
            i++;
        }
        satcountview.setText(""+i);
        layout.redraw(bear,alimuth,elevation,snr, widthp,heightp, i);
          layout.invalidate();
    }
```

```
    private void getLocation()
    {

        // 查找到服务信息，位置数据标准类
        Criteria criteria = new Criteria();
        // 查询精度:高
        criteria.setAccuracy(Criteria.ACCURACY_FINE);
        // 是否查询海拔:是
        criteria.setAltitudeRequired(true);
        // 是否查询方位角:是
        criteria.setBearingRequired(true);
        // 是否允许付费
        criteria.setCostAllowed(true);
        // 电量要求:低
        criteria.setPowerRequirement(Criteria.POWER_LOW);
        // 是否查询速度:是
        criteria.setSpeedRequired(true);

        provider = locationManager.getBestProvider(criteria, true);


        // 获取GPS信息，获取位置提供者provider中的位置信息
        location = locationManager.getLastKnownLocation(provider);
        // 通过GPS获取位置
        updateToNewLocation(location);
        // 设置监听器，自动更新的最小时间为间隔N秒（1秒为1*1000毫秒，这样写主要为了方便）或最小位
        // 移变化超过N米
        // 实时获取位置提供者provider中的数据，一旦发生位置变化，立即通知应用程序
        locationManager.requestLocationUpdates(provider, 1000, 0,locationListener);
        // 监听卫星
        locationManager.addGpsStatusListener(statusListener);
        }

    private void updateToNewLocation(Location location)
    {

        if (location != null)
        {
            bear = location.getBearing();
            double  latitude = location.getLatitude();       //维度
            double longitude= location.getLongitude();       //经度
            float GpsSpeed = location.getSpeed();            //速度
            long GpsTime = location.getTime();               //时间
            Date date = new Date(GpsTime);

            DateFormat df = new SimpleDateFormat("yyyy-MM-dd HH:mm:ss");

            double GpsAlt = location.getAltitude();          //海拔
            latitudeview.setText("" + latitude);
            longitudeview.setText("" + longitude);
            speedview.setText(""+GpsSpeed);
            timeview.setText(""+df.format(date));
            altitudeview.setText(""+GpsAlt);
            bearingview.setText(""+bear);

            }
        else
        {
            errorview.setText("无法获取地理信息");
            }
        }
}
```

本实例在模拟器中的执行效果如图 7-3 所示。

▲图 7-3　在模拟器中的执行效果

7.3 在 Android 设备中使用地图

在 Android 设备中可以直接使用 Google 地图，可以用地图的形式显示位置信息。接下来将详细讲解在 Android 设备中使用 Google 地图的方法。

7.3.1　准备工作

Google 地图给人们的生活带来了极大的方便，例如可以通过 Google 地图查找商户信息、查看地图和获取行车路线等。Android 平台也提供了一个 map 包（com.google.android.maps），通过其中的 MapView 就能够方便地利用 Google 地图的资源来进行编程。在使用前需要预先进行如下必要的设置。

1. 添加 maps.jar

在 Android SDK 中，以 JAR 库的形式提供了和 Google Map 有关的 API，此 JAR 库位与“android-sdk-windows\add-ons\google_apis-4”目录下。要把 maps.jar 添加到项目中，可以在项目属性中的“Android”栏中指定使用包含 Google API 的 Target 作为项目的构建目标。如图 7-5 所示。

2. 将地图嵌入到应用

通过使用 MapActivity 和 MapView 控件，可以轻松地将地图嵌入到应用程序当中。在此步骤中，需要将 Google API 添加到构建路径中。方法是在图 7-4 所示界面中选择“Java Build Path”，然后在 Target Name 中选中相应的 Google API，设置项目中包含 Google API。如图 7-5 所示。

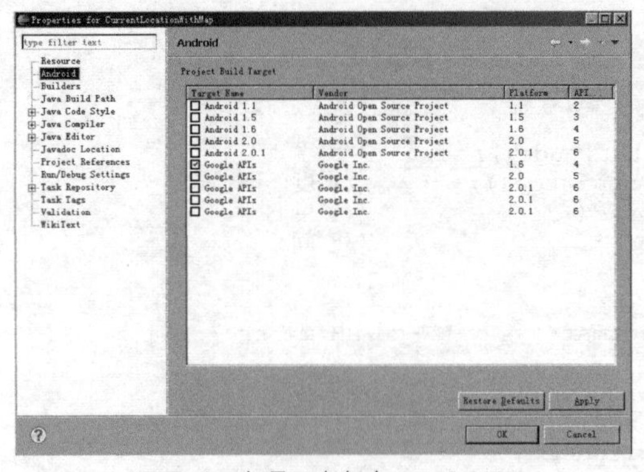

▲图 7-4　在项目中包含 Google API

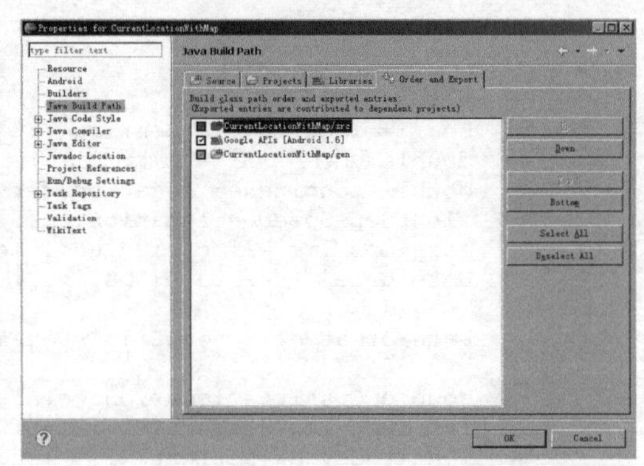

▲图 7-5　将 Google API 添加到构建路径

3. 获取 Map API 密钥

在利用 MapView 之前，必须要先申请一个 Android Map API Key。具体步骤如下所示。

第 1 步：找到 debug.keystore 文件，通常位于如下目录。

C:\Documents and Settings\机器的当前用户\Local Settings\Application Data\Android

第 2 步：获取 MD5 指纹，运行 cmd.exe，执行如下命令获取 MD5 指纹。

```
>keytool -list -alias androiddebugkey -keystore "debug.keystore 的路径" -storepass android
-keypass android
```

例如这里机器输入如下命令。

```
keytool -list -alias androiddebugkey -keystore "C:\Documents and Settings\
Administrator\.android\debug.keystore" -storepass android -keypass android
```

此时系统会提示输入 keystore 密码，这时候输入 android，系统就会输出申请到的 MD5 认证指纹。如图 7-6 所示。

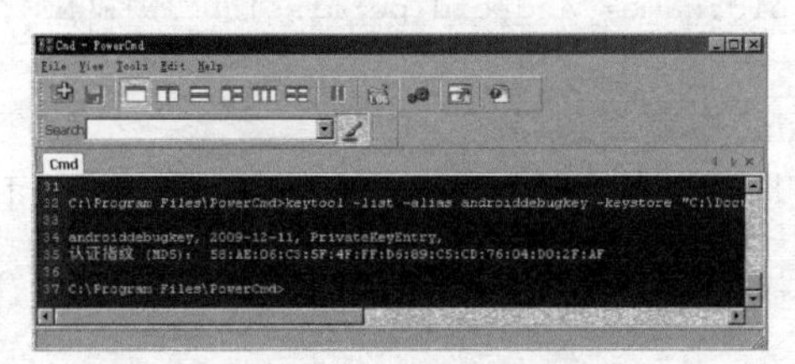

▲图 7-6　获取的认证指纹

注意　因为在 CMD 中不能直接复制、粘贴使用 CMD 命令，这样很影响编程效率，所以这里使用了第三方软件 PowerCmd 来代替机器中自带的 CMD 工具。

第 3 步：申请 Android map 的 API Key。

打开浏览器，输入下面的网址：http://code.google.com/intl/zh-CN/android/maps-api-signup.html。如图 7-8 所示。

在 google 的 android map API Key 申请页面上输入图 7-7 中得到的 MD5 认证指纹，单击“Generate API Key”按钮后即可转到新画面，得到申请到的 API Key。如图 7-8 所示。

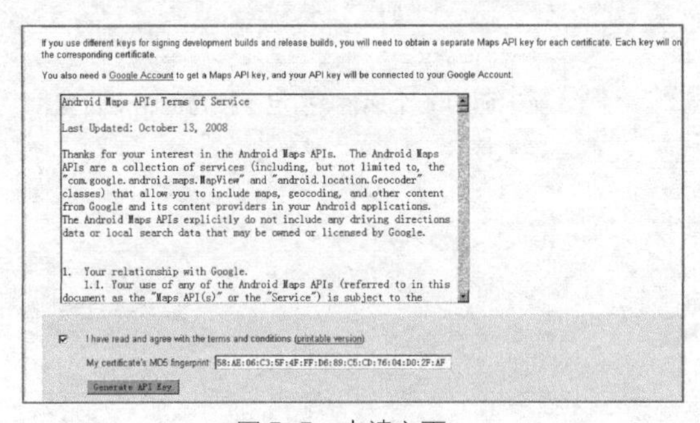

▲图 7-7　申请主页

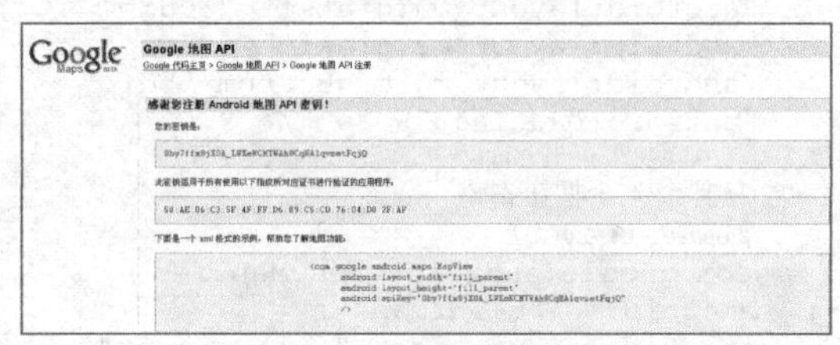

▲图 7-8　得到的 API Key

至此，就成功地获取了一个 API Key。

7.3.2　使用 Map API 密钥的基本流程

通过上一节的讲解，已经申请到了一个 Android Map API Key，下面开始讲解使用 Map API 密钥实现编程的基本流程。

第 1 步：在 AndroidManifest.xml 中声明权限。

在 Anroid 系统中，如果程序执行需要读取到安全敏感的项目，那么必须在 AndroidManifest.xml 中声明相关权限请求，例如这个地图程序需要从网络读取相关数据。所以必须声明 android.permission.INTERNET 权限。具体方法是在文件 AndroidManifest.xml 中添加如下代码。

```
<uses-permission android:name="android.permission.INTERNET" />
```

另外，因为 maps 类不是 Android 启动的默认类，所以还需要在文件 AndroidManifest.xml 的 application 标签中申明要用 maps 类。

```
<uses-library android:name="com.google.android.maps" />
```

下面是基本的 AndroidManifest.xml 文件代码。

```
<manifest xmlns:android="http://schemas.android.com/apk/res/android"
```

```
    <application android:icon="@drawable/icon" android:label="@string/app_name">
     <uses-library android:name="com.google.android.maps" />
    </application>
  <uses-permission android:name="android.permission.INTERNET" />
</manifest>
```

第2步：在文件main.xml中完成布局。

假设要显示杭州的卫星地图，并在地图上方有5个按钮，分别可以放大地图、缩小地图或者切换显示模式（卫星、交通、街景）。即整个界面主要由两个部分组成，上面是一排5个按钮，下面是Map View。

在 Android 中的 LinearLayout 是可以互相嵌套的，在此可以把上面 5 个按钮放在一个子LinearLayout里边（子LinearLayout的指定可以由android:addStatesFromChildren="true"实现），然后再把这个子LinearLayout加到外面的父LinearLayout里边。具体实现如下。

```
*为了简化篇幅,去掉一些不是重点说明的属性
<LinearLayout xmlns:android="http://schemas.android.com/apk/res/android"
 android:orientation="vertical" android:layout_width="fill_parent"
 android:layout_height="fill_parent">

 <LinearLayout android:layout_width="fill_parent"
  android:addStatesFromChildren="true"                /*说明是子 Layout
  android:gravity="center_vertical"                   /*这个子 Layout 里边的按钮是横向排列的
  >

  <Button android:id="@+id/ZoomOut"
   android:text="放大"
   android:layout_width="wrap_content"
   android:layout_height="wrap_content"
   android:layout_marginTop="5dip"                    /*下面的 4 个属性,指定了按钮的相对位置
   android:layout_marginLeft="30dip"
   android:layout_marginRight="5dip"
   android:layout_marginBottom="5dip"
   android:padding="5dip" />

  /*其余 4 个按钮省略
 </LinearLayout>
 <com.google.android.maps.MapView
  android:id="@+id/map"
  android:layout_width="fill_parent"
  android:layout_height="fill_parent"
  android:enabled="true"
  android:clickable="true"
  android:apiKey="在此输入上一节申请的 API Key"          /*必须加上上一节申请的 API Key
 />

</LinearLayout>
```

第3步：首先完成主Java程序代码，主文件的这个类必须继承MapActivity。

```
public class Mapapp extends MapActivity {
```

接下来看onCreate()函数，其核心代码如下所示。

```
 public void onCreate(Bundle icicle) {
 //取得地图 View
  myMapView = (MapView) findViewById(R.id.map);
  //设置为卫星模式
  myMapView.setSatellite(true);
  //地图初始化的点:杭州
  GeoPoint p = new GeoPoint((int) (30.27 * 1000000),
    (int) (120.16 * 1000000));
  //取得地图 View 的控制
  MapController mc = myMapView.getController();
  //定位到杭州
  mc.animateTo(p);
  //设置初始化倍数
```

```
    mc.setZoom(DEFAULT_ZOOM_LEVEL);
}
```

然后编写缩放按钮的处理代码，具体如下所示。

```
btnZoomIn.setOnClickListener(new View.OnClickListener() {
 public void onClick(View view) {
   myMapView.getController().setZoom(myMapView.getZoomLevel() - 1);
 });
```

地图模式的切换由下面代码实现。

```
btnSatellite.setOnClickListener(new View.OnClickListener() {
public void onClick(View view) {
 myMapView.setSatellite(true);                    //卫星模式为 true
 myMapView.setTraffic(false);                     //交通模式为 false
 myMapView.setStreetView(false);                  //街景模式为 false
 }
});
```

到此为止，就完成了第 1 个使用 Map API 的应用程序。

7.3.3　实战演练——在 Android 设备中使用谷歌地图实现定位

题目	目的	源码路径
实例 7-3	在 Android 设备中使用谷歌地图实现定位	\daima\7\LocationMap

本实例的功能是在 Android 设备中使用谷歌地图实现定位功能，具体实现流程如下所示。

（1）在布局文件 main.xml 中插入了两个 Button，分别实现对地图的“放大”和“缩小”，然后，通过 ToggleButton 用于控制是否显示卫星地图，最后，设置申请的 API Key。具体代码如下所示。

```
<?xml version="1.0" encoding="utf-8"?>
<LinearLayout xmlns:android="http://schemas.android.com/apk/res/android"
    android:orientation="vertical"
    android:layout_width="fill_parent"
    android:layout_height="fill_parent"
    >
<TextView
    android:id="@+id/myLocationText"
    android:layout_width="fill_parent"
    android:layout_height="wrap_content"
    />
<LinearLayout
    android:orientation="horizontal"
    android:layout_width="fill_parent"
    android:layout_height="wrap_content" >
    <Button
        android:id="@+id/in"
        android:layout_width="fill_parent"
        android:layout_height="wrap_content"
        android:layout_weight="1"
        android:text="放大地图" />
    <Button
        android:id="@+id/out"
        android:layout_width="fill_parent"
        android:layout_height="wrap_content"
        android:layout_weight="1"
        android:text="缩小地图" />
</LinearLayout>
<ToggleButton
    android:id="@+id/switchMap"
    android:layout_width="wrap_content"
    android:layout_height="wrap_content"
    android:textOff="卫星开关"
    android:textOn="卫星开关"/>
<com.google.android.maps.MapView
    android:id="@+id/myMapView"
```

```
    android:layout_width="fill_parent"
    android:layout_height="fill_parent"
    android:clickable="true"
    android:apiKey="0by7ffx8jX0A_LWXeKCMTWAh8CqHAlqvzetFqjQ"
     />
</LinearLayout>
```

（2）在文件 AndroidManifest.xml 中分别声明 android.permission.INTERNET 和 INTERNET 权限。具体代码如下所示。

```
<?xml version="1.0" encoding="utf-8"?>
<manifest xmlns:android="http://schemas.android.com/apk/res/android"
      package="com.UserCurrentLocationMap"
      android:versionCode="1"
      android:versionName="1.0.0">
    <application android:icon="@drawable/icon" android:label="@string/app_name">
        <activity android:name=".UserCurrentLocationMap"
                  android:label="@string/app_name">
            <intent-filter>
                <action android:name="android.intent.action.MAIN" />
                <category android:name="android.intent.category.LAUNCHER" />
            </intent-filter>
        </activity>
        <uses-library android:name="com.google.android.maps"/>
    </application>
    <uses-permission android:name="android.permission.INTERNET"/>
    <uses-permission android:name="android.permission.ACCESS_FINE_LOCATION"/>
</manifest>
```

（3）编写主程序文件 CurrentLocationWithMap.java，具体实现流程如下所示。

- 通过方法 onCreate()将 MapView 绘制到屏幕上。因为 MapView 只能继承自 MapActivity 中的活动，所以必须用方法 onCreate 将 MapView 绘制到屏幕上，并同时覆盖方法 isRouteDisplayed()，它表示是否需要在地图上绘制导航线路。主要代码如下所示。

```
package com.UserCurrentLocationMap;
..............................................
public class CurrentLocationWithMap extends MapActivity {
    MapView map;

    MapController ctrlMap;
    Button inBtn;
    Button outBtn;
    ToggleButton switchMap;
        @Override
    protected boolean isRouteDisplayed() {
        return false;
    }
```

- 定义方法 onCreate()。首先引入主布局 main.xml，并通过方法 findViewById()获得 MapView 对象的引用，接着调用 getOverlays()方法获取其 Overylay 链表，并将构建好的 MyLocationOverlay 对象添加到链表中去。其中 MyLocationOverlay 对象调用的 enableMyLocation()方法表示尝试通过位置服务来获取当前的位置。具体代码如下所示。

```
@Override
public void onCreate(Bundle savedInstanceState) {
    super.onCreate(savedInstanceState);
    setContentView(R.layout.main);

    map = (MapView)findViewById(R.id.myMapView);
    List<Overlay> overlays = map.getOverlays();
    MyLocationOverlay myLocation = new MyLocationOverlay(this,map);
    myLocation.enableMyLocation();
    overlays.add(myLocation);
```

- 为“放大”和“缩小”这两个按钮设置处理程序。首先通过方法 getController()获取 MapView

的 MapController 对象，然后在“放大”和“缩小”两个按钮单击事件监听器的回放方法里，根据按钮的不同实现对 MapView 的缩放。具体代码如下所示。

```
ctrlMap = map.getController();
inBtn = (Button)findViewById(R.id.in);
outBtn = (Button)findViewById(R.id.out);
OnClickListener listener = new OnClickListener() {
    @Override
    public void onClick(View v) {
        switch (v.getId()) {
        case R.id.in:                                   /*如果是缩放*/
            ctrlMap.zoomIn();
            break;
        case R.id.out:                                  /*如果是放大*/
            ctrlMap.zoomOut();
            break;
        default:
            break;
        }
    }
};
inBtn.setOnClickListener(listener);
outBtn.setOnClickListener(listener);
```

- 通过方法 onCheckedChanged()获取是否选择了 switchMap，如果选择了则显示卫星地图。首先通过方法 findViewById 获取对应 id 的 ToggleButton 对象的引用，然后调用 setOnChecked ChangeListener()方法，设置对事件监听器选中的事件进行处理。根据 ToggleButton 是否被选中，进而通过 setSatellite()方法启用或禁用卫星地图功能。具体代码如下所示。

```
switchMap = (ToggleButton)findViewById(R.id.switchMap);
switchMap.setOnCheckedChangeListener(new OnCheckedChangeListener() {
    @Override
    public void onCheckedChanged(CompoundButton cBtn, boolean isChecked) {
        if (isChecked == true) {
            map.setSatellite(true);
        } else {
            map.setSatellite(false);
        }
    }
});
```

- 通过 LocationManager 获取当前的位置，然后通过 getBestProvider()方法来获取和查询条件，最后设置更新位置信息的最小间隔为两 s，位移变化在 10m 以上。具体代码如下所示。

```
    LocationManager locationManager;
    String context = Context.LOCATION_SERVICE;
    locationManager = (LocationManager)getSystemService(context);
    //String provider = LocationManager.GPS_PROVIDER;

    Criteria criteria = new Criteria();
    criteria.setAccuracy(Criteria.ACCURACY_FINE);
    criteria.setAltitudeRequired(false);
    criteria.setBearingRequired(false);
    criteria.setCostAllowed(true);
    criteria.setPowerRequirement(Criteria.POWER_LOW);
    String provider = locationManager.getBestProvider(criteria, true);

    Location location = locationManager.getLastKnownLocation(provider);
    updateWithNewLocation(location);
    locationManager.requestLocationUpdates(provider, 2000, 10,
            locationListener);
}
```

- 设置回调方法何时被调用。具体代码如下所示。

```
private final LocationListener locationListener = new LocationListener() {
    public void onLocationChanged(Location location) {
```

```
        updateWithNewLocation(location);
        }
        public void onProviderDisabled(String provider){
        updateWithNewLocation(null);
        }
        public void onProviderEnabled(String provider){ }
        public void onStatusChanged(String provider, int status,
        Bundle extras){ }
    };
```

● 定义方法 updateWithNewLocation(Location location)来显示地理信息和地图信息。具体代码如下所示。

```
        private void updateWithNewLocation(Location location) {
            String latLongString;
            TextView myLocationText;
            myLocationText = (TextView)findViewById(R.id.myLocationText);
            if (location != null) {
                double lat = location.getLatitude();
                double lng = location.getLongitude();
                latLongString = "纬度是:" + lat + "\n经度是:" + lng;

                ctrlMap.animateTo(new GeoPoint((int)(lat*1E6),(int)(lng*1E6)));
            } else {
                latLongString = "获取失败";
            }
            myLocationText.setText("当前的位置是:\n" +
            latLongString);

        }
    }
```

至此，整个实例全部介绍完毕，在图 7-9 中选定一个经度和维度位置后，可以显示此位置的定位信息，并且定位信息分别以文字和地图形式显示出来。如图 7-10 所示。

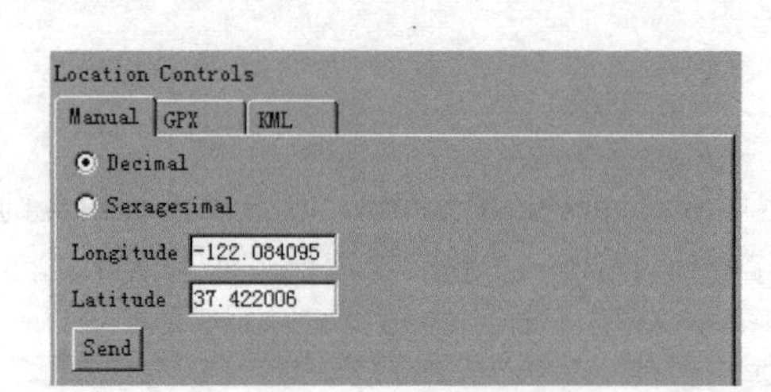

▲图 7-9 指定位置

▲图 7-10 显示对应信息

单击“放大地图”和“缩小地图”按钮后，能控制地图的大小显示，如图 7-11 所示。打开卫星地图后，可以显示此位置范围对应的卫星地图，如图 7-12 所示。

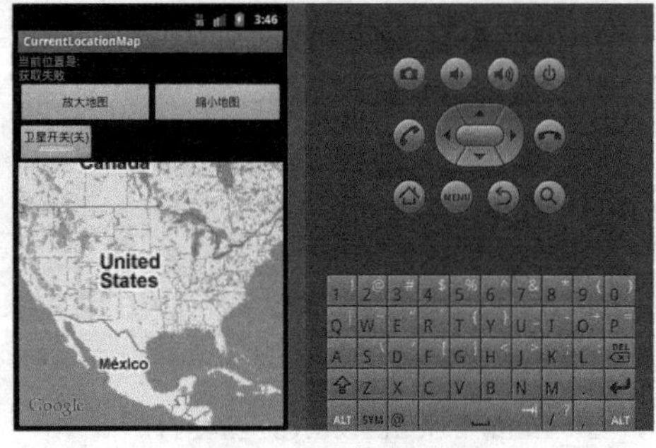

▲图 7-11 放大后效果

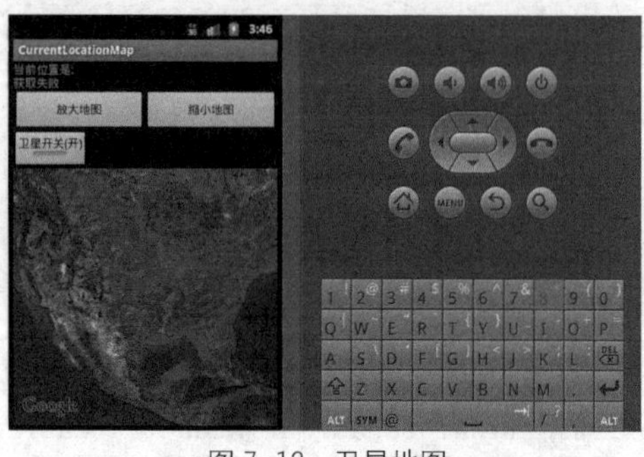

▲图 7-12 卫星地图

7.4 在 Android 设备中实现地址查询

Google 提供了一个 Geocoder 服务，在非商用情况下，使用 Geocoder 可以反查 Address 地址对象服务。这一功能在设备中比较常见，例如在户外手表中可以通过此功能方便地查询某个位置信息。

题目	目的	源码路径
实例 7-4	在 Android 设备中实现地址查询	\daima\7\dicha

本实例的功能是通过使用 Geocoder 实现地址反查功能，具体实现流程如下所示。

（1）编写布局文件 main.xml，主要代码如下所示。

```
<TextView
  android:id="@+id/myTextView1"
  android:layout_width="fill_parent"
  android:layout_height="wrap_content"
  android:textColor="@drawable/blue"
  android:text="@string/hello"/>
<EditText
  android:id="@+id/myEditText1"
  android:layout_width="wrap_content"
  android:layout_height="wrap_content"
  android:text="" />
<LinearLayout
  android:orientation="horizontal"
  android:layout_width="wrap_content"
  android:layout_height="wrap_content"
>
<Button
  android:id="@+id/myButton1"
  android:layout_width="wrap_content"
  android:layout_height="wrap_content"
  android:text="@string/str_button1"/>
<Button
  android:id="@+id/myButton2"
  android:layout_width="wrap_content"
  android:layout_height="wrap_content"
  android:text="@string/str_button2"/>
<Button
  android:id="@+id/myButton3"
  android:layout_width="wrap_content"
  android:layout_height="wrap_content"
  android:text="@string/str_button3"/>
</LinearLayout>
<com.google.android.maps.MapView
  android:id="@+id/myMapView1"
  android:layout_width="fill_parent"
  android:layout_height="fill_parent"
  android:enabled="true"
  android:clickable="true"
  android:apiKey="0by7ffx8jX0A_LWXeKCMTWAh8CqHAlqvzetFqjQ" />
```

（2）编写文件 dicha.java，在此文件中通过地址来获取 GeoPoint()方法，在自定义函数 getGeoByAddress()中传入唯一的值字符串的方式传入地址，使用方法 Geocoder.getFromLocationName()来获取从 Google 服务器找到的结果。文件 dicha.java 的主要代码如下所示。

```
protected void onCreate(Bundle icicle)
{
  super.onCreate(icicle);
  setContentView(R.layout.main);
  mEditText01 = (EditText)findViewById(R.id.myEditText1);
  mEditText01.setText
  (
    getResources().getText(R.string.str_default_address).toString()
```

```
    );
    /* 创建 MapView 对象 */
    mMapView01 = (MapView)findViewById(R.id.myMapView1);
    mMapController01 = mMapView01.getController();
    // 设置 MapView 的显示选项（卫星、街道）
    mMapView01.setSatellite(true);
    mMapView01.setStreetView(true);
    mButton01 = (Button)findViewById(R.id.myButton1);
    mButton01.setOnClickListener(new Button.OnClickListener()
    {
      @Override
      public void onClick(View v)
      {
        if(mEditText01.getText().toString()!="")
        {
          refreshMapViewByGeoPoint
          (
            getGeoByAddress
            (
              mEditText01.getText().toString()
            ),mMapView01,intZoomLevel,true
          );
        }
      }
    });
    /* 放大 */
    mButton02 = (Button)findViewById(R.id.myButton2);
    mButton02.setOnClickListener(new Button.OnClickListener()
    {
      @Override
      public void onClick(View v)
      {
        // TODO Auto-generated method stub
        intZoomLevel++;
        if(intZoomLevel>mMapView01.getMaxZoomLevel())
        {
          intZoomLevel = mMapView01.getMaxZoomLevel();
        }
        mMapController01.setZoom(intZoomLevel);
      }
    });
    /* 缩小 */
    mButton03 = (Button)findViewById(R.id.myButton3);
    mButton03.setOnClickListener(new Button.OnClickListener()
    {
      @Override
      public void onClick(View v)
      {
        intZoomLevel--;
        if(intZoomLevel<1)
        {
          intZoomLevel = 1;
        }
        mMapController01.setZoom(intZoomLevel);
      }
    });
    /* 初次查询地点 */
    refreshMapViewByGeoPoint
    (
      getGeoByAddress
      (
          getResources().getText(R.string.str_default_address).toString()
      ),mMapView01,intZoomLevel,true
    );
  }
  private GeoPoint getGeoByAddress(String strSearchAddress)
  {
    GeoPoint gp = null;
    try
```

```
    {
      if(strSearchAddress!="")
      {
        Geocoder mGeocoder01 = new Geocoder(dicha.this, Locale.getDefault());
        List<Address> lstAddress = mGeocoder01.getFromLocationName(strSearchAddress, 1);
        if (!lstAddress.isEmpty())
        {
          for (int i = 0; i < lstAddress.size(); ++i)
          {
            Address adsLocation = lstAddress.get(i);
            Log.i(TAG, "Address found = " + adsLocation.toString());
          }
          Address adsLocation = lstAddress.get(0);
          double geoLatitude = adsLocation.getLatitude()*1E6;
          double geoLongitude = adsLocation.getLongitude()*1E6;
          gp = new GeoPoint((int) geoLatitude, (int) geoLongitude);
        }
        else
        {
          Log.i(TAG, "Address GeoPoint NOT Found.");
        }
      }
    }
    catch (Exception e)
    {
      e.printStackTrace();
    }
    return gp;
  }
  public static void refreshMapViewByGeoPoint(GeoPoint gp, MapView mv, int zoomLevel,
boolean bIfSatellite)
  {
    try
    {
      mv.displayZoomControls(true);
      /* 取得MapView的MapController */
      MapController mc = mv.getController();
      /* 移至该地理坐标地址 */
      mc.animateTo(gp);

      /* 放大地图层级 */
      mc.setZoom(zoomLevel);
      /* 设置MapView的显示选项（卫星、街道）*/
      if(bIfSatellite)
      {
        mv.setSatellite(true);
        mv.setStreetView(true);
      }
      else
      {
        mv.setSatellite(false);
      }
    }
    catch(Exception e)
    {
      e.printStackTrace();
    }
  }

  @Override
  protected boolean isRouteDisplayed()
  {
    // TODO Auto-generated method stub
    return false;
  }
}
```

执行后能够实现地址反查处理。

注意

所谓 Geocoder 是指根据某个 key 寻找地理位置的坐标，这个 key 可以是地址。近年来 Google Map API 做了很多改进，已经自动整合了中文地址的定位功能，不再需要以前的糅合了。以前只有 6 个国家提供了街道级别的定位，Google 从不让人失望。当然每日 50000 次查询的 limit 还在，但是一般也足够了，何况还有客户端的 built-in cache 改善用户体验。

这个 key 还可以是 IP 地址，例如做一个 library 可以读取本地数据库转换 IP 为坐标，这个坐标只是城市中心的坐标，不过一般也够用了。同时 library 也将提供基于 http request 的公共服务，地址定位和 IP 定位现在都可以很容易实现了。

7.5 在 Android 设备中实现路径导航

路径导航功能在户外类的设备中比较常见，例如户外手表。作为喜欢户外露营和远程徒步的人群来说，通过此功能可以记录自己的行走轨迹，以供其他户外爱好人员提供线路参考。

题目	目的	源码路径
实例 7-5	在 Android 设备中实现路径导航	\daima\7\ludao

在本实例中，通过 Directions Route 实现了路径导航功能。首先调用 getLocationProvider()获取当前 Location 位置，以取得当前所在位置的地理坐标。并通过提供的 EditText Widget 来让用户输入将要前往的地址，通过地址来反复取得目的地的地理坐标。本实例的实现文件是 ludao.java，具体实现流程如下所示。

（1）创建 LocationManager 对象以获取系统 LOCATION 本地服务，然后设置使用 MapView 控件显示选项（卫星、街道）。实例说明代码如下所示。

```
public class ludao extends MapActivity
{
  private TextView mTextView01;
  private LocationManager mLocationManager01;
  private String strLocationProvider = "";
  private Location mLocation01;
  private MapController mMapController01;
  private MapView mMapView01;
  private Button mButton01,mButton02,mButton03;
  private EditText mEditText01;
  private int intZoomLevel=0;
  private GeoPoint fromGeoPoint, toGeoPoint;

  @Override
  protected void onCreate(Bundle icicle)
  {
    // TODO Auto-generated method stub
    super.onCreate(icicle);
    setContentView(R.layout.main);

    mTextView01 = (TextView)findViewById(R.id.myTextView1);

    mEditText01 = (EditText)findViewById(R.id.myEditText1);
    mEditText01.setText
    (
      getResources().getText
      (R.string.str_default_address).toString()
    );
    /* 创建 MapView 对象 */
    mMapView01 = (MapView)findViewById(R.id.myMapView1);
    mMapController01 = mMapView01.getController();
    // 设置 MapView 的显示选项（卫星、街道）
```

```
    mMapView01.setSatellite(true);
    mMapView01.setStreetView(true);
    // 放大的层级
    intZoomLevel = 15;
    mMapController01.setZoom(intZoomLevel);
    /* 创建 LocationManager 对象取得系统 LOCATION 服务 */
    mLocationManager01 =
    (LocationManager)getSystemService(Context.LOCATION_SERVICE);
    /*
     * 自定义函数，访问 Location Provider,
     * 并将之存储在 strLocationProvider 当中
     */
    getLocationProvider();
    /* 传入 Location 对象，显示 MapView */
    fromGeoPoint = getGeoByLocation(mLocation01);
    refreshMapViewByGeoPoint(fromGeoPoint,
                    mMapView01, intZoomLevel);
    /* 创建 LocationManager 对象，监听
     * Location 更改时事件，更新 MapView*/
    mLocationManager01.requestLocationUpdates
    (strLocationProvider, 2000, 10, mLocationListener01);
```

（2）定义单击 mButton01 按钮的处理事件，先获取用户要前往地址的 GeoPoint 对象，传入路径规划所需要的地标地址。实例说明代码如下所示。

```
mButton01 = (Button)findViewById(R.id.myButton1);
mButton01.setOnClickListener(new Button.OnClickListener()
{
  @Override
  public void onClick(View v)
  {
    // TODO Auto-generated method stub
    if(mEditText01.getText().toString()!="")
    {
      /* 取得 User 要前往地址的 GeoPoint 对象 */
      toGeoPoint =
      getGeoByAddress(mEditText01.getText().toString());

      /* 路径规划 Intent */
      Intent intent = new Intent();
      intent.setAction(android.content.Intent.ACTION_VIEW);

      /* 传入路径规划所需要的地标地址 */
      intent.setData
      (
        Uri.parse("http://maps.google.com/maps?f=d&saddr="+
        GeoPointToString(fromGeoPoint)+
        "&daddr="+GeoPointToString(toGeoPoint)+
        "&hl=cn" +
        "")
      );
      startActivity(intent);
    }
  }
});
```

（3）定义单击 mButton02 按钮的处理事件，单击后实现地图放大处理。具体代码如下所示。

```
/* 放大地图 */
mButton02 = (Button)findViewById(R.id.myButton2);
mButton02.setOnClickListener(new Button.OnClickListener()
{
  @Override
  public void onClick(View v)
  {
    // TODO Auto-generated method stub
    intZoomLevel++;
    if(intZoomLevel>mMapView01.getMaxZoomLevel())
    {
```

```
    intZoomLevel = mMapView01.getMaxZoomLevel();
  }
  mMapController01.setZoom(intZoomLevel);
 }
});
```

（4）定义单击 mButton03 按钮的处理事件，单击实现地图缩小处理。具体代码如下所示。

```
/* 缩小地图 */
mButton03 = (Button)findViewById(R.id.myButton3);
mButton03.setOnClickListener(new Button.OnClickListener()
{
  @Override
  public void onClick(View v)
  {
    // TODO Auto-generated method stub
    intZoomLevel--;
    if(intZoomLevel<1)
    {
      intZoomLevel = 1;
    }
    mMapController01.setZoom(intZoomLevel);
  }
 });
}
```

（5）捕捉手机 GPS 坐标更新时的事件，当手机收到位置更改时将 location 传入 getMyLocation 对象。具体代码如下所示。

```
/* 捕捉当手机 GPS 坐标更新时的事件 */
public final LocationListener mLocationListener01 =
new LocationListener()
{
  @Override
  public void onLocationChanged(Location location)
  {
    /* 当手机收到位置更改时，将 location 传入 getMyLocation */
    mLocation01 = location;
    fromGeoPoint = getGeoByLocation(location);
    refreshMapViewByGeoPoint(fromGeoPoint,
          mMapView01, intZoomLevel);
  }
};
```

（6）定义方法 getGeoByLocation()，设置当传入 Location 对象时取回其 GeoPoint 对象。具体代码如下所示。

```
/* 传入 Location 对象，取回其 GeoPoint 对象 */
private GeoPoint getGeoByLocation(Location location)
{
  GeoPoint gp = null;
  try
  {
    /* 如果 Location 存在 */
    if (location != null)
    {
      double geoLatitude = location.getLatitude()*1E6;
      double geoLongitude = location.getLongitude()*1E6;
      gp = new GeoPoint((int) geoLatitude, (int) geoLongitude);
    }
  }
  catch(Exception e)
  {
    e.printStackTrace();
  }
  return gp;
}
```

（7）定义方法 getGeoByAddress()，设置当输入地址时获取其 GeoPoint 对象。具体代码如下所示。

```
/* 输入地址，获得其 GeoPoint 对象 */
private GeoPoint getGeoByAddress(String strSearchAddress)
{
  GeoPoint gp = null;
  try
  {
    if(strSearchAddress!="")
    {
      Geocoder mGeocoder01 = new Geocoder
      (ludao.this, Locale.getDefault());
      List<Address> lstAddress = mGeocoder01.getFromLocationName
                        (strSearchAddress, 1);
      if (!lstAddress.isEmpty())
      {
        Address adsLocation = lstAddress.get(0);
        double geoLatitude = adsLocation.getLatitude()*1E6;
        double geoLongitude = adsLocation.getLongitude()*1E6;
        gp = new GeoPoint((int) geoLatitude, (int) geoLongitude);
      }
    }
  }
  catch (Exception e)
  {
    e.printStackTrace();
  }
  return gp;
}
```

（8）定义方法 refreshMapViewByGeoPoint()和 refreshMapViewByCode()，用于分别传入 geoPoint 更新 MapView 里的谷歌地图和传入经纬度更新 MapView 里的谷歌地图。具体代码如下所示。

```
/* 传入 geoPoint 更新 MapView 里的 Google Map */
public static void refreshMapViewByGeoPoint
(GeoPoint gp, MapView mapview, int zoomLevel)
{
  try
  {
    mapview.displayZoomControls(true);
    MapController myMC = mapview.getController();
    myMC.animateTo(gp);
    myMC.setZoom(zoomLevel);
    mapview.setSatellite(false);
  }
  catch(Exception e)
  {
    e.printStackTrace();
  }
}
/* 传入经纬度更新 MapView 里的 Google Map */
public static void refreshMapViewByCode
(double latitude, double longitude,
    MapView mapview, int zoomLevel)
{
  try
  {
    GeoPoint p = new GeoPoint((int) latitude, (int) longitude);
    mapview.displayZoomControls(true);
    MapController myMC = mapview.getController();
    myMC.animateTo(p);
    myMC.setZoom(zoomLevel);
    mapview.setSatellite(false);
  }
  catch(Exception e)
  {
    e.printStackTrace();
  }
}
```

（9）定义方法 GeoPointToString()将 GeoPoint 中的经纬度以“String,String”的格式返回。具体代码如下所示。

```
/* 将 GeoPoint 里的经纬度以 String,String 返回 */
private String GeoPointToString(GeoPoint gp)
{
  String strReturn="";
  try
  {
    /* 当 Location 存在 */
    if (gp != null)
    {
      double geoLatitude = (int)gp.getLatitudeE6()/1E6;
      double geoLongitude = (int)gp.getLongitudeE6()/1E6;
      strReturn = String.valueOf(geoLatitude)+","+
        String.valueOf(geoLongitude);
    }
  }
  catch(Exception e)
  {
    e.printStackTrace();
  }
  return strReturn;
}
```

执行后的效果如图 7-13 所示；单击“开始规划路径”按钮后弹出选择对话框，如图 7-14 所示。在此选择“Maps”后弹出规划界面，如图 7-15 所示；在图 7-15 所示界面中，在第 1 个文本框中设置出发位置，例如“Beijing”，在第 2 个文本框中设置目的地位置，例如“Tianjin”，选择汽车前往标志按钮，如图 7-16 所示；单击按钮“Go”，系统将实现北京出发，目的地到天津的线路规划，产生线路规划图。

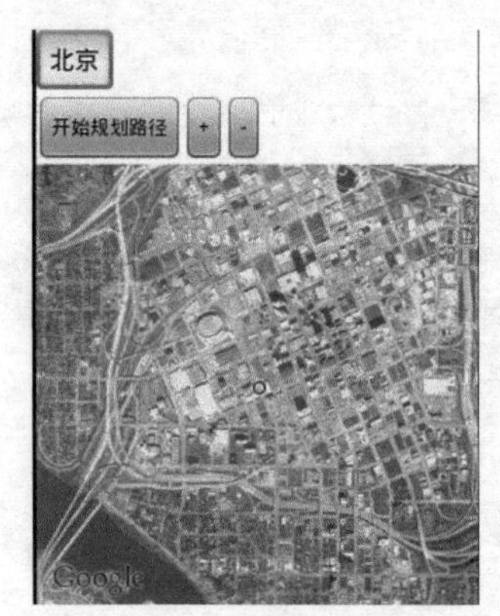

▲图 7-13　执行效果

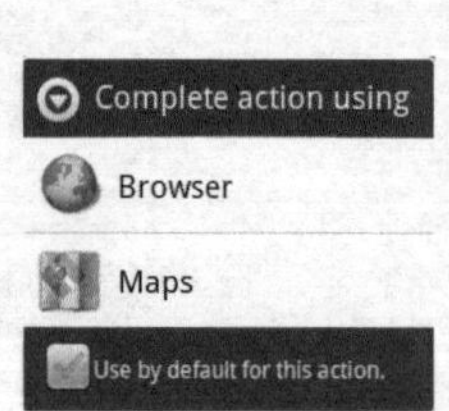

▲图 7-14　选择对话框

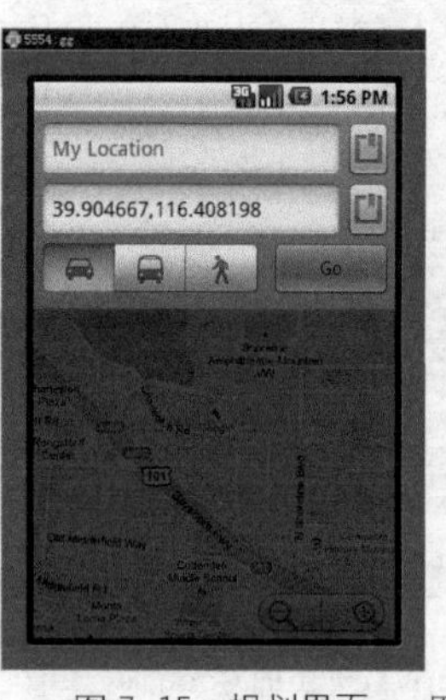

▲图 7-15　规划界面

▲图 7-16　设置出发地和目的地

单击图中的“Show on map”后将会在地图中显示行走线路。

第 8 章　光线传感器详解

在 Android 系统中支持多种传感器（Sensor），传感器系统可以让智能手机的功能更加丰富多彩。Android 系统支持多种传感器，有的传感器已经在 Android 的框架中使用，大多数传感器由应用程序来使用，使用传感器可以开发出包括游戏在内的很多新奇的应用。在 Android 系统中支持的传感器包括加速度传感器（Accelerometer）、姿态传感器（Orientation）、磁场传感器（Magnetic Field）和光传感器（Light）等。本章将详细讲解在 Android 设备中使用光线传感器的基本知识。

8.1 光线传感器基础

光线传感器的好处是可以根据手机所处环境的光线来调节手机屏幕的亮度和键盘灯。例如在光线充足的地方屏幕会很亮，键盘灯就会关闭。相反如果在暗处，键盘灯就会亮，屏幕较暗（与屏幕亮度的设置也有关系），这样既保护了眼睛又节省了电量。光线传感器在进入睡眠模式时候会发出蓝色周期性闪动的光，非常美观。在本节的内容中，将详细讲解 Android 系统光线传感器的基本知识。

8.1.1 光线传感器介绍

在 Android 设备中，光线传感器通常位于前摄像头旁边的一个小点，如果在光线充足的情况下（室外或者是灯光充足的室内），大约在 2～3 秒之后键盘灯会自动熄灭，即使再操作机器键盘灯也不会亮，除非到了光线比较暗的地方才会自动地亮起来。如果在光线充足的情况下用手将光线感应器遮上，在 2～3 秒后键盘灯会自动亮起来，在此过程中光线感应器起到了一个节电的功能。

要想在 Android 设备中监听光线传感器，需要掌握如下所示的监听方法。

（1）registerListenr(SensorListenerlistenr,int sensors,int rate)：已过时。

（2）registerListenr(SensorListenerlistenr,int sensors)：已过时。

（3）registerListenr(SensorEventListenerlistenr,Sensor sensors,int rate)。

（4）registerListenr(SensorEventListenerlistenr,Sensor sensors,int rate,Handlerhandler)：因为 SensorListener 已经过时，所以相应的注册方法也过时了。

在上述方法中，各个参数的具体说明如下所示。

- Listener：相应监听器的引用。
- Sensor：相应的感应器引用。
- Rate：感应器的反应速度，这个必须是系统提供的 4 个常量之一。
  - SENSOR_DELAY_NORMAL：匹配屏幕方向的变化。
  - SENSOR_DELAY_UI：匹配用户接口。
  - SENSOR_DELAY_GAME：匹配游戏。
  - SENSOR_DELAY_FASTEST.：匹配所能达到的最快。

开发光传感器应用时需要监测 SENSOR_LIGHT，例如下面的代码。

```
private SensorListener mySensorListener = new SensorListener(){
@Override
 public void onAccuracyChanged(int sensor, int accuracy) {}//重写onAccuracyChanged方法
 @Override
 public void onSensorChanged(int sensor, float[] values) { //重写onSensorChanged方法
    if(sensor == SensorManager.SENSOR_LIGHT){                 //只检查光强度的变化
       myTextView1.setText("光的强度为："+values[0]);          //将光的强度显示到TextView
           }
        }
};
@Override
protected void onResume() {                                //重写的onResume方法
       mySensorManager.registerListener(                    //注册监听
              mySensorListener,                             //监听器SensorListener对象
              SensorManager.SENSOR_LIGHT,                   //传感器的类型为光的强度
              SensorManager.SENSOR_DELAY_UI                 //频率
              );
       super.onResume();
}
```

在上述代码中，通过 if 语句判断是否为光的强度改变事件。在代码中只对光强度改变事件进行处理，将得到的光强度显示在屏幕中。光传感器只得到一个数据，而并不像其他传感器那样得到的是 *X*、*Y*、*Z* 3 个方向上的分量。

在注册监听时，通过传入"SensorManager.SENSOR_LIGHT"来通知系统只注册光传感器。

8.1.2　在 Android 中使用光线传感器的方法

在 Android 设备中，使用光线传感器的基本流程如下所示。

（1）通过一个 SensorManager 来管理各种感应器，要想获得这个管理器的引用，必须通过如下所示的代码来实现。

```
(SensorManager)getSystemService(Context.SENSOR_SERVICE);
```

（2）在 Android 系统中，所有的感应器都属于 Sensor 类的一个实例，并没有继续细分下去，所以 Android 对于感应器的处理几乎是一模一样的。既然都是 Sensor 类，那么怎么获得相应的感应器呢？这时就需要通过 SensorManager 来获得，可以通过如下所示的代码来确定要获得的感应器类型。

```
sensorManager.getDefaultSensor(Sensor.TYPE_LIGHT);
```

通过上述代码获得了光线感应器的引用。

（3）在获得相应的传感器的引用后可以用来感应光线强度的变化，此时需要通过监听传感器的方式来获得变化，监听功能通过前面介绍的监听方法实现。Android 提供了两个监听方式，一个是 SensorEventListener，另一个是 SensorListener，后者已经在 Android API 上显示过时了。

（4）在 Android 中注册传感器后，此时就说明启用了传感器。使用感应器是相当耗电的，这也是为什么传感器没有被广泛应用的主要原因，所以必须在不需要它的时候及时关掉。在 Android 中通过如下所示的注销方法来关闭。

- unregisterListener(SensorEventListenerlistener)；
- unregisterListener(SensorEventListenerlistener,Sensor sensor)。

（5）使用 SensorEventListener 来具体实现，在 Android 设备中有下面两种方式实现这个监听器的方法。

- onAccuracyChanged(Sensor sensor, int accuracy)：是反应速度变化的方法，也就是 rate 变化时的方法；

- onSensorChanged(SensorEvent event)：是传感器值变化的相应的方法。

读者需要注意的是，上述两个方法会同时响应。也就是说，当感应器发生变化时，这两个方法会一起被调用。上述方法中的 accuracy 的值是 4 个常量，对应的整数如下所示。

- SENSOR_DELAY_NORMAL：3。
- SENSOR_DELAY_UI：2。
- SENSOR_DELAY_GAME：1。
- SENSOR_DELAY_FASTEST：0。

而类 SensorEvent 有 4 个成员变量，具体说明如下所示。

- Accuracy：精确值。
- Sensor：发生变化的感应器。
- Timestamp：发生的时间，单位是纳秒。
- Values：发生变化后的值，这个是一个长度为 3 的数组。

光线传感器只需要 values[0]的值，其他两个都为 0。而 values[0]就是开发光线传感器所需要的，单位是 lux 照度单位。

8.2 实战演练——获取设备中光线传感器的值

在接下来的实例中，将演示在 Android 设备中使用光线传感器的方法。

实例	功能	源码路径
实例 8-1	获取设备中光线传感器的值	\daima\8\Sensor

本实例的功能是获取设备中光线传感器的值，具体实现流程如下所示。

（1）编写布局文件 activity_main.xml，具体实现代码如下所示。

```
<RelativeLayout xmlns:android="http://schemas.android.com/apk/res/android"
   xmlns:tools="http://schemas.android.com/tools"
   android:layout_width="match_parent"
   android:layout_height="match_parent"
   android:paddingBottom="@dimen/activity_vertical_margin"
   android:paddingLeft="@dimen/activity_horizontal_margin"
   android:paddingRight="@dimen/activity_horizontal_margin"
   android:paddingTop="@dimen/activity_vertical_margin"
   tools:context=".MainActivity" >
   <TextView
      android:layout_width="wrap_content"
      android:layout_height="wrap_content"
      android:text="@string/hello_world" />
</RelativeLayout>
```

（2）编写程序文件 MainActivity.java，具体实现代码如下所示。

```
package com.example.sensor;

import android.hardware.Sensor;
import android.hardware.SensorEvent;
import android.hardware.SensorEventListener;
import android.hardware.SensorListener;
import android.hardware.SensorManager;
import android.os.Bundle;
import android.renderscript.Sampler.Value;
import android.app.Activity;
import android.view.Menu;
import android.widget.TextView;

public class MainActivity extends Activity implements SensorEventListener  {
```

```
    private SensorManager sensor;
    private TextView text;
    @Override
    protected void onCreate(Bundle savedInstanceState) {
        super.onCreate(savedInstanceState);
        setContentView(R.layout.activity_main);
        sensor = (SensorManager)getSystemService(SENSOR_SERVICE);
        text = (TextView)findViewById(R.id.textView1);
    }

    @Override
    public boolean onCreateOptionsMenu(Menu menu) {
        // Inflate the menu; this adds items to the action bar if it is present
        getMenuInflater().inflate(R.menu.activity_main, menu);
        return true;
    }

    @Override
    protected void onPause() {
        // TODO Auto-generated method stub
        sensor.unregisterListener(this);
        super.onPause();
    }

    @Override
    protected void onResume() {
        // TODO Auto-generated method stub
        sensor.registerListener(this,sensor.getDefaultSensor(Sensor.TYPE_LIGHT),
        SensorManager.SENSOR_DELAY_GAME);
        super.onResume();
    }

    @Override
    protected void onStop() {
        // TODO Auto-generated method stub
        sensor.unregisterListener(this);
        super.onStop();
    }


    @Override
    public void onAccuracyChanged(Sensor sensor, int accuracy) {
        // TODO Auto-generated method stub

    }

    @Override
    public void onSensorChanged(SensorEvent event) {
        // TODO Auto-generated method stub
        float[] values = event.values;
        int sensorType = event.sensor.TYPE_LIGHT;
        if(sensorType==Sensor.TYPE_LIGHT)
        {
            text.setText(String.valueOf(values[0]));
        }
    }
}
```

在真机中执行后，将会显示设备中光线传感器的值。

8.3 实战演练——显示设备中光线传感器的强度

在接下来的实例中，将演示在 Android 设备中显示设备中光线传感器强度的方法。

实例	功能	源码路径
实例 8-2	显示设备中光线传感器的强度	\daima\8\qiang

本实例的功能是显示设备中光线传感器的强度值，具体实现流程如下所示。

（1）在 Eclipse 工程中引入两个开源开发包，如图 8-1 所示。

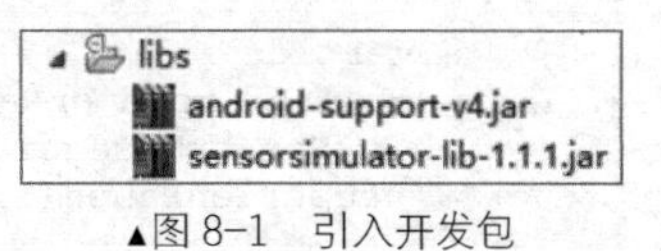

▲图 8-1 引入开发包

（2）编写布局文件 main.xml，具体实现代码如下所示。

```
<LinearLayout xmlns:android="http://schemas.android.com/apk/res/android"
   xmlns:tools="http://schemas.android.com/tools"
   android:layout_width="match_parent"
   android:layout_height="match_parent"
   android:orientation="vertical" >

   <TextView
      android:id="@+id/myTextView1"
      android:layout_width="wrap_content"
      android:layout_height="wrap_content"
    />

</LinearLayout>
```

（3）编写值文件 string.xml，具体实现代码如下所示。

```
<resources>
   <string name="app_name">Sample</string>
   <string name="title">光传感器</string>
   <string name="hello_world">Hello world!</string>
   <string name="menu_settings">Settings</string>
</resources>
```

（4）编写主程序文件 MainActivity.java，具体实现代码如下所示。

```
package com.example.qiang;
import org.openintents.sensorsimulator.hardware.Sensor;
import android.app.Activity;
import android.hardware.SensorManager;
import android.os.Bundle;
import android.widget.TextView;

public  class MainActivity extends Activity implements  android.hardware.SensorEvent
Listener {

   private TextView myTextView1;

   private SensorManager mySensorManager;

   @Override
   public void onCreate(Bundle savedInstanceState) {
      super.onCreate(savedInstanceState);
      setContentView(R.layout.main);
      myTextView1 = (TextView) findViewById(R.id.myTextView1);
      mySensorManager = (SensorManager) getSystemService(SENSOR_SERVICE);

   }

   @Override
   protected void onResume() {
      mySensorManager.registerListener(
              this,
              mySensorManager.getDefaultSensor(Sensor.TYPE_LIGHT),
              SensorManager.SENSOR_DELAY_GAME
              );
      super.onResume();
   }

   @Override
```

```
    protected void onStop() {
        // TODO Auto-generated method stub
        mySensorManager.unregisterListener(this);
        super.onStop();
    }
    @Override
    protected void onPause() {
        mySensorManager.unregisterListener(this);
        super.onPause();
    }

    @Override
    public void onAccuracyChanged(android.hardware.Sensor sensor, int accuracy) {
        // TODO Auto-generated method stub

    }

    @Override
    public void onSensorChanged(android.hardware.SensorEvent event) {
        // TODO Auto-generated method stub
        float[] values = event.values;
        int sensorType = event.sensor.TYPE_LIGHT;
        if (sensorType == Sensor.TYPE_LIGHT) {
            myTextView1.setText("当前光的强度为: "+values[0]);
        }
    }

}
```

8.4 实战演练——显示设备名称和光线强度

在接下来的实例中，将演示在 Android 设备中显示设备名称和光线强度的方法。

实例	功能	源码路径
实例 8-3	显示设备名称和光线强度	\daima\8\union

本实例的功能是显示设备名称、电量和光线强度值，具体实现流程如下所示。

（1）编写布局文件 main.xml，具体实现代码如下所示。

```
<?xml version="1.0" encoding="utf-8"?>
<LinearLayout xmlns:android="http://schemas.android.com/apk/res/android"
    android:layout_width="fill_parent"
    android:layout_height="fill_parent"
    android:orientation="vertical" >

    <TextView
        android:id="@+id/textView1"
        android:layout_width="wrap_content"
        android:layout_height="wrap_content"
        android:textSize="20dip"
        android:text="TextView" />

    <TextView
        android:id="@+id/textView2"
        android:layout_width="wrap_content"
        android:layout_height="wrap_content"
         android:textSize="20dip"
        android:text="TextView" />
</LinearLayout>
```

（2）编写主程序文件 mainActivity.java，具体实现代码如下所示。

```
package com.union;

import android.app.Activity;
import android.hardware.Sensor;
```

```
import android.hardware.SensorEvent;
import android.hardware.SensorEventListener;
import android.hardware.SensorManager;
import android.os.Bundle;
import android.widget.TextView;

public class mainActivity extends Activity {
   SensorManager mySm;
   Sensor myS;
   TextView tv;
   TextView tv1;
   @Override
   public void onCreate(Bundle savedInstanceState) {
      super.onCreate(savedInstanceState);
      setContentView(R.layout.main);
      mySm=(SensorManager)this.getSystemService(SENSOR_SERVICE);
      myS=mySm.getDefaultSensor(Sensor.TYPE_LIGHT);
      StringBuffer str=new StringBuffer();
      str.append("\n 名称: ");
      str.append(myS.getName());
      str.append("\n 耗电量(mA): ");
      str.append(myS.getPower());
      str.append("\n 类型编号: ");
      str.append(myS.getType());
      str.append("\n 版本: ");
      str.append(myS.getVersion());
      str.append("\n 最大测量范围:");
      str.append(myS.getMaximumRange());
      tv=(TextView)this.findViewById(R.id.textView2);
      tv.setText(str);

      tv1=(TextView)this.findViewById(R.id.textView1);
   }

   private SensorEventListener mySel=new SensorEventListener(){
      @Override
      public void onAccuracyChanged(Sensor sensor, int accuracy) {}

      @Override
      public void onSensorChanged(SensorEvent event) {
         float[] value=event.values;
         tv1.setText("光照强度为: "+value[0]);
      }
   };

   @Override
   protected void onResume() {
      super.onResume();
      mySm.registerListener(
            mySel,
            myS,
            SensorManager.SENSOR_DELAY_NORMAL);
   }
   @Override
   protected void onPause() {
      super.onPause();
      mySm.unregisterListener(mySel);
   }
}
```

在真机中执行上述实例代码后，会在屏幕中显示设备的名称、耗电量、类型编号、版本、最大测量范围和光线强度。

8.5 实战演练——智能楼宇灯光控制系统

在接下来的实例中，将演示在 Android 设备中开发一个智能楼宇灯光控制系统的方法。本实例是一个商业项目，需要编写驱动程序和底层蓝牙控制程序。在本节的内容中，将只讲解 Android 应用程序的实现过程。本实例对广大读者来说，为智能家居系统开发提供了很好的参考资料和素材。

实例	功能	源码路径
实例 8-4	显示设备名称和光线强度	\daima\8\light

8.5.1 布局文件

（1）编写主布局文件 main.xml，设置屏幕中间动态显示内容，屏幕底部为固定的界面，具体实现代码如下所示。

```
<RelativeLayout xmlns:android="http://schemas.android.com/apk/res/android"
    xmlns:custom="http://www.javaeye.com/custom"
    android:orientation="vertical"
    android:layout_width="fill_parent"
    android:layout_height="fill_parent"
    android:id="@+id/title_relativeLayout"
    android:background="@drawable/one">
    <!-- 中间动态显示界面 -->
     <ViewFlipper android:id="@+id/fliper"
        android:layout_width="fill_parent"
        android:layout_height="fill_parent"
        android:layout_below="@id/title_relativeLayout"
        android:background="#0000"
        android:layout_marginBottom="50.0dip">

    </ViewFlipper>

        <!-- 底部为固定的界面 -->
    <RelativeLayout android:orientation="horizontal"
        android:layout_width="fill_parent"
        android:layout_height="wrap_content"
        android:layout_alignParentBottom="true"
        style="@android:style/ButtonBar"
        android:background="@drawable/title_background">
    <ImageButton android:background="#0000" android:layout_width="wrap_content"
android:src="@drawable/add" android:id="@+id/imageButton1" android:layout_height=
"wrap_content" android:layout_alignParentBottom="true" android:layout_alignParent
Left="true"></ImageButton>
    <ImageButton android:layout_height="wrap_content" android:background="#0000"
android:id="@+id/imageButton2" android:layout_width="wrap_content" android:src=
"@drawable/menu" android:layout_alignTop="@+id/imageButton1" android:layout_align
ParentRight="true"></ImageButton>

</RelativeLayout>
</RelativeLayout>
```

（2）编写文件 bluetooth.xml，通过 Button 控件、ImageView 控件和 ToggleButton 控件实现蓝牙控制界面，具体实现代码如下所示。

```
<LinearLayout xmlns:android="http://schemas.android.com/apk/res/android"
    android:orientation="vertical" android:layout_width="fill_parent"
    android:layout_height="fill_parent"
    android:background="@drawable/one">
    <RelativeLayout android:layout_width="match_parent" android:layout_height="match_
parent" android:id="@+id/relativeLayout1">
        <ListView android:id="@+id/lvDevices" android:layout_height="wrap_content"
android:layout_width="match_parent" android:layout_alignParentLeft="true" android:
layout_alignParentBottom="true" android:layout_below="@+id/btnExit"></ListView>
        <Button android:text="Native Bluetooth visibility" android:layout_height="wrap_
content" android:layout_width="wrap_content" android:textSize="25sp" android:id="@+id/
btnDis" android:layout_alignParentTop="true" android:layout_alignRight="@+id/image
View1" android:layout_marginRight="64dp" android:layout_marginTop="315dp"></Button>
        <Button android:layout_width="wrap_content" android:textSize="25sp" android:id=
"@+id/btnExit" android:layout_height="wrap_content" android:text="Return to home menu"
android:layout_below="@+id/btnSearch" android:layout_alignRight="@+id/btnSearch"
```

```
android:layout_marginTop="37dp" android:layout_alignLeft="@+id/btnDis"></Button>
        <Button android:layout_width="110dip" android:textSize="25sp" android:id=
"@+id/btnSearch" android:layout_height="50dip" android:text="Search Equipment"
android:layout_alignBottom="@+id/btnDis" android:layout_alignLeft="@+id/tbtnSwitch">
</Button>
        <ToggleButton android:textOff="Off" android:layout_width="110dip" android:
layout_height="50dip" android:textOn="On" android:id="@+id/tbtnSwitch" android:
textSize="23sp" android:text="OFF" android:layout_alignBottom="@+id/imageView1"
android:layout_alignRight="@+id/imageView2" android:layout_marginRight="52dp" android:
layout_marginBottom="23dp"></ToggleButton>
        <ImageView android:layout_height="wrap_content" android:layout_width="wrap_
content" android:id="@+id/imageView1" android:src="@drawable/lanya" android:layout_
above="@+id/btnSearch" android:layout_toLeftOf="@+id/tbtnSwitch" android:layout_
marginRight="41dp"></ImageView>
        <ImageView android:layout_height="wrap_content" android:layout_width="wrap_
content" android:id="@+id/imageView2" android:src="@drawable/sousuo" android:layout_
above="@+id/tbtnSwitch" android:layout_centerHorizontal="true"></ImageView>
    </RelativeLayout>
</LinearLayout>
```

（3）编写文件 company.xml，功能是显示公司介绍信息，具体实现代码如下所示。

```
<LinearLayout
  xmlns:android="http://schemas.android.com/apk/res/android"
  android:orientation="vertical"
  android:layout_width="match_parent"
  android:background="@drawable/bac"
  android:layout_height="match_parent">
    <RelativeLayout android:id="@+id/relativeLayout1" android:layout_width="match_
parent" android:layout_height="match_parent">
        <ImageView android:src="@drawable/company" android:id="@+id/imageView1" android:
layout_height="wrap_content" android:layout_width="wrap_content" android:layout_
alignParentBottom="true" android:layout_centerHorizontal="true"></ImageView>
        <ImageButton android:src="@drawable/back" android:background="#0000" android:
layout_height="wrap_content" android:id="@+id/back" android:layout_width="wrap_content"
android:layout_alignParentBottom="true" android:layout_alignParentRight="true">
</ImageButton>
        <ImageView android:src="@drawable/ctitle" android:id="@+id/imageView2" android:
layout_height="wrap_content" android:layout_width="wrap_content" android:layout_
alignParentTop="true" android:layout_alignRight="@+id/imageView1" android:layout_
marginRight="22dp"></ImageView>
    </RelativeLayout>
</LinearLayout>
```

（4）编写文件 dialog.xml，功能是实现一个系统功能介绍对话框效果，具体实现代码如下所示。

```
<LinearLayout
  xmlns:android="http://schemas.android.com/apk/res/android"
  android:orientation="vertical"
  android:layout_width="match_parent"
  android:layout_height="match_parent"
  android:id="@+id/dialog" android:weightSum="1">
    <RelativeLayout android:id="@+id/relativeLayout1" android:layout_width="match_
parent" android:layout_height="wrap_content" android:layout_weight="1.09">
        <TextView android:text="@string/intr1" android:layout_width="wrap_content"
android:layout_height="wrap_content" android:id="@+id/textView2" android:text
Appearance="?android:attr/textAppearanceLarge" android:layout_below="@+id/textView1"
android:layout_alignLeft="@+id/textView1" android:layout_marginLeft="24dp"></TextView>
        <TextView android:text="@string/intr0" android:layout_width="wrap_content"
android:layout_height="wrap_content" android:id="@+id/textView1" android:text
Appearance="?android:attr/textAppearanceLarge" android:layout_alignParentTop="true"
android:layout_alignParentLeft="true" android:layout_marginLeft="50dp" android:
layout_marginTop="53dp"></TextView>
        <TextView android:text="@string/intr2" android:layout_width="wrap_content"
android:layout_height="wrap_content" android:id="@+id/textView3" android:text
Appearance="?android:attr/textAppearanceLarge" android:layout_below="@+id/textView2"
android:layout_alignLeft="@+id/textView1"></TextView>
        <TextView android:text="@string/intr3" android:layout_width="wrap_content"
android:layout_height="wrap_content" android:id="@+id/textView4" android:text
Appearance="?android:attr/textAppearanceLarge" android:layout_below="@+id/textView3"
```

```
android:layout_alignLeft="@+id/textView2"></TextView>
        <ImageView android:id="@+id/imageView1" android:layout_width="wrap_content"
android:layout_height="wrap_content" android:src="@drawable/main_add_enable" android:
layout_below="@+id/textView4" android:layout_alignLeft="@+id/textView3" android:
layout_marginTop="37dp"></ImageView>
        <TextView android:text="@string/intr4" android:layout_width="wrap_content"
android:layout_height="wrap_content" android:id="@+id/textView5" android:text
Appearance="?android:attr/textAppearanceLarge" android:layout_alignBottom="@+id/
imageView1" android:layout_toRightOf="@+id/imageView1" android:layout_marginLeft=
"34dp"></TextView>
        <ImageView android:id="@+id/imageView2" android:layout_width="wrap_content"
android:layout_height="wrap_content" android:src="@drawable/main_menu_enable"
android:layout_below="@+id/imageView1" android:layout_alignLeft="@+id/imageView1"
android:layout_marginTop="24dp"></ImageView>
        <TextView android:text="@string/intr5" android:layout_width="wrap_content"
android:layout_height="wrap_content" android:id="@+id/textView6" android:text
Appearance="?android:attr/textAppearanceLarge" android:layout_alignBottom="@+id/
imageView2" android:layout_alignLeft="@+id/textView5"></TextView>
        <ImageView android:id="@+id/imageView3" android:layout_width="wrap_content"
android:layout_height="wrap_content" android:src="@drawable/main_back_enable" android:
layout_below="@+id/imageView2" android:layout_alignLeft="@+id/imageView2" android:
layout_marginTop="30dp"></ImageView>
        <TextView android:text="@string/intr6" android:layout_width="wrap_content"
android:layout_height="wrap_content" android:id="@+id/textView7" android:
textAppearance="?android:attr/textAppearanceLarge" android:layout_alignBottom=
"@+id/imageView3" android:layout_alignLeft="@+id/textView6"></TextView>
        <TextView android:text="@string/intr7" android:textColor="#00ff00" android:
layout_width="wrap_content" android:layout_height="wrap_content" android:id="@+id/
textView8" android:textAppearance="?android:attr/textAppearanceLarge" android:layout_
below="@+id/imageView3" android:layout_alignLeft="@+id/imageView3" android:layout_
marginTop="52dp"></TextView>
        <TextView android:text="@string/intr8" android:textColor="#00ff00" android:
layout_width="wrap_content" android:layout_height="wrap_content" android:id="@+id/
textView9" android:textAppearance="?android:attr/textAppearanceLarge" android:layout_
below="@+id/textView8" android:layout_alignLeft="@+id/textView8" android:layout_
marginTop="47dp"></TextView>
        <TextView android:text="@string/intr9" android:textColor="#00ff00" android:
layout_width="wrap_content" android:layout_height="wrap_content" android:id="@+id/
textView10" android:textAppearance="?android:attr/textAppearanceLarge" android:
layout_below="@+id/textView9" android:layout_alignLeft="@+id/textView9" android:
layout_marginTop="45dp"></TextView>
    </RelativeLayout>
</LinearLayout>
```

（5）编写文件 first.xml，功能是通过单选按钮列表实现第 1 路调光设置界面，具体实现代码如下所示。

```
<LinearLayout xmlns:android="http://schemas.android.com/apk/res/android"
    android:layout_width="fill_parent"
    android:layout_height="fill_parent"
    android:background="#0000"
    android:weightSum="1" android:orientation="vertical">
    <ImageButton
    android:layout_width="wrap_content"
    android:layout_height="wrap_content"
    android:src="@drawable/help_2"
    android:background="#0000"
    android:id="@+id/imageButton1"
    android:layout_gravity="right"></ImageButton>

    <TextView
    android:id="@+id/textview1"
    android:layout_height="wrap_content"
    android:layout_width="wrap_content"
    android:layout_alignParentTop="true"
    android:layout_alignParentRight="true"
    android:layout_marginRight="76dp"
    android:layout_marginTop="15dp"></TextView>
```

```
<RelativeLayout
android:id="@+id/relativeLayout1"
android:layout_width="match_parent" android:layout_height="696dp" android:gravity=
"left">
    <RadioGroup android:layout_height="wrap_content" android:id="@+id/radioGroup1"
android:layout_width="wrap_content" android:layout_alignParentTop="true" android:
layout_toRightOf="@+id/imageButton2" android:layout_marginLeft="18dp" android:layout_
marginTop="32dp">
    </RadioGroup>
    <ImageButton android:id="@+id/imageButton5" android:layout_height="wrap_
content" android:background="#0000" android:layout_width="wrap_content" android:src=
"@drawable/button2_c" android:layout_alignTop="@+id/imageButton3" android:layout_
alignLeft="@+id/imageButton4"></ImageButton>
    <RadioGroup android:layout_height="wrap_content" android:id="@+id/radioGroup3"
android:layout_width="wrap_content" android:layout_alignTop="@+id/radioGroup2"
android:layout_toRightOf="@+id/radioGroup1" android:layout_marginLeft="45dp">
        <RadioButton android:text="0%" android:layout_height="wrap_content"
android:id="@+id/radio21" android:layout_width="wrap_content" ></RadioButton>
        <RadioButton android:text="20%" android:layout_height="wrap_content"
android:id="@+id/radio22" android:layout_width="wrap_content"></RadioButton>
        <RadioButton android:text="40%" android:layout_height="wrap_content"
android:id="@+id/radio23" android:layout_width="wrap_content"></RadioButton>
        <RadioButton android:text="60%" android:layout_height="wrap_content"
android:id="@+id/radio24" android:layout_width="wrap_content"></RadioButton>
        <RadioButton android:text="80%" android:layout_height="wrap_content"
android:id="@+id/radio25" android:layout_width="wrap_content"></RadioButton>
        <RadioButton android:text="100%" android:layout_height="wrap_content"
android:id="@+id/radio26" android:layout_width="wrap_content"></RadioButton>
    </RadioGroup>
    <ImageButton android:id="@+id/imageButton4" android:layout_height="wrap_content"
android:background="#0000" android:layout_width="wrap_content" android:src=
"@drawable/button2_o" android:layout_alignTop="@+id/imageButton2" android:layout_to
RightOf="@+id/radioGroup1" android:layout_marginLeft="24dp"></ImageButton>
    <RadioGroup android:layout_height="wrap_content" android:id="@+id/radioGroup4"
android:layout_width="wrap_content" android:layout_alignTop="@+id/radioGroup3"
android:layout_toRightOf="@+id/imageButton4" android:layout_marginLeft="76dp">
        <RadioButton android:text="0%" android:layout_height="wrap_content"
android:id="@+id/radio31" android:layout_width="wrap_content"></RadioButton>
        <RadioButton android:text="20%" android:layout_height="wrap_content"
android:id="@+id/radio32" android:layout_width="wrap_content"></RadioButton>
        <RadioButton android:text="40%" android:layout_height="wrap_content"
android:id="@+id/radio33" android:layout_width="wrap_content"></RadioButton>
        <RadioButton android:text="60%" android:layout_width="wrap_content"
android:id="@+id/radio34" android:layout_height="wrap_content" android:layout_below=
"@+id/radioGroup4" android:layout_alignLeft="@+id/radioGroup4"></RadioButton>
        <RadioButton android:text="80%" android:layout_width="wrap_content"
android:id="@+id/radio35" android:layout_height="wrap_content" android:layout_below=
"@+id/radioButton4" android:layout_alignLeft="@+id/radioButton4"></RadioButton>
        <RadioButton android:text="100%" android:layout_width="wrap_content"
android:id="@+id/radio36" android:layout_height="wrap_content" android:layout_below=
"@+id/radioButton5" android:layout_alignLeft="@+id/radioButton5"></RadioButton>
</RadioGroup>
            <TextView android:layout_height="wrap_content" android:text="第 3 路调光"
android:textColor="#ffffff" android:id="@+id/textView4" android:layout_width="wrap_
content" android:layout_alignTop="@+id/textView2" android:layout_alignLeft="@+id/
radioGroup4"></TextView>
        <ImageButton android:layout_width="wrap_content" android:layout_height=
"wrap_content" android:src="@drawable/button3_c" android:background="#0000" android:
id="@+id/imageButton7" android:layout_alignTop="@+id/imageButton5" android:layout_
alignLeft="@+id/imageButton6"></ImageButton>
        <ImageButton android:layout_width="wrap_content" android:layout_
height="wrap_content" android:src="@drawable/button3_o" android:background="#0000"
android:id="@+id/imageButton6" android:layout_alignTop="@+id/imageButton4" android:
layout_toRightOf="@+id/imageButton4" android:layout_marginLeft="56dp"></ImageButton>
        <TextView android:layout_width="wrap_content" android:text="第 4 路调光"
android:layout_height="wrap_content" android:id="@+id/textView5" android:textColor=
"#ffffff" android:layout_alignTop="@+id/textView4" android:layout_alignRight="@+id/
radioGroup5"></TextView>
        <ImageButton android:src="@drawable/button4_c" android:layout_height=
```

```
"wrap_content" android:id="@+id/imageButton9" android:background="#0000" android:
layout_width="wrap_content" android:layout_alignTop="@+id/imageButton7" android:
layout_alignLeft="@+id/imageButton8"></ImageButton>
        <ImageButton android:src="@drawable/split" android:layout_height="wrap_
content" android:id="@+id/imageButton10" android:background="#0000" android:layout_
width="1000dip" android:layout_alignTop="@+id/radioGroup2" android:layout_alignRight=
"@+id/radioGroup4" android:layout_marginTop="47dp"></ImageButton>
        <ImageButton android:src="@drawable/split" android:layout_height="wrap_
content" android:id="@+id/imageButton11" android:background="#0000" android:layout_
width="wrap_content" android:layout_alignTop="@+id/imageButton10" android:layout_
alignRight="@+id/radioGroup5"></ImageButton>
        <ImageButton android:background="#0000" android:layout_height=
"wrap_content" android:src="@drawable/split" android:id="@+id/imageButton12" android:
layout_width="wrap_content" android:layout_below="@+id/imageButton10" android:layout_
alignLeft="@+id/radioGroup2" android:layout_marginTop="35dp"></ImageButton>
        <ImageButton android:background="#0000" android:layout_height="wrap_
content" android:src="@drawable/split" android:id="@+id/imageButton15" android:layout_
width="wrap_content" android:layout_alignTop="@+id/imageButton12" android:layout_
alignRight="@+id/radioGroup5"></ImageButton>
        <ImageButton android:background="#0000" android:layout_height="wrap_
content" android:src="@drawable/split" android:id="@+id/imageButton13" android:layout_
width="wrap_content" android:layout_below="@+id/imageButton12" android:layout_align
Right="@+id/imageButton4" android:layout_marginTop="43dp"></ImageButton>
        <ImageButton android:background="#0000" android:layout_height="wrap_
content" android:src="@drawable/split" android:id="@+id/imageButton17" android:layout_
width="wrap_content" android:layout_alignTop="@+id/imageButton13" android:layout_
alignRight="@+id/radioGroup5"></ImageButton>
        <ImageButton android:background="#0000" android:layout_height="wrap_
content" android:src="@drawable/split" android:id="@+id/imageButton18" android:layout_
width="wrap_content" android:layout_below="@+id/imageButton13" android:layout_
alignLeft="@+id/imageButton13" android:layout_marginTop="39dp"></ImageButton>
        <ImageButton android:background="#0000" android:layout_height="wrap_
content" android:src="@drawable/split" android:id="@+id/imageButton19" android:layout_
width="wrap_content" android:layout_alignTop="@+id/imageButton18" android:layout_
alignRight="@+id/radioGroup5"></ImageButton>
        <ImageButton android:id="@+id/imageButton3" android:layout_width="wrap_
content" android:src="@drawable/button1_c" android:layout_height="wrap_content"
android:background="#0000" android:layout_below="@+id/imageButton2" android:layout_
alignLeft="@+id/imageButton2" android:layout_marginTop="30dp"></ImageButton>
        <RadioGroup android:layout_width="wrap_content" android:id="@+id/
radioGroup2" android:layout_height="wrap_content" android:layout_below="@+id/radio
Group1" android:layout_alignParentLeft="true" android:layout_marginLeft="179dp"
android:layout_marginTop="105dp">
            <RadioButton android:layout_height="wrap_content" android:layout_width=
"wrap_content" android:text="0%" android:id="@+id/radio11"></RadioButton>
            <RadioButton android:layout_height="wrap_content" android:layout_width=
"wrap_content" android:text="20%" android:id="@+id/radio12"></RadioButton>
            <RadioButton android:layout_height="wrap_content" android:layout_width=
"wrap_content" android:text="40%" android:id="@+id/radio13"></RadioButton>
            <RadioButton android:layout_height="wrap_content" android:layout_width=
"wrap_content" android:text="60%" android:id="@+id/radio14"></RadioButton>
            <RadioButton android:layout_height="wrap_content" android:layout_width=
"wrap_content" android:text="80%" android:id="@+id/radio15"></RadioButton>
            <RadioButton android:layout_height="wrap_content" android:layout_width=
"wrap_content" android:text="100%" android:id="@+id/radio16"></RadioButton>
        </RadioGroup>
        <TextView android:text="第 2 路调光" android:textColor="#ffffff" android:
layout_width="wrap_content" android:id="@+id/textView2" android:layout_height="wrap_
content" android:layout_alignTop="@+id/textView3" android:layout_alignLeft="@+id/
radioGroup3"></TextView>
        <TextView android:text="第 1 路调光" android:textColor="#ffffff" android:
layout_width="wrap_content" android:id="@+id/textView3" android:layout_height="wrap_
content" android:layout_below="@+id/radioGroup1" android:layout_alignLeft="@+id/
radioGroup2" android:layout_marginTop="44dp"></TextView>
        <TextView android:text=" 一" android:layout_width="wrap_content" android:
id="@+id/textView7" android:textSize="30sp" android:layout_height="wrap_content"
android:layout_below="@+id/imageButton11" android:layout_alignRight="@+id/textView6">
</TextView>
        <TextView android:text="调" android:layout_width="wrap_content"
```

```
android:id="@+id/textView8" android:textSize="30sp" android:layout_height="wrap_
content" android:layout_below="@+id/imageButton15" android:layout_alignRight="@+id/
textView7"></TextView>
        <TextView android:text="光" android:layout_width="wrap_content"
android:id="@+id/textView9" android:textSize="30sp" android:layout_height="wrap_
content" android:layout_below="@+id/imageButton17" android:layout_alignRight="@+id/
textView8"></TextView>
        <TextView android:text="界" android:layout_width="wrap_content"
android:id="@+id/textView10" android:textSize="30sp" android:layout_height="wrap_
content" android:layout_below="@+id/imageButton19" android:layout_alignRight="@+id/
textView9"></TextView>
        <ImageButton android:id="@+id/imageButton8" android:layout_width="wrap_
content" android:src="@drawable/button4_o" android:layout_height="wrap_content"
android:background="#0000" android:layout_alignTop="@+id/imageButton6" android:
layout_toRightOf="@+id/imageButton6" android:layout_marginLeft="39dp"></ImageButton>
        <RadioGroup android:layout_width="wrap_content" android:id="@+id/
radioGroup5" android:layout_height="wrap_content" android:layout_alignTop="@+id/
radioGroup4" android:layout_alignRight="@+id/imageButton8">
            <RadioButton android:layout_height="wrap_content" android:layout_width=
"wrap_content" android:text="0%" android:id="@+id/radio41"></RadioButton>
            <RadioButton android:layout_height="wrap_content" android:layout_width=
"wrap_content" android:text="20%" android:id="@+id/radio42"></RadioButton>
            <RadioButton android:layout_height="wrap_content" android:layout_width=
"wrap_content" android:text="40%" android:id="@+id/radio43"></RadioButton>
            <RadioButton android:layout_height="wrap_content" android:layout_width=
"wrap_content" android:text="60%" android:id="@+id/radio44"></RadioButton>
            <RadioButton android:layout_height="wrap_content" android:layout_width=
"wrap_content" android:text="80%" android:id="@+id/radio45"></RadioButton>
            <RadioButton android:layout_height="wrap_content" android:layout_width=
"wrap_content" android:text="100%" android:id="@+id/radio46"></RadioButton>
        </RadioGroup>
        <ImageButton android:id="@+id/imageButton14" android:layout_width="1000dip"
android:src="@drawable/line" android:layout_height="5dip" android:layout_below="@+id/
radioGroup3"></ImageButton>
        <ImageButton android:layout_width="5dip" android:src="@drawable/line"
android:layout_height="1000dip" android:id="@+id/imageButton16" android:layout_
alignParentTop="true" android:layout_toRightOf="@+id/textView5" android:layout_margin
Left="46dp"></ImageButton>
        <TextView android:textSize="30sp" android:layout_width="wrap_content"
android:text="第" android:id="@+id/textView6" android:layout_height="wrap_content"
android:layout_alignTop="@+id/radioGroup5" android:layout_toRightOf="@+id/image
Button16" android:layout_marginLeft="24dp"></TextView>
        <ImageButton android:layout_width="wrap_content" android:src="@drawable/
button1_add_x" android:layout_height="wrap_content" android:background="#0000"
android:id="@+id/imageButton2" android:layout_below="@+id/imageButton14" android:
layout_alignRight="@+id/radioGroup2" android:layout_marginTop="56dp"></ImageButton>
        <TextView android:id="@+id/textView11" android:text="面"
android:textSize="30sp" android:layout_height="wrap_content" android:layout_width=
"wrap_content" android:layout_above="@+id/imageButton14" android:layout_alignLeft=
"@+id/textView10"></TextView>
        <ImageButton android:layout_height="wrap_content" android:background=
"#0000" android:id="@+id/btnopen" android:layout_width="wrap_content" android:src=
"@drawable/allop" android:layout_alignTop="@+id/imageButton8" android:layout_align
Left="@+id/textView11"></ImageButton>
        <ImageButton android:layout_height="wrap_content" android:id="@+id/
btnclose" android:layout_width="wrap_content" android:src="@drawable/allcl" android:
background="#0000" android:layout_alignTop="@+id/imageButton9" android:layout_
alignLeft="@+id/btnopen"></ImageButton>


    </RelativeLayout>

</LinearLayout>
```

（6）编写文件 home.xml，此文件是系统执行后进入的主界面，具体实现代码如下所示。

```
<LinearLayout
  xmlns:android="http://schemas.android.com/apk/res/android"
  android:orientation="vertical"
  android:layout_width="match_parent"
```

```
  android:background="@drawable/bac"
  android:layout_height="match_parent" android:weightSum="1">
    <RelativeLayout android:id="@+id/relativeLayout1" android:layout_width="match_
parent" android:layout_height="682dp">
        <ImageView android:layout_alignParentLeft="true" android:id="@+id/imageView2"
android:src="@drawable/enter1" android:layout_height="wrap_content" android:layout_
width="wrap_content" android:layout_below="@+id/imageView1"></ImageView>
        <ImageView android:id="@+id/imageView4" android:src="@drawable/enter2" android:
layout_height="wrap_content" android:layout_width="wrap_content" android:layout_
below="@+id/imageView2" android:layout_alignParentLeft="true" android:layout_
marginLeft="48dp" android:layout_marginTop="86dp"></ImageView>
        <ImageView android:id="@+id/imageView5" android:src="@drawable/enter3" android:
layout_height="wrap_content" android:layout_width="wrap_content" android:layout_
alignParentBottom="true" android:layout_alignLeft="@+id/imageView4" android:layout_
marginBottom="66dp"></ImageView>
        <ImageView android:id="@+id/imageView7" android:src="@drawable/split" android:
layout_height="wrap_content" android:layout_width="wrap_content" android:layout_
alignBottom="@+id/imageView5" android:layout_alignLeft="@+id/imageView6"></ImageView>
        <ImageButton android:id="@+id/Enterp" android:background="#0000" android:
src="@drawable/enter0" android:layout_height="wrap_content" android:layout_width=
"wrap_content" android:layout_alignBottom="@+id/imageView6" android:layout_alignLeft=
"@+id/Enterr"></ImageButton>
        <ImageButton android:id="@+id/Enterc" android:background="#0000" android:src=
"@drawable/enter0" android:layout_height="wrap_content" android:layout_width="wrap_
content" android:layout_above="@+id/imageView7" android:layout_alignLeft="@+id/
Enterp"></ImageButton>
        <ImageView android:id="@+id/imageView6" android:src="@drawable/split" android:
layout_height="wrap_content" android:layout_width="wrap_content" android:layout_
below="@+id/imageView4" android:layout_alignLeft="@+id/imageView3"></ImageView>
        <ImageView android:id="@+id/imageView1" android:src="@drawable/logo" android:
layout_height="wrap_content" android:layout_width="wrap_content" android:layout_
alignParentTop="true" android:layout_centerHorizontal="true"></ImageView>
        <ImageButton android:background="#0000" android:id="@+id/Enterr" android:
src="@drawable/enter0" android:layout_height="wrap_content" android:layout_width=
"wrap_content" android:layout_alignBottom="@+id/imageView3" android:layout_toRightOf=
"@+id/imageView1" android:layout_marginLeft="63dp"></ImageButton>
        <ImageView android:id="@+id/imageView3" android:src="@drawable/split" android:
layout_height="wrap_content" android:layout_width="wrap_content" android:layout_
below="@+id/imageView2" android:layout_alignRight="@+id/imageView1" android:layout_
marginRight="87dp"></ImageView>

</RelativeLayout>
    <RelativeLayout android:layout_weight="0.95" android:layout_height="wrap_content"
android:id="@+id/relativeLayout2" android:layout_width="match_parent"></RelativeLayout>
    <RelativeLayout android:layout_width="fill_parent"
        android:layout_height="wrap_content"
        android:layout_alignParentBottom="true"
        style="@android:style/ButtonBar"
        android:background="@drawable/title_background">
        <ImageButton android:layout_width="wrap_content" android:layout_height="wrap_
content" android:background="#0000" android:id="@+id/back" android:src="@drawable/
back" android:layout_centerHorizontal="true" android:layout_alignTop="@+id/
imageButton3"></ImageButton>></RelativeLayout>
</LinearLayout>
```

（7）编写文件 lightstandard.xml，功能是显示不同房间的照明亮度参考值，具体实现代码如下所示。

```
<LinearLayout
  xmlns:android="http://schemas.android.com/apk/res/android"
  android:orientation="vertical"
  android:layout_width="match_parent"
  android:layout_height="match_parent"
  android:background="@drawable/bac"
  android:id="@+id/stand" android:weightSum="1">
    <RelativeLayout   android:id="@+id/relativeLayout1"   android:layout_width="match_
parent" android:layout_height="wrap_content" android:layout_weight="1.09">
        <ImageView android:layout_height="wrap_content" android:src="@drawable/
twitter1" android:id="@+id/imageView1" android:background="#0000" android:layout_
```

```
width="wrap_content" android:layout_alignParentTop="true" android:layout_alignRight=
"@+id/imageView2" android:layout_marginRight="119dp"></ImageView>
        <ImageView android:layout_height="wrap_content" android:src="@drawable/stand1"
android:id="@+id/imageView2" android:layout_width="wrap_content" android:layout_
below="@+id/imageView1" android:layout_alignParentRight="true" android:layout_
marginRight="173dp"></ImageView>
    </RelativeLayout>
</LinearLayout>
```

（8）编写文件 product.xml，功能是显示本产品的详细介绍信息，具体实现代码如下所示。

```
<LinearLayout
  xmlns:android="http://schemas.android.com/apk/res/android"
  android:orientation="vertical"
  android:background="@drawable/bac"
  android:layout_width="match_parent"
  android:layout_height="match_parent">
    <RelativeLayout android:id="@+id/relativeLayout1" android:layout_height="match_
parent" android:layout_width="match_parent">
        <ImageButton android:src="@drawable/back" android:background="#0000"
android:layout_height="wrap_content" android:id="@+id/back" android:layout_width=
"wrap_content" android:layout_alignParentBottom="true" android:layout_alignParent
Right="true"></ImageButton>
        <ImageView android:src="@drawable/product" android:id="@+id/imageView1"
android:layout_height="wrap_content" android:layout_width="wrap_content" android:
layout_above="@+id/back" android:layout_alignParentLeft="true" android:layout_
marginLeft="117dp" android:layout_marginBottom="31dp"></ImageView>
        <ImageView android:src="@drawable/title" android:id="@+id/imageView2" android:
layout_height="wrap_content" android:layout_width="wrap_content" android:layout_
alignParentTop="true" android:layout_alignLeft="@+id/imageView1" android:layout_
marginLeft="59dp" android:layout_marginTop="42dp"></ImageView>
    </RelativeLayout>
</LinearLayout>
```

（9）编写文件 second.xml，功能是通过单选按钮列表实现一个 5 路调光设置界面效果，具体实现代码如下所示。

```
<LinearLayout xmlns:android="http://schemas.android.com/apk/res/android"
    android:layout_width="fill_parent"
    android:layout_height="fill_parent"
    android:background="#0000"
    android:weightSum="1" android:orientation="vertical">
    <ImageButton
    android:layout_width="wrap_content"
    android:layout_height="wrap_content"
    android:src="@drawable/help_2"
    android:background="#0000"
    android:id="@+id/imageButton1"
    android:layout_gravity="right"></ImageButton>

    <TextView
    android:id="@+id/textview1"
    android:layout_height="wrap_content"
    android:layout_width="wrap_content"
    android:layout_alignParentTop="true"
    android:layout_alignParentRight="true"
    android:layout_marginRight="76dp"
    android:layout_marginTop="15dp"></TextView>
    <RelativeLayout
    android:id="@+id/relativeLayout1"
    android:layout_width="match_parent" android:layout_height="696dp" android:gravity=
"left">
        <RadioGroup android:layout_height="wrap_content" android:id="@+id/radioGroup1"
android:layout_width="wrap_content" android:layout_alignParentTop="true" android:
layout_toRightOf="@+id/imageButton2" android:layout_marginLeft="18dp" android:layout_
marginTop="32dp">
        </RadioGroup>
        <ImageButton android:layout_width="wrap_content" android:id="@+id/
```

```
imageButton3" android:background="#0000" android:layout_height="wrap_content"
android:src="@drawable/button5_c" android:layout_below="@+id/imageButton2"
android:layout_alignLeft="@+id/imageButton2"
android:layout_marginTop="26dp"></ImageButton>
      <ImageButton android:id="@+id/imageButton5" android:layout_height="wrap_
content" android:background="#0000" android:layout_width="wrap_content" android:src=
"@drawable/button6_c" android:layout_alignTop="@+id/imageButton3" android:layout_
alignLeft="@+id/imageButton4"></ImageButton>
      <RadioGroup android:layout_height="wrap_content" android:id="@+id/radioGroup3"
android:layout_width="wrap_content" android:layout_alignTop="@+id/radioGroup2"
android:layout_toRightOf="@+id/radioGroup1" android:layout_marginLeft="45dp">
         <RadioButton android:text="0%" android:layout_height="wrap_content"
android:id="@+id/radio21" android:layout_width="wrap_content" ></RadioButton>
         <RadioButton android:text="20%" android:layout_height="wrap_content"
android:id="@+id/radio22" android:layout_width="wrap_content"></RadioButton>
         <RadioButton android:text="40%" android:layout_height="wrap_content"
android:id="@+id/radio23" android:layout_width="wrap_content"></RadioButton>
         <RadioButton android:text="60%" android:layout_height="wrap_content"
android:id="@+id/radio24" android:layout_width="wrap_content"></RadioButton>
         <RadioButton android:text="80%" android:layout_height="wrap_content"
android:id="@+id/radio25" android:layout_width="wrap_content"></RadioButton>
         <RadioButton android:text="100%" android:layout_height="wrap_content"
android:id="@+id/radio26" android:layout_width="wrap_content"></RadioButton>
      </RadioGroup>
      <ImageButton android:id="@+id/imageButton4" android:layout_height="wrap_
content" android:background="#0000" android:layout_width="wrap_content" android:src=
"@drawable/button6_o" android:layout_alignTop="@+id/imageButton2" android:layout_
toRightOf="@+id/radioGroup1" android:layout_marginLeft="24dp"></ImageButton>
      <RadioGroup android:layout_height="wrap_content" android:id="@+id/radioGroup4"
android:layout_width="wrap_content" android:layout_alignTop="@+id/radioGroup3"
android:layout_toRightOf="@+id/imageButton4" android:layout_marginLeft="76dp">
         <RadioButton android:text="0%" android:layout_height="wrap_content"
android:id="@+id/radio31" android:layout_width="wrap_content"></RadioButton>
         <RadioButton android:text="20%" android:layout_height="wrap_content"
android:id="@+id/radio32" android:layout_width="wrap_content"></RadioButton>
         <RadioButton android:text="40%" android:layout_height="wrap_content"
android:id="@+id/radio33" android:layout_width="wrap_content"></RadioButton>
         <RadioButton android:text="60%" android:layout_width="wrap_content"
android:id="@+id/radio34" android:layout_height="wrap_content" android:layout_below=
"@+id/radioGroup4" android:layout_alignLeft="@+id/radioGroup4"></RadioButton>
         <RadioButton android:text="80%" android:layout_width="wrap_content"
android:id="@+id/radio35" android:layout_height="wrap_content" android:layout_below=
"@+id/radioButton4" android:layout_alignLeft="@+id/radioButton4"></RadioButton>
         <RadioButton android:text="100%" android:layout_width="wrap_content"
android:id="@+id/radio36" android:layout_height="wrap_content" android:layout_below=
"@+id/radioButton5" android:layout_alignLeft="@+id/radioButton5"></RadioButton>
</RadioGroup>
            <TextView android:layout_height="wrap_content" android:text="第 7 路调光"
android:textColor="#ffffff" android:id="@+id/textView4" android:layout_width="wrap_
content" android:layout_alignTop="@+id/textView2" android:layout_alignLeft="@+id/
radioGroup4"></TextView>
         <ImageButton android:layout_width="wrap_content" android:layout_height=
"wrap_content" android:src="@drawable/button7_c" android:background="#0000" android:
id="@+id/imageButton7" android:layout_alignTop="@+id/imageButton5" android:layout_
alignLeft="@+id/imageButton6"></ImageButton>
         <ImageButton android:layout_width="wrap_content" android:layout_height=
"wrap_content" android:src="@drawable/button7_o" android:background="#0000"
android:id="@+id/imageButton6" android:layout_alignTop="@+id/imageButton4" android:
layout_toRightOf="@+id/imageButton4" android:layout_marginLeft="56dp"></ImageButton>
         <TextView android:layout_height="wrap_content" android:text="第 8 路调光"
android:textColor="#ffffff" android:id="@+id/textView5" android:layout_width=
"wrap_content" android:layout_alignTop="@+id/textView4" android:layout_alignLeft=
"@+id/radioGroup5"></TextView>
         <ImageButton android:layout_width="wrap_content" android:layout_height=
"wrap_content" android:src="@drawable/button8_c" android:background="#0000"
android:id="@+id/imageButton9" android:layout_alignTop="@+id/imageButton7" android:
layout_alignLeft="@+id/imageButton8"></ImageButton>
         <RadioGroup android:id="@+id/radioGroup2" android:layout_width="wrap_
content" android:layout_height="wrap_content" android:layout_below="@+id/radioGroup1"
```

```
android:layout_alignParentLeft="true" android:layout_marginLeft="207dp" android:
layout_marginTop="84dp">
            <RadioButton android:text="0%" android:id="@+id/radio11" android:layout_
width="wrap_content" android:layout_height="wrap_content"></RadioButton>
            <RadioButton android:text="20%" android:id="@+id/radio12" android:
layout_width="wrap_content" android:layout_height="wrap_content"></RadioButton>
            <RadioButton android:text="40%" android:id="@+id/radio13" android:
layout_width="wrap_content" android:layout_height="wrap_content"></RadioButton>
            <RadioButton android:text="60%" android:id="@+id/radio14" android:
layout_width="wrap_content" android:layout_height="wrap_content"></RadioButton>
            <RadioButton android:text="80%" android:id="@+id/radio15" android:
layout_width="wrap_content" android:layout_height="wrap_content"></RadioButton>
            <RadioButton android:text="100%" android:id="@+id/radio16" android:
layout_width="wrap_content" android:layout_height="wrap_content"></RadioButton>
        </RadioGroup>
        <TextView android:textColor="#ffffff" android:layout_width="wrap_content"
android:text="第 5 路调光" android:id="@+id/textView3" android:layout_height="wrap_
content" android:layout_below="@+id/radioGroup1" android:layout_alignLeft="@+id/
radioGroup2" android:layout_marginTop="31dp"></TextView>
        <ImageButton android:layout_width="wrap_content" android:src="@drawable/
button5_o" android:layout_height="wrap_content" android:background="#0000" android:
id="@+id/imageButton2" android:layout_below="@+id/radioGroup2" android:layout_
alignRight="@+id/radioGroup2" android:layout_marginTop="76dp"></ImageButton>
        <TextView android:textColor="#ffffff" android:layout_width="wrap_content"
android:text="第 6 路调光" android:id="@+id/textView2" android:layout_height="wrap_
content" android:layout_alignTop="@+id/textView3" android:layout_alignLeft="@+id/
radioGroup3"></TextView>
        <RadioGroup android:id="@+id/radioGroup5" android:layout_width="wrap_
content" android:layout_height="wrap_content" android:layout_alignTop="@+id/
radioGroup4" android:layout_toRightOf="@+id/imageButton6" android:layout_marginLeft=
"59dp">
            <RadioButton android:text="0%" android:id="@+id/radio41" android:layout_
width="wrap_content" android:layout_height="wrap_content"></RadioButton>
            <RadioButton android:text="20%" android:id="@+id/radio42" android:
layout_width="wrap_content" android:layout_height="wrap_content"></RadioButton>
            <RadioButton android:text="40%" android:id="@+id/radio43" android:
layout_width="wrap_content" android:layout_height="wrap_content"></RadioButton>
            <RadioButton android:text="60%" android:id="@+id/radio44" android:
layout_width="wrap_content" android:layout_height="wrap_content"></RadioButton>
            <RadioButton android:text="80%" android:id="@+id/radio45" android:
layout_width="wrap_content" android:layout_height="wrap_content"></RadioButton>
            <RadioButton android:text="100%" android:id="@+id/radio46" android:
layout_width="wrap_content" android:layout_height="wrap_content"></RadioButton>
        </RadioGroup>
        <ImageButton android:layout_width="wrap_content" android:src="@drawable/
button8_o" android:layout_height="wrap_content" android:background="#0000" android:id=
"@+id/imageButton8" android:layout_alignTop="@+id/imageButton6" android:layout_
toRightOf="@+id/imageButton6" android:layout_marginLeft="39dp"></ImageButton>
        <ImageButton android:layout_width="wrap_content" android:src="@drawable/
split" android:layout_height="wrap_content" android:background="#0000" android:id=
"@+id/imageButton10" android:layout_alignTop="@+id/radioGroup3" android:layout_
alignRight="@+id/radioGroup3" android:layout_marginTop="42dp"></ImageButton>
        <ImageButton android:layout_width="wrap_content" android:src="@drawable/
split" android:layout_height="wrap_content" android:background="#0000" android:id=
"@+id/imageButton11" android:layout_below="@+id/imageButton10" android:layout_
alignRight="@+id/radioGroup3" android:layout_marginTop="43dp"></ImageButton>
        <ImageButton android:layout_width="wrap_content" android:src="@drawable/
split" android:layout_height="wrap_content" android:background="#0000" android:id=
"@+id/imageButton12" android:layout_below="@+id/imageButton11" android:layout_align
Right="@+id/textView2" android:layout_marginTop="42dp"></ImageButton>
        <ImageButton android:layout_width="wrap_content" android:src="@drawable/
split" android:layout_height="wrap_content" android:background="#0000" android:id=
"@+id/imageButton13" android:layout_alignBottom="@+id/radioGroup3" android:layout_
alignRight="@+id/radioGroup3" android:layout_marginBottom="46dp"></ImageButton>
        <ImageButton android:layout_width="wrap_content" android:src="@drawable/
split" android:layout_height="wrap_content" android:background="#0000" android:id=
"@+id/imageButton14" android:layout_alignTop="@+id/imageButton10" android:layout_
alignRight="@+id/textView5"></ImageButton>
        <ImageButton android:layout_width="wrap_content" android:src="@drawable/
```

```
split" android:layout_height="wrap_content" android:background="#0000" android:id=
"@+id/imageButton15" android:layout_alignTop="@+id/imageButton11" android:layout_
alignRight="@+id/radioGroup5"></ImageButton>
        <ImageButton android:layout_width="wrap_content" android:src="@drawable/
split" android:layout_height="wrap_content" android:background="#0000" android:id=
"@+id/imageButton16" android:layout_alignTop="@+id/imageButton12" android:layout_a
lignRight="@+id/radioGroup5"></ImageButton>
        <ImageButton android:layout_width="wrap_content" android:src="@drawable/
split" android:layout_height="wrap_content" android:background="#0000" android:id=
"@+id/imageButton17" android:layout_alignTop="@+id/imageButton13" android:layout_
alignRight="@+id/radioGroup5"></ImageButton>
        <ImageButton android:layout_width="1000dip" android:src="@drawable/line"
android:layout_height="5dip" android:id="@+id/imageButton18" android:layout_below=
"@+id/radioGroup3" android:layout_alignParentLeft="true" android:layout_marginTop=
"29dp"></ImageButton>
        <ImageButton android:layout_width="5dip" android:src="@drawable/line"
android:layout_height="800dip" android:id="@+id/imageButton19" android:layout_
alignParentTop="true" android:layout_above="@+id/imageButton9" android:layout_
toRightOf="@+id/imageButton9"></ImageButton>
        <TextView android:layout_width="wrap_content" android:text="    二"
android:textSize="30sp" android:id="@+id/textView7" android:layout_height="wrap_
content" android:layout_below="@+id/imageButton14" android:layout_alignRight="@+id/
textView6"></TextView>
        <TextView android:layout_width="wrap_content" android:text="    调"
android:textSize="30sp" android:id="@+id/textView8" android:layout_height="wrap_
content" android:layout_below="@+id/imageButton15" android:layout_alignRight="@+id/
textView7"></TextView>
        <TextView android:layout_width="wrap_content" android:text="    光"
android:textSize="30sp" android:id="@+id/textView9" android:layout_height="wrap_
content" android:layout_below="@+id/imageButton16" android:layout_alignRight="@+id/
textView8"></TextView>
        <TextView android:layout_height="wrap_content" android:text="    界"
android:textSize="30sp" android:id="@+id/textView10" android:layout_width="wrap_
content" android:layout_above="@+id/imageButton17" android:layout_alignRight="@+id/
textView9"></TextView>
        <TextView android:layout_height="wrap_content" android:text="    面"
android:textSize="30sp" android:id="@+id/textView11" android:layout_width="wrap_
content" android:layout_below="@+id/imageButton17" android:layout_alignRight="@+id/
textView10"></TextView>
        <TextView android:layout_height="wrap_content" android:text="    第"
android:textSize="30sp" android:id="@+id/textView6" android:layout_width=
"wrap_content" android:layout_above="@+id/textView7" android:layout_toRightOf=
"@+id/imageButton19"></TextView>


    </RelativeLayout>

</LinearLayout>
```

8.5.2　实现程序文件

（1）编写文件Main.java，功能是响应用户按键处理事件，根据用户触摸的选项进入对应的模式，通过动画效果过渡进入HOME界面。文件Main.java的具体实现代码如下所示。

```
public class Main extends ActivityGroup implements OnGestureListener,OnTouchListener {
    //声明ViewFlipper对象
    private ViewFlipper m_ViewFlipper;
    //声明GestureDetector对象
    private GestureDetector m_GestureDetector;
    //声明LocalActivityManager对象
    private LocalActivityManager m_ActivityManager;
    private static int FLING_MIN_DISTANCE = 100;
    private static int FLING_MIN_VELOCITY = 200;
    //定义自定义图片加文字按钮ImageButton对象
    //private ImageButton mButton1;

    //单选按键部分1
    private String[] areas = new String[]{"一般模式", "会议模式", "视频模式","迎接模式", "
```

```
返回"};
    private RadioOnClick radioOnClick = new RadioOnClick(4);
    @SuppressWarnings("unused")
    private ListView RadioListView;
    private ImageButton imagebtn,imagebutton;
    OutputStream tmpOut = null;
    Timer timer=new Timer();
    @Override
    public void onCreate(Bundle savedInstanceState)
    {
        super.onCreate(savedInstanceState);
        requestWindowFeature(Window.FEATURE_NO_TITLE);
        //设置内容试图
        setContentView(R.layout.main);
        getWindow().setFlags(WindowManager.LayoutParams.FLAG_FULLSCREEN, WindowManager.
        LayoutParams.FLAG_FULLSCREEN);
        imagebutton=(ImageButton)findViewById(R.id.imageButton1);
        imagebutton.setOnClickListener(new ImageButton.OnClickListener()
        {
            @Override
            public void onClick(View v)
            {
                /*Intent intent=new Intent();
                intent.setClass(Main.this, stand.class);
                startActivity(intent);
                Main.this.finish();
                overridePendingTransition(R.anim.zoomin,R.anim.zoomout);*/
                LayoutInflater inflater = getLayoutInflater();
                  View layout = inflater.inflate(R.layout.lightstandard,
                  (ViewGroup) findViewById(R.id.stand));

                  new AlertDialog.Builder(Main.this)
                  .setTitle("Ihe nationl atandard of illumination")
                  .setView(layout)
                  .setPositiveButton("Return", null)
                  //.setNegativeButton("取消", null)
                  .show();
            }
        }
                );
        //单选按键部分 2
        imagebtn=(ImageButton)findViewById(R.id.imageButton2);
        imagebtn.setOnClickListener(new RadioClickListener());
        //构建 ViewFlipper 对象
        m_ViewFlipper = (ViewFlipper) findViewById(R.id.fliper);
        //获取 Activity 消息
        m_ActivityManager = getLocalActivityManager();
        //注册一个用于手势识别的类
        m_GestureDetector = new GestureDetector(this);
        //添加视图，指定每个视图对应的 Activity
        m_ViewFlipper.addView((m_ActivityManager.startActivity("", new Intent(Main.
        this,Firstpage.class)).getDecorView()),0);
        m_ViewFlipper.addView((m_ActivityManager.startActivity("", new Intent(Main.
        this,Secondpage.class)).getDecorView()),1);
        //给 ViewFlipper 设置一个 listener
        m_ViewFlipper.setOnTouchListener(this);
        //默认为正在播放页面并设置图标
        //设置相应元素索引显示的子视图
        m_ViewFlipper.setDisplayedChild(0);
        //允许长按住 ViewFlipper，这样才能识别拖动等手势
        m_ViewFlipper.setLongClickable(true);
        //监听
        /** Called when the activity is first created. */
        /*取得 Button 对象*/
        //返回按钮按键事件
       /* btnbac = (ImageButton) findViewById(R.id.btnback);
       btnbac.setOnClickListener(new ImageButton.OnClickListener(){
        @Override
        public void onClick(View v)
```

```
    {
       // TODO Auto-generated method stub
        Intent intent=new Intent();
        intent.setClass(Main.this, home.class);
        startActivity(intent);
        Main.this.finish();
        overridePendingTransition(R.anim.zoomin,R.anim.zoomout);
        /*Intent intent = new Intent();
        intent.setClass(Main.this, home.class);
        intent.setFlags(Intent.FLAG_ACTIVITY_CLEAR_TOP);  //注意本行的 FLAG 设置
        startActivity(intent);

    } });*/}
@Override
public boolean onKeyDown(int keyCode, KeyEvent event) {
    //按下键盘上返回按钮
    if(keyCode == KeyEvent.KEYCODE_BACK){
        new AlertDialog.Builder(this)
            //.setIcon(R.drawable.services)
        .setTitle(R.string.app_about)
          /*设置弹出窗口的图式*/
         // .setIcon(R.drawable.hot)
          /*设置弹出窗口的信息*/
          .setMessage(R.string.app_about_msg)
          .setPositiveButton(R.string.str_ok,
          new DialogInterface.OnClickListener()
        {
         public void onClick(DialogInterface dialoginterface, int i)
         {
          finish();/*关闭窗口*/
         }
         }
        )
          /*设置弹出窗口的返回事件*/
         .setNegativeButton(R.string.str_no,
          new DialogInterface.OnClickListener()
          {
         public void onClick(DialogInterface dialoginterface, int i)
         {
         }
          })
        .show();
        //setResult(11,this.getIntent());
        return true;
    }else{
        return super.onKeyDown(keyCode, event);
    }
}
@Override
protected void onDestroy() {
    super.onDestroy();

    android.os.Process.killProcess(android.os.Process.myPid());
    System.exit(0);
    overridePendingTransition(R.anim.zoomin,R.anim.zoomout);
    //或者下面这种方式



}
/**
 * 定义从右侧进入的动画效果
 * @return
 */
public Animation inFromRightAnimation()
{
    Animation inFromRight = new TranslateAnimation(
            Animation.RELATIVE_TO_PARENT, +1.0f,
```

```
                Animation.RELATIVE_TO_PARENT, 0.0f,
                Animation.RELATIVE_TO_PARENT, 0.0f,
                Animation.RELATIVE_TO_PARENT, 0.0f);
        inFromRight.setDuration(500);
        inFromRight.setInterpolator(new AccelerateInterpolator());
        return inFromRight;
    }
    /**
     * 定义从左侧退出的动画效果
     * @return
     */
    public Animation outToLeftAnimation()
    {
        Animation outtoLeft = new TranslateAnimation(
                Animation.RELATIVE_TO_PARENT, 0.0f,
                Animation.RELATIVE_TO_PARENT, -1.0f,
                Animation.RELATIVE_TO_PARENT, 0.0f,
                Animation.RELATIVE_TO_PARENT, 0.0f);
        outtoLeft.setDuration(500);
        outtoLeft.setInterpolator(new AccelerateInterpolator());
        return outtoLeft;
    }
    /**
     * 定义从左侧进入的动画效果
     * @return
     */
    public Animation inFromLeftAnimation()
    {
        Animation inFromLeft = new TranslateAnimation(
                Animation.RELATIVE_TO_PARENT, -1.0f,
                Animation.RELATIVE_TO_PARENT, 0.0f,
                Animation.RELATIVE_TO_PARENT, 0.0f,
                Animation.RELATIVE_TO_PARENT, 0.0f);
        inFromLeft.setDuration(500);
        inFromLeft.setInterpolator(new AccelerateInterpolator());
        return inFromLeft;
    }
    /**
     * 定义从右侧退出时的动画效果
     * @return
     */
    public Animation outToRightAnimation()
    {
        Animation outtoRight = new TranslateAnimation(
                Animation.RELATIVE_TO_PARENT, 0.0f,
                Animation.RELATIVE_TO_PARENT, +1.0f,
                Animation.RELATIVE_TO_PARENT, 0.0f,
                Animation.RELATIVE_TO_PARENT, 0.0f);
        outtoRight.setDuration(500);
        outtoRight.setInterpolator(new AccelerateInterpolator());
        return outtoRight;
    }

    @Override
    public boolean onDown(MotionEvent e) {
        return false;
    }
    @Override
    public boolean onFling(MotionEvent e1, MotionEvent e2, float velocityX,
            float velocityY) {
        //当向左侧滑动的时候
        if(e1.getX()-e2.getX()>FLING_MIN_DISTANCE && Math.abs(velocityX)>FLING_MIN_VELOCITY)
        {
            //设置View进入屏幕的时候使用的动画
            m_ViewFlipper.setInAnimation(inFromRightAnimation());
            //设置View退出屏幕的时候使用的动画
            m_ViewFlipper.setOutAnimation(outToLeftAnimation());
            //下一个页面
            m_ViewFlipper.showNext();
```

```
            //获取相应元素索引显示的子视图
        }
        //当向右侧滑动的时候
        else if(e2.getX()-e1.getX()>FLING_MIN_DISTANCE && Math.abs(velocityX)>FLING_
        MIN_VELOCITY)
        {
            //设置View进入屏幕时候使用的动画
            m_ViewFlipper.setInAnimation(inFromLeftAnimation());
            //设置View退出屏幕时候使用的动画
            m_ViewFlipper.setOutAnimation(outToRightAnimation());
            //上一个页面
            m_ViewFlipper.showPrevious();
            //获取相应元素索引显示的子视图
        }
        return false;
    }
    @Override
    public void onLongPress(MotionEvent e) {

    }
    @Override
    public boolean onScroll(MotionEvent e1, MotionEvent e2, float distanceX,
            float distanceY) {

        return false;
    }
    @Override
    public void onShowPress(MotionEvent e) {

    }
    @Override
    public boolean onSingleTapUp(MotionEvent e) {
        return false;
    }
    @Override
    public boolean onTouch(View v, MotionEvent event) {
        //一定要将触屏事件交给手势识别类去处理（自己处理会很麻烦的）
        return m_GestureDetector.onTouchEvent(event);
    }

    //单选按键部分3
    class RadioClickListener implements OnClickListener {
      @Override
      public void onClick(View v) {
       AlertDialog ad =new AlertDialog.Builder(Main.this).setTitle("选择模式")
       .setSingleChoiceItems(areas,radioOnClick.getIndex(),radioOnClick).create();
       RadioListView=ad.getListView();
       ad.show();
    }
    }

    /**
     * 单击单选框事件
     *
     */
    class RadioOnClick implements DialogInterface.OnClickListener{
  private int index;

  public RadioOnClick(int index){
   this.index = index;
  }
  public void setIndex(int index){
   this.index=index;
  }
  public int getIndex(){
   return index;
  }

  public void onClick(DialogInterface dialog, int whichButton){
```

```
 setIndex(whichButton);
// Toast.makeText(Main.this, "您选择了： " + areas[index], Toast.LENGTH_LONG).show();
 switch (index)
 {
//一般模式下
   case 0:

   {
     //第 1 个子节点的亮度
     send(0x01,0x5a);

   //第 2 个子节点的亮度
       send(0x02,0x5a);

   //第 3 个子节点的亮度
       send(0x03,0x5a);

  //第 4 个子节点的亮度
       send(0x04,0x5a);
     Toast.makeText(Main.this,"您选择了： "+areas[index],Toast.LENGTH_LONG).show();
      dialog.dismiss();
 }
 break;


// 会议模式下
   case 1:
   {
       //第 1 路子节点
       send(0x01,0x23);

       //第 2 路子节点
       send(0x02,0x23);

      //第 3 路子节点
       send(0x03,0x23);

       //第 4 路子节点
       send(0x04,0x23);
       Toast.makeText(Main.this,"您选择了： "+areas[index],Toast.LENGTH_LONG).show();
       dialog.dismiss();
 }  break;


 //视频模式下
   case 2:
   {
       //第 1 路子节点
       send(0x01,0xc8);

       //第 2 路子节点
       send(0x02,0xc8);

      //第 3 路子节点
       send(0x03,0xc8);

       //第 4 路子节点
       send(0x04,0xc8);

       Toast.makeText(Main.this,"您选择了： "+areas[index],Toast.LENGTH_LONG).show();
       dialog.dismiss();
 }  break;


 //迎接模式下
   case 3:
   {
       //第 1 路子节点
       send(0x01,0x23);
```

```
        //第 2 路子节点
        TimerTask timerTask = new TimerTask() {
            @Override
            public void run()
            {
                    // 你要干的活
                send(0x02,0x23);
            }
    };
    timer.schedule(timerTask, 1000 * 3);
     //第 3 路子节点
    TimerTask timerTask1 = new TimerTask() {
        @Override
        public void run() {

                // 你要干的活
            send(0x03,0x23);

        }
    };
           timer.schedule(timerTask1, 1000 * 6); //两秒后执行

        //第 4 路子节点
     TimerTask timerTask2 = new TimerTask() {
          @Override
          public void run() {
                      // 你要干的活
              send(0x04,0x23);

          }};
          timer.schedule(timerTask2, 1000 * 9);
        Toast.makeText(Main.this,"您选择了: "+areas[index],Toast.LENGTH_LONG).show();
         dialog.dismiss();
  }   break;

  //返回
    case 4:

    {//Toast.makeText(Main.this,"您选择了:  "+areas[index],Toast.LENGTH_LONG).show();
     dialog.dismiss();  }
}


}

private void  send(int Room,int Grade){

    try {

          //String strpass="@@UP....";
         byte[] byteone=new byte[8];//=strpass.getBytes("US-ASCII");
             byteone[0]=(byte)0xf5;
             byteone[1]=(byte)0x5f;
             byteone[2]=(byte)0x00;
             byteone[3]=(byte)Room;
             byteone[4]=(byte)0x03;
             byteone[5]=(byte)0x00;
             byteone[6]=(byte)Grade;
             byteone[7]=(byte)0x06;

          //byte[] bytekai=strpass.getBytes("US-ASCII");
       tmpOut = BluetoothMain.btSocket.getOutputStream();
       tmpOut.write(byteone);}
    catch (IOException e) {
       Log.e("BluetoothReadService", "temp sockets not created", e);
   }
    }
}}
```

（2）编写文件 home.java，功能是监听用户触摸单击的选项来执行对应的处理程序，来到对应的 Activity 界面。文件 home.java 的具体实现代码如下所示。

```
public class home extends Activity {
    private ImageButton enterr,enterp,enterc,back;
    /** Called when the activity is first created. */
    @Override
    public void onCreate(Bundle savedInstanceState) {
        super.onCreate(savedInstanceState);
        requestWindowFeature(Window.FEATURE_NO_TITLE);
        setContentView(R.layout.home);
        getWindow().setFlags(WindowManager.LayoutParams.FLAG_FULLSCREEN,
        WindowManager.LayoutParams.FLAG_FULLSCREEN);
        enterr=(ImageButton) findViewById (R.id.Enterr);
        enterp=(ImageButton) findViewById (R.id.Enterp);
        enterc=(ImageButton) findViewById (R.id.Enterc);
        back=(ImageButton) findViewById (R.id.back);
        enterr.setOnClickListener(new Button.OnClickListener()
        {

            @Override
            public void onClick(View v) {
                // TODO Auto-generated method stub
                Intent intent = new Intent();
                intent.setClass(home.this, BluetoothMain.class);
                intent.setFlags(Intent.FLAG_ACTIVITY_CLEAR_TOP);  //注意本行的 FLAG 设置
                startActivity(intent);
                 //home.this.finish();
                overridePendingTransition(R.anim.zoomin,R.anim.zoomout);


            }

        }
                );

        enterp.setOnClickListener(new Button.OnClickListener()
        {
            @Override
            public void onClick(View v) {
                // TODO Auto-generated method stub
                Intent intent=new Intent();
                intent.setClass(home.this, product.class);
                startActivity(intent);
                home.this.finish();
                overridePendingTransition(R.anim.zoomin,R.anim.zoomout);

        }}
                );

        enterc.setOnClickListener(new Button.OnClickListener()

        {

            @Override
            public void onClick(View v) {
                // TODO Auto-generated method stub
                Intent intent=new Intent();
                intent.setClass(home.this, company.class);
                startActivity(intent);
                home.this.finish();
                overridePendingTransition(R.anim.zoomin,R.anim.zoomout);

        }}
                );

       back.setOnClickListener(new Button.OnClickListener()
        {
```

```
            @Override
            public void onClick(View v) {

                AlertDialog.Builder alertbBuilder=new AlertDialog.Builder(home.this);
                 //.setIcon(R.drawable.services)
                   alertbBuilder.setTitle(R.string.app_about)
                 /*设置弹出窗口的图式*/
                 //.setIcon(R.drawable.hot)
                 /*设置弹出窗口的信息*/
                 .setMessage(R.string.app_about_msg)
                 .setPositiveButton(R.string.str_ok,
                 new DialogInterface.OnClickListener()
                {
                public void onClick(DialogInterface dialoginterface, int i)
                 {
                  finish();/*关闭窗口*/
                 }
                }
               )
                 /*设置弹出窗口的返回事件*/
                 .setNegativeButton(R.string.str_no,
                  new DialogInterface.OnClickListener()
                  {
                public void onClick(DialogInterface dialoginterface, int i)
                {
                }
                 })
                .show();

            }
        });}
}
```

（3）编写文件 BluetoothMain.java，功能是根据用户的设置选项来设置系统的蓝牙参数，实现控制本系统蓝牙设备的功能。文件 BluetoothMain.java 的具体实现代码如下所示。

```
public class BluetoothMain extends Activity {
    static final String SPP_UUID = "00001101-0000-1000-8000-00805F9B34FB";

    Button btnSearch, btnDis, btnExit;
    ToggleButton tbtnSwitch;
    ListView lvBTDevices;
    ArrayAdapter<String> adtDevices;
    List<String> lstDevices = new ArrayList<String>();

    BluetoothAdapter btAdapt;
    public static BluetoothSocket btSocket;

    @Override
    public void onCreate(Bundle savedInstanceState) {
        super.onCreate(savedInstanceState);
        setContentView(R.layout.bluetooth);
        // Button 设置
        btnSearch = (Button) this.findViewById(R.id.btnSearch);
        btnSearch.setOnClickListener(new ClickEvent());
        btnExit = (Button) this.findViewById(R.id.btnExit);
        btnExit.setOnClickListener(new ClickEvent());
        btnDis = (Button) this.findViewById(R.id.btnDis);
        btnDis.setOnClickListener(new ClickEvent());

        // ToogleButton 设置
        tbtnSwitch = (ToggleButton) this.findViewById(R.id.tbtnSwitch);
        tbtnSwitch.setOnClickListener(new ClickEvent());

        // ListView 及其数据源适配器
        lvBTDevices = (ListView) this.findViewById(R.id.lvDevices);
        adtDevices = new ArrayAdapter<String>(BluetoothMain.this,
                android.R.layout.simple_list_item_1, lstDevices);
        lvBTDevices.setAdapter(adtDevices);
```

```
        //lvBTDevices.setOnItemClickListener(new ItemClickEvent());

        // 初始化本机蓝牙功能，读取蓝牙状态并显示
        btAdapt = BluetoothAdapter.getDefaultAdapter();
        if (btAdapt.getState() == BluetoothAdapter.STATE_OFF)
            tbtnSwitch.setChecked(false);
        else if (btAdapt.getState() == BluetoothAdapter.STATE_ON)
            tbtnSwitch.setChecked(true);

        // 注册 Receiver 来获取蓝牙设备相关的结果
        IntentFilter intent = new IntentFilter();
        intent.addAction(BluetoothDevice.ACTION_FOUND);// 用 BroadcastReceiver 来取得搜索结果
        intent.addAction(BluetoothDevice.ACTION_BOND_STATE_CHANGED);
        intent.addAction(BluetoothAdapter.ACTION_SCAN_MODE_CHANGED);
        intent.addAction(BluetoothAdapter.ACTION_STATE_CHANGED);
        registerReceiver(searchDevices, intent);

        if (btAdapt.getState() == BluetoothAdapter.STATE_OFF) {// 如果蓝牙还没开启
            Toast.makeText(BluetoothMain.this, "Bluetooth is openning, Just a minute ,
            please", 1000).show();
            btAdapt.enable();
            tbtnSwitch.setChecked(true);
        }
        setTitle("The Bluetooth address: " + btAdapt.getAddress());
        lstDevices.clear();
        btAdapt.startDiscovery();

    }


    private BroadcastReceiver searchDevices = new BroadcastReceiver() {

        public void onReceive(Context context, Intent intent) {
            String action = intent.getAction();
            Bundle b = intent.getExtras();
            Object[] lstName = b.keySet().toArray();

            // 显示所有收到的消息及其细节
            for (int i = 0; i < lstName.length; i++) {
                String keyName = lstName[i].toString();
                Log.e(keyName, String.valueOf(b.get(keyName)));
            }

            //搜索设备时，取得设备的 MAC 地址
            if (BluetoothDevice.ACTION_FOUND.equals(action)) {
                BluetoothDevice device = intent
                        .getParcelableExtra(BluetoothDevice.EXTRA_DEVICE);
                String str= device.getName() + "|" + device.getAddress();
                String str1 = device.getAddress();
            //if(str1.equals("00:11:08:01:06:78"))  //用来判断是否为所需要的蓝牙
            //{

                if (lstDevices.indexOf(str) == -1)// 防止重复添加
                    lstDevices.add(str); // 获取设备名称和 MAC 地址
                adtDevices.notifyDataSetChanged();
                /* 添加判断，如果为已配对或者已存 MAC 地址的设备
                 * 自动进行连接，并跳转到另一个 Activity
                 * */
                if(true){   //未添加条件
                    btAdapt.cancelDiscovery();
                    //String str = lstDevices.get();
                    //String[] values = str.split("\\|");
                    //String address=values[1];
                    //Log.e("address",values[1]);
                    UUID uuid = UUID.fromString(SPP_UUID);
                    BluetoothDevice btDev = btAdapt.getRemoteDevice(device.getAddress());
                    try {
                        btSocket = btDev
```

```
                    .createRfcommSocketToServiceRecord(uuid);
                btSocket.connect();

                Intent intent1 = new Intent();
                intent1.setClass(BluetoothMain.this, Main.class);
                startActivity(intent1);
                overridePendingTransition(R.anim.zoomin,R.anim.zoomout);
            } catch (IOException e) {
                // TODO Auto-generated catch block
                e.printStackTrace();
            }
        }


    //}
    }
  }
};

@Override
protected void onDestroy() {
    this.unregisterReceiver(searchDevices);
    super.onDestroy();
    android.os.Process.killProcess(android.os.Process.myPid());
}
class ClickEvent implements View.OnClickListener {
    @Override
    public void onClick(View v) {
        if (v == btnSearch)// 搜索蓝牙设备，在 BroadcastReceiver 显示结果
        {

          if (btAdapt.getState() == BluetoothAdapter.STATE_OFF) {// 如果蓝牙还没开启
              Toast.makeText(BluetoothMain.this, "请先打开蓝牙", 1000).show();
              return;
          }

          setTitle("本机蓝牙地址：" + btAdapt.getAddress());
          lstDevices.clear();
          btAdapt.startDiscovery();
        } else if (v == tbtnSwitch) {// 本机蓝牙启动/关闭
          if (btAdapt.getState() == BluetoothAdapter.STATE_OFF){
              btAdapt.enable();
              tbtnSwitch.setChecked(true);
          }

          else if (btAdapt.getState() == BluetoothAdapter.STATE_ON){
              btAdapt.disable();
              tbtnSwitch.setChecked(false);
          }
        } else if (v == btnDis)// 本机可以被搜索
        {
          Intent discoverableIntent = new Intent(
                  BluetoothAdapter.ACTION_REQUEST_DISCOVERABLE);
          discoverableIntent.putExtra(
                  BluetoothAdapter.EXTRA_DISCOVERABLE_DURATION, 300);
          startActivity(discoverableIntent);
        } else if (v == btnExit) {
          try {
              if (btSocket != null)
                  btSocket.close();
          } catch (IOException e) {
              e.printStackTrace();
          }
         if (btAdapt.getState() == BluetoothAdapter.STATE_ON){
              btAdapt.disable();
         }
          BluetoothMain.this.finish();
        }

    }
}
```

（4）编写文件 Firstpage.java，功能是监听用户在第 1 路调光单选按钮列表中的选择值，根据这个值来控制光线的亮度。文件 Firstpage.java 的具体实现代码如下所示。

```
public class Firstpage extends Activity {
    private ImageButton imagebutton;
    private ImageButton ibutton1_o,ibutton1_c,ibutton2_o,ibutton2_c,ibutton3_o,
    ibutton3_c,ibutton4_o,ibutton4_c,btnallop,btnallcl;
    RadioGroup radiogroup0,radiogroup1,radiogroup2,radiogroup3;
    RadioButton radio1,radio2,radio3,radio4,radio5,radio6;
    RadioButton radio11,radio12,radio13,radio14,radio15,radio16;
    RadioButton radio21,radio22,radio23,radio24,radio25,radio26;
    RadioButton radio31,radio32,radio33,radio34,radio35,radio36;
    OutputStream tmpOut = null;
    @Override
    protected void onCreate(Bundle savedInstanceState) {
        super.onCreate(savedInstanceState);
        setContentView(R.layout.first);
        //全开按钮
        btnallop=(ImageButton) findViewById(R.id.btnopen);
        btnallop.setOnClickListener(new Button.OnClickListener(){
            @Override
            public void onClick(View v)
            {
                //第 1 路灯亮
                send(0x01,0x23);

                //第 2 路灯亮
                send(0x02,0x23);

                //第 3 路灯亮
                send(0x03,0x23);

                //第 4 路灯亮
                send(0x04,0x23);


            }});

        //全关按钮

        btnallcl=(ImageButton) findViewById(R.id.btnclose);
        btnallcl.setOnClickListener(new Button.OnClickListener(){
            @Override
            public void onClick(View v)
            {
                //第 1 路灯亮
                send(0x01,0xff);

                //第 2 路灯亮
                send(0x02,0xff);
                //第 3 路灯亮
                send(0x03,0xff);
                //第 4 路灯亮
                send(0x04,0xff);

            }
            });
        //产品介绍按钮
        imagebutton=(ImageButton) findViewById(R.id.imageButton1);
        imagebutton.setOnClickListener(new Button.OnClickListener(){
            @Override
            public void onClick(View v)
            {
                 LayoutInflater inflater = getLayoutInflater();
                View layout = inflater.inflate(R.layout.dialog,
                (ViewGroup) findViewById(R.id.dialog));

                 new AlertDialog.Builder(Firstpage.this)
                 .setTitle("Introduce")
```

```
                .setView(layout)
                .setPositiveButton("Return", null)
                //.setNegativeButton("取消", null)
                .show();
            } });

    //第 1 子节点的占空比
        radiogroup0=(RadioGroup)findViewById(R.id.radioGroup2);
        radio1=(RadioButton)findViewById(R.id.radio11);
        radio2=(RadioButton)findViewById(R.id.radio12);
        radio3=(RadioButton)findViewById(R.id.radio13);
        radio4=(RadioButton)findViewById(R.id.radio14);
        radio5=(RadioButton)findViewById(R.id.radio15);
        radio6=(RadioButton)findViewById(R.id.radio16);
        radiogroup0.setOnCheckedChangeListener(new RadioGroup.OnCheckedChangeListener() {

            @Override
            public void onCheckedChanged(RadioGroup group, int checkedId) {
                // TODO Auto-generated method stub

                switch (checkedId)
                {
                case R.id.radio11:
                     /*DisplayToast("灯光亮度 0%! ");*/
                    Toast.makeText(Firstpage.this, "第 1 路灯光亮度为 0%! ",
                            Toast.LENGTH_SHORT).show();
                    {
                    send(0x01,0xff);

                }
                    break;
                case R.id.radio12:
                   // DisplayToast("灯光亮度 20%! ");
                    Toast.makeText(Firstpage.this, "第 1 路灯光亮度为 20%! ",
                              Toast.LENGTH_SHORT).show();
                    {
                    send(0x01,0xc8);

                 }
                  break;
                case R.id.radio13:
                  //DisplayToast("灯光亮度 40%! ");
                    Toast.makeText(Firstpage.this, "第 1 路灯光亮度为 40%! ",
                              Toast.LENGTH_SHORT).show();
                    {
                    send(0x01,0x91);

                  }
                  break;
                case R.id.radio14:
                    //DisplayToast("灯光亮度 60%! ");
                    Toast.makeText(Firstpage.this, "第 1 路灯光亮度为 60%! ",
                              Toast.LENGTH_SHORT).show();
                    {
                    send(0x01,0x5a);

                  }
                  break;
                case R.id.radio15:
                   // DisplayToast("灯光亮度 80%! ");
                    Toast.makeText(Firstpage.this, "第 1 路灯光亮度为 80%! ",
                              Toast.LENGTH_SHORT).show();
                    {
                    send(0x01,0x23);

                  }
                  break;
                default:
                   // DisplayToast("灯光亮度 100%! ");
```

```
            Toast.makeText(Firstpage.this, "第 1 路灯光亮度为 100%! ",
                       Toast.LENGTH_SHORT).show();
        {
            send(0x01,0x00);

        }
          break;
        }

      }
  });
//第 2 子节点的占空比
  radiogroup1=(RadioGroup)findViewById(R.id.radioGroup3);
  radio11=(RadioButton)findViewById(R.id.radio21);
  radio12=(RadioButton)findViewById(R.id.radio22);
  radio13=(RadioButton)findViewById(R.id.radio23);
  radio14=(RadioButton)findViewById(R.id.radio24);
  radio15=(RadioButton)findViewById(R.id.radio25);
  radio16=(RadioButton)findViewById(R.id.radio26);
  radiogroup1.setOnCheckedChangeListener(new RadioGroup.OnCheckedChangeListener() {

      @Override
      public void onCheckedChanged(RadioGroup group, int checkedId) {
          // TODO Auto-generated method stub

          switch (checkedId)
          {
          case R.id.radio21:
                /*DisplayToast("灯光亮度 0%! ");*/
              Toast.makeText(Firstpage.this, "第 2 路灯光亮度为 0%! ",
                      Toast.LENGTH_SHORT).show();
          {
              send(0x02,0xff);

        }
              break;
          case R.id.radio22:
             // DisplayToast("灯光亮度 20%! ");
              Toast.makeText(Firstpage.this, "第 2 路灯光亮度为 20%! ",
                         Toast.LENGTH_SHORT).show();
          {
              send(0x02,0xc8);

        }
            break;
          case R.id.radio23:
             //DisplayToast("灯光亮度 40%! ");
              Toast.makeText(Firstpage.this, "第 3 路灯光亮度为 40%! ",
                         Toast.LENGTH_SHORT).show();
          {
              send(0x02,0x91);

          }
            break;
          case R.id.radio24:
              //DisplayToast("灯光亮度 60%! ");
              Toast.makeText(Firstpage.this, "第 2 路灯光亮度为 60%! ",
                         Toast.LENGTH_SHORT).show();
          {
              send(0x02,0x5a);

          }
            break;
          case R.id.radio25:
             // DisplayToast("灯光亮度 80%! ");
              Toast.makeText(Firstpage.this, "第 2 路灯光亮度为 80%! ",
                         Toast.LENGTH_SHORT).show();
          {
              send(0x02,0x23);
```

```
                }
                  break;
                default:
                   // DisplayToast("灯光亮度 100%! ");
                    Toast.makeText(Firstpage.this, "第 2 路灯光亮度为 100%! ",
                                 Toast.LENGTH_SHORT).show();
                {
                    send(0x02,0x00);

                }
                  break;
                }

            }
        });

  //第 3 子节点的占空比
    radiogroup2=(RadioGroup)findViewById(R.id.radioGroup4);
    radio21=(RadioButton)findViewById(R.id.radio31);
    radio22=(RadioButton)findViewById(R.id.radio32);
    radio23=(RadioButton)findViewById(R.id.radio33);
    radio24=(RadioButton)findViewById(R.id.radio34);
    radio25=(RadioButton)findViewById(R.id.radio35);
    radio26=(RadioButton)findViewById(R.id.radio36);
    radiogroup2.setOnCheckedChangeListener(new RadioGroup.OnCheckedChangeListener() {

        @Override
        public void onCheckedChanged(RadioGroup group, int checkedId) {
            // TODO Auto-generated method stub

            switch (checkedId)
            {
            case R.id.radio31:
                  /*DisplayToast("灯光亮度 0%! ");*/
                    Toast.makeText(Firstpage.this, "第 3 路灯光亮度为 0%! ",
                              Toast.LENGTH_SHORT).show();
            {
                    send(0x03,0xff);

              }
                    break;
            case R.id.radio32:
                 // DisplayToast("灯光亮度 20%! ");
                    Toast.makeText(Firstpage.this, "第 3 路灯光亮度为 20%! ",
                                 Toast.LENGTH_SHORT).show();
            {
                    send(0x03,0xc8);

              }
                  break;
            case R.id.radio33:
                 //DisplayToast("灯光亮度 40%! ");
                    Toast.makeText(Firstpage.this, "第 3 路灯光亮度为 40%! ",
                                 Toast.LENGTH_SHORT).show();
            {
                    send(0x03,0x91);

              }
                  break;
            case R.id.radio34:
                  //DisplayToast("灯光亮度 60%! ");
                    Toast.makeText(Firstpage.this, "第 3 路灯光亮度为 60%! ",
                                 Toast.LENGTH_SHORT).show();
            {
                    send(0x03,0x5a);

            }
                  break;
```

```
          case R.id.radio35:
            // DisplayToast("灯光亮度 80%! ");
             Toast.makeText(Firstpage.this, "第 3 路灯光亮度为 80%! ",
                        Toast.LENGTH_SHORT).show();
          {
             send(0x03,0x23);

         }
           break;
          default:
            // DisplayToast("灯光亮度 100%! ");
             Toast.makeText(Firstpage.this, "第 3 路灯光亮度为 100%! ",
                        Toast.LENGTH_SHORT).show();
          {
             send(0x03,0x00);

          }
           break;
          }

      }
  });   //第 4 路

//第 4 子节点的占空比
  radiogroup3=(RadioGroup)findViewById(R.id.radioGroup5);
  radio31=(RadioButton)findViewById(R.id.radio41);
  radio32=(RadioButton)findViewById(R.id.radio42);
  radio33=(RadioButton)findViewById(R.id.radio43);
  radio34=(RadioButton)findViewById(R.id.radio44);
  radio35=(RadioButton)findViewById(R.id.radio45);
  radio36=(RadioButton)findViewById(R.id.radio46);
  radiogroup3.setOnCheckedChangeListener(new RadioGroup.OnCheckedChangeListener() {

      @Override
      public void onCheckedChanged(RadioGroup group, int checkedId) {
          // TODO Auto-generated method stub

          switch (checkedId)
          {
          case R.id.radio41:
               /*DisplayToast("灯光亮度 0%! ");*/
             Toast.makeText(Firstpage.this, "第 4 路灯光亮度为 0%! ",
                     Toast.LENGTH_SHORT).show();
          {
             send(0x04,0xff);

        }
             break;
          case R.id.radio42:
            // DisplayToast("灯光亮度 20%! ");
             Toast.makeText(Firstpage.this, "第 4 路灯光亮度为 20%! ",
                        Toast.LENGTH_SHORT).show();
          {
             send(0x04,0xc8);

         }
           break;
          case R.id.radio43:
            //DisplayToast("灯光亮度 40%! ");
             Toast.makeText(Firstpage.this, "第 4 路灯光亮度为 40%! ",
                        Toast.LENGTH_SHORT).show();
          {
             send(0x04,0x91);

          }
           break;
          case R.id.radio44:
             //DisplayToast("灯光亮度 60%! ");
             Toast.makeText(Firstpage.this, "第 4 路灯光亮度为 60%! ",
```

```
                    Toast.LENGTH_SHORT).show();
            {
              send(0x04,0x5a);

            }
              break;
            case R.id.radio45:
              // DisplayToast("灯光亮度 80%! ");
              Toast.makeText(Firstpage.this, "第 4 路灯光亮度为 80%! ",
                    Toast.LENGTH_SHORT).show();
            {
              send(0x04,0x23);

            }
              break;
            default:
              // DisplayToast("灯光亮度 100%! ");
              Toast.makeText(Firstpage.this, "第 4 路灯光亮度为 100%! ",
                    Toast.LENGTH_SHORT).show();
            {
              send(0x04,0x00);

            }
              break;
            }

        }
    });


    ibutton1_o=(ImageButton) findViewById(R.id.imageButton2);
    ibutton1_o.setOnClickListener(new ClickEventKey());
    ibutton1_c=(ImageButton) findViewById(R.id.imageButton3);
    ibutton1_c.setOnClickListener(new ClickEventKey());
    ibutton2_o=(ImageButton) findViewById(R.id.imageButton4);
    ibutton2_o.setOnClickListener(new ClickEventKey());
    ibutton2_c=(ImageButton) findViewById(R.id.imageButton5);
    ibutton2_c.setOnClickListener(new ClickEventKey());
    ibutton3_o=(ImageButton) findViewById(R.id.imageButton6);
    ibutton3_o.setOnClickListener(new ClickEventKey());
    ibutton3_c=(ImageButton) findViewById(R.id.imageButton7);
    ibutton3_c.setOnClickListener(new ClickEventKey());
    ibutton4_o=(ImageButton) findViewById(R.id.imageButton8);
    ibutton4_o.setOnClickListener(new ClickEventKey());
    ibutton4_c=(ImageButton) findViewById(R.id.imageButton9);
    ibutton4_c.setOnClickListener(new ClickEventKey());
}


class ClickEventKey implements View.OnClickListener {
    @Override
    public void onClick(View v) {
    //    if (v == ibutton1_o)
        switch (v.getId())

        {
        case R.id.imageButton2:
        {
            send(0x01,0x23);
        }

        break;
        case R.id.imageButton3:

        {
            send(0x01,0xff);
```

```
                }

                break;
                case R.id.imageButton4:
                {
                    send(0x02,0x23);
                }

                break;
                case R.id.imageButton5:
                {
                    send(0x02,0xff);
                }

                break;
                 case R.id.imageButton6:
                {
                    send(0x03,0x23);
                }

                break;
                 case R.id.imageButton7:
                {
                    send(0x03,0xff);
                }

                break;
                 case R.id.imageButton8:
                {
                    send(0x04,0x23);
                }

                break;
                 case R.id.imageButton9:
                {
                    send(0x04,0xff);
                }

                break;
            }
                }
            }
private void  send(int Room,int Grade){

        try {

                //String strpass="@@UP....";
            byte[] byteone=new byte[8];//=strpass.getBytes("US-ASCII");
                byteone[0]=(byte)0xf5;
                byteone[1]=(byte)0x5f;
                byteone[2]=(byte)0x00;
                byteone[3]=(byte)Room;
                byteone[4]=(byte)0x03;
                byteone[5]=(byte)0x00;
                byteone[6]=(byte)Grade;
                byteone[7]=(byte)0x06;

                //byte[] bytekai=strpass.getBytes("US-ASCII");
            tmpOut = BluetoothMain.btSocket.getOutputStream();
            tmpOut.write(byteone);}
          catch (IOException e) {
            Log.e("BluetoothReadService", "temp sockets not created", e);
        }
      }
    }
```

（5）编写文件Secondpage.java，功能是监听用户在第5路调光单选按钮列表中的选择值，根据这个值来控制光线的亮度。文件Secondpage.java的具体实现代码如下所示。

```
public class Secondpage extends Activity {
    private ImageButton imagebutton;
    private ImageButton ibutton5_o,ibutton5_c,ibutton6_o,ibutton6_c,ibutton7_o,
    ibutton7_c,ibutton8_o,ibutton8_c;
    RadioGroup radiogroup0,radiogroup1,radiogroup2,radiogroup3;
    RadioButton radio1,radio2,radio3,radio4,radio5,radio6;
    RadioButton radio11,radio12,radio13,radio14,radio15,radio16;
    RadioButton radio21,radio22,radio23,radio24,radio25,radio26;
    RadioButton radio31,radio32,radio33,radio34,radio35,radio36;
    OutputStream tmpOut = null;
    @Override
    protected void onCreate(Bundle savedInstanceState) {
        super.onCreate(savedInstanceState);
        setContentView(R.layout.second);
        imagebutton=(ImageButton) findViewById(R.id.imageButton1);
        imagebutton.setOnClickListener(new Button.OnClickListener(){
            @Override
            public void onClick(View v)
            {
                LayoutInflater inflater = getLayoutInflater();
                View layout = inflater.inflate(R.layout.dialog,
                (ViewGroup) findViewById(R.id.dialog));

                 new AlertDialog.Builder(Secondpage.this)
                 .setTitle("Introduce")
                 .setView(layout)
                 .setPositiveButton("Return", null)
                 //.setNegativeButton("取消", null)
                 .show();
                 } });

        radiogroup0=(RadioGroup)findViewById(R.id.radioGroup2);
        radio1=(RadioButton)findViewById(R.id.radio11);
        radio2=(RadioButton)findViewById(R.id.radio12);
        radio3=(RadioButton)findViewById(R.id.radio13);
        radio4=(RadioButton)findViewById(R.id.radio14);
        radio5=(RadioButton)findViewById(R.id.radio15);
        radio6=(RadioButton)findViewById(R.id.radio16);
        radiogroup0.setOnCheckedChangeListener(new RadioGroup.OnCheckedChangeListener() {

            @Override
            public void onCheckedChanged(RadioGroup group, int checkedId) {
                // TODO Auto-generated method stub

                switch (checkedId)
                {
                case R.id.radio11:
                    /*DisplayToast("灯光亮度 0%! ");*/
                  Toast.makeText(Secondpage.this, "第 5 路灯光亮度为 0%! ",
                          Toast.LENGTH_SHORT).show();
                {
                  send(0x05,0xff);
                }
                  break;
                case R.id.radio12:
                   // DisplayToast("灯光亮度 20%! ");
                    Toast.makeText(Secondpage.this, "第 5 路灯光亮度为 20%! ",
                            Toast.LENGTH_SHORT).show();
                {
                   send(0x05,0xc8);
                }
                 break;
                case R.id.radio13:
                   //DisplayToast("灯光亮度 40%! ");
                    Toast.makeText(Secondpage.this, "第 5 路灯光亮度为 40%! ",
                            Toast.LENGTH_SHORT).show();
                {
                   send(0x05,0x91);
                 }
```

```
            break;
          case R.id.radio14:
            //DisplayToast("灯光亮度 60%! ");
            Toast.makeText(Secondpage.this, "第 5 路灯光亮度为 60%! ",
                       Toast.LENGTH_SHORT).show();
          {
            send(0x05,0x5a);
          }
            break;
          case R.id.radio15:
           // DisplayToast("灯光亮度 80%! ");
            Toast.makeText(Secondpage.this, "第 5 路灯光亮度为 80%! ",
                       Toast.LENGTH_SHORT).show();
          {
            send(0x05,0x23);
          }
            break;
          default:
           // DisplayToast("灯光亮度 100%! ");
            Toast.makeText(Secondpage.this, "第 1 路灯光亮度为 100%! ",
                       Toast.LENGTH_SHORT).show();
          {
            send(0x05,0x00);
          }
            break;
          }

      }
  });

  radiogroup1=(RadioGroup)findViewById(R.id.radioGroup3);
  radio11=(RadioButton)findViewById(R.id.radio21);
  radio12=(RadioButton)findViewById(R.id.radio22);
  radio13=(RadioButton)findViewById(R.id.radio23);
  radio14=(RadioButton)findViewById(R.id.radio24);
  radio15=(RadioButton)findViewById(R.id.radio25);
  radio16=(RadioButton)findViewById(R.id.radio26);
  radiogroup1.setOnCheckedChangeListener(new RadioGroup.OnCheckedChangeListener() {

      @Override
      public void onCheckedChanged(RadioGroup group, int checkedId) {
          // TODO Auto-generated method stub

          switch (checkedId)
          {
          case R.id.radio21:
               /*DisplayToast("灯光亮度 0%! ");*/
            Toast.makeText(Secondpage.this, "第 6 路灯光亮度为 0%! ",
                     Toast.LENGTH_SHORT).show();
          {
            send(0x06,0xff);
            }
            break;
          case R.id.radio22:
           // DisplayToast("灯光亮度 20%! ");
            Toast.makeText(Secondpage.this, "第 6 路灯光亮度为 20%! ",
                       Toast.LENGTH_SHORT).show();
          {
            send(0x06,0xc8);
            }
            break;
          case R.id.radio23:
           //DisplayToast("灯光亮度 40%! ");
            Toast.makeText(Secondpage.this, "第 6 路灯光亮度为 40%! ",
                       Toast.LENGTH_SHORT).show();
          {
            send(0x06,0x91);
            }
            break;
```

```
            case R.id.radio24:
                //DisplayToast("灯光亮度 60%! ");
                Toast.makeText(Secondpage.this, "第 6 路灯光亮度为 60%! ",
                        Toast.LENGTH_SHORT).show();
            {
                send(0x06,0x5a);
              }
              break;
            case R.id.radio25:
              // DisplayToast("灯光亮度 80%! ");
                Toast.makeText(Secondpage.this, "第 6 路灯光亮度为 80%! ",
                        Toast.LENGTH_SHORT).show();
            {
                send(0x06,0x23);
            }
              break;
            default:
              // DisplayToast("灯光亮度 100%! ");
                Toast.makeText(Secondpage.this, "第 6 路灯光亮度为 100%! ",
                        Toast.LENGTH_SHORT).show();
            {
                send(0x06,0x00);
            }
              break;
            }

        }
    });


    radiogroup2=(RadioGroup)findViewById(R.id.radioGroup4);
    radio21=(RadioButton)findViewById(R.id.radio31);
    radio22=(RadioButton)findViewById(R.id.radio32);
    radio23=(RadioButton)findViewById(R.id.radio33);
    radio24=(RadioButton)findViewById(R.id.radio34);
    radio25=(RadioButton)findViewById(R.id.radio35);
    radio26=(RadioButton)findViewById(R.id.radio36);
    radiogroup2.setOnCheckedChangeListener(new RadioGroup.OnCheckedChangeListener() {

        @Override
        public void onCheckedChanged(RadioGroup group, int checkedId) {
            // TODO Auto-generated method stub

            switch (checkedId)
            {
            case R.id.radio31:
                  /*DisplayToast("灯光亮度 0%! ");*/
                Toast.makeText(Secondpage.this, "第 7 路灯光亮度为 0%! ",
                        Toast.LENGTH_SHORT).show();
            {

                send(0x07,0xff);
              }
                break;
            case R.id.radio32:
              // DisplayToast("灯光亮度 20%! ");
                Toast.makeText(Secondpage.this, "第 7 路灯光亮度为 20%! ",
                        Toast.LENGTH_SHORT).show();
            {
                send(0x07,0xc8);
              }
              break;
            case R.id.radio33:
              //DisplayToast("灯光亮度 40%! ");
                Toast.makeText(Secondpage.this, "第 7 路灯光亮度为 40%! ",
                        Toast.LENGTH_SHORT).show();
            {
                send(0x07,0x91);
              }
```

```
        break;
      case R.id.radio34:
         //DisplayToast("灯光亮度 60%! ");
         Toast.makeText(Secondpage.this, "第 7 路灯光亮度为 60%! ",
                  Toast.LENGTH_SHORT).show();
      {
         send(0x07,0x5a);
        }
        break;
      case R.id.radio35:
        // DisplayToast("灯光亮度 80%! ");
         Toast.makeText(Secondpage.this, "第 7 路灯光亮度为 80%! ",
                  Toast.LENGTH_SHORT).show();
      {
         send(0x07,0x23);
        }
        break;
      default:
        // DisplayToast("灯光亮度 100%! ");
         Toast.makeText(Secondpage.this, "第 7 路灯光亮度为 100%! ",
                  Toast.LENGTH_SHORT).show();
      {
         send(0x07,0x00);
      }
        break;
      }

    }
});  //第 8 路


radiogroup3=(RadioGroup)findViewById(R.id.radioGroup5);
radio31=(RadioButton)findViewById(R.id.radio41);
radio32=(RadioButton)findViewById(R.id.radio42);
radio33=(RadioButton)findViewById(R.id.radio43);
radio34=(RadioButton)findViewById(R.id.radio44);
radio35=(RadioButton)findViewById(R.id.radio45);
radio36=(RadioButton)findViewById(R.id.radio46);
radiogroup3.setOnCheckedChangeListener(new RadioGroup.OnCheckedChangeListener() {

    @Override
    public void onCheckedChanged(RadioGroup group, int checkedId) {
      // TODO Auto-generated method stub

      switch (checkedId)
      {
      case R.id.radio41:
            /*DisplayToast("灯光亮度 0%! ");*/
         Toast.makeText(Secondpage.this, "第 8 路灯光亮度为 0%! ",
               Toast.LENGTH_SHORT).show();
      {
         send(0x08,0xff);
      }
         break;
      case R.id.radio42:
        // DisplayToast("灯光亮度 20%! ");
         Toast.makeText(Secondpage.this, "第 8 路灯光亮度为 20%! ",
                  Toast.LENGTH_SHORT).show();
      {
         send(0x08,0xc8);
        }
        break;
      case R.id.radio43:
        //DisplayToast("灯光亮度 40%! ");
         Toast.makeText(Secondpage.this, "第 8 路灯光亮度为 40%! ",
                  Toast.LENGTH_SHORT).show();
      {
         send(0x08,0x91);
        }
```

```
                break;
              case R.id.radio44:
                 //DisplayToast("灯光亮度 60%! ");
                 Toast.makeText(Secondpage.this, "第 8 路灯光亮度为 60%! ",
                         Toast.LENGTH_SHORT).show();
              {
                 send(0x08,0x5a);
                }
                break;
              case R.id.radio45:
                 // DisplayToast("灯光亮度 80%! ");
                 Toast.makeText(Secondpage.this, "第 8 路灯光亮度为 80%! ",
                         Toast.LENGTH_SHORT).show();
              {
                 send(0x08,0x23);
                }
                break;
              default:
                 // DisplayToast("灯光亮度 100%! ");
                 Toast.makeText(Secondpage.this, "第 8 路灯光亮度为 100%! ",
                         Toast.LENGTH_SHORT).show();
              {
                 send(0x08,0x00);
                }
                break;
              }

          }
      });

      ibutton5_o=(ImageButton) findViewById(R.id.imageButton2);
      ibutton5_o.setOnClickListener(new ClickEventKey());
      ibutton5_c=(ImageButton) findViewById(R.id.imageButton3);
      ibutton5_c.setOnClickListener(new ClickEventKey());
      ibutton6_o=(ImageButton) findViewById(R.id.imageButton4);
      ibutton6_o.setOnClickListener(new ClickEventKey());
      ibutton6_c=(ImageButton) findViewById(R.id.imageButton5);
      ibutton6_c.setOnClickListener(new ClickEventKey());
      ibutton7_o=(ImageButton) findViewById(R.id.imageButton6);
      ibutton7_o.setOnClickListener(new ClickEventKey());
      ibutton7_c=(ImageButton) findViewById(R.id.imageButton7);
      ibutton7_c.setOnClickListener(new ClickEventKey());
      ibutton8_o=(ImageButton) findViewById(R.id.imageButton8);
      ibutton8_o.setOnClickListener(new ClickEventKey());
      ibutton8_c=(ImageButton) findViewById(R.id.imageButton9);
      ibutton8_c.setOnClickListener(new ClickEventKey());
  }

  class ClickEventKey implements View.OnClickListener {
      @Override
      public void onClick(View v) {
      //    if (v == ibutton1_o)
          switch (v.getId())

          {
          case R.id.imageButton2:
          {
              send(0x05,0x23);
             }

          break;
          case R.id.imageButton3:
             {
              send(0x05,0xff);
          }

          break;
          case R.id.imageButton4:
          {
```

```
            send(0x06,0x23);
          }

        break;
        case R.id.imageButton5:
        {
            send(0x06,0xff);
        }

        break;
         case R.id.imageButton6:
        {
            send(0x07,0x23);
        }

        break;
         case R.id.imageButton7:
        {
            send(0x07,0xff);
        }

        break;
         case R.id.imageButton8:
        {
            send(0x08,0x23);
        }

        break;
         case R.id.imageButton9:
        {
            send(0x08,0xff);
        }

        break;
        }
        }
    }
private void  send(int Room,int Grade){

        try {

            //String strpass="@@UP....";
         byte[] byteone=new byte[8];//=strpass.getBytes("US-ASCII");
             byteone[0]=(byte)0xf5;
             byteone[1]=(byte)0x5f;
             byteone[2]=(byte)0x00;
             byteone[3]=(byte)Room;
             byteone[4]=(byte)0x03;
             byteone[5]=(byte)0x00;
             byteone[6]=(byte)Grade;
             byteone[7]=(byte)0x06;

            //byte[] bytekai=strpass.getBytes("US-ASCII");
          tmpOut = BluetoothMain.btSocket.getOutputStream();
          tmpOut.write(byteone);}
       catch (IOException e) {
          Log.e("BluetoothReadService", "temp sockets not created", e);
      }
  }
}
```

到此为止，本实例的主要功能模块的实现过程介绍完毕。有关蓝牙方面的知识，将在本书后面的章节中进行讲解。本实例执行后的效果读者自行演示。

第9章　接近警报传感器详解

本章将详细讲解在Android设备中使用接近警报技术的基本知识。

9.1　类 Geocoder 详解

在Android系统中，可以使用LocationManager来设置接近警报功能。此功能和本书前面讲解的地图定位功能类似，但是可以在设备进入或离开某一个指定区域时发送通知应用，而并不是在新位置时才发送通知程序。本节将详细讲解在Android系统中实现接近警报应用的方法。

9.1.1　类 Geocoder 基础

在现实世界中，地图和定位服务通常使用经纬度来精确地指出地理位置。在Android 系统中，提供了地理编码类Geocoder来转换经纬度和现实世界的地址。地理编码是一个街道、地址或者其他位置（经度、纬度）转化为坐标的过程。反向地理编码是将坐标转换为地址（经度、纬度）的过程。一组反向地理编码结果间可能会有所差异。例如在一个结果中可能包含最临近建筑的完整街道地址，而另一个可能只包含城市名称和邮政编码。Geocoder要求的后端服务并没有包含在基本的Android框架中。如果没有此后端服务，执行Geocoder的查询方法将返回一个空列表。使用isPresent()方法，以确定Geocoder是否能够正常执行。

在Android系统中，类Geocoder的继承关系如下所示。

```
public final class Geocoder extends Object
java.lang.Object
android.location.Geocoder
```

9.1.2　公共构造器

在Android 系统中，类Geocoder包含了如下所示的公共构造器。

（1）public Geocoder(Context context, Local local)：功能是根据给定的语言环境构造一个Geocoder对象。各个参数的具体说明如下所示。

- context：当前的上下文对象。
- local the：当前语言环境。

（2）public Geocoder(Context context)：功能是根据给定的系统默认语言环境构造一个Geocoder对象。参数context表示当前的上下文对象。

9.1.3　公共方法

在Android系统中，类Geocoder包含了如下所示的公共方法。

（1）public List<Address> getFromLocation(double latitude, double longitude, int maxResults)：功

能是根据给定的经纬度返回一个描述此区域的地址数组。返回的地址将根据构造器提供的语言环境进行本地化。

返回值：一组地址对象，如果没找到匹配项，或者后台服务无效的话则返回 null 或者空序列。也可能通过网络获取，返回结果是一个最好的估计值，但不能保证其完全正确。

各个参数的具体说明如下所示。

- latitude：纬度。
- longitude：经度。
- maxResults：要返回的最大结果数，推荐 1～5。

包含的异常如下所示。

- IllegalArgumentException：纬度小于-90 或者大于 90。
- IllegalArgumentException：经度小于-180 或者大于 180。
- IOException：如果没有网络或者 IO 错误。

（2）public List<Address> getFromLocationName(String locationName, int maxResults, double lowerLeftLatitude, double lowerLeftLongitude, double upperRightLatitude, double upperRightLongitude)：功能是返回一个由给定的位置名称参数所描述的地址数组。名称参数可以是一个位置名称，例如“Dalvik, Iceland”，也可以是一个地址，例如“1600 Amphitheatre Parkway, Mountain View, CA”，也可以是一个机场代号，例如“SFO”，……，返回的地址将根据构造器提供的语言环境进行本地化。

也可以指定一个搜索边界框，该边界框由左下方坐标经纬度和右上方坐标经纬度确定。

返回值是一组地址对象，如果没找到匹配项，或者后台服务无效的话则返回 null 或者空序列。也有可能是通过网络获取。返回结果是一个最好的估计值，但不能保证其完全正确。通过 UI 主线程的后台线程来调用这个方法可能更加有用。

各个参数的具体说明如下所示。

- locationName：用户提供的位置描述。
- maxResults：要返回的最大结果数，推荐 1～5。
- lowerLeftLatitude：左下角纬度，用来设定矩形范围。
- lowerLeftLongitude：左下角经度，用来设定矩形范围。
- upperRightLatitude：右上角纬度，用来设定矩形范围。
- upperRightLongitude：右上角经度，用来设定矩形范围。

（3）public List<Address> getFromLocationName(String locationName, int maxResults)：功能是返回一个由给定的位置名称参数所描述的地址数组。名称参数可以是一个位置名称，例如“Dalvik, Iceland”，也可以是一个地址，例如“1600 Amphitheatre Parkway, Mountain View, CA”，也可以是一个机场代号，例如“SFO”，……，返回的地址将根据构造器提供的语言环境进行本地化。在现实应用中，通过 UI 主线程的后台线程来调用这个方法可能会更加有用。

返回值是一组地址对象，如果没找到匹配项，或者后台服务无效的话则返回 null 或者空序列。也有可能是通过网络获取。返回结果是一个最好的估计值，但不能保证其完全正确。

各个参数的具体说明如下所示。

- locationName：用户提供的位置描述。
- maxResults：要返回的最大结果数，推荐 1～5。

包含的异常如下所示。

- IllegalArgumentException：如果位置描述为空。
- IOException：如果没有网络或者 IO 错误。

（4）public static boolean isPresent()：如果 Geocoder 的方法 getFromLocation 和方法

getFromLcationName 都实现，则返回 true。当没有网络连接时，这些方法仍然可能返回空值或者空序列。

9.1.4　Geocoder 的主要功能

在 Android 系统中，类 Geocoder 的主要功能如下所示。

1. 设置模拟器以支持定位服务

GPS 数据格式有 GPX 和 KML 两种，其中 GPX 是一个 XML 格式文件，为应用软件设计的通用的 GPS 数据格式，可以用来描述路点、轨迹和路程。而 KML 是基于 XML（eXtensible MarkupLanguage，可扩展标记语言）语法标准的一种标记语言（markup language），采用标记结构，含有嵌套的元素和属性。由 Google 旗下的 Keyhole 公司发展并维护，用来表达地理标记。

LBS 是 Location-Based Service 的缩写，是一个总称，用来描述用于找到设备当前位置的不同技术。主要包含如下所示的两个元素。

- locationManager：用于提供 LBS 的钩子 hook，获得当前位置，跟踪移动，设置移入和移出指定区域的接近警报。
- LocationProviders：其中的每一个都代表不同的用于确定设备当前位置的位置发现技术，两个常用的是 Providers GPS_PROVIDER 和 NETWORK_PROVIDER。

```
String providerName =LocationManager.GPS_PROVIDER;
LocationProvidergpsProvider;
gpsProvider =locationManager.getProvider(providerName);
```

在 Eclipse 开发环境中，依次单击“DDMS”｜“Location Controls”可以设置位置变化数据以在模拟器中测试应用程序，如图 9-1 所示。使用 ManualTab，可以指定特定的纬度/经度对。另外，KML 和 GPX 可以载入 KML 和 GPX 文件。一旦加载，可以跳转到特定的航点（位置）或顺序播放每个位置。

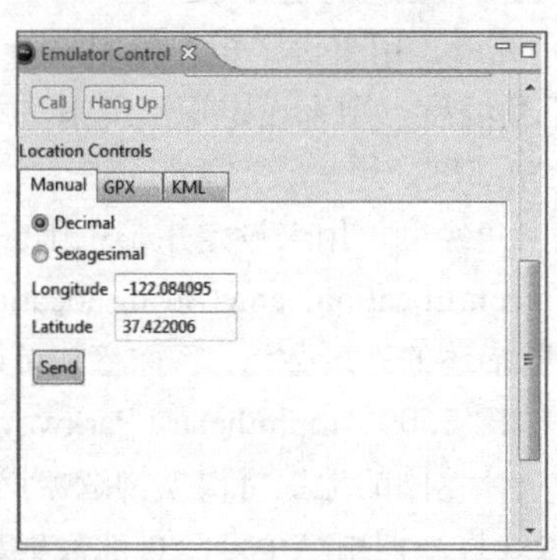

▲图 9-1　单击“DDMS”｜“Location Controls”设置位置变化数据

也可以用类 Criteria 设置符合要求的 provider 的条件查询（精度=精确/粗略、能耗=高/中/低、成本、返回海拔、速度、方位的能力），例如下面的代码。

```
Criteria criteria = newCriteria();
criteria.setAccuracy(Criteria.ACCURACY_COARSE);
criteria.setPowerRequirement(Criteria.POWER_LOW);
criteria.setAltitudeRequired(false);
criteria.setBearingRequired(false);
criteria.setSpeedRequired(false);
criteria.setCostAllowed(true);
String bestProvider = locationManager.getBestProvider(criteria, true);
//或者用 getProviders 返回所有可能匹配的 Provider
```

```
List<String>matchingProviders = locationManager.getProviders(criteria,false);
```

在使用 LocationManager 前，需要将 uses-permission 加载到 manifest 文件中以支持对 LBS 硬件的访问。GPS 需要 finepermission 权限，Network 需要 coarsepermission 权限。

```
<uses-permissionandroid:name="android.permission.ACCESS_FINE_LOCATION"/>
<uses-permissionandroid:name="android.permission.ACCESS_COARSE_LOCATION"/>
```

使用 GetLastKnownLocation 方法可以获得最新的位置。

```
String provider =LocationManager.GPS_PROVIDER;
Location location = locationManager.getLastKnownLocation(provider);
```

2. 跟踪运动（TrackingMovement）

- 可以使用 requestLocationUpdates 方法取得最新的位置变化，为优化性能可以指定位置变化的最小时间（毫秒）和最小距离（米）。当超出最小时间和距离值时，Location Listener 将触发 onLocationChanged 事件。

```
locationManager.requestLocationUpdates(provider,t, distance,myLocationListener);
```

- 使用 RomoveUpdates 方法停止位置更新。
- 大多数 GPS 硬件都非常明显地消耗电能。

3. 邻近警告（ProximityAlerts）

通过邻近警告功能让运用程序设置触发器，当用户在地理位置上移动或超出设定距离时触发。

- 可以使用 PendingIntent 定义 Proximity Alert 触发时广播的 Intent。
- 为了处理 proximityalert，需要创建 BroadcastReceiver，并重写 onReceive 方法。例如下面的代码。

```
public classProximityIntentReceiver extends BroadcastReceiver {
        @Override
        public voidonReceive (Context context, Intent intent) {
        String key =LocationManager.KEY_PROXIMITY_ENTERING;
        Booleanentering = intent.getBooleanExtra(key, false);
        [ ...perform proximity alert actions ... ]
        }
        }
```

- 如果想要想启动监听，需要注册 Receriver。

```
IntentFilter filter =new IntentFilter(TREASURE_PROXIMITY_ALERT);
registerReceiver(newProximityIntentReceiver(), filter);
```

9.1.5 地理编码和地理反编码

在智能设备地图操作应用中，地理编码（Geocoding）和地理反编码（Reverse Geocoding）是最常见操作之一。其中前者表示通过街道地址请求空间坐标，后者表示通过空间坐标请求街道地址。通俗地说，两者就是街道地址与经纬度的转换。例如前者输入查询“上海市 1239 号”会得到(31.285207060526762, 121.50546412914991)，而后者则表示上述反过程的查询工作。

在实际的 Android 应用开发过程中，地图相关的操作对于地理编码与地理反编码的使用都是十分普遍的。Android 在 MapView 控件中对于这两者都进行了封装，因此开发者可以方便地利用 Google Map Service 对两者进行查询。具体开发步骤如下所示。

（1）申请 MapKey，任何地图的显示都需要申请一个 MapKey，具体步骤请参考本书的第 11 章。

（2）建立一个基于 Google APIs 的程序，并且在文件 AndroidManifest.xml 中加入地图 API 的支持。

```
<?xml version="1.0" encoding="utf-8"?>
<manifest xmlns:android="http://schemas.android.com/apk/res/android"
     package="net.learn2develop.GoogleMaps"
     android:versionCode="1"
     android:versionName="1.0.0">
   <application android:icon="@drawable/icon" android:label="@string/app_name">
   <uses-library android:name="com.google.android.maps" />
      <activity android:name=".MapsActivity"
              android:label="@string/app_name">
          <intent-filter>
              <action android:name="android.intent.action.MAIN" />
              <category android:name="android.intent.category.LAUNCHER" />
          </intent-filter>
      </activity>
   </application>

   <uses-permission android:name="android.permission.INTERNET" />
</manifest>
</xml>
```

（3）在主 Layout 文件中加入对于地图的引用显示，在里面需要加入刚才申请的 MapKey，否则无法正常显示地图。

```
<?xml version="1.0" encoding="utf-8"?>
<RelativeLayout xmlns:android="http://schemas.android.com/apk/res/android"
   android:layout_width="fill_parent"
   android:layout_height="fill_parent">
   <com.google.android.maps.MapView
      android:id="@+id/mapView"
      android:layout_width="fill_parent"
      android:layout_height="fill_parent"
      android:enabled="true"
      android:clickable="true"
      android:apiKey="MapKey"
      />

</RelativeLayout>
```

（4）在主 Activity 编写代码以支持对地图的显示。

```
import com.google.android.maps.MapActivity;
import com.google.android.maps.MapView;
import android.os.Bundle;

public class MapsActivity extends MapActivity
{
   /** Called when the activity is first created. */
   @Override
   public void onCreate(Bundle savedInstanceState)
   {
      super.onCreate(savedInstanceState);
      setContentView(R.layout.main);
   MapView mapView = (MapView) findViewById(R.id.mapView);
   mapView.setBuiltInZoomControls(true);
   }
}
```

此时执行后可以在模拟器中显示一幅地图。

（5）向程序中加入地理编码与地理反编码的功能，其中实现地理编码的演示代码如下所示。

```
Geocoder geoCoder = new Geocoder(this, Locale.getDefault());
try {
   List<Address> addresses = geoCoder.getFromLocationName(
      "上海市1239号", 5);
   String add = "";
   if (addresses.size() > 0) {
      p = new GeoPoint(
            (int) (addresses.get(0).getLatitude() * 1E6),
            (int) (addresses.get(0).getLongitude() * 1E6));
```

```
        mc.animateTo(p);
        mapView.invalidate();
      }
  } catch (IOException e) {
    e.printStackTrace();
  }
```

下面的代码是演示地理反编码的过程，其中 MapOverlay 为地图图层上的叠加图层，用于标识的显示以及单击事件的捕捉。

```
class MapOverlay extends com.google.android.maps.Overlay
    {
        @Override
        public boolean draw(Canvas canvas, MapView mapView,
        boolean shadow, long when)
        {
          //...
        }

        @Override
        public boolean onTouchEvent(MotionEvent event, MapView mapView)
        {
            //---when user lifts his finger---
            if (event.getAction() == 1) {
                GeoPoint p = mapView.getProjection().fromPixels(
                    (int) event.getX(),
                    (int) event.getY());

                Geocoder geoCoder = new Geocoder(
                    getBaseContext(), Locale.getDefault());
                try {
                    List<Address> addresses = geoCoder.getFromLocation(
                        p.getLatitudeE6()  / 1E6,
                        p.getLongitudeE6() / 1E6, 1);

                    String add = "";
                    if (addresses.size() > 0)
                    {
                        for (int i=0; i<addresses.get(0).getMaxAddressLineIndex();
                            i++)
                           add += addresses.get(0).getAddressLine(i) + "/n";
                    }

                    Toast.makeText(getBaseContext(), add, Toast.LENGTH_SHORT).show();
                }
                catch (IOException e) {
                    e.printStackTrace();
                }
                return true;
            }
            else
                return false;
        }
    }
```

此时执行后就会实现地址的反编码查询。

9.2 实战演练——在设备地图中快速查询某个位置

实例	功能	源码路径
实例 9-1	在设备地图中快速查询某个位置	\daima\9\fast

在本实例中，插入一个输入框控件和按钮控件，当在输入框中输入一个地址并单击“开始查询”按钮后，会在地图中显示这个位置的信息。本实例的具体实现流程如下所示。

（1）编写文件 mymap.xml，在里面添加 Map 密钥以实现对地图的引用。主要代码如下所示。

```
<com.google.android.maps.MapView
            android:id="@+id/mapview_mymap_display"
            android:layout_width="fill_parent"
            android:layout_height="fill_parent"
            android:apiKey="0NFa8R5kt6KmenQdcxhItm2rcaSZaNhOe3WZQTw"
            />
```

（2）编写文件 MyMap.java，获取用户在文本框中输入的地址，并根据这个地址进行查询操作。文件 MyMap.java 的主要代码如下所示。

```
public class MyMap extends MapActivity{//程序列表中添加联网的权限还需要加一个类库

   MapView mapview;
   private MapController mapcontroller;
   private GeoPoint geopoint;
   protected String addressname;

   protected void onCreate(Bundle savedInstanceState) {
      // TODO Auto-generated method stub
      super.onCreate(savedInstanceState);
      setContentView(R.layout.mymap);
      //用于显示地图上的一个 ViewGroup
      mapview=(MapView)findViewById(R.id.mapview_mymap_display);


      Bundle bundle=getIntent().getExtras();
      Log.d("MyMap_Oncreate_bundle",bundle+"");
      addressname=bundle.getString("address");
      Log.d("MyMap_oncreate",addressname);
      //使得这个 view 可以获得单击事件
      mapview.setClickable(true);
      //是否可以设置自动缩放设置
      mapview.setBuiltInZoomControls(true);

      //获取控制缩放的操作对象
      mapcontroller=mapview.getController();
      //通过系统默认区域设置进行地图的定位
      Geocoder geocoder=new Geocoder(this);

      mapview.setTraffic(true);
      try
      {
            List<Address> addresses=geocoder.getFromLocationName(addressname,2);
         Log.d("MyMap_oncreate_addressname3",addressname);
         geopoint = new GeoPoint(
               (int) (addresses.get(0).getLatitude() * 1E6),
               (int) (addresses.get(0).getLongitude() * 1E6));
            MyOverlay myoverlay=new MyOverlay();

            mapview.getOverlays().add(myoverlay);
            mapcontroller.setZoom(20);
            mapcontroller.animateTo(geopoint);
      }
      catch(Exception e)
      {
         e.printStackTrace();
      }
   }
   @Override
   protected boolean isRouteDisplayed() {        return false;
   }

   class MyOverlay extends Overlay
   {
      @Override
      public boolean draw(Canvas canvas, MapView mapview, boolean shadow, long when) {
         // TODO Auto-generated method stub
```

```
            Paint paint=new Paint();
            Point screenPoint=new Point();
            //经纬度坐标和屏幕像素坐标的一个映射
            mapview.getProjection().toPixels(geopoint, screenPoint);
            //并且这个映射可以把地理上的经纬度转换在屏幕上的像素点
            Bitmap bitmap=BitmapFactory.decodeResource(getResources(),R.drawable.flag1);
            canvas.drawBitmap(bitmap,screenPoint.x,screenPoint.y, paint);
            canvas.drawText(addressname,screenPoint.x,screenPoint.y, paint);
            return super.draw(canvas, mapview, shadow, when);
        }
    }
}
```

执行后的效果如图 9-2 所示，输入一个地址并单击“开始查询”按钮后会在地图中显示此位置。

在本实例中，在 Activity 的 OnCreate 方法中设置 MapView 的各个属性，设置了是否可以获得单击事件（setClickable()方法），设置地图缩放尺度（setBuiltInZoomControls(true)），设置了地图的视图模式，谷歌地图一共有 3 种视图模式。

- 街道视图：mapview.setStreetView()。
- 卫星视图：mapview.setSatelite()。
- 一般地图：mapview.setTraffic()。

▲图 9-2 执行效果

对地图的操作是通过对一个 MapController 对象的操作，该对象是通过 MapView.getController()方法获取的。在使地图显示某一个地点时，则是 MapController.animateTo()方法，参数是一个 GeoPoint 类型，经度和维度的一个组合。个人感觉类似于坐标值，并且可以通过 MapController.setZoom 来设置放大的倍数，其中数值越大，地图越详细。类 GeoCoder 是处理地理编码的一个类，根据输入的地点可以获取一个和此地点相关的 Address 类的集合。

方法 getFromLLocationName()有两个参数，一个是输入的地点，另一个是获取的地点的个数。通过继承 OverLay 类也可为地图设置一个图标图层。

最后不要忘记在列表中添加访问 Internet 的权限。

```
<uses-permission android:name="android:permission.Internet"/>
```

并且还需要为应用添加类库。

```
<uses- library android:name="com.google.android.maps"/>
```

9.3 实战演练——接近某个位置时实现自动提醒

实例	功能	源码路径
实例 9-2	接近某个位置时实现自动提醒	https://github.com/gast-lib/gast-lib

本实例的功能是，当设备接近某个位置时实现自动提醒。本实例源码是开源代码，来源于下面地址，读者可以自行登录并下载。

```
https://github.com/gast-lib/gast-lib
```

本实例的具体实现流程如下所示。

（1）编写主界面 GeocodeActivity 用于从用户处获取目标位置的信息，并使用 LocationManager 对象对位置进行编码处理。主界面 GeocodeActivity 的布局文件是 geocode.xml，功能是供用户选择一个目标位置信息，具体实现代码如下所示。

```
<RelativeLayout xmlns:android="http://schemas.android.com/apk/res/android"
    android:layout_width="match_parent"
    android:layout_height="match_parent"
    android:orientation="vertical" >
```

```
    <TextView android:id="@+id/enterLocationLabel"
        android:layout_height="wrap_content"
        android:layout_width="wrap_content"
        android:text="@string/enterLocationLabel"
        android:layout_alignParentTop="true" />

    <EditText android:id="@+id/enterLocationValue"
        android:layout_height="wrap_content"
        android:layout_width="match_parent"
        android:layout_below="@id/enterLocationLabel"
        android:text="springfield" />

    <Button android:id="@+id/lookupLocationButton"
        android:layout_width="match_parent"
        android:layout_height="wrap_content"
        android:text="@string/lookupLocationButton"
        android:layout_below="@id/enterLocationValue"
        android:onClick="onLookupLocationClick" />

    <Button android:id="@+id/okButton"
        android:layout_width="match_parent"
        android:layout_height="wrap_content"
        android:layout_alignParentBottom="true"
        android:text="@android:string/ok"
        android:onClick="onOkClick" />
    <ListView android:id="@android:id/list"
        android:layout_width="match_parent"
        android:layout_height="match_parent"
        android:drawSelectorOnTop="false"
        android:layout_above="@id/okButton"
        android:layout_below="@id/lookupLocationButton"
        android:choiceMode="singleChoice" />
</RelativeLayout>
```

主界面 GeocodeActivity 的程序文件是 GeocodeActivity.java，功能是对设备中提供的地址实现地理编码和反向地理编码处理。当单击“Lookup Location”按钮时会运行 onLookupLocationClick 方法，在此方法中调用了类 Geocode。文件 GeocodeActivity.java 的具体实现代码如下所示。

```
public class GeocodeActivity extends ListActivity
{
    private static final String TAG = "GeocodeActivity";
    private static final int MAX_ADDRESSES = 30;

    @Override
    protected void onCreate(Bundle savedInstanceState)
    {
        super.onCreate(savedInstanceState);

        setContentView(R.layout.geocode);
    }

    public void onLookupLocationClick(View view)
    {
        if (Geocoder.isPresent())
        {
            EditText addressText = (EditText) findViewById(R.id.enterLocationValue);

            try
            {
                List<Address> addressList = new Geocoder(this).getFromLocationName
                (addressText.getText().toString(), MAX_ADDRESSES);

                List<AddressWrapper>addressWrapperList = new ArrayList<Address Wrapper>();

                for (Address address : addressList)
                {
                    addressWrapperList.add(new AddressWrapper(address));
                }
```

```
            setListAdapter(new ArrayAdapter<AddressWrapper>(this, android.R.layout.
            simple_list_item_single_choice, addressWrapperList));
        }
        catch (IOException e)
        {
            Log.e(TAG, "Could not geocode address", e);

            new AlertDialog.Builder(this)
                .setMessage(R.string.geocodeErrorMessage)
                .setTitle(R.string.geocodeErrorTitle)
                .setPositiveButton(android.R.string.ok, new DialogInterface.
                OnClickListener()
                {
                    @Override
                    public void onClick(DialogInterface dialog, int which)
                    {
                        dialog.dismiss();
                    }
                }).show();
        }
    }
}

public void onOkClick(View view)
{
    ListView listView = getListView();

    Intent intent = getIntent();
    if (listView.getCheckedItemPosition() != ListView.INVALID_POSITION)
    {
        AddressWrapper addressWrapper = (AddressWrapper)listView.getItemAtPosition
        (listView.getCheckedItemPosition());

        intent.putExtra("name", addressWrapper.toString());
        intent.putExtra("latitude", addressWrapper.getAddress().getLatitude());
        intent.putExtra("longitude", addressWrapper.getAddress().getLongitude());
    }

    this.setResult(RESULT_OK, intent);
    finish();
}

private static class AddressWrapper
{
    private Address address;

    public AddressWrapper(Address address)
    {
        this.address = address;
    }

    @Override
    public String toString()
    {
        StringBuilder stringBuilder = new StringBuilder();

        for (int i = 0; i < address.getMaxAddressLineIndex(); i++)
        {
            stringBuilder.append(address.getAddressLine(i));

            if ((i + 1) < address.getMaxAddressLineIndex())
            {
                stringBuilder.append(", ");
            }
        }

        return stringBuilder.toString();
    }
```

```
        public Address getAddress()
        {
            return address;
        }
    }
}
```

在上述代码中，调用方法 isPresent()检验当前设备是否支持地理编码和反向地理编码功能。方法 fromLocationName 能够解析输入的位置字符串和标志性建筑的坐标，这一功能是通过网络查询来实现的。得到查询结果后，会在下方使用 ListView 列表进行展示。

（2）开始实现接近警报设置界面，此功能通过 LocationManager 实现。接近警报设置界面的布局文件是 proximity_alert.xml，提供了两个单选按钮供用户选择设置类型，并且可以设置允许的接近读取半径值。文件 proximity_alert.xml 的具体实现代码如下所示。

```
<RelativeLayout
 xmlns:android="http://schemas.android.com/apk/res/android"
 android:orientation="vertical"
 android:layout_width="match_parent"
 android:layout_height="match_parent">

   <TextView android:id="@+id/locationLabel"
      android:layout_width="wrap_content"
      android:layout_height="wrap_content"
      android:text="@string/locationLabel"
      android:layout_alignParentTop="true"
      android:layout_alignParentLeft="true"
      style="@style/apptext" />

   <TextView android:id="@+id/locationValue"
      android:layout_width="wrap_content"
      android:layout_height="wrap_content"
      android:layout_alignTop="@id/locationLabel"
      android:layout_toRightOf="@id/locationLabel"
      android:text="@string/none"
      android:paddingLeft="5dip"
      style="@style/apptext" />

   <TextView android:id="@+id/latitudeLabel"
      android:layout_width="wrap_content"
      android:layout_height="wrap_content"
      android:text="@string/latitudeLabel"
      android:layout_below="@id/locationValue"
      android:layout_alignParentLeft="true"
      style="@style/apptext" />

   <TextView android:id="@+id/latitudeValue"
      android:layout_width="wrap_content"
      android:layout_height="wrap_content"
      android:layout_alignTop="@id/latitudeLabel"
      android:layout_toRightOf="@id/latitudeLabel"
      android:text="@string/none"
      android:paddingLeft="5dip"
      style="@style/apptext" />

   <TextView android:id="@+id/longitudeLabel"
      android:layout_width="wrap_content"
      android:layout_height="wrap_content"
      android:text="@string/longitudeLabel"
      android:layout_below="@id/latitudeLabel"
      android:layout_alignParentLeft="true"
      style="@style/apptext" />

   <TextView android:id="@+id/longitudeValue"
      android:layout_width="wrap_content"
      android:layout_height="wrap_content"
```

```
        android:layout_alignTop="@id/longitudeLabel"
        android:layout_toRightOf="@id/longitudeLabel"
        android:text="@string/none"
        android:paddingLeft="5dip"
        style="@style/apptext" />

    <TextView android:id="@+id/radiusLabel"
        android:layout_width="wrap_content"
        android:layout_height="wrap_content"
        android:text="@string/radiusLabel"
        android:layout_below="@id/longitudeLabel"
        android:layout_alignParentLeft="true"
        style="@style/apptext" />

    <EditText android:id="@+id/radiusValue"
        android:layout_width="wrap_content"
        android:layout_height="wrap_content"
        android:layout_alignTop="@id/radiusLabel"
        android:layout_toRightOf="@id/radiusLabel"
        android:text="10"
        android:paddingLeft="5dip"
        style="@style/apptext"
        android:inputType="number" />

    <Button
        android:layout_height="wrap_content"
        android:layout_width="match_parent"
        android:text="@string/setLocation"
        android:onClick="onSetLocationClick"
        android:layout_below="@id/radiusValue" />

    <LinearLayout android:id="@+id/buttons"
        android:layout_width="match_parent"
        android:layout_height="wrap_content"
        android:orientation="horizontal"
        android:layout_alignParentBottom="true" >

        <Button android:id="@+id/setProximityAlert"
            android:layout_height="wrap_content"
            android:layout_width="match_parent"
            android:text="@string/setProximityAlert"
            android:onClick="onSetProximityAlertClick"
            android:enabled="false"
            android:layout_weight="1" />

        <Button android:id="@+id/clearProximityAlert"
            android:layout_height="wrap_content"
            android:layout_width="match_parent"
            android:text="@string/clearProximityAlert"
            android:onClick="onClearProximityAlertClick"
            android:enabled="false"
            android:layout_weight="1" />
    </LinearLayout>

    <RadioGroup android:id="@+id/proximityTypeRadioGroup"
        android:layout_height="wrap_content"
        android:layout_width="match_parent"
        android:orientation="horizontal"
        android:layout_above="@id/buttons" >

        <RadioButton android:id="@+id/androidProximityAlert"
            android:layout_height="wrap_content"
            android:layout_width="wrap_content"
            android:text="@string/androidProximityAlertTypeLabel"
            android:layout_weight="1"
            style="@style/apptext" />
        <RadioButton android:id="@+id/customProximityAlert"
            android:layout_height="wrap_content"
            android:layout_width="wrap_content"
```

```
            android:text="@string/customProximityAlertTypeLabel"
            android:layout_weight="1"
            style="@style/apptext" />
    </RadioGroup>

    <TextView
        android:layout_above="@id/proximityTypeRadioGroup"
        style="@style/apptext"
        android:text="Select Proximity Alert Type" />
</RelativeLayout>
```

接近警报设置界面的 Activity 是 ProximityAlertActivity.java，具体实现流程如下所示。

- 首先通过 onSetProximityAlertClick 读取输入的半径值，当来到预定目标的这一半径范围之内时会发出警报。
- 读取半径值后调用 locationManager.addProximityAlert 来传递坐标、半径、失效和广播 Intent 等参数。
- 编写方法 onClearProximityAlertClick 用于处理单击“Clear Proximity Alert”按钮的动作，这样可以在不需要时关闭接近警报功能，这样做的好处是可以避免这个接近警报过期。

```
public class ProximityAlertActivity extends Activity
{
    private static final String USE_ANDROID_PROXIMITY_TYPE_KEY = "useAndroidProximity
    TypeKey";

    private LocationManager locationManager;
    private PendingIntent pendingIntent;
    private SharedPreferences preferences;
    private RadioButton androidProximityTypeRadioButton;
    private Button setProximityAlert;
    private Button clearProximityAlert;
    private double latitude = Double.MAX_VALUE;
    private double longitude = Double.MAX_VALUE;

    @Override
    protected void onCreate(Bundle savedInstanceState)
    {
        super.onCreate(savedInstanceState);
        setContentView(R.layout.proximity_alert);

        locationManager = (LocationManager) getSystemService(LOCATION_SERVICE);

        pendingIntent = ProximityPendingIntentFactory.createPendingIntent(this);

        preferences = getPreferences(MODE_PRIVATE);
        androidProximityTypeRadioButton =
                (RadioButton)findViewById(R.id.androidProximityAlert);

        setProximityAlert = (Button) findViewById(R.id.setProximityAlert);
        clearProximityAlert = (Button) findViewById(R.id.clearProximityAlert);
    }

    @Override
    protected void onResume()
    {
        super.onResume();

        if (preferences.getBoolean(USE_ANDROID_PROXIMITY_TYPE_KEY, true))
        {
            androidProximityTypeRadioButton.setChecked(true);
        }
        else
        {
            ((RadioButton)findViewById(R.id.customProximityAlert)).setChecked(true);
        }
    }
```

```
@Override
protected void onPause()
{
    super.onPause();

    locationManager.removeProximityAlert(pendingIntent);
    preferences.edit().putBoolean(USE_ANDROID_PROXIMITY_TYPE_KEY, androidProximity
    TypeRadioButton.isChecked()).commit();
}

public void onSetProximityAlertClick(View view)
{
    EditText radiusView = (EditText)findViewById(R.id.radiusValue);
    int radius =
            Integer.parseInt(radiusView.getText().toString());

    if (androidProximityTypeRadioButton.isChecked())
    {
        locationManager.addProximityAlert(latitude,
                                          longitude,
                                          radius,
                                          -1,
                                          pendingIntent);
    }
    else
    {
        Criteria criteria = new Criteria();
        criteria.setAccuracy(Criteria.ACCURACY_COARSE);
        Intent intent = new Intent(this, ProximityAlertService.class);
        intent.putExtra(ProximityAlertService.LATITUDE_INTENT_KEY, latitude);
        intent.putExtra(ProximityAlertService.LONGITUDE_INTENT_KEY, longitude);
        intent.putExtra(ProximityAlertService.RADIUS_INTENT_KEY, (float)radius);
        startService(intent);
    }

    setProximityAlert.setEnabled(false);
    clearProximityAlert.setEnabled(true);
}

public void onClearProximityAlertClick(View view)
{
    if (androidProximityTypeRadioButton.isChecked())
    {
        locationManager.removeProximityAlert(pendingIntent);
    }

    setProximityAlert.setEnabled(true);
    clearProximityAlert.setEnabled(false);
}

public void onSetLocationClick(View view)
{
    startActivityForResult(new Intent(this, GeocodeActivity.class), 1);
}
@Override
protected void onActivityResult(int requestCode, int resultCode, Intent data)
{
    super.onActivityResult(requestCode, resultCode, data);

    if (resultCode == RESULT_OK
            && data != null
            && data.hasExtra("name")
            && data.hasExtra("latitude")
            && data.hasExtra("longitude"))
    {
        latitude = data.getDoubleExtra("latitude", Double.MAX_VALUE);
        longitude = data.getDoubleExtra("longitude", Double.MAX_VALUE);
```

```
            ((TextView)findViewById(R.id.locationValue)).setText(data.getStringExtra
            ("name"));
            ((TextView)findViewById(R.id.latitudeValue)).setText(String.valueOf
            (latitude));
            ((TextView)findViewById(R.id.longitudeValue)).setText(String.valueOf
            (longitude));

            setProximityAlert.setEnabled(true);
            clearProximityAlert.setEnabled(false);
        }
    }
}
```

（3）开始实现发送接近警报响应信息模块。当设备进入或离开预定坐标的指定半径区域时，接近警报会广播发送一个 Intent。上述功能是通过文件 ProximityAlertBroadcastReceiver.java 实现的，主要代码如下所示。

```
public class ProximityAlertBroadcastReceiver extends LocationBroadcastReceiver
{
    private static final int NOTIFICATION_ID = 9999;

    @Override
    public void onEnteringProximity(Context context)
    {
        displayNotification(context, "Entering Proximity");
    }

    @Override
    public void onExitingProximity(Context context)
    {
        displayNotification(context, "Exiting Proximity");
    }

    private void displayNotification(Context context, String message)
    {
        NotificationManager notificationManager = (NotificationManager)context.
        getSystemService(Context.NOTIFICATION_SERVICE);

        PendingIntent pi = PendingIntent.getActivity(context, 0, new Intent(), 0);

        Notification notification = new Notification(R.drawable.icon, message,
        System.currentTimeMillis());
        notification.setLatestEventInfo(context, "GAST", "Proximity Alert", pi);

        notificationManager.notify(NOTIFICATION_ID, notification);
    }
}
```

当设备进入和离开预定义区域时，会发送一个通知信息。

（4）因为设备的电池通常是有限的，而接近警报功能需要长时间开启，所以很费电。为了解决上述局限性，在接近警报设置界面中提供了两个单选按钮供用户选择设置类型，其中“Custom”类型可供用户自定义设置。编写文件 ProximityAlertActivity.java，实现了对默认接近警报的优化，不但限制了对 GPS 的使用，而且降低了请求位置更新功能的频率。设置在最后才会将服务注册到网络提供者，以获取位置更新功能，这样可以长时间使用网络提供者，并且只有在网络提供者无法对设备和目标区域之间的距离做出准确计算的情况下才会启用 GPS。当计算完两者距离之后，需要通过服务来确定是否要广播一个 Intent 来发出接近警报。如果不需要，则取消当前位置的更新请求，并重新计算最短距离，并用新的距离来注册位置更新请求。文件 ProximityAlertActivity.java 的具体实现代码如下所示。

```
public class ProximityAlertActivity extends Activity
{
    private static final String USE_ANDROID_PROXIMITY_TYPE_KEY = "useAndroidProximity
TypeKey";
```

```
private LocationManager locationManager;
private PendingIntent pendingIntent;
private SharedPreferences preferences;
private RadioButton androidProximityTypeRadioButton;
private Button setProximityAlert;
private Button clearProximityAlert;
private double latitude = Double.MAX_VALUE;
private double longitude = Double.MAX_VALUE;

@Override
protected void onCreate(Bundle savedInstanceState)
{
    super.onCreate(savedInstanceState);
    setContentView(R.layout.proximity_alert);

    locationManager = (LocationManager) getSystemService(LOCATION_SERVICE);

    pendingIntent = ProximityPendingIntentFactory.createPendingIntent(this);

    preferences = getPreferences(MODE_PRIVATE);
    androidProximityTypeRadioButton =
            (RadioButton)findViewById(R.id.androidProximityAlert);

    setProximityAlert = (Button) findViewById(R.id.setProximityAlert);
    clearProximityAlert = (Button) findViewById(R.id.clearProximityAlert);
}

@Override
protected void onResume()
{
    super.onResume();

    if (preferences.getBoolean(USE_ANDROID_PROXIMITY_TYPE_KEY, true))
    {
        androidProximityTypeRadioButton.setChecked(true);
    }
    else
    {
        ((RadioButton)findViewById(R.id.customProximityAlert)).setChecked(true);
    }
}

@Override
protected void onPause()
{
    super.onPause();

    locationManager.removeProximityAlert(pendingIntent);
    preferences.edit().putBoolean(USE_ANDROID_PROXIMITY_TYPE_KEY, androidProximity
    TypeRadioButton.isChecked()).commit();
}

public void onSetProximityAlertClick(View view)
{
    EditText radiusView = (EditText)findViewById(R.id.radiusValue);
    int radius =
            Integer.parseInt(radiusView.getText().toString());

    if (androidProximityTypeRadioButton.isChecked())
    {
        locationManager.addProximityAlert(latitude,
                                          longitude,
                                          radius,
                                          -1,
                                          pendingIntent);
    }
    else
    {
```

```
            Criteria criteria = new Criteria();
            criteria.setAccuracy(Criteria.ACCURACY_COARSE);
            Intent intent = new Intent(this, ProximityAlertService.class);
            intent.putExtra(ProximityAlertService.LATITUDE_INTENT_KEY, latitude);
            intent.putExtra(ProximityAlertService.LONGITUDE_INTENT_KEY, longitude);
            intent.putExtra(ProximityAlertService.RADIUS_INTENT_KEY, (float)radius);
            startService(intent);
        }

        setProximityAlert.setEnabled(false);
        clearProximityAlert.setEnabled(true);
    }

    public void onClearProximityAlertClick(View view)
    {
        if (androidProximityTypeRadioButton.isChecked())
        {
            locationManager.removeProximityAlert(pendingIntent);
        }

        setProximityAlert.setEnabled(true);
        clearProximityAlert.setEnabled(false);
    }

    public void onSetLocationClick(View view)
    {
        startActivityForResult(new Intent(this, GeocodeActivity.class), 1);
    }

    @Override
    protected void onActivityResult(int requestCode, int resultCode, Intent data)
    {
        super.onActivityResult(requestCode, resultCode, data);

        if (resultCode == RESULT_OK
                && data != null
                && data.hasExtra("name")
                && data.hasExtra("latitude")
                && data.hasExtra("longitude"))
        {
            latitude = data.getDoubleExtra("latitude", Double.MAX_VALUE);
            longitude = data.getDoubleExtra("longitude", Double.MAX_VALUE);

            ((TextView)findViewById(R.id.locationValue)).setText(data.getStringExtra
            ("name"));
            ((TextView)findViewById(R.id.latitudeValue)).setText(String.valueOf
            (latitude));
            ((TextView)findViewById(R.id.longitudeValue)).setText(String.valueOf
            (longitude));

            setProximityAlert.setEnabled(true);
            clearProximityAlert.setEnabled(false);
        }
    }
}
```

到此为止，整个实例介绍完毕。对于本实例的具体实现源码，读者可以参阅开源站点的源码。

第10章　磁场传感器详解

在 Android 设备中，经常需要检测设备的方向，例如设备的朝向和移动方向。在 Android 系统中，通常使用重力传感器、加速度传感器、磁场传感器和旋转矢量传感器来检测设备的方向。本章将详细讲解在 Android 设备中使用磁场传感器检测设备方向的基本知识，为读者步入本书后面知识的学习打下基础。

10.1 磁场传感器基础

磁场传感器是可以将各种磁场及其变化的量转变成电信号输出的装置。自然界和人类社会生活的许多地方都存在磁场或与磁场相关的信息。在本节的内容中，将详细讲解在 Android 设备中使用磁场传感器的基本知识。

10.1.1　什么是磁场传感器

磁场传感器是利用人工设置的永久磁体产生的磁场，可以作为许多种信息的载体，被广泛用于探测、采集、存储、转换、复现和监控各种磁场和磁场中承载的各种信息的任务。在当今的信息社会中，磁场传感器已成为信息技术和信息产业中不可缺少的基础元件。目前，人们已研制出利用各种物理、化学和生物效应的磁场传感器，并已在科研、生产和社会生活的各个方面得到广泛应用，承担起探究种种信息的任务。

在现实市面中，最早的磁场传感器是伴随测磁仪器的进步而逐步发展的。在众多的测试磁场方法中，大多都是将磁场信息变成电信号进行测量。在测磁仪器中“探头”或“取样装置”就是磁场传感器。随着信息产业、工业自动化、交通运输、电力电子技术、办公自动化、家用电器、医疗仪器等的飞速发展和电子计算机应用的普及，需用大量的传感器将需要进行测量和控制的非电参量，转换成可与计算机兼容的信号，作为它们的输入信号，这就给磁场传感器的快速发展提供了机会，形成了相当可观的磁场传感器产业。

10.1.2　磁场传感器的分类

在现实应用中，磁场传感器的主要分类如下所示。

1. 薄膜磁致电阻传感器

铁磁性物质在磁化过程中，它的电阻值沿磁化方向将增加，并将达到饱和的现象称为磁阻效应。薄膜磁阻元件是利用薄膜工艺和微细加工技术，将 NiPe、NiCo 合金用真空蒸镇或溯射工艺沉积到硅片或铁氧体基片上，通过微细加工技术制成一定形状的磁咀图形，形成三端式、四端式以及多端式器件。例如，BMber 结构桥式电路磁阻元件具有灵敏度高、工作频率特性好、温度稳定性好、结构简单、体积小等特点。可制成高密度磁咀磁头、磁性编码器、磁阻位移传感器和磁阻电流传感器等。

2. 磁阻敏感器

将物质在磁场中电阻发生变化的现象称为磁阻效应。对于铁、钴、镍及其合金等强磁性金属，当外加磁场平行于磁体内部磁化方向时，电阻几乎不随外加磁场变化；当外加磁场偏离金属的内磁化方向时，此类金属的电阻值将减小，这就是强磁金属的各向异性磁阻效应。

3. 电涡流式传感器

近年来，国内外正发展建立在电涡流效应原理上的传感器，即电涡流式传感器。这种传感器不但具有测量线性范围大、灵敏度高、结构简单、抗干扰能力强、不受油污等介质的影响等优点，而且又具有无损、非接触测量的特点，目前正广泛地应用于工业各部门中的位移、尺寸、厚度、振动、转速、压力、电导率、温度、波面等测量以及探测全属材料和加工件表面裂纹及缺陷。

4. 磁性液体加速度传感器

磁性液体作为一种新型的纳米功能材料，一经问世便走到科学技术发展的前沿，目前科学家们已经将这种新型功能材料应用到广阔的领域中。以此为基础的磁性液体传感器技术也引起了国际技术领域广泛的关注。

5. 磁性液体水平传感器

磁性液体传感器的研究与应用起源于美国。早在 1983 年美国新墨西哥州阿尔帕克基应用技术公司就与美国空军签订了合同，那时便开始研制基于磁性液体动力学原理（MHD）的主动式和被动式传感器。磁性液体传感器应用领域很广，既可以应用到民用上，也可以应用到军工上，有其他传感器所代替不了的功能。正因如此，国外比较早地意识到开发磁性液体传感器的意义，并且已开始了研究和生产，并将该种传感器应用到航空、航天、宇航站等尖端军事领域。美国、法国、德国、俄罗斯、日本和罗马尼亚等国家已开始利用磁性液体来制作各种传感器。磁性液体水平传感器对控制机器人工作状态、使太阳能栅板保持朝向太阳、使抛物天线持续朝向通信系统中的人造卫星等方面有着重要的应用。目前国外已研制出单轴、双轴和三轴等类型的磁性液体水平传感器，而我国有关磁性液体传感器的研究尚处于实验和探索阶段。

10.2 Android 系统中的磁场传感器

在 Android 系统中，磁场传感器 TYPE_MAGNETIC_FIELD，单位是 uT（微特斯拉），能够测量设备周围 3 个物理轴（x，y，z）的磁场。在 Android 设备中，磁场传感器主要用于感应周围的磁感应强度，在注册监听器后主要用于捕获如下 3 个参数：

- values[0]；
- values[l]；
- values[2]。

上述 3 个参数分别代表磁感应强度在空间坐标系中 3 个方向轴上的分量。所有数据的单位都为 uT，即微特斯拉。

在 Android 系统中，磁场传感器主要包含了如下所示的公共方法。

- int getFifoMaxEventCount()：返回该传感器可以处理事件的最大值。如果该值为 0，表示当前模式不支持此传感器。

- int getFifoReservedEventCount()：保留传感器在批处理模式中 FIFO 的事件数，给出了一个在保证可以分批事件的最小值。
- float getMaximumRange()：传感器单元的最大范围。
- int getMinDelay()：最小延迟。
- String getName()：获取传感器的名称。
- float getPower()：获取传感器电量。
- float getResolution()：获得传感器的分辨率。
- int getType()：获取传感器的类型。
- String getVendor()：获取传感器的供应商字符串。
- Int etVersion()：获取该传感器模块版本。
- String toString()：返回一个对当前传感器的字符串描述。

10.3 实战演练——获取磁场传感器的 3 个分量

在接下来的实例中，将演示在 Android 设备中使用磁场传感器的方法。

实例	功能	源码路径
实例 10-1	使用磁场传感器	\daima\10\cichang

本实例的实现文件是 cichangLI.java，在此文件中定义了监听器类对象和注册监听的方法。主要代码如下所示。

```
public class cichangLI extends Activity {
    TextView myTextView1;//x 方向磁场分量
    TextView myTextView2;//y 方向磁场分量
    TextView myTextView3;//z 方向磁场分量
    //SensorManager mySensorManager;//引用 SensorManager 对象
    SensorManagerSimulator mySensorManager;//声明 SensorManagerSimulator 对象,调试时用
    @Override
    public void onCreate(Bundle savedInstanceState) {//重写 onCreate 方法
        super.onCreate(savedInstanceState);
        setContentView(R.layout.main);//当前的用户界面
        myTextView1 = (TextView) findViewById(R.id.myTextView1);//得到 myTextView1 引用
        myTextView2 = (TextView) findViewById(R.id.myTextView2);//得到 myTextView2 引用
        myTextView3 = (TextView) findViewById(R.id.myTextView3);//得到 myTextView3 引用
        //调试时用
        mySensorManager = SensorManagerSimulator.getSystemService(this, SENSOR_SERVICE);
        mySensorManager.connectSimulator();                //连接 Simulator 服务器
    }
    @SuppressWarnings("deprecation")
    private SensorListener mySensorListener = new SensorListener(){
        @Override
        public void onAccuracyChanged(int sensor,int accuracy){}//重写 onAccuracyChanged 方法
        @Override
        public void onSensorChanged(int sensor,float[]values){//重写 onSensorChanged 方法
            if(sensor == SensorManager.SENSOR_MAGNETIC_FIELD){//检查磁场的变化
                myTextView1.setText("x 方向的磁场分量为："+values[0]);  //数据显示在 TextView
                myTextView2.setText("y 方向的磁场分量为："+values[1]);  //数据显示在 TextView
                myTextView3.setText("z 方向的磁场分量为："+values[2]);  //数据显示在 TextView
            }
        }
    };
    @Override
    protected void onResume() {//重写的 onResume 方法
        mySensorManager.registerListener(//注册监听
                mySensorListener, //监听器 SensorListener 对象
                SensorManager.SENSOR_MAGNETIC_FIELD,//传感器的类型为加速度
                SensorManager.SENSOR_DELAY_UI//传感器事件传递的频度
                );
```

```
        super.onResume();
    }
    @Override
    protected void onPause() {//重写 onPause 方法
        //取消注册监听器
        mySensorManager.unregisterListener((SensorEventListener) mySensorListener);
        super.onPause();
    }
}
```

因为本实例比较简单，是根据 SensorSimulator 中附带的开源代码改编的，所以在此不再进行详细介绍，读者只需阅读本书源程序即可。

10.4 实战演练——演示常用传感器的基本用法

在接下来的实例中，将演示在 Android 设备中使用常用传感器的基本方法。

实例	功能	源码路径
实例 10-2	使用磁场传感器	\daima\10\HelloSensor

10.4.1　实现布局文件

布局文件 main.xml 的功能是使用 ListView 控件列表显示常用的传感器类型，具体实现代码如下所示。

```
<LinearLayout xmlns:android="http://schemas.android.com/apk/res/android"
    android:orientation="vertical"
    android:layout_width="fill_parent"
    android:layout_height="fill_parent">
    <ListView android:id="@+id/ListView01"
            android:layout_width="wrap_content"
            android:layout_height="wrap_content"/>
</LinearLayout>
```

10.4.2　实现程序文件

主 Activity 的实现文件是 HelloSensor.java，功能是响应用户选择的传感器来执行对应的处理程序，具体实现代码如下所示。

```
public class HelloSensor extends Activity {
    //5 个范例的菜单名称和应用程序 Class
    private Object[] activities = {
        "Compass", CompassDemo.class,
        "Orientation", OrientationDemo.class,
        "Accelerometer", AccelerometerDemo.class,
        "Magnetic Field", MagneticFieldDemo.class,
        "Temperature", TemperatureDemo.class,
    };
    //HelloSensor 主程式
    @Override
    public void onCreate(Bundle savedInstanceState) {
        super.onCreate(savedInstanceState);
        setContentView(R.layout.main);
        //建立 5 个范例菜单名称的数组 list
        CharSequence[] list = new CharSequence[activities.length / 2];
        for (int i = 0; i < list.length; i++) {
            list[i] = (String)activities[i * 2];
        }
        //将 5 个范例菜单名称放置在 listView 中
        ArrayAdapter<CharSequence> adapter = new ArrayAdapter<CharSequence>(this,
        android.R.layout.simple_list_item_1, list);
        ListView listView = (ListView)findViewById(R.id.ListView01);
        listView.setAdapter(adapter);
```

```
        //按下菜单名称指向相关的应用程序 Class
        listView.setOnItemClickListener(new OnItemClickListener() {
            public void onItemClick(AdapterView<?>parent, View view, int position, long id) {
                Intent intent = new Intent(HelloSensor.this, (Class<?>)activities[position
                * 2 + 1]);
                startActivity(intent);
            }
        });
    }
}
```

当在主 Activity 界面选“CompassDemo”选项后，会执行文件 CompassDemo.java 启动罗盘传感器，具体实现代码如下所示。

```
public class CompassDemo extends Activity implements SensorEventListener {
    private SensorManager sensorManager;
    private MySurfaceView view;
    private Object[] orientation = {
            "Rotate Z-axis Orientation", "Rotate X-axis Orientation","Rotate Y-axis
            Orientation",
    };
    //CompassDemo 主程序
    @Override
    public void onCreate(Bundle savedInstanceState) {
        super.onCreate(savedInstanceState);
        sensorManager = (SensorManager)getSystemService(SENSOR_SERVICE);
        view = new MySurfaceView(this);
        setContentView(view);
    }
    @Override
    protected void onResume() {
        super.onResume();
        List<Sensor> sensors = sensorManager.getSensorList(Sensor.TYPE_ORIENTATION);
        if (sensors.size() > 0) {
            sensorManager.registerListener(this, sensors.get(0), SensorManager.SENSOR_
            DELAY_NORMAL);
        }
    }
    @Override
    protected void onPause() {
        super.onPause();
        sensorManager.unregisterListener(this);
    }
    public void onAccuracyChanged(Sensor sensor, int accuracy) {
    }
    public void onSensorChanged(SensorEvent event) {
        view.onValueChanged(event.values);
    }
    class MySurfaceView extends SurfaceView implements SurfaceHolder.Callback {
        private Bitmap bitmap,curBitmap;
        private float x, y, z, delta=-8;
        private int curWidth, curHeight;
        public MySurfaceView(Context context) {
            super(context);
            getHolder().addCallback(this);
            bitmap = BitmapFactory.decodeResource(getResources(), R.drawable.compass);
        }
        public void surfaceChanged(SurfaceHolder holder, int format, int width, int height) {
            x = getWidth()/2;
            y = getHeight()/2;
            onValueChanged(new float[3]);
        }
        public void surfaceCreated(SurfaceHolder holder) {
        }
        public void surfaceDestroyed(SurfaceHolder holder) {
        }
        @SuppressWarnings("static-access")
        void onValueChanged(float[] values) {
            Canvas canvas = getHolder().lockCanvas();
            if (canvas != null) {
```

```
                Paint paint = new Paint();
                paint.setAntiAlias(true);
                paint.setColor(Color.BLUE);
                paint.setTextSize(24);
                canvas.drawColor(Color.WHITE);
                canvas.save();
                Matrix matrix = new Matrix();
                curWidth = (int) (bitmap.getWidth()* 1);
                curHeight =  (int) (bitmap.getHeight()* 1);
                curBitmap = bitmap.createScaledBitmap(bitmap, curWidth, curHeight, false);
                matrix.setRotate(-values[0]+delta, x , y );
                canvas.setMatrix(matrix);
                canvas.drawBitmap(curBitmap, x-curWidth/2, y-curHeight/2, null);
                canvas.restore();
                for (int i = 0; i < values.length; i++) {
                    canvas.drawText(orientation[i] + ": "  + values[i], 0, paint.
                    getTextSize() * (i + 1), paint);
                }
                getHolder().unlockCanvasAndPost(canvas);
            }
        }
    }
}
```

当在主 Activity 界面选择“AccelerometerDemo”选项后，会执行文件 AccelerometerDemo.java 启动加速计传感器，具体实现代码如下所示。

```
public class AccelerometerDemo extends Activity implements SensorEventListener {
    private SensorManager sensorManager;
    private MySurfaceView view;
    private Object[] accelerometer = {
            "X-axis Accelerometer", "Y-axis Accelerometer","Z-axis Accelerometer",
    };
    //AccelerometerDemo 主程序
    @Override
    public void onCreate(Bundle savedInstanceState) {
        super.onCreate(savedInstanceState);
        sensorManager = (SensorManager)getSystemService(SENSOR_SERVICE);
        view = new MySurfaceView(this);
        setContentView(view);
    }
    @Override
    protected void onResume() {
        super.onResume();
        List<Sensor> sensors = sensorManager.getSensorList(Sensor.TYPE_ACCELEROMETER);
        if (sensors.size() > 0) {
            sensorManager.registerListener(this, sensors.get(0), SensorManager.SENSOR_
            DELAY_FASTEST);
        }
    }
    @Override
    protected void onPause() {
        super.onPause();
        sensorManager.unregisterListener(this);
    }
    public void onAccuracyChanged(Sensor sensor, int accuracy) {
    }
    public void onSensorChanged(SensorEvent event) {
        view.onValueChanged(event.values);
    }
    class MySurfaceView extends SurfaceView implements SurfaceHolder.Callback {
        private Bitmap bitmap, curBitmap;
        private float x, y, z;
        private int curWidth, curHeight;
        public MySurfaceView(Context context) {
            super(context);
            getHolder().addCallback(this);
            bitmap = BitmapFactory.decodeResource(getResources(), R.drawable.android);
        }
```

```
        public void surfaceChanged(SurfaceHolder holder, int format, int width, int height) {
            x = (getWidth() - bitmap.getWidth()) / 2;
            y = (getHeight() - bitmap.getHeight()) / 2;
            onValueChanged(new float[3]);
        }
        public void surfaceCreated(SurfaceHolder holder) {
        }
        public void surfaceDestroyed(SurfaceHolder holder) {
        }
        @SuppressWarnings("static-access")
        void onValueChanged(float[] values) {
            z = 2 + values[2]/5;
            curWidth = (int) (bitmap.getWidth()* z);
            curHeight =  (int) (bitmap.getHeight()* z);
            curBitmap = bitmap.createScaledBitmap(bitmap, curWidth, curHeight, false);
            x = (getWidth() - curWidth) / 2;
            y = (getHeight() - curHeight) / 2;
            x -= values[0]*10;
            y += values[1]*10;
            Canvas canvas = getHolder().lockCanvas();
            if (canvas != null) {
                Paint paint = new Paint();
                paint.setAntiAlias(true);
                paint.setColor(Color.BLUE);
                paint.setTextSize(24);
                canvas.drawColor(Color.WHITE);
                canvas.drawBitmap(curBitmap, x, y, null);
                for (int i = 0; i < values.length; i++) {
                    canvas.drawText(accelerometer[i] + ": " + values[i], 0, paint.
                    getTextSize() * (i + 1), paint);
                }
                getHolder().unlockCanvasAndPost(canvas);
            }
        }
    }
}
```

当在主 Activity 界面选择“Magnetic Field”选项后，会执行文件 MagneticFieldDemo.java 启动磁场传感器，具体实现代码如下所示。

```
public class MagneticFieldDemo extends Activity implements SensorEventListener {
    private SensorManager sensorManager;
    private MySurfaceView view;
    private float max;
    private Object[] magnetic = {
        "X-axis Maganetic Field", "Y-axis Maganetic Field","Z-axis Maganetic Field",
    };
    //MagneticFieldDemo 主程序
    @Override
    public void onCreate(Bundle savedInstanceState) {
        super.onCreate(savedInstanceState);
        sensorManager = (SensorManager)getSystemService(SENSOR_SERVICE);
        view = new MySurfaceView(this);
        setContentView(view);
    }
    @Override
    protected void onResume() {
        super.onResume();
        List<Sensor> sensors = sensorManager.getSensorList(Sensor.TYPE_MAGNETIC_FIELD);
        if (sensors.size() > 0) {
            Sensor sensor = sensors.get(0);
            max = sensor.getMaximumRange();
            sensorManager.registerListener(this, sensor, SensorManager.SENSOR_DELAY_NORMAL);
        }
    }
    @Override
    protected void onPause() {
        super.onPause();
        sensorManager.unregisterListener(this);
```

```
    }
    public void onAccuracyChanged(Sensor sensor, int accuracy) {
    }
    public void onSensorChanged(SensorEvent event) {
        view.onValueChanged(event.values);
    }
    class MySurfaceView extends SurfaceView implements SurfaceHolder.Callback {
        public MySurfaceView(Context context) {
            super(context);
            getHolder().addCallback(this);
        }
        public void surfaceChanged(SurfaceHolder holder, int format, int width, int height) {
            onValueChanged(new float[3]);
        }
        public void surfaceCreated(SurfaceHolder holder) {
        }
        public void surfaceDestroyed(SurfaceHolder holder) {
        }
        void onValueChanged(float[] values) {
            Canvas canvas = getHolder().lockCanvas();
            if (canvas != null) {
                Paint paint = new Paint();
                paint.setAntiAlias(true);
                paint.setColor(Color.BLUE);
                paint.setTextSize(24);
                canvas.drawColor(Color.WHITE);
                for (int i = 0; i < values.length; i++) {
                    canvas.drawText(magnetic[i] + ": " + values[i], 0, paint.getTextSize()
                    * (i + 1), paint);
                }
                canvas.drawText("max: " + max, 0, paint.getTextSize() * 5, paint);
                getHolder().unlockCanvasAndPost(canvas);
            }
        }
    }
}
```

当在主 Activity 界面选择“Orientation”选项后，会执行文件 Magnet OrientationDemo.java 启动方向传感器，具体实现代码如下所示。

```
public class OrientationDemo extends Activity implements SensorEventListener {
    private SensorManager sensorManager;
    private MySurfaceView view;
    private Object[] orientation = {
            "Rotate Z-axis Orientation", "Rotate X-axis Orientation","Rotate Y-axis
            Orientation",
    };
    //OrientationDemo 主程序
    @Override
    public void onCreate(Bundle savedInstanceState) {
        super.onCreate(savedInstanceState);
        sensorManager = (SensorManager)getSystemService(SENSOR_SERVICE);
        view = new MySurfaceView(this);
        setContentView(view);
    }
    @Override
    protected void onResume() {
        super.onResume();
        List<Sensor> sensors = sensorManager.getSensorList(Sensor.TYPE_ORIENTATION);
        if (sensors.size() > 0) {
            sensorManager.registerListener(this, sensors.get(0), SensorManager.SENSOR_
            DELAY_NORMAL);
        }
    }
    @Override
    protected void onPause() {
        super.onPause();
        sensorManager.unregisterListener(this);
    }
```

```
    public void onAccuracyChanged(Sensor sensor, int accuracy) {
    }
    public void onSensorChanged(SensorEvent event) {
        view.onValueChanged(event.values);
    }
    class MySurfaceView extends SurfaceView implements SurfaceHolder.Callback {
        private Bitmap bitmap, bitmap1, bitmap2,curBitmap;
        private float x, y, z;
        private float x1=130, y1=160;
        private int curWidth, curHeight;
        public MySurfaceView(Context context) {
            super(context);
            getHolder().addCallback(this);
            bitmap = BitmapFactory.decodeResource(getResources(), R.drawable. androidplate);
            bitmap1 = BitmapFactory.decodeResource(getResources(), R.drawable. androidheight);
            bitmap2 = BitmapFactory.decodeResource(getResources(), R.drawable. androidwidth);
        }
        public void surfaceChanged(SurfaceHolder holder, int format, int width, int height) {
            x = getWidth()/2;
            y = getHeight()/2;
            onValueChanged(new float[3]);
        }
        public void surfaceCreated(SurfaceHolder holder) {
        }
        public void surfaceDestroyed(SurfaceHolder holder) {
        }
        @SuppressWarnings("static-access")
        void onValueChanged(float[] values) {
            Canvas canvas = getHolder().lockCanvas();
            if (canvas != null) {
                Paint paint = new Paint();
                paint.setAntiAlias(true);
                paint.setColor(Color.BLUE);
                paint.setTextSize(24);
                canvas.drawColor(Color.WHITE);
                canvas.save();
                Matrix matrix = new Matrix();
                curWidth = (int) (bitmap.getWidth()* 1);
                curHeight =  (int) (bitmap.getHeight()* 1);
                curBitmap = bitmap.createScaledBitmap(bitmap, curWidth, curHeight, false);
                matrix.setRotate(-values[0], x , y );
                canvas.setMatrix(matrix);
                canvas.drawBitmap(curBitmap, x-curWidth/2, y-curHeight/2, null);
                matrix.setRotate(values[1], x-x1 , y+y1 );
                canvas.setMatrix(matrix);
                canvas.drawBitmap(bitmap1, x-x1-bitmap1.getWidth()/2, y+y1-bitmap1.
                getHeight()/2, null);
                matrix.setRotate(-values[2], x+x1 , y+y1 );
                canvas.setMatrix(matrix);
                canvas.drawBitmap(bitmap2, x+x1-bitmap2.getWidth()/2, y+y1-bitmap2.
                getHeight()/2, null);
                canvas.restore();
                for (int i = 0; i < values.length; i++) {
                    canvas.drawText(orientation[i] + ": " + values[i], 0, paint.
                    getTextSize() * (i + 1), paint);
                }
                getHolder().unlockCanvasAndPost(canvas);
            }
        }
    }
}
```

当在主 Activity 界面选择“Temperature”选项后，会执行文件 TemperatureDemo.java 启动温度传感器，具体实现代码如下所示。

```
public class TemperatureDemo extends Activity implements SensorEventListener {
    private SensorManager sensorManager;
    private MySurfaceView view;
    private String vendor = "UNKNOWN";
```

```
    private String name = "UNKNOWN";
    private int version = 0;
    private Object[] temperature = {
            "Temperature", "No Data","No Data",
    };
    //TemperatureDemo 主程序
    @Override
    public void onCreate(Bundle savedInstanceState) {
        super.onCreate(savedInstanceState);
        sensorManager = (SensorManager)getSystemService(SENSOR_SERVICE);
        view = new MySurfaceView(this);
        setContentView(view);
    }
    @Override
    protected void onResume() {
        super.onResume();
        List<Sensor> sensors = sensorManager.getSensorList(Sensor.TYPE_TEMPERATURE);
        if (sensors.size() > 0) {
            Sensor sensor = sensors.get(0);
            vendor = sensor.getVendor();
            name = sensor.getName();
            version = sensor.getVersion();
            sensorManager.registerListener(this, sensor, SensorManager.SENSOR_DELAY_NORMAL);
        }
    }
    @Override
    protected void onPause() {
        super.onPause();
        sensorManager.unregisterListener(this);
    }
    public void onAccuracyChanged(Sensor sensor, int accuracy) {
    }
    public void onSensorChanged(SensorEvent event) {
        view.onValueChanged(event.values);
    }
    class MySurfaceView extends SurfaceView implements SurfaceHolder.Callback {
        public MySurfaceView(Context context) {
            super(context);
            getHolder().addCallback(this);
        }
        public void surfaceChanged(SurfaceHolder holder, int format, int width, int height) {
            onValueChanged(new float[3]);
        }
        public void surfaceCreated(SurfaceHolder holder) {
        }
        public void surfaceDestroyed(SurfaceHolder holder) {
        }
        void onValueChanged(float[] values) {
            Canvas canvas = getHolder().lockCanvas();
            if (canvas != null) {
                Paint paint = new Paint();
                paint.setAntiAlias(true);
                paint.setColor(Color.BLUE);
                paint.setTextSize(24);
                canvas.drawColor(Color.WHITE);
                for (int i = 0; i < values.length; i++) {
                    canvas.drawText(temperature[i] + ": " + values[i], 0, paint.
                    getTextSize() * (i + 1), paint);
                }
                canvas.drawText("Vender: "+vendor, 0, paint.getTextSize() * 5, paint);
                canvas.drawText("Name: "+name, 0, paint.getTextSize() * 6, paint);
                canvas.drawText("Version: "+ version, 0, paint.getTextSize() * 7, paint);
                getHolder().unlockCanvasAndPost(canvas);
            }
        }
    }
}
```

到此为止，整个实例介绍完毕，执行后的效果如图 10-1 所示。

▲图 10-1　执行效果

第 11 章　加速度传感器详解

传统意义上的加速度传感器是一种能够测量加速力的电子设备。加速力是指当物体在加速过程中作用在物体上的力，就好比地球引力，也就是重力。加速力既可以是个常量，也可以是变量。加速度计有两种，其中一种是角加速度计，是由陀螺仪（角速度传感器）的改进的，另一种就是线加速度计。在本章的内容中，将详细讲解在 Android 设备中使用加速度传感器检测设备方向的基本知识，为读者步入本书后面知识的学习打下基础。

11.1 加速度传感器基础

在现实应用中，加速度传感器可以帮助机器人了解它现在身处的环境，能够分辨出是在爬山，还是在走下坡，或是否摔倒等。一个好的程序员能够使用加速度传感器来分辨出上述情形，加速度传感器甚至可以用来分析发动机的振动。在本节的内容中，将简要讲解加速度传感器的基础性知识。

11.1.1　加速度传感器的分类

在实际应用过程中，可以将加速度传感器分为如下所示的 4 类。

1. 压电式

压电式加速度传感器又称压电加速度计。它也属于惯性式传感器。压电式加速度传感器的原理是利用压电陶瓷或石英晶体的压电效应，在加速度计受振时，质量块加在压电元件上的力也随之变化。当被测振动频率远低于加速度计的固有频率时，则力的变化与被测加速度成正比。

2. 压阻式

基于世界领先的 MEMS 硅微加工技术，压阻式加速度传感器具有体积小、低功耗等特点，易于集成在各种模拟和数字电路中，广泛应用于汽车碰撞实验、测试仪器和设备振动监测等领域。加速度传感器网为客户提供压阻式加速度传感器/压阻加速度计各品牌的型号、参数、原理、价格和接线图等信息。

3. 电容式

电容式加速度传感器是基于电容原理的极距变化型的电容传感器。电容式加速度传感器/电容式加速度计是对比较通用的加速度传感器。在某些领域无可替代，如安全气囊、手机移动设备等。电容式加速度传感器/电容式加速度计采用了微机电系统（MEMS）工艺，在大量生产时变得经济，从而保证了较低的成本。

4. 伺服式

伺服式加速度传感器是一种闭环测试系统，具有动态性能好、动态范围大和线性度好等特点。其工作原理，传感器的振动系统由“*m-k*”系统组成，与一般加速度计相同，但质量 *m* 上还接着一个电磁线圈，当基座上有加速度输入时，质量块偏离平衡位置，该位移大小由位移传感器检测出来，经伺服放大器放大后转换为电流输出，该电流流过电磁线圈，在永久磁铁的磁场中产生电磁恢复力，使质量块保持在仪表壳体中原来的平衡位置上，所以伺服加速度传感器在闭环状态下工作。由于有反馈作用，增强了抗干扰的能力，提高了测量精度，扩大了测量范围，伺服加速度测量技术广泛地应用于惯性导航和惯性制导系统中，在高精度的振动测量和标定中也有应用。

11.1.2　加速度传感器的主要应用领域

在计算机领域中，加速度传感器可以测量牵引力产生的加速度。例如在 IBM Thinkpad 笔记本电脑中就内置了加速度传感器，能够动态地监测出笔记本在使用中的振动。根据这些振动数据，系统会智能地选择关闭硬盘还是让其继续运行，这样可以由于振动产生最大程度的保护。所以，加速度传感器主要应用在手柄振动/摇晃、仪器仪表、汽车制动启动、地震、报警系统、玩具、结构物、环境监视、工程测振、地质勘探、铁路、桥梁、大坝的振动测试与分析，还有鼠标、高层建筑结构动态特性和安全保卫振动侦察上。在接下来的内容中，将详细讲解加速度传感器的主要应用领域。

1. 汽车安全

加速度传感器主要用于汽车安全气囊、防抱死系统和牵引控制系统等安全性能方面。在汽车安全应用中，加速度计的快速反应非常重要。安全气囊应在什么时候弹出要迅速确定，所以加速度计必须在瞬间做出反应。通过采用可迅速达到稳定状态而不是振动不止的传感器设计可以缩短器件的反应时间。其中，压阻式加速度传感器由于在汽车工业中的广泛应用而得到迅速发展。

2. 游戏控制

加速度传感器可以检测上下左右的倾角的变化，因此通过前后倾斜手持设备来实现对游戏中物体的前后左右的方向控制，就变得很简单。

3. 图像自动翻转

用加速度传感器检测手持设备的旋转动作及方向，实现所要显示图像的转正。

4. 电子指南针倾斜校正

磁传感器是通过测量磁通量的大小来确定方向的。当磁传感器发生倾斜时，通过磁传感器的地磁通量将发生变化，从而使方向指向产生误差。因此，如果不带倾斜校正的电子指南针，需要用户水平放置。而利用加速度传感器可以测量倾角的这一原理，可以对电子指南针的倾斜进行补偿。

5. GPS 导航系统死角的补偿

GPS 系统是通过接收 3 颗呈 120 度分布的卫星信号来最终确定物体的方位的。在一些特殊的场合和地貌，如隧道、高楼林立、丛林地带，GPS 信号会变弱甚至完全失去，这也就是所谓的死角。而通过加装加速度传感器及以前所通用的惯性导航，便可以进行系统死区的测量。对加速度传感器进行一次积分，就变成了单位时间里的速度变化量，从而测出在死区内物体的移动。

6. 计步器功能

加速度传感器可以检测交流信号以及物体的振动，人在走动的时候会产生一定规律性的振动，而加速度传感器可以检测振动的过零点，从而计算出走或跑的步数，进而计算出人所移动的位移。并且利用一定的公式可以计算出卡路里的消耗。

7. 防手抖功能

用加速度传感器检测手持设备的振动/晃动幅度，当振动/晃动幅度过大时锁住照相快门，使所拍摄的图像永远是清晰的。

8. 闪信功能

通过挥动手持设备实现在空中显示文字，用户可以自己编写显示的文字。这个闪信功能是利用人们的视觉残留现象，用加速度传感器检测挥动的周期，实现所显示文字的准确定位。

9. 硬盘保护

利用加速度传感器检测自由落体状态，从而对迷你硬盘实施必要的保护。硬盘在读取数据时，磁头与盘片之间的间距很小，因此，外界的轻微振动就会对硬盘产生很严重的后果，使数据丢失。而利用加速度传感器可以检测自由落体状态。当检测到自由落体状态时，让磁头复位，以减少硬盘的受损程度。

10. 设备或终端姿态检测

加速度传感器和陀螺仪通常称为惯性传感器，常用于各种设备或终端中实现姿态检测、运动检测等，很适合玩体感游戏的人群。加速度传感器利用重力加速度，可以用于检测设备的倾斜角度，但是它会受到运动加速度的影响，使倾角测量不够准确，所以通常需要利用陀螺仪和磁传感器补偿。同时磁传感器测量方位角时，也是利用地磁场，当系统中电流变化或周围有导磁材料以及当设备倾斜时，测量出的方位角也不准确，这时需要用加速度传感器（倾角传感器）和陀螺仪进行补偿。

11. 智能产品

加速度传感器在微信功能中的创新功能突破了电子产品的千篇一律，这个功能的实现来源传感器的方向、加速表、光线、磁场、临近性和温度等参数的特性。这个原理是手机里面集成的加速度传感器导致的，它能够分别测量 *X*、*Y*、*Z* 三个方面的加速度值，*X* 方向值的大小代表手机水平移动，*Y* 方向值的大小代表手机垂直移动，*Z* 方向值的大小代表手机的空间垂直方向，天空的方向为正，地球的方向为负，然后把相关的加速度值传输给操作系统，通过判断其大小变化，就能知道同时玩微信的朋友。

11.2 Android 系统中的加速度传感器

在 Android 系统中，加速度传感器是 TYPE_ACCELEROMETER，单位是 m/s^2，能够测量应用于设备 *x*、*y*、*z* 轴上的加速度，又叫作 G-sensor。在开发过程中，通过 Android 的加速度传感器可以取得 *x*、*y*、*z* 3 个轴的加速度。在 Android 系统中，在类 SensorManager 中定义了很多星体的重力加速度值。如表 11-1 所示。

表 11-1　　类 SensorManager 被定义的各新星体的重力加速度值　　单位：m/s^2

常 量 名	说　明	实际的值
GRAVITY_DEATH_STAR_1	死亡星	3.5303614E-7
GRAVITY_EARTH	地球	9.80665
GRAVITY_JUPITER	木星	23.12
GRAVITY_MARS	火星	3.71
GRAVITY_MERCURY	水星	3.7
GRAVITY_MOON	月亮	1.6
GRAVITY_NEPTUNE	海王星	11.0
GRAVITY_PLUTO	冥王星	0.6
GRAVITY_SATURN	土星	8.96
GRAVITY_SUN	太阳	275.0
GRAVITY_THE_ISLAND	岛屿星	4.815162
GRAVITY_URANUS	天王星	8.69
GRAVITY_VENUS	金星	8.87

通常来说，从加速度传感器获取的值，拿手机等智能设备的人的手振动或放在摇晃的场所的时候，受振动影响设备的值增幅变化是存在的。手的摇动、轻微振动的影响是属于长波形式，去掉这种长波干扰的影响，可以取得高精度的值。去掉长波这种过滤机制叫作 Low-Pass Filter。Low-Pass Filter 机制有如下所示的 3 种封装方法。

- 从抽样数据中取得中间值的方法。
- 最近取得加速度的值每个很少变化的方法。
- 从抽样数据中取得中间值的方法。

在 Android 应用中，有时需要获取瞬间加速度值，例如类似计步器、作用力测定的应用开发的时候，如果想要检测出加速度急剧的变化。此时的处理和 Low-Pass Filter 处理相反，需要去掉短周波的影响，这样可以取得数据。像这种去掉短周波的影响的过滤器叫作 High-Pass Filter。

11.2.1　实战演练——获取 *x*、*y*、*z* 轴的加速度值

在接下来的实例中，将演示在 Android 中使用加速度传感器的方法。

实例	功能	源码路径
实例 11-1	获取 *x*、*y*、*z* 轴的加速度值	\daima\11\jiaS

本实例的具体实现流程如下所示。

（1）编写布局文件 main.xml，主要代码如下所示。

```
<?xml version="1.0" encoding="utf-8"?>              <!-- 声明 xml 的版本以及编码格式 -->
<LinearLayout xmlns:android="http://schemas.android.com/apk/res/android"
    android:orientation="vertical"
    android:layout_width="fill_parent"
    android:layout_height="fill_parent">              <!-- 添加一个垂直的线性布局 -->
    <TextView
        android:id="@+id/title"
        android:gravity="center_horizontal"
        android:textSize="20px"
        android:layout_width="fill_parent"
        android:layout_height="wrap_content"
        android:text="@string/title"/>                <!-- 添加一个 TextView 控件 -->
    <TextView
```

```
        android:id="@+id/myTextView1"
        android:textSize="18px"
        android:layout_width="fill_parent"
        android:layout_height="wrap_content"
        android:text="@string/myTextView1"/>        <!-- 添加一个 TextView 控件 -->
    <TextView
        android:id="@+id/myTextView2"
        android:textSize="18px"
        android:layout_width="fill_parent"
        android:layout_height="wrap_content"
        android:text="@string/myTextView2"/>        <!-- 添加一个 TextView 控件 -->
    <TextView
        android:id="@+id/myTextView3"
        android:textSize="18px"
        android:layout_width="fill_parent"
        android:layout_height="wrap_content"
        android:text="@string/myTextView3"/>        <!-- 添加一个 TextView 控件 -->
</LinearLayout>
```

（2）编写主程序文件 jiaSLI.java，此文件的具体实现流程如下所示。

- 声明 3 个 TextView 的引用，分别用于显示 3 个方向上的加速度。
- 声明对 SensorManager 对象的引用，此处因使用的是 SensorSimulator 工具来模拟传感器。
- 设置当前的用户界面，然后得到 XML 文件中配置的各个控件的引用。
- 初始化 SensorManager 对象，同样因为调试的原因用专用代码替代。
- 初始化监听器类，并重写了该类中的两个方法。
- 在 onSensorChanged 方法中只处理加速度的变化，并将得到的数值显示到 TextView 中。
- 重写类 Activity 的 onResume()方法，在该方法中为 SensorManager 添加监听，还需要重写 onPause()方法，在方法中取消注册的监听器。

文件 jiaSLI.java 的主要实现代码如下所示。

```
public class jiaSCH extends Activity {
    TextView myTextView1;//x 方向加速度
    TextView myTextView2;//y 方向加速度
    TextView myTextView3;//z 方向加速度
    //SensorManager mySensorManager;//SensorManager 对象引用
    SensorManagerSimulator mySensorManager;//声明 SensorManagerSimulator 对象,调试时用
    @Override
    public void onCreate(Bundle savedInstanceState) {//重写 onCreate 方法
        super.onCreate(savedInstanceState);
        setContentView(R.layout.main);//设置当前的用户界面
        myTextView1 = (TextView) findViewById(R.id.myTextView1);//得到 myTextView1 引用
        myTextView2 = (TextView) findViewById(R.id.myTextView2);//得到 myTextView2 引用
        myTextView3 = (TextView) findViewById(R.id.myTextView3);//得到 myTextView3 引用
        //调试时用
        mySensorManager = SensorManagerSimulator.getSystemService(this, SENSOR_SERVICE);
        mySensorManager.connectSimulator();                    //连接 Simulator 服务器
    }
    private SensorListener mySensorListener = new SensorListener(){
        @Override
        public void onAccuracyChanged(int sensor, int accuracy){}//重写 onAccuracyChanged 方法
        @Override
        public void onSensorChanged(int sensor,float[] values){//重写 onSensorChanged 方法
            if(sensor == SensorManager.SENSOR_ACCELEROMETER){//只检查加速度的变化
                myTextView1.setText("x 方向上的加速度为"+values[0]);//将提取的 x 数据显示到 TextView
                myTextView2.setText("y 方向上的加速度为"+values[1]);//将提取的 y 数据显示到 TextView
                myTextView3.setText("z 方向上的加速度为"+values[2]);//将提取的 z 数据显示到 TextView
            }
        }
    };
    @Override
    protected void onResume() {//重写的 onResume 方法
        mySensorManager.registerListener(//注册监听
                mySensorListener, //监听 SensorListener 对象
                SensorManager.SENSOR_ACCELEROMETER,//设置传感器的类型为加速度
```

```
            SensorManager.SENSOR_DELAY_UI//用传感器事件传递频率
            );
        super.onResume();
    }
    @Override
    protected void onPause() {//重写 onPause 方法
        mySensorManager.unregisterListener(mySensorListener);//取消注册监听器
        super.onPause();
    }
}
```

（3）为了调试本实例代码，需要为该程序添加网络权限，因为 SensorSimulator 安装在 Android 模拟器中的客户端需要和桌面端的服务器进行通信。

11.2.2　实战演练——实现控件的抖动效果

本实例不是基于传感器实现的，本实例的目的是演示不使用传感器实现控件抖动效果的方法。

实例	功能	源码路径
实例 11-2	获取 *x*、*y*、*z* 轴的加速度值	\daima\11\ShakeDemo

本实例的布局文件是 main.xml，具体实现代码如下所示。

```
<RelativeLayout xmlns:android="http://schemas.android.com/apk/res/android"
    android:id="@+id/bg"
    android:layout_width="fill_parent"
    android:layout_height="fill_parent"
    android:background="@drawable/bg"
    android:orientation="vertical" >

    <TextView
        android:id="@+id/text"
        android:layout_width="fill_parent"
        android:layout_height="wrap_content"
        android:gravity="center"
        android:text="@string/mShake"
        android:textColor="#000000" />

    <EditText
        android:id="@+id/passWd"
        android:layout_width="fill_parent"
        android:layout_height="wrap_content"
        android:layout_below="@+id/text"
        android:clickable="true"
        android:password="true"
        android:singleLine="true" />

    <Button
        android:id="@+id/shake_x"
        android:layout_width="wrap_content"
        android:layout_height="wrap_content"
        android:layout_below="@+id/passWd"
        android:text="X 轴抖动" />

    <Button
        android:id="@+id/shake_y"
        android:layout_width="wrap_content"
        android:layout_height="wrap_content"
        android:layout_alignBaseline="@+id/shake_x"
        android:layout_alignBottom="@+id/shake_x"
        android:layout_alignParentRight="true"
        android:text="Y 轴抖动" />

    <Button
        android:id="@+id/shake"
        android:layout_width="wrap_content"
        android:layout_height="wrap_content"
```

```
        android:layout_alignBaseline="@+id/shake_x"
        android:layout_alignBottom="@+id/shake_x"
        android:layout_centerHorizontal="true"
        android:text="抖动" />

    <ImageView
        android:id="@+id/image"
        android:layout_width="wrap_content"
        android:layout_height="wrap_content"
        android:layout_centerInParent="true"
        android:src="@drawable/icon" />

</RelativeLayout>
```

程序文件 ShakeDemoActivity.java 的具体实现代码如下所示。

```
public class ShakeDemoActivity extends Activity {
    /** Called when the activity is first created. */
    RelativeLayout mLayout;
    Button shake_x;
    Button shake;
    Button shake_y;
    Context mContext;
    EditText passWd;
    private ImageView image;

    public void onCreate(Bundle savedInstanceState) {
        super.onCreate(savedInstanceState);
        setContentView(R.layout.main);
        mContext = this;
        findView();

        shake.setOnClickListener(new OnClickListener() {

            public void onClick(View v) {
                Animation shake = AnimationUtils.loadAnimation(
                        ShakeDemoActivity.this, R.anim.anim);
                shake.reset();
                shake.setFillAfter(true);
                image.startAnimation(shake);
            }
        });
        shake_x.setOnClickListener(new OnClickListener() {

            public void onClick(View v) {
                Animation shakeAnim = AnimationUtils.loadAnimation(mContext,
                        R.anim.shake_x);
                passWd.startAnimation(shakeAnim);
            }
        });

        shake_y.setOnClickListener(new OnClickListener() {

            public void onClick(View v) {
                Animation shakeAnim = AnimationUtils.loadAnimation(mContext,
                        R.anim.shake_y);
                passWd.startAnimation(shakeAnim);
            }
        });
    }

    public void findView() {
        mLayout = (RelativeLayout) findViewById(R.id.bg);
        passWd = (EditText) findViewById(R.id.passWd);
        shake_x = (Button) findViewById(R.id.shake_x);
        shake = (Button) findViewById(R.id.shake);
        shake_y = (Button) findViewById(R.id.shake_y);
        image = (ImageView) findViewById(R.id.image);
    }
}
```

执行后会使按钮控件分别在 *X* 轴和 *Y* 轴实现抖动效果。

11.2.3　实战演练——实现仿微信“摇一摇”效果

本实例的功能是实现仿微信的“摇一摇”效果。

实例	功能	源码路径
实例 11-3	在设备中实现仿微信“摇一摇”效果	\daima\11\Shake

本实例的布局文件是 shake.xml，在界面上方设置了一个图片，在下方设置了按钮控件和文本控件，具体实现代码如下所示。

```
<LinearLayout xmlns:android="http://schemas.android.com/apk/res/android"
   android:layout_width="fill_parent"
   android:layout_height="fill_parent"
   android:orientation="vertical" >

   <RelativeLayout
      android:layout_width="fill_parent"
      android:layout_height="fill_parent"
      android:layout_centerInParent="true" >

      <ImageView
         android:id="@+id/shakeBg"
         android:layout_width="wrap_content"
         android:layout_height="wrap_content"
         android:layout_centerInParent="true"
         android:src="@drawable/shake_all" />

      <LinearLayout
         android:layout_width="fill_parent"
         android:layout_height="wrap_content"
         android:layout_centerInParent="true"
         android:orientation="vertical" >

         <RelativeLayout
            android:id="@+id/shakeImgUp"
            android:layout_width="fill_parent"
            android:layout_height="190dp"
            android:background="#111111">
            <ImageView
               android:layout_width="wrap_content"
               android:layout_height="wrap_content"
               android:layout_alignParentBottom="true"
               android:layout_centerHorizontal="true"
               android:src="@drawable/shake_up"
                />
         </RelativeLayout>
         <RelativeLayout
            android:id="@+id/shakeImgDown"
            android:layout_width="fill_parent"
            android:layout_height="190dp"
            android:background="#111111">
            <ImageView
               android:layout_width="wrap_content"
               android:layout_height="wrap_content"
               android:layout_centerHorizontal="true"
               android:src="@drawable/shake_down"
                />
         </RelativeLayout>
      </LinearLayout>
   </RelativeLayout>

   <RelativeLayout
      android:id="@+id/shake_title_bar"
        android:layout_width="fill_parent"
```

```
        android:layout_height="45dp"
        android:background="@drawable/title_bar"
        android:gravity="center_vertical"  >
            <Button
              android:layout_width="70dp"
              android:layout_height="wrap_content"
              android:layout_centerVertical="true"
              android:text="返回"
              android:textSize="14sp"
              android:textColor="#fff"
              android:onClick="shake_activity_back"
                android:background="@drawable/title_btn_back"/>
            <TextView
              android:layout_width="wrap_content"
              android:layout_height="wrap_content"
              android:text="摇一摇"
              android:layout_centerInParent="true"
              android:textSize="20sp"
              android:textColor="#ffffff" />
          <ImageButton
              android:layout_width="67dp"
              android:layout_height="wrap_content"
              android:layout_alignParentRight="true"
              android:layout_centerVertical="true"
              android:layout_marginRight="5dp"
              android:src="@drawable/mm_title_btn_menu"
              android:background="@drawable/title_btn_right"
              android:onClick="linshi"
              />
    </RelativeLayout>


    <SlidingDrawer
        android:id="@+id/slidingDrawer1"
        android:layout_width="match_parent"
        android:layout_height="match_parent"
        android:content="@+id/content"
        android:handle="@+id/handle" >
        <Button
            android:id="@+id/handle"
            android:layout_width="wrap_content"
            android:layout_height="wrap_content"

            android:background="@drawable/shake_report_dragger_up" />
        <LinearLayout
            android:id="@+id/content"
            android:layout_width="match_parent"
            android:layout_height="match_parent"
            android:background="#f9f9f9" >
            <ImageView
                android:layout_width="match_parent"
                android:layout_height="wrap_content"
                android:scaleType="fitXY"
                android:src="@drawable/shake_line_up" />
        </LinearLayout>
    </SlidingDrawer>
</LinearLayout>
```

程序文件 shakeActivity.java 实现了主 Activity，核心功能是监听设备的摇动方向，定义了摇一摇动画，并设置摇动过程中的振动效果。文件 shakeActivity.java 的主要实现代码如下所示。

```
public class shakeActivity extends Activity{

    ShakeListener mShakeListener = null;
    Vibrator mVibrator;
    private RelativeLayout mImgUp;
    private RelativeLayout mImgDn;
    private RelativeLayout mTitle;
```

```
    private SlidingDrawer mDrawer;
    private Button mDrawerBtn;

    @Override
    public void onCreate(Bundle savedInstanceState) {
        // TODO Auto-generated method stub
        super.onCreate(savedInstanceState);
        setContentView(R.layout.shake);
        //drawerSet ();//设置 drawer 监听切换按钮的方向

        mVibrator = (Vibrator)getApplication().getSystemService(VIBRATOR_SERVICE);

        mImgUp = (RelativeLayout) findViewById(R.id.shakeImgUp);
        mImgDn = (RelativeLayout) findViewById(R.id.shakeImgDown);
        mTitle = (RelativeLayout) findViewById(R.id.shake_title_bar);

        mDrawer = (SlidingDrawer) findViewById(R.id.slidingDrawer1);
        mDrawerBtn = (Button) findViewById(R.id.handle);
        mDrawer.setOnDrawerOpenListener(new OnDrawerOpenListener()
        {   public void onDrawerOpened()
            {
                mDrawerBtn.setBackgroundDrawable(getResources().getDrawable(R.drawable.
                shake_down));
                TranslateAnimation titleup = new TranslateAnimation(Animation.
                RELATIVE_TO_SELF,0f,Animation.RELATIVE_TO_SELF,0f,Animation.RELATIVE_
                TO_SELF,0f,Animation.RELATIVE_TO_SELF,-1.0f);
                titleup.setDuration(200);
                titleup.setFillAfter(true);
                mTitle.startAnimation(titleup);
            }
        });
         /* 设定 SlidingDrawer 被关闭的事件处理 */
        mDrawer.setOnDrawerCloseListener(new OnDrawerCloseListener()
        {   public void onDrawerClosed()
            {
                mDrawerBtn.setBackgroundDrawable(getResources().getDrawable(R.drawable.
                shake_report_dragger_up));
                TranslateAnimation titledn = new TranslateAnimation(Animation.RELATIVE_
                TO_SELF,0f,Animation.RELATIVE_TO_SELF,0f,Animation.RELATIVE_TO_SELF,
                -1.0f,Animation.RELATIVE_TO_SELF,0f);
                titledn.setDuration(200);
                titledn.setFillAfter(false);
                mTitle.startAnimation(titledn);
            }
        });

        mShakeListener = new ShakeListener(this);
        mShakeListener.setOnShakeListener(new OnShakeListener() {
            public void onShake() {
                //Toast.makeText(getApplicationContext(), "抱歉，暂时没有找到在同一时刻摇一摇
                //的人。\n 再试一次吧！", Toast.LENGTH_SHORT).show();
                startAnim();  //开始摇一摇手掌动画
                mShakeListener.stop();
                startVibrato(); //开始振动
                new Handler().postDelayed(new Runnable(){
                    @Override
                    public void run(){
                        //Toast.makeText(getApplicationContext(), "抱歉，暂时没有找到\n 在同
                        //一时刻摇一摇的人。\n 再试一次吧！", 500).setGravity
                        //(Gravity.CENTER,0,0).show();
                        Toast mtoast;
                        mtoast = Toast.makeText(getApplicationContext(),
                                "抱歉，暂时没有找到\n 在同一时刻摇一摇的人。\n 再试一次吧！", 10);
                              //mtoast.setGravity(Gravity.CENTER, 0, 0);
                              mtoast.show();
                              mVibrator.cancel();
                              mShakeListener.start();
                    }
                }, 2000);
```

```
            }
        });
    }
    public void startAnim () {   //定义摇一摇动画
        AnimationSet animup = new AnimationSet(true);
        TranslateAnimation mytranslateanimup0 = new TranslateAnimation(Animation.
        RELATIVE_TO_SELF,0f,Animation.RELATIVE_TO_SELF,0f,Animation.RELATIVE_TO_SELF,
        0f,Animation.RELATIVE_TO_SELF,-0.5f);
        mytranslateanimup0.setDuration(1000);
        TranslateAnimation mytranslateanimup1 = new TranslateAnimation(Animation.
        RELATIVE_TO_SELF,0f,Animation.RELATIVE_TO_SELF,0f,Animation.RELATIVE_TO_SELF,
        0f,Animation.RELATIVE_TO_SELF,+0.5f);
        mytranslateanimup1.setDuration(1000);
        mytranslateanimup1.setStartOffset(1000);
        animup.addAnimation(mytranslateanimup0);
        animup.addAnimation(mytranslateanimup1);
        mImgUp.startAnimation(animup);

        AnimationSet animdn = new AnimationSet(true);
        TranslateAnimation mytranslateanimdn0 = new TranslateAnimation(Animation.
        RELATIVE_TO_SELF,0f,Animation.RELATIVE_TO_SELF,0f,Animation.RELATIVE_TO_SELF,
        0f,Animation.RELATIVE_TO_SELF,+0.5f);
        mytranslateanimdn0.setDuration(1000);
        TranslateAnimation mytranslateanimdn1 = new TranslateAnimation(Animation.
        RELATIVE_TO_SELF,0f,Animation.RELATIVE_TO_SELF,0f,Animation.RELATIVE_TO_SELF,
        0f,Animation.RELATIVE_TO_SELF,-0.5f);
        mytranslateanimdn1.setDuration(1000);
        mytranslateanimdn1.setStartOffset(1000);
        animdn.addAnimation(mytranslateanimdn0);
        animdn.addAnimation(mytranslateanimdn1);
        mImgDn.startAnimation(animdn);
    }
    public void startVibrato(){        //定义振动
        mVibrator.vibrate( new long[]{500,200,500,200}, -1); //第1个{}里面是节奏数组，第
        //2个参数是重复次数，-1为不重复，非-1则从pattern的指定下标开始重复
    }

    public void shake_activity_back(View v) {     //标题栏返回按钮
        this.finish();
    }
    public void linshi(View v) {     //标题栏
        startAnim();
    }
    @Override
    protected void onDestroy() {
        super.onDestroy();
        if (mShakeListener != null) {
            mShakeListener.stop();
        }
    }
}
```

上述代码实现了与设备的交互控制——振动，振动是一种提醒或替换铃声的事件，通过上述代码可以了解到如何触发手机振动事件，虽然振动是手机默认的模式，但通过程序的辅助，可以做更精密的控制，例如振动周期、持续时间等。在设置振动（Vibration）事件中，必须要知道命令其振动的时间长短、振动事件的周期等。而在Android里设置的数值都是以毫秒（1000毫秒=1秒）来做计算的，所以在做设置时需要注意，如果设置的时间值太小的话，会感觉不出来。要想让设备乖乖的振动，需要创建Vibrator对象，通过调用Vibrate方法来达到振动的目的，在Vibrator的构造器中有4个参数，前3个的值是设置振动的大小，在这边可以把数值改成一大一小，这样就可以明显感觉出振动的差异，而最后一个值是设置振动的时间。

程序文件ShakeListener.java的功能是为设备实现了一个设备摇晃的监听器，这一功能是通过传感器实现的。文件ShakeListener.java的主要实现代码如下所示。

```
public class ShakeListener implements SensorEventListener {
```

```
// 速度阈值，当摇晃速度达到此值后产生作用
private static final int SPEED_SHRESHOLD = 3000;
// 两次检测的时间间隔
private static final int UPTATE_INTERVAL_TIME = 70;
// 传感器管理器
private SensorManager sensorManager;
// 传感器
private Sensor sensor;
// 重力感应监听器
private OnShakeListener onShakeListener;
// 上下文
private Context mContext;
// 手机上一个位置重力感应坐标
private float lastX;
private float lastY;
private float lastZ;
// 上次检测时间
private long lastUpdateTime;

// 构造器
public ShakeListener(Context c) {
    // 获得监听对象
    mContext = c;
    start();
}

// 开始
public void start() {
    // 获得传感器管理器
    sensorManager = (SensorManager) mContext
            .getSystemService(Context.SENSOR_SERVICE);
    if (sensorManager != null) {
        // 获得重力传感器
        sensor = sensorManager.getDefaultSensor(Sensor.TYPE_ACCELEROMETER);
    }
    // 注册
    if (sensor != null) {
        sensorManager.registerListener(this, sensor,
                SensorManager.SENSOR_DELAY_GAME);
    }

}

// 停止检测
public void stop() {
    sensorManager.unregisterListener(this);
}

// 设置重力感应监听器
public void setOnShakeListener(OnShakeListener listener) {
    onShakeListener = listener;
}

// 重力感应监听器感应获得变化数据
public void onSensorChanged(SensorEvent event) {
    // 现在检测时间
    long currentUpdateTime = System.currentTimeMillis();
    // 两次检测的时间间隔
    long timeInterval = currentUpdateTime - lastUpdateTime;
    // 判断是否达到了检测时间间隔
    if (timeInterval < UPTATE_INTERVAL_TIME)
        return;
    // 将现在的时间变成 last 时间
    lastUpdateTime = currentUpdateTime;

    // 获得 x、y、z 的坐标
    float x = event.values[0];
    float y = event.values[1];
    float z = event.values[2];
```

```
        // 获得x、y、z的变化值
        float deltaX = x - lastX;
        float deltaY = y - lastY;
        float deltaZ = z - lastZ;

        // 将现在的坐标变成last坐标
        lastX = x;
        lastY = y;
        lastZ = z;

        double speed = Math.sqrt(deltaX * deltaX + deltaY * deltaY + deltaZ
                * deltaZ)
                / timeInterval * 10000;
        Log.v("thelog", "===========log===================");
        // 达到速度阀值，发出提示
        if (speed >= SPEED_SHRESHOLD) {
            onShakeListener.onShake();
        }
    }

    public void onAccuracyChanged(Sensor sensor, int accuracy) {

    }

    // 摇晃监听接口
    public interface OnShakeListener {
        public void onShake();
    }

}
```

在真机中执行后将会实现和微信“摇一摇”类似的效果，如图11-1所示。

▲图11-1 “摇一摇”效果

11.3 线性加速度传感器详解

线性加速度传感器是加速度传感器的一种，单独分开讲的原因是为了和螺旋仪传感器分开。陀螺仪是测角速度的，加速度是测线性加速度的。其中前者利用了惯性原理，而后者利用了力平衡原理。在本节的内容中，将详细讲解在Android设备中使用线性加速度传感器的基本知识，为读者步入本书后面知识的学习打下基础。

11.3.1 线性加速度传感器的原理

线性加速度传感器利用了惯性原理：A（加速度）=F（惯性力）/M（质量），只需要测量F即可。怎么测量F？通过电磁力去平衡这个力就可以得到F对应于电流的关系，只需要用实验去标定这个比例系数就行了。多数加速度传感器是根据压电效应的原理来工作的。所谓的压电效应就是“对于不存在对称中心的异极晶体加在晶体上的外力除了使晶体发生形变以外，还将改变晶体的极化状态，在晶体内部建立电场，这种由于机械力作用使介质发生极化的现象称为正压电效应”。

一般加速度传感器就是利用了其内部的由于加速度造成的晶体变形这个特性。由于这个变形会产生电压，只需要计算出产生电压和所施加的加速度之间的关系，就可以将加速度转化成电压输出。当然，还有很多其他方法来制作加速度传感器，例如压阻技术、电容效应、热气泡效应、光效应，但是其最基本的原理都是由于加速度某个介质产生变形，通过测量其变形量并用相关电路转化成电压输出。每种技术都有各自的优势和问题。

压阻式加速度传感器由于在汽车工业中的广泛应用而发展最快。由于安全性越来越成为汽车制造商的卖点，这种附加系统也越来越多。压阻式加速度传感器 2000 年的市场规模约为 4.2 亿美元，根据有关调查，预计其市值将按年平均 4.1%速度增长，至 2007 年达到 5.6 亿美元。这其中，欧洲市场的速度最快，因为欧洲是许多安全气囊和汽车生产企业的所在地。

压电技术主要在工业上用于防止机器故障，使用这种传感器可以检测机器潜在的故障以达到自保护以及避免对工人产生意外伤害，这种传感器具有用户，尤其是质量行业的用户所追求的可重复性、稳定性和自生性。但是在许多新的应用领域，很多用户尚无使用这类传感器的意识，销售商冒险进入这种尚待开发的市场会麻烦多多，因为终端用户对由于使用这种传感器而带来的问题和解决方法都认识不多。如果这些问题能够得到解决，将会促进压电传感器得到更快的发展。2002 年压电传感器市值为 3 亿美元，预计其年增长率将达到 4.9%，到 2007 年达到 4.2 亿美元。

使用加速度传感器有时会碰到低频场合测量时输出信号出现失真的情况，用多种测量判断方法一时找不出故障出现的原因，经过分析总结，导致测量结果失真的因素主要是系统低频响应差、系统低频信噪比差和外界环境对测量信号的影响。 所以，只要出现加速度传感器低频测量信号失真情况，对比以上 3 点观察是哪个因素造成的，有针对性的进行解决即可。

在 Android 系统中，线性加速度传感器的类型是 TYPE_LINEAR_ACCELERATION，单位是 m/s^2，能够获取加速度传感器去除重力的影响得到数据。

11.3.2　实战演练——测试小球的运动

本实例的功能是，在屏幕中设置一个小球，当设备运动时小球也会随之运行，并在屏幕上方显示当前小球在 *X* 轴、*Y* 轴和 *Z* 轴的重力值。

实例	功能	源码路径
实例 11-4	在测试小球的运动	\daima\11\Accelerometer

1. 实现布局文件

编写布局文件 main.xml，在界面上方设置了一个背景图片，并使用文本框来显示当前小球在 *X* 轴、*Y* 轴和 *Z* 轴的重力值。文件 main.xml 的具体实现代码如下所示。

```
<LinearLayout xmlns:android="http://schemas.android.com/apk/res/android"
   android:layout_width="fill_parent"
   android:layout_height="fill_parent"
   android:orientation="vertical" >

   <RelativeLayout
      android:id="@+android:id/ball_top"
      android:layout_width="fill_parent"
      android:layout_height="56dp"
      android:orientation="vertical" >

   <TextView
      android:id="@+android:id/ball_prompt"
      android:layout_width="fill_parent"
      android:layout_height="wrap_content"
      android:text="Sensor: 0, 0, 0" />

   <TextView
      android:id="@+id/tv3"
      android:layout_width="wrap_content"
      android:layout_height="wrap_content"
      android:layout_alignParentBottom="true"
      android:layout_alignParentRight="true"
      android:layout_marginRight="32dp"
      android:text="上下"
      android:textAppearance="?android:attr/textAppearanceLarge" />
```

```
    <TextView
        android:id="@+id/tv1"
        android:layout_width="wrap_content"
        android:layout_height="wrap_content"
        android:layout_alignParentBottom="true"
        android:layout_alignParentLeft="true"
        android:layout_marginLeft="29dp"
        android:text="左右"
        android:textAppearance="?android:attr/textAppearanceLarge" />

    <TextView
        android:id="@+id/tv2"
        android:layout_width="wrap_content"
        android:layout_height="wrap_content"
        android:layout_alignParentBottom="true"
        android:layout_centerHorizontal="true"
        android:text="前后"
        android:textAppearance="?android:attr/textAppearanceLarge" />
     </RelativeLayout>
    <RelativeLayout
        android:id="@+android:id/ball_container"
        android:layout_width="fill_parent"
        android:layout_height="fill_parent"
        android:background="@drawable/bg"
        android:orientation="vertical" >
        <cn.accelerometer.view.BallView
            xmlns:android="http://schemas.android.com/apk/res/android"
            android:id="@+android:id/ball"
            android:layout_width="wrap_content"
            android:layout_height="wrap_content"
            android:layout_alignParentLeft="true"
            android:layout_alignParentTop="true"
            android:scaleType="center"
            android:src="@drawable/ball" />
    </RelativeLayout>
</LinearLayout>
```

2. 实现程序文件

编写文件 BallView.java 实现小球视图，具体实现代码如下所示。

```
public class BallView extends ImageView {

    public BallView(Context context) {
        super(context);
    }

    public BallView(Context context, AttributeSet attrs) {
        super(context, attrs);
    }

    public BallView(Context context, AttributeSet attrs, int defStyle) {
        super(context, attrs, defStyle);
    }

    public void moveTo(int x, int y) {//加一个 TextView，球就不能移动了
        super.setFrame(x, y, x + getWidth(), y + getHeight());//绘制视图，由左上角与右下角
        //确定视图矩形位置
    }
}
```

编写文件 AccelerometerActivity.java，功能是监听传感器的运动轨迹，获取小球在 X 轴、Y 轴和 Z 轴的重力值。文件 AccelerometerActivity.java 的具体实现代码如下所示。

```
public class AccelerometerActivity extends Activity {
    private static final float MAX_ACCELEROMETER = 9.81f;
    private SensorManager sensorManager;
    private BallView ball;
    private boolean success = false;
```

```
    private boolean init = false;
    private int container_width = 0;
    private int container_height = 0;
    private int ball_width = 0;
    private int ball_height = 0;
    private TextView prompt;
    private TextView tv1;
    private TextView tv2;
    private TextView tv3;
    /*
    private int a=0;
    private int b=0;
     */
    @Override
    public void onCreate(Bundle savedInstanceState) {
        super.onCreate(savedInstanceState);
        setContentView(R.layout.main);
        //获取感应器管理器
        sensorManager = (SensorManager) getSystemService(SENSOR_SERVICE);
        prompt = (TextView) findViewById(R.id.ball_prompt);
        tv1 = (TextView)findViewById(R.id.tv1);
        tv2 = (TextView)findViewById(R.id.tv2);
        tv3 = (TextView)findViewById(R.id.tv3);
    }

    @Override
    public void onWindowFocusChanged(boolean hasFocus) {//ball_container控件显示出来后才
//能获取其宽和高，所以通过此方法得到其宽高
        super.onWindowFocusChanged(hasFocus);
        if(hasFocus && !init){
            View container = findViewById(R.id.ball_container);
            container_width = container.getWidth();
            container_height = container.getHeight();
            ball = (BallView) findViewById(R.id.ball);
            ball_width = ball.getWidth();
            ball_height = ball.getHeight();
            moveTo(0f, 0f);
            init = true;
        }
    }

    @Override
    protected void onResume() {
        Sensor sensor = sensorManager.getDefaultSensor(Sensor.TYPE_ACCELEROMETER);
        //获取重力加速度感应器
        success = sensorManager.registerListener(listener, sensor, SensorManager.SENSOR_
        DELAY_GAME);//注册listener，第3个参数是检测的精确度
        super.onResume();
    }

    @Override
    protected void onPause() {
        if(success) sensorManager.unregisterListener(listener);
        super.onPause();
    }

    private SensorEventListener listener = new SensorEventListener() {
        @Override
        public void onSensorChanged(SensorEvent event) {


            /*
            Button bt1= (Button)findViewById(R.id.bt1);//测试图片移动
            bt1.setOnClickListener(new OnClickListener() {
                @Override
                public void onClick(View paramView) {
                    moveTo(0,++b);
                    //if(b>50)
```

```
                //    b=0;
            }
        });

        Button bt2= (Button)findViewById(R.id.bt2);
        bt2.setOnClickListener(new OnClickListener() {
            @Override
            public void onClick(View paramView) {
                moveTo(0,--b);
                //if(b>50)
                //    b=0;
            }
        });
        */

        if (!init) return ;
        float x = event.values[SensorManager.DATA_X];
        float y = event.values[SensorManager.DATA_Y];
        float z = event.values[SensorManager.DATA_Z];
        prompt.setText("X=" + x + ",Y= " + y + ", Z=" + z);
        //当重力x、y为0时，球处于中心位置，以y为轴心（固定不动），转动手机，x会在（0~9.81）
        //之间变化，负号代表方向
        moveTo(-x, y);//x方向取反


        if(x>0){
            tv1.setTextColor(Color.WHITE);
            tv1.setText("向左");
        }

        if(x<0){
            tv1.setTextColor(Color.CYAN);
            tv1.setText("向右");
        }

        if(y>0){
            tv2.setTextColor(Color.WHITE);
            tv2.setText("向后");
        }

        if(y<0){
            tv2.setTextColor(Color.RED);
            tv2.setText("向前");
        }

        if(z>0){
            tv3.setTextColor(Color.WHITE);
            tv3.setText("向上");
        }

        if(z<0){
            tv3.setTextColor(Color.YELLOW);
            tv3.setText("向下");
        }


    }
    @Override
    public void onAccuracyChanged(Sensor sensor, int accuracy) {
    }
};

private void moveTo(float x, float y) {


    int max_x = (container_width - ball_width) / 2;//在x轴可移动的最大值
    int max_y = (container_height - ball_height) / 2;//在y轴可移动的最大值
    //手机沿x、y轴垂直摆放时，自由落体加速度最大为9.81，当手机沿x、y轴成某个角度摆放时，变量x
    //和y即为该角度的加速度
```

```
        float percentageX = x / MAX_ACCELEROMETER;//得到当前加速度的比率，如果手机沿 x 轴垂直
        摆放，比率为 100%，即球在 x 轴上移动到最大值
        float percentageY = y / MAX_ACCELEROMETER;


        int pixel_x = (int) (max_x * percentageX);//得到 x 轴偏移量
        int pixel_y = (int) (max_y * percentageY);//得到 y 轴偏移量
        //以球在中心位置的坐标为参考点，加上偏移量，得到球的对应位置，然后移动球到该位置

        int x3=max_x + pixel_x;//屏幕中心位置+x 轴偏移
        int y3=max_y + pixel_y;//屏幕中心位置+y 轴偏移

        ball.moveTo(x3, y3);

    }
}
```

执行后会在屏幕上方显示当前滚动小球在 x 轴、y 轴和 z 轴的重力值，如图 11-2 所示。

▲图 11-2　执行效果

第 12 章　方向传感器详解

在 Android 设备中，经常需要检测设备的方向，例如设备的朝向和移动方向。在 Android 系统中，通常使用重力传感器、加速度传感器、磁场传感器和旋转矢量传感器来检测设备的方向。本章将详细讲解在 Android 设备中使用方向传感器检测设备方向的基本知识，为读者步入本书后面知识的学习打下基础。

12.1 方向传感器基础

方向传感器也被称为姿态传感器，与加速度传感器不同，方向传感器主要用于感应设备方位的变化。本节将详细讲解方向传感器的基本知识。

12.1.1　方向传感器必备知识

在现实世界中，方向传感器通过对力敏感的传感器，感受手机等设备在变换姿势时的重心变化，使手机等设备光标变化位置从而实现选择的功能。方向传感器运用了欧拉角的知识，欧拉角的基本思想是将角位移分解为绕 3 个互相垂直轴的 3 个旋转组成的序列。其实，任意 3 个轴和任意顺序都可以，但最有意义的是使用笛卡儿坐标系并按一定的顺序所组成的旋转序列。

在学习欧拉角知识之前先介绍几种不同概念的坐标系，以便于读者理解欧拉角知识。

（1）世界坐标系

世界坐标系是一个特殊的坐标系，建立了描述其他坐标系所需要的参考框架。能够用世界坐标系描述其他坐标系的位置，而不能用更大的、外部的坐标系来描述世界坐标系。例如，“向西”“向东”等词汇就是世界坐标系中的描述词汇。

（2）物体坐标系

物体坐标系是和特定物体相关联的坐标系，每个物体都有它们独立的坐标系。当物体移动或改变方向时，和该物体相关联的坐标系将随之移动或改变方向。例如，“向左”“向右”等词汇就是物体坐标系中的描述词汇。

（3）摄像机坐标系

摄像机坐标系是和观察者密切相关的坐标系。在摄像机坐标系中，摄像机在原点，x 轴向右，z 轴向前（朝向屏幕内或摄像机方向），y 轴向上（不是世界的上方而是摄像机本身的上方）。

（4）惯性坐标系

惯性坐标系是为了简化世界坐标系到物体坐标系的转换而引入的一种新的坐标系。惯性坐标系的原点和物体坐标系的原点重合，但惯性坐标系的轴平行于世界坐标系的轴。

在欧拉角中，表示一个物体的方位用“Yaw-Pitch-Roll”约定。在这个系统中，一个方位被定义为一个 Yaw 角、一个 Pitch 角和一个 Ron 角。欧拉角的基本思想是让物体开始于“标准”方位，目的是使物体坐标轴和惯性坐标轴对齐。在标准方位上，让物体做 Yaw、Pitch 和 Roll 旋转，最后物体到达我们想要描述的方位。

（5）Yaw 轴

Yaw 轴是 3 个方向轴中唯一不变的轴，其方向总是竖直向上，和世界坐标系中的 z 轴是等同的，也就是重力加速度 g 的反方向。如图 12-5 所示，为在图 12-4 的基础上手机绕 Yaw 轴旋转的效果。

（6）Pitch 轴

Pitch 轴方向依赖于手机沿 Yaw 轴的转动情况，即当手机沿 Yaw 转过一定的角度后，Pitch 轴也相应围绕 Yaw 轴转动相同的角度。Pitch 轴的位置依赖于手机沿 Yaw 轴转过的角度，好比 Yaw 轴和 Pitch 轴是两根焊死在一起成 90 度。

12.1.2　Android 中的方向传感器

在 Android 系统中，方向传感器的类型是 TYPE_ORIENTATION，用于测量设备围绕 3 个物理轴（x，y，z）的旋转角度，在新版本中已经使用 SensorManager.getOrientation()替代。Android 系统中的方向传感器在生活中的典型应用例子是指南针，接下来先来简单介绍一下传感器中 3 个参数 X、Y、Z 的含义，如图 12-1 所示。

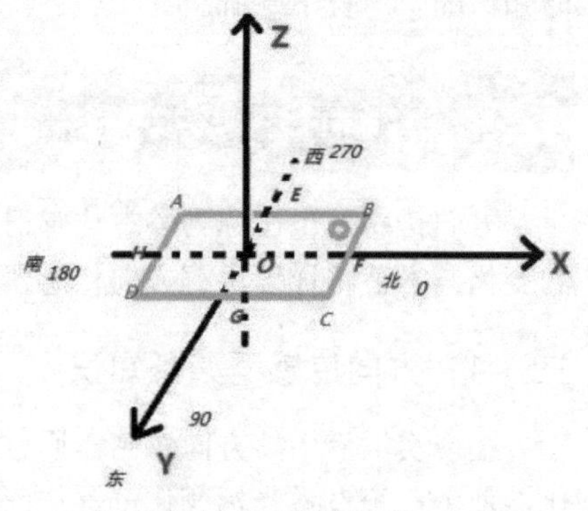

▲图 12-1　参数 X、Y、Z

如图 12-1 所示，矩形 ABCD 表示一个手机，带有小圈那一头是手机头部，各个部分的具体说明如下所示。

传感器中的 X：如上图所示，规定 X 正半轴为北，手机头部指向 OF 方向，此时 X 的值为 0。如果手机头部指向 OG 方向，此时 X 值为 90，指向 OH 方向，X 值为 180，指向 OE，X 值为 270。

传感器中的 Y：现在将手机沿着 BC 轴慢慢向上抬起，即手机头部不动，尾部慢慢向上翘起来，直到 AD 跑到 BC 右边并落在 XOY 平面上，Y 的值将从 0～180 之间变动，如果手机沿着 AD 轴慢慢向上抬起，即手机尾部不动，直到 BC 跑到 AD 左边并且落在 XOY 平面上，Y 的值将从 0～−180 之间变动，这就是方向传感器中 Y 的含义。

传感器中的 Z：现在将手机沿着 AB 轴慢慢向上抬起，即手机左边框不动，右边框慢慢向上翘起来，直到 CD 跑到 AB 右边并落在 XOY 平面上，Z 的值将从 0～180 之间变动，如果手机沿着 CD 轴慢慢向上抬起，即手机右边框不动，直到 AB 跑到 CD 左边并且落在 XOY 平面上，Z 的值将从 0～−180 之间变动，这就是方向传感器中 Z 的含义。

12.2　实战演练——测试当前设备的 3 个方向值

在接下来的实例中，将演示在 Android 中使用方向传感器的方法。

实例	功能	源码路径
实例 12-1	使用姿态传感器测试当前设备的 3 个方向值	\daima\12\zitai

12.2.1　实现布局文件

编写布局文件 main.xml，主要代码如下所示。

```
<TextView
    android:id="@+id/title"
```

```
        android:gravity="center_horizontal"
        android:textSize="20px"
        android:layout_width="fill_parent"
        android:layout_height="wrap_content"
        android:text="@string/title"/><!-- 添加一个 TextView 控件 -->
    <TextView
        android:id="@+id/myTextView1"
        android:textSize="18px"
        android:layout_width="fill_parent"
        android:layout_height="wrap_content"
        android:text="@string/myTextView1"/><!-- 添加一个 TextView 控件 -->
    <TextView
        android:id="@+id/myTextView2"
        android:textSize="18px"
        android:layout_width="fill_parent"
        android:layout_height="wrap_content"
        android:text="@string/myTextView2"/><!-- 添加一个 TextView 控件 -->
    <TextView
        android:id="@+id/myTextView3"
        android:textSize="18px"
        android:layout_width="fill_parent"
        android:layout_height="wrap_content"
        android:text="@string/myTextView3"/><!-- 添加一个 TextView 控件 -->
```

12.2.2 实现主程序文件

编写主程序文件zitaiLI.java，此文件的具体实现流程如下所示。

- 声明3个分别用来显示Yaw、Pitch、Roll的TextView的引用。
- 声明SensorManager引用，而因调试原因用SensorManagerSimulator代替。
- 重写Activity的onCreate方法，在方法中设置当前的用户界面，然后得到各个控件的引用，并初始化SensorManager或SensorManagerSimulator。
- 初始化监听器类的对象，在重写的onSensorChanged方法中只对姿态SENSOR_ORIENTATION变化进行处理，将3个姿态值显示到TextView中。
- 重写Activity的onResume方法，在方法中为SensorManager注册监听，此处传入的传感器类型为SENSOR_ORIENTATION，表示只读取姿态数据。

文件zitaiCH.java的主要代码如下所示。

```
public class zitaiLI extends Activity {
    TextView myTextView1;
    TextView myTextView2;
    TextView myTextView3;
    //声明 SensorManagerSimulator 对象,调试时用
    SensorManagerSimulator mySensorManager;
    @Override
    public void onCreate(Bundle savedInstanceState) {          //重写 onCreate 方法
        super.onCreate(savedInstanceState);
        setContentView(R.layout.main);                        //设置用户界面
        myTextView1 = (TextView) findViewById(R.id.myTextView1);    //myTextView1 的引用
        myTextView2 = (TextView) findViewById(R.id.myTextView2);    //myTextView2 的引用
        myTextView3 = (TextView) findViewById(R.id.myTextView3);    //myTextView3 的引用
        //mySensorManager =
        //    (SensorManager)getSystemService(SENSOR_SERVICE);    //获得 SensorManager
        //调试时用
        mySensorManager = SensorManagerSimulator.getSystemService(this, SENSOR_SERVICE);
        mySensorManager.connectSimulator();                    //与 Simulator 服务器连接
    }
    private SensorListener mySensorListener = new SensorListener(){
        @Override
        public void onAccuracyChanged(int sensor, int accuracy) {}//重写 onAccuracyChanged 方法
        @Override
        public void onSensorChanged(int sensor,float[]values){//重写 onSensorChanged 方法
            if(sensor == SensorManager.SENSOR_ORIENTATION){       //检查姿态变化
                myTextView1.setText("Yaw 为: "+values[0]);         //TextView 数据显示
```

```
                myTextView2.setText("Pitch 为: "+values[1]);          //TextView 数据显示
                myTextView3.setText("Roll 为: "+values[2]);           //TextView 数据显示
            }
        }
    };
    @Override
    protected void onResume() {                                    //重写的 onResume 方法
        mySensorManager.registerListener(                          //注册监听
                mySensorListener,                                  //监听器 SensorListener 对象
                SensorManager.SENSOR_ORIENTATION,                  //姿态传感器的类型
                SensorManager.SENSOR_DELAY_UI                      //传感器事件传递频度
                );
        super.onResume();
    }
    @Override
    protected void onPause() {                                        //重写 onPause 方法
        mySensorManager.unregisterListener(mySensorListener);          //取消注册监听器
        super.onPause();
    }
}
```

因为此实例比较简单，是根据 SensorSimulator 中附带的开源代码改编的，所以在此不再进行详细介绍，读者只需阅读本书附带程序中的源码即可。

12.3 实战演练——开发一个指南针程序

在接下来的实例中，将演示在 Android 中使用方向传感器开发指南针应用程序的方法。在本实例中首先准备一张指南针素材图片，该图片上方向指南针指向北方。接下来开发一个检测方向的传感器，传感器程序可以检测到手机顶部绕 Z 转过的多少度。在实例中需添加了一张图片，并让图片总是反转方向传感器返回的第一个角度值。

实例	功能	源码路径
实例 12-2	在设备中开发指南针程序	\daima\12\zhinanzhen

12.3.1 实现布局文件

编写布局文件 main.xml，功能是插入准备好的素材图片，主要实现代码如下所示。

```
<LinearLayout xmlns:android="http://schemas.android.com/apk/res/android"
    android:orientation="vertical"
    android:layout_width="fill_parent"
    android:layout_height="fill_parent"
    android:background="#fff"
    >
<ImageView
    android:id="@+id/znzImage"
    android:layout_width="fill_parent"
    android:layout_height="fill_parent"
    android:scaleType="fitCenter"
    android:src="@drawable/compass" />
</LinearLayout>
```

12.3.2 实现程序文件

编写程序文件 Zhinanzheng.java，使用传感器获取设备的旋转角度值，并根据这个值返回指南针的角度。文件 Zhinanzheng.java 的具体实现代码如下所示。

```
public class Zhinanzheng extends Activity implements SensorEventListener{

    ImageView image;  //指南针图片
    float currentDegree = 0f; //指南针图片转过的角度
```

```
    SensorManager mSensorManager; //管理器

    /** Called when the activity is first created. */
    @Override
    public void onCreate(Bundle savedInstanceState) {
        super.onCreate(savedInstanceState);
        setContentView(R.layout.main);

        image = (ImageView)findViewById(R.id.znzImage);
        mSensorManager = (SensorManager)getSystemService(SENSOR_SERVICE); //获取管理服务
    }

    @Override
    protected void onResume(){
        super.onResume();
        //注册监听器
        mSensorManager.registerListener(this
                , mSensorManager.getDefaultSensor(Sensor.TYPE_ORIENTATION), SensorManager.
                SENSOR_DELAY_GAME);
    }

    //取消注册
    @Override
    protected void onPause(){
        mSensorManager.unregisterListener(this);
        super.onPause();

    }

    @Override
    protected void onStop(){
        mSensorManager.unregisterListener(this);
        super.onStop();

    }

    //传感器值改变
    @Override
    public void onAccuracyChanged(Sensor sensor, int accuracy) {
        // TODO Auto-generated method stub


    }
    //精度改变
    @Override
    public void onSensorChanged(SensorEvent event) {
        // TODO Auto-generated method stub
        //获取触发event的传感器类型
        int sensorType = event.sensor.getType();

        switch(sensorType){
        case Sensor.TYPE_ORIENTATION:
            float degree = event.values[0]; //获取z转过的角度
            //创建旋转动画
            RotateAnimation ra = new RotateAnimation
(currentDegree,-degree,Animation. RELATIVE_TO_SELF,0.5f
                ,Animation.RELATIVE_TO_SELF,0.5f);
         ra.setDuration(100);//动画持续时间
         image.startAnimation(ra);
         currentDegree = -degree;
         break;

        }
    }
}
```

执行后的效果如图12-2所示。

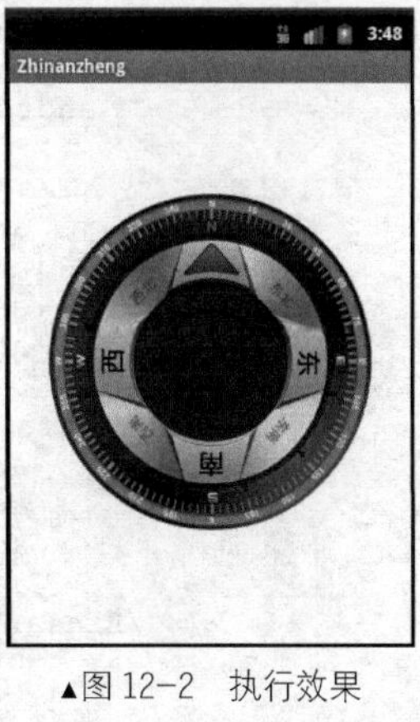

▲图12-2 执行效果

12.4 开发一个具有定位功能的指南针

在接下来的实例中，将演示在 Android 设备中使用方向传感器实现具有定位功能指南针的方法。在本实例中，通过 handler 设置每隔 20 毫秒后执行一次，以检测方向的变化值，达到更新指南针旋转的功能。在实例界面中自定义了一个 ImageView，并增加了一个旋转图片的方法。为了实现指南针的动画效果，特意使用了加速插入器功能。在 Android 系统中提供了如下所示的插入器。

- AccelerateInterpolator：动画从开始到结束，变化率是一个加速的过程。
- DecelerateInterpolator：动画从开始到结束，变化率是一个减速的过程。
- CycleInterpolator：动画从开始到结束，变化率是循环给定次数的正弦曲线。
- AccelerateDecelerateInterpolator：动画从开始到结束，变化率是先加速后减速的过程。
- LinearInterpolator：动画从开始到结束，变化率是线性变化。

实例	功能	源码路径
实例 12-3	开发一个具有定位功能的指南针	\daima\12\Compass

12.4.1 实现布局文件

编写界面的布局文件 main.xml，具体代码如下所示。

```
<FrameLayout xmlns:android="http://schemas.android.com/apk/res/android"
   android:layout_width="fill_parent"
   android:layout_height="fill_parent" >

   <LinearLayout
      android:id="@+id/view_compass"
      android:layout_width="fill_parent"
      android:layout_height="fill_parent"
      android:orientation="vertical" >

      <FrameLayout
         android:layout_width="fill_parent"
         android:layout_height="0.0dip"
         android:layout_weight="1.0"
         android:background="@drawable/background_compass"
         android:gravity="center" >

         <!--
         <LinearLayout
            android:id="@id/pressure_altitude"
            android:layout_width="fill_parent"
            android:layout_height="wrap_content"
            android:layout_gravity="center"
            android:gravity="center"
            android:orientation="horizontal" >

            <ImageView
               android:id="@+id/pressure_show"
               android:layout_width="wrap_content"
               android:layout_height="wrap_content"
               android:background="@drawable/pressure_bg" />

            <ImageView
               android:id="@+id/altitude_show"
               android:layout_width="wrap_content"
               android:layout_height="wrap_content"
               android:background="@drawable/altitude_bg" />
         </LinearLayout>
         -->
```

```
            <LinearLayout
                android:id="@+id/layout_direction"
                android:layout_width="wrap_content"
                android:layout_height="wrap_content"
                android:layout_gravity="center_horizontal"
                android:layout_marginTop="@dimen/direction_margin_top"
                android:textColor="@android:color/white"
                android:textSize="@dimen/text_direction_size" />

            <net.micode.compass.CompassView
                android:id="@+id/compass_pointer"
                android:layout_width="wrap_content"
                android:layout_height="wrap_content"
                android:layout_gravity="center"
                android:layout_marginTop="@dimen/compass_margin_top"
                android:src="@drawable/compass" />

            <LinearLayout
                android:id="@+id/layout_angle"
                android:layout_width="wrap_content"
                android:layout_height="wrap_content"
                android:layout_gravity="center"
                android:layout_marginLeft="2.0dip"
                android:layout_marginTop="@dimen/degree_text_margin_top"
                android:orientation="horizontal" />
        </FrameLayout>

        <LinearLayout
            android:id="@+id/bottom_unit"
            android:layout_width="fill_parent"
            android:layout_height="wrap_content"
            android:background="@drawable/background_bottom"
            android:gravity="center_vertical"
            android:orientation="vertical" >

            <TextView
                android:id="@+id/textview_location_latitude_degree"
                android:layout_width="fill_parent"
                android:layout_height="wrap_content"
                android:gravity="center_vertical"
                android:paddingLeft="10dip"
                android:paddingRight="10dip"
                android:text="@string/cannot_get_location"
                android:textColor="@android:color/white" />

            <TextView
                android:id="@+id/textview_location_longitude_degree"
                android:layout_width="fill_parent"
                android:layout_height="wrap_content"
                android:gravity="center_vertical"
                android:paddingLeft="10dip"
                android:paddingRight="10dip"
                android:text="@string/cannot_get_location"
                android:textColor="@android:color/white" />
        </LinearLayout>
    </LinearLayout>

    <FrameLayout
        android:id="@+id/view_guide"
        android:layout_width="fill_parent"
        android:layout_height="fill_parent"
        android:visibility="gone" >

        <ImageView
            android:id="@+id/guide_description"
            android:layout_width="fill_parent"
            android:layout_height="fill_parent"
            android:layout_gravity="center"
            android:background="@drawable/guide" />
```

```
        <ImageView
            android:id="@+id/guide_animation"
            android:layout_width="wrap_content"
            android:layout_height="wrap_content"
            android:layout_gravity="top|center"
            android:layout_marginTop="195.0dip"
            android:src="@drawable/calibrate_animation" />
    </FrameLayout>

</FrameLayout>
```

12.4.2　实现程序文件

编写文件 CompassView.java，自定义一个继承于 ImageView 的 View 类 CompassView，增加一个通用的旋转图片资源的方法。文件 CompassView.java 的具体实现代码如下所示。

```
/**
 * 自定义一个 View 继承 ImageView，增加一个通用的旋转图片资源的方法
 *
 */
public class CompassView extends ImageView {
    private float mDirection;// 方向旋转浮点数
    private Drawable compass;// 图片资源

    //3 个构造器
    public CompassView(Context context) {
        super(context);
        mDirection = 0.0f;// 默认不旋转
        compass = null;
    }

    public CompassView(Context context, AttributeSet attrs) {
        super(context, attrs);
        mDirection = 0.0f;
        compass = null;
    }

    public CompassView(Context context, AttributeSet attrs, int defStyle) {
        super(context, attrs, defStyle);
        mDirection = 0.0f;
        compass = null;
    }

    @Override
    protected void onDraw(Canvas canvas) {
        if (compass == null) {
            compass = getDrawable();// 获取当前 view 的图片资源
            compass.setBounds(0, 0, getWidth(), getHeight());// 图片资源在 view 的位置，此处
            //相当于充满 view
        }

        canvas.save();
        canvas.rotate(mDirection, getWidth() / 2, getHeight() / 2);// 绕图片中心点旋转
        compass.draw(canvas);// 把旋转后的图片画在 view 上，即保持旋转后的样子
        canvas.restore();// 保存一下
    }

    /**
     * 自定义更新方向的方法
     *
     * @param direction
     *            传入的方向
     */
    public void updateDirection(float direction) {
        mDirection = direction;
        invalidate();// 重新刷新一下，更新方向
    }
}
```

编写文件 CompassActivity.java，实现传感器的旋转判断处理，并获取旋转角度以返回指南针的具体方位。文件 CompassActivity.java 的具体实现代码如下所示。

```
public class CompassActivity extends Activity {
   private static final int EXIT_TIME = 2000;// 两次按返回键的间隔判断
   private final float MAX_ROATE_DEGREE = 1.0f;// 最多旋转一周，即 360°
   private SensorManager mSensorManager;// 传感器管理对象
   private Sensor mOrientationSensor;// 传感器对象
   // private LocationManager mLocationManager;// 位置管理对象
   // private String mLocationProvider;// 位置提供者名称，GPS 设备还是网络
   private float mDirection;// 当前浮点方向
   private float mTargetDirection;// 目标浮点方向
   private AccelerateInterpolator mInterpolator;// 动画从开始到结束，变化率是一个加速的过程，
   //就是一个动画速率
   protected final Handler mHandler = new Handler();
   private boolean mStopDrawing;// 是否停止指南针旋转的标志位
   private boolean mChinease;// 系统当前是否使用中文
   private long firstExitTime = 0L;// 用来保存第一次按返回键的时间

   LocationApplication application;
   View mCompassView;
   CompassView mPointer;// 指南针 view
   // TextView mLocationTextView;// 显示位置的 view
   TextView mLatitudeTV;// 纬度
   TextView mLongitudeTV;// 经度
   LinearLayout mDirectionLayout;// 显示方向（东南西北）的 view
   LinearLayout mAngleLayout;// 显示方向度数的 view
   View mViewGuide;
   ImageView mGuideAnimation;
   protected Handler invisiableHandler = new Handler() {
      public void handleMessage(Message msg) {
         mViewGuide.setVisibility(View.GONE);
      }
   };

   public void onWindowFocusChanged(boolean hasFocus) {
      AnimationDrawable anim = (AnimationDrawable) mGuideAnimation
            .getDrawable();
      anim.start();
   }

   // 这个是更新指南针旋转的线程，handler 的灵活使用，每 20 毫秒检测方向变化值，对应更新指南针旋转
   protected Runnable mCompassViewUpdater = new Runnable() {
      @Override
      public void run() {
         if (mPointer != null && !mStopDrawing) {
            if (mDirection != mTargetDirection) {

               // calculate the short routine
               float to = mTargetDirection;
               if (to - mDirection > 180) {
                  to -= 360;
               } else if (to - mDirection < -180) {
                  to += 360;
               }

               // limit the max speed to MAX_ROTATE_DEGREE
               float distance = to - mDirection;
               if (Math.abs(distance) > MAX_ROATE_DEGREE) {
                  distance = distance > 0 ? MAX_ROATE_DEGREE
                        : (-1.0f * MAX_ROATE_DEGREE);
               }

               // need to slow down if the distance is short
               mDirection = normalizeDegree(mDirection
                     + ((to - mDirection) * mInterpolator
                        .getInterpolation(Math.abs(distance)>MAX_ROATE_DEGREE ? 0.4f
                              : 0.3f)));// 用了一个加速动画去旋转图片，很细致
               mPointer.updateDirection(mDirection);// 更新指南针旋转
```

```
            }

            updateDirection();// 更新方向值

            mHandler.postDelayed(mCompassViewUpdater, 20);// 20 毫秒后重新执行自己，比定时器好
        }
    }
};

@Override
protected void onCreate(Bundle savedInstanceState) {
    super.onCreate(savedInstanceState);
    application = LocationApplication.getInstance();
    setContentView(R.layout.main);
    initResources();// 初始化 view
    initServices();// 初始化传感器和位置服务
    application.mTv = mLatitudeTV;
    application.mAddress = mLongitudeTV;

    if (application.mData != null)
        mLatitudeTV.setText(application.mData);
    if (application.address != null)
        mLatitudeTV.setText(application.address);
    application.mLocationClient.start();
}

@Override
public void onBackPressed() {// 覆盖返回键
    long curTime = System.currentTimeMillis();
    if (curTime - firstExitTime < EXIT_TIME) {// 两次按返回键的时间小于 2 秒就退出应用
        finish();
    } else {
        Toast.makeText(this, "再按一次退出", Toast.LENGTH_SHORT).show();
        firstExitTime = curTime;
    }
}

@Override
protected void onResume() {// 在恢复的生命周期里判断、启动位置更新服务和传感器服务
    super.onResume();
    // if (mLocationProvider != null) {
    // updateLocation(mLocationManager
    // .getLastKnownLocation(mLocationProvider));
    // mLocationManager.requestLocationUpdates(mLocationProvider, 2000,
    // 10, mLocationListener);// 2 秒或者距离变化 10 米时更新一次地理位置
    // }
    // else {
    // mLocationTextView.setText(R.string.cannot_get_location);
    // }
    if (mOrientationSensor != null) {
        mSensorManager.registerListener(mOrientationSensorEventListener,
                mOrientationSensor, SensorManager.SENSOR_DELAY_GAME);
    } else {
        // Toast.makeText(this, R.string.cannot_get_sensor,
        // Toast.LENGTH_SHORT)
        // .show();
    }
    mStopDrawing = false;
    mHandler.postDelayed(mCompassViewUpdater, 20);// 20 毫秒执行一次更新指南针图片旋转
}

@Override
protected void onPause() {// 在暂停的生命周期里注销传感器服务和位置更新服务
    super.onPause();
    mStopDrawing = true;
    if (mOrientationSensor != null) {
        mSensorManager.unregisterListener(mOrientationSensorEventListener);
    }
    // if (mLocationProvider != null) {
```

```
        // mLocationManager.removeUpdates(mLocationListener);
        // }
    }

    // 初始化view
    private void initResources() {
        mViewGuide = findViewById(R.id.view_guide);
        mViewGuide.setVisibility(View.VISIBLE);
        invisiableHandler.sendMessageDelayed(new Message(), 3000);
        mGuideAnimation = (ImageView) findViewById(R.id.guide_animation);
        mDirection = 0.0f;// 初始化起始方向
        mTargetDirection = 0.0f;// 初始化目标方向
        mInterpolator = new AccelerateInterpolator();// 实例化加速动画对象
        mStopDrawing = true;
        mChinease = TextUtils.equals(Locale.getDefault().getLanguage(), "zh");
        // 判断系统当前使用的语言是否为中文

        mCompassView = findViewById(R.id.view_compass);
        // 实际上是一个LinearLayout，装指南针ImageView和位置TextView
        mPointer = (CompassView) findViewById(R.id.compass_pointer);// 自定义的指南针view
        // mLocationTextView = (TextView)
        // findViewById(R.id.textview_location);// 显示位置信息的TextView
        mLongitudeTV = (TextView) findViewById(R.id.textview_location_longitude_ degree);
        mLatitudeTV = (TextView) findViewById(R.id.textview_location_latitude_degree);
        mDirectionLayout = (LinearLayout) findViewById(R.id.layout_direction);
        // 顶部显示方向名称（东南西北）的LinearLayout
        mAngleLayout = (LinearLayout) findViewById(R.id.layout_angle);
        // 顶部显示方向具体度数的LinearLayout

        // mPointer.setImageResource(mChinease ? R.drawable.compass_cn
        // : R.drawable.compass);// 如果系统使用中文，就用中文的指南针图片
    }

    // 初始化传感器和位置服务
    private void initServices() {
        // sensor manager
        mSensorManager = (SensorManager) getSystemService(Context.SENSOR_SERVICE);
        mOrientationSensor = mSensorManager.getSensorList(
                Sensor.TYPE_ORIENTATION).get(0);
        // Log.i("way", mOrientationSensor.getName());

        // location manager
        // mLocationManager = (LocationManager)
        // getSystemService(Context.LOCATION_SERVICE);
        // Criteria criteria = new Criteria();// 条件对象，即指定条件过滤获得LocationProvider
        // criteria.setAccuracy(Criteria.ACCURACY_FINE);// 较高精度
        // criteria.setAltitudeRequired(false);// 是否需要高度信息
        // criteria.setBearingRequired(false);// 是否需要方向信息
        // criteria.setCostAllowed(true);// 是否产生费用
        // criteria.setPowerRequirement(Criteria.POWER_LOW);// 设置低电耗
        // mLocationProvider = mLocationManager.getBestProvider(criteria,
        // true);// 获取条件最好的Provider

    }

    // 更新顶部方向显示的方法
    private void updateDirection() {
        LayoutParams lp = new LayoutParams(LayoutParams.WRAP_CONTENT,
                LayoutParams.WRAP_CONTENT);
        // 先移除layout中所有的view
        mDirectionLayout.removeAllViews();
        mAngleLayout.removeAllViews();

        // 下面是根据mTargetDirection，做方向名称图片的处理
        ImageView east = null;
        ImageView west = null;
        ImageView south = null;
        ImageView north = null;
        float direction = normalizeDegree(mTargetDirection * -1.0f);
```

```
if (direction > 22.5f && direction < 157.5f) {
    // east
    east = new ImageView(this);
    east.setImageResource(mChinease ? R.drawable.e_cn : R.drawable.e);
    east.setLayoutParams(lp);
} else if (direction > 202.5f && direction < 337.5f) {
    // west
    west = new ImageView(this);
    west.setImageResource(mChinease ? R.drawable.w_cn : R.drawable.w);
    west.setLayoutParams(lp);
}

if (direction > 112.5f && direction < 247.5f) {
    // south
    south = new ImageView(this);
    south.setImageResource(mChinease ? R.drawable.s_cn : R.drawable.s);
    south.setLayoutParams(lp);
} else if (direction < 67.5 || direction > 292.5f) {
    // north
    north = new ImageView(this);
    north.setImageResource(mChinease ? R.drawable.n_cn : R.drawable.n);
    north.setLayoutParams(lp);
}
// 下面是根据系统使用语言，更换对应的语言图片资源
if (mChinease) {
    // east/west should be before north/south
    if (east != null) {
        mDirectionLayout.addView(east);
    }
    if (west != null) {
        mDirectionLayout.addView(west);
    }
    if (south != null) {
        mDirectionLayout.addView(south);
    }
    if (north != null) {
        mDirectionLayout.addView(north);
    }
} else {
    // north/south should be before east/west
    if (south != null) {
        mDirectionLayout.addView(south);
    }
    if (north != null) {
        mDirectionLayout.addView(north);
    }
    if (east != null) {
        mDirectionLayout.addView(east);
    }
    if (west != null) {
        mDirectionLayout.addView(west);
    }
}
// 下面是根据方向度数显示度数图片数字
int direction2 = (int) direction;
boolean show = false;
if (direction2 >= 100) {
    mAngleLayout.addView(getNumberImage(direction2 / 100));
    direction2 %= 100;
    show = true;
}
if (direction2 >= 10 || show) {
    mAngleLayout.addView(getNumberImage(direction2 / 10));
    direction2 %= 10;
}
mAngleLayout.addView(getNumberImage(direction2));
// 下面是增加一个° 的图片
ImageView degreeImageView = new ImageView(this);
degreeImageView.setImageResource(R.drawable.degree);
```

```
    degreeImageView.setLayoutParams(lp);
    mAngleLayout.addView(degreeImageView);
}

// 获取方向度数对应的图片，返回 ImageView
private ImageView getNumberImage(int number) {
    ImageView image = new ImageView(this);
    LayoutParams lp = new LayoutParams(LayoutParams.WRAP_CONTENT,
            LayoutParams.WRAP_CONTENT);
    switch (number) {
    case 0:
        image.setImageResource(R.drawable.number_0);
        break;
    case 1:
        image.setImageResource(R.drawable.number_1);
        break;
    case 2:
        image.setImageResource(R.drawable.number_2);
        break;
    case 3:
        image.setImageResource(R.drawable.number_3);
        break;
    case 4:
        image.setImageResource(R.drawable.number_4);
        break;
    case 5:
        image.setImageResource(R.drawable.number_5);
        break;
    case 6:
        image.setImageResource(R.drawable.number_6);
        break;
    case 7:
        image.setImageResource(R.drawable.number_7);
        break;
    case 8:
        image.setImageResource(R.drawable.number_8);
        break;
    case 9:
        image.setImageResource(R.drawable.number_9);
        break;
    }
    image.setLayoutParams(lp);
    return image;
}

// 更新位置显示
private void updateLocation(Location location) {
    if (location == null) {
        // mLocationTextView.setText(R.string.getting_location);
        return;
    } else {
        // StringBuilder sb = new StringBuilder();
        double latitude = location.getLatitude();
        double longitude = location.getLongitude();
        String latitudeStr;
        String longitudeStr;
        if (latitude >= 0.0f) {
            latitudeStr = getString(R.string.location_north,
                    getLocationString(latitude));
        } else {
            latitudeStr = getString(R.string.location_south,
                    getLocationString(-1.0 * latitude));
        }

        // sb.append("    ");

        if (longitude >= 0.0f) {
            longitudeStr = getString(R.string.location_east,
                    getLocationString(longitude));
```

```
            } else {
                longitudeStr = getString(R.string.location_west,
                        getLocationString(-1.0 * longitude));
            }
            mLatitudeTV.setText(latitudeStr);
            mLongitudeTV.setText(longitudeStr);
            // mLocationTextView.setText(sb.toString());//
            // 显示经纬度，其实还可以进行反向编译，显示具体地址
        }
    }

    // 把经纬度转换成度分秒显示
    private String getLocationString(double input) {
        int du = (int) input;
        int fen = (((int) ((input - du) * 3600))) / 60;
        int miao = (((int) ((input - du) * 3600))) % 60;
        return String.valueOf(du) + "° " + String.valueOf(fen) + "′ "
                + String.valueOf(miao) + "″ ";
    }

    // 方向传感器变化监听
    private  SensorEventListener  mOrientationSensorEventListener  =  new  SensorEvent
Listener() {

        @Override
        public void onSensorChanged(SensorEvent event) {
            float direction = event.values[mSensorManager.DATA_X] * -1.0f;
            mTargetDirection = normalizeDegree(direction);// 赋值给全局变量，让指南针旋转
            // Log.i("way", event.values[mSensorManager.DATA_Y] + "");
        }

        @Override
        public void onAccuracyChanged(Sensor sensor, int accuracy) {
        }
    };

    // 调整方向传感器获取的值
    private float normalizeDegree(float degree) {
        return (degree + 720) % 360;
    }

    // 位置信息更新监听
    // LocationListener mLocationListener = new LocationListener() {
    //
    // @Override
    // public void onStatusChanged(String provider, int status, Bundle extras) {
    // if (status != LocationProvider.OUT_OF_SERVICE) {
    // updateLocation(mLocationManager
    // .getLastKnownLocation(mLocationProvider));
    // }
    // // else {
    // // mLocationTextView.setText(R.string.cannot_get_location);
    // // }
    // }
    //
    // @Override
    // public void onProviderEnabled(String provider) {
    // }
    //
    // @Override
    // public void onProviderDisabled(String provider) {
    // }
    //
    // @Override
    // public void onLocationChanged(Location location) {
    // updateLocation(location);// 更新位置
    // }
    // };
}
```

编写文件 LocationApplication.java，功能是获取并输出当前设备的具体经度和纬度位置，具体实现代码如下所示。

```
public class LocationApplication extends Application {
    private static LocationApplication instance;
    public LocationClient mLocationClient = null;
    public String mData;
    public String address;
    public MyLocationListenner myListener = new MyLocationListenner();
    public TextView mTv;
    public TextView mAddress;
    public static LocationApplication getInstance() {
        return instance;
    }

    @Override
    public void onCreate() {
        // TODO Auto-generated method stub
        instance = this;
        mLocationClient = new LocationClient(getApplicationContext());
        mLocationClient.registerLocationListener(myListener);
        setLocationOption();
        super.onCreate();
    }

    // 设置相关参数
    public void setLocationOption() {
        LocationClientOption option = new LocationClientOption();
        option.setProdName("Compass");
        option.setOpenGps(true); // 打开 gps
        option.setCoorType("bd09ll"); // 设置坐标类型
        option.setAddrType("all"); // 设置地址信息，仅设置为“all”时有地址信息，默认无地址信息
        option.setScanSpan(5 * 60 * 1000);// 设置定位模式，小于1秒则一次定位;大于等于1秒则定时定位
        option.setPriority(LocationClientOption.GpsFirst);
        mLocationClient.setLocOption(option);
    }


    /**
     *监听函数，有新位置的时候，格式化成字符串，输出到屏幕中
     */
    public class MyLocationListenner implements BDLocationListener {

        @Override
        public void onReceiveLocation(BDLocation location) {
            // TODO Auto-generated method stub
            if (location == null)
                return;
            StringBuffer sb = new StringBuffer(256);
            // sb.append("时间: ");
            // sb.append(location.getTime());
            sb.append("纬度 : ");
            sb.append(location.getLatitude() + "°");
            sb.append(", 经度 : ");
            sb.append(location.getLongitude() + "°");
//              sb.append(", 精度 : ");
//              sb.append(location.getRadius() + " 米");
            mData = sb.toString();
            if(mTv != null) mTv.setText(sb);

            if (location.getLocType() == BDLocation.TypeGpsLocation) {
                address = "速度 : " + location.getSpeed();
                // sb.append("\n?;? : ");
                // sb.append(location.getSpeed());
            } else if (location.getLocType() == BDLocation.TypeNetWorkLocation) {
                // sb.append("\n? ?_ : ");
                // sb.append(location.getAddrStr());
                address = "地址 : " + location.getAddrStr();
            }
```

```
            if(mAddress !=null) mAddress.setText(address);
//              logMsg(sb.toString());
        }

        @Override
        public void onReceivePoi(BDLocation poiLocation) {
            // TODO Auto-generated method stub

        }

    }
}
```

到此为止，整个实例介绍完毕，执行后的效果如图 12-3 所示。

▲图 12-3　执行效果

第 13 章　陀螺仪传感器详解

陀螺仪传感器是一个基于自由空间移动和手势的定位和控制系统。例如我们可以在假想的平面上移动鼠标，屏幕上的光标就会随之跟着移动，并且可以绕着链接画圈和点击按键。又比如当我们正在演讲或离开桌子时，这些操作都能够很方便地实现。陀螺仪传感器已经被广泛运用于手机、平板等移动便携设备上，在将来其他设备也会陆续使用陀螺仪传感器。本章将详细讲解在Android 设备中使用陀螺仪传感器的基本知识，为读者步入本书后面知识的学习打下基础。

13.1 陀螺仪传感器基础

陀螺仪的原理是，当一个旋转物体的旋转轴所指的方向在不受外力影响时是不会改变的。根据这个道理，可以用陀螺仪来保持方向。然后用多种方法读取轴所指示的方向，并自动将数据信号传给控制系统。在现实生活中，骑自行车便是利用了这个原理。轮子转得越快越不容易倒，因为车轴有一股保持水平的力量。现代陀螺仪可以精确地确定运动物体的方位，在现代航空、航海、航天和国防工业中广泛使用一种惯性导航仪器。传统的惯性陀螺仪主要部分有机械式的陀螺仪，而机械式的陀螺仪对工艺结构的要求很高。20 世纪 70 年代提出了现代光纤陀螺仪的基本设想，到 80 年代以后，光纤陀螺仪得到了非常迅速的发展，激光谐振陀螺仪也有了很大的发展。光纤陀螺仪结构紧凑，灵敏度高，工作可靠，在很多领域已经完全取代了机械式的传统陀螺仪，成为现代导航仪器中的关键部件。与光纤陀螺仪同时发展的还有环式激光陀螺仪。

根据框架的数目和支承的形式以及附件的性质进行划分，陀螺仪传感器的主要类型如下所示。

（1）二自由度陀螺仪：只有一个框架，使转子自转轴具有一个转动自由度。根据二自由度陀螺仪中所使用的反作用力矩的性质，可以把这种陀螺仪分为如下所示的 3 种类型。

- 积分陀螺仪（它使用的反作用力矩是阻尼力矩）。
- 速率陀螺仪（它使用的反作力矩是弹性力矩）。
- 无约束陀螺（它仅有惯性反作用力矩）。

另外，除了机、电框架式陀螺仪外还出现了某些新型陀螺仪，例如静电式自由转子陀螺仪、挠性陀螺仪和激光陀螺仪等。

（2）三自由度陀螺仪：具有内、外两个框架，使转子自转轴具有两个转动自由度。在没有任何力矩装置时，它就是一个自由陀螺仪。

在当前技术水平条件下，陀螺仪传感器主要用于如下两个领域。

（1）国防工业

陀螺仪传感器原本是运用到直升机模型上的，现在它已经被广泛运用于手机这类移动便携设备上。不仅如此，现代陀螺仪是一种能够精确地确定运动物体的方位的仪器，所以陀螺仪传感器是现代航空、航海、航天和国防工业应用中必不可少的控制装置。陀螺仪传感器是法国的物理学家莱昂·傅科在研究地球自转时命名的，到如今一直是航空和航海上航行姿态及速率等最方便实用的参考仪表。

（2）开门报警器

陀螺仪传感器可以测量开门的角度，当门被打开一个角度后会发出报警声，或者结合 GPRS 模块发送短信以提醒门被打开了。另外，陀螺仪传感器集成了加速度传感器的功能，当门被打开的瞬间，将产生一定的加速度值，陀螺仪传感器会测量到这个加速度值，达到预设的门槛值后，将发出报警声，或者结合 GPRS 模块发送短信以提醒门被打开了。报警器内还可以集成雷达感应测量功能，当有人进入房间内移动时就会被雷达测量到。双重保险提醒防盗，可靠性高，误报率低，非常适合重要场合的防盗报警。

13.2 Android 中的陀螺仪传感器

在 Android 系统中，陀螺仪传感器的类型是 TYPE_GYROSCOPE，单位是 rad/s，能够测量设备 x、y、z 三轴的角加速度数据。Android 中的陀螺仪传感器又名为 Gyro-sensor 角速度器，利用内部震动机械结构侦测物体转动所产生的角速度，进而计算出物体移动的角度，侦测水平改变的状态，但无法计算移动的激烈程度。在本节的内容中，将详细讲解 Android 中的陀螺仪传感器的基本知识。

13.2.1　陀螺仪传感器和加速度传感器的对比

在 Android 的传感器系统中，陀螺仪传感器和加速度传感器非常类似，两者的区别如下所示。

- 加速度传感器：用于测量加速度，借助一个三轴加速度计可以测得一个固定平台相对地球表面的运动方向，但是一旦平台运动起来，情况就会变得复杂的多。如果平台做自由落体，加速度计测得的加速度值为零。如果平台朝某个方向做加速度运动，各个轴向加速度值会含有重力产生的加速度值，使得无法获得真正的加速度值。例如，安装在 60 度横滚角飞机上的三轴加速度计会测得 2G 的垂直加速度值，而事实上飞机相对地区表面是 60 度的倾角。因此，单独使用加速度计无法使飞机保持一个固定的航向。
- 陀螺仪传感器：用于测量机体围绕某个轴向的旋转角速率值。当使用陀螺仪测量飞机机体轴向的旋转角速率时，如果飞机在旋转，测得的值为非零值，飞机不旋转时，测量的值为零。因此，在 60 度横滚角的飞机上的陀螺仪测得的横滚角速率值为零，同样在飞机做水平直线飞行时的角速率值为零。可以通过角速率值的时间积分来估计当前的横滚角度，前提是没有误差的累积。陀螺仪测量的值会随时间漂移，经过几分钟甚至几秒钟定会累积出额外的误差来，而最终会导致对飞机当前相对水平面横滚角度完全错误的认知。因此，单独使用陀螺仪也无法保持飞机的特定航向。

综上所述，加速度传感器在较长时间的测量值（确定飞机航向）是正确的，而在较短时间内由于信号噪声的存在而有误差。陀螺仪传感器则在较短时间内比较准确，而较长时间会由于漂移而存有误差。因此，需要两者（相互调整）来确保航向的正确。

13.2.2　智能设备中的陀螺仪传感器

在穿戴设备中，三自由度陀螺仪是一个可以识别设备，能够相对于地面绕 X、Y、Z 轴转动角度的感应器（自己的理解，不够严谨）。无论是可设备，还是智能手机、平板电脑，通过使用陀螺仪传感器可以实现很多好玩的应用，例如说指南针。

在实际开发过程中，可以用一个磁场感应器（Magnetic Sensor）来实现陀螺仪。磁场感应器是用来测量磁场感应强度的。一个 3 轴的磁 sensor IC 可以得到当前环境下 X、Y 和 Z 方向上的磁场感应强度，对于 Android 中间层来说就是读取该感应器测量到的这 3 个值。当需要时，上报给上层应用程序。磁感应强度的单位是 T（特斯拉）或者是 Gs（高斯），1T 等于 10000Gs。

在了解陀螺仪之前，需要先了解 Android 系统定义坐标系的方法，如下所示的文件中进行了定义。

```
/hardware/libhardware/include/hardware/sensors.h
```

在上述文件 sensors.h 中，有一个如下图 13-1 所示的效果图。

图 13-1 表示设备的正上方是 Y 轴方向，右边是 X 轴方向，垂直设备屏幕平面向上的是 Z 轴方向，这个很重要。因为应用程序就是根据这样的定义来写的，所以我们报给应用的数据要跟这个定义符合。还需要清楚磁 Sensor 芯片贴在板上的坐标系。我们从芯片读出数据后要把芯片的坐标系转换为设备的实际坐标系。除非芯片贴在板上刚好跟设备的 X、Y、Z 轴方向刚好一致。

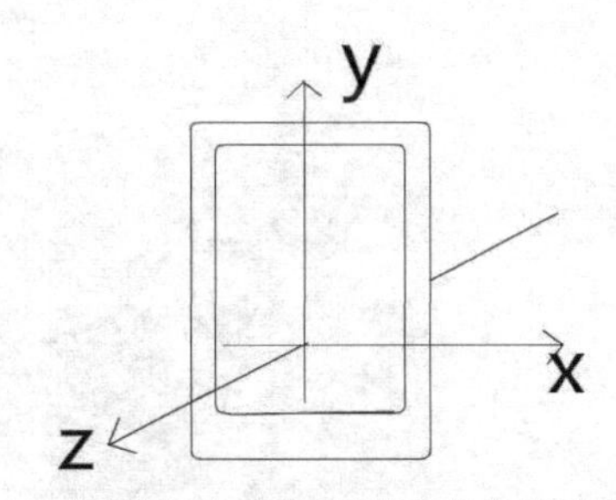

▲图 13-1 Android 系统定义的坐标系

陀螺仪的实现是根据磁场感应强度的 3 个值计算出另外 3 个值。当需要时可以计算出这 3 个值上报给应用程序，这样就实现了陀螺仪的功能。在接下来的内容中，将详细讲解这 3 个值具体含义和计算方法。

（1）Azimuth 方位角：即绕 Z 轴转动的角度，0 度=正北。假设 Y 轴指向地磁正北方，直升机正前方的方向如下图 13-2 所示。

90 度=正东时如图 13-3 所示。

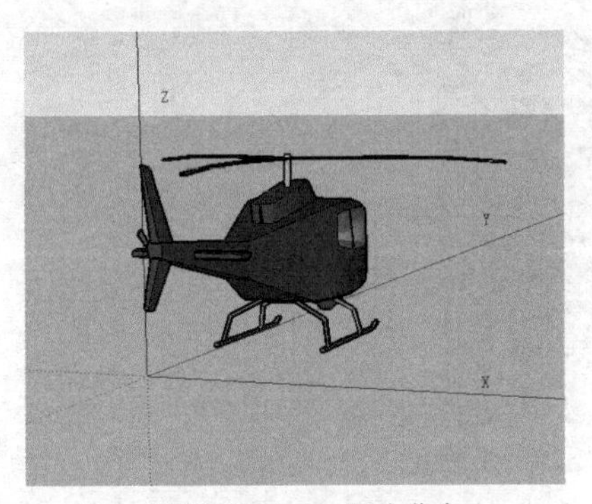
▲图 13-2 Azimuth 方位角

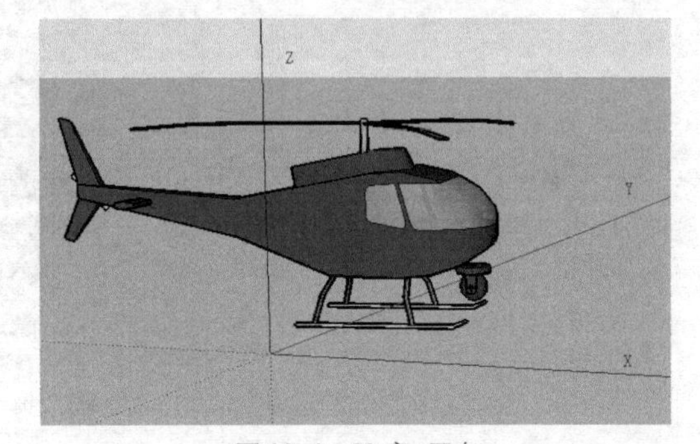
▲图 13-3 90 度=正东

180 度=正南时如图 13-4 所示。

270 度=正西时如图 13-5 所示。

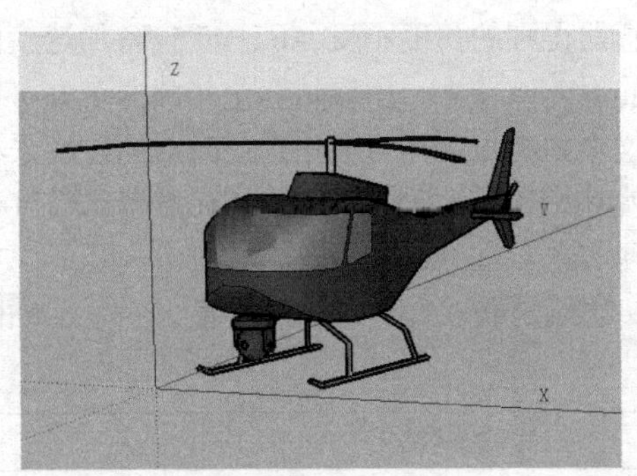
▲图 13-4 180 度=正南

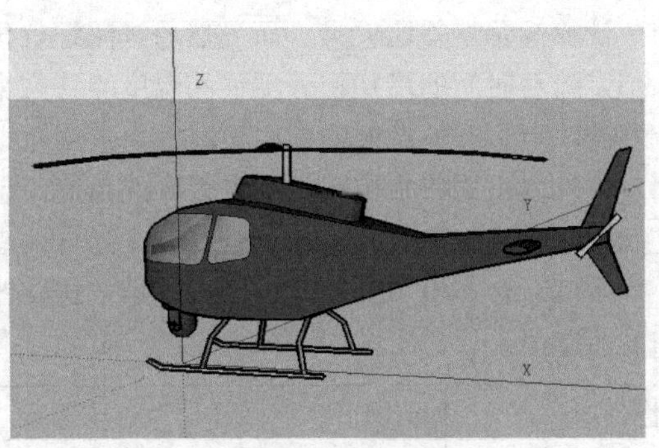
▲图 13-5 270 度=正西

在这种情况下，通过计算 *X* 和 *Y* 方向的磁感应强度的反正切的方式就可以得到方位角。要想实现指南针，只需要用这个值即可（不考虑设备非水平的情况）。

（2）Pitch 仰俯：绕 X 轴转动的角度（–180<=pitch<=180），如果设备水平放置，前方向下俯就是正，如图 13-6 所示。

前方向上仰就是负值，如图 13-7 所示。

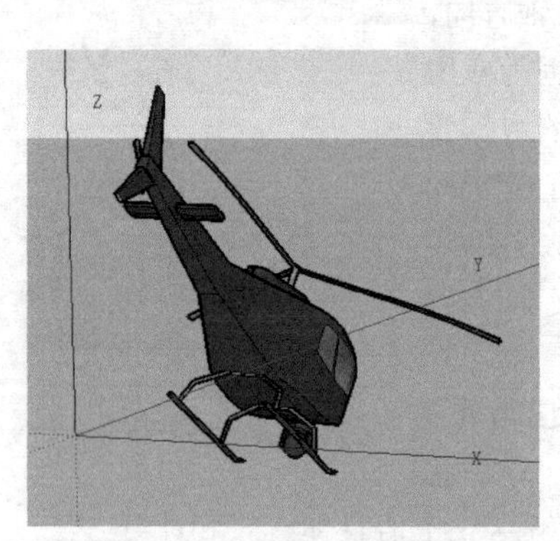

▲图 13-6　前方向下俯就是正

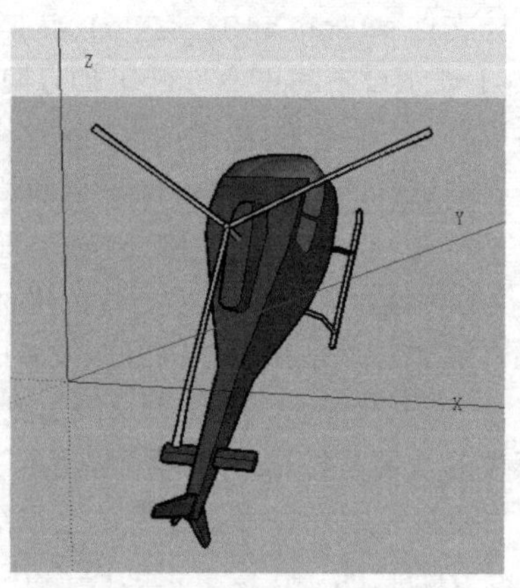

▲图 13-7　前方向上仰就是负值

在这种情况下，计算磁 Sensor 的 *Y* 和 *Z* 反正切就可以得到此角度值。如图 13-8 所示。

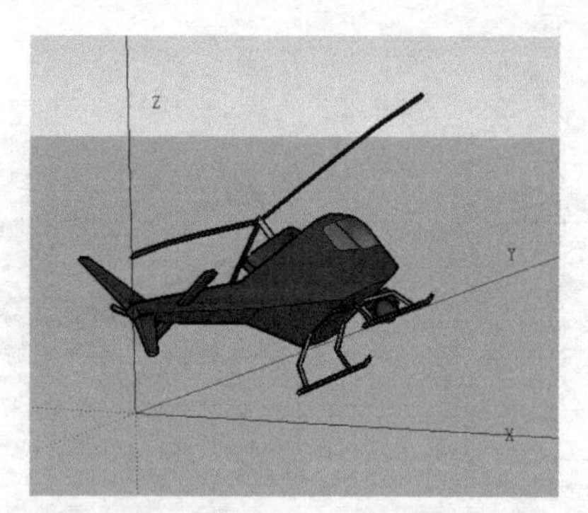

▲图 13-8　计算磁 Sensor 的 *Y* 和 *Z* 反正切

13.3　实战演练——联合使用互补滤波器和陀螺仪传感器

在现实应用中，加速度传感器对 Android 设备（其他设备也是如此）的加速度比较敏感，取瞬时值计算倾角误差比较大。而虽然通过陀螺仪传感器得到的角度不受 Android 设备加速度的影响，但是随着时间的增加积分漂移和温度漂移带来的误差比较大。所以这两个传感器正好可以弥补相互的缺点。互补滤波就是在短时间内采用陀螺仪得到的角度作为最优，定时对加速度采样来的角度进行取平均值来校正陀螺仪得到的角度。也就是说，在短时间内用陀螺传感器比较准确，而在长时间时用加速度传感器比较准确，这样两者便实现了互补。

本实例联合使用了互补滤波器和陀螺仪传感器，功能是获取并显示当前设备 *X*、*Y*、*Z* 轴的坐标值和角度值。

实例	功能	源码路径
实例 13-1	联合使用互补滤波器和陀螺仪传感器	\daima\13\FusedGyroscope

13.3.1 实现布局文件

本实例的布局文件是 activity_fused_gyroscope.xml，功能是调用文本控件显示获取的 *X*、*Y* 和 *Z* 的值，具体实现代码如下所示。

```
<LinearLayout xmlns:android="http://schemas.android.com/apk/res/android"
    xmlns:tools="http://schemas.android.com/tools"
    android:layout_width="match_parent"
    android:layout_height="match_parent"
    android:orientation="vertical"
    tools:context=".GyroscopeActivity" >

    <RelativeLayout
        android:id="@+id/layout_calibrated_header"
        android:layout_width="match_parent"
        android:layout_height="wrap_content"
        android:layout_marginTop="5dp" >

        <RelativeLayout
            android:layout_width="wrap_content"
            android:layout_height="wrap_content"
            android:layout_centerHorizontal="true" >

            <TextView
                android:id="@+id/label_calibrated_filter_name"
                android:layout_width="wrap_content"
                android:layout_height="wrap_content"
                android:fontFamily="sans-serif-condensed"
                android:text="@string/sensor_calibrated_name"
                android:textAppearance="?android:attr/textAppearanceSmall"
                android:textColor="@color/dark_orange" />

            <TextView
                android:id="@+id/label_calibrated_filter_description"
                android:layout_width="wrap_content"
                android:layout_height="wrap_content"
                android:layout_toRightOf="@+id/label_calibrated_filter_name"
                android:fontFamily="sans-serif-condensed"
                android:text="@string/sensor_label"
                android:textAppearance="?android:attr/textAppearanceSmall" />
        </RelativeLayout>
    </RelativeLayout>

    <RelativeLayout
        android:id="@+id/layout_calibrated_statistics"
        android:layout_width="match_parent"
        android:layout_height="wrap_content" >

        <RelativeLayout
            android:layout_width="wrap_content"
            android:layout_height="wrap_content"
            android:layout_centerHorizontal="true"
            android:layout_centerVertical="true" >

            <TableLayout
                android:id="@+id/table_calibrated_statistics_left"
                android:layout_width="wrap_content"
                android:layout_height="wrap_content" >

                <TableRow
                    android:id="@+id/table_calibrated_statistics_left_row_0"
                    android:layout_width="fill_parent"
                    android:layout_height="wrap_content"
                    android:padding="2dip" >

                    <RelativeLayout
                        android:layout_width="wrap_content"
                        android:layout_height="wrap_content"
```

```
        android:layout_marginRight="20dp"
        android:layout_weight="1" >

        <TextView
            android:id="@+id/label_x_axis_calibrated"
            android:layout_width="wrap_content"
            android:layout_height="wrap_content"
            android:layout_alignParentLeft="true"
            android:fontFamily="sans-serif-condensed"
            android:text="@string/label_x_axis"
            android:textAppearance="?android:attr/textAppearanceMedium" />

        <TextView
            android:id="@+id/value_x_axis_calibrated"
            android:layout_width="wrap_content"
            android:layout_height="wrap_content"
            android:layout_toRightOf="@+id/label_x_axis_calibrated"
            android:fontFamily="sans-serif-condensed"
            android:text="@string/value_default"
            android:textAppearance="?android:attr/textAppearanceMedium" />
    </RelativeLayout>

    <RelativeLayout
        android:layout_width="fill_parent"
        android:layout_height="wrap_content"
        android:layout_weight="1" >

        <TextView
            android:id="@+id/label_y_axis_calibrated"
            android:layout_width="wrap_content"
            android:layout_height="wrap_content"
            android:layout_alignParentLeft="true"
            android:fontFamily="sans-serif-condensed"
            android:text="@string/label_y_axis"
            android:textAppearance="?android:attr/textAppearanceMedium" />

        <TextView
            android:id="@+id/value_y_axis_calibrated"
            android:layout_width="wrap_content"
            android:layout_height="wrap_content"
            android:layout_toRightOf="@+id/label_y_axis_calibrated"
            android:fontFamily="sans-serif-condensed"
            android:text="@string/value_default"
            android:textAppearance="?android:attr/textAppearanceMedium" />
    </RelativeLayout>

    <RelativeLayout
        android:layout_width="fill_parent"
        android:layout_height="wrap_content"
        android:layout_marginLeft="20dp"
        android:layout_weight="1" >

        <TextView
            android:id="@+id/label_z_axis_calibrated"
            android:layout_width="wrap_content"
            android:layout_height="wrap_content"
            android:layout_alignParentLeft="true"
            android:fontFamily="sans-serif-condensed"
            android:text="@string/label_z_axis"
            android:textAppearance="?android:attr/textAppearanceMedium" />

        <TextView
            android:id="@+id/value_z_axis_calibrated"
            android:layout_width="wrap_content"
            android:layout_height="wrap_content"
            android:layout_toRightOf="@+id/label_z_axis_calibrated"
            android:fontFamily="sans-serif-condensed"
            android:text="@string/value_default"
            android:textAppearance="?android:attr/textAppearanceMedium" />
```

```
            </RelativeLayout>
          </TableRow>
        </TableLayout>
      </RelativeLayout>
    </RelativeLayout>

    <View
        android:layout_width="fill_parent"
        android:layout_height="1dp"
        android:background="@android:color/darker_gray" />

    <RelativeLayout
        android:layout_width="match_parent"
        android:layout_height="wrap_content"
        android:layout_marginTop="5dp" >

        <com.kircherelectronics.fusedgyroscopeexplorer.gauge.flat.GaugeBearingFlat
            android:id="@+id/gauge_bearing_calibrated"
            android:layout_width="140dp"
            android:layout_height="140dp"
            android:layout_alignParentLeft="true"
            android:layout_marginLeft="20dp" />

        <com.kircherelectronics.fusedgyroscopeexplorer.gauge.flat.GaugeRotationFlat
            android:id="@+id/gauge_tilt_calibrated"
            android:layout_width="140dp"
            android:layout_height="140dp"
            android:layout_alignParentRight="true"
            android:layout_marginRight="20dp" />
    </RelativeLayout>

    <RelativeLayout
        android:id="@+id/layout_raw_header"
        android:layout_width="match_parent"
        android:layout_height="wrap_content"
        android:layout_marginTop="20dp" >

        <RelativeLayout
            android:layout_width="wrap_content"
            android:layout_height="wrap_content"
            android:layout_centerHorizontal="true" >

            <TextView
                android:id="@+id/label_raw_filter_name"
                android:layout_width="wrap_content"
                android:layout_height="wrap_content"
                android:fontFamily="sans-serif-condensed"
                android:text="@string/sensor_uncalibrated_name"
                android:textAppearance="?android:attr/textAppearanceSmall"
                android:textColor="@color/dark_orange" />

            <TextView
                android:id="@+id/label_raw_filter_description"
                android:layout_width="wrap_content"
                android:layout_height="wrap_content"
                android:layout_toRightOf="@+id/label_raw_filter_name"
                android:fontFamily="sans-serif-condensed"
                android:text="@string/sensor_label"
                android:textAppearance="?android:attr/textAppearanceSmall" />
        </RelativeLayout>
    </RelativeLayout>

    <RelativeLayout
        android:id="@+id/layout_raw_statistics"
        android:layout_width="match_parent"
        android:layout_height="wrap_content" >

        <RelativeLayout
            android:layout_width="wrap_content"
```

```
        android:layout_height="wrap_content"
        android:layout_centerHorizontal="true"
        android:layout_centerVertical="true" >

        <TableLayout
            android:id="@+id/table_raw_statistics_left"
            android:layout_width="wrap_content"
            android:layout_height="wrap_content" >

            <TableRow
                android:id="@+id/table_raw_statistics_left_row_0"
                android:layout_width="fill_parent"
                android:layout_height="wrap_content"
                android:padding="2dip" >

                <RelativeLayout
                    android:layout_width="wrap_content"
                    android:layout_height="wrap_content"
                    android:layout_marginRight="20dp"
                    android:layout_weight="1" >

                    <TextView
                        android:id="@+id/label_x_axis_raw"
                        android:layout_width="wrap_content"
                        android:layout_height="wrap_content"
                        android:layout_alignParentLeft="true"
                        android:fontFamily="sans-serif-condensed"
                        android:text="@string/label_x_axis"
                        android:textAppearance="?android:attr/textAppearanceMedium" />

                    <TextView
                        android:id="@+id/value_x_axis_raw"
                        android:layout_width="wrap_content"
                        android:layout_height="wrap_content"
                        android:layout_toRightOf="@+id/label_x_axis_raw"
                        android:fontFamily="sans-serif-condensed"
                        android:text="@string/value_default"
                        android:textAppearance="?android:attr/textAppearanceMedium" />
                </RelativeLayout>

                <RelativeLayout
                    android:layout_width="fill_parent"
                    android:layout_height="wrap_content"
                    android:layout_weight="1" >

                    <TextView
                        android:id="@+id/label_y_axis_raw"
                        android:layout_width="wrap_content"
                        android:layout_height="wrap_content"
                        android:layout_alignParentLeft="true"
                        android:fontFamily="sans-serif-condensed"
                        android:text="@string/label_y_axis"
                        android:textAppearance="?android:attr/textAppearanceMedium" />

                    <TextView
                        android:id="@+id/value_y_axis_raw"
                        android:layout_width="wrap_content"
                        android:layout_height="wrap_content"
                        android:layout_toRightOf="@+id/label_y_axis_raw"
                        android:fontFamily="sans-serif-condensed"
                        android:text="@string/value_default"
                        android:textAppearance="?android:attr/textAppearanceMedium" />
                </RelativeLayout>

                <RelativeLayout
                    android:layout_width="fill_parent"
                    android:layout_height="wrap_content"
                    android:layout_marginLeft="20dp"
                    android:layout_weight="1" >
```

```
                    <TextView
                        android:id="@+id/label_z_axis_raw"
                        android:layout_width="wrap_content"
                        android:layout_height="wrap_content"
                        android:layout_alignParentLeft="true"
                        android:fontFamily="sans-serif-condensed"
                        android:text="@string/label_z_axis"
                        android:textAppearance="?android:attr/textAppearanceMedium" />

                    <TextView
                        android:id="@+id/value_z_axis_raw"
                        android:layout_width="wrap_content"
                        android:layout_height="wrap_content"
                        android:layout_toRightOf="@+id/label_z_axis_raw"
                        android:fontFamily="sans-serif-condensed"
                        android:text="@string/value_default"
                        android:textAppearance="?android:attr/textAppearanceMedium" />
                </RelativeLayout>
            </TableRow>
        </TableLayout>
    </RelativeLayout>
</RelativeLayout>

<View
    android:layout_width="fill_parent"
    android:layout_height="1dp"
    android:background="@android:color/darker_gray" />

<RelativeLayout
    android:layout_width="match_parent"
    android:layout_height="wrap_content"
    android:layout_marginTop="5dp" >

    <com.kircherelectronics.fusedgyroscopeexplorer.gauge.flat.GaugeBearingFlat
        android:id="@+id/gauge_bearing_raw"
        android:layout_width="140dp"
        android:layout_height="140dp"
        android:layout_alignParentLeft="true"
        android:layout_marginLeft="20dp" />

    <com.kircherelectronics.fusedgyroscopeexplorer.gauge.flat.GaugeRotationFlat
        android:id="@+id/gauge_tilt_raw"
        android:layout_width="140dp"
        android:layout_height="140dp"
        android:layout_alignParentRight="true"
        android:layout_marginRight="20dp" />
</RelativeLayout>

<RelativeLayout
    android:layout_width="match_parent"
    android:layout_height="fill_parent" >

    <RelativeLayout
        android:layout_width="match_parent"
        android:layout_height="wrap_content"
        android:layout_above="@+id/color_bar" >

        <RelativeLayout
            android:layout_width="wrap_content"
            android:layout_height="wrap_content"
            android:layout_centerHorizontal="true" >

            <ImageView
                android:id="@+id/image_developer_icon"
                android:layout_width="wrap_content"
                android:layout_height="wrap_content"
                android:layout_marginRight="4dp"
                android:src="@drawable/ke_icon" />
```

```
                <TextView
                    android:id="@+id/label_developer_description"
                    android:layout_width="wrap_content"
                    android:layout_height="wrap_content"
                    android:layout_toRightOf="@+id/image_developer_icon"
                    android:fontFamily="sans-serif-condensed"
                    android:text="@string/developer_url"
                    android:textAppearance="?android:attr/textAppearanceSmall" />
            </RelativeLayout>
        </RelativeLayout>

        <RelativeLayout
            android:id="@+id/color_bar"
            android:layout_width="match_parent"
            android:layout_height="wrap_content"
            android:layout_alignParentBottom="true"
            android:layout_marginBottom="5dp"
            android:layout_marginTop="5dp" >

            <RelativeLayout
                android:layout_width="wrap_content"
                android:layout_height="wrap_content"
                android:layout_centerHorizontal="true" >

                <ImageView
                    android:id="@+id/image_color_bar"
                    android:layout_width="wrap_content"
                    android:layout_height="wrap_content"
                    android:src="@drawable/color_bar" />
            </RelativeLayout>
        </RelativeLayout>
    </RelativeLayout>

</LinearLayout>
```

13.3.2　实现主 Activity 文件

本实例的主 Activity 文件是 FusedGyroscopeActivity.java，功能是使用一个包含 Android 陀螺仪的传感器来测试设备的旋转。在此文件中定义了监听用户在菜单项中的选择操作事件处理程序，并定义了监听各种传感器发生变化后的处理方法。文件 FusedGyroscopeActivity.java 的具体实现代码如下所示。

```
public class FusedGyroscopeActivity extends Activity implements
        GyroscopeSensorObserver, GravitySensorObserver, MagneticSensorObserver,
        FusedGyroscopeSensorObserver
{

    public static final float EPSILON = 0.000000001f;

    private static final String tag = FusedGyroscopeActivity.class
            .getSimpleName();
    private static final float NS2S = 1.0f / 1000000000.0f;
    private static final int MEAN_FILTER_WINDOW = 10;
    private static final int MIN_SAMPLE_COUNT = 30;

    private boolean hasInitialOrientation = false;
    private boolean stateInitializedAndroid = false;

    // 虽然运行在 UI 线程，但是易于使用
    private GaugeBearingFlat gaugeBearingFused;
    private GaugeBearingFlat gaugeBearingAndroid;
    private GaugeRotationFlat gaugeTiltFused;
    private GaugeRotationFlat gaugeTiltAndroid;

    private DecimalFormat df;
```

```
// 标准方法
private float[] currentRotationMatrixAndroid;
private float[] deltaRotationMatrixAndroid;
private float[] deltaRotationVectorAndroid;
private float[] gyroscopeOrientationAndroid;

//加速度计和基于旋转矩阵磁强计
private float[] initialRotationMatrix;

//加速度矢量
private float[] gravity;

//磁场矢量
private float[] magnetic;

private int accelerationSampleCount = 0;
private int magneticSampleCount = 0;

private long timestampOldCalibrated = 0;

private MeanFilter gravityFilter;
private MeanFilter magneticFilter;

private FusedGyroscopeSensor fusedGyroscopeSensor;
private GravitySensor gravitySensor;
private GyroscopeSensor gyroscopeSensor;
private MagneticSensor magneticSensor;

private TextView xAxisAndroid;
private TextView yAxisAndroid;
private TextView zAxisAndroid;

private TextView xAxisFused;
private TextView yAxisFused;
private TextView zAxisFused;
@Override
protected void onCreate(Bundle savedInstanceState)
{
    super.onCreate(savedInstanceState);

    setContentView(R.layout.activity_fused_gyroscope);

    initUI();
    initMaths();
    initSensors();
    initFilters();
};

@Override
public boolean onCreateOptionsMenu(Menu menu)
{
    //将菜单项目添加到动作条
    getMenuInflater().inflate(R.menu.fused_gyroscope, menu);
    return true;
}

/**
 * 根据选择菜单项的 ID 执行对应事件处理程序
 * */
@Override
public boolean onOptionsItemSelected(MenuItem item)
{
    switch (item.getItemId())
    {

    // Reset everything
    case R.id.action_reset:
        reset();
```

```
            restart();
            return true;

        default:
            return super.onOptionsItemSelected(item);
        }
    }

    public void onStart()
    {
        super.onStart();

        restart();
    }

    public void onPause()
    {
        super.onPause();

        reset();
    }
//处理重力传感器的方法
    @Override
    public void onGravitySensorChanged(float[] gravity, long timeStamp)
    {
        // Get a local copy of the raw magnetic values from the device sensor.
        System.arraycopy(gravity, 0, this.gravity, 0, gravity.length);

        // Use a mean filter to smooth the sensor inputs
        this.gravity = gravityFilter.filterFloat(this.gravity);

        // Count the number of samples received.
        accelerationSampleCount++;

        // Only determine the initial orientation after the acceleration sensor
        // and magnetic sensor have had enough time to be smoothed by the mean
        // filters. Also, only do this if the orientation hasn't already been
        // determined since we only need it once.
        if (accelerationSampleCount > MIN_SAMPLE_COUNT
                && magneticSampleCount > MIN_SAMPLE_COUNT
                && !hasInitialOrientation)
        {
            calculateOrientation();
        }
    }
//处理陀螺传感器改变时的方法
    @Override
    public void onGyroscopeSensorChanged(float[] gyroscope, long timestamp)
    {
        // don't start until first accelerometer/magnetometer orientation has
        // been acquired
        if (!hasInitialOrientation)
        {
            return;
        }

        //基于矩阵旋转的陀螺初始化
        if (!stateInitializedAndroid)
        {
            currentRotationMatrixAndroid = matrixMultiplication(
                    currentRotationMatrixAndroid, initialRotationMatrix);

            stateInitializedAndroid = true;
        }

        // This timestep's delta rotation to be multiplied by the current
        // rotation after computing it from the gyro sample data.
        if (timestampOldCalibrated != 0 && stateInitializedAndroid)
        {
```

```
            final float dT = (timestamp - timestampOldCalibrated) * NS2S;

            // Axis of the rotation sample, not normalized yet.
            float axisX = gyroscope[0];
            float axisY = gyroscope[1];
            float axisZ = gyroscope[2];

            // Calculate the angular speed of the sample
            float omegaMagnitude = (float) Math.sqrt(axisX * axisX + axisY
                    * axisY + axisZ * axisZ);

            // Normalize the rotation vector if it's big enough to get the axis
            if (omegaMagnitude > EPSILON)
            {
                axisX /= omegaMagnitude;
                axisY /= omegaMagnitude;
                axisZ /= omegaMagnitude;
            }

            // Integrate around this axis with the angular speed by the timestep
            // in order to get a delta rotation from this sample over the
            // timestep. We will convert this axis-angle representation of the
            // delta rotation into a quaternion before turning it into the
            // rotation matrix.
            float thetaOverTwo = omegaMagnitude * dT / 2.0f;

            float sinThetaOverTwo = (float) Math.sin(thetaOverTwo);
            float cosThetaOverTwo = (float) Math.cos(thetaOverTwo);

            deltaRotationVectorAndroid[0] = sinThetaOverTwo * axisX;
            deltaRotationVectorAndroid[1] = sinThetaOverTwo * axisY;
            deltaRotationVectorAndroid[2] = sinThetaOverTwo * axisZ;
            deltaRotationVectorAndroid[3] = cosThetaOverTwo;

            SensorManager.getRotationMatrixFromVector(
                    deltaRotationMatrixAndroid, deltaRotationVectorAndroid);

            currentRotationMatrixAndroid = matrixMultiplication(
                    currentRotationMatrixAndroid, deltaRotationMatrixAndroid);

            SensorManager.getOrientation(currentRotationMatrixAndroid,
                    gyroscopeOrientationAndroid);
        }

        timestampOldCalibrated = timestamp;

        gaugeBearingAndroid.updateBearing(gyroscopeOrientationAndroid[0]);
        gaugeTiltAndroid.updateRotation(gyroscopeOrientationAndroid);

        xAxisAndroid.setText(df.format(gyroscopeOrientationAndroid[0]));
        yAxisAndroid.setText(df.format(gyroscopeOrientationAndroid[1]));
        zAxisAndroid.setText(df.format(gyroscopeOrientationAndroid[2]));
    }

//角速度传感器改变时的处理方法
@Override
    public void onAngularVelocitySensorChanged(float[] angularVelocity,
            long timeStamp)
    {
        gaugeBearingFused.updateBearing(angularVelocity[0]);
        gaugeTiltFused.updateRotation(angularVelocity);

        xAxisFused.setText(df.format(angularVelocity[0]));
        yAxisFused.setText(df.format(angularVelocity[1]));
        zAxisFused.setText(df.format(angularVelocity[2]));
    }

//磁场传感器改变时的处理方法
@Override
```

```
public void onMagneticSensorChanged(float[] magnetic, long timeStamp)
{
    // Get a local copy of the raw magnetic values from the device sensor.
    System.arraycopy(magnetic, 0, this.magnetic, 0, magnetic.length);

    // Use a mean filter to smooth the sensor inputs
    this.magnetic = magneticFilter.filterFloat(this.magnetic);

    // Count the number of samples received.
    magneticSampleCount++;
}

/**
 * 计算方向
 */
private void calculateOrientation()
{
    hasInitialOrientation = SensorManager.getRotationMatrix(
            initialRotationMatrix, null, gravity, magnetic);

    // Remove the sensor observers since they are no longer required.
    if (hasInitialOrientation)
    {
        gravitySensor.removeGravityObserver(this);
        magneticSensor.removeMagneticObserver(this);
    }
}

/**
 *初始化均值滤波器
 */
private void initFilters()
{
    gravityFilter = new MeanFilter();
    gravityFilter.setWindowSize(MEAN_FILTER_WINDOW);

    magneticFilter = new MeanFilter();
    magneticFilter.setWindowSize(MEAN_FILTER_WINDOW);
}

/**
 *初始化前面方法所需的数据结构
 */
private void initMaths()
{
    gravity = new float[3];
    magnetic = new float[3];

    initialRotationMatrix = new float[9];

    deltaRotationVectorAndroid = new float[4];
    deltaRotationMatrixAndroid = new float[9];
    currentRotationMatrixAndroid = new float[9];
    gyroscopeOrientationAndroid = new float[3];

    // Initialize the current rotation matrix as an identity matrix...
    currentRotationMatrixAndroid[0] = 1.0f;
    currentRotationMatrixAndroid[4] = 1.0f;
    currentRotationMatrixAndroid[8] = 1.0f;
}

/**
 * 初始化传感器
 */
private void initSensors()
{
    fusedGyroscopeSensor = new FusedGyroscopeSensor();
    gravitySensor = new GravitySensor(this);
    magneticSensor = new MagneticSensor(this);
```

```
    gyroscopeSensor = new GyroscopeSensor(this);
}

/**
 * 初始化 UI
 */
private void initUI()
{
    // Get a decimal formatter for the text views
    df = new DecimalFormat("#.##");

    // Initialize the raw (uncalibrated) text views
    xAxisAndroid = (TextView) this.findViewById(R.id.value_x_axis_raw);
    yAxisAndroid = (TextView) this.findViewById(R.id.value_y_axis_raw);
    zAxisAndroid = (TextView) this.findViewById(R.id.value_z_axis_raw);

    // Initialize the calibrated text views
    xAxisFused = (TextView) this.findViewById(R.id.value_x_axis_calibrated);
    yAxisFused = (TextView) this.findViewById(R.id.value_y_axis_calibrated);
    zAxisFused = (TextView) this.findViewById(R.id.value_z_axis_calibrated);

    // Initialize the raw (uncalibrated) gauge views
    gaugeBearingAndroid = (GaugeBearingFlat) findViewById(R.id.gauge_bearing_raw);
    gaugeTiltAndroid = (GaugeRotationFlat) findViewById(R.id.gauge_tilt_raw);

    // Initialize the calibrated gauges views
    gaugeBearingFused = (GaugeBearingFlat) findViewById(R.id.gauge_bearing_calibrated);
    gaugeTiltFused = (GaugeRotationFlat) findViewById(R.id.gauge_tilt_calibrated);
}

/**
 *实现矩阵乘法算法
 *
 * @param a
 * @param b
 * @return a*b
 */
private float[] matrixMultiplication(float[] a, float[] b)
{
    float[] result = new float[9];

    result[0] = a[0] * b[0] + a[1] * b[3] + a[2] * b[6];
    result[1] = a[0] * b[1] + a[1] * b[4] + a[2] * b[7];
    result[2] = a[0] * b[2] + a[1] * b[5] + a[2] * b[8];

    result[3] = a[3] * b[0] + a[4] * b[3] + a[5] * b[6];
    result[4] = a[3] * b[1] + a[4] * b[4] + a[5] * b[7];
    result[5] = a[3] * b[2] + a[4] * b[5] + a[5] * b[8];

    result[6] = a[6] * b[0] + a[7] * b[3] + a[8] * b[6];
    result[7] = a[6] * b[1] + a[7] * b[4] + a[8] * b[7];
    result[8] = a[6] * b[2] + a[7] * b[5] + a[8] * b[8];

    return result;
}

/**
 *重新启动系统的处理方法
 */
private void restart()
{
    gravitySensor.registerGravityObserver(this);
    magneticSensor.registerMagneticObserver(this);
    gyroscopeSensor.registerGyroscopeObserver(this);

    gravitySensor.registerGravityObserver(fusedGyroscopeSensor);
    magneticSensor.registerMagneticObserver(fusedGyroscopeSensor);
    gyroscopeSensor.registerGyroscopeObserver(fusedGyroscopeSensor);
```

```
        fusedGyroscopeSensor.registerObserver(this);
    }

    /**
     *单击"RESET"按钮时的处理方法
     */
    private void reset()
    {
        gravitySensor.removeGravityObserver(this);
        magneticSensor.removeMagneticObserver(this);
        gyroscopeSensor.removeGyroscopeObserver(this);

        gravitySensor.removeGravityObserver(fusedGyroscopeSensor);
        magneticSensor.removeMagneticObserver(fusedGyroscopeSensor);
        gyroscopeSensor.removeGyroscopeObserver(fusedGyroscopeSensor);

        fusedGyroscopeSensor.removeObserver(this);

        initMaths();

        accelerationSampleCount = 0;
        magneticSampleCount = 0;

        hasInitialOrientation = false;
        stateInitializedAndroid = false;
    }
}
```

13.3.3　实现均值滤波器

编写文件 MeanFilter.java，功能是实现一种均值滤波器算法，原理是以平滑的数据点为基础计算平均值。文件 MeanFilter.java 的具体实现代码如下所示。

```
public class MeanFilter
{
    // 均值滤波器滚动窗口的大小
    private int filterWindow = 30;

    private boolean dataInit;

    private ArrayList<LinkedList<Number>> dataLists;

    /**
     *初始化一个新的 meanfilter 对象
     */
    public MeanFilter()
    {
        dataLists = new ArrayList<LinkedList<Number>>();
        dataInit = false;
    }

    /**
     *筛选数据
     *
     * @param iterator
     *          contains input the data.
     * @return the filtered output data.
     */
    public float[] filterFloat(float[] data)
    {
        for (int i = 0; i < data.length; i++)
        {
            // Initialize the data structures for the data set.
            if (!dataInit)
            {
                dataLists.add(new LinkedList<Number>());
            }
```

```
            dataLists.get(i).addLast(data[i]);

            if (dataLists.get(i).size() > filterWindow)
            {
                dataLists.get(i).removeFirst();
            }
        }

        dataInit = true;

        float[] means = new float[dataLists.size()];

        for (int i = 0; i < dataLists.size(); i++)
        {
            means[i] = (float) getMean(dataLists.get(i));
        }

        return means;
    }

    /**
     *获得的数据集的平均值
     *
     * @param data
     *          the data set.
     * @return the mean of the data set.
     */
    private float getMean(List<Number> data)
    {
        float m = 0;
        float count = 0;

        for (int i = 0; i < data.size(); i++)
        {
            m += data.get(i).floatValue();
            count++;
        }

        if (count != 0)
        {
            m = m / count;
        }

        return m;
    }

    public void setWindowSize(int size)
    {
        this.filterWindow = size;
    }
}
```

13.3.4 测量各个平面的值

编写文件 GaugeBearingFlat.java，功能是绘制一个指针样式的测量表来测试模拟器的轴承值。文件 GaugeBearingFlat.java 的具体实现代码如下所示。

```
public final class GaugeBearingFlat extends View
{
    private static final String tag = GaugeBearingFlat.class.getSimpleName();
    //绘图工具
    private RectF rimRect;
    private Paint rimPaint;

    private RectF faceRect;
    private Paint facePaint;

    // 外圆矩形
    private RectF rimOuterTopRect;
```

```
    private RectF rimOuterBottomRect;
    private RectF rimOuterLeftRect;
    private RectF rimOuterRightRect;

    //表面静态位图
    private Bitmap hand;
    private Paint handPaint;
    private Path handPath;

    private Paint backgroundPaint;
    // end drawing tools

    private Bitmap background; // holds the cached static part

    private Canvas handCanvas;

    // the one in the top center (12 o'clock)
    private static final int centerDegree = 0;
    private static final int minDegrees = 0;
    private static final int maxDegrees = 360;

    // hand dynamics -- all are angular expressed in F degrees
    private boolean handInitialized = false;
    private float handPosition = centerDegree;
    private float handTarget = centerDegree;
    private float handVelocity = 0.0f;
    private float handAcceleration = 0.0f;
    private long lastHandMoveTime = -1L;

    private int unitsOfMeasure = UnitsOfMeasure.DEGREES;

    /**
     *创建一个新的实例
     *
     * @param context
     */
    public GaugeBearingFlat(Context context)
    {
        super(context);
        init();
    }

    /**
     *创建一个新的实例
     *
     * @param context
     * @param attrs
     */
    public GaugeBearingFlat(Context context, AttributeSet attrs)
    {
        super(context, attrs);
        init();
    }

    /**
     * Create a new instance.
     *
     * @param context
     * @param attrs
     * @param defStyle
     */
    public GaugeBearingFlat(Context context, AttributeSet attrs, int defStyle)
    {
        super(context, attrs, defStyle);
        init();
    }

    public int getUnitsOfMeasure()
    {
```

```
    return unitsOfMeasure;
}

public void setUnitsOfMeasure(int unitsOfMeasure)
{
    this.unitsOfMeasure = unitsOfMeasure;
}

/**
 *更新设备的轴承
 *
 * @param azimuth
 */
public void updateBearing(float azimuth)
{
    // Adjust the range: 0 < range <= 360 (from: -180 < range <=
    // 180)
    azimuth = (float) (Math.toDegrees(azimuth) + 360) % 360;

    setHandTarget(azimuth);
}

/**
 *运行实例，此处可以看作是 ondraw()
 */
protected void onDraw(Canvas canvas)
{
    drawBackground(canvas);
    drawHand(canvas);
    canvas.restore();
    moveHand();
}
@Override
protected void onRestoreInstanceState(Parcelable state)
{
    Bundle bundle = (Bundle) state;
    Parcelable superState = bundle.getParcelable("superState");
    super.onRestoreInstanceState(superState);
    handInitialized = bundle.getBoolean("handInitialized");
    handPosition = bundle.getFloat("handPosition");
    handTarget = bundle.getFloat("handTarget");
    handVelocity = bundle.getFloat("handVelocity");
    handAcceleration = bundle.getFloat("handAcceleration");
    lastHandMoveTime = bundle.getLong("lastHandMoveTime");
}

@Override
protected Parcelable onSaveInstanceState()
{
    Parcelable superState = super.onSaveInstanceState();

    Bundle state = new Bundle();
    state.putParcelable("superState", superState);
    state.putBoolean("handInitialized", handInitialized);
    state.putFloat("handPosition", handPosition);
    state.putFloat("handTarget", handTarget);
    state.putFloat("handVelocity", handVelocity);
    state.putFloat("handAcceleration", handAcceleration);
    state.putLong("lastHandMoveTime", lastHandMoveTime);
    return state;
}
/**
 *初始化绘图工具
 */
private void initDrawingTools()
{

    rimRect = new RectF(0.1f, 0.1f, 0.9f, 0.9f);
```

```
    // the linear gradient is a bit skewed for realism
    rimPaint = new Paint();
    rimPaint.setAntiAlias(true);
    rimPaint.setFlags(Paint.ANTI_ALIAS_FLAG);
    rimPaint.setColor(Color.rgb(255, 255, 255));

    float rimSize = 0.03f;
    faceRect = new RectF();
    faceRect.set(rimRect.left + rimSize, rimRect.top + rimSize,
            rimRect.right - rimSize, rimRect.bottom - rimSize);

    rimOuterTopRect = new RectF(0.46f, 0.076f, 0.54f, 0.11f);

    rimOuterBottomRect = new RectF(0.46f, 0.89f, 0.54f, 0.924f);

    rimOuterLeftRect = new RectF(0.076f, 0.46f, 0.11f, 0.54f);

    rimOuterRightRect = new RectF(0.89f, 0.46f, 0.924f, 0.54f);

    facePaint = new Paint();
    facePaint.setStyle(Paint.Style.FILL);
    facePaint.setFlags(Paint.ANTI_ALIAS_FLAG);
    facePaint.setAntiAlias(true);

    handPaint = new Paint();
    handPaint.setAntiAlias(true);
    handPaint.setFlags(Paint.ANTI_ALIAS_FLAG);
    handPaint.setColor(Color.WHITE);
    handPaint.setStyle(Paint.Style.FILL);

    handPath = new Path();
    handPath.moveTo(0.5f, 0.5f + 0.32f);
    handPath.lineTo(0.5f - 0.02f, 0.5f + 0.32f - 0.32f);

    handPath.lineTo(0.5f, 0.5f - 0.32f);
    handPath.lineTo(0.5f + 0.02f, 0.5f + 0.32f - 0.32f);
    handPath.lineTo(0.5f, 0.5f + 0.32f);
    handPath.addCircle(0.5f, 0.5f, 0.025f, Path.Direction.CW);

    backgroundPaint = new Paint();
    backgroundPaint.setFilterBitmap(true);
}

@Override
protected void onMeasure(int widthMeasureSpec, int heightMeasureSpec)
{
    int widthMode = MeasureSpec.getMode(widthMeasureSpec);
    int widthSize = MeasureSpec.getSize(widthMeasureSpec);

    int heightMode = MeasureSpec.getMode(heightMeasureSpec);
    int heightSize = MeasureSpec.getSize(heightMeasureSpec);

    int chosenWidth = chooseDimension(widthMode, widthSize);
    int chosenHeight = chooseDimension(heightMode, heightSize);

    int chosenDimension = Math.min(chosenWidth, chosenHeight);

    setMeasuredDimension(chosenDimension, chosenDimension);
}

/**
 *选择视图的尺寸
 *
 * @param mode
 * @param size
 * @return
 */
private int chooseDimension(int mode, int size)
{
```

```
    if (mode == MeasureSpec.AT_MOST || mode == MeasureSpec.EXACTLY)
    {
        return size;
    }
    else
    { // (mode == MeasureSpec.UNSPECIFIED)
        return getPreferredSize();
    }
}
/**
 *如果没有指定的大小
 * @return
 */
private int getPreferredSize()
{
    return 300;
}
/**
 *绘制边缘的表
 *
 * @param canvas
 */
private void drawRim(Canvas canvas)
{
    // first, draw the metallic body
    canvas.drawOval(rimRect, rimPaint);

    // top rect
    canvas.drawRect(rimOuterTopRect, rimPaint);
    // bottom rect
    canvas.drawRect(rimOuterBottomRect, rimPaint);
    // left rect
    canvas.drawRect(rimOuterLeftRect, rimPaint);
    // right rect
    canvas.drawRect(rimOuterRightRect, rimPaint);
}

/**
 *绘制表面
 *
 * @param canvas
 */
private void drawFace(Canvas canvas)
{
    canvas.drawOval(faceRect, facePaint);
}
/**
 *转换成角度
 *
 * @param degree
 * @return
 */
private float degreeToAngle(float degree)
{
    return degree;
}
private void drawHand(Canvas canvas)
{
    // *Bug Notice* We draw the hand with a bitmap and a new canvas because
    // canvas.drawPath() doesn't work. This seems to be related to devices
    // with hardware acceleration enabled.

    // free the old bitmap
    if (hand != null)
    {
        hand.recycle();
    }

    hand = Bitmap.createBitmap(getWidth(), getHeight(),
```

```
            Bitmap.Config.ARGB_8888);
        handCanvas = new Canvas(hand);
        float scale = (float) getWidth();
        handCanvas.scale(scale, scale);

        if (handInitialized)
        {
            float handAngle = degreeToAngle(handPosition);
            handCanvas.save(Canvas.MATRIX_SAVE_FLAG);
            handCanvas.rotate(handAngle, 0.5f, 0.5f);
            handCanvas.drawPath(handPath, handPaint);
        }
        else
        {
            float handAngle = degreeToAngle(0);
            handCanvas.save(Canvas.MATRIX_SAVE_FLAG);
            handCanvas.rotate(handAngle, 0.5f, 0.5f);
            handCanvas.drawPath(handPath, handPaint);
        }

        canvas.drawBitmap(hand, 0, 0, backgroundPaint);
    }

    /**
     *绘制仪表背景
     *
     * @param canvas
     */
    private void drawBackground(Canvas canvas)
    {
        if (background == null)
        {
            Log.w(tag, "Background not created");
        }
        else
        {
            canvas.drawBitmap(background, 0, 0, backgroundPaint);
        }
    }

    @Override
    protected void onSizeChanged(int w, int h, int oldw, int oldh)
    {
        Log.d(tag, "Size changed to " + w + "x" + h);

        regenerateBackground();
    }

    /**
     * 生成屏幕背景变化时的图像，缓存以后无需生成即可使用
     */
    private void regenerateBackground()
    {
        // free the old bitmap
        if (background != null)
        {
            background.recycle();
        }

        background = Bitmap.createBitmap(getWidth(), getHeight(),
            Bitmap.Config.ARGB_8888);
        Canvas backgroundCanvas = new Canvas(background);
        float scale = (float) getWidth();
        backgroundCanvas.scale(scale, scale);

        drawRim(backgroundCanvas);
        drawFace(backgroundCanvas);
    }
    /**
```

```
     * Indicate where the hand should be moved to.
     *
     * @param bearing
     */
    private void setHandTarget(float bearing)
    {
        if (bearing < minDegrees)
        {
            bearing = minDegrees;
        }
        else if (bearing > maxDegrees)
        {
            bearing = maxDegrees;
        }

        handTarget = bearing;
        handInitialized = true;

        invalidate();
    }

}
```

编写文件 GaugeRotationFlat.java，功能是根据从传感器获取的值显示 *X*、*Y*、*Z* 三个空间的旋转测量值。文件 GaugeRotationFlat.java 的具体实现代码如下所示。

```
public final class GaugeRotationFlat extends View
{
    private static final String TAG = GaugeRotationFlat.class.getSimpleName();

    // drawing tools
    private RectF rimOuterRect;
    private RectF rimTopRect;
    private RectF rimBottomRect;
    private RectF rimLeftRect;
    private RectF rimRightRect;
    private Paint rimOuterPaint;
    private Bitmap bezel;
    private Bitmap face;

    private Canvas faceCanvas;
    private float[] rotation = new float[3];
    private RectF earthRect;
    private RectF faceRect;
    private RectF rimRect;
    private RectF skyRect;
    private RectF skyBackgroundRect;
    private Paint arrowPaint;
    private Paint backgroundPaint;
    private Paint earthPaint;
    private Paint gaugeGuidePaint;
    private Paint numericPaint;
    private Paint rimCirclePaint;
    private Paint rimPaint;
    private Paint rimShadowPaint;
    private Paint scalePaint;
    private Paint skyPaint;
    private Paint smallTickPaint;
    private Paint thickScalePaint;

    public GaugeRotationFlat(Context context)
    {
        super(context);

        initDrawingTools();
    }
    public GaugeRotationFlat(Context context, AttributeSet attrs)
    {
        super(context, attrs);
```

```
        initDrawingTools();
    }

    public GaugeRotationFlat(Context context, AttributeSet attrs, int defStyle)
    {
        super(context, attrs, defStyle);

        initDrawingTools();
    }
    public void updateRotation(float[] rotation)
    {
        this.rotation = rotation;

        this.invalidate();
    }

    private void initDrawingTools()
    {
        // Rectangle for the rim of the gauge bezel
        rimRect = new RectF(0.12f, 0.12f, 0.88f, 0.88f);

        // Paint for the rim of the gauge bezel
        rimPaint = new Paint();
        rimPaint.setFlags(Paint.ANTI_ALIAS_FLAG);
        // The linear gradient is a bit skewed for realism
        rimPaint.setShader(new LinearGradient(0.40f, 0.0f, 0.60f, 1.0f, Color
                .rgb(0, 0, 0), Color.rgb(0, 0, 0), Shader.TileMode.CLAMP));

        float rimOuterSize = -0.04f;
        rimOuterRect = new RectF();
        rimOuterRect.set(rimRect.left + rimOuterSize, rimRect.top
                + rimOuterSize, rimRect.right - rimOuterSize, rimRect.bottom
                - rimOuterSize);

        // still a work in progress changing the rimOuterSize will not
        // dynamically
        // change the small rectangles to the appropriate size.
        rimTopRect = new RectF(0.5f, 0.106f, 0.5f, 0.06f);
        rimTopRect.set(rimTopRect.left + rimOuterSize, rimTopRect.top
                + rimOuterSize, rimTopRect.right - rimOuterSize,
                rimTopRect.bottom - rimOuterSize);

        rimBottomRect = new RectF(0.5f, 0.94f, 0.5f, 0.894f);
        rimBottomRect.set(rimBottomRect.left + rimOuterSize, rimBottomRect.top
                + rimOuterSize, rimBottomRect.right - rimOuterSize,
                rimBottomRect.bottom - rimOuterSize);

        rimLeftRect = new RectF(0.106f, 0.5f, 0.06f, 0.5f);
        rimLeftRect.set(rimLeftRect.left + rimOuterSize, rimLeftRect.top
                + rimOuterSize, rimLeftRect.right - rimOuterSize,
                rimLeftRect.bottom - rimOuterSize);

        rimRightRect = new RectF(0.94f, 0.5f, 0.894f, 0.5f);
        rimRightRect.set(rimRightRect.left + rimOuterSize, rimRightRect.top
                + rimOuterSize, rimRightRect.right - rimOuterSize,
                rimRightRect.bottom - rimOuterSize);

        rimOuterPaint = new Paint();
        rimOuterPaint.setFlags(Paint.ANTI_ALIAS_FLAG);
        rimOuterPaint.setColor(Color.rgb(255, 255, 255));

        // Paint for the outer circle of the gauge bezel
        rimCirclePaint = new Paint();
        rimCirclePaint.setAntiAlias(true);
        rimCirclePaint.setStyle(Paint.Style.STROKE);
        rimCirclePaint.setColor(Color.argb(0x4f, 0x33, 0x36, 0x33));
        rimCirclePaint.setStrokeWidth(0.005f);

        float rimSize = 0.02f;
```

```
faceRect = new RectF();
faceRect.set(rimRect.left + rimSize, rimRect.top + rimSize,
      rimRect.right - rimSize, rimRect.bottom - rimSize);

earthRect = new RectF();
earthRect.set(rimRect.left + rimSize, rimRect.top + rimSize,
      rimRect.right - rimSize, rimRect.bottom - rimSize);

skyRect = new RectF();
skyRect.set(rimRect.left + rimSize, rimRect.top + rimSize,
      rimRect.right - rimSize, rimRect.bottom - rimSize);

skyBackgroundRect = new RectF();
skyBackgroundRect.set(rimRect.left + rimSize, rimRect.top + rimSize,
      rimRect.right - rimSize, rimRect.bottom - rimSize);

// now set to black
skyPaint = new Paint();
skyPaint.setFlags(Paint.ANTI_ALIAS_FLAG);
skyPaint.setShader(new LinearGradient(0.40f, 0.0f, 0.60f, 1.0f, Color
      .rgb(0, 0, 0), Color.rgb(0, 0, 0), Shader.TileMode.CLAMP));

// now set to white
earthPaint = new Paint();
earthPaint.setFlags(Paint.ANTI_ALIAS_FLAG);
earthPaint.setShader(new LinearGradient(0.40f, 0.0f, 0.60f, 1.0f, Color
      .rgb(238, 238, 238), Color.rgb(238, 238, 238),
      Shader.TileMode.CLAMP));

rimShadowPaint = new Paint();
rimShadowPaint.setShader(new RadialGradient(0.5f, 0.5f, faceRect
      .width() / 2.0f, new int[]
{ 0x00000000, 0x00000500, 0x50000500 }, new float[]
{ 0.96f, 0.96f, 0.99f }, Shader.TileMode.MIRROR));
rimShadowPaint.setStyle(Paint.Style.FILL);

scalePaint = new Paint();
scalePaint.setStyle(Paint.Style.STROKE);
scalePaint.setColor(Color.WHITE);
scalePaint.setStrokeWidth(0.005f);
scalePaint.setAntiAlias(true);

arrowPaint = new Paint();
arrowPaint.setStyle(Paint.Style.FILL_AND_STROKE);
arrowPaint.setColor(Color.RED);
arrowPaint.setStrokeWidth(0.005f);
arrowPaint.setAntiAlias(true);

thickScalePaint = new Paint();
thickScalePaint.setStyle(Paint.Style.STROKE);
thickScalePaint.setColor(Color.WHITE);
thickScalePaint.setStrokeWidth(0.008f);
thickScalePaint.setAntiAlias(true);

numericPaint = new Paint();
numericPaint.setTextSize(0.035f);
numericPaint.setColor(Color.WHITE);
numericPaint.setTypeface(Typeface.DEFAULT);
numericPaint.setAntiAlias(true);
numericPaint.setTextAlign(Paint.Align.CENTER);
numericPaint.setLinearText(true);

smallTickPaint = new Paint();
smallTickPaint.setStyle(Paint.Style.STROKE);
smallTickPaint.setColor(Color.argb(100, 255, 255, 255));
smallTickPaint.setStrokeWidth(0.005f);
smallTickPaint.setAntiAlias(true);

gaugeGuidePaint = new Paint();
```

```
        gaugeGuidePaint.setFilterBitmap(true);
        BitmapFactory.Options options = new BitmapFactory.Options();
        options.inPreferredConfig = Bitmap.Config.ARGB_8888;

        backgroundPaint = new Paint();
        backgroundPaint.setFilterBitmap(true);
    }

    @Override
    protected void onMeasure(int widthMeasureSpec, int heightMeasureSpec)
    {
        int widthMode = MeasureSpec.getMode(widthMeasureSpec);
        int widthSize = MeasureSpec.getSize(widthMeasureSpec);
        int heightMode = MeasureSpec.getMode(heightMeasureSpec);
        int heightSize = MeasureSpec.getSize(heightMeasureSpec);
        int chosenWidth = chooseDimension(widthMode, widthSize);
        int chosenHeight = chooseDimension(heightMode, heightSize);
        int chosenDimension = Math.min(chosenWidth, chosenHeight);
        setMeasuredDimension(chosenDimension, chosenDimension);
    }
    private int chooseDimension(int mode, int size)
    {
        if (mode == MeasureSpec.AT_MOST || mode == MeasureSpec.EXACTLY)
        {
            return size;
        }
        else
        { // (mode == MeasureSpec.UNSPECIFIED)
            return getPreferredSize();
        }
    }
    // in case there is no size specified
    private int getPreferredSize()
    {
        return 300;
    }
    private void drawRim(Canvas canvas)
    {
        // First draw the most back rim
        canvas.drawOval(rimOuterRect, rimOuterPaint);
        // Then draw the small black line
        canvas.drawOval(rimRect, rimPaint);
        // now the outer rim circle
        // canvas.drawOval(rimRect, rimCirclePaint);

        canvas.drawRect(rimTopRect, rimOuterPaint);
        // bottom rect
        canvas.drawRect(rimBottomRect, rimOuterPaint);
        // left rect
        canvas.drawRect(rimLeftRect, rimOuterPaint);
        // right rect
        canvas.drawRect(rimRightRect, rimOuterPaint);
    }
    private void drawFace(Canvas canvas)
    {
        // free the old bitmap
        if (face != null)
        {
            face.recycle();
        }
        face = Bitmap.createBitmap(getWidth(), getHeight(),
                Bitmap.Config.ARGB_8888);
        faceCanvas = new Canvas(face);
        float scale = (float) getWidth();
        faceCanvas.scale(scale, scale);
        float rimSize = 0.02f;
        float radius = ((rimRect.left + rimSize) + (rimRect.right - rimSize)) / 2;
        float aPos = rotation[1];
        if (aPos < 0)
```

```
    {
        aPos = 0;
    }
    float aNeg = rotation[1];
    if (aNeg > 0)
    {
        aNeg = 0;
    }
    float radiusSquaredNeg = (float) Math.pow(radius, 2);
    float aSquaredNeg = (float) Math.pow(aNeg, 2);
    float bNeg = (float) Math.sqrt(radiusSquaredNeg - aSquaredNeg);
    float radiusSquaredPos = (float) Math.pow(radius, 2);
    float aSquaredPos = (float) Math.pow(aPos, 2);
        float bPos = (float) Math.sqrt(radiusSquaredPos - aSquaredPos);
    skyBackgroundRect.set(rimRect.left + rimSize, rimRect.top + rimSize,
          rimRect.right - rimSize, rimRect.bottom - rimSize);
    if (aPos == 0)
    {
        faceCanvas.drawArc(skyBackgroundRect,
              (float) (0 + (rotation[2] * (180 / Math.PI))), 360, true,
              skyPaint);
    }
    if (aNeg== 0)
    {
        faceCanvas.drawArc(skyBackgroundRect,
              (float) (0 + (rotation[2] * (180 / Math.PI))), 360, true,
              earthPaint);
    }
    skyRect.set((rimRect.left + rimSize) + (radius - bPos) / 2, rimRect.top
          + rimSize, (rimRect.right - rimSize) - (radius - bPos) / 2,
          (rimRect.bottom - rimSize) - (aPos));
    faceCanvas.drawArc(skyRect,
          (float) (0 + (rotation[2] * (180 / Math.PI))), -180, true,
          skyPaint);
    earthRect.set((rimRect.left + rimSize) + (radius - bNeg) / 2,
          (rimRect.top + rimSize) - (aNeg), (rimRect.right - rimSize)
              - (radius - bNeg) / 2, rimRect.bottom - rimSize);
    faceCanvas.drawArc(earthRect,
          (float) (0 + (rotation[2] * (180 / Math.PI))), 180, true,
          earthPaint);
    canvas.drawBitmap(face, 0, 0, backgroundPaint);
}
private void drawBezel(Canvas canvas)
{
    if (bezel == null)
    {
        Log.w(TAG, "Bezel not created");
    }
    else
    {
        canvas.drawBitmap(bezel, 0, 0, backgroundPaint);
    }
}
@Override
protected void onSizeChanged(int w, int h, int oldw, int oldh)
{
    Log.d(TAG, "Size changed to " + w + "x" + h);
    regenerateBezel();
}
private void regenerateBezel()
{
    // free the old bitmap
    if (bezel != null)
    {
        bezel.recycle();
    }
    bezel = Bitmap.createBitmap(getWidth(), getHeight(),
          Bitmap.Config.ARGB_8888);
    Canvas bezelCanvas = new Canvas(bezel);
```

```
        float scale = (float) getWidth();
        bezelCanvas.scale(scale, scale);
        drawRim(bezelCanvas);
    }
    @Override
    protected void onDraw(Canvas canvas)
    {
        drawBezel(canvas);
        drawFace(canvas);
        canvas.restore();
    }
}
```

13.3.5　传感器处理

编写文件 FusedGyroscopeSensor.java，功能是联合使用陀螺传感器、磁场传感器和重力传感器实现精确的测量值。文件 FusedGyroscopeSensor.java 的具体实现代码如下所示。

```
public class FusedGyroscopeSensor implements GyroscopeSensorObserver,
        MagneticSensorObserver, GravitySensorObserver
{
    private static final String tag = FusedGyroscopeSensor.class
            .getSimpleName();
    public static final float FILTER_COEFFICIENT = 0.5f;
    public static final float EPSILON = 0.000000001f;
    private static final float NS2S = 1.0f / 1000000000.0f;
    private ArrayList<FusedGyroscopeSensorObserver> observersAngularVelocity;
    private boolean hasOrientation = false;
    private float dT = 0;
    private float omegaMagnitude = 0;
    private float thetaOverTwo = 0;
    private float sinThetaOverTwo = 0;
    private float cosThetaOverTwo = 0;
    private float[] gravity = new float[]
    { 0, 0, 0 };
    private float[] gyroscope = new float[3];
    private float[] gyroMatrix = new float[9];
    private float[] gyroOrientation = new float[3];
    private float[] magnetic = new float[3];
    private float[] orientation = new float[3];
    private float[] fusedOrientation = new float[3];
    private float[] rotationMatrix = new float[9];
    private float[] absoluteFrameOrientation = new float[3];
    private float[] deltaVector = new float[4];
    private float[] deltaMatrix = new float[9];
    private long timeStamp;

    private boolean initState = false;
    private MeanFilter meanFilterAcceleration;
    private MeanFilter meanFilterMagnetic;
    /**
     * Initialize a singleton instance.
     */
    public FusedGyroscopeSensor()
    {
        super();
        observersAngularVelocity = new ArrayList<FusedGyroscopeSensorObserver>();
        meanFilterAcceleration = new MeanFilter();
        meanFilterAcceleration.setWindowSize(10);
        meanFilterMagnetic = new MeanFilter();
        meanFilterMagnetic.setWindowSize(10);
        gyroOrientation[0] = 0.0f;
        gyroOrientation[1] = 0.0f;
        gyroOrientation[2] = 0.0f;
        // Initialize gyroMatrix with identity matrix
        gyroMatrix[0] = 1.0f;
        gyroMatrix[1] = 0.0f;
        gyroMatrix[2] = 0.0f;
        gyroMatrix[3] = 0.0f;
```

```
    gyroMatrix[4] = 1.0f;
    gyroMatrix[5] = 0.0f;
    gyroMatrix[6] = 0.0f;
    gyroMatrix[7] = 0.0f;
    gyroMatrix[8] = 1.0f;
}
public void notifyObservers()
{
    System.arraycopy(gyroOrientation, 0, absoluteFrameOrientation, 0, 3);
    for (FusedGyroscopeSensorObserver g : observersAngularVelocity)
    {
        g.onAngularVelocitySensorChanged(absoluteFrameOrientation,
                timeStamp);
    }
}
public void registerObserver(FusedGyroscopeSensorObserver g)
{
    observersAngularVelocity.add(g);
}

public void removeObserver(FusedGyroscopeSensorObserver g)
{
    int i = observersAngularVelocity.indexOf(g);
    if (i >= 0)
    {
        observersAngularVelocity.remove(i);
    }
}

/**
 * Calculates orientation angles from accelerometer and magnetometer output.
 */
private void calculateOrientation()
{
    if (SensorManager.getRotationMatrix(rotationMatrix, null, gravity,
            magnetic))
    {
        SensorManager.getOrientation(rotationMatrix, orientation);

        hasOrientation = true;
    }
}

/**
 * Get the rotation matrix from the current orientation. Android Sensor
 * Manager does not provide a method to transform the orientation into a
 * rotation matrix, only the orientation from a rotation matrix. The basic
 * rotations can be found in Wikipedia with the caveat that the rotations
 * are *transposed* relative to what is required for this method.
 */
private float[] getRotationMatrixFromOrientation(float[] orientation)
{
    float[] xM = new float[9];
    float[] yM = new float[9];
    float[] zM = new float[9];

    float sinX = (float) Math.sin(orientation[1]);
    float cosX = (float) Math.cos(orientation[1]);
    float sinY = (float) Math.sin(orientation[2]);
    float cosY = (float) Math.cos(orientation[2]);
    float sinZ = (float) Math.sin(orientation[0]);
    float cosZ = (float) Math.cos(orientation[0]);

    // rotation about x-axis (pitch)
    xM[0] = 1.0f;
    xM[1] = 0.0f;
    xM[2] = 0.0f;
    xM[3] = 0.0f;
    xM[4] = cosX;
    xM[5] = sinX;
```

```
        xM[6] = 0.0f;
        xM[7] = -sinX;
        xM[8] = cosX;

        // rotation about y-axis (roll)
        yM[0] = cosY;
        yM[1] = 0.0f;
        yM[2] = sinY;
        yM[3] = 0.0f;
        yM[4] = 1.0f;
        yM[5] = 0.0f;
        yM[6] = -sinY;
        yM[7] = 0.0f;
        yM[8] = cosY;

        // rotation about z-axis (azimuth)
        zM[0] = cosZ;
        zM[1] = sinZ;
        zM[2] = 0.0f;
        zM[3] = -sinZ;
        zM[4] = cosZ;
        zM[5] = 0.0f;
        zM[6] = 0.0f;
        zM[7] = 0.0f;
        zM[8] = 1.0f;

        // Build the composite rotation... rotation order is y, x, z (roll,
        // pitch, azimuth)
        float[] resultMatrix = matrixMultiplication(xM, yM);
        resultMatrix = matrixMultiplication(zM, resultMatrix);
        return resultMatrix;
    }

    /**
     * Calculates a rotation vector from the gyroscope angular speed values.
     */
    private void getRotationVectorFromGyro(float timeFactor)
    {

        // Calculate the angular speed of the sample
        omegaMagnitude = (float) Math.sqrt(Math.pow(gyroscope[0], 2)
              + Math.pow(gyroscope[1], 2) + Math.pow(gyroscope[2], 2));

        // Normalize the rotation vector if it's big enough to get the axis
        if (omegaMagnitude > EPSILON)
        {
            gyroscope[0] /= omegaMagnitude;
            gyroscope[1] /= omegaMagnitude;
            gyroscope[2] /= omegaMagnitude;
        }
        thetaOverTwo = omegaMagnitude * timeFactor;
        sinThetaOverTwo = (float) Math.sin(thetaOverTwo);
        cosThetaOverTwo = (float) Math.cos(thetaOverTwo);

        deltaVector[0] = sinThetaOverTwo * gyroscope[0];
        deltaVector[1] = sinThetaOverTwo * gyroscope[1];
        deltaVector[2] = sinThetaOverTwo * gyroscope[2];
        deltaVector[3] = cosThetaOverTwo;
    }

    /**
     * Multiply A by B.
     */
    private float[] matrixMultiplication(float[] A, float[] B)
    {
        float[] result = new float[9];

        result[0] = A[0] * B[0] + A[1] * B[3] + A[2] * B[6];
        result[1] = A[0] * B[1] + A[1] * B[4] + A[2] * B[7];
        result[2] = A[0] * B[2] + A[1] * B[5] + A[2] * B[8];
```

```
        result[3] = A[3] * B[0] + A[4] * B[3] + A[5] * B[6];
        result[4] = A[3] * B[1] + A[4] * B[4] + A[5] * B[7];
        result[5] = A[3] * B[2] + A[4] * B[5] + A[5] * B[8];

        result[6] = A[6] * B[0] + A[7] * B[3] + A[8] * B[6];
        result[7] = A[6] * B[1] + A[7] * B[4] + A[8] * B[7];
        result[8] = A[6] * B[2] + A[7] * B[5] + A[8] * B[8];

        return result;
    }

    /**
     * Calculate the fused orientation.
     */
    private void calculateFusedOrientation()
    {
        float oneMinusCoeff = (1.0f - FILTER_COEFFICIENT);
        // azimuth
        if (gyroOrientation[0] < -0.5 * Math.PI && orientation[0] > 0.0)
        {
            fusedOrientation[0] = (float) (FILTER_COEFFICIENT
                    * (gyroOrientation[0] + 2.0 * Math.PI) + oneMinusCoeff
                    * orientation[0]);
            fusedOrientation[0] -= (fusedOrientation[0] > Math.PI) ? 2.0 * Math.PI
                    : 0;
        }
        else if (orientation[0] < -0.5 * Math.PI && gyroOrientation[0] > 0.0)
        {
            fusedOrientation[0] = (float) (FILTER_COEFFICIENT
                    * gyroOrientation[0] + oneMinusCoeff
                    * (orientation[0] + 2.0 * Math.PI));
            fusedOrientation[0] -= (fusedOrientation[0] > Math.PI) ? 2.0 * Math.PI
                    : 0;
        }
        else
        {
            fusedOrientation[0] = FILTER_COEFFICIENT * gyroOrientation[0]
                    + oneMinusCoeff * orientation[0];
        }

        // pitch
        if (gyroOrientation[1] < -0.5 * Math.PI && orientation[1] > 0.0)
        {
            fusedOrientation[1] = (float) (FILTER_COEFFICIENT
                    * (gyroOrientation[1] + 2.0 * Math.PI) + oneMinusCoeff
                    * orientation[1]);
            fusedOrientation[1] -= (fusedOrientation[1] > Math.PI) ? 2.0 * Math.PI
                    : 0;
        }
        else if (orientation[1] < -0.5 * Math.PI && gyroOrientation[1] > 0.0)
        {
            fusedOrientation[1] = (float) (FILTER_COEFFICIENT
                    * gyroOrientation[1] + oneMinusCoeff
                    * (orientation[1] + 2.0 * Math.PI));
            fusedOrientation[1] -= (fusedOrientation[1] > Math.PI) ? 2.0 * Math.PI
                    : 0;
        }
        else
        {
            fusedOrientation[1] = FILTER_COEFFICIENT * gyroOrientation[1]
                    + oneMinusCoeff * orientation[1];
        }

        // roll
        if (gyroOrientation[2] < -0.5 * Math.PI && orientation[2] > 0.0)
        {
            fusedOrientation[2] = (float) (FILTER_COEFFICIENT
                    * (gyroOrientation[2] + 2.0 * Math.PI) + oneMinusCoeff
                    * orientation[2]);
```

```
            fusedOrientation[2] -= (fusedOrientation[2] > Math.PI) ? 2.0 * Math.PI
                    : 0;
        }
        else if (orientation[2] < -0.5 * Math.PI && gyroOrientation[2] > 0.0)
        {
            fusedOrientation[2] = (float) (FILTER_COEFFICIENT
                    * gyroOrientation[2] + oneMinusCoeff
                    * (orientation[2] + 2.0 * Math.PI));
            fusedOrientation[2] -= (fusedOrientation[2] > Math.PI) ? 2.0 * Math.PI
                    : 0;
        }
        else
        {
            fusedOrientation[2] = FILTER_COEFFICIENT * gyroOrientation[2]
                    + oneMinusCoeff * orientation[2];
        }

        // overwrite gyro matrix and orientation with fused orientation
        // to comensate gyro drift
        gyroMatrix = getRotationMatrixFromOrientation(fusedOrientation);

        System.arraycopy(fusedOrientation, 0, gyroOrientation, 0, 3);

        notifyObservers();
    }

    @Override
    public void onMagneticSensorChanged(float[] magnetic, long timeStamp)
    {
        // Get a local copy of the raw magnetic values from the device sensor.
        System.arraycopy(magnetic, 0, this.magnetic, 0, magnetic.length);

        this.magnetic = meanFilterMagnetic.filterFloat(this.magnetic);
    }

    @Override
    public void onGravitySensorChanged(float[] gravity, long timeStamp)
    {
        System.arraycopy(gravity, 0, this.gravity, 0, gravity.length);

        this.gravity = meanFilterAcceleration.filterFloat(this.gravity);

        calculateOrientation();
    }

    @Override
    public void onGyroscopeSensorChanged(float[] gyroscope, long timeStamp)
    {
        // don't start until first accelerometer/magnetometer orientation has
        // been acquired
        if (!hasOrientation)
        {
            return;
        }
        if (!initState)
        {
            gyroMatrix = matrixMultiplication(gyroMatrix, rotationMatrix);
            initState = true;
        }

        if (this.timeStamp != 0)
        {
            dT = (timeStamp - this.timeStamp) * NS2S;

            System.arraycopy(gyroscope, 0, this.gyroscope, 0, 3);
            getRotationVectorFromGyro(dT / 2.0f);
        }
        this.timeStamp = timeStamp;
        SensorManager.getRotationMatrixFromVector(deltaMatrix, deltaVector);
        gyroMatrix = matrixMultiplication(gyroMatrix, deltaMatrix);
```

```
        SensorManager.getOrientation(gyroMatrix, gyroOrientation);

        calculateFusedOrientation();
    }
}
```

编写文件 GravitySensor.java，功能是使用重力传感器测试具体数值，在此类中通过一个观察者模式测量了加速度的值。文件 GravitySensor.java 的具体实现代码如下所示。

```
public class GravitySensor implements SensorEventListener
{
    /*
     * Developer Note: Quaternions are used for the internal representations of
     * the rotations which prevents the polar anomalies associated with Gimbal
     * lock when using Euler angles for the rotations.
     */
    private static final String tag = GravitySensor.class.getSimpleName();
    // Keep track of observers.
    private ArrayList<GravitySensorObserver> observersAcceleration;
    // Keep track of the application mode. Vehicle Mode occurs when the device
    // is in the Landscape orientation and the sensors are rotated to face the
    // -Z-Axis (along the axis of the camera).
    private boolean vehicleMode = false;
    // We need the Context to register for Sensor Events.
    private Context context;
    // Keep a local copy of the acceleration values that are copied from the
    // sensor event.
    private float[] gravity = new float[3];
    // The time stamp of the most recent Sensor Event.
    private long timeStamp = 0;
    // Quaternion data structures to rotate a matrix from the absolute Android
    // orientation to the orientation that the device is actually in. This is
    // needed because the the device sensors orientation is fixed in hardware.
    // Also remember the many algorithms require a NED orientation which is not
    // the same as the absolute Android orientation. Do not confuse this
    // rotation with a rotation into absolute earth frame!
    private Rotation yQuaternion;
    private Rotation xQuaternion;
    private Rotation rotationQuaternion;
    // We need the SensorManager to register for Sensor Events.
    private SensorManager sensorManager;
    // The vectors that will be rotated when the application is in Vehicle Mode.
    private Vector3D vIn;
    private Vector3D vOut;
    /**
     * Initialize the state.
     */
    public GravitySensor(Context context)
    {
        super();
        this.context = context;
        initQuaternionRotations();
        observersAcceleration = new ArrayList<GravitySensorObserver>();
        sensorManager = (SensorManager) this.context
                .getSystemService(Context.SENSOR_SERVICE);
    }
    /**
     * Register for Sensor.TYPE_GRAVITY measurements.
     */
    public void registerGravityObserver(GravitySensorObserver observer)
    {
        // If there are currently no observers, but one has just requested to be
        // registered, register to listen for sensor events from the device.
        if (observersAcceleration.size() == 0)
        {
            sensorManager.registerListener(this,
                    sensorManager.getDefaultSensor(Sensor.TYPE_GRAVITY),
                    SensorManager.SENSOR_DELAY_FASTEST);
        }
```

```
        // Only register the observer if it is not already registered.
        int i = observersAcceleration.indexOf(observer);
        if (i == -1)
        {
            observersAcceleration.add(observer);
        }
    }
    /**
     * Remove Sensor.TYPE_GRAVITY measurements.
     */
    public void removeGravityObserver(GravitySensorObserver observer)
    {
        int i = observersAcceleration.indexOf(observer);
        if (i >= 0)
        {
            observersAcceleration.remove(i);
        }
        // If there are no observers, then don't listen for Sensor Events.
        if (observersAcceleration.size() == 0)
        {
            sensorManager.unregisterListener(this);
        }
    }
    @Override
    public void onAccuracyChanged(Sensor sensor, int accuracy)
    {
        // Do nothing.
    }
    @Override
    public void onSensorChanged(SensorEvent event)
    {
        if (event.sensor.getType() == Sensor.TYPE_GRAVITY)
        {
            System.arraycopy(event.values, 0, gravity, 0, event.values.length);
            timeStamp = event.timestamp;
            if (vehicleMode)
            {
                gravity = quaternionToDeviceVehicleMode(gravity);
            }
            notifyGravityObserver();
        }
    }
    /**
     * Vehicle mode occurs when the device is put into the landscape
     * orientation. On Android phones, the positive Y-Axis of the sensors faces
     * towards the top of the device. In vehicle mode, we want the sensors to
     * face the negative Z-Axis so it is aligned with the camera of the device.
     */
    public void setVehicleMode(boolean vehicleMode)
    {
        this.vehicleMode = vehicleMode;
    }

    /**
     * To avoid anomalies at the poles with Euler angles and Gimbal lock,
     * quaternions are used instead.
     */
    private void initQuaternionRotations()
    {
        // Rotate by 90 degrees or pi/2 radians.
        double rotation = Math.PI / 2;
        // Create the rotation around the x-axis
        Vector3D xV = new Vector3D(1, 0, 0);
        xQuaternion = new Rotation(xV, rotation);
        // Create the rotation around the y-axis
        Vector3D yV = new Vector3D(0, 1, 0);
        yQuaternion = new Rotation(yV, -rotation);
        // Create the composite rotation.
        rotationQuaternion = yQuaternion.applyTo(xQuaternion);
```

```
    }
    /**
     * Notify observers with new measurements.
     */
    private void notifyGravityObserver()
    {
        // The iterator is a work around for a concurrency exception...
        GravitySensorObserver observer;
        ArrayList<GravitySensorObserver> notificationList = new ArrayList<GravitySensor
        Observer>(observersAcceleration);
        for (Iterator<GravitySensorObserver> iterator = notificationList
                .iterator(); iterator.hasNext();)
        {
            observer = iterator.next();
            observer.onGravitySensorChanged(this.gravity, this.timeStamp);
        }
    }
    /**
     * Orient the measurements from the absolute Android device rotation into
     * the current device orientation. Note that the rotation is different based
     * on the current rotation of the device relative to the absolute Android
     * rotation. Do not confuse this with a rotation into absolute earth frame,
     * or the NED orientation that the algorithm assumes.
     */
    private float[] quaternionToDeviceVehicleMode(float[] matrix)
    {
        vIn = new Vector3D(matrix[0], matrix[1], matrix[2]);
        vOut = rotationQuaternion.applyTo(vIn);
        float[] rotation =
        { (float) vOut.getX(), (float) vOut.getY(), (float) vOut.getZ() };
        return rotation;
    }
}
```

编写文件 GyroscopeSensor.java，功能是使用陀螺仪传感器测试具体数值，在此类中通过一个观察者模式测量了旋转角度的值。文件 GyroscopeSensor.java 的具体实现代码如下所示。

```
public class GyroscopeSensor implements SensorEventListener
{
    /*
     * Developer Note: Quaternions are used for the internal representations of
     * the rotations which prevents the polar anomalies associated with Gimbal
     * lock when using Euler angles for the rotations.
     */

    private static final String tag = GyroscopeSensor.class.getSimpleName();

    // Keep track of observers.
    private ArrayList<GyroscopeSensorObserver> observersGyroscope;

    // Keep track of the application mode. Vehicle Mode occurs when the device
    // is in the Landscape orientation and the sensors are rotated to face the
    // -Z-Axis (along the axis of the camera).
    private boolean vehicleMode = false;

    // We need the Context to register for Sensor Events.
    private Context context;

    // Keep a local copy of the rotation values that are copied from the
    // sensor event.
    private float[] gyroscope = new float[3];

    // The time stamp of the most recent Sensor Event.
    private long timeStamp = 0;

    // Quaternion data structures to rotate a matrix from the absolute Android
    // orientation to the orientation that the device is actually in. This is
    // needed because the the device sensors orientation is fixed in hardware.
    // Also remember the many algorithms require a NED orientation which is not
```

```
// the same as the absolute Android orientation. Do not confuse this
// rotation with a rotation into absolute earth frame!
private Rotation yQuaternion;
private Rotation xQuaternion;
private Rotation rotationQuaternion;

// We need the SensorManager to register for Sensor Events.
private SensorManager sensorManager;

// The vectors that will be rotated when the application is in Vehicle Mode.
private Vector3D vIn;
private Vector3D vOut;
/**
 * Initialize the state.
 */
public GyroscopeSensor(Context context)
{
    super();
    this.context = context;
    initQuaternionRotations();
    observersGyroscope = new ArrayList<GyroscopeSensorObserver>();
    sensorManager = (SensorManager) this.context
        .getSystemService(Context.SENSOR_SERVICE);
}
/**
 * Register for Sensor.TYPE_GYROSCOPE measurements.
 */
public void registerGyroscopeObserver(GyroscopeSensorObserver observer)
{
    if (observersGyroscope.size() == 0)
    {
        sensorManager.registerListener(this,
            sensorManager.getDefaultSensor(Sensor.TYPE_GYROSCOPE),
            SensorManager.SENSOR_DELAY_FASTEST);
    }

    // Only register the observer if it is not already registered.
    int i = observersGyroscope.indexOf(observer);
    if (i == -1)
    {
        observersGyroscope.add(observer);
    }
}
/**
 * Remove Sensor.TYPE_GYROSCOPE measurements.
 */
public void removeGyroscopeObserver(GyroscopeSensorObserver observer)
{
    int i = observersGyroscope.indexOf(observer);
    if (i >= 0)
    {
        observersGyroscope.remove(i);
    }
    // If there are no observers, then don't listen for Sensor Events.
    if (observersGyroscope.size() == 0)
    {
        sensorManager.unregisterListener(this);
    }
}
@Override
public void onSensorChanged(SensorEvent event)
{
    if (event.sensor.getType() == Sensor.TYPE_GYROSCOPE)
    {
        System.arraycopy(event.values, 0, this.gyroscope, 0,
            event.values.length);
        this.timeStamp = event.timestamp;
        if (vehicleMode)
        {
```

```
            this.gyroscope = quaternionToDeviceVehicleMode(this.gyroscope);
        }
        notifyGyroscopeObserver();
    }
}
/**
 * Vehicle mode occurs when the device is put into the landscape
 * orientation. On Android phones, the positive Y-Axis of the sensors faces
 * towards the top of the device. In vehicle mode, we want the sensors to
 * face the negative Z-Axis so it is aligned with the camera of the device.
 */
public void setVehicleMode(boolean vehicleMode)
{
    this.vehicleMode = vehicleMode;
}

/**
 * To avoid anomalies at the poles with Euler angles and Gimbal lock,
 * quaternions are used instead.
 */
private void initQuaternionRotations()
{
    // Rotate by 90 degrees or pi/2 radians.
    double rotation = Math.PI / 2;
    // Create the rotation around the x-axis
    Vector3D xV = new Vector3D(1, 0, 0);
    xQuaternion = new Rotation(xV, rotation);
    // Create the rotation around the y-axis
    Vector3D yV = new Vector3D(0, 1, 0);
    yQuaternion = new Rotation(yV, -rotation);
    // Create the composite rotation.
    rotationQuaternion = yQuaternion.applyTo(xQuaternion);
}

/**
 * Notify observers with new measurements.
 */
private void notifyGyroscopeObserver()
{
    for (GyroscopeSensorObserver a : observersGyroscope)
    {
        a.onGyroscopeSensorChanged(this.gyroscope, this.timeStamp);
    }
}

/**
 * Orient the measurements from the absolute Android device rotation into
 * the current device orientation. Note that the rotation is different based
 * on the current rotation of the device relative to the absolute Android
 * rotation. Do not confuse this with a rotation into absolute earth frame,
 * or the NED orientation that the algorithm assumes.
 */
private float[] quaternionToDeviceVehicleMode(float[] matrix)
{
    vIn = new Vector3D(matrix[0], matrix[1], matrix[2]);
    vOut = rotationQuaternion.applyTo(vIn);
    float[] rotation =
    { (float) vOut.getX(), (float) vOut.getY(), (float) vOut.getZ() };
    return rotation;
}
}
```

编写文件 MagneticSensor.java，功能是使用磁场传感器测试具体数值，在此类中通过一个观察者模式测量了磁场强度的值。文件 MagneticSensor.java 的具体实现代码如下所示。

```
public class MagneticSensor implements SensorEventListener
{
    /*
     * Developer Note: Quaternions are used for the internal representations of
```

```
 * the rotations which prevents the polar anomalies associated with Gimbal
 * lock when using Euler angles for the rotations.
 */
private static final String tag = MagneticSensor.class.getSimpleName();
private ArrayList<MagneticSensorObserver> observersMagnetic;
private boolean vehicleMode = false;
private Context context;
private float[] magnetic = new float[3];
private long timeStamp = 0;
private Rotation yQuaternion;
private Rotation xQuaternion;
private Rotation rotationQuaternion;
private SensorManager sensorManager;
private Vector3D vIn;
private Vector3D vOut;
/**
 * Initialize the state.
 */
public MagneticSensor(Context context)
{
    super();
    this.context = context;
    initQuaternionRotations();
    observersMagnetic = new ArrayList<MagneticSensorObserver>();
    sensorManager = (SensorManager) this.context
            .getSystemService(Context.SENSOR_SERVICE);
}
/**
 * Register for Sensor.TYPE_MAGNETIC measurements.
 */
public void registerMagneticObserver(MagneticSensorObserver observer)
{
    if (observersMagnetic.size() == 0)
    {
        sensorManager.registerListener(this,
                sensorManager.getDefaultSensor(Sensor.TYPE_MAGNETIC_FIELD),
                SensorManager.SENSOR_DELAY_FASTEST);
    }
    // Only register the observer if it is not already registered.
    int i = observersMagnetic.indexOf(observer);
    if (i == -1)
    {
        observersMagnetic.add(observer);
    }
}
/**
 * Remove Sensor.TYPE_MAGNETIC measurements.
 *          The observer to be removed.
 */
public void removeMagneticObserver(MagneticSensorObserver observer)
{
    int i = observersMagnetic.indexOf(observer);
    if (i >= 0)
    {
        observersMagnetic.remove(i);
    }
    // If there are no observers, then don't listen for Sensor Events.
    if (observersMagnetic.size() == 0)
    {
        sensorManager.unregisterListener(this);
    }
}
@Override
public void onSensorChanged(SensorEvent event)
{
    if (event.sensor.getType() == Sensor.TYPE_MAGNETIC_FIELD)
    {
        System.arraycopy(event.values, 0, magnetic, 0, event.values.length);
        timeStamp = event.timestamp;
```

```
            if (vehicleMode)
            {
                this.magnetic = quaternionToDeviceVehicleMode(this.magnetic);
            }
            notifyMagneticObserver();
        }
    }
    /**
     * Vehicle mode occurs when the device is put into the landscape
     * orientation. On Android phones, the positive Y-Axis of the sensors faces
     * towards the top of the device. In vehicle mode, we want the sensors to
     * face the negative Z-Axis so it is aligned with the camera of the device.
     */
    public void setVehicleMode(boolean vehicleMode)
    {
        this.vehicleMode = vehicleMode;
    }
    /**
     * To avoid anomalies at the poles with Euler angles and Gimbal lock,
     * quaternions are used instead.
     */
    private void initQuaternionRotations()
    {
        // Rotate by 90 degrees or pi/2 radians.
        double rotation = Math.PI / 2;
        // Create the rotation around the x-axis
        Vector3D xV = new Vector3D(1, 0, 0);
        xQuaternion = new Rotation(xV, rotation);
        // Create the rotation around the y-axis
        Vector3D yV = new Vector3D(0, 1, 0);
        yQuaternion = new Rotation(yV, -rotation);
        // Create the composite rotation.
        rotationQuaternion = yQuaternion.applyTo(xQuaternion);
    }
    /**
     * Notify observers with new measurements.
     */
    private void notifyMagneticObserver()
    {
        // The iterator is a work around for a concurrency exception... Not the best work
        // around, but it works for now.
        MagneticSensorObserver observer;
        ArrayList<MagneticSensorObserver> notificationList = new ArrayList
        <MagneticSensorObserver>(
                observersMagnetic);
        for (Iterator<MagneticSensorObserver> iterator = notificationList
                .iterator(); iterator.hasNext();)
        {
            observer = iterator.next();
            observer.onMagneticSensorChanged(this.magnetic, this.timeStamp);
        }
    }
    /**
     * Orient the measurements from the absolute Android device rotation into
     * the current device orientation. Note that the rotation is different based
     * on the current rotation of the device relative to the absolute Android
     * rotation. Do not confuse this with a rotation into absolute earth frame,
     * or the NED orientation that the algorithm assumes.
     */
    private float[] quaternionToDeviceVehicleMode(float[] matrix)
    {
        vIn = new Vector3D(matrix[0], matrix[1], matrix[2]);
        vOut = rotationQuaternion.applyTo(vIn);
        float[] rotation =
        { (float) vOut.getX(), (float) vOut.getY(), (float) vOut.getZ() };
        return rotation;
    }
}
```

到此为止，整个实例介绍完毕，执行后的效果读者自行演示。

第 14 章　旋转向量传感器详解

旋转向量传感器也被称为旋转矢量传感器，简称 RV-sensor。旋转矢量代表设备的方向，是一个将坐标轴和角度混合计算得到的数据。本章将详细讲解在 Android 设备中使用旋转向量传感器的基本知识，为读者步入本书后面知识的学习打下基础。

14.1 Android 中的旋转向量传感器

在 Android 系统中，旋转向量传感器的值是 TYPE_ROTATION_VECTOR，旋转矢量代表设备的方向，是一个将坐标轴和角度混合计算得到的数据。对 Android 旋转向量传感器的具体说明如表 14-1 所示。

表 14-1　Android 旋转向量传感器的具体说明

传 感 器	传感器事件数据	说　明	测量单位
TYPE_ROTATION_VECTOR	SensorEvent.values[0]	旋转向量沿 x 轴的部分（x * sin(θ/2)）	无
	SensorEvent.values[1]	旋转向量沿 y 轴的部分（y * sin(θ/2)）	
	SensorEvent.values[2]	旋转向量沿 z 轴的部分（z * sin(θ/2)）	
	SensorEvent.values[3]	旋转向量的数值部分（(cos(θ/2))1	

由表 14-1 可知，RV-sensor 能够输出如下所示的 3 个数据：

- x*sin(θ/2)；
- y*sin(θ/2)；
- z*sin(θ/2)。

则 sin(theta/2)表示 RV 的数量级，RV 的方向与轴旋转的方向相同，这样 RV 的 3 个数值与 cos(theta/2)组成一个四元组。而 RV 的数据没有单位，使用的坐标系与加速度相同。例如下面的演示代码。

```
sensors_event_t.data[0] = x*sin(theta/2)
sensors_event_t.data[1] = y*sin(theta/2)
sensors_event_t.data[2] = z*sin(theta/2)
sensors_event_t.data[3] =  cos(theta/2)
```

GV、LA 和 RV 的数值没有物理传感器可以直接给出，需要 G-sensor、O-sensor 和 Gyro-sensor 经过算法计算后得出。

由此可见，旋转向量代表了设备的方位，这个方位结果由角度和坐标轴信息组成，在里面包含了设备围绕坐标轴（x、y、z）旋转的角度θ。 例如下面的代码演示了获取默认的旋转向量传感器的方法。

```
private SensorManager mSensorManager;
private Sensor mSensor;
```

```
...
mSensorManager = (SensorManager) getSystemService(Context.SENSOR_SERVICE);
mSensor = mSensorManager.getDefaultSensor(Sensor.TYPE_ROTATION_VECTOR);
```

在 Android 系统中，旋转向量的 3 个元素等于四元组的后 3 个部分（cos(θ /2)、x*sin(θ /2)、y*sin(θ /2)、z*sin(θ /2)），没有单位。*X*、*Y*、*Z* 轴的具体定义与加速度传感器的相同。旋转向量传感器的坐标系如图 14-1 所示。

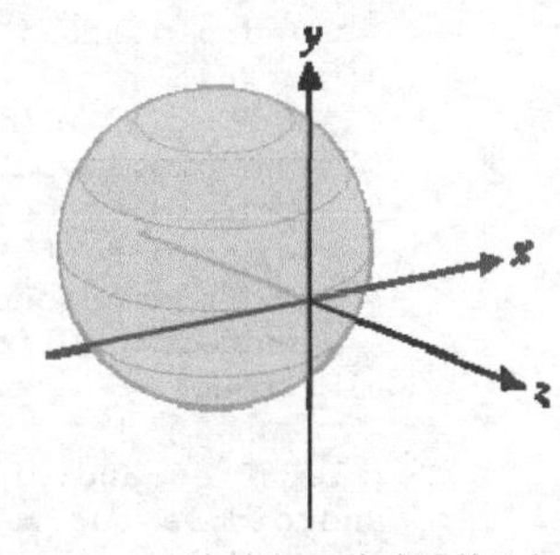

▲图 14-1 旋转向量传感器的坐标系

上述坐标系具有如下所示的特点。

- *X*：定义为向量积 Y×Z。它是以设备当前位置为切点的地球切线，方向朝东。
- *Y*：以设备当前位置为切点的地球切线，指向地磁北极。
- *Z*：与地平面垂直，指向天空。

14.2 实战演练——确定设备当前的具体方向

本实例联合使用了旋转向量传感器、磁场传感器、重力传感器和加速度传感器，功能是获取当前设备的方向。

实例	功能	源码路径
实例 14-1	确定设备当前的具体方向	https://github.com/gast-lib/gast-lib

本实例的功能是，当设备接近某个位置时实现自动提醒。本实例源码是开源代码，来源于如下地址，读者可以自行登录并下载。

```
https://github.com/gast-lib/gast-lib/
```

14.2.1 实现主 Activity

本实例的主 Activity 是 DetermineOrientationActivity，通过布局文件 determine_orientation.xml 实现布局，在屏幕中提供一组单选按钮供用户选择所需要的传感器，并在屏幕下方显示具体的显示传感器返回的数据。布局文件 determine_orientation.xml 的具体实现代码如下所示。

```
<RelativeLayout xmlns:android="http://schemas.android.com/apk/res/android"
    android:layout_width="match_parent"
    android:layout_height="match_parent"
    android:orientation="vertical" >

    <RadioGroup android:id="@+id/sensorSelector"
        android:layout_width="match_parent"
        android:layout_height="wrap_content"
        android:layout_alignParentTop="true" >

        <RadioButton android:id="@+id/gravitySensor"
            android:layout_width="match_parent"
            android:layout_height="wrap_content"
            android:text="@string/gravitySensorLabel"
            android:checked="true"
            android:onClick="onSensorSelectorClick" />

        <RadioButton android:id="@+id/accelerometerMagnetometer"
            android:layout_width="match_parent"
            android:layout_height="wrap_content"
            android:text="@string/accelerometerMagnetometerLabel"
            android:checked="false"
            android:onClick="onSensorSelectorClick" />
```

```
    <RadioButton android:id="@+id/gravityMagnetometer"
        android:layout_width="match_parent"
        android:layout_height="wrap_content"
        android:text="@string/gravityMagnetometerLabel"
        android:checked="false"
        android:onClick="onSensorSelectorClick" />

    <RadioButton android:id="@+id/rotationVector"
        android:layout_width="match_parent"
        android:layout_height="wrap_content"
        android:text="@string/rotationVectorLabel"
        android:checked="false"
        android:onClick="onSensorSelectorClick" />
</RadioGroup>

<ToggleButton android:id="@+id/ttsNotificationsToggleButton"
    android:layout_width="wrap_content"
    android:layout_height="wrap_content"
    android:text="@string/speakOrientationLabel"
    android:checked="true"
    android:layout_below="@id/sensorSelector"
    android:textOn="@string/ttsNotificationsOn"
    android:textOff="@string/ttsNotificationsOff"
    android:onClick="onTtsNotificationsToggleButtonClicked" />

<TextView android:id="@+id/selectedSensorLabel"
    android:layout_width="wrap_content"
    android:layout_height="wrap_content"
    android:text="@string/selectedSensorLabel"
    android:layout_below="@id/ttsNotificationsToggleButton"
    android:layout_marginRight="5dip" />

<TextView android:id="@+id/selectedSensorValue"
    android:layout_width="wrap_content"
    android:layout_height="wrap_content"
    android:layout_toRightOf="@id/selectedSensorLabel"
    android:layout_alignTop="@id/selectedSensorLabel"
    android:layout_alignBottom="@id/selectedSensorLabel" />

<TextView android:id="@+id/orientationLabel"
    android:layout_width="wrap_content"
    android:layout_height="wrap_content"
    android:text="@string/orientationLabel"
    android:layout_below="@id/selectedSensorValue"
    android:layout_marginRight="5dip" />

<TextView android:id="@+id/orientationValue"
    android:layout_width="wrap_content"
    android:layout_height="wrap_content"
    android:layout_toRightOf="@id/orientationLabel"
    android:layout_alignTop="@id/orientationLabel"
    android:layout_alignBottom="@id/orientationLabel" />

<TextView android:id="@+id/sensorXLabel"
    android:layout_width="wrap_content"
    android:layout_height="wrap_content"
    android:layout_below="@id/orientationValue"
    android:layout_marginRight="5dip" />

<TextView android:id="@+id/sensorXValue"
    android:layout_width="wrap_content"
    android:layout_height="wrap_content"
    android:layout_toRightOf="@id/sensorXLabel"
    android:layout_alignTop="@id/sensorXLabel"
    android:layout_alignBottom="@id/sensorXLabel" />

<TextView android:id="@+id/sensorYLabel"
    android:layout_width="wrap_content"
    android:layout_height="wrap_content"
```

```
        android:layout_below="@id/sensorXLabel"
        android:layout_marginRight="5dip" />

    <TextView android:id="@+id/sensorYValue"
        android:layout_width="wrap_content"
        android:layout_height="wrap_content"
        android:layout_toRightOf="@id/sensorYLabel"
        android:layout_alignTop="@id/sensorYLabel"
        android:layout_alignBottom="@id/sensorYLabel" />

    <TextView android:id="@+id/sensorZLabel"
        android:layout_width="wrap_content"
        android:layout_height="wrap_content"
        android:layout_below="@id/sensorYLabel"
        android:layout_marginRight="5dip" />

    <TextView android:id="@+id/sensorZValue"
        android:layout_width="wrap_content"
        android:layout_height="wrap_content"
        android:layout_toRightOf="@id/sensorZLabel"
        android:layout_alignTop="@id/sensorZLabel"
        android:layout_alignBottom="@id/sensorZLabel" />
</RelativeLayout>
```

主 Activity 程序文件 DetermineOrientationActivity.java 的功能是，获取 SensorManager 的引用，根据用户单选按钮的选择来注册这个传感器，然后调用这个传感器来获取数据。假如选择的是重力传感器，则会注册重力传感器，然后获取数组 SensorEvent.Values 中的 *X*、*Y* 和 *Z* 轴上的重力大小。当选择使用旋转向量传感器时，会调用方法 determineOrientation(rotationMatrix)根据给出的旋转矩阵计算出具体的方向。当使用者确定了设备的具体朝向时，会使用文本转语音的功能来提醒用户。本实例的语音提醒功能是通过 TTS 机制实现的，有关语音提醒方面的知识将在本书后面的章节中进行详细讲解。文件 DetermineOrientationActivity.java 的具体实现代码如下所示。

```
protected void onCreate(Bundle savedInstanceState)
{
    super.onCreate(savedInstanceState);
    super.setContentView(R.layout.determine_orientation);

    // Keep the screen on so that changes in orientation can be easily
    // observed
    getWindow().addFlags(WindowManager.LayoutParams.FLAG_KEEP_SCREEN_ON);

    // Set up stream to use for Text-To-Speech
    ttsParams = new HashMap<String, String>();
    ttsParams.put(Engine.KEY_PARAM_STREAM, String.valueOf(TTS_STREAM));

    // Set the volume control to use the same stream as TTS which allows
    // the user to easily adjust the TTS volume
    this.setVolumeControlStream(TTS_STREAM);

    // Get a reference to the sensor service
    sensorManager = (SensorManager) getSystemService(SENSOR_SERVICE);

    // Initialize references to the UI views that will be updated in the
    // code
    sensorSelector = (RadioGroup) findViewById(R.id.sensorSelector);
    selectedSensorValue = (TextView) findViewById(R.id.selectedSensorValue);
    orientationValue = (TextView) findViewById(R.id.orientationValue);
    sensorXLabel = (TextView) findViewById(R.id.sensorXLabel);
    sensorXValue = (TextView) findViewById(R.id.sensorXValue);
    sensorYLabel = (TextView) findViewById(R.id.sensorYLabel);
    sensorYValue = (TextView) findViewById(R.id.sensorYValue);
    sensorZLabel = (TextView) findViewById(R.id.sensorZLabel);
    sensorZValue = (TextView) findViewById(R.id.sensorZValue);
    ttsNotificationsToggleButton =
            (ToggleButton) findViewById(R.id.ttsNotificationsToggleButton);
```

```
        // Retrieve stored preferences
        preferences = getPreferences(MODE_PRIVATE);
        ttsNotifications =
                preferences.getBoolean(TTS_NOTIFICATION_PREFERENCES_KEY, true);
    }

    @Override
    protected void onResume()
    {
        super.onResume();

        ttsNotificationsToggleButton.setChecked(ttsNotifications);
        updateSelectedSensor();
    }

    @Override
    protected void onPause()
    {
        super.onPause();

        // Unregister updates from sensors
        sensorManager.unregisterListener(this);

        // Shutdown TTS facility
        if (tts != null)
        {
            tts.shutdown();
        }
    }

    @Override
    public void onSensorChanged(SensorEvent event)
    {
        float[] rotationMatrix;

        switch (event.sensor.getType())
        {
            case Sensor.TYPE_GRAVITY:
                sensorXLabel.setText(R.string.xAxisLabel);
                sensorXValue.setText(String.valueOf(event.values[0]));

                sensorYLabel.setText(R.string.yAxisLabel);
                sensorYValue.setText(String.valueOf(event.values[1]));

                sensorZLabel.setText(R.string.zAxisLabel);
                sensorZValue.setText(String.valueOf(event.values[2]));

                sensorYLabel.setVisibility(View.VISIBLE);
                sensorYValue.setVisibility(View.VISIBLE);
                sensorZLabel.setVisibility(View.VISIBLE);
                sensorZValue.setVisibility(View.VISIBLE);

                if (selectedSensorId == R.id.gravitySensor)
                {
                    if (event.values[2] >= GRAVITY_THRESHOLD)
                    {
                        onFaceUp();
                    }
                    else if (event.values[2] <= (GRAVITY_THRESHOLD * -1))
                    {
                        onFaceDown();
                    }
                }
                else
                {
                    accelerationValues = event.values.clone();
                    rotationMatrix = generateRotationMatrix();

                    if (rotationMatrix != null)
```

```
                {
                    determineOrientation(rotationMatrix);
                }
            }

            break;
        case Sensor.TYPE_ACCELEROMETER:
            accelerationValues = event.values.clone();
            rotationMatrix = generateRotationMatrix();

            if (rotationMatrix != null)
            {
                determineOrientation(rotationMatrix);
            }
            break;
        case Sensor.TYPE_MAGNETIC_FIELD:
            magneticValues = event.values.clone();
            rotationMatrix = generateRotationMatrix();

            if (rotationMatrix != null)
            {
                determineOrientation(rotationMatrix);
            }
            break;
        case Sensor.TYPE_ROTATION_VECTOR:

            rotationMatrix = new float[16];
            SensorManager.getRotationMatrixFromVector(rotationMatrix,
                    event.values);
            determineOrientation(rotationMatrix);
            break;
    }
}

@Override
public void onAccuracyChanged(Sensor sensor, int accuracy)
{
    Log.d(TAG,
            String.format("Accuracy for sensor %s = %d",
            sensor.getName(), accuracy));
}

/**
 * Generates a rotation matrix using the member data stored in
 * accelerationValues and magneticValues.
 *
 * @return The rotation matrix returned from
 * {@link android.hardware.SensorManager#getRotationMatrix(float[], float[],
 float[], float[])}
 * or <code>null</code> if either <code>accelerationValues</code> or
 * <code>magneticValues</code> is null.
 */
private float[] generateRotationMatrix()
{
    float[] rotationMatrix = null;

    if (accelerationValues != null && magneticValues != null)
    {
        rotationMatrix = new float[16];
        boolean rotationMatrixGenerated;
        rotationMatrixGenerated =
                SensorManager.getRotationMatrix(rotationMatrix,
                null,
                accelerationValues,
                magneticValues);

        if (!rotationMatrixGenerated)
        {
            Log.w(TAG, getString(R.string.rotationMatrixGenFailureMessage));
```

```
            rotationMatrix = null;
        }
    }

    return rotationMatrix;
}

/**
 * Uses the last read accelerometer and gravity values to determine if the
 * device is face up or face down.
 *
 * @param rotationMatrix The rotation matrix to use if the orientation
 * calculation
 */
private void determineOrientation(float[] rotationMatrix)
{
    float[] orientationValues = new float[3];
    SensorManager.getOrientation(rotationMatrix, orientationValues);

    double azimuth = Math.toDegrees(orientationValues[0]);
    double pitch = Math.toDegrees(orientationValues[1]);
    double roll = Math.toDegrees(orientationValues[2]);

    sensorXLabel.setText(R.string.azimuthLabel);
    sensorXValue.setText(String.valueOf(azimuth));

    sensorYLabel.setText(R.string.pitchLabel);
    sensorYValue.setText(String.valueOf(pitch));

    sensorZLabel.setText(R.string.rollLabel);
    sensorZValue.setText(String.valueOf(roll));

    sensorYLabel.setVisibility(View.VISIBLE);
    sensorYValue.setVisibility(View.VISIBLE);
    sensorZLabel.setVisibility(View.VISIBLE);
    sensorZValue.setVisibility(View.VISIBLE);

    if (pitch <= 10)
    {
        if (Math.abs(roll) >= 170)
        {
            onFaceDown();
        }
        else if (Math.abs(roll) <= 10)
        {
            onFaceUp();
        }
    }
}

/**
 * Handler for device being face up.
 */
private void onFaceUp()
{
    if (!isFaceUp)
    {
        if (tts != null && ttsNotificationsToggleButton.isChecked())
        {
            tts.speak(getString(R.string.faceUpText),
                    TextToSpeech.QUEUE_FLUSH,
                    ttsParams);
        }

        orientationValue.setText(R.string.faceUpText);
        isFaceUp = true;
    }
}
```

```
/**
 * Handler for device being face down.
 */
private void onFaceDown()
{
    if (isFaceUp)
    {
        if (tts != null && ttsNotificationsToggleButton.isChecked())
        {
            tts.speak(getString(R.string.faceDownText),
                    TextToSpeech.QUEUE_FLUSH,
                    ttsParams);
        }

        orientationValue.setText(R.string.faceDownText);
        isFaceUp = false;
    }
}

/**
 * Updates the views for when the selected sensor is changed
 */
private void updateSelectedSensor()
{
    // Clear any current registrations
    sensorManager.unregisterListener(this);

    // Determine which radio button is currently selected and enable the
    // appropriate sensors
    selectedSensorId = sensorSelector.getCheckedRadioButtonId();
    if (selectedSensorId == R.id.accelerometerMagnetometer)
    {
        sensorManager.registerListener(this,
                sensorManager.getDefaultSensor(Sensor.TYPE_ACCELEROMETER),
                RATE);

        sensorManager.registerListener(this,
                sensorManager.getDefaultSensor(Sensor.TYPE_MAGNETIC_FIELD),
                RATE);
    }
    else if (selectedSensorId == R.id.gravityMagnetometer)
    {
        sensorManager.registerListener(this,
                sensorManager.getDefaultSensor(Sensor.TYPE_GRAVITY),
                RATE);

        sensorManager.registerListener(this,
                sensorManager.getDefaultSensor(Sensor.TYPE_MAGNETIC_FIELD),
                RATE);
    }
    else if ((selectedSensorId == R.id.gravitySensor))
    {
        sensorManager.registerListener(this,
                sensorManager.getDefaultSensor(Sensor.TYPE_GRAVITY),
                RATE);
    }
    else
    {
        sensorManager.registerListener(this,
                sensorManager.getDefaultSensor(Sensor.TYPE_ROTATION_VECTOR),
                RATE);
    }

    // Update the label with the currently selected sensor
    RadioButton selectedSensorRadioButton =
            (RadioButton) findViewById(selectedSensorId);
    selectedSensorValue.setText(selectedSensorRadioButton.getText());
}
```

```
    /**
     * Handles click event for the sensor selector.
     *
     * @param view The view that was clicked
     */
    public void onSensorSelectorClick(View view)
    {
        updateSelectedSensor();
    }

    /**
     * Handles click event for the TTS toggle button.
     *
     * @param view The view for the toggle button
     */
    public void onTtsNotificationsToggleButtonClicked(View view)
    {
        ttsNotifications = ((ToggleButton) view).isChecked();
        preferences.edit()
            .putBoolean(TTS_NOTIFICATION_PREFERENCES_KEY, ttsNotifications)
            .commit();
    }

    @Override
    public void onSuccessfulInit(TextToSpeech tts)
    {
        super.onSuccessfulInit(tts);
        this.tts = tts;
    }

    @Override
    protected void receiveWhatWasHeard(List<String> heard, float[] confidenceScores)
    {
        // no-op
    }
}
```

14.2.2 获取设备的旋转向量

编写文件 NorthFinder.java，首先获取设备的旋转向量，并将旋转向量的坐标映射到摄像头的轴上。如果取消了对方法 remapCoordinateSystem 的调用，则将当前设备指向北方，而并不是将后置摄像头指向北方。除此之外，在此文件中还使用 OpenGL 改变了屏幕的颜色，当后置摄像头指向北方时（允许误差 20 度以内），将屏幕颜色从红色变为绿色。文件 NorthFinder.java 的具体实现代码如下所示。

```
public class NorthFinder extends Activity implements SensorEventListener
{
    private static final int ANGLE = 20;

    private TextView tv;
    private GLSurfaceView mGLSurfaceView;
    private MyRenderer mRenderer;
    private SensorManager mSensorManager;
    private Sensor mRotVectSensor;
    private float[] orientationVals = new float[3];

    private final float[] mRotationMatrix = new float[16];

    @Override
    protected void onCreate(Bundle savedInstanceState)
    {
        super.onCreate(savedInstanceState);

        setContentView(R.layout.sensors_north_main);
```

```
        mRenderer = new MyRenderer();
        mGLSurfaceView = (GLSurfaceView) findViewById(R.id.glsurfaceview);
        mGLSurfaceView.setRenderer(mRenderer);

        tv = (TextView) findViewById(R.id.tv);

        mSensorManager = (SensorManager) getSystemService(SENSOR_SERVICE);
        mRotVectSensor =
                mSensorManager.getDefaultSensor(Sensor.TYPE_ROTATION_VECTOR);

    }

    @Override
    protected void onResume()
    {
        super.onResume();
        mSensorManager.registerListener(this, mRotVectSensor, 10000);
    }

    @Override
    protected void onPause()
    {
        super.onPause();
        mSensorManager.unregisterListener(this);
    }

    @Override
    public void onSensorChanged(SensorEvent event)
    {
        // It is good practice to check that we received the proper sensor event
        if (event.sensor.getType() == Sensor.TYPE_ROTATION_VECTOR)
        {
            // Convert the rotation-vector to a 4x4 matrix.
            SensorManager.getRotationMatrixFromVector(mRotationMatrix,
                    event.values);
            SensorManager
                    .remapCoordinateSystem(mRotationMatrix,
                            SensorManager.AXIS_X, SensorManager.AXIS_Z,
                            mRotationMatrix);
            SensorManager.getOrientation(mRotationMatrix, orientationVals);

            // Optionally convert the result from radians to degrees
            orientationVals[0] = (float) Math.toDegrees(orientationVals[0]);
            orientationVals[1] = (float) Math.toDegrees(orientationVals[1]);
            orientationVals[2] = (float) Math.toDegrees(orientationVals[2]);

            tv.setText(" Yaw: " + orientationVals[0] + "\n Pitch: "
                    + orientationVals[1] + "\n Roll (not used): "
                    + orientationVals[2]);

        }
    }

    @Override
    public void onAccuracyChanged(Sensor sensor, int accuracy)
    {
        // no-op
    }

    class MyRenderer implements GLSurfaceView.Renderer
    {
        public void onDrawFrame(GL10 gl)
        {
            // Clear screen
            gl.glClear(GL10.GL_COLOR_BUFFER_BIT);

            // Detect if the device is pointing within +/- ANGLE of north
            if (orientationVals[0] < ANGLE && orientationVals[0] > -ANGLE
                    && orientationVals[1] < ANGLE
```

```
                && orientationVals[1] > -ANGLE)
            {
                gl.glClearColor(0, 1, 0, 1); // Make background green
            }
            else
            {
                gl.glClearColor(1, 0, 0, 1); // Make background red
            }
        }

        @Override
        public void onSurfaceChanged(GL10 gl, int width, int height)
        {
            // no-op
        }

        @Override
        public void onSurfaceCreated(GL10 gl, EGLConfig config)
        {
            // no-op
        }
    }
}
```

到此为止，整个实例的核心代码介绍完毕。为节省本书篇幅，其余的代码将不再进行详细讲解。

第 15 章　距离传感器详解

在 Android 设备应用程序开发过程中，经常需要检测设备的运动数据，例如设备的运动速率和运动距离等。这些数据对于健身类设备来说，都是十分重要的数据，例如健身手表可以及时测试晨练的运动距离和速率。在 Android 系统中，通常使用加速度传感器、线性加速度传感器和距离传感器来检测设备的运动数据。本章将详细讲解在 Android 设备中检测运动数据的基本知识，为读者步入本书后面知识的学习打下基础。

15.1 距离传感器基础

在 Android 系统中，需要使用加速度传感器、线性加速度传感器和距离传感器来检测设备的运动数据。本书在前几章已经详细讲解了加速度传感器和线性加速度传感器的知识。本节将详细讲解 Android 系统距离传感器的基本知识。

15.1.1 距离传感器介绍

在当前的技术条件下，距离传感器是指利用“飞行时间法”（flying time）的原理来实现测量距离，以实现检测物体的距离的一种传感器。“飞行时间法”是通过发射特别短的光脉冲，并测量此光脉冲从发射到被物体反射回来的时间，通过测时间间隔来计算与物体之间的距离。

在现实世界中，距离传感器在智能手机中的应用比较常见。一般触屏智能手机在默认设置下，都会有一个延时锁屏的设置，就是在一段时间内，如果手机检测不到任何操作，就会进入锁屏状态。这样是有一定好处的。手机作为移动终端的一种，追求低功耗是设计的目标之一。延时锁屏既可以避免不必要的能量消耗，又能保证不丢失重要信息。另外，在使用触屏手机设备时，当接电话的时候距离传感器会起作用，当脸靠近屏幕时屏幕灯会熄灭，并自动锁屏，这样可以防止脸误操作。当脸离开屏幕时屏幕灯会自动开启，并且自动解锁。

除了被广泛应用于手机设备之外，距离传感器还被用于野外环境（山体情况、峡谷深度等）、飞机高度检测、矿井深度、物料高度测量等领域。在野外应用领域中，主要用于检测山体情况和峡谷深度等。而对飞机高度测量功能是通过检测飞机在起飞和降落时距离地面的高度，并将结果实时显示在控制面板上实现的。也可以使用距离传感器测量物料各点高度，用于计算物料的体积。在现实应用中，用于飞机高度和物料高度的距离传感器有 LDM301 系列，用于野外应用的距离传感器有 LDM4x 系列。

在当前的可设备应用中，距离传感器被应用于智能皮带中。在皮带扣里嵌入了距离传感器，当把皮带调整至合适宽度、卡好皮带扣后，如果皮带在 10 秒钟内没有重新解开，传感器就会自动生成本次的腰围数据。皮带与皮带扣连接处的其中一枚铆钉将被数据传输装置所替代。当你将智能手机放在铆钉处保持两秒钟静止，手机里的自我健康管理 App 会被自动激活，并获取本次腰围数据。

15.1.2　Android 系统中的距离传感器

在 Android 系统中，距离传感器也被称为 P-Sensor，值是 TYPE_PROXIMITY，单位是 cm，能够测量某个对象到屏幕的距离。可以在打电话时判断人耳到电话屏幕距离，以关闭屏幕而达到省电功能。

P-Sensor 主要用于在通话过程中防止用户误操作屏幕，接下来以通话过程为例来讲解电话程序对 P-Sensor 的操作流程。

（1）在启动电话程序的时候，在“.java”文件中新建了一个 P-Sensor 的 WakeLock 对象。代码如下所示。

```
mProximityWakeLock =  pm.newWakeLock(
PowerManager.PROXIMITY_SCREEN_OFF_WAKE_LOCK, LOG_TAG
);
```

对象 wackLock 的功能是请求控制屏幕的点亮或熄灭。

（2）在电话状态发生改变时，例如接通了电话，调用“.java”文件中的方法根据当前电话的状态来决定是否打开 P-Sensor。如果在通话过程中，电话是 OFF-HOOK 状态时打开 P-Sensor。演示代码如下所示。

```
if (!mProximityWakeLock.isHeld()) {
                    if (DBG) Log.d(LOG_TAG, "updateProximitySensorMode: acquiring...");
                    mProximityWakeLock.acquire();
                }
```

在上述代码中，mProximityWakeLock.acquire()会调用到另外的方法打开 P-Senso，这个另外的方法会判断当前手机有没有 P-Sensor。如果有的话，就会向 SensorManager 注册一个 P-Sensor 监听器。这样当 P-Sensor 检测到手机和人体距离发生改变时，就会调用服务监听器进行处理。同样，当电话挂断时，电话模块会去调用方法取消 P-Sensor 监听器。

在 Android 系统中，PowerManagerService 中 P-Sensor 监听器会进行实时监听工作，当 P-Sensor 检测到距离有变化时就会进行监听。具体监听过程的代码如下所示。

```
SensorEventListener mProximityListener = new SensorEventListener() {
        public void onSensorChanged(SensorEvent event) {
            long milliseconds = SystemClock.elapsedRealtime();
            synchronized (mLocks) {
                float distance = event.values[0];  //检测到手机和人体的距离
                long timeSinceLastEvent = milliseconds - mLastProximityEventTime;
                //这次检测和上次检测的时间差
                mLastProximityEventTime = milliseconds;  //更新上一次检测的时间
                mHandler.removeCallbacks(mProximityTask);
                boolean proximityTaskQueued = false;

                // compare against getMaximumRange to support sensors that only return 0 or 1
                boolean active = (distance >= 0.0 && distance < PROXIMITY_THRESHOLD &&
                        distance < mProximitySensor.getMaximumRange());  //如果距离小于某一
                        //个距离阈值，默认是 5.0f，说明手机和脸部距离贴近，应该要熄灭屏幕

                if (mDebugProximitySensor) {
                    Slog.d(TAG, "mProximityListener.onSensorChanged active: " + active);
                }
                if (timeSinceLastEvent < PROXIMITY_SENSOR_DELAY) {
                    // enforce delaying atleast PROXIMITY_SENSOR_DELAY before processing
                    mProximityPendingValue = (active ? 1 : 0);
                    mHandler.postDelayed(mProximityTask, PROXIMITY_SENSOR_DELAY -
                    timeSinceLastEvent);
                    proximityTaskQueued = true;
                } else {
                    // process the value immediately
                    mProximityPendingValue = -1;
                    proximityChangedLocked(active);   //熄灭屏幕操作
```

```
            }

            // update mProximityPartialLock state
            boolean held = mProximityPartialLock.isHeld();
            if (!held && proximityTaskQueued) {
               // hold wakelock until mProximityTask runs
               mProximityPartialLock.acquire();
            } else if (held && !proximityTaskQueued) {
               mProximityPartialLock.release();
            }
         }
      }

      public void onAccuracyChanged(Sensor sensor, int accuracy) {
         // ignore
      }
   };
```

由上述代码可知，在监听时会首先通过“float distance = event.values[0];”获取变化的距离。例如，如果发现检测这次距离变化和上次距离变化时间差小于系统设置的阈值，则不会去熄灭屏幕。过于频繁的操作系统会忽略掉。如果感觉P-Sensor不够灵敏，就可以修改如下所示的系统默认值。

```
private static final int PROXIMITY_SENSOR_DELAY = 1000;
```

将上述值改小后就会发现P-Sensor会变得灵敏很多。

如果P-Sensor检测到这次距离变化小于系统默认值，并且这次是一次正常的变化，那么需要通过如下代码熄灭屏幕。

```
proximityChangedLocked(active);
```

此处会判断P-Sensor是否可以用，如果不可用则返回，并忽略这次距离变化。

```
if (!mProximitySensorEnabled) {
          Slog.d(TAG, "Ignoring proximity change after sensor is disabled");
          return;
      }
```

如果一切都满足，则调用如下代码灭灯。

```
goToSleepLocked(SystemClock.uptimeMillis(),
             WindowManagerPolicy.OFF_BECAUSE_OF_PROX_SENSOR);
```

15.2 实战演练——使用距离传感器实现自动锁屏功能

在当前的Android智能设备中，特别是触屏智能手机设备，在默认设置下都会有一个延时锁屏的功能，通过此功能设置如果在一段时间内设置手机检测不到任何操作，手机会自动进入锁屏状态。通过延时锁屏功能，不但可以避免不必要的能量消耗，而且又能保证不丢失重要的信息。

实例	功能	源码路径
实例15-1	获取设备中光线传感器的值	\daima\15\AutoLock

本实例的功能是，使用距离传感器实现自动锁屏功能，具体实现流程如下所示。

（1）编写布局文件activity_main.xml，功能是在屏幕中分别设置“启动服务”“停止服务”和“退出”3个按钮，具体实现代码如下所示。

```
<RelativeLayout xmlns:android="http://schemas.android.com/apk/res/android"
   xmlns:tools="http://schemas.android.com/tools"
   android:layout_width="match_parent"
   android:layout_height="match_parent"
   android:paddingBottom="@dimen/activity_vertical_margin"
   android:paddingLeft="@dimen/activity_horizontal_margin"
   android:paddingRight="@dimen/activity_horizontal_margin"
```

```
    android:paddingTop="@dimen/activity_vertical_margin"
    tools:context=".MainActivity" >

    <TextView
        android:id="@+id/title_tv"
        android:layout_centerHorizontal="true"
        android:layout_width="wrap_content"
        android:layout_height="wrap_content"
        android:textSize="20sp"
        android:text="@string/title" />

     <Button
        android:id="@+id/start"
        android:layout_below="@id/title_tv"
        android:layout_centerHorizontal="true"
        android:layout_width="fill_parent"
        android:layout_height="wrap_content"
        android:textSize="20sp"
        android:text="@string/start" />

     <Button
        android:id="@+id/stop"
        android:layout_below="@id/start"
        android:layout_centerHorizontal="true"
        android:layout_width="fill_parent"
        android:layout_height="wrap_content"
        android:textSize="20sp"
        android:text="@string/stop" />

     <Button
        android:id="@+id/exit"
        android:layout_below="@id/stop"
        android:layout_centerHorizontal="true"
        android:layout_width="fill_parent"
        android:layout_height="wrap_content"
        android:textSize="20sp"
        android:text="@string/exit" />

     <TextView
        android:id="@+id/sensortitle_tv"
        android:layout_below="@id/exit"
        android:layout_centerHorizontal="true"
        android:layout_width="wrap_content"
        android:layout_height="wrap_content"
        android:textSize="20sp"
        android:text="@string/sensorinfo" />

      <TextView
        android:id="@+id/sensorinfo_tv"
        android:layout_below="@id/sensortitle_tv"
        android:layout_centerHorizontal="true"
        android:layout_width="fill_parent"
        android:layout_height="wrap_content"
        android:textSize="20sp"
        android:text="@string/sensorinfo" />
</RelativeLayout>
```

（2）编写文件 MainActivity.java，在启动时显示传感器名和版本号，并根据用户的按钮操作执行对应的事件处理程序。文件 MainActivity.java 的具体实现代码如下所示。

```
public class MainActivity extends Activity {

    private Button start;
    private Button stop;
    private Button exit;
    private TextView sensorinfo_tv;
    private Intent intent;
    private SensorManager sm = null;
    private Sensor promixty = null;
```

```
@Override
protected void onCreate(Bundle savedInstanceState) {
    super.onCreate(savedInstanceState);
    setContentView(R.layout.activity_main);
    if (null == sm) {
        sm = (SensorManager) getSystemService(SENSOR_SERVICE); // 获取传感器管理类
        promixty = sm.getDefaultSensor(Sensor.TYPE_PROXIMITY); // 获取距离传感器
    }
    String sensorInfo;
    if (null != promixty) {
        sensorInfo = "传感器名称：" + promixty.getName() + "\n"
                + "  设备版本：" + promixty.getVersion() + "\n" + "  供应商："
                + promixty.getVendor() + "\n";
    }
    else {
        sensorInfo = "无法获取距离传感器信息，可能是您的手机不支持该传感器。";
    }
    initUI();
    intent = new Intent("org.hq.autoLockService");
    start.setOnClickListener( new OnClickListener() {
        @Override
        public void onClick(View v) {
            //开启服务
            start();
        }
    });
    stop.setOnClickListener(new OnClickListener() {
        @Override
        public void onClick(View v) {
            // 停止服务
            stop();
        }
    });
    exit.setOnClickListener( new OnClickListener() {

        @Override
        public void onClick(View v) {
            // 结束本次Activity
            finish();
        }
    });
    sensorinfo_tv.setText(sensorInfo);
}

private void initUI(){
    start = (Button) super.findViewById(R.id.start);
    stop = (Button) super.findViewById(R.id.stop);
    exit = (Button) super.findViewById(R.id.exit);
    sensorinfo_tv = (TextView) super.findViewById(R.id.sensorinfo_tv);
}

private void start(){
    Bundle bundle = new Bundle();
    bundle.putInt("distance", 3);
    bundle.putBoolean("activited", true);
    intent.putExtras(bundle);
    startService(intent);                              // startService
    //结束本次Activity
    //finish();
}
//终止服务
private void stop(){
    stopService(intent);
    Toast.makeText(this, "已停止后台服务。", Toast.LENGTH_SHORT).show();
}

@Override
```

```
    public boolean onCreateOptionsMenu(Menu menu) {
        // Inflate the menu; this adds items to the action bar if it is present.
        getMenuInflater().inflate(R.menu.main, menu);
        return true;
    }

}
```

（3）编写文件 AutoLockService.java 实现自动锁屏服务，通过举例传感器来监听距离，自动进入锁屏状态。文件 AutoLockService.java 的具体实现代码如下所示。

```
public class AutoLockService extends Service implements SensorEventListener {

    private SensorManager sm = null;
    private Sensor promixty = null;
    // 默认启用锁屏
    private static boolean ACTIVITED = true;
    // 锁屏距离（单位：厘米）
    private static int LOCK_DIST = 3;

    @Override
    public IBinder onBind(Intent intent) {
        // TODO Auto-generated method stub
        return null;
    }

    @Override
    public void onCreate() {
        super.onCreate();
        if (null == sm) {
            sm = (SensorManager) getSystemService(SENSOR_SERVICE); // 获取传感器管理类
            promixty = sm.getDefaultSensor(Sensor.TYPE_PROXIMITY); // 获取距离传感器
        }
        // 显示距离传感器信息
        if (null != promixty) {
            Toast.makeText(this, "已创建后台服务。", Toast.LENGTH_SHORT).show();
        } else {
            Toast.makeText(this, "无法找到距离传感器", Toast.LENGTH_SHORT).show();
        }
    }

    @Override
    public void onDestroy() {
        super.onDestroy();
        if( null != sm ){
            //撤销监听器
            sm.unregisterListener(this);
        }
    }

    @Override
    public void onStart(Intent intent, int startId) {
        if (intent != null) {
            Bundle bundle = intent.getExtras();
            Toast.makeText(this, "后台服务已启动。", Toast.LENGTH_SHORT).show();
            // 从 Intent 中获取设置参数
            if (bundle != null) {
                int dist = bundle.getInt("distance");
                ACTIVITED = bundle.getBoolean("activited");
                if (dist > 0 && dist < 9) {
                    LOCK_DIST = dist;
                }
            }
            // 注册监听器
            sm.registerListener(this, promixty,SensorManager.SENSOR_DELAY_NORMAL);
        }
    }

    //监听精度变化
```

```
    @Override
    public void onAccuracyChanged(Sensor sensor, int accuracy) {
        Toast.makeText(this, "距离传感器promixty精度变为" + accuracy,
                Toast.LENGTH_SHORT).show();
    }

    @Override
    public void onSensorChanged(SensorEvent event) {
        if (event.values[0] < LOCK_DIST) // 距离小于5，锁屏
        {
            if (ACTIVITED) {
                lockScreen();
            }
        }
    }

    // 跳至锁屏页面
    private void lockScreen() {
        Intent intent = new Intent();
        //在Activity之外启动，要加上FLAG_ACTIVITY_NEW_TASK flag
        intent.setFlags( Intent.FLAG_ACTIVITY_NEW_TASK );
        intent.setClass(this, LockScreen.Controller.class);
        startActivity(intent);
    }
}
```

（4）编写文件LockScreen.java，功能是实现锁屏功能，在锁屏之前需要先获取锁屏权限。文件LockScreen.java的具体实现代码如下所示。

```
public class LockScreen extends DeviceAdminReceiver {
    static final int RESULT_ENABLE = 1;

    public static class Controller extends Activity {

        DevicePolicyManager mDPM;
        ComponentName mDeviceAdminSample;

        @Override
        protected void onCreate(Bundle savedInstanceState) {
            super.onCreate(savedInstanceState);

            // 首先我们要获得android设备管理代理
            mDPM = (DevicePolicyManager) getSystemService(Context.DEVICE_POLICY_SERVICE);

            // LockScreen 继承自 DeviceAdminReceiver
            mDeviceAdminSample = new ComponentName(Controller.this,
                    LockScreen.class);
            // 得到当前设备管理器有没有激活
            boolean active = mDPM.isAdminActive(mDeviceAdminSample);
            if (!active) {
                // 如果没有激活，则提示用户激活（第一次运行程序时）
                getAdmin();
            } else {
                // 如果已经激活，则执行立即锁屏
                mDPM.lockNow();
            }
            // killMyself ，锁屏之后立即kill掉我们的Activity，避免资源的浪费
            //android.os.Process.killProcess(android.os.Process.myPid());
            finish();
        }

        //获取锁屏权限
        public void getAdmin() {
            // Launch the activity to have the user enable our admin.
            Intent intent = new Intent(
                    DevicePolicyManager.ACTION_ADD_DEVICE_ADMIN);
            intent.putExtra(DevicePolicyManager.EXTRA_DEVICE_ADMIN,
                    mDeviceAdminSample);
            intent.putExtra(DevicePolicyManager.EXTRA_ADD_EXPLANATION,
```

```
                "欢迎您的使用！在第一次使用时，请授予该程序锁屏权限。");
            startActivityForResult(intent, RESULT_ENABLE);
        }
    }
}
```

（5）在文件 AndroidManifest.xml 中声明权限，特别是需要注册一个广播接收者，具体实现代码如下所示。

```
<!-- 注册锁屏 Activity -->
<activity android:name="org.lock.LockScreen$Controller" >
</activity>

<service
    android:name="org.lock.AutoLockService"
    android:permission="android.permission.BIND_ACCESSIBILITY_SERVICE"
    android:enabled="true" >
    <intent-filter>
        <action android:name="org.lock.autoLockService" />
    </intent-filter>
</service>
<receiver
    android:name="org.lock.LockScreen"
    android:permission="android.permission.BIND_DEVICE_ADMIN" >
    <meta-data
        android:name="android.app.device_admin"
        android:resource="@xml/device_admin_sample" />

    <intent-filter>
        <action  android:name="android.app.action.DEVICE_ADMIN_
ENABLED" />
    </intent-filter>
</receiver>
```

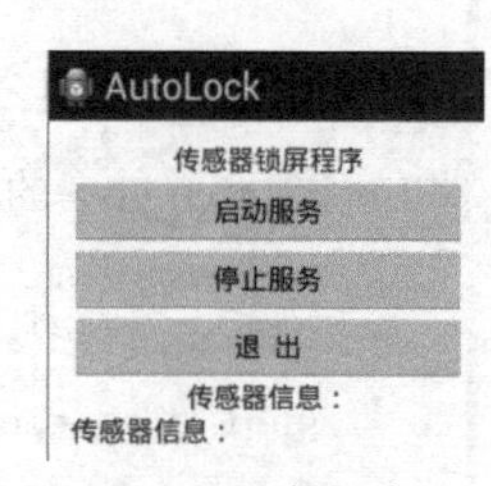

▲图 15-1　执行效果

到此为止，整个实例介绍完毕，执行后的效果如图 15-1 所示。

15.3 实战演练——根据设备的距离实现自动锁屏功能

在下面的实例中，使用类 PowerManager 来控制电源状态的，通过使用该类提供的 API 可以控制电池的待机时间。在确实需要使用类 PowerManager 时，需要尽可能地使用最低级别的 WakeLocks 锁。并且确保使用完后释放它。我们可以通过使用 context.getSystemService(Context.POWER_SERVICE)的方式来获得 PowerManager 的实例。

实例	功能	源码路径
实例 15-2	根据设备的距离实现自动锁屏功能	\daima\15\lLock

本实例的功能是根据设备的距离实现自动锁屏功能，核心文件是 MyService.java，设置当贴近手机时申请设备电源锁定，当远离手机时释放设备电源锁。文件 MyService.java 的具体实现代码如下所示。

```
public class MyService extends Service {

    private SensorManager mManager;

    private Sensor mSensor = null;

    private SensorEventListener mListener = null;

    private PowerManager localPowerManager = null;
    private PowerManager.WakeLock localWakeLock = null;

    @Override
```

```
public void onCreate() {

    // 获取系统服务 POWER_SERVICE，返回一个 PowerManager 对象
    localPowerManager = (PowerManager) getSystemService(Context.POWER_SERVICE);
    // 获取 PowerManager.WakeLock 对象,后面的参数|表示同时传入两个值,最后的是 LogCat 里用的 Tag
    localWakeLock = this.localPowerManager.newWakeLock(32, "MyPower");

    // 获取系统服务 SENSOR_SERVICE，返回一个 SensorManager 对象
    mManager = (SensorManager) getSystemService(Context.SENSOR_SERVICE);
    // 获取距离感应器对象
    mSensor = mManager.getDefaultSensor(Sensor.TYPE_PROXIMITY);
    // 注册感应器事件
    mListener = new SensorEventListener() {

        @Override
        public void onSensorChanged(SensorEvent event) {

            float[] its = event.values;

            if (its != null

            && event.sensor.getType() == Sensor.TYPE_PROXIMITY) {

                System.out.println("its[0]:" + its[0]);

                // 经过测试，当手贴近距离感应器的时候 its[0]返回值为 0.0，当手离开时返回 1.0
                if (its[0] == 0.0) {// 贴近手机

                    System.out.println("手放上去了...");

                    if (localWakeLock.isHeld()) {
                        return;
                    } else
                        localWakeLock.acquire();// 申请设备电源锁

                } else {// 远离手机

                    System.out.println("手拿开了...");

                    if (localWakeLock.isHeld()) {
                        return;
                    } else
                        localWakeLock.setReferenceCounted(false);
                    localWakeLock.release(); // 释放设备电源锁
                }
            }
        }

        @Override
        public void onAccuracyChanged(Sensor sensor, int accuracy) {

        }
    };
}

@Override
public int onStartCommand(Intent intent, int flags, int startId) {

    // 注册监听
    mManager.registerListener(mListener, mSensor,
            SensorManager.SENSOR_DELAY_GAME);

    return super.onStartCommand(intent, flags, startId);
}

@Override
public void onDestroy() {

    // 取消监听
```

```
        mManager.unregisterListener(mListener);

        super.onDestroy();
    }

    @Override
    public IBinder onBind(Intent intent) {
        return null;
    }

}
```

通过上述代码，当系统锁屏或黑屏时会广播如下两个消息：

- ACTION_SCREEN_OFF；
- ACTION_SCREEN_ON。

此时可以在文件 MyReceiver.java 中自定义一个 BroadcastReceiver 来接收这个广播，并且做相应的处理。具体处理方法是在黑屏的时候停止 Service，然后在屏幕亮时再启动 Service。文件 MyReceiver.java 的具体实现代码如下所示。

```
public class MyReceiver extends BroadcastReceiver {

    @Override
    public void onReceive(Context context, Intent intent) {

        String action = intent.getAction();

        if (action.equals("android.intent.action.SCREEN_OFF")) {
        } else if (action.equals("android.intent.action.SCREEN_ON")) {
        }
    }
}
```

最后需要在文件 AndroidManifest.xml 中声明如下所示的权限。

```
<uses-permissionandroid:name="android.permission.DEVICE_POWER"/>
<uses-permission android:name="android.permission.WAKE_LOCK"/>
```

到此为止，整个实例介绍完毕，在真机中执行后会实现预期的效果。

15.4 实战演练——绘制运动曲线

本实例联合使用了线性加速度传感器、磁场传感器、重力传感器和加速度传感器，功能是在当前设备中绘制运动曲线。

实例	功能	源码路径
实例 15-3	在设备中绘制运动曲线	https://github.com/gast-lib/gast-lib

本实例源码是开源代码，来源于如下地址，读者可以自行登录并下载。

```
https://github.com/gast-lib/gast-lib/
```

15.4.1 实现布局文件

编写布局文件 determine_movement.xml，在屏幕上方提供了单选按钮供用户选择传感器类型，提供复选框用户设置是否开启语音功能和高通滤波器。在屏幕下方使用 Androidplot 插件实现了一个绘制曲线界面，通过这个界面可以绘制运动曲线。Androidplot 插件是一个开源工具，本项目使用的压缩包是 Androidplot-core-0.5.0-release.jar，我们只需将这个压缩包包含进我们的 Eclipse 工程即可。文件 determine_movement.xml 的具体实现代码如下所示。

```
<RelativeLayout xmlns:android="http://schemas.android.com/apk/res/android"
    android:layout_width="match_parent"
    android:layout_height="match_parent"
    android:orientation="vertical" >

    <TextView android:id="@+id/sensorSelectorLabel"
        style="@style/apptext"
        android:text="@string/selectSensorSectionHeading"
        android:layout_alignParentTop="true"/>

    <RadioGroup android:id="@+id/sensorSelector"
        android:layout_width="match_parent"
        android:layout_height="wrap_content"
        android:layout_below="@id/sensorSelectorLabel">

        <RadioButton android:id="@+id/accelerometer"
            android:layout_width="match_parent"
            android:layout_height="wrap_content"
            android:text="@string/accelerometerLabel"
            android:checked="true"
            android:onClick="onSensorSelectorClick" />

        <RadioButton android:id="@+id/linearAcceleration"
            android:layout_width="match_parent"
            android:layout_height="wrap_content"
            android:text="@string/linearAccelerationSensorLabel"
            android:checked="false"
            android:onClick="onSensorSelectorClick" />
    </RadioGroup>

    <TextView android:id="@+id/optionsLabel"
        style="@style/apptext"
        android:text="@string/optionsSectionHeading"
        android:layout_below="@id/sensorSelector" />

    <CheckBox android:id="@+id/ttsNotificationsCheckBox"
        android:layout_width="wrap_content"
        android:layout_height="wrap_content"
        android:text="@string/speakMovementDetectedLabel"
        android:checked="@bool/useTtsDefaultValue"
        android:layout_below="@id/optionsLabel"
        android:onClick="onTtsNotificationsCheckBoxClicked" />

    <CheckBox android:id="@+id/highPassFilterCheckBox"
        android:layout_width="wrap_content"
        android:layout_height="wrap_content"
        android:text="@string/enableHighPassFilterLabel"
        android:checked="@bool/useHighPassFilterDefaultValue"
        android:layout_below="@id/ttsNotificationsCheckBox"
        android:onClick="onHighPassFilterCheckBoxClicked" />

    <ToggleButton android:id="@+id/readAccelerationDataToggleButton"
        android:layout_width="match_parent"
        android:layout_height="wrap_content"
        android:checked="false"
        android:layout_alignParentBottom="true"
        android:textOn="@string/readingAccelerationData"
        android:textOff="@string/startReadingAccelerationData"
        android:onClick="onReadAccelerationDataToggleButtonClicked" />

    <com.androidplot.xy.XYPlot android:id="@+id/XYPlot"
        android:layout_width="match_parent"
        android:layout_height="200dp"
        title="Acceleration Plot"
        android:layout_below="@id/highPassFilterCheckBox"
        android:layout_above="@id/readAccelerationDataToggleButton"/>
</RelativeLayout>
```

15.4.2　实现 Activity 程序文件

编写主 Activity 文件 DetermineMovementActivity.java，功能是获取用户在单选按钮中选择的传感器类型，获取用户在复选框中选中的选项，监听用户单击“Start Reading Acceleration Data”按钮，单击此按钮后将进行绘制曲线操作。文件 DetermineMovementActivity.java 的具体实现代码如下所示。

```
public class DetermineMovementActivity extends SpeechRecognizingAndSpeakingActivity
{
    private static final String TAG = "DetermineMovementActivity";
    private static final int RATE = SensorManager.SENSOR_DELAY_NORMAL;
    private static final String USE_HIGH_PASS_FILTER_PREFERENCE_KEY =
            "USE_HIGH_PASS_FILTER_PREFERENCE_KEY";
    private static final String USE_TTS_NOTIFICATION_PREFERENCE_KEY =
            "USE_TTS_NOTIFICATION_PREFERENCE_KEY";
    private static final String SELECTED_SENSOR_TYPE_PREFERENCE_KEY =
            "SELECTED_SENSOR_TYPE_PREFERENCE_KEY";
    private static final int TTS_STREAM = AudioManager.STREAM_NOTIFICATION;

    private SensorManager sensorManager;
    private TextToSpeech tts;
    private RadioGroup sensorSelector;
    private int selectedSensorType;
    private boolean readingAccelerationData;
    private SharedPreferences preferences;
    private AccelerationEventListener accelerometerListener;
    private AccelerationEventListener linearAccelerationListener;
    private boolean useTtsNotification;
    private boolean useHighPassFilter;
    private XYPlot xyPlot;
    private CheckBox ttsNotificationsCheckBox;
    private CheckBox highPassFilterCheckBox;
    private HashMap<String, String> ttsParams;

    @Override
    protected void onCreate(Bundle savedInstanceState)
    {
        super.onCreate(savedInstanceState);
        super.setContentView(R.layout.determine_movement);

        getWindow().addFlags(WindowManager.LayoutParams.FLAG_KEEP_SCREEN_ON);

        ttsParams = new HashMap<String, String>();
        ttsParams.put(Engine.KEY_PARAM_STREAM, String.valueOf(TTS_STREAM));

        this.setVolumeControlStream(TTS_STREAM);

        sensorSelector = (RadioGroup)findViewById(R.id.sensorSelector);
        ttsNotificationsCheckBox = (CheckBox)findViewById(R.id.ttsNotificationsCheckBox);
        highPassFilterCheckBox = (CheckBox)findViewById(R.id.highPassFilterCheckBox);

        sensorManager = (SensorManager) getSystemService(SENSOR_SERVICE);

        readingAccelerationData = false;

        preferences = getPreferences(MODE_PRIVATE);

        useHighPassFilter =
                getResources().getBoolean(R.bool.useHighPassFilterDefaultValue);
        useHighPassFilter =
                preferences.getBoolean(USE_HIGH_PASS_FILTER_PREFERENCE_KEY,
                                       useHighPassFilter);
        ((CheckBox)findViewById(R.id.highPassFilterCheckBox)).setChecked
(useHighPassFilter);

        useTtsNotification =
                getResources().getBoolean(R.bool.useTtsDefaultValue);
```

```
        useTtsNotification =
                preferences.getBoolean(USE_TTS_NOTIFICATION_PREFERENCE_KEY,
                            useTtsNotification);
        ((CheckBox)findViewById(R.id.ttsNotificationsCheckBox)).setChecked
(useTtsNotification);

        selectedSensorType =
                preferences.getInt(SELECTED_SENSOR_TYPE_PREFERENCE_KEY,
                        Sensor.TYPE_ACCELEROMETER);

        if (selectedSensorType == Sensor.TYPE_ACCELEROMETER)
        {
            ((RadioButton)findViewById(R.id.accelerometer)).setChecked(true);
        }
        else
        {
            ((RadioButton)findViewById(R.id.linearAcceleration)).setChecked(true);
        }

        xyPlot = (XYPlot)findViewById(R.id.XYPlot);
        xyPlot.setDomainLabel("Elapsed Time (ms)");
        xyPlot.setRangeLabel("Acceleration (m/sec^2)");
        xyPlot.setBorderPaint(null);
        xyPlot.disableAllMarkup();
        xyPlot.setRangeBoundaries(-10, 10, BoundaryMode.FIXED);
    }

    @Override
    protected void onPause()
    {
        super.onPause();
        stopReadingAccelerationData();

        if (tts != null)
        {
            tts.shutdown();
        }
    }

    @Override
    public void onSuccessfulInit(TextToSpeech tts)
    {
        super.onSuccessfulInit(tts);
        this.tts = tts;
    }

    public void onSensorSelectorClick(View view)
    {
        int selectedSensorId = sensorSelector.getCheckedRadioButtonId();
        if (selectedSensorId == R.id.accelerometer)
        {
            selectedSensorType = Sensor.TYPE_ACCELEROMETER;
        }
        else if (selectedSensorId == R.id.linearAcceleration)
        {
            selectedSensorType = Sensor.TYPE_LINEAR_ACCELERATION;
        }

        preferences
            .edit()
            .putInt(SELECTED_SENSOR_TYPE_PREFERENCE_KEY, selectedSensorType)
            .commit();
    }
//下面是开关按钮处理程序
    public void onReadAccelerationDataToggleButtonClicked(View view)
    {
        ToggleButton toggleButton = (ToggleButton)view;

        if (toggleButton.isChecked())
```

```
        {
           startReadingAccelerationData();//开始绘制曲线操作
        }
        else
        {
           stopReadingAccelerationData();//停止绘制曲线操作
        }
     }
  //下面代码实现具体的绘制功能
     private void startReadingAccelerationData()
     {
        if (!readingAccelerationData)
        {
           // Clear any plot that may already exist on the chart
           xyPlot.clear();
           xyPlot.redraw();

           // Disable UI components so they cannot be changed while plotting
           // sensor data
           for (int i = 0; i < sensorSelector.getChildCount(); i++)
           {
              sensorSelector.getChildAt(i).setEnabled(false);
           }
           ttsNotificationsCheckBox.setEnabled(false);
           highPassFilterCheckBox.setEnabled(false);

           // Data files are stored on the external cache directory so they can
           // be pulled off of the device by the user
           File accelerometerDataFile =
                 new File(getExternalCacheDir(), "accelerometer.csv");
           File linearAccelerationDataFile =
                 new File(getExternalCacheDir(), "linearAcceleration.csv");

           if (selectedSensorType == Sensor.TYPE_ACCELEROMETER)
           {
              xyPlot.setTitle("Sensor.TYPE_ACCELEROMETER");
              accelerometerListener =
                    new AccelerationEventListener(xyPlot,
                          useHighPassFilter,
                          accelerometerDataFile,
                          (useTtsNotification ? tts : null),
                          ttsParams,
                          getString(R.string.movementDetectedText));

              linearAccelerationListener =
                    new AccelerationEventListener(null,
                          useHighPassFilter,
                          linearAccelerationDataFile,
                          (useTtsNotification ? tts : null),
                          ttsParams,
                          getString(R.string.movementDetectedText));
           }
           else
           {
              xyPlot.setTitle("Sensor.TYPE_LINEAR_ACCELERATION");
              accelerometerListener =
                    new AccelerationEventListener(null,
                          useHighPassFilter,
                          accelerometerDataFile,
                          (useTtsNotification ? tts : null),
                          ttsParams,
                          getString(R.string.movementDetectedText));

              linearAccelerationListener =
                    new AccelerationEventListener(xyPlot,
                          useHighPassFilter,
                          linearAccelerationDataFile,
                          (useTtsNotification ? tts : null),
                          ttsParams,
```

```
                    getString(R.string.movementDetectedText));
        }

        sensorManager.registerListener(accelerometerListener,
                sensorManager.getDefaultSensor(Sensor.TYPE_ACCELEROMETER),
                RATE);

        sensorManager.registerListener(linearAccelerationListener,
                sensorManager.getDefaultSensor(Sensor.TYPE_LINEAR_ACCELERATION),
                RATE);

        readingAccelerationData = true;

        Log.d(TAG, "Started reading acceleration data");
    }
}
//下面的代码取消绘制功能
private void stopReadingAccelerationData()
{
    if (readingAccelerationData)
    {
        // Re-enable sensor and options UI views
        for (int i = 0; i < sensorSelector.getChildCount(); i++)
        {
            sensorSelector.getChildAt(i).setEnabled(true);
        }
        ttsNotificationsCheckBox.setEnabled(true);
        highPassFilterCheckBox.setEnabled(true);

        sensorManager.unregisterListener(accelerometerListener);
        sensorManager.unregisterListener(linearAccelerationListener);

        // Tell listeners to clean up after themselves
        accelerometerListener.stop();
        linearAccelerationListener.stop();

        readingAccelerationData = false;

        Log.d(TAG, "Stopped reading acceleration data");
    }
}

public void onTtsNotificationsCheckBoxClicked(View view)
{
    useTtsNotification = ((CheckBox)view).isChecked();
    preferences
        .edit()
        .putBoolean(USE_TTS_NOTIFICATION_PREFERENCE_KEY, useTtsNotification)
        .commit();
}
public void onHighPassFilterCheckBoxClicked(View view)
{
   useHighPassFilter = ((CheckBox)view).isChecked();
    preferences
        .edit()
        .putBoolean(USE_HIGH_PASS_FILTER_PREFERENCE_KEY, useHighPassFilter)
        .commit();
}
@Override
protected void receiveWhatWasHeard(List<String> heard, float[] confidenceScores)
{
    // no-op
}
}
```

上述代码的核心是绘制方法 startReadingAccelerationData，此方法创建了两个 Acceleration EventListener 对象，并且都通过 registerListener 进行了注册工作。因为加速度和线性加速度都拥有一个侦听器，这两个传感器会同时获得加速度数据，并将数据写入到两个不同的 CSV 文件中。

尽管如此，最终只有一个传感器的数据被绘制到图表中。

15.4.3　实现监听事件处理

编写文件 AccelerationEventListener.java 实现监听处理，获取并处理加速度的数据，根据过滤数据计算总的加速度值，并将得到的数据写入到 CSV 文件，最后将数据绘制在图表中以实现检测运动的目的。文件 AccelerationEventListener.java 的具体实现代码如下所示。

```
public class AccelerationEventListener implements SensorEventListener
{
    private static final String TAG = "AccelerationEventListener";
    private static final char CSV_DELIM = ',';
    private static final int THRESHHOLD = 2;
    private static final String CSV_HEADER =
          "X Axis,Y Axis,Z Axis,Acceleration,Time";
    private static final float ALPHA = 0.8f;
    private static final int HIGH_PASS_MINIMUM = 10;
    private static final int MAX_SERIES_SIZE = 30;
    private static final int CHART_REFRESH = 125;

    private PrintWriter printWriter;
    private long startTime;
    private float[] gravity;
    private int highPassCount;
    private SimpleXYSeries xAxisSeries;
    private SimpleXYSeries yAxisSeries;
    private SimpleXYSeries zAxisSeries;
    private SimpleXYSeries accelerationSeries;
    private XYPlot xyPlot;
    private long lastChartRefresh;
    private boolean useHighPassFilter;
    private TextToSpeech tts;
    private HashMap<String, String> ttsParams;
    private String movementText;

    public AccelerationEventListener(XYPlot xyPlot,
                                     boolean useHighPassFilter,
                                     File dataFile,
                                     TextToSpeech tts,
                                     HashMap<String, String> ttsParams,
                                     String movementText)
    {
        this.xyPlot = xyPlot;
        this.useHighPassFilter = useHighPassFilter;
        this.tts = tts;
        this.ttsParams = ttsParams;
        this.movementText = movementText;

        xAxisSeries = new SimpleXYSeries("X Axis");
        yAxisSeries = new SimpleXYSeries("Y Axis");
        zAxisSeries = new SimpleXYSeries("Z Axis");
        accelerationSeries = new SimpleXYSeries("Acceleration");

        gravity = new float[3];
        startTime = SystemClock.uptimeMillis();
        highPassCount = 0;

        try
        {
            printWriter =
                    new PrintWriter(new BufferedWriter(new FileWriter(dataFile)));

            printWriter.println(CSV_HEADER);
        }
        catch (IOException e)
        {
            Log.e(TAG, "Could not open CSV file(s)", e);
```

```
    }

    if (xyPlot != null)
    {
        xyPlot.addSeries(xAxisSeries,
                LineAndPointRenderer.class,
                new LineAndPointFormatter(Color.RED, null, null));
        xyPlot.addSeries(yAxisSeries,
                LineAndPointRenderer.class,
                new LineAndPointFormatter(Color.GREEN, null, null));
        xyPlot.addSeries(zAxisSeries,
                LineAndPointRenderer.class,
                new LineAndPointFormatter(Color.BLUE, null, null));
        xyPlot.addSeries(accelerationSeries,
                LineAndPointRenderer.class,
                new LineAndPointFormatter(Color.CYAN, null, null));
    }
}

private void writeSensorEvent(PrintWriter printWriter,
                              float x,
                              float y,
                              float z,
                              double acceleration,
                              long eventTime)
{
    if (printWriter != null)
    {
        StringBuffer sb = new StringBuffer()
            .append(x).append(CSV_DELIM)
            .append(y).append(CSV_DELIM)
            .append(z).append(CSV_DELIM)
            .append(acceleration).append(CSV_DELIM)
            .append((eventTime / 1000000) - startTime);

        printWriter.println(sb.toString());
        if (printWriter.checkError())
        {
            Log.w(TAG, "Error writing sensor event data");
        }
    }
}

@Override
public void onSensorChanged(SensorEvent event)
{
    float[] values = event.values.clone();

    // Pass values through high-pass filter if enabled
    if (useHighPassFilter)
    {
        values = highPass(values[0],
                values[1],
                values[2]);
    }

    // Ignore data if the high-pass filter is enabled, has not yet received
    // some data to set it
    if (!useHighPassFilter || (++highPassCount >= HIGH_PASS_MINIMUM))
    {
        double sumOfSquares = (values[0] * values[0])
                + (values[1] * values[1])
                + (values[2] * values[2]);
        double acceleration = Math.sqrt(sumOfSquares);

        // Write to data file
        writeSensorEvent(printWriter,
                values[0],
                values[1],
```

```
                        values[2],
                        acceleration,
                        event.timestamp);

        // If the plot is null, the sensor is not active. Do not plot the
        // data or used the data to determine if the device is moving
        if (xyPlot != null)
        {
            long current = SystemClock.uptimeMillis();

            // Limit how much the chart gets updated
            if ((current - lastChartRefresh) >= CHART_REFRESH)
            {
                long timestamp = (event.timestamp / 1000000) - startTime;

                // Plot data
                addDataPoint(xAxisSeries, timestamp, values[0]);
                addDataPoint(yAxisSeries, timestamp, values[1]);
                addDataPoint(zAxisSeries, timestamp, values[2]);
                addDataPoint(accelerationSeries, timestamp, acceleration);

                xyPlot.redraw();

                lastChartRefresh = current;
            }

            // A "movement" is only triggered of the total acceleration is
            // above a threshold
            if (acceleration > THRESHHOLD)
            {
                Log.i(TAG, "Movement detected");

                if (tts != null)
                {
                    tts.speak(movementText,
                            TextToSpeech.QUEUE_FLUSH,
                            ttsParams);
                }
            }
        }
    }
}

private void addDataPoint(SimpleXYSeries series,
                          Number timestamp,
                          Number value)
{
    if (series.size() == MAX_SERIES_SIZE)
    {
        series.removeFirst();
    }

    series.addLast(timestamp, value);
}

/**
 * This method derived from the Android documentation and is available under
 * the Apache 2.0 license.
 *
 * @see http://developer.android.com/reference/android/hardware/SensorEvent.html
 */
private float[] highPass(float x, float y, float z)
{
    float[] filteredValues = new float[3];

    gravity[0] = ALPHA * gravity[0] + (1 - ALPHA) * x;
    gravity[1] = ALPHA * gravity[1] + (1 - ALPHA) * y;
```

```
        gravity[2] = ALPHA * gravity[2] + (1 - ALPHA) * z;

        filteredValues[0] = x - gravity[0];
        filteredValues[1] = y - gravity[1];
        filteredValues[2] = z - gravity[2];

        return filteredValues;
    }

    public void stop()
    {
        if (printWriter != null)
        {
            printWriter.close();
        }

        if (printWriter.checkError())
        {
            Log.e(TAG, "Error closing writer");
        }
    }
    @Override
    public void onAccuracyChanged(Sensor sensor, int accuracy)
    {
        // no-op
    }
}
```

在上述代码中，方法 highPass(float x, float y, float z)实现了高通滤波功能，上述解决方案是谷歌提供的，读者可以直接使用或者对其进行升级后再使用。到此为止，整个实例介绍完毕，执行后的效果如图 15-2 所示。

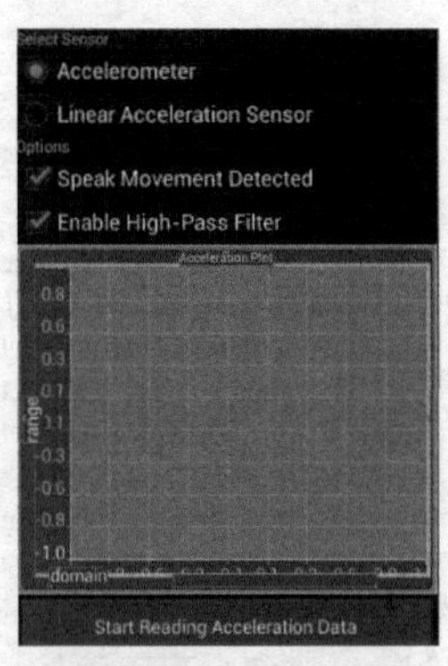

▲图 15-2　执行效果

15.5 实战演练——开发一个健身计步器

随着人们生活水平的日益提高，高血压、高血脂和高血糖的发病者愈发频繁，影响了人们的生活质量。随着 2008 奥运会在北京的成功举办，全民健身潮已经日益深入民心。在运动的过程中拥有一款科学的设备——计步器，是一件既新潮又健身科学的事情。本节将详细讲解开发一个 Android 设备计步器的基本知识，为读者步入现实工作岗位打下基础。

实例	功能	源码路径
实例 15-4	在设备中开发一个计步器	\daima\15\step

15.5.1　系统功能模块介绍

本章计步器系统的功能是，统计穿戴者徒步行走的步数和距离，计算行走速度和消耗的热量。并且还具有启动和关闭功能，还可以对整个系统进行灵活设置。本章计步器系统的构成模块结构如图 15-3 所示。

15.5.2　系统主界面

系统主界面是运行程序后首先呈现在用户面前的界面。在本节的内容中，将详细讲解系统主界面的具体实现流程。

1. 布局文件

本系统主界面的布局文件是 main.xml，具体实现代码如下所示。

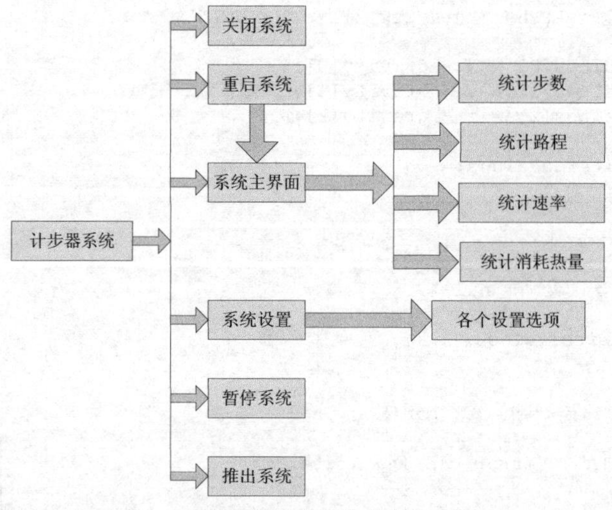

▲图 15-3　系统构成模块

```
<?xml version="1.0" encoding="utf-8"?>
<LinearLayout xmlns:android="http://schemas.android.com/apk/res/android"
    android:orientation="vertical"
    android:layout_width="fill_parent"
    android:layout_height="fill_parent"
    android:padding="@dimen/margin"
    android:background="@color/screen_background">

    <LinearLayout android:id="@+id/row_1"
        android:orientation="horizontal"
        android:layout_width="fill_parent"
        android:layout_height="wrap_content"
        android:paddingBottom="@dimen/row_spacing">

        <LinearLayout android:id="@+id/box_steps"
            android:orientation="vertical"
            android:layout_width="fill_parent"
            android:layout_height="wrap_content"
            android:gravity="center_horizontal"
            android:paddingRight="@dimen/margin"
            android:layout_weight="1">

            <TextView android:id="@+id/step_value"
                android:textSize="@dimen/value"
                android:layout_width="fill_parent"
                android:layout_height="wrap_content"
                android:gravity="center_horizontal"
                android:background="@color/display_background"
                android:paddingLeft="@dimen/padding"
                android:paddingRight="@dimen/padding"
                android:paddingTop="@dimen/padding"
                android:text=""/>
            <TextView android:id="@+id/step_units"
                android:gravity="center_horizontal"
                android:layout_width="fill_parent"
                android:layout_height="wrap_content"
                android:textSize="@dimen/units"
                android:text="@string/steps"
                android:background="@color/display_background"
                android:paddingBottom="@dimen/padding"/>

        </LinearLayout>

        <LinearLayout android:id="@+id/box_distance"
            android:orientation="vertical"
            android:layout_width="fill_parent"
```

```
        android:layout_height="wrap_content"
        android:gravity="center_horizontal"
        android:layout_weight="1">

        <TextView android:id="@+id/distance_value"
            android:textSize="@dimen/value"
            android:layout_width="fill_parent"
            android:layout_height="wrap_content"
            android:gravity="center_horizontal"
            android:background="@color/display_background"
            android:paddingTop="@dimen/padding"
            android:paddingRight="@dimen/padding"
            android:paddingLeft="@dimen/padding"
            android:text=""/>
        <TextView android:id="@+id/distance_units"
            android:gravity="center_horizontal"
            android:layout_width="fill_parent"
            android:layout_height="wrap_content"
            android:textSize="@dimen/units"
            android:text="@string/kilometers"
            android:background="@color/display_background"
            android:paddingBottom="@dimen/padding"/>

    </LinearLayout>
</LinearLayout>

<LinearLayout android:id="@+id/row_2"
    android:orientation="horizontal"
    android:layout_width="fill_parent"
    android:layout_height="wrap_content"
    android:paddingBottom="@dimen/row_spacing">

    <LinearLayout android:id="@+id/box_pace"
        android:orientation="vertical"
        android:layout_height="wrap_content"
        android:paddingRight="@dimen/margin"
        android:layout_width="fill_parent"
        android:layout_weight="1">

        <TextView android:id="@+id/pace_value"
            android:gravity="center_horizontal"
            android:layout_width="fill_parent"
            android:layout_height="wrap_content"
            android:background="@color/display_background"
            android:textSize="@dimen/small_value"
            android:paddingLeft="@dimen/padding"
            android:paddingRight="@dimen/padding"
            android:paddingTop="@dimen/padding"
            android:text=""/>
        <TextView android:id="@+id/pace_units"
            android:gravity="center_horizontal"
            android:layout_width="fill_parent"
            android:layout_height="wrap_content"
            android:textSize="@dimen/units"
            android:text="@string/steps_per_minute"
            android:paddingBottom="@dimen/padding"
            android:background="@color/display_background"/>

    </LinearLayout>

    <LinearLayout android:id="@+id/box_speed"
        android:orientation="vertical"
        android:paddingRight="@dimen/margin"
        android:layout_height="wrap_content"
        android:layout_width="fill_parent"
        android:layout_weight="1">

        <TextView android:id="@+id/speed_value"
            android:gravity="center_horizontal"
```

```
            android:layout_width="fill_parent"
            android:layout_height="wrap_content"
            android:background="@color/display_background"
            android:textSize="@dimen/small_value"
            android:paddingLeft="@dimen/padding"
            android:paddingRight="@dimen/padding"
            android:paddingTop="@dimen/padding"
            android:text=""/>
        <TextView android:id="@+id/speed_units"
            android:gravity="center_horizontal"
            android:layout_width="fill_parent"
            android:layout_height="wrap_content"
            android:textSize="@dimen/units"
            android:text="@string/kilometers_per_hour"
            android:paddingBottom="@dimen/padding"
            android:background="@color/display_background"/>
    </LinearLayout>

    <LinearLayout android:id="@+id/box_calories"
        android:orientation="vertical"
        android:layout_height="wrap_content"
        android:layout_width="fill_parent"
        android:layout_weight="1">

        <TextView android:id="@+id/calories_value"
            android:gravity="center_horizontal"
            android:layout_width="fill_parent"
            android:layout_height="wrap_content"
            android:background="@color/display_background"
            android:textSize="@dimen/small_value"
            android:paddingLeft="@dimen/padding"
            android:paddingRight="@dimen/padding"
            android:paddingTop="@dimen/padding"
            android:text=""/>
        <TextView android:id="@+id/calories_units"
            android:gravity="center_horizontal"
            android:layout_width="fill_parent"
            android:layout_height="wrap_content"
            android:textSize="@dimen/units"
            android:text="@string/calories_burned"
            android:paddingBottom="@dimen/padding"
            android:background="@color/display_background"/>

    </LinearLayout>

</LinearLayout>

<!-- Desired pace/speed row -->
<LinearLayout
    android:id="@+id/desired_pace_control"
    android:paddingTop="@dimen/row_spacing"
    android:gravity="center_horizontal"
    android:orientation="horizontal"
    android:layout_width="fill_parent"
    android:layout_height="wrap_content"
    android:layout_weight="1">

    <!-- Button "-", for decrementing desired pace/speed -->
    <Button android:id="@+id/button_desired_pace_lower"
        android:text="-"
        android:textSize="@dimen/button_sign"
        android:layout_width="@dimen/button"
        android:layout_height="@dimen/button"/>

    <!-- Container for desired pace/speed -->
    <LinearLayout
        android:gravity="center_horizontal"
        android:orientation="vertical"
        android:layout_width="@dimen/desired_pace_width"
```

```
            android:layout_height="wrap_content">

            <TextView android:id="@+id/desired_pace_label"
                android:gravity="center_horizontal"
                android:layout_width="fill_parent"
                android:layout_height="wrap_content"
                android:text="@string/desired_pace"/>

            <!-- Current desired pace/speed -->
            <TextView android:id="@+id/desired_pace_value"
                android:gravity="center_horizontal"
                android:textSize="@dimen/desired_pace"
                android:layout_width="@dimen/desired_pace_width"
                android:layout_height="wrap_content"/>

        </LinearLayout>

        <!-- Button "+", for incrementing desired pace/speed -->
        <Button android:id="@+id/button_desired_pace_raise"
            android:text="+"
            android:textSize="@dimen/button_sign"
            android:layout_width="@dimen/button"
            android:layout_height="@dimen/button"/>

    </LinearLayout>

</LinearLayout>
```

通过上述代码，在屏幕中分别显示了穿戴者徒步行走的步数、行走距离、行走速度和消耗的热量等信息。

2. 系统主 Activity

本系统的主 Activity 是 Pedometer，实现文件是 Pedometer.java，具体实现流程如下所示。

（1）定义类 Pedometer，功能是定义系统中需要的变量和初始值，具体实现代码如下所示。

```
public class Pedometer extends Activity {
    private static final String TAG = "Pedometer";
    private SharedPreferences mSettings;
    private PedometerSettings mPedometerSettings;
    private Utils mUtils;

    private TextView mStepValueView;
    private TextView mPaceValueView;
    private TextView mDistanceValueView;
    private TextView mSpeedValueView;
    private TextView mCaloriesValueView;
    TextView mDesiredPaceView;
    private int mStepValue;//mStepValueView 的值
    private int mPaceValue;//mPaceValueView 的值
    private float mDistanceValue;//mDistanceValueView 的值
    private float mSpeedValue;//mSpeedValueView 的值
    private int mCaloriesValue;//mCaloriesValueView 的值
    private float mDesiredPaceOrSpeed;//
    private int mMaintain;//是否是爬山
    private boolean mIsMetric;//公制和米制切换标志
    private float mMaintainInc;//
    private boolean mQuitting = false; //
    private boolean mIsRunning;//程序是否运行的标志位
```

（2）编写函数 onStart()用于启动计步器，具体代码如下所示。

```
//开始函数，重写该函数，加入日志。
@Override
protected void onStart() {
    Log.i(TAG, "[ACTIVITY] onStart");
    super.onStart();
}
```

（3）编写函数 onResume()用于恢复计步器系统，具体代码如下所示。

```
//重写回复函数
@Override
protected void onResume() {
    Log.i(TAG, "[ACTIVITY] onResume");
    super.onResume();

    mSettings = PreferenceManager.getDefaultSharedPreferences(this);
    mPedometerSettings = new PedometerSettings(mSettings);

    mUtils.setSpeak(mSettings.getBoolean("speak", false));

    // Read from preferences if the service was running on the last onPause
    mIsRunning = mPedometerSettings.isServiceRunning();

    // Start the service if this is considered to be an application start (last onPause
    // was long ago)
    if (!mIsRunning && mPedometerSettings.isNewStart()) {
        startStepService();
        bindStepService();
    }
    else if (mIsRunning) {
        bindStepService();
    }

    mPedometerSettings.clearServiceRunning();

    mStepValueView     = (TextView) findViewById(R.id.step_value);
    mPaceValueView     = (TextView) findViewById(R.id.pace_value);
    mDistanceValueView = (TextView) findViewById(R.id.distance_value);
    mSpeedValueView    = (TextView) findViewById(R.id.speed_value);
    mCaloriesValueView = (TextView) findViewById(R.id.calories_value);
    mDesiredPaceView   = (TextView) findViewById(R.id.desired_pace_value);

    mIsMetric = mPedometerSettings.isMetric();
    ((TextView) findViewById(R.id.distance_units)).setText(getString(
            mIsMetric
            ? R.string.kilometers
            : R.string.miles
    ));
    ((TextView) findViewById(R.id.speed_units)).setText(getString(
            mIsMetric
            ? R.string.kilometers_per_hour
            : R.string.miles_per_hour
    ));

    mMaintain = mPedometerSettings.getMaintainOption();
    ((LinearLayout) this.findViewById(R.id.desired_pace_control)).setVisibility(
            mMaintain != PedometerSettings.M_NONE
            ? View.VISIBLE
            : View.GONE
        );
    if (mMaintain == PedometerSettings.M_PACE) {
        mMaintainInc = 5f;
        mDesiredPaceOrSpeed = (float)mPedometerSettings.getDesiredPace();
    }
    else
    if (mMaintain == PedometerSettings.M_SPEED) {
        mDesiredPaceOrSpeed = mPedometerSettings.getDesiredSpeed();
        mMaintainInc = 0.1f;
    }
    Button button1 = (Button) findViewById(R.id.button_desired_pace_lower);
    button1.setOnClickListener(new View.OnClickListener() {
        public void onClick(View v) {
            mDesiredPaceOrSpeed -= mMaintainInc;
            mDesiredPaceOrSpeed = Math.round(mDesiredPaceOrSpeed * 10) / 10f;
            displayDesiredPaceOrSpeed();
            setDesiredPaceOrSpeed(mDesiredPaceOrSpeed);
```

```
        }
    });
    Button button2 = (Button) findViewById(R.id.button_desired_pace_raise);
    button2.setOnClickListener(new View.OnClickListener() {
        public void onClick(View v) {
            mDesiredPaceOrSpeed += mMaintainInc;
            mDesiredPaceOrSpeed = Math.round(mDesiredPaceOrSpeed * 10) / 10f;
            displayDesiredPaceOrSpeed();
            setDesiredPaceOrSpeed(mDesiredPaceOrSpeed);
        }
    });
    if (mMaintain != PedometerSettings.M_NONE) {
        ((TextView) findViewById(R.id.desired_pace_label)).setText(
                mMaintain == PedometerSettings.M_PACE
                ? R.string.desired_pace
                : R.string.desired_speed
        );
    }


    displayDesiredPaceOrSpeed();
}
```

（4）编写函数 displayDesiredPaceOrSpeed()，用于显示想要的步伐节奏或速率，具体实现代码如下所示。

```
private void displayDesiredPaceOrSpeed() {
    if (mMaintain == PedometerSettings.M_PACE) {
        mDesiredPaceView.setText("" + (int)mDesiredPaceOrSpeed);
    }
    else {
        mDesiredPaceView.setText("" + mDesiredPaceOrSpeed);
    }
}
```

（5）定义函数 onPause()用于暂停计步器，具体实现代码如下所示。

```
@Override
protected void onPause() {
    Log.i(TAG, "[ACTIVITY] onPause");
    if (mIsRunning) {
        unbindStepService();
    }
    if (mQuitting) {
        mPedometerSettings.saveServiceRunningWithNullTimestamp(mIsRunning);
    }
    else {
        mPedometerSettings.saveServiceRunningWithTimestamp(mIsRunning);
    }

    super.onPause();
    savePaceSetting();
}
```

（6）定义函数 onStop()用于停止计步器系统，具体实现代码如下所示。

```
@Override
protected void onStop() {
    Log.i(TAG, "[ACTIVITY] onStop");
    super.onStop();
}
```

（7）定义函数 onDestroy()用于退出整个计步器系统，具体实现代码如下所示。

```
protected void onDestroy() {
    Log.i(TAG, "[ACTIVITY] onDestroy");
    super.onDestroy();
}
```

（8）编写函数 onRestart()用于重启计步器，具体代码如下所示。

```
protected void onRestart() {
    Log.i(TAG, "[ACTIVITY] onRestart");
    super.onDestroy();
}
```

（9）编写函数 setDesiredPaceOrSpeed，用于设置想要的步伐节奏或速率，具体实现代码如下所示。

```
private void setDesiredPaceOrSpeed(float desiredPaceOrSpeed) {
    if (mService != null) {
        if (mMaintain == PedometerSettings.M_PACE) {
            mService.setDesiredPace((int)desiredPaceOrSpeed);
        }
        else
        if (mMaintain == PedometerSettings.M_SPEED) {
            mService.setDesiredSpeed(desiredPaceOrSpeed);
        }
    }
}

private void savePaceSetting() {
    mPedometerSettings.savePaceOrSpeedSetting(mMaintain, mDesiredPaceOrSpeed);
}
```

（10）编写函数 startStepService()，用于启动计步服务，具体实现代码如下所示。

```
private void startStepService() {
    if (! mIsRunning) {
        Log.i(TAG, "[SERVICE] Start");
        mIsRunning = true;
        startService(new Intent(Pedometer.this,
                StepService.class));
    }
}
```

（11）编写函数 bindStepService()，用于绑定计步服务，具体实现代码如下所示。

```
private void bindStepService() {
    Log.i(TAG, "[SERVICE] Bind");
    bindService(new Intent(Pedometer.this,
            StepService.class), mConnection, Context.BIND_AUTO_CREATE + Context.BIND_
DEBUG_UNBIND);
}
```

（12）编写函数 unbindStepService()，用于解除对当前计步服务的绑定，具体实现代码如下所示。

```
private void unbindStepService() {
    Log.i(TAG, "[SERVICE] Unbind");
    unbindService(mConnection);
}
```

（13）编写函数 stopStepService()，用于停止当前的计步服务，具体实现代码如下所示。

```
private void stopStepService() {
    Log.i(TAG, "[SERVICE] Stop");
    if (mService != null) {
        Log.i(TAG, "[SERVICE] stopService");
        stopService(new Intent(Pedometer.this,
              StepService.class));
    }
    mIsRunning = false;
}
```

（14）编写函数 resetValues ()，用于重置当前计步器中的各个数值，具体实现代码如下所示。

```
private void resetValues(boolean updateDisplay) {
    if (mService != null && mIsRunning) {
        mService.resetValues();
    }
    else {
        mStepValueView.setText("0");
```

```
        mPaceValueView.setText("0");
        mDistanceValueView.setText("0");
        mSpeedValueView.setText("0");
        mCaloriesValueView.setText("0");
        SharedPreferences state = getSharedPreferences("state", 0);
        SharedPreferences.Editor stateEditor = state.edit();
        if (updateDisplay) {
            stateEditor.putInt("steps", 0);
            stateEditor.putInt("pace", 0);
            stateEditor.putFloat("distance", 0);
            stateEditor.putFloat("speed", 0);
            stateEditor.putFloat("calories", 0);
            stateEditor.commit();
        }
    }
}
```

（15）编写函数 onPrepareOptionsMenu(Menu menu)，功能是创建系统菜单中的各个选项，具体实现代码如下所示。

```
public boolean onPrepareOptionsMenu(Menu menu) {
    menu.clear();
    if (mIsRunning) {
        menu.add(0, MENU_PAUSE, 0, R.string.pause)
        .setIcon(android.R.drawable.ic_media_pause)
        .setShortcut('1', 'p');
    }
    else {
        menu.add(0, MENU_RESUME, 0, R.string.resume)
        .setIcon(android.R.drawable.ic_media_play)
        .setShortcut('1', 'p');
    }
    menu.add(0, MENU_RESET, 0, R.string.reset)
    .setIcon(android.R.drawable.ic_menu_close_clear_cancel)
    .setShortcut('2', 'r');
    menu.add(0, MENU_SETTINGS, 0, R.string.settings)
    .setIcon(android.R.drawable.ic_menu_preferences)
    .setShortcut('8', 's')
    .setIntent(new Intent(this, Settings.class));
    menu.add(0, MENU_QUIT, 0, R.string.quit)
    .setIcon(android.R.drawable.ic_lock_power_off)
    .setShortcut('9', 'q');
    return true;
}
```

（16）编写函数 onOptionsItemSelected(MenuItem item)，用于根据用户选择的选项进行处理，具体实现代码如下所示。

```
public boolean onOptionsItemSelected(MenuItem item) {
    switch (item.getItemId()) {
        case MENU_PAUSE:
            unbindStepService();
            stopStepService();
            return true;
        case MENU_RESUME:
            startStepService();
            bindStepService();
            return true;
        case MENU_RESET:
            resetValues(true);
            return true;
        case MENU_QUIT:
            resetValues(false);
            unbindStepService();
            stopStepService();
            mQuitting = true;
            finish();
            return true;
    }
```

```
    return false;
}
```

（17）通过如下代码在屏幕文本框中显示统计数据的变化状况，具体实现代码如下所示。

```
// TODO: unite all into 1 type of message
private StepService.ICallback mCallback = new StepService.ICallback() {
    public void stepsChanged(int value) {
        mHandler.sendMessage(mHandler.obtainMessage(STEPS_MSG, value, 0));
    }
    public void paceChanged(int value) {
        mHandler.sendMessage(mHandler.obtainMessage(PACE_MSG, value, 0));
    }
    public void distanceChanged(float value) {
        mHandler.sendMessage(mHandler.obtainMessage(DISTANCE_MSG, (int)(value*1000), 0));
    }
    public void speedChanged(float value) {
        mHandler.sendMessage(mHandler.obtainMessage(SPEED_MSG, (int)(value*1000), 0));
    }
    public void caloriesChanged(float value) {
        mHandler.sendMessage(mHandler.obtainMessage(CALORIES_MSG, (int)(value), 0));
    }
};

private static final int STEPS_MSG = 1;
private static final int PACE_MSG = 2;
private static final int DISTANCE_MSG = 3;
private static final int SPEED_MSG = 4;
private static final int CALORIES_MSG = 5;
```

（18）编写函数 handleMessage(Message msg)，在文本框中显示具体统计的数值，具体实现代码如下所示。

```
    private Handler mHandler = new Handler() {
        @Override public void handleMessage(Message msg) {
            switch (msg.what) {
                case STEPS_MSG:
                    mStepValue = (int)msg.arg1;
                    mStepValueView.setText("" + mStepValue);
                    break;
                case PACE_MSG:
                    mPaceValue = msg.arg1;
                    if (mPaceValue <= 0) {
                        mPaceValueView.setText("0");
                    }
                    else {
                        mPaceValueView.setText("" + (int)mPaceValue);
                    }
                    break;
                case DISTANCE_MSG:
                    mDistanceValue = ((int)msg.arg1)/1000f;
                    if (mDistanceValue <= 0) {
                        mDistanceValueView.setText("0");
                    }
                    else {
                        mDistanceValueView.setText(
                                ("" + (mDistanceValue + 0.000001f)).substring(0, 5)
                        );
                    }
                    break;
                case SPEED_MSG:
                    mSpeedValue = ((int)msg.arg1)/1000f;
                    if (mSpeedValue <= 0) {
                        mSpeedValueView.setText("0");
                    }
                    else {
                        mSpeedValueView.setText(
                                ("" + (mSpeedValue + 0.000001f)).substring(0, 4)
                        );
```

```
                }
                break;
            case CALORIES_MSG:
                mCaloriesValue = msg.arg1;
                if (mCaloriesValue <= 0) {
                    mCaloriesValueView.setText("0");
                }
                else {
                    mCaloriesValueView.setText("" + (int)mCaloriesValue);
                }
                break;
            default:
                super.handleMessage(msg);
        }
    }
};
}
```

到此为止，整个系统主界面介绍完毕，执行后的效果如图15-4所示。

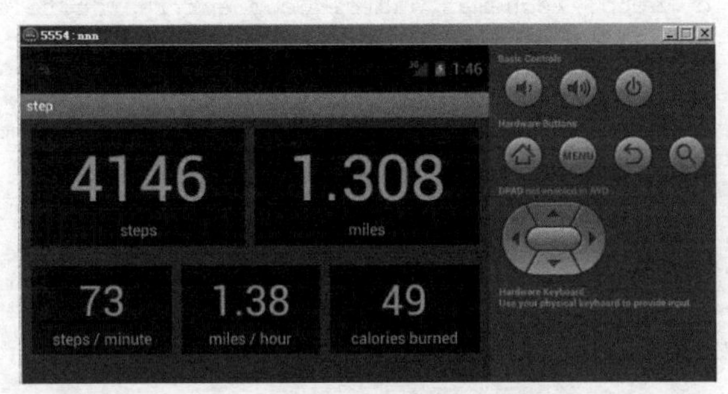

▲图15-4 系统主界面的执行效果

15.5.3 系统设置模块

在系统主界面中，当按下MENU按钮后会在屏幕下方弹出图15-5所示的界面。

单击图15-5中的选项后会进入系统设置界面，如图15-6所示。

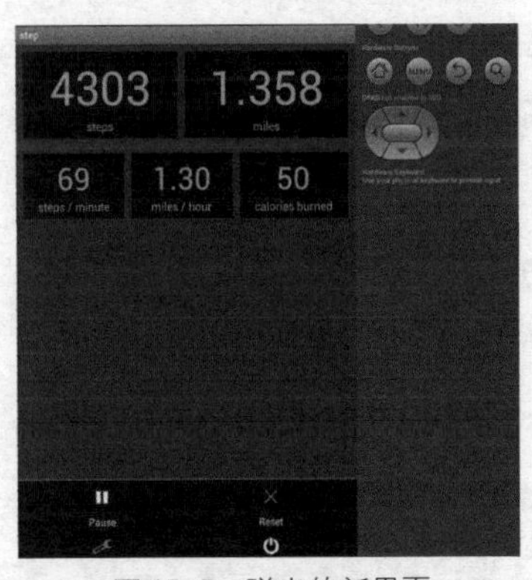

▲图15-5 弹出的新界面

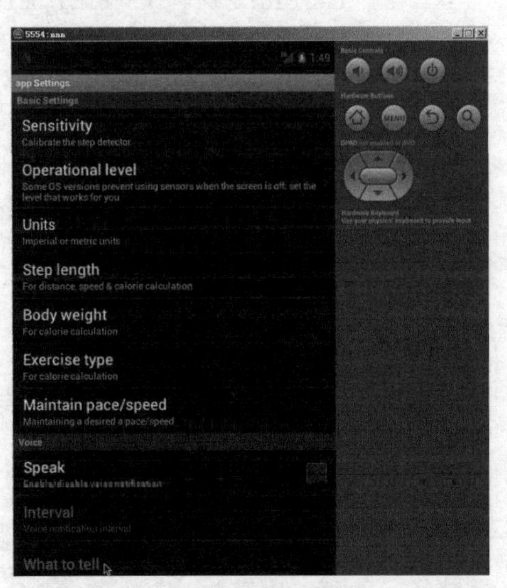

▲图15-6 系统设置界面

在图15-6所示的界面中，可以对整个系统的参数进行设置。在本节的内容中，将详细讲解系统设置模块的具体实现过程。

1. 系统设置 Activity

本系统设置 Activity 的实现文件是 Settings.java，功能是载入系统设置界面的布局文件 preferences.xml，具体实现代码如下所示。

```
public class Settings extends PreferenceActivity {
   /** Called when the activity is first created. */
   @Override
   public void onCreate(Bundle savedInstanceState) {
       super.onCreate(savedInstanceState);

       addPreferencesFromResource(R.xml.preferences);
   }
}
```

系统设置界面的布局文件是 preferences.xml，具体实现代码如下所示。

```
<?xml version="1.0" encoding="utf-8"?>
<PreferenceScreen
     xmlns:android="http://schemas.android.com/apk/res/android">
   <PreferenceCategory
         android:title="@string/steps_settings_title">
       <ListPreference
             android:key="sensitivity"
             android:title="@string/sensitivity_setting"
             android:summary="@string/sensitivity_setting_details"
             android:entries="@array/sensitivity_preference"
             android:entryValues="@array/sensitivity_preference_values"
             android:dialogTitle="@string/sensitivity_setting_title"
             android:defaultValue="30" />
       <ListPreference
             android:key="operation_level"
             android:title="@string/operation_level_setting"
             android:summary="@string/operation_level_setting_details"
             android:entries="@array/operation_level_preference"
             android:entryValues="@array/operation_level_preference_values"
             android:dialogTitle="@string/operation_level_setting_title"
             android:defaultValue="30" />
       <ListPreference
             android:key="units"
             android:title="@string/units_setting"
             android:summary="@string/units_setting_details"
             android:entries="@array/units_preference"
             android:entryValues="@array/units_preference_values"
             android:dialogTitle="@string/units_setting_title"
             android:defaultValue="imperial" />
       <name.step.preferences.StepLengthPreference
             android:key="step_length"
             android:title="@string/step_length_setting"
             android:summary="@string/step_length_setting_details"
             android:dialogTitle="@string/step_length_setting_title"
             android:defaultValue="20" />
       <name.step.preferences.BodyWeightPreference
             android:key="body_weight"
             android:title="@string/body_weight_setting"
             android:summary="@string/body_weight_setting_details"
             android:dialogTitle="@string/body_weight_setting_title"
             android:defaultValue="50" />
       <ListPreference
             android:key="exercise_type"
             android:title="@string/exercise_type_setting"
             android:summary="@string/exercise_type_setting_details"
             android:entries="@array/exercise_type_preference"
             android:entryValues="@array/exercise_type_preference_values"
             android:dialogTitle="@string/exercise_type_setting_title"
             android:defaultValue="running" />
       <ListPreference
             android:key="maintain"
             android:title="@string/maintain_setting"
```

```
            android:summary="@string/maintain_setting_details"
            android:entries="@array/maintain_preference"
            android:entryValues="@array/maintain_preference_values"
            android:dialogTitle="@string/maintain_setting_title"
            android:defaultValue="none" />
    </PreferenceCategory>
    <PreferenceCategory
        android:title="@string/voice_settings_title">
        <CheckBoxPreference
            android:key="speak"
            android:title="@string/voice_setting"
            android:summary="@string/voice_setting_details"
            android:defaultValue="false" />
        <ListPreference
            android:key="speaking_interval"
            android:title="@string/speaking_interval_setting"
            android:summary="@string/speaking_interval_setting_details"
            android:entries="@array/speaking_interval_preference"
            android:entryValues="@array/speaking_interval_preference_values"
            android:dependency="speak"
            android:defaultValue="1" />
        <PreferenceScreen
            android:key="tell_what"
            android:title="@string/tell_what"
            android:dependency="speak">
            <PreferenceCategory
                android:title="@string/tell_what">
                <CheckBoxPreference
                    android:key="tell_steps"
                    android:title="@string/tell_steps_setting"
                    android:summary="@string/tell_steps_setting_details"
                    android:defaultValue="false" />
                <CheckBoxPreference
                    android:key="tell_pace"
                    android:title="@string/tell_pace_setting"
                    android:summary="@string/tell_pace_setting_details"
                    android:defaultValue="false" />
                <CheckBoxPreference
                    android:key="tell_distance"
                    android:title="@string/tell_distance_setting"
                    android:summary="@string/tell_distance_setting_details"
                    android:defaultValue="false" />
                <CheckBoxPreference
                    android:key="tell_speed"
                    android:title="@string/tell_speed_setting"
                    android:summary="@string/tell_speed_setting_details"
                    android:defaultValue="false" />
                <CheckBoxPreference
                    android:key="tell_calories"
                    android:title="@string/tell_calories_setting"
                    android:summary="@string/tell_calories_setting_details"
                    android:defaultValue="false" />
                <CheckBoxPreference
                    android:key="tell_fasterslower"
                    android:title="@string/tell_fasterslower_setting"
                    android:summary="@string/tell_fasterslower_setting_details"
                    android:defaultValue="false" />
            </PreferenceCategory>
        </PreferenceScreen>
    </PreferenceCategory>
</PreferenceScreen>
```

2. 获取各个设置值

编写文件 PedometerSettings.java，功能是定义各种系统参数的设置函数，获取图 15-4 所示的各个设置值。文件 PedometerSettings.java 的具体实现代码如下所示。

```
public class PedometerSettings {
    SharedPreferences mSettings;
```

```
public static int M_NONE = 1;
public static int M_PACE = 2;
public static int M_SPEED = 3;

public PedometerSettings(SharedPreferences settings) {
    mSettings = settings;
}
//公制和米制切换标志：imperial 公制，metric 米制
public boolean isMetric() {
    return mSettings.getString("units", "imperial").equals("metric");
}

//取步长
public float getStepLength() {
    try {
        return Float.valueOf(mSettings.getString("step_length", "20").trim());
    }
    catch (NumberFormatException e) {
        // TODO: reset value, & notify user somehow
        return 0f;
    }
}
//取体重
public float getBodyWeight() {
    try {
        return Float.valueOf(mSettings.getString("body_weight", "50").trim());
    }
    catch (NumberFormatException e) {
        // TODO: reset value, & notify user somehow
        return 0f;
    }
}
//判断是走路还是跑步
public boolean isRunning() {
    return mSettings.getString("exercise_type", "running").equals("running");
}
//登山还是平时走路
public int getMaintainOption() {
    String p = mSettings.getString("maintain", "none");
    return
        p.equals("none") ? M_NONE : (
            p.equals("pace") ? M_PACE : (
                    p.equals("speed") ? M_SPEED : (0)
            )
        );
}
//-------------------------------------------------------------------
// Desired pace & speed:
// these can not be set in the preference activity, only on the main
// screen if "maintain" is set to "pace" or "speed"

public int getDesiredPace() {
    return mSettings.getInt("desired_pace", 180); // steps/minute
}

public float getDesiredSpeed() {
    return mSettings.getFloat("desired_speed", 4f); // km/h or mph
}

public void savePaceOrSpeedSetting(int maintain, float desiredPaceOrSpeed) {
    SharedPreferences.Editor editor = mSettings.edit();
    if (maintain == M_PACE) {
        editor.putInt("desired_pace", (int)desiredPaceOrSpeed);
    }
    else
    if (maintain == M_SPEED) {
        editor.putFloat("desired_speed", desiredPaceOrSpeed);
    }
```

```
        editor.commit();
    }

    //-------------------------------------------------------------------
    // Speaking:

    public boolean shouldSpeak() {
        return mSettings.getBoolean("speak", false);
    }

    public float getSpeakingInterval() {
        try {
            return Float.valueOf(mSettings.getString("speaking_interval", "1"));
        }
        catch (NumberFormatException e) {
            // This could not happen as the value is selected from a list.
            return 1;
        }
    }

    public boolean shouldTellSteps() {
        return mSettings.getBoolean("speak", false)
        && mSettings.getBoolean("tell_steps", false);
    }

    public boolean shouldTellPace() {
        return mSettings.getBoolean("speak", false)
        && mSettings.getBoolean("tell_pace", false);
    }

    public boolean shouldTellDistance() {
        return mSettings.getBoolean("speak", false)
        && mSettings.getBoolean("tell_distance", false);
    }

    public boolean shouldTellSpeed() {
        return mSettings.getBoolean("speak", false)
        && mSettings.getBoolean("tell_speed", false);
    }

    public boolean shouldTellCalories() {
        return mSettings.getBoolean("speak", false)
        && mSettings.getBoolean("tell_calories", false);
    }

    public boolean shouldTellFasterslower() {
        return mSettings.getBoolean("speak", false)
        && mSettings.getBoolean("tell_fasterslower", false);
    }

    public boolean wakeAggressively() {
        return   mSettings.getString("operation_level",   "run_in_background").equals
("wake_up");
    }

    public boolean keepScreenOn() {
        return   mSettings.getString("operation_level",   "run_in_background").equals
("keep_screen_on");
    }

    //
    // Internal

    public void saveServiceRunningWithTimestamp(boolean running) {
        SharedPreferences.Editor editor = mSettings.edit();
        editor.putBoolean("service_running", running);
        editor.putLong("last_seen", Utils.currentTimeInMillis());
        editor.commit();
    }
```

```
    public void saveServiceRunningWithNullTimestamp(boolean running) {
        SharedPreferences.Editor editor = mSettings.edit();
        editor.putBoolean("service_running", running);
        editor.putLong("last_seen", 0);
        editor.commit();
    }
    public void clearServiceRunning() {
        SharedPreferences.Editor editor = mSettings.edit();
        editor.putBoolean("service_running", false);
        editor.putLong("last_seen", 0);
        editor.commit();
    }
    public boolean isServiceRunning() {
        return mSettings.getBoolean("service_running", false);
    }
    public boolean isNewStart() {
        // activity last paused more than 10 minutes ago
        return  mSettings.getLong("last_seen",  0)  <  Utils.currentTimeInMillis()  -
1000*60*10;
    }
}
```

3. 系统服务设置

编写文件 StepService.java，功能是设置计步器系统的服务功能。这里的服务是指与用户进行交互的后台服务。文件 StepService.java 的具体实现流程如下所示。

（1）定义服务类 StepService，定义系统需要的变量并设置初始值，具体实现代码如下所示。

```
public class StepService extends Service {
    private static final String TAG = "name.bagi.levente.pedometer.StepService";
    private SharedPreferences mSettings;
    private PedometerSettings mPedometerSettings;
    private SharedPreferences mState;
    private SharedPreferences.Editor mStateEditor;
    private Utils mUtils;
    private SensorManager mSensorManager;
    private Sensor mSensor;
    private StepDetector mStepDetector;
    // private StepBuzzer mStepBuzzer; // used for debugging
    private StepDisplayer mStepDisplayer;
    private PaceNotifier mPaceNotifier;
    private DistanceNotifier mDistanceNotifier;
    private SpeedNotifier mSpeedNotifier;
    private CaloriesNotifier mCaloriesNotifier;
    private SpeakingTimer mSpeakingTimer;

    private PowerManager.WakeLock wakeLock;
    private NotificationManager mNM;

    private int mSteps;
    private int mPace;
    private float mDistance;
    private float mSpeed;
    private float mCalories;

    public class StepBinder extends Binder {
        StepService getService() {
            return StepService.this;
        }
    }
```

（2）编写函数 onCreate()，载入系统初始设置的系统参数值，并且开启检测功能。具体实现代码如下所示。

```
public void onCreate() {
    Log.i(TAG, "[SERVICE] onCreate");
```

```
    super.onCreate();

    mNM = (NotificationManager)getSystemService(NOTIFICATION_SERVICE);
    showNotification();

    // Load settings
    mSettings = PreferenceManager.getDefaultSharedPreferences(this);
    mPedometerSettings = new PedometerSettings(mSettings);
    mState = getSharedPreferences("state", 0);

    mUtils = Utils.getInstance();
    mUtils.setService(this);
    mUtils.initTTS();

    acquireWakeLock();

    // 开启检测功能
    mStepDetector = new StepDetector();
    mSensorManager = (SensorManager) getSystemService(SENSOR_SERVICE);
    registerDetector();

    // 注册 ACTION_SCREEN_OFF ，将电话进入待机模式
    IntentFilter filter = new IntentFilter(Intent.ACTION_SCREEN_OFF);
    registerReceiver(mReceiver, filter);

    mStepDisplayer = new StepDisplayer(mPedometerSettings, mUtils);
    mStepDisplayer.setSteps(mSteps = mState.getInt("steps", 0));
    mStepDisplayer.addListener(mStepListener);
    mStepDetector.addStepListener(mStepDisplayer);

    mPaceNotifier     = new PaceNotifier(mPedometerSettings, mUtils);
    mPaceNotifier.setPace(mPace = mState.getInt("pace", 0));
    mPaceNotifier.addListener(mPaceListener);
    mStepDetector.addStepListener(mPaceNotifier);

    mDistanceNotifier = new DistanceNotifier(mDistanceListener, mPedometerSettings,
mUtils);
    mDistanceNotifier.setDistance(mDistance = mState.getFloat("distance", 0));
    mStepDetector.addStepListener(mDistanceNotifier);

    mSpeedNotifier = new SpeedNotifier(mSpeedListener,  mPedometerSettings, mUtils);
    mSpeedNotifier.setSpeed(mSpeed = mState.getFloat("speed", 0));
    mPaceNotifier.addListener(mSpeedNotifier);

    mCaloriesNotifier = new CaloriesNotifier(mCaloriesListener, mPedometerSettings,
mUtils);
    mCaloriesNotifier.setCalories(mCalories = mState.getFloat("calories", 0));
    mStepDetector.addStepListener(mCaloriesNotifier);

    mSpeakingTimer = new SpeakingTimer(mPedometerSettings, mUtils);
    mSpeakingTimer.addListener(mStepDisplayer);
    mSpeakingTimer.addListener(mPaceNotifier);
    mSpeakingTimer.addListener(mDistanceNotifier);
    mSpeakingTimer.addListener(mSpeedNotifier);
    mSpeakingTimer.addListener(mCaloriesNotifier);
    mStepDetector.addStepListener(mSpeakingTimer);

    // 启动声音
    reloadSettings();

    // Tell the user we started.
    Toast.makeText(this, getText(R.string.started), Toast.LENGTH_SHORT).show();
}
```

（3）编写函数 onStart(Intent intent, int startId)以启动系统服务，具体实现代码如下所示。

```
public void onStart(Intent intent, int startId) {
    Log.i(TAG, "[SERVICE] onStart");
    super.onStart(intent, startId);
}
```

（4）编写函数 onDestroy()以销毁系统服务，具体实现代码如下所示。

```
public void onDestroy() {
    Log.i(TAG, "[SERVICE] onDestroy");
    mUtils.shutdownTTS();

    // 取消接收机制
    unregisterReceiver(mReceiver);
    unregisterDetector();

    mStateEditor = mState.edit();
    mStateEditor.putInt("steps", mSteps);
    mStateEditor.putInt("pace", mPace);
    mStateEditor.putFloat("distance", mDistance);
    mStateEditor.putFloat("speed", mSpeed);
    mStateEditor.putFloat("calories", mCalories);
    mStateEditor.commit();

    mNM.cancel(R.string.app_name);

    wakeLock.release();

    super.onDestroy();

    // 停止检测
    mSensorManager.unregisterListener(mStepDetector);

    // 通知用户停止
    Toast.makeText(this, getText(R.string.stopped), Toast.LENGTH_SHORT).show();
}
```

（5）编写函数 registerDetector()以注册探测服务，具体实现代码如下所示。

```
private void registerDetector() {
    mSensor = mSensorManager.getDefaultSensor(
        Sensor.TYPE_ACCELEROMETER /*|
        Sensor.TYPE_MAGNETIC_FIELD |
        Sensor.TYPE_ORIENTATION*/);
    mSensorManager.registerListener(mStepDetector,
        mSensor,
        SensorManager.SENSOR_DELAY_FASTEST);
}
```

（6）编写函数 unregisterDetector()以注销探测服务，具体实现代码如下所示。

```
private void unregisterDetector() {
    mSensorManager.unregisterListener(mStepDetector);
}
```

（7）编写函数 onBind(Intent intent)以绑定系统服务，具体实现代码如下所示。

```
public IBinder onBind(Intent intent) {
    Log.i(TAG, "[SERVICE] onBind");
    return mBinder;
}
private final IBinder mBinder = new StepBinder();

public interface ICallback {
    public void stepsChanged(int value);
    public void paceChanged(int value);
    public void distanceChanged(float value);
    public void speedChanged(float value);
    public void caloriesChanged(float value);
}

private ICallback mCallback;

public void registerCallback(ICallback cb) {
    mCallback = cb;
    //mStepDisplayer.passValue();
```

```
    //mPaceListener.passValue();
}

private int mDesiredPace;
private float mDesiredSpeed;
```

（8）编写函数 setDesiredPace(int desiredPace)，功能是重设用户修改后的每步距离值，具体实现代码如下所示。

```
public void setDesiredPace(int desiredPace) {
    mDesiredPace = desiredPace;
    if (mPaceNotifier != null) {
        mPaceNotifier.setDesiredPace(mDesiredPace);
    }
}
```

（9）编写函数 setDesiredSpeed(float desiredSpeed)，功能是重设用户修改后的速率值，具体实现代码如下所示。

```
public void setDesiredSpeed(float desiredSpeed) {
    mDesiredSpeed = desiredSpeed;
    if (mSpeedNotifier != null) {
        mSpeedNotifier.setDesiredSpeed(mDesiredSpeed);
    }
}
```

（10）编写函数 reloadSettings()，功能是重新载入系统设置值，具体实现代码如下所示。

```
public void reloadSettings() {
    mSettings = PreferenceManager.getDefaultSharedPreferences(this);

    if (mStepDetector != null) {
        mStepDetector.setSensitivity(
                Float.valueOf(mSettings.getString("sensitivity", "10"))
        );
    }

    if (mStepDisplayer    != null) mStepDisplayer.reloadSettings();
    if (mPaceNotifier     != null) mPaceNotifier.reloadSettings();
    if (mDistanceNotifier != null) mDistanceNotifier.reloadSettings();
    if (mSpeedNotifier    != null) mSpeedNotifier.reloadSettings();
    if (mCaloriesNotifier != null) mCaloriesNotifier.reloadSettings();
    if (mSpeakingTimer    != null) mSpeakingTimer.reloadSettings();
}
```

（11）编写函数 resetValues()，功能是重新设置值，具体实现代码如下所示。

```
public void resetValues() {
    mStepDisplayer.setSteps(0);
    mPaceNotifier.setPace(0);
    mDistanceNotifier.setDistance(0);
    mSpeedNotifier.setSpeed(0);
    mCaloriesNotifier.setCalories(0);
}
```

（12）通过如下所示的代码从 PaceNotifie 获取前进速率值。

```
private StepDisplayer.Listener mStepListener = new StepDisplayer.Listener() {
    public void stepsChanged(int value) {
        mSteps = value;
        passValue();
    }
    public void passValue() {
        if (mCallback != null) {
            mCallback.stepsChanged(mSteps);
        }
    }
};
```

（13）通过如下所示的代码从 PaceNotifier 获取步长值。

```
private PaceNotifier.Listener mPaceListener = new PaceNotifier.Listener() {
    public void paceChanged(int value) {
        mPace = value;
        passValue();
    }
    public void passValue() {
        if (mCallback != null) {
            mCallback.paceChanged(mPace);
        }
    }
};
```

（14）通过如下所示的代码从 DistanceNotifier 获取距离值。

```
private DistanceNotifier.Listener mDistanceListener = new DistanceNotifier.Listener() {
    public void valueChanged(float value) {
        mDistance = value;
        passValue();
    }
    public void passValue() {
        if (mCallback != null) {
            mCallback.distanceChanged(mDistance);
        }
    }
};
```

（15）从如下所示的代码从 SpeedNotifier 获取速率值。

```
private SpeedNotifier.Listener mSpeedListener = new SpeedNotifier.Listener() {
    public void valueChanged(float value) {
        mSpeed = value;
        passValue();
    }
    public void passValue() {
        if (mCallback != null) {
            mCallback.speedChanged(mSpeed);
        }
    }
};
```

（16）通过如下所示的代码从 CaloriesNotifier 获取热量值。

```
private CaloriesNotifier.Listener mCaloriesListener = new CaloriesNotifier.Listener() {
    public void valueChanged(float value) {
        mCalories = value;
        passValue();
    }
    public void passValue() {
        if (mCallback != null) {
            mCallback.caloriesChanged(mCalories);
        }
    }
};
```

（17）编写函数 showNotification()，功能是显示一个“该服务正在运行”的通知，具体实现代码如下所示。

```
private void showNotification() {
    CharSequence text = getText(R.string.app_name);
    Notification notification = new Notification(R.drawable.ic_notification, null,
            System.currentTimeMillis());
    notification.flags = Notification.FLAG_NO_CLEAR | Notification.FLAG_ONGOING_EVENT;
    Intent pedometerIntent = new Intent();
    pedometerIntent.setComponent(new ComponentName(this, Pedometer.class));
    pedometerIntent.addFlags(Intent.FLAG_ACTIVITY_CLEAR_TOP);
    PendingIntent contentIntent = PendingIntent.getActivity(this, 0,
            pedometerIntent, 0);
    notification.setLatestEventInfo(this, text,
            getText(R.string.notification_subtitle), contentIntent);
```

```
    mNM.notify(R.string.app_name, notification);
}
```

（18）通过如下所示的代码处理广播接收机制 ACTION_SCREEN_OFF。

```
private BroadcastReceiver mReceiver = new BroadcastReceiver() {
    @Override
    public void onReceive(Context context, Intent intent) {
        // Check action just to be on the safe side.
        if (intent.getAction().equals(Intent.ACTION_SCREEN_OFF)) {
            // Unregisters the listener and registers it again.
            StepService.this.unregisterDetector();
            StepService.this.registerDetector();
            if (mPedometerSettings.wakeAggressively()) {
                wakeLock.release();
                acquireWakeLock();
            }
        }
    }
};
```

（19）编写函数 acquireWakeLock()获取唤醒锁，具体实现代码如下所示。

```
    private void acquireWakeLock() {
        PowerManager pm = (PowerManager) getSystemService(Context.POWER_SERVICE);
        int wakeFlags;
        if (mPedometerSettings.wakeAggressively()) {
            wakeFlags=PowerManager.SCREEN_DIM_WAKE_LOCK|PowerManager.ACQUIRE_CAUSES_WAKEUP;
        }
        else if (mPedometerSettings.keepScreenOn()) {
            wakeFlags = PowerManager.SCREEN_DIM_WAKE_LOCK;
        }
        else {
            wakeFlags = PowerManager.PARTIAL_WAKE_LOCK;
        }
        wakeLock = pm.newWakeLock(wakeFlags, TAG);
        wakeLock.acquire();
    }
}
```

4. 获取并显示热量

编写文件 CaloriesNotifier.java，功能是获取并显示行走过程中所消耗的近似热量值，热量的单位是卡路里。文件 CaloriesNotifier.java 的具体实现代码如下所示。

```
public class CaloriesNotifier implements StepListener, SpeakingTimer.Listener {

    public interface Listener {
        public void valueChanged(float value);
        public void passValue();
    }
    private Listener mListener;

    private static double METRIC_RUNNING_FACTOR = 1.02784823;
    private static double IMPERIAL_RUNNING_FACTOR = 0.75031498;

    private static double METRIC_WALKING_FACTOR = 0.708;
    private static double IMPERIAL_WALKING_FACTOR = 0.517;

    private double mCalories = 0;

    PedometerSettings mSettings;
    Utils mUtils;

    boolean mIsMetric;
    boolean mIsRunning;
    float mStepLength;
    float mBodyWeight;

    public CaloriesNotifier(Listener listener, PedometerSettings settings, Utils utils) {
```

```
        mListener = listener;
        mUtils = utils;
        mSettings = settings;
        reloadSettings();
    }
    public void setCalories(float calories) {
        mCalories = calories;
        notifyListener();
    }
    public void reloadSettings() {
        mIsMetric = mSettings.isMetric();
        mIsRunning = mSettings.isRunning();
        mStepLength = mSettings.getStepLength();
        mBodyWeight = mSettings.getBodyWeight();
        notifyListener();
    }
    public void resetValues() {
        mCalories = 0;
    }

    public void isMetric(boolean isMetric) {
        mIsMetric = isMetric;
    }
    public void setStepLength(float stepLength) {
        mStepLength = stepLength;
    }

    public void onStep() {

        if (mIsMetric) {
            mCalories +=
                (mBodyWeight * (mIsRunning ? METRIC_RUNNING_FACTOR : METRIC_WALKING_FACTOR))
                // Distance:
                * mStepLength // centimeters
                / 100000.0; // centimeters/kilometer
        }
        else {
            mCalories +=
                (mBodyWeight*(mIsRunning?IMPERIAL_RUNNING_FACTOR:IMPERIAL_WALKING_FACTOR))
                // Distance:
                * mStepLength // inches
                / 63360.0; // inches/mile
        }

        notifyListener();
    }

    private void notifyListener() {
        mListener.valueChanged((float)mCalories);
    }

    public void passValue() {

    }

    public void speak() {
        if (mSettings.shouldTellCalories()) {
            if (mCalories > 0) {
                mUtils.say("" + (int)mCalories + " calories burned");
            }
        }

    }
}
```

5. 显示行走距离

编写文件 DistanceNotifier.java，功能是获取并显示行走的距离，具体实现代码如下所示。

```
public class DistanceNotifier implements StepListener, SpeakingTimer.Listener {

    public interface Listener {
        public void valueChanged(float value);
        public void passValue();
    }
    private Listener mListener;

    float mDistance = 0;

    PedometerSettings mSettings;
    Utils mUtils;

    boolean mIsMetric;
    float mStepLength;

    public DistanceNotifier(Listener listener, PedometerSettings settings, Utils utils) {
        mListener = listener;
        mUtils = utils;
        mSettings = settings;
        reloadSettings();
    }
    public void setDistance(float distance) {
        mDistance = distance;
        notifyListener();
    }

    public void reloadSettings() {
        mIsMetric = mSettings.isMetric();
        mStepLength = mSettings.getStepLength();
        notifyListener();
    }

    public void onStep() {

        if (mIsMetric) {
            mDistance += (float)(// kilometers
                mStepLength // centimeters
                / 100000.0); // centimeters/kilometer
        }
        else {
            mDistance += (float)(// miles
                mStepLength // inches
                / 63360.0); // inches/mile
        }

        notifyListener();
    }

    private void notifyListener() {
        mListener.valueChanged(mDistance);
    }

    public void passValue() {
        // Callback of StepListener - Not implemented
    }

    public void speak() {
        if (mSettings.shouldTellDistance()) {
            if (mDistance >= .001f) {
                mUtils.say(("" + (mDistance + 0.000001f)).substring(0, 4) + (mIsMetric ?
" kilometers" : " miles"));
                // TODO: format numbers (no "." at the end)
            }
        }
    }
}
```

6. 获取并显示步伐速率

编写文件 PaceNotifier.java，功能是获取并显示步伐速率，提示每分钟多少步。我们可以为系统设置步伐速率，系统会根据用户设置的值提示快还是慢。文件 PaceNotifier.java 的具体实现代码如下所示。

```
public class PaceNotifier implements StepListener, SpeakingTimer.Listener {

    public interface Listener {
        public void paceChanged(int value);
        public void passValue();
    }
    private ArrayList<Listener> mListeners = new ArrayList<Listener>();

    int mCounter = 0;

    private long mLastStepTime = 0;
    private long[] mLastStepDeltas = {-1, -1, -1, -1};
    private int mLastStepDeltasIndex = 0;
    private long mPace = 0;

    PedometerSettings mSettings;
    Utils mUtils;

    /** Desired pace, adjusted by the user */
    int mDesiredPace;

    /** Should we speak? */
    boolean mShouldTellFasterslower;

    /** When did the TTS speak last time */
    private long mSpokenAt = 0;

    public PaceNotifier(PedometerSettings settings, Utils utils) {
        mUtils = utils;
        mSettings = settings;
        mDesiredPace = mSettings.getDesiredPace();
        reloadSettings();
    }
    public void setPace(int pace) {
        mPace = pace;
        int avg = (int)(60*1000.0 / mPace);
        for (int i = 0; i < mLastStepDeltas.length; i++) {
            mLastStepDeltas[i] = avg;
        }
        notifyListener();
    }
    public void reloadSettings() {
        mShouldTellFasterslower =
            mSettings.shouldTellFasterslower()
            && mSettings.getMaintainOption() == PedometerSettings.M_PACE;
        notifyListener();
    }

    public void addListener(Listener l) {
        mListeners.add(l);
    }

    public void setDesiredPace(int desiredPace) {
        mDesiredPace = desiredPace;
    }

    public void onStep() {
        long thisStepTime = System.currentTimeMillis();
        mCounter ++;

        // Calculate pace based on last x steps
        if (mLastStepTime > 0) {
```

```
        long delta = thisStepTime - mLastStepTime;

        mLastStepDeltas[mLastStepDeltasIndex] = delta;
        mLastStepDeltasIndex = (mLastStepDeltasIndex + 1) % mLastStepDeltas.length;

        long sum = 0;
        boolean isMeaningfull = true;
        for (int i = 0; i < mLastStepDeltas.length; i++) {
            if (mLastStepDeltas[i] < 0) {
                isMeaningfull = false;
                break;
            }
            sum += mLastStepDeltas[i];
        }
        if (isMeaningfull && sum > 0) {
            long avg = sum / mLastStepDeltas.length;
            mPace = 60*1000 / avg;

            // TODO: remove duplication. This also exists in SpeedNotifier
            if (mShouldTellFasterslower && !mUtils.isSpeakingEnabled()) {
                if (thisStepTime - mSpokenAt > 3000 && !mUtils.isSpeakingNow()) {
                    float little = 0.10f;
                    float normal = 0.30f;
                    float much = 0.50f;

                    boolean spoken = true;
                    if (mPace < mDesiredPace * (1 - much)) {
                        mUtils.say("much faster!");
                    }
                    else
                    if (mPace > mDesiredPace * (1 + much)) {
                        mUtils.say("much slower!");
                    }
                    else
                    if (mPace < mDesiredPace * (1 - normal)) {
                        mUtils.say("faster!");
                    }
                    else
                    if (mPace > mDesiredPace * (1 + normal)) {
                        mUtils.say("slower!");
                    }
                    else
                    if (mPace < mDesiredPace * (1 - little)) {
                        mUtils.say("a little faster!");
                    }
                    else
                    if (mPace > mDesiredPace * (1 + little)) {
                        mUtils.say("a little slower!");
                    }
                    else {
                        spoken = false;
                    }
                    if (spoken) {
                        mSpokenAt = thisStepTime;
                    }
                }
            }
        }
        else {
            mPace = -1;
        }
    }
    mLastStepTime = thisStepTime;
    notifyListener();
}

private void notifyListener() {
    for (Listener listener : mListeners) {
        listener.paceChanged((int)mPace);
```

```
        }
    }

    public void passValue() {
        // Not used
    }

    //-----------------------------------------------------
    // Speaking

    public void speak() {
        if (mSettings.shouldTellPace()) {
            if (mPace > 0) {
                mUtils.say(mPace + " steps per minute");
            }
        }
    }
}
```

7. 获取并显示行走速率

编写文件 SpeedNotifier.java，功能是获取并显示行走速率，提示每小时走多少公里。我们可以为系统设置行走速率，系统会根据用户设置的值提示快还是慢。文件 SpeedNotifier.java 的具体实现代码如下所示。

```
public class SpeedNotifier implements PaceNotifier.Listener, SpeakingTimer.Listener {
    public interface Listener {
        public void valueChanged(float value);
        public void passValue();
    }
    private Listener mListener;

    int mCounter = 0;
    float mSpeed = 0;

    boolean mIsMetric;
    float mStepLength;

    PedometerSettings mSettings;
    Utils mUtils;

    /** Desired speed, adjusted by the user */
    float mDesiredSpeed;

    /** Should we speak? */
    boolean mShouldTellFasterslower;
    boolean mShouldTellSpeed;

    /** When did the TTS speak last time */
    private long mSpokenAt = 0;

    public SpeedNotifier(Listener listener, PedometerSettings settings, Utils utils) {
        mListener = listener;
        mUtils = utils;
        mSettings = settings;
        mDesiredSpeed = mSettings.getDesiredSpeed();
        reloadSettings();
    }
    public void setSpeed(float speed) {
        mSpeed = speed;
        notifyListener();
    }
    public void reloadSettings() {
        mIsMetric = mSettings.isMetric();
        mStepLength = mSettings.getStepLength();
        mShouldTellSpeed = mSettings.shouldTellSpeed();
        mShouldTellFasterslower =
            mSettings.shouldTellFasterslower()
```

```
            && mSettings.getMaintainOption() == PedometerSettings.M_SPEED;
        notifyListener();
    }
    public void setDesiredSpeed(float desiredSpeed) {
        mDesiredSpeed = desiredSpeed;
    }

    private void notifyListener() {
        mListener.valueChanged(mSpeed);
    }

    public void paceChanged(int value) {
        if (mIsMetric) {
            mSpeed = // kilometers / hour
                value * mStepLength // centimeters / minute
                / 100000f * 60f; // centimeters/kilometer
        }
        else {
            mSpeed = // miles / hour
                value * mStepLength // inches / minute
                / 63360f * 60f; // inches/mile
        }
        tellFasterSlower();
        notifyListener();
    }

    /**
     * Say slower/faster, if needed.
     */
    private void tellFasterSlower() {
        if (mShouldTellFasterslower && mUtils.isSpeakingEnabled()) {
            long now = System.currentTimeMillis();
            if (now - mSpokenAt > 3000 && !mUtils.isSpeakingNow()) {
                float little = 0.10f;
                float normal = 0.30f;
                float much = 0.50f;

                boolean spoken = true;
                if (mSpeed < mDesiredSpeed * (1 - much)) {
                    mUtils.say("much faster!");
                }
                else
                if (mSpeed > mDesiredSpeed * (1 + much)) {
                    mUtils.say("much slower!");
                }
                else
                if (mSpeed < mDesiredSpeed * (1 - normal)) {
                    mUtils.say("faster!");
                }
                else
                if (mSpeed > mDesiredSpeed * (1 + normal)) {
                    mUtils.say("slower!");
                }
                else
                if (mSpeed < mDesiredSpeed * (1 - little)) {
                    mUtils.say("a little faster!");
                }
                else
                if (mSpeed > mDesiredSpeed * (1 + little)) {
                    mUtils.say("a little slower!");
                }
                else {
                    spoken = false;
                }
                if (spoken) {
                    mSpokenAt = now;
                }
            }
        }
```

```
    }

    public void passValue() {
        // Not used
    }

    public void speak() {
        if (mSettings.shouldTellSpeed()) {
            if (mSpeed >= .01f) {
                mUtils.say(("" + (mSpeed + 0.000001f)).substring(0, 4) + (mIsMetric ? " 
kilometers per hour" : " miles per hour"));
            }
        }
    }
}
```

到此为止，本系统的主要功能模块全部介绍完毕。为了节省篇幅，本书只对重点和难点内容进行了剖析，其他部分的具体实现请读者参阅本书程序中附带的源码。

第16章　气压传感器详解

在Android设备开发应用过程中，通常需要使用设备来感知当前所处环境的信息，例如气压、GPS、海拔、湿度和温度。在Android系统中，专门提供了压传感器、海拔传感器、湿度传感器和温度传感器来支持上述功能。本章将详细讲解在Android设备中使用气压传感器详解的基本知识，为读者步入本书后面知识的学习打下基础。

16.1　气压传感器基础

在现实应用中，气压传感器主要用于测量气体的绝对压强，主要适用于与气体压强相关的物理实验，如气体定律等，也可以在生物和化学实验中测量干燥、无腐蚀性的气体压强。本节将详细讲解Android系统中气压传感器的基本知识。

16.1.1　什么是气压传感器

气压传感器的原理比较简单，其主要的传感元件是一个对气压传感器内的强弱敏感的薄膜和一个顶针开控制，电路方面它连接了一个柔性电阻器。当被测气体的压力强降低或升高时，这个薄膜变形带动顶针，同时该电阻器的阻值将会改变。电阻器的阻值发生变化。从传感元件取得0～5V的信号电压，经过A/D转换由数据采集器接收，然后数据采集器以适当的形式把结果传送给计算机。

在现实应用中，很多气压传感器的主要部件为变容式硅膜盒。当该变容硅膜盒外界大气压力发生变化时顶针动作，单晶硅膜盒随着发生弹性变形，从而引起硅膜盒平行板电容器电容量的变化来控制气压传感器。

国标GB7665-87对传感器的定义是："能感受规定的被测量并按照一定的规律转换成可用信号的器件或装置，通常由敏感元件和转换元件组成"。而气压传感器是由一种检测装置，能感受到被测量的信息，并能将检测感受到的信息，按一定规律变换成为电信号或其他所需形式的信息输出，以满足信息的传输、处理、存储、显示、记录和控制等要求，是实现自动化检测和控制的首要环节。

16.1.2　气压传感器在智能手机中的应用

随着智能手机设备的发展，气压传感器得到了大力的普及。气压传感器首次在智能手机上使用是在Galaxy Nexus上，而之后推出的一些Android旗舰手机里也包含了这一传感器，像Galaxy SIII、Galaxy Note2也都有。对于喜欢登山的人来说，都会非常关心自己所处的高度。海拔高度的测量方法，一般常用的有两种方式，一是通过GPS全球定位系统，二是通过测出大气压，然后根据气压值计算出海拔高度。由于受到技术和其他方面原因的限制，GPS计算海拔高度一般误差都会有十米左右，而且在树林里或者是在悬崖下面时，有时候甚至接收不到GPS卫星信号。同时当用户处于楼宇内时，内置感应器可能会无法接收到 GPS 信号，从而不能够识别地理位置。配合气压传感器、加速计、陀螺仪等就能够实现精确定位。这样当你在商场购物时，你能够更好地找到目标商品。

另外在汽车导航领域中，经常会有人抱怨在高架桥里导航常常会出错。比如在高架桥上时，GPS 说右转，而实际上右边根本没有右转出口，这主要是 GPS 无法判断你是在桥上还是桥下而造成的错误导航。一般高架桥上下两层的高度会有几米到十几米的距离，而 GPS 的误差可能会有几十米，所以发生上面的事情也就可以理解了。此时如果在手机中增加一个气压传感器就不一样了，它的精度可以做到 1 米的误差，这样就可以很好地辅助 GPS 来测量出所处的高度，错误导航的问题也就容易解决了。

气压的方式可选择的范围会广些，而且可以把成本可以控制在比较低的水平。另外像 Galaxy Nexus 等手机的气压传感器还包括温度传感器，它可以捕捉到温度来对结果进行修正，以增加测量结果的精度。所以在手机原有 GPS 的基础上再增加气压传感器的功能，可以让三维定位更加精准。

在 Android 系统中，气压传感器的类型是 TYPE_PRESSURE，单位是 hPa（百帕斯卡），能够返回当前环境下的压强。

16.2 实战演练——开发一个 Android 气压计系统

在接下来的实例中，将详细讲解开发一个 Web 版气压测试系统的方法。

实例	功能	源码路径
实例 16-1	开发一个 Android 气压计系统	\daima\16\cordova-plugin-barometer

16.2.1 编写插件调用文件

编写插件调用文件 plugin.xml，具体代码如下所示。

```
<plugin xmlns="http://apache.org/cordova/ns/plugins/1.0"
        id="org.dartlang.phonegap.barometer"
    version="0.0.2">

    <name>Device Barometer</name>
    <description>Cordova Device Barometer Plugin</description>
    <license>LICENSE</license>
    <keywords>cordova,device,barometer</keywords>
    <repo>https://github.com/zanderso/cordova-plugin-barometer</repo>
    <issue></issue>

    <js-module src="www/Pressure.js" name="Pressure">
        <clobbers target="Pressure" />
    </js-module>

    <js-module src="www/barometer.js" name="barometer">
        <clobbers target="navigator.barometer" />
    </js-module>

    <!-- android -->
    <platform name="android">

        <config-file target="res/xml/config.xml" parent="/*">
            <feature name="Barometer">
                <param  name="android-package"  value="org.dartlang.phonegap.barometer.
BarometerListener"/>
            </feature>
        </config-file>

        <source-file   src="src/android/BarometerListener.java"   target-dir="src/org/
```

```
dartlang/phonegap/barometer" />

    </platform>
</plugin>
```

16.2.2 编写 Cordova 插件文件

编写 Cordova 插件文件 barometer.js 设置可以访问气压计的数据，具体实现代码如下所示。

```
var argscheck = require('cordova/argscheck'),
    utils = require("cordova/utils"),
    exec = require("cordova/exec"),
    Acceleration = require('./Pressure');

// Is the barometer sensor running?
var running = false;

// Keeps reference to watchPressure calls.
var timers = {};

// Array of listeners; used to keep track of when we should call start and stop.
var listeners = [];

// Last returned pressure object from native
var pressure = null;

// 告知本地开始测试获取数据
function start() {
    exec(function(a) {
        var tempListeners = listeners.slice(0);
        pressure = new Pressure(a.val, a.timestamp);
        for (var i = 0, l = tempListeners.length; i < l; i++) {
            tempListeners[i].win(pressure);
        }
    }, function(e) {
        var tempListeners = listeners.slice(0);
        for (var i = 0, l = tempListeners.length; i < l; i++) {
            tempListeners[i].fail(e);
        }
    }, "Barometer", "start", []);
    running = true;
}

// 告知本地停止获取数据
function stop() {
    exec(null, null, "Barometer", "stop", []);
    running = false;
}

//在监听数组中添加回调
function createCallbackPair(win, fail) {
    return {win:win, fail:fail};
}

// Removes a win/fail listener pair from the listeners array
function removeListeners(l) {
    var idx = listeners.indexOf(l);
    if (idx > -1) {
        listeners.splice(idx, 1);
        if (listeners.length === 0) {
            stop();
        }
    }
}

var barometer = {
    /**
     * 异步获取当前的气压
     *
```

```
 * @param {Function} successCallback   The function to call when the pressure data
 is available
 * @param {Function} errorCallback    The function to call when there is an error
 getting the pressure data. (OPTIONAL)
 * @param {BarometerOptions} options   The options for getting the barometer data
 such as frequency. (OPTIONAL)
 */
getCurrentPressure: function(successCallback, errorCallback, options) {
    argscheck.checkArgs('fFO', 'barometer.getCurrentPressure', arguments);

    var p;
    var win = function(a) {
        removeListeners(p);
        successCallback(a);
    };
    var fail = function(e) {
        removeListeners(p);
        errorCallback && errorCallback(e);
    };

    p = createCallbackPair(win, fail);
    listeners.push(p);

    if (!running) {
        start();
    }
},

/**
 *在一个指定的时间间隔内不间断地异步获取气压值
 *
 * @param {Function} successCallback
 * @param {Function} errorCallback    The function to call when there is an error
 getting the pressure data. (OPTIONAL)
 * @param {BarometerOptions} options   The options for getting the barometer data
 such as frequency. (OPTIONAL)
 * @return String    The watch id that must be passed to #clearWatch to stop watching.
 */
watchPressure: function(successCallback, errorCallback, options) {
    argscheck.checkArgs('fFO', 'barometer.watchPressure', arguments);
    // Default interval (10 sec)
    var frequency = (options && options.frequency && typeof options.frequency ==
    'number') ? options.frequency : 10000;

    // 读取压力数值
    var id = utils.createUUID();

    var p = createCallbackPair(function(){}, function(e) {
        removeListeners(p);
        errorCallback && errorCallback(e);
    });
    listeners.push(p);

    timers[id] = {
        timer:window.setInterval(function() {
            if (pressure) {
                successCallback(pressure);
            }
        }, frequency),
        listeners:p
    };

    if (running) {
        // If we're already running then immediately invoke the success callback
        // but only if we have retrieved a value, sample code does not check for null ...
        if (pressure) {
            successCallback(pressure);
        }
    } else {
```

```
            start();
        }

        return id;
    },

    /**
     *清除指定的气压表
     *
     * @param {String} id     The id of the watch returned from #watchPressure.
     */
    clearWatch: function(id) {
        // Stop javascript timer & remove from timer list
        if (id && timers[id]) {
            window.clearInterval(timers[id].timer);
            removeListeners(timers[id].listeners);
            delete timers[id];
        }
    }
};
module.exports = barometer;
```

16.2.3 定义每个时间点的压力值

编写文件 Pressure.js 定义每个时间点的压力值，具体代码如下所示。

```
var Pressure = function(val, timestamp) {
    this.val = val;
    this.timestamp = timestamp || (new Date()).getTime();
};

module.exports = Pressure;
```

16.2.4 监听传感器传来的和存储的新压力值

在 Android 平台下下编写 Java 程序文件 BarometerListener.java，功能是监听传感器传来的和存储的新压力值，具体实现代码如下所示。

```
/**
 * 监听传感器传来的和存储的新压力值
 */
public class BarometerListener extends CordovaPlugin implements SensorEventListener {

    public static int STOPPED = 0;
    public static int STARTING = 1;
    public static int RUNNING = 2;
    public static int ERROR_FAILED_TO_START = 3;

    private float pressure;  // 最近的压力值
    private long timestamp;  //最近时间值
    private int status;  // 监听级别
    private int accuracy = SensorManager.SENSOR_STATUS_UNRELIABLE;

    private SensorManager sensorManager;  // SensorManager 对象
    private Sensor mSensor;  // 传递气压传感器

    private CallbackContext callbackContext;  //跟踪 JS 回调的上下文
.

    private Handler mainHandler=null;
    private Runnable mainRunnable = new Runnable() {
        public void run() {
            BarometerListener.this.timeout();
        }
    };

    /**
     * 创建一个气压监听
```

```
 */
public BarometerListener() {
    this.pressure = 0;
    this.timestamp = 0;
    this.setStatus(BarometerListener.STOPPED);
 }

/**
 *初始化处理，得到和活动相关的路径
 *
 * @param cordova The context of the main Activity.
 * @param webView The associated CordovaWebView.
 */
@Override
public void initialize(CordovaInterface cordova, CordovaWebView webView) {
    super.initialize(cordova, webView);
    this.sensorManager = (SensorManager) cordova.getActivity().getSystemService
    (Context.SENSOR_SERVICE);
}

/**
 *执行获取请求
 *
 * @param action         The action to execute.
 * @param args           The exec() arguments.
 * @param callbackId     The callback id used when calling back into JavaScript.
 * @return               Whether the action was valid.
 */
public boolean execute(String action, JSONArray args, CallbackContext callbackContext) {
    if (action.equals("start")) {
        this.callbackContext = callbackContext;
        if (this.status != BarometerListener.RUNNING) {
            // If not running, then this is an async call, so don't worry about waiting
            // We drop the callback onto our stack, call start, and let start and the
            // sensor callback fire off the callback down the road
            this.start();
        }
    }
    else if (action.equals("stop")) {
        if (this.status == BarometerListener.RUNNING) {
            this.stop();
        }
    } else {
      // Unsupported action
        return false;
    }

    PluginResult result = new PluginResult(PluginResult.Status.NO_RESULT, "");
    result.setKeepCallback(true);
    callbackContext.sendPluginResult(result);
    return true;
}

/**
 *当气压传感器被关闭时则停止监听器
 */
public void onDestroy() {
    this.stop();
}

//--------------------------------------------------------------------------
// 下面是本地方法
//--------------------------------------------------------------------------
//
/**
 *开始监听传感器
 *
 * @return          status of listener
```

```
*/
private int start() {
    //如果已经开始或运行则返回
    if ((this.status == BarometerListener.RUNNING) || (this.status ==
    BarometerListener.STARTING)) {
        return this.status;
    }

    this.setStatus(BarometerListener.STARTING);

    // 从传感器获取气压
    List<Sensor> list = this.sensorManager.getSensorList(Sensor.TYPE_PRESSURE);

    // If found, then register as listener
    if ((list != null) && (list.size() > 0)) {
      this.mSensor = list.get(0);
      this.sensorManager.registerListener(this, this.mSensor, SensorManager.SENSOR_
      DELAY_UI);
      this.setStatus(BarometerListener.STARTING);
    } else {
      this.setStatus(BarometerListener.ERROR_FAILED_TO_START);
      this.fail(BarometerListener.ERROR_FAILED_TO_START, "No sensors found to
      register barometer listening to.");
      return this.status;
    }

    //设置一个超时回调的主线程
    stopTimeout();
    mainHandler = new Handler(Looper.getMainLooper());
    mainHandler.postDelayed(mainRunnable, 2000);

    return this.status;
}
private void stopTimeout() {
    if(mainHandler!=null){
        mainHandler.removeCallbacks(mainRunnable);
    }
}
/**
 *停止监听
 */
private void stop() {
    stopTimeout();
    if (this.status != BarometerListener.STOPPED) {
        this.sensorManager.unregisterListener(this);
    }
    this.setStatus(BarometerListener.STOPPED);
    this.accuracy = SensorManager.SENSOR_STATUS_UNRELIABLE;
}

/**
 *如果传感器还没有开始则返回一个错误
 *
 * Called two seconds after starting the listener.
 */
private void timeout() {
    if (this.status == BarometerListener.STARTING) {
        this.setStatus(BarometerListener.ERROR_FAILED_TO_START);
        this.fail(BarometerListener.ERROR_FAILED_TO_START, "Barometer could not be
        started.");
    }
}

/**
 * Called when the accuracy of the sensor has changed.
 *
 * @param sensor
 * @param accuracy
 */
```

```
public void onAccuracyChanged(Sensor sensor, int accuracy) {
    // Only look at barometer events
    if (sensor.getType() != Sensor.TYPE_PRESSURE) {
        return;
    }

    // If not running, then just return
    if (this.status == BarometerListener.STOPPED) {
        return;
    }
    this.accuracy = accuracy;
}

/**
 *传感器监听事件
 *
 * @param SensorEvent event
 */
public void onSensorChanged(SensorEvent event) {
    // Only look at barometer events
    if (event.sensor.getType() != Sensor.TYPE_PRESSURE) {
        return;
    }

    // If not running, then just return
    if (this.status == BarometerListener.STOPPED) {
        return;
    }
    this.setStatus(BarometerListener.RUNNING);

    if (this.accuracy >= SensorManager.SENSOR_STATUS_ACCURACY_MEDIUM) {

        // Save time that event was received
        this.timestamp = System.currentTimeMillis();
        this.pressure = event.values[0];

        this.win();
    }
}

/**
 * Called when the view navigates.
 */
@Override
public void onReset() {
    if (this.status == BarometerListener.RUNNING) {
        this.stop();
    }
}

// 发送一个错误到 JS 文件
private void fail(int code, String message) {
    // Error object
    JSONObject errorObj = new JSONObject();
    try {
        errorObj.put("code", code);
        errorObj.put("message", message);
    } catch (JSONException e) {
        e.printStackTrace();
    }
    PluginResult err = new PluginResult(PluginResult.Status.ERROR, errorObj);
    err.setKeepCallback(true);
    callbackContext.sendPluginResult(err);
}

private void win() {
    // Success return object
    PluginResult result = new PluginResult(PluginResult.Status.OK, this.get
    PressureJSON());
```

```
        result.setKeepCallback(true);
        callbackContext.sendPluginResult(result);
    }

    private void setStatus(int status) {
        this.status = status;
    }
    private JSONObject getPressureJSON() {
        JSONObject r = new JSONObject();
        try {
            r.put("val", this.pressure);
            r.put("timestamp", this.timestamp);
        } catch (JSONException e) {
            e.printStackTrace();
        }
        return r;
    }
}
```

到此为止，整个实例介绍完毕，这样便成功构建了一个简单的气压计模型。读者可以以此为基础进行拓展，设计一个自己喜欢的气压计程序。

16.3 实战演练——获取当前相对海拔和绝对海拔的数据

在接下来的实例中，将演示在 Android 可设备中联合使用 GPS 和气压传感器的方法。

实例	功能	源码路径
实例 16-2	获取当前相对海拔和绝对海拔的数据	https://github.com/gast-lib/gast-lib

本实例源码来源于如下地址，读者可以自行免费获得。

```
https://github.com/gast-lib/gast-lib
```

16.3.1 实现布局文件

编写布局文件 determine_altitude.xml，功能是分别通过文本框控件在上方显示 GPS 相对海拔信息和绝对海拔高度信息，并且也通过文本框控件显示相对气压值和绝对气压值，在下方设置了一个“Mark Starting Altitude”按钮用于标记开始时的高度，这样用当前海拔减去开始高度就会得到相对海拔高度。布局文件 determine_altitude.xml 的具体实现代码如下所示。

```
<RelativeLayout xmlns:android="http://schemas.android.com/apk/res/android"
    android:layout_width="match_parent"
    android:layout_height="match_parent"
    android:orientation="vertical" >

    <!-- GPS Altitude -->
    <TextView android:id="@+id/gpsAltitudeSectionHeading"
        style="@style/apptext"
        android:text="@string/gpsAltitudeLabel"
        android:layout_alignParentTop="true" />

    <TextView android:id="@+id/gpsAltitudeSectionDivider"
        style="@style/line_separator"
        android:layout_below="@id/gpsAltitudeSectionHeading" />

    <TextView android:id="@+id/gpsAltitudeLabel"
        android:layout_width="wrap_content"
        android:layout_height="wrap_content"
        android:text="@string/altitudeLabel"
        android:layout_below="@id/gpsAltitudeSectionDivider"/>

    <TextView android:id="@+id/gpsAltitude"
        android:layout_width="wrap_content"
```

```
        android:layout_height="wrap_content"
        android:layout_alignTop="@id/gpsAltitudeLabel"
        android:layout_alignBottom="@id/gpsAltitudeLabel"
        android:layout_toRightOf="@id/gpsAltitudeLabel"
        android:text="@string/notAvailable"/>

    <TextView android:id="@+id/gpsRelativeAltitudeLabel"
        android:layout_width="wrap_content"
        android:layout_height="wrap_content"
        android:text="@string/relativeAltitudeLabel"
        android:layout_below="@id/gpsAltitude"/>

    <TextView android:id="@+id/gpsRelativeAltitude"
        android:layout_width="wrap_content"
        android:layout_height="wrap_content"
        android:layout_alignTop="@id/gpsRelativeAltitudeLabel"
        android:layout_alignBottom="@id/gpsRelativeAltitudeLabel"
        android:layout_toRightOf="@id/gpsRelativeAltitudeLabel"
        android:text="@string/notAvailable"/>

    <!-- Standard Pressure Barometer Altitude -->
    <TextView android:id="@+id/barometerAltitudeSectionHeading"
        style="@style/apptext"
        android:text="@string/barometerAltitudeLabel"
        android:layout_below="@id/gpsRelativeAltitudeLabel"
        android:layout_marginTop="10dip" />

    <TextView android:id="@+id/barometerAltitudeSectionDivider"
        style="@style/line_separator"
        android:layout_below="@id/barometerAltitudeSectionHeading" />

    <TextView android:id="@+id/barometerAltitudeLabel"
        android:layout_width="wrap_content"
        android:layout_height="wrap_content"
        android:text="@string/altitudeLabel"
        android:layout_below="@id/barometerAltitudeSectionDivider"/>

    <TextView android:id="@+id/barometerAltitude"
        android:layout_width="wrap_content"
        android:layout_height="wrap_content"
        android:layout_alignTop="@id/barometerAltitudeLabel"
        android:layout_alignBottom="@id/barometerAltitudeLabel"
        android:layout_toRightOf="@id/barometerAltitudeLabel"/>

    <TextView android:id="@+id/barometerRelativeAltitudeLabel"
        android:layout_width="wrap_content"
        android:layout_height="wrap_content"
        android:text="@string/relativeAltitudeLabel"
        android:layout_below="@id/barometerAltitude"/>

    <TextView android:id="@+id/barometerRelativeAltitude"
        android:layout_width="wrap_content"
        android:layout_height="wrap_content"
        android:layout_alignTop="@id/barometerRelativeAltitudeLabel"
        android:layout_alignBottom="@id/barometerRelativeAltitudeLabel"
        android:layout_toRightOf="@id/barometerRelativeAltitudeLabel"/>

    <TextView android:id="@+id/standardPressureLabel"
        android:layout_width="wrap_content"
        android:layout_height="wrap_content"
        android:text="@string/standardPressureLabel"
        android:layout_below="@id/barometerRelativeAltitude"/>

    <TextView android:id="@+id/standardPressure"
        android:layout_width="wrap_content"
        android:layout_height="wrap_content"
        android:layout_alignTop="@id/standardPressureLabel"
        android:layout_alignBottom="@id/standardPressureLabel"
        android:layout_toRightOf="@id/standardPressureLabel"
```

```
        android:text="@string/notAvailable"/>

    <!-- MSLP Barometer Altitude -->
    <TextView android:id="@+id/mslpBarometerAltitudeSectionHeading"
        style="@style/apptext"
        android:text="@string/mslpBarometerAltitudeLabel"
        android:layout_below="@id/standardPressureLabel"
        android:layout_marginTop="10dip" />

    <TextView android:id="@+id/mslpBarometerAltitudeSectionDivider"
        style="@style/line_separator"
        android:layout_below="@id/mslpBarometerAltitudeSectionHeading" />

    <TextView android:id="@+id/mslpBarometerAltitudeLabel"
        android:layout_width="wrap_content"
        android:layout_height="wrap_content"
        android:text="@string/altitudeLabel"
        android:layout_below="@id/mslpBarometerAltitudeSectionDivider"/>

    <TextView android:id="@+id/mslpBarometerAltitude"
        android:layout_width="wrap_content"
        android:layout_height="wrap_content"
        android:layout_alignTop="@id/mslpBarometerAltitudeLabel"
        android:layout_alignBottom="@id/mslpBarometerAltitudeLabel"
        android:layout_toRightOf="@id/mslpBarometerAltitudeLabel"
        android:text="@string/notAvailable"/>

    <TextView android:id="@+id/mslpRelativeAltitudeLabel"
        android:layout_width="wrap_content"
        android:layout_height="wrap_content"
        android:text="@string/relativeAltitudeLabel"
        android:layout_below="@id/mslpBarometerAltitude"/>

    <TextView android:id="@+id/mslpBarometerRelativeAltitude"
        android:layout_width="wrap_content"
        android:layout_height="wrap_content"
        android:layout_alignTop="@id/mslpRelativeAltitudeLabel"
        android:layout_alignBottom="@id/mslpRelativeAltitudeLabel"
        android:layout_toRightOf="@id/mslpRelativeAltitudeLabel"
        android:text="@string/notAvailable"/>

    <TextView android:id="@+id/mslpLabel"
        android:layout_width="wrap_content"
        android:layout_height="wrap_content"
        android:text="@string/mslpLabel"
        android:layout_below="@id/mslpBarometerRelativeAltitude"/>

    <TextView android:id="@+id/mslp"
        android:layout_width="wrap_content"
        android:layout_height="wrap_content"
        android:layout_alignTop="@id/mslpLabel"
        android:layout_alignBottom="@id/mslpLabel"
        android:layout_toRightOf="@id/mslpLabel"
        android:text="@string/notAvailable"/>

    <ToggleButton
        android:layout_width="match_parent"
        android:layout_height="wrap_content"
        android:layout_below="@id/mslpLabel"
        android:onClick="onToggleClick"
        android:textOff="Mark Starting Altitude"
        android:textOn="Compute Relative Altitude" />
</RelativeLayout>
```

16.3.2 实现主 Activity

本实例 Activity 程序文件是 DetermineAltitudeActivity.java，具体实现流程如下所示。

- 定义系统需要的成员变量和常量，具体代码如下所示。

```
public class DetermineAltitudeActivity extends Activity implements SensorEventListener,
LocationListener
{
    private static final String TAG = "DetermineAltitudeActivity";
    private static final int TIMEOUT = 1000; //1 second
    private static final long NS_TO_MS_CONVERSION = (long)1E6;

    // System services
    private SensorManager sensorManager;
    private LocationManager locationManager;

    // UI Views
    private TextView gpsAltitudeView;
    private TextView gpsRelativeAltitude;
    private TextView barometerAltitudeView;
    private TextView barometerRelativeAltitude;
    private TextView mslpBarometerAltitudeView;
    private TextView mslpBarometerRelativeAltitude;
    private TextView mslpView;

    // Member state
    private Float mslp;
    private long lastGpsAltitudeTimestamp = -1;
    private long lastBarometerAltitudeTimestamp = -1;
    private float bestLocationAccuracy = -1;
    private float currentBarometerValue;
    private float lastBarometerValue;
    private double lastGpsAltitude;
    private double currentGpsAltitude;
    private boolean webServiceFetching;
    private long lastErrorMessageTimestamp = -1;
```

- 定义 onCreate 方法，通过此方法设置加载的布局文件是 determine_altitude.xml，并且获取了对被更新 UI 视图的引用，还有位置服务和传感器服务的句柄。具体实现代码如下所示。

```
@Override
protected void onCreate(Bundle savedInstanceState)
{
    super.onCreate(savedInstanceState);
    setContentView(R.layout.determine_altitude);
    getWindow().addFlags(WindowManager.LayoutParams.FLAG_KEEP_SCREEN_ON);

    sensorManager =
          (SensorManager) getSystemService(Context.SENSOR_SERVICE);
    locationManager = (LocationManager) getSystemService(LOCATION_SERVICE);

    gpsAltitudeView = (TextView) findViewById(R.id.gpsAltitude);

    gpsRelativeAltitude =
          (TextView) findViewById(R.id.gpsRelativeAltitude);

    barometerAltitudeView = (TextView) findViewById(R.id.barometerAltitude);
    barometerRelativeAltitude =
          (TextView) findViewById(R.id.barometerRelativeAltitude);
    mslpBarometerAltitudeView =
          (TextView) findViewById(R.id.mslpBarometerAltitude);
    mslpBarometerRelativeAltitude =
          (TextView) findViewById(R.id.mslpBarometerRelativeAltitude);
    mslpView = (TextView) findViewById(R.id.mslp);

    webServiceFetching = false;

    TextView standardPressure =
          (TextView)findViewById(R.id.standardPressure);
    String standardPressureString =
          String.valueOf(SensorManager.PRESSURE_STANDARD_ATMOSPHERE);
    standardPressure.setText(standardPressureString);
}
```

● 在方法 onResume()中注册 GPS 和气压传感器，通过遍历从 getProviders(true)返回的数组来注册已经激活位置提供者。

```
@Override
protected void onResume()
{
    super.onResume();

    List<String> enabledProviders = locationManager.getProviders(true);

    if (enabledProviders.isEmpty()
            || !enabledProviders.contains(LocationManager.GPS_PROVIDER))
    {
        Toast.makeText(this,
                R.string.gpsNotEnabledMessage,
                Toast.LENGTH_LONG).show();
    }
    else
    {
        // Register every location provider returned from LocationManager
        for (String provider : enabledProviders)
        {
            // Register for updates every minute
            locationManager.requestLocationUpdates(provider,
                    60000,  // minimum time of 60000 ms (1 minute)
                    0,      // Minimum distance of 0
                    this,
                    null);
        }
    }

    Sensor sensor = sensorManager.getDefaultSensor(Sensor.TYPE_PRESSURE);

    // Only make registration call if device has a pressure sensor
    if (sensor != null)
    {
        sensorManager.registerListener(this,
                sensor,
                SensorManager.SENSOR_DELAY_NORMAL);
    }
}
```

● 编写方法停止注册工作，具体实现代码如下所示。

```
@Override
protected void onPause()
{
    super.onPause();

    sensorManager.unregisterListener(this);
    locationManager.removeUpdates(this);
}
```

● 编写方法 onSensorChanged，功能是通过气压传感器获取海拔高度，将读取的气压数值传入到 SensorManager.getAltitude 方法，在此方法中封装了实现数值计算的公式。具体实现代码如下所示。

```
@Override
public void onSensorChanged(SensorEvent event)
{
    float altitude;
    currentBarometerValue = event.values[0];

    double currentTimestamp = event.timestamp / NS_TO_MS_CONVERSION;
    double elapsedTime = currentTimestamp - lastBarometerAltitudeTimestamp;
    if (lastBarometerAltitudeTimestamp == -1 || elapsedTime > TIMEOUT)
    {
        altitude =
```

```
            SensorManager
                .getAltitude(SensorManager.PRESSURE_STANDARD_ATMOSPHERE,
                    currentBarometerValue);
        barometerAltitudeView.setText(String.valueOf(altitude));

        if (mslp != null)
        {
            altitude = SensorManager.getAltitude(mslp,
                currentBarometerValue);
            mslpBarometerAltitudeView.setText(String.valueOf(altitude));
            mslpView.setText(String.valueOf(mslp));
        }

        lastBarometerAltitudeTimestamp = (long)currentTimestamp;
    }
}

@Override
public void onAccuracyChanged(Sensor sensor, int accuracy)
{
    // no-op
}
```

● 编写方法 onLocationChanged，功能是获取 GPS 提供的海拔高度。在此方法中，通过获取 location.getAltitude()方法的方式返回以米为单位的海拔高度。具体实现代码如下所示。

```
@Override
public void onLocationChanged(Location location)
{
    if (LocationManager.GPS_PROVIDER.equals(location.getProvider())
            && (lastGpsAltitudeTimestamp == -1
                || location.getTime() - lastGpsAltitudeTimestamp > TIMEOUT))
    {
        double altitude = location.getAltitude();
        gpsAltitudeView.setText(String.valueOf(altitude));
        lastGpsAltitudeTimestamp = location.getTime();
        currentGpsAltitude = altitude;
    }

    float accuracy = location.getAccuracy();
    boolean betterAccuracy = accuracy < bestLocationAccuracy;
    if (mslp == null  || (bestLocationAccuracy > -1 && betterAccuracy))
    {
        bestLocationAccuracy = accuracy;

        if (!webServiceFetching)
        {
            webServiceFetching = true;
            new MetarAsyncTask().execute(location.getLatitude(),
                    location.getLongitude());
        }
    }
}
```

● 编写方法 onToggleClick，功能是如果监听到获取相对海拔的高度，则减去记录的原始高度来得到相对海拔的值。具体实现代码如下所示。

```
public void onToggleClick(View view)
{
    if (((ToggleButton)view).isChecked())
    {
        lastGpsAltitude = currentGpsAltitude;
        lastBarometerValue = currentBarometerValue;
        gpsRelativeAltitude.setVisibility(View.INVISIBLE);
        barometerRelativeAltitude.setVisibility(View.INVISIBLE);

        if (mslp != null)
        {
            mslpBarometerRelativeAltitude.setVisibility(View.INVISIBLE);
```

```
        }
    }
    else
    {
        double delta;

        delta = currentGpsAltitude - lastGpsAltitude;
        gpsRelativeAltitude.setText(String.valueOf(delta));
        gpsRelativeAltitude.setVisibility(View.VISIBLE);

        delta = SensorManager
                .getAltitude(SensorManager.PRESSURE_STANDARD_ATMOSPHERE,
                    currentBarometerValue)
            - SensorManager
                .getAltitude(SensorManager.PRESSURE_STANDARD_ATMOSPHERE,
                    lastBarometerValue);

        barometerRelativeAltitude.setText(String.valueOf(delta));
        barometerRelativeAltitude.setVisibility(View.VISIBLE);

        if (mslp != null)
        {
            delta = SensorManager.getAltitude(mslp, currentBarometerValue)
                    - SensorManager.getAltitude(mslp, lastBarometerValue);
            mslpBarometerRelativeAltitude.setText(String.valueOf(delta));
            mslpBarometerRelativeAltitude.setVisibility(View.VISIBLE);
        }
    }
}
```

- 编写类 MetarAsyncTask，功能是访问远程天气 Web 服务器，并获取 MSLP 数据的类。执行完毕后，Web 服务器会返回变量 response 中的响应。具体实现代码如下所示。

```
private class MetarAsyncTask extends AsyncTask<Number, Void, Float>
{
    private static final String WS_URL =
            "http://ws.geonames.org/findNearByWeatherJSON";
    private static final String SLP_STRING = "slp";

    @Override
    protected Float doInBackground(Number... params)
    {
        Float mslp = null;
        HttpURLConnection urlConnection = null;

        try
        {
            // Generate URL with parameters for web service
            Uri uri =
                    Uri.parse(WS_URL)
                    .buildUpon()
                    .appendQueryParameter("lat", String.valueOf(params[0]))
                    .appendQueryParameter("lng", String.valueOf(params[1]))
                    .build();

            // Connect to web service
            URL url = new URL(uri.toString());
            urlConnection = (HttpURLConnection) url.openConnection();

            // Read web service response and convert to a string
            InputStream inputStream =
                    new BufferedInputStream(urlConnection.getInputStream());

            // Convert InputStream to String using a Scanner
            Scanner inputStreamScanner =
                    new Scanner(inputStream).useDelimiter("\\A");
            String response = inputStreamScanner.next();
            inputStreamScanner.close();
```

```
            Log.d(TAG, "Web Service Response -> " + response);

            JSONObject json = new JSONObject(response);

            String observation =
                  json
                     .getJSONObject("weatherObservation")
                     .getString("observation");

            // Split on whitespace
            String[] values = observation.split("\\s");

            // Iterate of METAR string until SLP string is found
            String slpString = null;
            for (int i = 1; i < values.length; i++)
            {
               String value = values[i];

               if (value.startsWith(SLP_STRING.toLowerCase())
                     || value.startsWith(SLP_STRING.toUpperCase()))
               {
                  slpString =
                        value.substring(SLP_STRING.length());
                  break;
               }
            }

            // Decode SLP string into numerical representation
            StringBuffer sb = new StringBuffer(slpString);

            sb.insert(sb.length() - 1, ".");

            float val1 = Float.parseFloat("10" + sb);
            float val2 = Float.parseFloat("09" + sb);

            mslp =
                  (Math.abs((1000 - val1)) < Math.abs((1000 - val2)))
                     ? val1
                     : val2;
         }
         catch (Exception e)
         {
            Log.e(TAG, "Could not communicate with web service", e);
         }
         finally
         {
            if (urlConnection != null)
            {
               urlConnection.disconnect();
            }
         }

         return mslp;
      }

      @Override
      protected void onPostExecute(Float result)
      {
         long uptime = SystemClock.uptimeMillis();

         if (result == null
               && (lastErrorMessageTimestamp == -1
                  || ((uptime - lastErrorMessageTimestamp) > 30000)))
         {
            Toast.makeText(DetermineAltitudeActivity.this,
                  R.string.webServiceConnectionFailureMessage,
                  Toast.LENGTH_LONG).show();

            lastErrorMessageTimestamp = uptime;
```

```
            }
            else
            {
                DetermineAltitudeActivity.this.mslp = result;
            }

            DetermineAltitudeActivity.this.webServiceFetching = false;
        }
    }
}
```

到此为止，整个实例介绍完毕，执行后会获取当前设备所处位置的相对海拔和绝对海拔的数据。如图 16-1 所示。

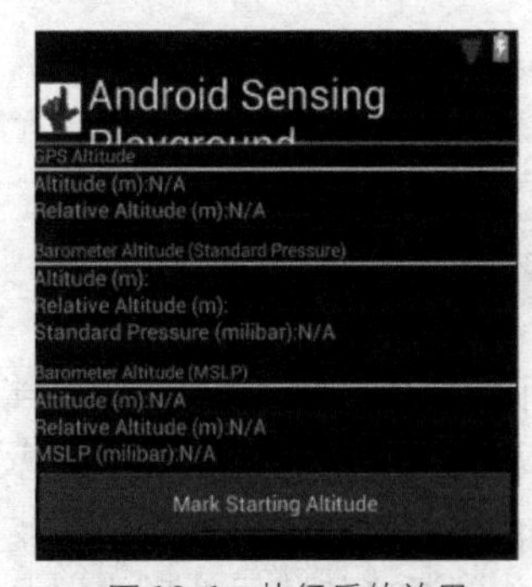

▲图 16-1　执行后的效果

第 17 章 温度传感器详解

温度传感器（temperature transducer）是指能感受温度并转换成可用输出信号的传感器。本章将详细讲解在 Android 设备中使温度传感器的基本知识，为读者步入本书后面知识的学习打下基础。

17.1 温度传感器基础

温度传感器从 17 世纪初人们开始利用温度进行测量。在半导体技术的支持下，本世纪相继开发了半导体热电偶传感器、PN 结温度传感器和集成温度传感器。与之相应，根据波与物质的相互作用规律，相继开发了声学温度传感器、红外传感器和微波传感器。温度传感器是五花八门的各种传感器中最为常用的一种，现代的温度传感器外形非常得小，这样更加让它广泛应用在生产实践的各个领域中，也为人们的生活提供了无数的便利和功能。

温度传感器有 4 种主要类型：热电偶、热敏电阻、电阻温度检测器（RTD）和 IC 温度传感器。IC 温度传感器又包括模拟输出和数字输出两种类型。在现实世界中，温度传感器是温度测量仪表的核心部分，品种繁多。按测量方式可以分为接触式和非接触式两大类，按照传感器材料及电子元件特性分为热电阻和热电偶两类。

在当前的技术水平条件下，温度传感器的主要原理如下所示。

（1）金属膨胀原理设计的传感器

金属在环境温度变化后会产生一个相应的延伸，因此传感器可以以不同方式对这种反应进行信号转换。

（2）双金属片式传感器

双金属片由两片不同膨胀系数的金属贴在一起而组成，随着温度变化，材料 A 比另外一种金属膨胀程度要高，引起金属片弯曲。弯曲的曲率可以转换成一个输出信号。

（3）双金属杆和金属管传感器

随着温度升高，金属管（材料 A）长度增加，而不膨胀钢杆（金属 B）的长度并不增加，这样由于位置的改变，金属管的线性膨胀就可以进行传递。反过来，这种线性膨胀可以转换成一个输出信号。

（4）液体和气体的变形曲线设计的传感器

在温度变化时，液体和气体同样会相应产生体积的变化。

综上所述，多种类型的结构可以把这种膨胀的变化转换成位置的变化，这样产生位置的变化可以输出为电位计、感应偏差、挡流板等形式的结果。

17.2 Android 系统中的温度传感器

在 Android 系统中，早期版本的温度传感器值是 TYPE_TEMPERATURE，在新版本中被 TYPE_

AMBIENT_TEMPERATURE 替换。Android 温度传感器的单位是℃，能够测量并返回当前的温度。

在 Android 内核平台中自带了大量的传感器源码，读者可以在 Rexsee 的开源社区 http://www.rexsee.com/找到相关的原生代码。其中使用温度传感器相关的原生代码如下所示。

```
package rexsee.sensor;

import rexsee.core.browser.JavascriptInterface;
import rexsee.core.browser.RexseeBrowser;
import android.content.Context;
import android.hardware.Sensor;
import android.hardware.SensorEvent;
import android.hardware.SensorEventListener;
import android.hardware.SensorManager;

public class RexseeSensorTemperature implements JavascriptInterface {

      private static final String INTERFACE_NAME = "Temperature";
      @Override
      public String getInterfaceName() {
            return mBrowser.application.resources.prefix + INTERFACE_NAME;
      }
      @Override
      public JavascriptInterface getInheritInterface(RexseeBrowser childBrowser) {
            return this;
      }
      @Override
      public JavascriptInterface getNewInterface(RexseeBrowser childBrowser) {
            return new RexseeSensorTemperature(childBrowser);
      }

      public static final String EVENT_ONTEMPERATURECHANGED = "onTemperatureChanged";

      private final Context mContext;
      private final RexseeBrowser mBrowser;
      private final SensorManager mSensorManager;
      private final SensorEventListener mSensorListener;
      private final Sensor mSensor;

      private int mRate = SensorManager.SENSOR_DELAY_NORMAL;
      private int mCycle = 100; //milliseconds
      private int mEventCycle = 100; //milliseconds
      private float mAccuracy = 0;

      private long lastUpdate = -1;
      private long lastEvent = -1;

      private float value = -999f;

      public RexseeSensorTemperature(RexseeBrowser browser) {
            mContext = browser.getContext();
            mBrowser = browser;
            browser.eventList.add(EVENT_ONTEMPERATURECHANGED);

            mSensorManager = (SensorManager) mContext.getSystemService(Context.
            SENSOR_SERVICE);

            mSensor = mSensorManager.getDefaultSensor(Sensor.TYPE_TEMPERATURE);

            mSensorListener = new SensorEventListener() {
                  @Override
                  public void onAccuracyChanged(Sensor sensor, int accuracy) {
                  }
                  @Override
                  public void onSensorChanged(SensorEvent event) {
                        if (event.sensor.getType() != Sensor.TYPE_TEMPERATURE) return;
                        long curTime = System.currentTimeMillis();
                        if (lastUpdate == -1 || (curTime - lastUpdate) > mCycle) {
                              lastUpdate = curTime;
```

```
                                        float lastValue = value;
                                        value = event.values[SensorManager.DATA_X];
                                        if (lastEvent == -1 || (curTime - lastEvent) >
mEventCycle) {
                                                if (Math.abs(value - lastValue) > mAccuracy) {
                                                        lastEvent = curTime;
                                                        mBrowser.eventList.run(EVENT_
ONTEMPERATURECHANGED);
                                                }
                                        }
                                }
                        }
                };

        }

        public String getLastKnownValue() {
                return (value == -999) ? "null" : String.valueOf(value);
        }

        public void setRate(String rate) {
                mRate = SensorRate.getInt(rate);
        }
        public String getRate() {
                return SensorRate.getString(mRate);
        }
        public void setCycle(int milliseconds) {
                mCycle = milliseconds;
        }
        public int getCycle() {
                return mCycle;
        }
        public void setEventCycle(int milliseconds) {
                mEventCycle = milliseconds;
        }
        public int getEventCycle() {
                return mEventCycle;
        }
        public void setAccuracy(float value) {
                mAccuracy = Math.abs(value);
        }
        public float getAccuracy() {
                return mAccuracy;
        }

        public boolean isReady() {
                return (mSensor == null) ? false : true;
        }
        public void start() {
                if (isReady()) {
                        mSensorManager.registerListener(mSensorListener, mSensor, mRate);
                } else {
                        mBrowser.exception(getInterfaceName(),"Temperature sensor is not found.");
                }
        }
        public void stop() {
                if (isReady()) {
                        mSensorManager.unregisterListener(mSensorListener);
                }
        }

}
```

17.3 实战演练——让 Android 设备变为温度计

在接下来的实例中，将详细讲解在 Android 设备中实现温度计功能的方法。

实例	功能	源码路径
实例 17-1	开发一个 Android 温度计系统	\daima\17\wendu

17.3.1 实现布局文件

编写布局文件 main.xml，通过 TextView 控件显示获取的温度值，具体实现代码如下所示。

```
<LinearLayout xmlns:android="http://schemas.android.com/apk/res/android"
   android:orientation="vertical"
   android:layout_width="fill_parent"
   android:layout_height="fill_parent"><!-- 添加一个垂直的线性布局 -->
   <TextView
      android:id="@+id/title"
      android:gravity="center_horizontal"
      android:textSize="20px"
      android:layout_width="fill_parent"
      android:layout_height="wrap_content"
      android:text="@string/title"/><!-- 添加一个 TextView 控件 -->
   <TextView
      android:id="@+id/myTextView1"
      android:textSize="18px"
      android:layout_width="fill_parent"
      android:layout_height="wrap_content"
      android:text="@string/myTextView1"/><!-- 添加一个 TextView 控件 -->
</LinearLayout>
```

17.3.2 检测温度传感器的温度变化

为该项目添加 SensorSimulator 工具的的 JAR 包，编写文件 activity.java 来检测温度传感器的温度变化，并设置在 onSensorChanged 方法中只对 SENSOR_TEMPERATURE 即温度变化进行检测。文件 activity.java 的具体实现代码如下所示。

```
import org.openintents.sensorsimulator.hardware.SensorManagerSimulator;

import wendu.R;
import android.app.Activity;
import android.hardware.SensorListener;
import android.hardware.SensorManager;
import android.os.Bundle;
import android.widget.TextView;
public class activity extends Activity {
   TextView myTextView1;//当前温度
   //SensorManager mySensorManager;//SensorManager 对象引用
   SensorManagerSimulator mySensorManager;//声明 SensorManagerSimulator 对象,调试时用
   @Override
   public void onCreate(Bundle savedInstanceState) {//重写 onCreate 方法
      super.onCreate(savedInstanceState);
      setContentView(R.layout.main);//设置当前的用户界面
      myTextView1 = (TextView) findViewById(R.id.myTextView1);//得到 myTextView1 的引用
      //mySensorManager = (SensorManager)getSystemService(SENSOR_SERVICE);
      //获得 SensorManager
      //调试时用
      mySensorManager = SensorManagerSimulator.getSystemService(this, SENSOR_SERVICE);
      mySensorManager.connectSimulator();                //与 Simulator 连接
   }
   private SensorListener mySensorListener = new SensorListener(){
      @Override
      public void onAccuracyChanged(int sensor, int accuracy) {}
      //重写 onAccuracyChanged 方法
      @Override
      public void onSensorChanged(int sensor,float[]values){//重写 onSensorChanged 方法
         if(sensor == SensorManager.SENSOR_TEMPERATURE){//只检查温度的变化
            myTextView1.setText("当前的温度为: "+values[0]);  //将当前温度显示到 TextView
         }
      }
   };
```

```
    @Override
    protected void onResume() {//重写的 onResume 方法
        mySensorManager.registerListener(//注册监听
                mySensorListener, //监听器 SensorListener 对象
                SensorManager.SENSOR_TEMPERATURE,//传感器的类型为温度
                SensorManager.SENSOR_DELAY_UI//传感器事件传递的频度
                );
        super.onResume();
    }
    @Override
    protected void onPause() {//重写 onPause 方法
        mySensorManager.unregisterListener(mySensorListener);//取消注册监听器
        super.onPause();
    }
}
```

到此为止，整个实例介绍完毕。运行电脑端的 sensorsimulator 工具，调整工具中的参数使其模拟温度传感器。在“Environment Sensors”中选择“Temperature”，在“Quick Setting”面板中对温度进行快速设置，在传感器模拟器中会显示当前所有的模拟参数。运行后会显示获取的温度值。

17.4 实战演练——电池温度测试仪

在接下来的实例中，将详细讲解测试 Android 设备的电源温度的方法。本实例在现实中可能不会有市场效益，但是演示了测试电源温度的方法，对初学者来说很具有指导意义。

实例	功能	源码路径
实例 17-2	开发一个 Android 温度计系统	\daima\17\ThermoRecord

17.4.1 实现布局文件

编写布局文件 main.xml，功能是通过按钮控件来控制测温功能，具体实现代码如下所示。

```
<?xml version="1.0" encoding="utf-8"?>
<LinearLayout xmlns:android="http://schemas.android.com/apk/res/android"
    xmlns:mako = "http://schemas.android.com/apk/res/com.mako"
    android:orientation="horizontal"
    android:layout_width="match_parent"
    android:layout_height="match_parent"
    >
    <LinearLayout
        android:orientation="vertical"
        android:layout_height="match_parent"
        android:layout_width="62dp"
        android:gravity="bottom|center_horizontal"
        >
        <ZoomControls android:id = "@+id/zoomControl"
            android:rotation="90"
            android:layout_width="wrap_content"
            android:layout_height="wrap_content"
            android:gravity="center_horizontal"
            android:orientation="vertical" />
        <com.mako.RecCheckButton android:id="@+id/recCheckButton"
            android:layout_width="wrap_content"
            android:layout_height="wrap_content" />
    </LinearLayout>
    <com.mako.TemperatureDataView
        android:layout_width="match_parent"
        android:layout_height="match_parent" />
    <!-- <com.mako.FrogListView
        android:layout_width="match_parent"
        android:layout_height="match_parent" /> -->
</LinearLayout>
```

17.4.2 实现程序文件

编写文件 mainact.java，功能是监听用户在主屏幕的操作，根据操作来执行对应的响应事件处理函数。文件 mainact.java 的具体实现代码如下所示。

```
public class mainact extends Activity{
   protected ArrayList<ArrayList<Datum>> temperaturesBaseRecordings; //interspersed
with empty sections;
   protected LinkedList<SummaryLayer> summaries;
   public void beginRecording(){
   }

   public void haltRecording(){
   }

   @Override
   public void onCreate(Bundle savedInstanceState){
      super.onCreate(savedInstanceState);
      setContentView(R.layout.main);
      //access the base db and update the summary db accordingly;
      //bind with the service;
   }

   @Override
   public void onRestoreInstanceState(Bundle state){
      super.onRestoreInstanceState(state);
   }
   @Override
   public void onSaveInstanceState(Bundle state){
      super.onSaveInstanceState(state);
   }
   @Override
   public void onResume(){
      super.onResume();
   }
   @Override
   public void onRestart(){
      super.onRestart();
   }
   @Override
   public void onDestroy(){
      super.onDestroy();
      //unbind from the service?
   }
}
```

编写文件 Datum.java 实现基准数据处理，具体实现代码如下所示。

```
public class Datum implements Parcelable {
   public float temperature;
   public static final int
    CONTAMINATED_BY_ENGAGEMENT=1 /*that is, if it's being used, Likely to draw more power
and raise battery temp*/,
    CONTAMINATED_BY_CHARGING=2,
    CONTAMINATED_BY_EMPTIES=4, /*this kind of contamination is not really much of a
contamination though, just a roughening. These do not occur in base level datums, just
at the edges of contigious recordings in higher summarylayers*/
    NO_DATA = 5; //means this is a meaningless datum, this only occurs in particular places,
so you shouldn't normally worry about it;
   public int flags;
   public Datum(float temp, int flag){temperature = temp; flags = flag;}

   public int describeContents(){return 0;}
   public void writeToParcel(Parcel out, int flaggings){
      out.writeInt(flags);
      out.writeFloat(temperature);
   }
   public static final Parcelable.Creator<Datum> CREATOR = new Parcelable.Creator
<Datum>() {
```

```
        public Datum createFromParcel(Parcel in){
            float tempor = in.readFloat();
            int flagings = in.readInt();
            return new Datum(tempor, flagings);
        }
        public Datum [] newArray(int size){
            return new Datum[size];
        }
    };
    @Override
    public Datum clone(){
        return new Datum(temperature, flags);
    }
}
```

编写文件 DatumView.java，功能是调用基准数据类 Datum 实现绘制基准视图界面功能，具体实现代码如下所示。

```
class DatumView extends LinearLayout{
    protected Datum v;
    protected TextView tv;
    protected FlagImage fi;
    public static final float defaultHeightInches = (float)0.4;
    protected static BitmapDrawable[] flagBitmaps = new BitmapDrawable[8];
    protected static BitmapDrawable nodatBitmap;
    public static Paint backGroundPaint;
    public static Paint haskPaint;
    public static Paint contaminatedEmpty;
    public static Paint contaminatedUsage;
    public static Paint contaminatedCharging;
    public static Path laef;
    protected class FlagImage extends ImageView{
        public FlagImage(Context cont, AttributeSet attrs, int defStyle){
            super(cont, attrs, defStyle); init(cont); }
        public FlagImage(Context cont, AttributeSet attrs){
            super(cont, attrs); init(cont); }
        public FlagImage(Context cont){
            super(cont); init(cont); }
        protected int flags;
        protected void init(Context cont){
            takeFlag(Datum.NO_DATA);
        }
        protected Bitmap paintBlank(int w, int h){
            Bitmap surf = Bitmap.createBitmap(w, h, Bitmap.Config.ARGB_8888);
            Canvas c = new Canvas(surf);
            c.drawPaint(backGroundPaint);
            //now draw the hask
            float haskrad = (h*8)/10;
            float  xo = w/2 - haskrad,   yo = h/2 - haskrad;
            Path haskPath = new Path();
            RectF arcRect = new RectF(xo,yo,xo + haskrad,yo + haskrad);
            float arcRad = 64; //180 means complete circle;
            haskPath.addArc(arcRect, -45 - arcRad/2, arcRad);
            haskPath.addArc(arcRect, -(45+180) + arcRad/2, -arcRad);
            haskPath.close();
            c.drawPath(haskPath, haskPaint);
            return surf;
        }
        protected Bitmap paintFlags(int w, int h, int flaggings){
            Bitmap surf = Bitmap.createBitmap(w, h, Bitmap.Config.ARGB_8888);
            Canvas c = new Canvas(surf);
            c.drawPaint(backGroundPaint);
            float separation = 2;
            float oneFlagSpan =
                (w <= 5*separation)?
                    0:
                    (h < 2*separation + (w - 5*separation)/3)?
                        h - 2*separation:
                        (w - 5*separation)/3;
```

```
            float fullWidth = oneFlagSpan*3 + 5*separation;
            float fullHeight = oneFlagSpan + 2*separation;
            Path mlaef = new Path(laef);
            Matrix transf = new Matrix(); transf.setRectToRect(
                new RectF(0,0,1,1),
                new RectF((h - fullHeight)/2 + separation,  (w - fullWidth)/2 + separation,
                separation+oneFlagSpan,  separation+oneFlagSpan),
                Matrix.ScaleToFit.CENTER);
            mlaef.transform(transf);
            if((flaggings & Datum.CONTAMINATED_BY_ENGAGEMENT) != 0) c.drawPath(mlaef,
            contaminatedUsage);
            mlaef.offset(oneFlagSpan + separation, 0);
            if((flaggings & Datum.CONTAMINATED_BY_CHARGING) != 0) c.drawPath(mlaef,
            contaminatedCharging);
            mlaef.offset(oneFlagSpan + separation, 0);
            if((flaggings & Datum.CONTAMINATED_BY_EMPTIES) != 0) c.drawPath(mlaef,
            contaminatedEmpty);
            return surf;
        }
        protected void installFlag(){
            if(flags == 0) return;
            if((flags & Datum.NO_DATA) != 0){
                if(nodatBitmap != null) setImageDrawable(nodatBitmap);
                else
                    setImageDrawable(( nodatBitmap = new BitmapDrawable(
                        getContext().getResources(),
                        paintBlank(getWidth(), getHeight())) ));
            }else{
                int flagIndex = flags&(7);
                if(flagBitmaps[flagIndex] != null) setImageDrawable(flagBitmaps[flagIndex]);
                else
                    setImageDrawable(( flagBitmaps[flagIndex] = new BitmapDrawable(
                        getContext().getResources(),
                        paintFlags(getWidth(), getHeight(), flags)) ));
            }
        }
        public void takeFlag(int inFlag){ //also serves as the image update function;
            flags = inFlag;
            if(getHeight() != 0 && getWidth() != 0){
                installFlag();
            }
        }
        @Override
        public void onSizeChanged(int oldw, int oldh, int w, int h){
            for(int i = 0; i < flagBitmaps.length; ++i) flagBitmaps[i] = null;
            nodatBitmap = null;
            takeFlag(flags);
        }
    }
    public DatumView(Context cont, AttributeSet attrs, int defStyle){
        super(cont, attrs, defStyle);
        init(cont);
    }
    public DatumView(Context cont, AttributeSet attrs){
        super(cont, attrs);
        init(cont);
    }
    public DatumView(Context cont){
        super(cont);
        init(cont);
    }
    protected void init(Context cont){
        Log.v("frog", "DatumView created");
        if(backGroundPaint == null){
            backGroundPaint = new Paint();
            backGroundPaint.setColor(0xffb0b0b0);
        }
        if(haskPaint == null){
            haskPaint = new Paint(Paint.ANTI_ALIAS_FLAG);
```

```
            haskPaint.setColor(0xffb0b0b0);
        }
        if(contaminatedCharging == null){
            contaminatedCharging = new Paint(Paint.ANTI_ALIAS_FLAG);
            contaminatedCharging.setColor(0xffffff00);
        }
        if(contaminatedUsage == null){
            contaminatedUsage = new Paint(Paint.ANTI_ALIAS_FLAG);
            contaminatedUsage.setColor(0xff00ff00);
        }
        if(contaminatedEmpty == null){
            contaminatedEmpty = new Paint(Paint.ANTI_ALIAS_FLAG);
            contaminatedEmpty.setColor(0xff00ffff);
        }
        if(laef == null){
            float cornerDeg = (float)0.1; //EG: a value of 1 would make the entire thing
            //an invisible diagonal sliver, 0 would make it a square.
            laef = new Path();
            laef.moveTo(0,cornerDeg);
            laef.lineTo(cornerDeg,0);
            laef.lineTo(1,0);
            laef.lineTo(1,1 - cornerDeg);
            laef.lineTo(1 - cornerDeg,1);
            laef.lineTo(0,1);
            laef.close();
        }
        fi = new FlagImage(cont);
        fi.takeFlag(Datum.NO_DATA);
        tv = new TextView(cont);
        LinearLayout.LayoutParams tvLParams = new LinearLayout.LayoutParams(0,
ViewGroup.LayoutParams.WRAP_CONTENT);    tvLParams.gravity = android.view.Gravity.LEFT;
tvLParams.weight = 1;
        LinearLayout.LayoutParams fiLParams = new LinearLayout.LayoutParams(0, ViewGroup.
LayoutParams.MATCH_PARENT,  0);    tvLParams.gravity  =  android.view.Gravity.RIGHT;
tvLParams.weight = 0;
        addView( tv,  tvLParams );
        addView( fi,  fiLParams );
    }
    protected void installDatum(Datum in){
        v=in;
        if(v == null || (v.flags & Datum.NO_DATA) != 0){
        }else{
            tv.setText(""+v.temperature);
            fi.takeFlag(v.flags);
        }
    }
}
```

编写文件 recording.java，功能是记录测试的数据，并将各个记录的值保存到 DB 数据库中。文件 recording.java 的具体实现代码如下所示。

```
public class recording extends Service{
    public static final long MEASURE_INTERVAL_IN_MILLISECONDS = 1000* 60* 5; //it's a bit
hard-coded /= Lame but I'm sure it'll ever matter;
    protected int currentTemperature;
    protected boolean isCharging;
    protected boolean isBeingUsed;
    protected ScheduledThreadPoolExecutor timer;
    protected ScheduledFuture<?> future;
    protected Runnable recordingTask = new Runnable(){
        @Override public void run(){
            //take our variables and push a protobuf datum to recordingStream;

            //if activity:thermo is bound, pass it the new datum.

        }
    };
    protected BroadcastReceiver batteryInfoReceiver = new BroadcastReceiver(){
        @Override
```

```
        public void onReceive(Context context, Intent intent){
            currentTemperature = intent.getIntExtra(BatteryManager.EXTRA_TEMPERATURE,0);
            isCharging = ((intent.getIntExtra(BatteryManager.EXTRA_STATUS,0) &
            BatteryManager.BATTERY_STATUS_CHARGING) != 0);
        }};
    protected BroadcastReceiver screenIsOnReciever = new BroadcastReceiver(){
        @Override
        public void onReceive(Context context, Intent intent){   isBeingUsed = true;   }};
    protected BroadcastReceiver screenIsOffReciever = new BroadcastReceiver(){
        @Override
        public void onReceive(Context context, Intent intent){ isBeingUsed = false; }};
    @Override public int onStartCommand(Intent intent, int flags, int startId){
        this.registerReceiver(batteryInfoReceiver, new IntentFilter(Intent.ACTION_
        BATTERY_CHANGED));
        this.registerReceiver(screenIsOnReciever,     new   IntentFilter(Intent.ACTION_
SCREEN_ON)); //using SCREEN_ON instead of USER_PRESENT as screen is likely to tax the
battery even without active use. I suppose the variable this binds to is a misnomer;
        this.registerReceiver(screenIsOffReciever,    new   IntentFilter(Intent.ACTION_
SCREEN_OFF));
        timer = new ScheduledThreadPoolExecutor(1);
        //only one.. No action at a distance for me, thanks =J
        return START_STICKY;
    }
    public void startRecording(){
        //open db
        //create db?
        //int initialDelay =
        future = timer.scheduleAtFixedRate(
            recordingTask,
            0,
            MEASURE_INTERVAL_IN_MILLISECONDS,
            java.util.concurrent.TimeUnit.MILLISECONDS );
    }
    public void haltRecording(){
        future.cancel(false); //will this actuall prevent repetition?
        //close db connection;
    }
    @Override
    public IBinder onBind(Intent intent){return null;}
}
```

编写文件 TemperatureDataView.java，功能是将存储记录的数字生成一个图形化的温度数据视图。文件 TemperatureDataView.java 的具体实现代码如下所示。

```
class TemperatureDataView extends ScrollView{
    protected RelativeLayout layout;
    protected View upperView;
    protected static final int upperViewID=0;
    protected View lowerView;
    protected static final int lowerViewID=1;
    protected int upperChild=-1, lowerChild=-1;
    protected int highestID=2; //the highest ID is available;
    protected int dbLength;
    protected boolean seesEnd;
    protected int[] viewsPageCycle; //the ints herein are of layout's indexes. It's a cyclic
    //array, the beginning and end are thought to be at viewsPageEye*pageSize. Index
    //increasing : going down the list.
    protected int viewsPageEye; //tells which page is the 'first' at this time.
    protected int basalRecord;
    //tells which record the base of the viewsPageCycle is looking at.
    protected int pageSize; //unit: datums.
    protected int nPages;
    protected boolean sizeGiven = false;
    protected ExecutorService pool;
    protected Future<ArrayList<Datum>> lowerAccess;
    protected Future<ArrayList<Datum>> upperAccess;
    public static final float defaultItemHeightInches = (float)0.4;
    protected int itemHeight;
    //targetDb
```

```
    protected RelativeLayout.LayoutParams defItemParams(){
        return new RelativeLayout.LayoutParams(ViewGroup.LayoutParams.MATCH_PARENT,
        itemHeight);
    }

    protected class AccessTask implements Callable<ArrayList<Datum>>{
        int first, n;
        Datum defdat = new Datum(0, Datum.CONTAMINATED_BY_ENGAGEMENT | Datum.
        CONTAMINATED_BY_CHARGING | Datum.CONTAMINATED_BY_EMPTIES);
        //DB
        public AccessTask(int firstID, int nItems){ //, DB db){
            first = firstID; n = nItems;
        }
        public ArrayList<Datum> call(){
            ArrayList<Datum> ret = new ArrayList<Datum>(n);
            //get from DB

            for(Datum dat : ret) dat = defdat.clone();
            return ret;
        }
    }

    protected void increaseUpperPadSize(){
        //remove the views in the highest page, adding their size to that of upperView;
        RelativeLayout.LayoutParams newUppersLayoutParams = new RelativeLayout.Layout
Params(ViewGroup.LayoutParams.MATCH_PARENT, upperView.getHeight()+pageSize*itemHeight);
        newUppersLayoutParams.addRule(RelativeLayout.ALIGN_PARENT_TOP);
        layout.updateViewLayout(upperView, newUppersLayoutParams);
    }
    protected void accessLower(){ //get from return value right before scrolling to the
lower page concerned[an asynchronous call is definately unnecessary, but the goal here
is not to ship but to learn];
        if(lowerAccess == null) //[otherwise a task is already assigned]
            lowerAccess = pool.submit(new AccessTask(basalRecord+nPages*pageSize, pageSize));
    }
    protected void finalizeShiftViewDown(){ //call accessLower some time before this is
    needed, then this will be instant;
        ArrayList<Datum> resultList;
        if(lowerAccess == null){
            accessLower();
        }
        try{
            resultList = lowerAccess.get();
        }catch(InterruptedException ex){throw new RuntimeException("db access thread
        interrupted?");}
        catch(ExecutionException ex){throw new RuntimeException("db access thread
        interrupted?");}

        increaseUpperPadSize();
        int viewsPageCycleIter = viewsPageEye*pageSize;
        int end = viewsPageCycleIter+pageSize;
        int previousLayoutID =
            (viewsPageEye == 0)?
                viewsPageCycle[nPages*pageSize - 1]:
                viewsPageCycle[viewsPageEye*pageSize -1];
        Iterator<Datum> iter = resultList.iterator();
        while(viewsPageCycleIter != end){
            layout.removeViewAt(viewsPageCycle[viewsPageCycleIter]);
            int newLayoutID = highestID++;
            viewsPageCycle[viewsPageCycleIter] = newLayoutID;
            RelativeLayout.LayoutParams params = defItemParams();
            params.addRule(RelativeLayout.BELOW, previousLayoutID);
            DatumView dv = new DatumView(getContext());
            dv.installDatum(iter.next());
            layout.addView(dv, newLayoutID, params);
            ++viewsPageCycleIter;
            previousLayoutID = newLayoutID;
        }
        /*//decrease the lower pad size and reattach it to previousLayoutID;
```

```
        //no. size only changes when dbLength changes.
        RelativeLayout.LayoutParams params = new RelativeLayout.LayoutParams(ViewGroup.
        LayoutParams.MATCH_PARENT, lowerView.getHeight()-pageSize*itemHeight);
        params.addRule(RelativeLayout.BELOW, previousLayoutID);
        layout.updateViewLayout(lowerView, params);*/
        ++viewsPageEye; if(viewsPageEye == nPages) viewsPageEye = 0;
        basalRecord+=pageSize;
        lowerAccess = null;
    }
    protected void acceptFirstSizing(){
        pageSize = this.getHeight()/itemHeight;
        nPages = 3; //may add more when the ScrollView expands. Fuck changing the page
        size. Too hard.
        assert pageSize > 0 : "it looks like size isn't assigned before initialization :(.
        just make pagesize big or something";
        viewsPageCycle = new int[nPages*pageSize];
        for(int i = 0; i < viewsPageCycle.length ; ++i) viewsPageCycle[i] = -1;
        viewsPageEye = 0;
        layout = new RelativeLayout(getContext());
        seesEnd = false;
        lowerView = new View(getContext());
        upperView = new View(getContext());
        RelativeLayout.LayoutParams params = new RelativeLayout.LayoutParams(ViewGroup.
        LayoutParams.MATCH_PARENT, 0);
        params.addRule(RelativeLayout.ALIGN_PARENT_TOP);
        layout.addView(upperView, upperViewID, params);
        dbLength = 50; //LIE
        params = new RelativeLayout.LayoutParams(ViewGroup.LayoutParams.MATCH_PARENT,
        dbLength*itemHeight);
        params.addRule(RelativeLayout.ALIGN_PARENT_TOP);
        layout.addView(lowerView, lowerViewID, params);
        addView(
            layout,
            new ViewGroup.LayoutParams(
                ViewGroup.LayoutParams.MATCH_PARENT,
                ViewGroup.LayoutParams.WRAP_CONTENT));
        if(dbLength*itemHeight < getHeight()) seesEnd=true;
        relocate(0);
        scrollTo(0,0);
    }
    protected void init(Context cont){
        pool = Executors.newFixedThreadPool(2);
        {//decide itemHeight:
        android.util.DisplayMetrics metrics = new android.util.DisplayMetrics();
        ((WindowManager)cont.getSystemService(Context.WINDOW_SERVICE)).
        getDefaultDisplay().getMetrics(metrics);
        itemHeight = (int)(defaultItemHeightInches * metrics.densityDpi);
        }
    }
    public TemperatureDataView(Context cont, AttributeSet attrs, int defStyle){
        super(cont, attrs, defStyle);
        init(cont);
    }
    public TemperatureDataView(Context cont, AttributeSet attrs){
        super(cont, attrs);
        init(cont);
    }
    public TemperatureDataView(Context cont){
        super(cont);
        init(cont);
    }
    protected void addDatumBegin(Datum v){
        DatumView view = new DatumView(getContext());
        view.installDatum(v);
        RelativeLayout.LayoutParams params = defItemParams();
        if(upperChild != -1) params.addRule(RelativeLayout.ABOVE, upperChild);
        else params.addRule(RelativeLayout.ALIGN_PARENT_TOP, upperChild);
        upperChild = highestID++;
        layout.addView(view, upperChild, params);
```

```
    }
    protected void addDatumEnd(Datum v){
        DatumView view = new DatumView(getContext());
        view.installDatum(v);
        RelativeLayout.LayoutParams params = defItemParams();
        if(lowerChild != -1) params.addRule(RelativeLayout.BELOW, lowerChild);
        else params.addRule(RelativeLayout.ALIGN_PARENT_BOTTOM, lowerChild);
        lowerChild = highestID++;
        layout.addView(view, lowerChild, params);
    }

    public void notifyFreshDatum(){
        ++dbLength;
        if(seesEnd){
            //get the datum and add it to the list

        }else{
            RelativeLayout.LayoutParams params = new RelativeLayout.LayoutParams
            (ViewGroup.LayoutParams.MATCH_PARENT, dbLength*itemHeight);
            params.addRule(RelativeLayout.ALIGN_PARENT_BOTTOM);
            layout.updateViewLayout(lowerView, params);
        }
    }

    @Override
    public void onSizeChanged(int oldw, int oldh, int w, int h){
        if(!sizeGiven){
            acceptFirstSizing();
            sizeGiven=true;
        }
        //change nPages

    }

    protected void relocate(int y){ //initializes cached Datums for the new location
        if(y < 0) y=0;
        //clear the cache
        for(int i = 0; i < nPages*pageSize; ++i){
            if(viewsPageCycle[i] != -1) layout.removeViewAt(viewsPageCycle[i]);
        }
        viewsPageEye = 0;
        basalRecord = (y/(itemHeight*pageSize) - itemHeight*(pageSize/2))*pageSize;
        highestID = 2;
        //add the new stuff
            //fetch the needed records
        ArrayList<Datum> data = (new AccessTask(basalRecord, nPages*pageSize)).call();
            //figure out whether and to what degree the data stretches before the beginning
            //of the layout[the portion before is ignorable];
        int validStart = (basalRecord < 0)? -basalRecord : 0;
                //change upperView accordingly
        RelativeLayout.LayoutParams upperLayoutParams;
        if(validStart == 0){
            upperLayoutParams = new RelativeLayout.LayoutParams(ViewGroup.LayoutParams.
            MATCH_PARENT, basalRecord*itemHeight);
        }else{
            upperLayoutParams = new RelativeLayout.LayoutParams(ViewGroup.LayoutParams.
            MATCH_PARENT, 0);
        }
        upperLayoutParams.addRule(RelativeLayout.ALIGN_PARENT_TOP);
        layout.updateViewLayout(upperView, upperLayoutParams);
            //dimension it
        int prevID = upperViewID;
        for(; validStart < data.size(); ++validStart){
            Datum cur = data.get(validStart);
            DatumView curv = new DatumView(getContext());
            if(cur != null) curv.installDatum(cur);
            int id = highestID++;
            RelativeLayout.LayoutParams params = defItemParams();
            params.addRule(RelativeLayout.BELOW, prevID);
```

```
            layout.addView(curv, id, params);
            viewsPageCycle[validStart] = id;
            prevID = id;
        }
        //adjust vars
        upperChild = 2;
        lowerChild = prevID;
        if(lowerAccess != null) lowerAccess.cancel(true);
        lowerAccess = null;
        if(upperAccess != null) upperAccess.cancel(true);
        upperAccess = null;
    }

    @Override
    public void scrollTo(int x, int y){
        int loc = layout.getScrollY();
        if(loc < y){ //is scrolling down
            int height = getHeight();
            if(y+height > (basalRecord+(int)(nPages*0.75)*pageSize)*itemHeight){
                if(y+height > (basalRecord+nPages*pageSize)*itemHeight){
                    finalizeShiftViewDown();
                }else{
                    accessLower();
                }
            }
        }
        super.scrollTo(x,y);
    }
    //public void selectDatabase() //pass two file handles, base and summaries. Summaries
    //may be null;
    //public void pushFreshDatum(Datum in) //not exactly sure how I'll do this.
}
```

到此为止，整个实例介绍完毕，在模拟器中的执行效果如图 17-1 所示，在真机中会显示获取的温度值。

▲图 17-1 执行效果

17.5 实战演练——测试温度、湿度、光照和压力

在接下来的实例中，将详细讲解在 Android 设备中将温度、湿度、光照和压力转换为可视化数据，从而读取到上述四种传感器的具体数值。本实例底层基于了 Arduino 开发版，测试 APP 是在 Android 设备中实现的。

实例	功能	源码路径
实例 17-3	将温度、湿度和光照转换为压力传感器节点	\daima\17\HTLPSensorNode

17.5.1 实现 Arduino 文件

编写 Arduino 文件 HTLPSensorNodeArduino.pde，读者在使用此文件时，请务必修改为自己设备的无线网络设置参数，具体实现代码如下所示。

```
#include <WiFly.h>
#include <Wire.h>
```

```
#define BMP085_ADDRESS 0x77  // I2C address of BMP085

//TCP/IP variables
Server server(80);
String requestMessage;

//SHT15 variables
int sht15TempCmd  = 0b00000011;
int sht15HumidCmd = 0b00000101;
int sht15DataPin  = 4;
int sht15ClockPin = 2;
int sht15TemperatureVal;
int sht15HumidityVal;
int ack;

//Photoresistor variables
int lightSensor = 0;
int lightADCReading;
double lightInputVoltage;
double lightResistance;

//BMP085 variables
const unsigned char OSS = 0;  // Oversampling Setting
// Calibration values
int ac1;
int ac2;
int ac3;
unsigned int ac4;
unsigned int ac5;
unsigned int ac6;
int b1;
int b2;
int mb;
int mc;
int md;
// b5 is calculated in bmp085GetTemperature(...), this variable is also used in
bmp085GetPressure(...)
// so ...Temperature(...) must be called before ...Pressure(...).
long b5;
int bmp085Temperature;

//Sensor result variables
int currentTemperatureInF;
int currentTemperatureInC;
int currentHumidityInPercent;
double currentLightInLux;
long currentPressure;

void setup() {
  Serial.begin(9600);
  SpiSerial.begin();

  //exit CMD mode if not already done
  SpiSerial.println("");
  SpiSerial.println("exit");
  delay(1000);

  //set into CMD mode
  SpiSerial.print("$$$");
  delay(1000);

  //set authorization level
  SpiSerial.println("set w a <insert>");
  delay(1000);

  //set passphrase
  SpiSerial.println("set w p <insert>");
  delay(1000);
```

```
  //set localport
  SpiSerial.println("set i l <insert>");
  delay(1000);

  //disable *HELLO* default message on connect
  SpiSerial.println("set comm remote 0");
  delay(1000);

  //join wifi network <ssid>
  SpiSerial.println("join <insert>");
  delay(5000);

  //exit CMD mode
  SpiSerial.println("exit");
  delay(1000);

  Wire.begin();
  bmp085Calibration();
}


void loop() {
  listenForClients();
}

int shiftIn(int dataPin, int clockPin, int numBits)
{
  int ret = 0;
  int i;

  for (i=0; i<numBits; ++i)
  {
    digitalWrite(clockPin, HIGH);
    delay(10);  // I don't know why I need this, but without it I don't get my 8 lsb of
temp
    ret = ret*2 + digitalRead(dataPin);
    digitalWrite(clockPin, LOW);
  }

  return(ret);
}

void sendCommandSHT(int command, int dataPin, int clockPin)
{
  // Transmission Start
  pinMode(dataPin, OUTPUT);
  pinMode(clockPin, OUTPUT);
  digitalWrite(dataPin, HIGH);
  digitalWrite(clockPin, HIGH);
  digitalWrite(dataPin, LOW);
  digitalWrite(clockPin, LOW);
  digitalWrite(clockPin, HIGH);
  digitalWrite(dataPin, HIGH);
  digitalWrite(clockPin, LOW);

  // The command (3 msb are address and must be 000, and last 5 bits are command)
  shiftOut(dataPin, clockPin, MSBFIRST, command);

  // Verify we get the correct ack
  digitalWrite(clockPin, HIGH);
  pinMode(dataPin, INPUT);
  ack = digitalRead(dataPin);
  if (ack != LOW){
  }
  digitalWrite(clockPin, LOW);
  ack = digitalRead(dataPin);
  if (ack != HIGH){
  }
```

```
}

void waitForResultSHT(int dataPin)
{
  int i;

  pinMode(dataPin, INPUT);

  for(i= 0; i < 200; ++i)
  {
    delay(5);
    ack = digitalRead(dataPin);

    if (ack == LOW)
      break;
  }

  if (ack == HIGH){
  }
  //Serial.println("Ack Error 2");
}

int getData16SHT(int dataPin, int clockPin)
{
  int val;

  // Get the most significant bits
  pinMode(dataPin, INPUT);
  pinMode(clockPin, OUTPUT);
  val = shiftIn(dataPin, clockPin, 8);
  val *= 256;

  // Send the required ack
  pinMode(dataPin, OUTPUT);
  digitalWrite(dataPin, HIGH);
  digitalWrite(dataPin, LOW);
  digitalWrite(clockPin, HIGH);
  digitalWrite(clockPin, LOW);

  // Get the least significant bits
  pinMode(dataPin, INPUT);
  val |= shiftIn(dataPin, clockPin, 8);

  return val;
}

void skipCrcSHT(int dataPin, int clockPin)
{
  // Skip acknowledge to end trans (no CRC)
  pinMode(dataPin, OUTPUT);
  pinMode(clockPin, OUTPUT);

  digitalWrite(dataPin, HIGH);
  digitalWrite(clockPin, HIGH);
  digitalWrite(clockPin, LOW);
}
void collectData() {
  bmp085Temperature = bmp085GetTemperature(bmp085ReadUT());
  currentPressure = bmp085GetPressure(bmp085ReadUP());

  sendCommandSHT(sht15TempCmd, sht15DataPin, sht15ClockPin);
  waitForResultSHT(sht15DataPin);
  sht15TemperatureVal = getData16SHT(sht15DataPin, sht15ClockPin);
  skipCrcSHT(sht15DataPin, sht15ClockPin);
  //Get more precise temperature by combining temperature readings of both sensors
  currentTemperatureInF = ((-40.2 + 0.018 * sht15TemperatureVal) + (((bmp085Temperature
/ 10) * 1.8) + 32)) / 2;
  currentTemperatureInC = ((-40.1 + 0.01 * sht15TemperatureVal) + (bmp085Temperature /
```

```
10)) / 2;

  sendCommandSHT(sht15HumidCmd, sht15DataPin, sht15ClockPin);
  waitForResultSHT(sht15DataPin);
  sht15HumidityVal = getData16SHT(sht15DataPin, sht15ClockPin);
  skipCrcSHT(sht15DataPin, sht15ClockPin);
  currentHumidityInPercent = -4.0 + 0.0405 * sht15HumidityVal + -0.0000028 *
sht15HumidityVal * sht15HumidityVal;

  lightADCReading = analogRead(lightSensor);
  // Calculating the voltage of the ADC for light
  lightInputVoltage = 5.0 * (lightADCReading / 1024.0);
  // Calculating the resistance of the photoresistor in the voltage divider
  lightResistance = (10.0 * 5.0) / lightInputVoltage - 10.0;
  // Calculating the intensity of light in lux
  currentLightInLux = 255.84 * pow(lightResistance, -10/9);
}

void listenForClients() {
  // listen for incoming clients
  Client client = server.available();
  if (client) {
    // an http request ends with a blank line
    boolean currentLineIsBlank = true;
    requestMessage = "";
    while (client.connected()) {
      if (client.available()) {
        char c = client.read();
        requestMessage += c;
        // if you've gotten to the end of the line (received a newline
        // character) and the line is blank, the http request has ended,
        // so you can send a reply
        if (c == '\n' && currentLineIsBlank) {
          //Serial.println("request in");
          //evaluate request message and send response
          Serial.println(requestMessage);
          if(requestMessage.indexOf("GET") != -1) {
            collectData();
            if(requestMessage.indexOf("/temperaturef") != -1) {
              client.println("HTTP/1.1 200 OK");
              client.println("Content-Type: application/json");
              client.println("");
              client.print("{\"temperatureInF\":");
              client.print(currentTemperatureInF);
              client.print("}");
            }
            else if(requestMessage.indexOf("/temperaturec") !=- 1) {
              client.println("HTTP/1.1 200 OK");
              client.println("Content-Type: application/json");
              client.println("");
              client.print("{\"temperatureInC\":");
              client.print(currentTemperatureInC);
              client.print("}");
            }
            else if(requestMessage.indexOf("/humidity") !=- 1) {
              client.println("HTTP/1.1 200 OK");
              client.println("Content-Type: application/json");
              client.println("");
              client.print("{\"humidityInPercent\":");
              client.print(currentHumidityInPercent);
              client.print("}");
            }
            else if(requestMessage.indexOf("/light") !=- 1) {
              client.println("HTTP/1.1 200 OK");
              client.println("Content-Type: application/json");
              client.println("");
              client.print("{\"lightInLux\":");
              client.print(currentLightInLux);
              client.print("}");
```

```
          }
          else if(requestMessage.indexOf("/pressure") !=- 1) {
            client.println("HTTP/1.1 200 OK");
            client.println("Content-Type: application/json");
            client.println("");
            client.print("{\"pressureInPa\":");
            client.print(currentPressure);
            client.print("}");
          }
          else if(requestMessage.indexOf("/sensors") !=- 1) {
            Serial.println("collect");
            client.println("HTTP/1.1 200 OK");
            client.println("Content-Type: application/json");
            client.println("");
            client.print("{\"temperatureInF\":");
            client.print(currentTemperatureInF);
            client.print(",\"temperatureInC\":");
            client.print(currentTemperatureInC);
            client.print(",\"humidityInPercent\":");
            client.print(currentHumidityInPercent);
            client.print(",\"lightInLux\":");
            client.print(currentLightInLux);
            client.print(",\"pressureInPa\":");
            client.print(currentPressure);
            client.print("}");
          }
          else {
            client.println("HTTP/1.1 404 Not Found");
            client.println("Content-Type: text/html");
            client.println("");
          }
        }
        else {
          client.println("HTTP/1.1 405 Method Not Allowed");
          client.println("Content-Type: text/html");
          client.println("");
        }
        break;
      }
      if (c == '\n') {
        // you're starting a new line
        currentLineIsBlank = true;
      }
      else if (c != '\r') {
        // you've gotten a character on the current line
        currentLineIsBlank = false;
      }
    }
  }
  // give the web browser time to receive the data
  delay(1);
  // close the connection:
  client.stop();
 }
}

// Stores all of the bmp085's calibration values into global variables
// Calibration values are required to calculate temp and pressure
// This function should be called at the beginning of the program
void bmp085Calibration()
{
  ac1 = bmp085ReadInt(0xAA);
  ac2 = bmp085ReadInt(0xAC);
  ac3 = bmp085ReadInt(0xAE);
  ac4 = bmp085ReadInt(0xB0);
  ac5 = bmp085ReadInt(0xB2);
  ac6 = bmp085ReadInt(0xB4);
  b1 = bmp085ReadInt(0xB6);
  b2 = bmp085ReadInt(0xB8);
```

```
  mb = bmp085ReadInt(0xBA);
  mc = bmp085ReadInt(0xBC);
  md = bmp085ReadInt(0xBE);
}

// Calculate temperature given ut.
// Value returned will be in units of 0.1 deg C
short bmp085GetTemperature(unsigned int ut)
{
  long x1, x2;

  x1 = (((long)ut - (long)ac6)*(long)ac5) >> 15;
  x2 = ((long)mc << 11)/(x1 + md);
  b5 = x1 + x2;

  return ((b5 + 8)>>4);
}

// Calculate pressure given up
// calibration values must be known
// b5 is also required so bmp085GetTemperature(...) must be called first.
// Value returned will be pressure in units of Pa.
long bmp085GetPressure(unsigned long up)
{
  long x1, x2, x3, b3, b6, p;
  unsigned long b4, b7;

  b6 = b5 - 4000;
  // Calculate B3
  x1 = (b2 * (b6 * b6)>>12)>>11;
  x2 = (ac2 * b6)>>11;
  x3 = x1 + x2;
  b3 = (((((long)ac1)*4 + x3)<<OSS) + 2)>>2;

  // Calculate B4
  x1 = (ac3 * b6)>>13;
  x2 = (b1 * ((b6 * b6)>>12))>>16;
  x3 = ((x1 + x2) + 2)>>2;
  b4 = (ac4 * (unsigned long)(x3 + 32768))>>15;

  b7 = ((unsigned long)(up - b3) * (50000>>OSS));
  if (b7 < 0x80000000)
    p = (b7<<1)/b4;
  else
    p = (b7/b4)<<1;

  x1 = (p>>8) * (p>>8);
  x1 = (x1 * 3038)>>16;
  x2 = (-7357 * p)>>16;
  p += (x1 + x2 + 3791)>>4;

  return p;
}

// Read 1 byte from the BMP085 at 'address'
char bmp085Read(unsigned char address)
{
  unsigned char data;

  Wire.beginTransmission(BMP085_ADDRESS);
  Wire.send(address);
  Wire.endTransmission();

  Wire.requestFrom(BMP085_ADDRESS, 1);
  while(!Wire.available())
    ;

  return Wire.receive();
}
```

```
// Read 2 bytes from the BMP085
// First byte will be from 'address'
// Second byte will be from 'address'+1
int bmp085ReadInt(unsigned char address)
{
  unsigned char msb, lsb;

  Wire.beginTransmission(BMP085_ADDRESS);
  Wire.send(address);
  Wire.endTransmission();

  Wire.requestFrom(BMP085_ADDRESS, 2);
  while(Wire.available()<2)
    ;
  msb = Wire.receive();
  lsb = Wire.receive();

  return (int) msb<<8 | lsb;
}

// Read the uncompensated temperature value
unsigned int bmp085ReadUT()
{
  unsigned int ut;

  // Write 0x2E into Register 0xF4
  // This requests a temperature reading
  Wire.beginTransmission(BMP085_ADDRESS);
  Wire.send(0xF4);
  Wire.send(0x2E);
  Wire.endTransmission();

  // Wait at least 4.5ms
  delay(5);

  // Read two bytes from registers 0xF6 and 0xF7
  ut = bmp085ReadInt(0xF6);
  return ut;
}

// Read the uncompensated pressure value
unsigned long bmp085ReadUP()
{
  unsigned char msb, lsb, xlsb;
  unsigned long up = 0;

  // Write 0x34+(OSS<<6) into register 0xF4
  // Request a pressure reading w/ oversampling setting
  Wire.beginTransmission(BMP085_ADDRESS);
  Wire.send(0xF4);
  Wire.send(0x34 + (OSS<<6));
  Wire.endTransmission();

  // Wait for conversion, delay time dependent on OSS
  delay(2 + (3<<OSS));

  // Read register 0xF6 (MSB), 0xF7 (LSB), and 0xF8 (XLSB)
  Wire.beginTransmission(BMP085_ADDRESS);
  Wire.send(0xF6);
  Wire.endTransmission();
  Wire.requestFrom(BMP085_ADDRESS, 3);

  // Wait for data to become available
  while(Wire.available() < 3)
    ;
  msb = Wire.receive();
  lsb = Wire.receive();
  xlsb = Wire.receive();
```

```
    up = (((unsigned long) msb << 16) | ((unsigned long) lsb << 8) | (unsigned long) xlsb)
>> (8-OSS);

    return up;
}
```

17.5.2 实现 Android APP

编写文件 main.xml，功能是通过文本控件显示读取压力、温度、湿度、亮度的数值。文件 main.xml 的具体实现代码如下所示。

```
<LinearLayout xmlns:android="http://schemas.android.com/apk/res/android"
    android:orientation="vertical" android:layout_width="fill_parent"
    android:layout_height="fill_parent" android:background="#ff000000">
    <DigitalClock android:id="@+id/DigitalClock"
        android:background="#ff000000" android:shadowColor="#FFFF00"
        android:shadowDx="2" android:shadowDy="2" android:shadowRadius="2"
        android:textAppearance="@style/textStyle" android:layout_gravity="center"
        android:layout_width="wrap_content" android:layout_height="wrap_content" />
    <LinearLayout android:id="@+id/innerLinearLayoutTemperature"
        android:background="#ff000000" android:orientation="horizontal"
        android:layout_width="wrap_content" android:layout_height="wrap_content">
        <ImageButton android:id="@+id/temperatureButton"
            android:src="@drawable/thermometer" android:background="#ff000000"
            android:layout_width="wrap_content" android:layout_height="wrap_content" />
        <TextView android:id="@+id/temperatureText"
            android:shadowColor="#FFFF00" android:shadowDx="1" android:shadowDy="1"
            android:shadowRadius="1" android:textAppearance="@style/textStyle"
            android:layout_width="wrap_content" android:layout_gravity="center"
            android:layout_height="wrap_content" />
    </LinearLayout>
    <LinearLayout android:id="@+id/innerLinearLayoutHumidity"
        android:background="#ff000000" android:orientation="horizontal"
        android:layout_width="wrap_content" android:layout_height="wrap_content">
        <ImageButton android:id="@+id/humidityButton" android:src="@drawable/humidity"
            android:background="#ff000000" android:layout_height="wrap_content"
            android:layout_width="wrap_content" />
        <TextView android:id="@+id/humidityText" android:shadowColor="#FFFF00"
            android:shadowDx="1" android:shadowDy="1" android:shadowRadius="1"
            android:textAppearance="@style/textStyle" android:layout_width="wrap_content"
            android:layout_gravity="center" android:layout_height="wrap_content" />
    </LinearLayout>
    <LinearLayout android:id="@+id/innerLinearLayoutLight"
        android:background="#ff000000" android:orientation="horizontal"
        android:layout_width="wrap_content" android:layout_height="wrap_content">
        <ImageButton android:id="@+id/lightButton" android:src="@drawable/light"
            android:background="#ff000000" android:layout_height="wrap_content"
            android:layout_width="wrap_content" />
        <TextView android:id="@+id/lightText" android:shadowColor="#FFFF00"
            android:shadowDx="1" android:shadowDy="1" android:shadowRadius="1"
            android:textAppearance="@style/textStyle" android:layout_width="wrap_content"
            android:layout_gravity="center" android:layout_height="wrap_content" />
    </LinearLayout>
    <LinearLayout android:id="@+id/innerLinearLayoutPressure"
        android:background="#ff000000" android:orientation="horizontal"
        android:layout_width="wrap_content" android:layout_height="wrap_content">
        <ImageButton android:id="@+id/pressureButton" android:src="@drawable/pressure"
            android:background="#ff000000" android:layout_height="wrap_content"
            android:layout_width="wrap_content" />
        <TextView android:id="@+id/pressureText" android:shadowColor="#FFFF00"
            android:shadowDx="1" android:shadowDy="1" android:shadowRadius="1"
            android:textAppearance="@style/textStyle" android:layout_width="wrap_content"
            android:layout_gravity="center" android:layout_height="wrap_content" />
    </LinearLayout>
</LinearLayout>
```

编写文件 SensorOverview.java，功能是测试指定目标传感器的温度、湿度、亮度和气压值，并将测试的数据显示出来。文件 SensorOverview.java 的具体实现代码如下所示。

```
public class SensorOverview extends Activity {
    /** Called when the activity is first created. */
    private TextView temperature;
    private TextView humidity;
    private TextView light;
    private TextView pressure;

    @Override
    public void onCreate(Bundle savedInstanceState) {
        super.onCreate(savedInstanceState);
        setContentView(R.layout.main);
        temperature = (TextView) findViewById(R.id.temperatureText);
        humidity = (TextView) findViewById(R.id.humidityText);
        light = (TextView) findViewById(R.id.lightText);
        pressure = (TextView) findViewById(R.id.pressureText);
        new Thread(new Runnable() {

            @Override
            public void run() {
                while (true) {
                    try {
                        HttpParams httpParameters = new BasicHttpParams();
                        HttpConnectionParams.setConnectionTimeout(
                                httpParameters, 3000);
                        HttpConnectionParams.setSoTimeout(httpParameters, 3000);
                        HttpClient hc = new DefaultHttpClient(httpParameters);
                        HttpGet get = new HttpGet(
                                "http://192.168.0.114/sensors/");
                        HttpResponse rp = hc.execute(get);
                        if (rp.getStatusLine().getStatusCode() == HttpStatus.SC_OK) {
                            String jsontext = EntityUtils.toString(rp
                                    .getEntity());
                            JSONObject entries = new JSONObject(jsontext);
                            String temperatureInCValue = entries
                                    .getString("temperatureInC");
                            String temperatureInFValue = entries
                            .getString("temperatureInF");
                            String humidityInPercentValue = entries
                                    .getString("humidityInPercent");
                            String lightInLuxValue = entries.getString("lightInLux");
                            String pressureInPaValue = entries.getString("pressureInPa");
                            Message message = new Message();
                            Bundle bundle = new Bundle();
                            bundle.putString("temperature", temperatureInCValue + "C " +
                            temperatureInFValue + "F");
                            bundle.putString("humidity", humidityInPercentValue + "%");
                            bundle.putString("light", lightInLuxValue + " lux");
                            bundle.putString("pressure", pressureInPaValue + " Pa");
                            message.setData(bundle);
                            handler.sendMessage(message);
                        }
                    } catch (Exception je) {

                    } finally {
                        try {
                            Thread.sleep(5000);
                        } catch (InterruptedException e) {
                            e.printStackTrace();
                        }
                    }
                }

            }
        }).start();
    }

    final Handler handler = new Handler() {
```

```
        @Override
        public void handleMessage(Message myMessage) {
            String temperatureMessage = myMessage.getData().getString(
                    "temperature");
            String humidityMessage = myMessage.getData().getString("humidity");
            String lightMessage = myMessage.getData().getString("light");
            String pressureMessage = myMessage.getData().getString("pressure");
            temperature.setText(temperatureMessage);
            humidity.setText(humidityMessage);
            light.setText(lightMessage);
            pressure.setText(pressureMessage);
        }
    };
}
```

到此为止，整个实例介绍完毕，执行后的效果如图 17-2 所示。

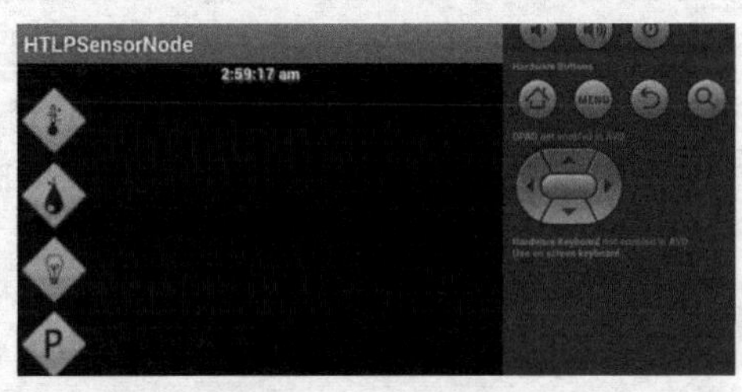

▲图 17-2　执行效果

第 18 章　湿度传感器详解

和测量重量、温度一样，选择湿度传感器首先要确定测量范围。除了气象、科研部门外，搞温度、湿度测控的一般不需要全湿程（0～100%RH）测量。在当今的信息时代，传感器技术与计算机技术、自动控制技术紧密结合着。测量的目的在于控制，测量范围与控制范围合称使用范围。当然，对不需要搞测控系统的应用者来说，直接选择通用型湿度仪就可以了。本章将详细讲解在 Android 设备中使湿度传感器的基本知识，为读者步入本书后面知识的学习打下基础。

18.1 湿度传感器基础

人类的生存和社会活动与湿度密切相关。随着现代化的实现，很难找出一个与湿度无关的领域来。由于应用领域不同，对湿度传感器的技术要求也不同。从制造角度来看，同是湿度传感器，材料、结构不同，工艺不同．其性能和技术指标有很大差异，因而价格也相差甚远。对使用者来说，选择湿度传感器时，首先要搞清楚需要什么样的传感器；自己的财力允许选购什么档次的产品，权衡好"需要与可能"的关系，不至于盲目行事。我们从与用户的来往中，发现有以下几个问题值得注意。

在现在实际应用过程中，生产厂商往往是分段给出其湿度传感器的精度的。如中、低湿段（0～80%RH）为±2%RH，而高湿段（80～100%RH）为±4%RH。而且此精度是在某一指定温度下（如25℃）的值。如果在不同温度下使用湿度传感器，其示值还要考虑温度漂移的影响。众所周知，相对湿度是温度的函数，温度严重地影响着指定空间内的相对湿度。温度每变化 0.1℃。将产生 0.5%RH 的湿度变化（误差）。使用场合如果难以做到恒温，则提出过高的测湿精度是不合适的。因为湿度随着温度的变化也飘忽不定的话，奢谈测湿精度将失去实际意义。所以控湿首先要控好温，这就是大量应用往往是温湿度一体化传感器而不单纯是湿度传感器的缘故。

在大多数情况下，如果没有精确的控温手段，或者被测空间是非密封的，±5%RH 的精度就足够了。对于要求精确控制恒温、恒湿的局部空间，或者需要随时跟踪记录湿度变化的场合，再选用±3%RH 以上精度的湿度传感器。与此相对应的温度传感器．其测温精度须满足±0.3℃以上，起码是±0.5℃的。而精度高于±2%RH 的要求恐怕连校准传感器的标准湿度发生器也难以做到，更何况传感器自身了。

在实际应用中，湿敏元件是最简单的湿度传感器。湿敏元件主要有电阻式、电容式两大类。其中湿敏电阻的特点是在基片上覆盖一层用感湿材料制成的膜，当空气中的水蒸气吸附在感湿膜上时，元件的电阻率和电阻值都发生变化，利用这一特性即可测量湿度。湿敏电容一般是用高分子薄膜电容制成的，常用的高分子材料有聚苯乙烯、聚酰亚胺、酪酸醋酸纤维等。当环境湿度发生改变时，湿敏电容的介电常数发生变化，使其电容量也发生变化，其电容变化量与相对湿度成正比。

电子式湿敏传感器的准确度可达 2～3%RH，这比干湿球测湿精度高。湿敏元件的线性度及抗污染性差，在检测环境湿度时，湿敏元件要长期暴露在待测环境中，很容易被污染而影响其测量精度及长期稳定性。这方面没有干湿球测湿方法好。

18.2 Android 系统中的湿度传感器

在 Android 系统中，湿度传感器的值是 TYPE_RELATIVE_HUMIDITY，单位是%，能够测量周围环境的相对湿度。Android 系统中的湿度与光线、气压、温度传感器的使用方式相同，可以从湿度传感器读取到相对湿度的原始数据。而且，如果设备同时提供了湿度传感器（TYPE_RELATIVE_HUMIDITY）和温度传感器（TYPE_AMBIENT_TEMPERATURE），那么就可以用这两个数据流来计算出结露点和绝对湿度。

（1）结露点

结露点是在固定的气压下，空气中所含的气态水达到饱和而凝结成液态水所需要降至的温度。以下给出了计算结露点温度的公式：

$$t_d(t,RH) = T_n \cdot \frac{\ln(RH/100\%) + m \cdot t/(T_n+t)}{m - [\ln(RH/100\%) + m \cdot t/(T_n+t)]}$$

在上述公式中，各个参数的具体说明如下所示。

- t_d = 结露点温度，单位是摄氏度 C。
- t = 当前温度，单位是摄氏度 C。
- RH = 当前相对湿度，单位是百分比（%）。
- m = 18.62。
- T_n = 243.12　单位是摄氏度 C。

（2）绝对湿度

绝对湿度是在一定体积的干燥空气中含有的水蒸气的质量。绝对湿度的计量单位是克/立方米。以下给出了计算绝对湿度的公式：

$$d_v(t,RH) = 218.7 \cdot \frac{(RH/100\%) \cdot A \cdot \exp(m \cdot t/(T_n+t)}{273.15 + t}$$

在上述公式中，各个参数的具体说明如下所示。

- d_v = 绝对湿度，单位是克/立方米。
- t = 当前温度，单位是摄氏度 C。
- RH = 当前相对湿度，单位是百分比（%）。
- m = 18.62。
- T_n = 243.12 单位是摄氏度 C。
- A = 6.112 hPa。

18.3 实战演练——获取远程湿度传感器的数据

在接下来的实例中，将通过一个演示实例来详细讲解使用湿度传感器的方法，功能是在 Android 设备中获取远程湿度传感器的数据。

实例	功能	源码路径
实例 18-1	获取远程湿度传感器的数据	\daima\18\sensor_app-master

18.3.1　编写布局文件

编写布局文件 activity_thread_async_task_main.xml，在界面中分别设置“Start”和“Stop”两个按钮，在下方使用 ProgressBar 控件显示获取的湿度值。文件 activity_thread_async_task_main.xml 的具体实现代码如下所示。

```
<RelativeLayout xmlns:android="http://schemas.android.com/apk/res/android"
   xmlns:tools="http://schemas.android.com/tools"
   android:layout_width="match_parent"
   android:layout_height="match_parent"
   android:onClick="updateUrl"
   android:paddingBottom="@dimen/activity_vertical_margin"
   android:paddingLeft="@dimen/activity_horizontal_margin"
   android:paddingRight="@dimen/activity_horizontal_margin"
   android:paddingTop="@dimen/activity_vertical_margin"
   tools:context=".ThreadAsyncTaskMainActivity" >

   <TextView
      android:id="@+id/textView1"
      android:layout_width="wrap_content"
      android:layout_height="wrap_content"
      android:text="@string/textView1"
      android:textSize="20sp" />

   <EditText
      android:id="@+id/editText"
      android:layout_width="wrap_content"
      android:layout_height="wrap_content"
      android:visibility="invisible" />

   <TableRow
      android:id="@+id/tableRow"
      android:layout_width="fill_parent"
      android:layout_height="wrap_content"
      android:layout_below="@id/editText" >

      <Button
         android:id="@+id/startButton"
         android:layout_width="fill_parent"
         android:layout_height="wrap_content"
         android:layout_weight="1"
         android:onClick="onClickStart"
         android:text="@string/startButton" />

      <Button
         android:id="@+id/stopButton"
         android:layout_width="fill_parent"
         android:layout_height="wrap_content"
         android:layout_weight="1"
         android:onClick="onClickStop"
         android:text="@string/stopButton" />

      <Button
         android:id="@+id/setButton"
         android:layout_width="fill_parent"
         android:layout_height="wrap_content"
         android:layout_weight="1"
         android:onClick="onClickSet"
         android:text="@string/setButton"
         android:visibility="invisible" />
   </TableRow>

   <ProgressBar
      android:id="@+id/progressBar1"
      style="?android:attr/progressBarStyleHorizontal"
      android:layout_width="wrap_content"
      android:layout_height="wrap_content"
      android:layout_alignLeft="@+id/tableRow"
```

```
        android:layout_alignRight="@+id/tableRow"
        android:layout_below="@+id/tableRow"
        android:layout_marginTop="40dp" />
</RelativeLayout>
```

18.3.2　监听用户触摸单击屏幕控件事件并处理

主 Activity 的实现文件是 ThreadAsyncTaskMainActivity.java，功能是监听用户触摸单击屏幕控件的事件，执行对应的处理方法。文件 ThreadAsyncTaskMainActivity.java 的具体实现代码如下所示。

```
public class ThreadAsyncTaskMainActivity extends Activity {
    private static final String URL_SENSOR = "http://lmi92.cnam.fr/ds2438/ds2438/";
    private Button startButton, stopButton, setButton;
    private ProgressBar progressBar;
    private EditText editText;
    private TextView textView;
    private String url;
    private long top, bot, acquitionTime = 3000; // 3 s en ms
    private WorkAsyncTask task;
    private HumiditySensorAbstract ds2438;

    @Override
    protected void onCreate(Bundle savedInstanceState) {
        super.onCreate(savedInstanceState);
        setContentView(R.layout.activity_thread_async_task_main);
        // 监听主屏幕的控件触摸事件
        startButton = (Button) findViewById(R.id.startButton);
        stopButton = (Button) findViewById(R.id.stopButton);
        setButton = (Button) findViewById(R.id.setButton);
        editText = (EditText) findViewById(R.id.editText);
        progressBar = (ProgressBar) findViewById(R.id.progressBar1);
        textView = (TextView) findViewById(R.id.textView1);
        startButton.setEnabled(true);
        stopButton.setEnabled(false);
        url = ThreadAsyncTaskMainActivity.loadURL_SENSOR(this);

    }

    /*
     * 单击“Start”按钮后调用 WorkAsyncTask()读取湿度
     */
    public void onClickStart(View v) {
        Toast.makeText(this, "Starting...", Toast.LENGTH_SHORT).show();
        task = new WorkAsyncTask();
        task.execute();

    }
    /*
     *单击“Stop”按钮后停止读取湿度工作
     */
    public void onClickStop(View v) {
        Toast.makeText(this, "Stopping...", Toast.LENGTH_SHORT).show();
        task.cancel(true);
        stopButton.setEnabled(false);
        startButton.setEnabled(true);
        textView.setText("Acquisition... Appuyez sur start");
    }
    public void onClickSet(View v) {
        url = editText.getText().toString();
        setButton.setVisibility(View.INVISIBLE);
        startButton.setEnabled(true);
        textView.setVisibility(View.VISIBLE);
        editText.setVisibility(View.INVISIBLE);
        ThreadAsyncTaskMainActivity.saveURL_SENSOR(this, url);
    }

    /*
```

```
     * 更新远程 URL 地址
     */
    public void updateUrl(View v) {
        setButton.setVisibility(View.VISIBLE);
        textView.setVisibility(View.INVISIBLE);
        editText.setVisibility(View.VISIBLE);
        editText.setText(url);
        stopButton.setEnabled(false);
        startButton.setEnabled(false);
    }

    /*
     * 保存远程传感器的 URL
     */
    private static void saveURL_SENSOR(Context context, String url) {
        SharedPreferences prefs = PreferenceManager
                .getDefaultSharedPreferences(context);
        Editor edit = prefs.edit();
        edit.putString("url_sensor", url);
        edit.commit();
    }

    /*
     *载入远程传感器的 URL
     */
    private static String loadURL_SENSOR(Context context) {
        SharedPreferences prefs = PreferenceManager
                .getDefaultSharedPreferences(context);
        return prefs.getString("url_sensor", URL_SENSOR);
    }

    private class WorkAsyncTask extends AsyncTask<Void, Float, Void> {

        @Override
        protected Void doInBackground(Void... params) {
            while (!isCancelled()) {

                try {
                    ds2438 = new HTTPHumiditySensor(url);
                    top = System.currentTimeMillis();
                    float taux = ds2438.value();
                    bot = System.currentTimeMillis();
                    long time = top - bot;
                    publishProgress(taux);
                    SystemClock.sleep(acquitionTime - time);
                } catch (Exception e) {

                    e.printStackTrace();
                }
            }
            return null;
        }

        @Override
        protected void onProgressUpdate(Float... values) {
            Time currentTime = new Time();
            currentTime.setToNow();
            String date = currentTime.format("%d.%m.%Y %H:%M:%S");
            // String date = currentTime.format("%H:%M:%S");
            textView.setText("[" + date + "] ds2438 : " + values[0]);
            progressBar.setProgress(values[0].intValue());
        }

        @Override
        protected void onPreExecute() {
            startButton.setEnabled(false);
            stopButton.setEnabled(true);
        }
```

```
        @Override
        protected void onCancelled() {
            progressBar.setProgress(0);

        }
    }
}
```

18.3.3 设置远程湿度传感器的初始 URL 地址

编写文件 HTTPHumiditySensor.java，功能是设置远程湿度传感器的初始 URL 地址，具体实现代码如下所示。

```
public class HTTPHumiditySensor extends HumiditySensorAbstract {
    // public final static String DEFAULT_HTTP_SENSOR =
    // "http://localhost:8999/ds2438/";
    public final static String DEFAULT_HTTP_SENSOR = "http://10.0.2.2:8999/ds2438/";
    public static final long ONE_MINUTE = 60L * 1000L;

    private static String urlSensor;

    private HTTPHumiditySensor() {
        this(DEFAULT_HTTP_SENSOR);
    }

    public HTTPHumiditySensor(String urlSensor) {
        this.urlSensor = urlSensor;
    }
    public float value() throws Exception {
        StringTokenizer st = new StringTokenizer(request(), "=");
        st.nextToken();
        float f = Float.parseFloat(st.nextToken()) * 10F;
        return ((int) f) / 10F;
    }

    public long minimalPeriod() {
        if (urlSensor.startsWith("http://localhost")
                || urlSensor.startsWith("http://10.0.2.2"))
            return 200L;
        else
            return 2000L; // ONE_MINUTE;
    }

    public String getUrl() {
        return HTTPHumiditySensor.urlSensor;
    }

    private String request() throws Exception {
        URL url = new URL(urlSensor);
        URLConnection connection = url.openConnection();
        BufferedReader in = new BufferedReader(new InputStreamReader(
                connection.getInputStream()), 128);
        String result = new String("");
        String inputLine = in.readLine();
        while (inputLine != null) {
            result = result + inputLine;
            inputLine = in.readLine();
        }
        in.close();
        return result;
        /****/
    }

}
```

到此为止，整个实例介绍完毕，在真机中执行后会获取并显示远程湿度传感器的湿度。

18.4 实战演练——开发一个湿度测试仪

在接下来的实例中，将通过一个演示实例来详细讲解使用湿度传感器开发一个湿度测试仪的方法。本实例可以在 Android 设备上运行，测试当前天气环境下的湿度。

实例	功能	源码路径
实例 18-2	获取远程湿度传感器的数据	\daima\18\humiditytracker-master

18.4.1　实现主界面

编写主界面布局文件 main.xml，功能是通过按钮控件在主界面中分别显示 6 个监测点的值，具体实现代码如下所示。

```
<LinearLayout xmlns:android="http://schemas.android.com/apk/res/android"
    android:orientation="vertical"
    android:layout_width="fill_parent"
    android:layout_height="fill_parent"
    android:padding="10dp"
    >
<Button
    android:id="@+id/reading"
    android:text="@string/takereading"
    android:layout_width="fill_parent"
    android:layout_height="wrap_content"
    android:layout_weight="0"
    style="@style/BigFont"
    android:paddingTop="15dp"
    android:paddingBottom="15dp"
    />
<TableLayout
    android:layout_width="fill_parent"
    android:layout_height="fill_parent"
    android:layout_weight="1"
    android:stretchColumns="0,1"
    >
    <TableRow>
        <Button
            android:id="@+id/stored1"
            android:textColor="@color/stored1"
            style="@style/StoredButton"
            />
        <Button
            android:id="@+id/stored2"
            android:textColor="@color/stored2"
            style="@style/StoredButton"
            />
    </TableRow>
    <TableRow>
        <Button
            android:id="@+id/stored3"
            android:textColor="@color/stored3"
            style="@style/StoredButton"
            />
        <Button
            android:id="@+id/stored4"
            android:textColor="@color/stored4"
            style="@style/StoredButton"
            />
    </TableRow>
    <TableRow>
        <Button
            android:id="@+id/stored5"
            android:textColor="@color/stored5"
            style="@style/StoredButton"
```

```
        />
    <Button
        android:id="@+id/stored6"
        android:textColor="@color/stored6"
        style="@style/StoredButton"
        />
  </TableRow>
</TableLayout>

<TextView
  android:text="@string/copyright"
  android:textSize="12dp"
  android:layout_width="fill_parent"
  android:layout_height="wrap_content"
  />

</LinearLayout>
```

主 Activity 的实现文件是 HumidityActivity.java，功能是监听用户触摸单击了屏幕中的哪一个按钮，并执行对应的处理事件函数。文件 HumidityActivity.java 的具体实现代码如下所示。

```
public class HumidityActivity extends Activity {
  public static final String EXTRA_HUMIDITY = "com.leafdigital.humidity.Humidity";
  public static final String EXTRA_TEMP = "com.leafdigital.humidity.Temp";
  public static final String EXTRA_HUMIDITY_ERROR = "com.leafdigital.humidity.
  HumidityError";
  public static final String EXTRA_TEMP_ERROR = "com.leafdigital.humidity.TempError";
  public static final String EXTRA_SAVEBANK = "com.leafdigital.humidity.SaveBank";

  public final static String PREFS_BANK_NAME = "bankName";

 /** Called when the activity is first created. */
 @Override
 public void onCreate(Bundle savedInstanceState)
 {
  super.onCreate(savedInstanceState);
  setContentView(R.layout.main);
    findViewById(R.id.reading).setOnClickListener(new OnClickListener()
  {
        @Override
        public void onClick(View v)
        {
          startActivityForResult(
              new Intent(HumidityActivity.this, TakeReadingActivity.class), 0);
        }
  });

  int[] ids = { R.id.stored1, R.id.stored2, R.id.stored3,
     R.id.stored4, R.id.stored5, R.id.stored6 };
  for(int i=0; i<6; i++)
  {
     final int storeNum = i;
     Button storedButton = (Button)findViewById(ids[i]);
        storedButton.setOnClickListener(new View.OnClickListener()
        {
          @Override
          public void onClick(View v)
          {
           Intent intent = new Intent(HumidityActivity.this, ShowRecordsActivity. class);
           intent.putExtra(EXTRA_SAVEBANK, storeNum);
           startActivity(intent);
          }
        });
  }
 }

 @Override
 protected void onResume()
 {
  int[] ids = { R.id.stored1, R.id.stored2, R.id.stored3,
```

```
        R.id.stored4, R.id.stored5, R.id.stored6 };
    for(int i=0; i<6; i++)
    {
        Button storedButton = (Button)findViewById(ids[i]);
        storedButton.setText(HumidityActivity.getBankName(this, i));
    }
      super.onResume();
  }

  @Override
  protected void onActivityResult(int requestCode, int resultCode, Intent data)
  {
      // Ignore anything but okay
        if(resultCode != RESULT_OK)
        {
            return;
        }

        if(data != null && data.getIntExtra(EXTRA_SAVEBANK, -1) != -1)
        {
            save(
                data.getIntExtra(EXTRA_SAVEBANK, -1),
                System.currentTimeMillis(),
                data.getDoubleExtra(EXTRA_TEMP, 0.0),
                data.getFloatExtra(EXTRA_TEMP_ERROR, 0.0f),
                data.getDoubleExtra(EXTRA_HUMIDITY, 0.0),
                data.getFloatExtra(EXTRA_HUMIDITY_ERROR, 0.0f));

            Intent intent = new Intent(HumidityActivity.this, ShowRecordsActivity.class);
          intent.putExtra(EXTRA_SAVEBANK, data.getIntExtra(EXTRA_SAVEBANK, -1));
          startActivity(intent);
        }
  }

  private void save(int bank, long time, double temp, float tempError,
      double humidity, float humidityError)
  {
      SQLiteDatabase db = new RecordsOpenHelper(this).getWritableDatabase();
      ContentValues values = new ContentValues();
      values.put("bank", bank);
      values.put("javatime", time);
      values.put("temperature", temp);
      values.put("humidity", humidity);
      values.put("temperature_error", tempError);
      values.put("humidity_error", humidityError);
      db.insert("readings", null, values);
      db.close();
  }

  public static String getBankName(Activity activity, int bank)
  {
      return activity.getSharedPreferences("com.leafdigital.humidity", MODE_PRIVATE).
getString(
          PREFS_BANK_NAME + bank, "" + (bank+1));
  }
}
```

系统主界面的执行效果如图 18-1 所示。

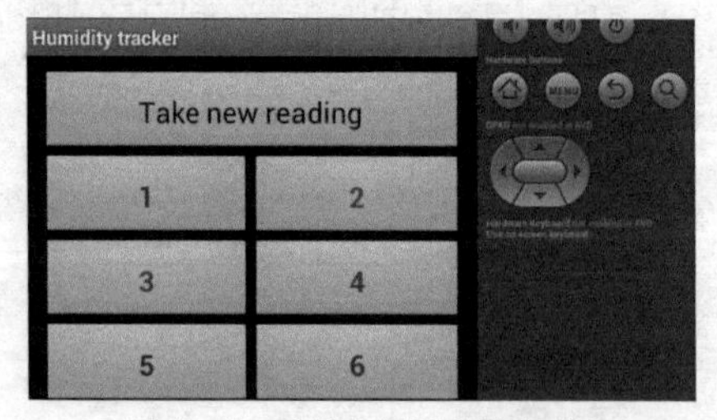

▲图 18-1　系统主界面的执行效果

18.4.2 设置具体值

当单击“Take new reading”按钮后会弹出设置具体值界面，此界面的布局文件是takereading.xml，在界面可以设置温度和湿度值，具体实现代码如下所示。

```
<LinearLayout xmlns:android="http://schemas.android.com/apk/res/android"
    android:orientation="vertical"
    android:layout_width="fill_parent"
    android:layout_height="fill_parent"
    android:padding="10dp"
    >

<TextView
    android:layout_width="fill_parent"
    android:layout_height="wrap_content"
    android:text="@string/temp"
    style="@style/BigFont"
    />

<LinearLayout
    android:orientation="horizontal"
    android:layout_width="fill_parent"
    android:layout_height="wrap_content"
    >

    <EditText
        android:layout_width="wrap_content"
        android:layout_height="wrap_content"
        android:id="@+id/temp"
        android:inputType="numberDecimal"
        android:width="100dp"
        />

    <TextView
        android:layout_width="wrap_content"
        android:layout_height="wrap_content"
        android:paddingLeft="5dp"
        android:text="@string/degreesc"
        style="@style/BigFont"
        />

    <TextView
        android:layout_width="wrap_content"
        android:layout_height="wrap_content"
        android:paddingLeft="10dp"
        android:text="@string/plusminus"
        style="@style/BigFont"
        />

    <EditText
        android:layout_width="wrap_content"
        android:layout_height="wrap_content"
        android:id="@+id/temperror"
        android:inputType="numberDecimal"
        android:width="60dp"
        />

</LinearLayout>

<TextView
    android:layout_width="fill_parent"
    android:layout_height="wrap_content"
    android:text="@string/humidity"
    android:paddingTop="15dp"
    style="@style/BigFont"
    />

<LinearLayout
```

```
    android:layout_width="fill_parent"
    android:layout_height="wrap_content"
    android:orientation="horizontal"
    android:paddingBottom="30dp"
    >

    <EditText
        android:layout_width="wrap_content"
        android:layout_height="wrap_content"
        android:id="@+id/humidity"
        android:inputType="numberDecimal"
        android:width="100dp"
        />

    <TextView
        android:layout_width="wrap_content"
        android:layout_height="wrap_content"
        android:paddingLeft="5dp"
        android:text="@string/percent"
        style="@style/BigFont"
        />

    <TextView
        android:layout_width="wrap_content"
        android:layout_height="wrap_content"
        android:paddingLeft="10dp"
        android:text="@string/plusminus"
        style="@style/BigFont"
        />

    <EditText
        android:layout_width="wrap_content"
        android:layout_height="wrap_content"
        android:id="@+id/humidityerror"
        android:inputType="numberDecimal"
        android:width="60dp"
        />

</LinearLayout>

<Button
    android:layout_width="fill_parent"
    android:layout_height="wrap_content"
    android:id="@+id/ok"
    android:text="@string/ok"
    android:enabled="false"
    style="@style/BigFont"
    />

</LinearLayout>
```

对应的程序文件是 TakeReadingActivity.java，功能是获取用户设置的温度值和湿度值进行保存，具体实现代码如下所示。

```
public class TakeReadingActivity extends Activity
{
    private final static String DECIMAL_REGEX = "[0-9]+(\\.[0-9]+)?";

    private final static String PREFS_TEMP_ERROR = "tempError",
        PREFS_HUMIDITY_ERROR = "humidityError";

    private Button ok;
    private EditText tempEdit, humidityEdit, tempErrorEdit, humidityErrorEdit;

    /**
     * Rounds a value for display. This rounds to two decimal places, but then
     * discards the decimal places if not used, so that it can display results
     * of the form 0.97, 0.9, and 0.
     * @param value Value
```

```
     * @return String
     */
    private static String roundValue(float value)
    {
        String s = String.format("%.2f", value);
        while(s.endsWith("0"))
        {
            s = s.substring(0, s.length() - 1);
        }
        if(s.endsWith("."))
        {
            s = s.substring(0, s.length() - 1);
        }
        return s;
    }

  /** Called when the activity is first created. */
  @Override
  public void onCreate(Bundle savedInstanceState)
  {
    super.onCreate(savedInstanceState);
    setContentView(R.layout.takereading);

    ok = (Button)findViewById(R.id.ok);
    tempEdit = (EditText)findViewById(R.id.temp);
    tempErrorEdit = (EditText)findViewById(R.id.temperror);
    humidityEdit = (EditText)findViewById(R.id.humidity);
    humidityErrorEdit = (EditText)findViewById(R.id.humidityerror);

    // Set default error values
    SharedPreferences prefs = getSharedPreferences("com.leafdigital.humidity",MODE_
PRIVATE);
    tempErrorEdit.setText(roundValue(prefs.getFloat(PREFS_TEMP_ERROR, 1.0f)));
    humidityErrorEdit.setText(roundValue(prefs.getFloat(PREFS_HUMIDITY_ERROR, 4.0f)));

    TextWatcher watcher = new TextWatcher()
    {
            @Override
            public void afterTextChanged(Editable arg0)
            {
                textChanged();
            }

            @Override
            public void beforeTextChanged(CharSequence arg0, int arg1, int arg2,
                int arg3)
            {
            }

            @Override
            public void onTextChanged(CharSequence arg0, int arg1, int arg2, int arg3)
            {
            }
    };
    tempEdit.addTextChangedListener(watcher);
    tempErrorEdit.addTextChangedListener(watcher);
    humidityEdit.addTextChangedListener(watcher);
    humidityErrorEdit.addTextChangedListener(watcher);

    ok.setOnClickListener(new OnClickListener()
        {
            @Override
            public void onClick(View v)
            {
              // User has entered a reading
              double temp = Double.parseDouble(tempEdit.getText().toString());
              double humidity = Double.parseDouble(humidityEdit.getText().toString());

              float tempError = Float.parseFloat(tempErrorEdit.getText().toString());
```

```
            float humidityError=Float.parseFloat(humidityErrorEdit.getText().toString());

            // Update preferences to track the error
            SharedPreferences prefs = getSharedPreferences("com.leafdigital.humidity",
            MODE_PRIVATE);
            if(prefs.getFloat(PREFS_TEMP_ERROR, 1.0f) != tempError
              || prefs.getFloat(PREFS_HUMIDITY_ERROR, 1.0f) != humidityError)
            {
              SharedPreferences.Editor editor = prefs.edit();
              editor.putFloat(PREFS_TEMP_ERROR, tempError);
              editor.putFloat(PREFS_HUMIDITY_ERROR, humidityError);
              editor.commit();
            }

             Intent intent = new Intent(TakeReadingActivity.this, ShowReadingActivity.class);
             intent.putExtra(EXTRA_TEMP, temp);
             intent.putExtra(EXTRA_TEMP_ERROR, tempError);
             intent.putExtra(EXTRA_HUMIDITY, humidity);
             intent.putExtra(EXTRA_HUMIDITY_ERROR, humidityError);

             startActivityForResult(intent, 0);
          }
      });
  }

  @Override
  protected void onActivityResult(int requestCode, int resultCode, Intent data)
  {
    // Pass result data through, and finish
      if(resultCode == RESULT_OK)
      {
        setResult(resultCode, data);
        finish();
      }
  }

  private void textChanged()
  {
    String tempString = tempEdit.getText().toString();
      String tempErrorString = tempErrorEdit.getText().toString();
      String humidityString = humidityEdit.getText().toString();
      String humidityErrorString = humidityErrorEdit.getText().toString();
      ok.setEnabled(
        tempString.matches(DECIMAL_REGEX)
        && tempErrorString.matches(DECIMAL_REGEX)
        && humidityString.matches(DECIMAL_REGEX)
        && humidityErrorString.matches(DECIMAL_REGEX)
        && Double.parseDouble(tempString) <= 40.0
        && Double.parseDouble(tempErrorString) <= 10.0
        && Double.parseDouble(humidityString) <= 100.0
        && Double.parseDouble(humidityString) >= 1.0
        && Double.parseDouble(humidityErrorString) <= 10.0);
  }
}
```

系统设置界面的执行效果如图 18-2 所示。

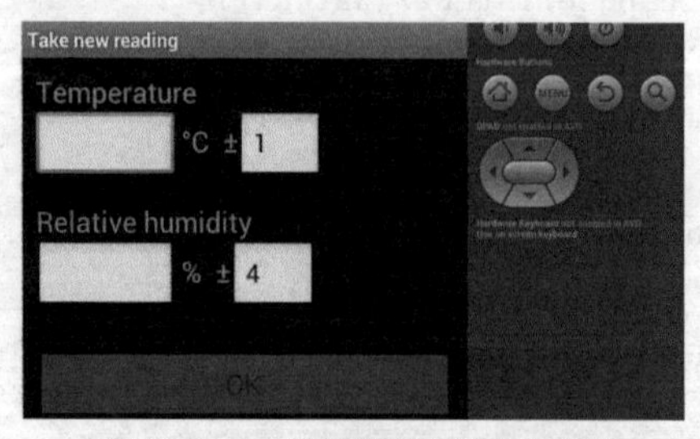

▲图 18-2　系统设置界面的执行效果

18.4.3 显示当前的值

布局文件showreading.xml的功能是，通过文本控件显示当前的温度、湿度、露点（大气中的湿空气由于温度下降，使所含的水蒸气达到饱和状态而开始凝结时的温度）和水密度值。文件showreading.xml的具体实现代码如下所示。

```
<LinearLayout xmlns:android="http://schemas.android.com/apk/res/android"
    android:orientation="vertical"
    android:layout_width="fill_parent"
    android:layout_height="fill_parent"
    android:padding="10dp"
    >

<LinearLayout
    android:orientation="horizontal"
    style="@style/ShowTableRow"
    >
    <TextView
        android:text="@string/temp"
        style="@style/ShowTableRowLabel"
        />
    <TextView
        android:id="@+id/showTempL"
        style="@style/ShowTableRowValue"
        />
    <TextView
        style="@style/ShowTableRowTo"
        />
    <TextView
        android:id="@+id/showTempH"
        style="@style/ShowTableRowValue2"
        />
    <TextView
        android:text="@string/degreesc"
        style="@style/ShowTableRowUnit"
        />
</LinearLayout>

<LinearLayout
    android:orientation="horizontal"
    style="@style/ShowTableRow"
    >
    <TextView
        android:text="@string/humidity"
        style="@style/ShowTableRowLabel"
        />
    <TextView
        android:id="@+id/showHumidityL"
        style="@style/ShowTableRowValue"
        />
    <TextView
        style="@style/ShowTableRowTo"
        />
    <TextView
        android:id="@+id/showHumidityH"
        style="@style/ShowTableRowValue2"
        />
    <TextView
        android:text="@string/percent"
        style="@style/ShowTableRowUnit"
        />
</LinearLayout>

<LinearLayout
    android:orientation="horizontal"
    style="@style/ShowTableRow"
    >
    <TextView
```

```
        android:text="@string/dewpoint"
        style="@style/ShowTableRowLabel"
        />
    <TextView
        android:id="@+id/showDewpointL"
        style="@style/ShowTableRowValue"
        />
    <TextView
        style="@style/ShowTableRowTo"
        />
    <TextView
        android:id="@+id/showDewpointH"
        style="@style/ShowTableRowValue2"
        />
    <TextView
        android:text="@string/degreesc"
        style="@style/ShowTableRowUnit"
        />
</LinearLayout>

<LinearLayout
    android:orientation="horizontal"
    style="@style/ShowTableRow"
    >
    <TextView
        android:text="@string/density"
        style="@style/ShowTableRowLabel"
        />
    <TextView
        android:id="@+id/showDensityL"
        style="@style/ShowTableRowValue"
        />
    <TextView
        style="@style/ShowTableRowTo"
        />
    <TextView
        android:id="@+id/showDensityH"
        style="@style/ShowTableRowValue2"
        />
    <TextView
        android:text="@string/gm3"
        style="@style/ShowTableRowUnit"
        />
</LinearLayout>

<LinearLayout
    android:paddingTop="15dp"
    android:layout_width="fill_parent"
    android:layout_height="wrap_content"
    >
    <Button
        android:layout_width="fill_parent"
        android:layout_height="wrap_content"
        android:id="@+id/ok"
        android:text="@string/ok"
        style="@style/BigFont"
        android:layout_weight="1"
        />
    <Button
        android:layout_width="fill_parent"
        android:layout_height="wrap_content"
        android:id="@+id/save"
        android:text="@string/save"
        style="@style/BigFont"
        android:layout_weight="1"
        />
</LinearLayout>
</LinearLayout>
```

程序文件 ShowReadingActivity.java 的功能是获取并显示当前温度、湿度、露点和水密度的值，具体实现代码如下所示。

```
public class ShowReadingActivity extends Activity
{

    /** Called when the activity is first created. */
    @Override
    public void onCreate(Bundle savedInstanceState)
    {
    super.onCreate(savedInstanceState);
    setContentView(R.layout.showreading);

    double temp = getIntent().getDoubleExtra(EXTRA_TEMP, 0.0);
    float tempError = getIntent().getFloatExtra(EXTRA_TEMP_ERROR, 0.0f);
    double humidity = getIntent().getDoubleExtra(EXTRA_HUMIDITY, 0.0);
    float humidityError = getIntent().getFloatExtra(EXTRA_HUMIDITY_ERROR, 0.0f);

    ClimateData data = new ClimateData(temp, tempError, humidity, humidityError);

        FieldValue value = data.getField(FIELD_TEMPERATURE);
    ((TextView)findViewById(R.id.showTempL)).setText(
        String.format("%.1f", value.getMin()));
    ((TextView)findViewById(R.id.showTempH)).setText(
        String.format("%.1f", value.getMax()));

        value = data.getField(FIELD_HUMIDITY);
    ((TextView)findViewById(R.id.showHumidityL)).setText(
        String.format("%.1f", value.getMin()));
    ((TextView)findViewById(R.id.showHumidityH)).setText(
        String.format("%.1f", value.getMax()));

        value = data.getField(FIELD_DEWPOINT);
    ((TextView)findViewById(R.id.showDewpointL)).setText(
        String.format("%.1f", value.getMin()));
    ((TextView)findViewById(R.id.showDewpointH)).setText(
        String.format("%.1f", value.getMax()));

        value = data.getField(FIELD_DENSITY);
    ((TextView)findViewById(R.id.showDensityL)).setText(
        String.format("%.2f", value.getMin()));
    ((TextView)findViewById(R.id.showDensityH)).setText(
        String.format("%.2f", value.getMax()));

    findViewById(R.id.ok).setOnClickListener(new OnClickListener()
        {
            @Override
            public void onClick(View v)
            {
              Intent intent = new Intent();
              intent.putExtra(EXTRA_SAVEBANK, -1);
                setResult(RESULT_OK, intent);
                finish();
            }
        });

    findViewById(R.id.save).setOnClickListener(new OnClickListener()
        {
            @Override
            public void onClick(View v)
            {
              // User has selected to save the reading
              Intent intent=new Intent(ShowReadingActivity.this,SaveReadingActivity.class);
              intent.putExtra(EXTRA_TEMP, getIntent().getDoubleExtra(EXTRA_TEMP, 0.0));
              intent.putExtra(EXTRA_TEMP_ERROR, getIntent().getFloatExtra(EXTRA_TEMP_
              ERROR, 0.0f));
              intent.putExtra(EXTRA_HUMIDITY,getIntent().getDoubleExtra(EXTRA_HUMIDITY,0.0));
```

```
            intent.putExtra(EXTRA_HUMIDITY_ERROR, getIntent().getFloatExtra(EXTRA_
            HUMIDITY_ERROR, 0.0f));
            startActivityForResult(intent, 0);
          }
      });
    }

    @Override
    protected void onActivityResult(int requestCode, int resultCode, Intent data)
    {
      // Pass result data through, and finish
        if(resultCode == RESULT_OK)
        {
          setResult(resultCode, data);
          finish();
        }
    }
}
```

系统显示当前值界面的执行效果如图 18-3 所示。

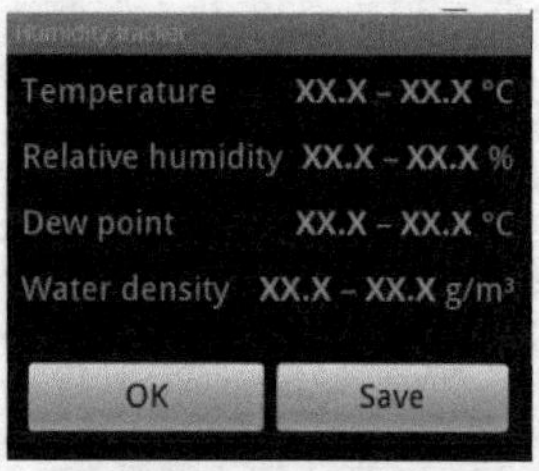

▲图 18-3 显示当前值界面的执行效果

18.4.4 保存当前数值

如果在当前值界面中单击“Save”按钮，系统会保存当前的测试数值。布局文件 savereading.xml 的功能是保存系统当前测试的湿度和温度值，具体实现代码如下所示。

```
<LinearLayout xmlns:android="http://schemas.android.com/apk/res/android"
    android:orientation="vertical"
    android:layout_width="fill_parent"
    android:layout_height="fill_parent"
    android:padding="10dp"
    >
<TableLayout
    android:layout_width="fill_parent"
    android:layout_height="fill_parent"
    android:layout_weight="1"
    android:stretchColumns="0,1"
    >
    <TableRow>
        <Button
            android:id="@+id/stored1"
            android:textColor="@color/stored1"
            style="@style/StoredButton"
            />
        <Button
            android:id="@+id/stored2"
            android:textColor="@color/stored2"
            style="@style/StoredButton"
            />
    </TableRow>
    <TableRow>
        <Button
            android:id="@+id/stored3"
            android:textColor="@color/stored3"
            style="@style/StoredButton"
            />
        <Button
            android:id="@+id/stored4"
            android:textColor="@color/stored4"
            style="@style/StoredButton"
            />
    </TableRow>
    <TableRow>
        <Button
            android:id="@+id/stored5"
            android:textColor="@color/stored5"
            style="@style/StoredButton"
            />
        <Button
```

```
            android:id="@+id/stored6"
            android:textColor="@color/stored6"
            style="@style/StoredButton"
            />
    </TableRow>
</TableLayout>
</LinearLayout>
```

程序文件 SaveReadingActivity.java 的功能是保存当前的值到数据库中，具体实现代码如下所示。

```
public class SaveReadingActivity extends Activity
{
    /** Called when the activity is first created. */
    @Override
    public void onCreate(Bundle savedInstanceState)
    {
        super.onCreate(savedInstanceState);
    setContentView(R.layout.savereading);

    int[] ids = { R.id.stored1, R.id.stored2, R.id.stored3,
        R.id.stored4, R.id.stored5, R.id.stored6 };
    for(int i=0; i<6; i++)
    {
        final int storeNum = i;
        Button storedButton = (Button)findViewById(ids[i]);
        storedButton.setText(HumidityActivity.getBankName(this, i));
            storedButton.setOnClickListener(new View.OnClickListener()
            {
                @Override
                public void onClick(View v)
                {
                  Intent intent = new Intent();
                  intent.putExtra(EXTRA_TEMP, getIntent().getDoubleExtra(EXTRA_TEMP, 0.0));
                  intent.putExtra(EXTRA_TEMP_ERROR, getIntent().getFloatExtra(EXTRA_
                  TEMP_ERROR, 0.0f));
                  intent.putExtra(EXTRA_HUMIDITY, getIntent().getDoubleExtra(EXTRA_
                  HUMIDITY, 0.0));
                  intent.putExtra(EXTRA_HUMIDITY_ERROR, getIntent().getFloatExtra(EXTRA_
                  HUMIDITY_ERROR, 0.0f));
                  intent.putExtra(EXTRA_SAVEBANK, storeNum);
                    setResult(RESULT_OK, intent);
                    finish();
                }
            });
    }
    }
}
```

18.4.5　图形化显示测试结果

布局文件 showrecords.xml 的功能是以图形化的方式显示当前的测试结果，具体实现代码如下所示。

```
<LinearLayout xmlns:android="http://schemas.android.com/apk/res/android"
    android:id="@+id/layout"
    android:orientation="vertical"
    android:layout_width="fill_parent"
    android:layout_height="fill_parent"
    android:padding="0dp"
    >
    <!-- Graph component added programatically -->
</LinearLayout>
```

程序文件 ShowRecordsActivity.java 的功能是监听用户的操作事件，根据用户的选项绘制图形化界面，以图形化方式显示当前的检测结果值。文件 ShowRecordsActivity.java 的具体实现代码如下所示。

```
public class ShowRecordsActivity extends Activity
{
    private int bank;
    private DataPoint[] data;
    private RecordsGraph graph;
    private int graphField;

    private int nearestPointIndex;

    private final static String PREFS_CURRENT_GRAPH = "currentGraph";

    public final static float[] MAX_RANGE = { 40.0f, 100.0f, 40.0f, 50.0f };
    public final static float[] RANGE_STEP = { 5.0f, 5.0f, 5.0f, 5.0f };
    public final static int[] FIELD_LABEL = { R.string.temp, R.string.humidity,
        R.string.dewpoint, R.string.density };
    public final static int[] FIELD_UNIT = { R.string.degreesc, R.string.percent,
        R.string.degreesc, R.string.gm3 };

    private final static float HIT_RADIUS = 48;
    private final static float GRAPH_PADDING = 10;

    private final static long ONE_DAY = 24L * 3600L * 1000L;

    private final static int DIALOG_POINT=1, DIALOG_NAME=2, DIALOG_DELETE=3;

    private static class DataPoint
    {
        long javaTime;
        ClimateData data;

        private DataPoint(long javaTime, ClimateData data)
        {
            this.javaTime = javaTime;
            this.data = data;
        }
    }

    private class RecordsGraph extends View
    {
        private float xScale, yScale, left, bottom, density;
        private long startTime;

        public RecordsGraph()
        {
            super(ShowRecordsActivity.this);
            setLayoutParams(new LinearLayout.LayoutParams(
                LayoutParams.FILL_PARENT, LayoutParams.FILL_PARENT, 1.0f));
        }

        @Override
        public boolean onTouchEvent(MotionEvent event)
        {
            switch(event.getAction())
            {
            case MotionEvent.ACTION_DOWN:
            case MotionEvent.ACTION_MOVE:
                updateNearestPoint(event.getX(), event.getY());
                return true;
            case MotionEvent.ACTION_UP:
                updateNearestPoint(event.getX(), event.getY());
                if(nearestPointIndex != -1)
                {
                    showDialog(DIALOG_POINT);
                }
                nearestPointIndex = -1;
                invalidate();
                return true;
            }
            return false;
```

```
}

private void updateNearestPoint(float x, float y)
{
    // Crap algorithm, but let's just try all the points within any kind of
    // range.
    float radius = HIT_RADIUS * density;
    float bestDiff = radius * radius;
    int bestIndex = -1;
    for(int i=0; i<data.length; i++)
    {
        // Get X point and check if within range
        float pointX = (float)(data[i].javaTime - startTime) * xScale + left;
        if (pointX > x + radius)
        {
            break;
        }
        if (pointX < x - radius)
        {
            continue;
        }

        // Find Y point
        FieldValue field = data[i].data.getField(graphField);
        float pointY = bottom - field.getMeasured() * yScale;

        // Work out difference (= distance squared)
        float diff = (pointY - y) * (pointY - y) + (pointX - x) * (pointX - x);
        if (diff < bestDiff)
        {
            bestDiff = diff;
            bestIndex = i;
        }
    }

    if(bestIndex != nearestPointIndex)
    {
        nearestPointIndex = bestIndex;
        invalidate();
    }
}

@Override
protected void onDraw(Canvas canvas)
{
    DisplayMetrics metrics = new DisplayMetrics();
    getWindowManager().getDefaultDisplay().getMetrics(metrics);
    density = metrics.density;
    float blobRadius = 2f * density;
    float padding = GRAPH_PADDING * density;

    // Clear to off-black
    canvas.drawColor(Color.argb(255, 30, 30, 30));

    // Get available width and height for graph
    float w = getWidth() - padding*2, h = getHeight() - padding*2;
    left = padding;
    bottom = getHeight() - padding;

    // Draw axes in grey
    Paint grey = new Paint(Paint.ANTI_ALIAS_FLAG);
    grey.setColor(Color.argb(255, 128, 128, 128));
    canvas.drawLine(left, bottom, left+w, bottom, grey);
    canvas.drawLine(left, bottom, left, bottom-h, grey);
    canvas.drawLine(left, bottom-h, left+ 1.5f*blobRadius, bottom-h, grey);

    // Work out Y scaling
    float actualMax = 0;
    for(int i=0; i<data.length; i++)
```

```
        {
            float value = data[i].data.getField(graphField).getMax();
            if(value > actualMax)
            {
                actualMax = value;
            }
        }
        float range = MAX_RANGE[graphField];
        if(data.length == 0)
        {
            actualMax = range;
        }
        while(range - RANGE_STEP[graphField] > actualMax)
        {
            range -= RANGE_STEP[graphField];
        }

        yScale = h / range;

        // Draw scale text
        String max = String.format("%.0f", range);
        grey.setTextSize(12f * density);
        Rect bounds = new Rect();
        grey.getTextBounds(max, 0, max.length(), bounds);
        canvas.drawText(max, left + blobRadius * 2 * density,
            bottom - h - bounds.top, grey);

        // If there is no data, stop here
        if(data.length == 0)
        {
            return;
        }

        // Work out X scaling
        startTime = data[0].javaTime;
        long timeRange = data[data.length - 1].javaTime - startTime;
        if(timeRange == 0)
        {
            timeRange = 1;
        }
        xScale = w / (float)timeRange;

        // Draw each point
        Paint white = new Paint(Paint.ANTI_ALIAS_FLAG);
        white.setColor(Color.argb(255, 255, 255, 255));
        Paint faintLine = new Paint(Paint.ANTI_ALIAS_FLAG);
        faintLine.setColor(Color.argb(160, 0, 255, 0));
        for(int i=0; i<data.length; i++)
        {
            float x = (float)(data[i].javaTime - startTime) * xScale;
            FieldValue field = data[i].data.getField(graphField);
            float yMin = field.getMin() * yScale, yMax = field.getMax() * yScale;
            float y = field.getMeasured() * yScale;
            if (i==nearestPointIndex)
            {
                Paint faint = new Paint(Paint.ANTI_ALIAS_FLAG);
                faint.setColor(Color.argb(64,255,255,255));
                canvas.drawCircle(x + left, bottom - y, blobRadius*20, faint);
            }

            canvas.drawLine(x + left, bottom - yMin, x + left, bottom - yMax, faintLine);
            canvas.drawCircle(x + left, bottom - y, blobRadius, white);
        }
    }

    @Override
    protected void onMeasure(int widthMeasureSpec, int heightMeasureSpec)
    {
        int w;
```

```
        if(MeasureSpec.getMode(widthMeasureSpec) == MeasureSpec.UNSPECIFIED)
        {
            w = 100;
        }
        else
        {
            w = MeasureSpec.getSize(widthMeasureSpec);
        }

        int h;
        if(MeasureSpec.getMode(heightMeasureSpec) == MeasureSpec.UNSPECIFIED)
        {
            h = 100;
        }
        else
        {
            h = MeasureSpec.getSize(heightMeasureSpec);
        }
        setMeasuredDimension(w, h);
    }
}

/** Called when the activity is first created. */
@Override
public void onCreate(Bundle savedInstanceState)
{
    super.onCreate(savedInstanceState);

    // Get current bank and set title
    bank = getIntent().getIntExtra(HumidityActivity.EXTRA_SAVEBANK, -1);

    loadData();

setContentView(R.layout.showrecords);
LinearLayout layout = (LinearLayout)findViewById(R.id.layout);

graph = new RecordsGraph();
layout.addView(graph, 0);

// Get default field
setField(getSharedPreferences("com.leafdigital.humidity", MODE_PRIVATE).getInt(
    PREFS_CURRENT_GRAPH, FIELD_HUMIDITY));
}

private void loadData()
{
    // Load records. (Note: This uses a writable database because it might need
    // to do a db version update, so it has to be able to write)
    SQLiteDatabase db = new RecordsOpenHelper(this).getWritableDatabase();
    Cursor cursor = db.query(
        "readings", new String[] { "javatime", "temperature", "temperature_error",
            "humidity", "humidity_error" },
        "bank = ?", new String[] { bank + "" }, null, null, "javatime");
    int count = cursor.getCount();
    data = new DataPoint[count];
    for(int i=0; i<count; i++)
    {
        cursor.moveToNext();
        data[i] = new DataPoint(cursor.getLong(0), new ClimateData(cursor.
        getFloat(1), cursor.getFloat(2),cursor.getFloat(3),cursor.getFloat(4)));
    }
    cursor.close();
    db.close();

    // Clear other fields that depend on data
    nearestPointIndex = -1;
}

private void setField(int field)
```

```
{
    this.graphField = field;
    setTitle(HumidityActivity.getBankName(this, bank) + " - " + getString(FIELD_LABEL
    [field]) + " (" + getString(FIELD_UNIT[field]) + ")");
    graph.invalidate();

    SharedPreferences prefs = getSharedPreferences("com.leafdigital.humidity",
    MODE_PRIVATE);
    if(prefs.getInt(PREFS_CURRENT_GRAPH, FIELD_HUMIDITY) != field)
    {
        if(field == FIELD_HUMIDITY)
        {
            prefs.edit().remove(PREFS_CURRENT_GRAPH).commit();
        }
        else
        {
            prefs.edit().putInt(PREFS_CURRENT_GRAPH, field).commit();
        }
    }
}

@Override
public boolean onCreateOptionsMenu(Menu menu)
{
    MenuInflater inflater = getMenuInflater();
    inflater.inflate(R.menu.showrecords, menu);
    return true;
}

@Override
public boolean onPrepareOptionsMenu(Menu menu)
{
    int id = -1;
    switch(graphField)
    {
    case FIELD_TEMPERATURE :
        id = R.id.temp;
        break;
    case FIELD_HUMIDITY :
        id = R.id.humidity;
        break;
    case FIELD_DEWPOINT :
        id = R.id.dewpoint;
        break;
    case FIELD_DENSITY :
        id = R.id.density;
        break;
    }
    menu.findItem(id).setChecked(true);

    menu.findItem(R.id.delete).setEnabled(data.length != 0);
    long now = System.currentTimeMillis();
    menu.findItem(R.id.deletebefore1d).setEnabled(data.length>0
        && now - data[0].javaTime > 1 * ONE_DAY);
    menu.findItem(R.id.deletebefore7d).setEnabled(data.length>0
        && now - data[0].javaTime > 7 * ONE_DAY);
    menu.findItem(R.id.deletebefore30d).setEnabled(data.length>0
        && now - data[0].javaTime > 30 * ONE_DAY);

    return super.onPrepareOptionsMenu(menu);
}

@Override
public boolean onOptionsItemSelected(MenuItem item)
{
    switch(item.getItemId())
    {
    case R.id.temp :
        setField(FIELD_TEMPERATURE);
```

```
            return true;
        case R.id.humidity :
            setField(FIELD_HUMIDITY);
            return true;
        case R.id.dewpoint :
            setField(FIELD_DEWPOINT);
            return true;
        case R.id.density :
            setField(FIELD_DENSITY);
            return true;
        case R.id.name :
            showDialog(DIALOG_NAME);
            return true;
        case R.id.deletelast :
            deleteData(data[data.length-1].javaTime-1, getString(R.string.deletelast),
            true);
            return true;
        case R.id.deletebefore1d :
            deleteData(System.currentTimeMillis() - ONE_DAY, getString(R.string.
            deletebefore1d), false);
            return true;
        case R.id.deletebefore7d :
            deleteData(System.currentTimeMillis() - 7 * ONE_DAY, getString(R.string.
            deletebefore7d), false);
            return true;
        case R.id.deletebefore30d :
            deleteData(System.currentTimeMillis() - 30 * ONE_DAY, getString(R.string.
            deletebefore30d), false);
            return true;
        case R.id.deleteall :
            deleteData(System.currentTimeMillis() + 1, getString(R.string.deleteall), false);
            return true;
        default:
            return super.onOptionsItemSelected(item);
        }
    }

    @Override
    protected Dialog onCreateDialog(int id)
    {
        // Create dialog
        final Dialog dialog = new Dialog(this);
        switch(id)
        {
        case DIALOG_POINT :
            dialog.setContentView(R.layout.pointdialog);
            View ok = dialog.findViewById(R.id.ok);
            ok.setOnClickListener(new OnClickListener()
            {
                @Override
                public void onClick(View v)
                {
                    dialog.cancel();
                }
            });
            break;

        case DIALOG_NAME :
            dialog.setContentView(R.layout.namedialog);
            dialog.setTitle(R.string.name);
            final EditText edit = (EditText)dialog.findViewById(R.id.name);
            final View okName = dialog.findViewById(R.id.ok);
            okName.setOnClickListener(new OnClickListener()
            {
                @Override
                public void onClick(View v)
                {
                    String name = edit.getText().toString();
                    SharedPreferences.Editor editor = getSharedPreferences("com.
```

```
                leafdigital.humidity",MODE_PRIVATE).edit();
                editor.putString(PREFS_BANK_NAME + bank, name);
                editor.commit();
                dialog.dismiss();
                setField(graphField);
            }
        });
        edit.addTextChangedListener(new TextWatcher()
        {
            @Override
            public void onTextChanged(CharSequence s, int start, int before, int count)
            {
            }

            @Override
            public void beforeTextChanged(CharSequence s, int start, int count, int after)
            {
            }

            @Override
            public void afterTextChanged(Editable s)
            {
                int length = edit.getText().length();
                okName.setEnabled(length <= 6 && length >= 1);
            }
        });
        break;

    case DIALOG_DELETE:
        AlertDialog.Builder builder = new AlertDialog.Builder(this);
        builder.setMessage(getString(R.string.confirmdelete));
        builder.setPositiveButton(getString(R.string.delete),
            new DialogInterface.OnClickListener()
            {
                @Override
                public void onClick(DialogInterface dialog, int which)
                {
                    confirmDelete();
                    dialog.dismiss();
                }
            });
        builder.setNegativeButton(getString(R.string.cancel),
            new DialogInterface.OnClickListener()
        {
            @Override
            public void onClick(DialogInterface dialog, int which)
            {
                dialog.cancel();
            }
        });
        builder.setTitle("Placeholder");
        return builder.create();
    }
    return dialog;
}

@Override
protected void onPrepareDialog(int id, Dialog dialog)
{
    switch(id)
    {
    case DIALOG_POINT :
        preparePointDialog(dialog);
        break;
    case DIALOG_NAME :
        prepareNameDialog(dialog);
        break;
    case DIALOG_DELETE :
        prepareDeleteDialog(dialog);
```

```
            break;
        }
    }

    private void preparePointDialog(Dialog dialog)
    {
        // Get data
        DataPoint point = data[nearestPointIndex];
        FieldValue field = point.data.getField(graphField);

        // Set title to date
        SimpleDateFormat date = new SimpleDateFormat("yyyy-MM-dd HH:mm");
        dialog.setTitle(date.format(new Date(point.javaTime)));

        // Set icon
        ImageView image = (ImageView)dialog.findViewById(R.id.icon);
        int icon = -1;
        switch(graphField)
        {
        case FIELD_TEMPERATURE:
            icon = R.drawable.menu_temp;
            break;
        case FIELD_HUMIDITY:
            icon = R.drawable.menu_humidity;
            break;
        case FIELD_DEWPOINT:
            icon = R.drawable.menu_dewpoint;
            break;
        case FIELD_DENSITY:
            icon = R.drawable.menu_density;
            break;
        }
        image.setImageResource(icon);

        // Set units
        for(int unitId : new int[] { R.id.unit1, R.id.unit2, R.id.unit3 })
        {
            ((TextView)dialog.findViewById(unitId)).setText(getString(FIELD_UNIT
            [graphField]));
        }

        // Set actual values
        String format = graphField == FIELD_DENSITY ? "%.2f" : "%.1f";
        ((TextView)dialog.findViewById(R.id.minValue)).setText(
            String.format(format, field.getMin()));
        ((TextView)dialog.findViewById(R.id.estValue)).setText(
            String.format(format, field.getMeasured()));
        ((TextView)dialog.findViewById(R.id.maxValue)).setText(
            String.format(format, field.getMax()));
    }

    private void prepareNameDialog(Dialog dialog)
    {
        ((EditText)dialog.findViewById(R.id.name)).setText(
            HumidityActivity.getBankName(this, bank));
        ((Button)dialog.findViewById(R.id.ok)).setEnabled(true);
    }

    private void prepareDeleteDialog(Dialog dialog)
    {
        dialog.setTitle(deleteMessage);
    }
    private long deleteTime;
    private boolean deleteAfter;
    private String deleteMessage;
    private void deleteData(long time, String message, boolean after)
    {
        // This is crappy, but I have to call showDislog and then use this stuff
        // later.
```

```
        this.deleteTime = time;
        this.deleteMessage = message;
        this.deleteAfter = after;
        showDialog(DIALOG_DELETE);
    }
    private void confirmDelete()
    {
        SQLiteDatabase db = new RecordsOpenHelper(this).getWritableDatabase();
        db.delete("readings", "javatime " + (deleteAfter ? ">" : "<")
            + " ? AND bank = ?", new String[] { deleteTime + "", bank + ""});
        db.close();
        loadData();
         graph.invalidate();
    }
}
```

当用户按钮设备中的“MENU”后，会在屏幕下方弹出 6 个操作选项，系统会以图形化界面效果显示检测结果，执行后效果如图 18-4 所示。

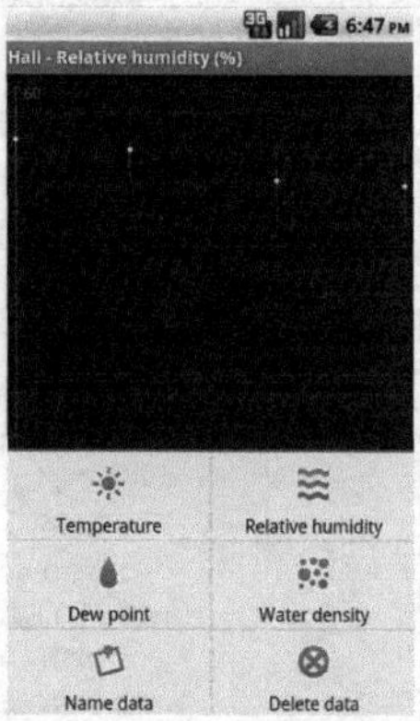

▲图 18-4　执行效果

18.4.6　湿度跟踪器

本系统检测湿度功能是通过开源代码湿度跟踪器实现的，实现文件 ClimateData.java 的具体实现代码如下所示。

```
public class ClimateData
{
    public final static int FIELD_TEMPERATURE=0, FIELD_HUMIDITY=1,
        FIELD_DEWPOINT=2, FIELD_DENSITY=3;

    private double[] measured = new double[4], min = new double[4], max = new double[4];

    public static class FieldValue
    {
        private float measured, min, max;

        private FieldValue(int field, double[] measured, double[] min, double[] max)
        {
            this.measured = (float)measured[field];
            this.min = (float)min[field];
            this.max = (float)max[field];
        }

        public float getMeasured()
        {
            return measured;
        }

        public float getMin()
        {
            return min;
        }

        public float getMax()
        {
            return max;
        }
    }

    public ClimateData(double temp, float tempError, double humidity, float
    humidityError)
    {
        measured[FIELD_TEMPERATURE] = temp;
        min[FIELD_TEMPERATURE] = Math.max(0, temp - tempError);
        max[FIELD_TEMPERATURE] = Math.min(40, temp + tempError);

        measured[FIELD_HUMIDITY] = humidity;
        min[FIELD_HUMIDITY] = Math.max(1, humidity - humidityError);
```

```
        max[FIELD_HUMIDITY] = Math.min(100, humidity + humidityError);

        measured[FIELD_DEWPOINT] = getDewpoint(measured);
        min[FIELD_DEWPOINT] = getDewpoint(min);
        max[FIELD_DEWPOINT] = getDewpoint(max);

        measured[FIELD_DENSITY] = getDensity(measured);
        min[FIELD_DENSITY] = getDensity(min);
        max[FIELD_DENSITY] = getDensity(max);
    }

  private static double getDewpoint(double[] values)
  {
     double temp = values[FIELD_TEMPERATURE];
     double humidity = values[FIELD_HUMIDITY];

       // Dewpoint valid up to 60 degrees, from
       // http://en.wikipedia.org/wiki/Dew_point#Calculating_the_dew_point
       double a = 17.271, b = 237.2;
       double gamma = ((a * temp) / (b + temp)) + Math.log(humidity/100.0);
       return b * gamma / (a - gamma);
  }

  private static double getDensity(double[] values)
  {
     double temp = values[FIELD_TEMPERATURE];
     double humidity = values[FIELD_HUMIDITY];

      // Approximate saturated vapour density valid up to 40 degrees, from
       // http://hyperphysics.phy-astr.gsu.edu/hbase/kinetic/relhum.html#c3
       double saturated = 5.018 + 0.32321 * temp + 0.0081847 * temp * temp
           + 0.00031243 * temp * temp * temp;
       return saturated * (humidity / 100.0);
  }

  public FieldValue getField(int field)
  {
     return new FieldValue(field, measured, min, max);
  }
}
```

到此为止，本实例的主要功能模块介绍完毕。和数据库操作相关的代码请读者参阅本书附带程序中的具体源码，为节省篇幅，在此不再进行详细讲解。

第 19 章　Android 蓝牙系统概述

蓝牙是一种支持设备短距离通信（一般 10m 内）的无线电技术，可以在包括移动电话、PDA、无线耳机、笔记本电脑、相关外设等众多设备之间进行无线信息交换。本章将首先讲解 Android 系统中蓝牙模块的底层源码和实现原理，为读者步入本书后面知识的学习打下基础。

19.1 蓝牙概述

蓝牙技术的数据传输速率为 1Mbps，采用时分双工传输方案实现全双工传输。本节将首先详细讲解蓝牙技术的发展历程，为读者步入本书后面知识的学习打下基础。

19.1.1 蓝牙技术的发展历程

蓝牙这一名称来自于 10 世纪的一位丹麦国王 Harald Blatand，Blatand 在英文里的意思可以被解释为 Bluetooth。因为国王喜欢吃蓝莓，牙龈每天都是蓝色的所以叫蓝牙。蓝牙的创始人是瑞典爱立信公司，爱立信早在 1994 年就已进行研发。1997 年，爱立信与其他设备生产商联系，并激发了他们对该项技术的浓厚兴趣。1998 年 2 月，5 个跨国大公司，包括爱立信、诺基亚、IBM、东芝及 Intel 组成了一个特殊兴趣小组（SIG），他们共同的目标是建立一个全球性的小范围无线通信技术，即现在的蓝牙。

Bluetooth 无线技术规格供我们全球的成员公司免费使用。许多行业的制造商都积极地在其产品中实施此技术，以减少使用零乱的电线，实现无缝连接、流传输立体声，传输数据或进行语音通信。Bluetooth 技术在 2.4 GHz 波段运行，该波段是一种无需申请许可证的工业、科技、医学（ISM）无线电波段。正因如此，使用 Bluetooth 技术不需要支付任何费用。但您必须向手机提供商注册使用 GSM 或 CDMA，除了设备费用外，您不需要为使用 Bluetooth 技术再支付任何费用。

Bluetooth 技术得到了空前广泛的应用，集成该技术的产品从手机、汽车到医疗设备，使用该技术的用户从消费者、工业市场到企业等等，不一而足。低功耗，小体积以及低成本的芯片解决方案使得 Bluetooth 技术甚至可以应用于极微小的设备中。

19.1.2 蓝牙的特点

Bluetooth 技术是一项即时技术，它不要求固定的基础设施，且易于安装和设置。您不需要电缆即可实现连接。新用户使用亦不费力，我们只需拥有 Bluetooth 品牌产品，检查可用的配置文件，将其连接至使用同一配置文件的另一 Bluetooth 设备即可。后续的 PIN 码流程就如同您在 ATM 机器上操作一样简单。外出时，您可以随身带上您的个人局域网（PAN），甚至可以与其他网络连接。

19.2 Android 系统中的蓝牙模块

Android 系统包含了对蓝牙网络协议栈的支持，这使得蓝牙设备能够无线连接其他蓝牙设备交换数据。Android 的应用程序框架提供了访问蓝牙功能的 APIs。这些 APIs 让应用程序能够无线连接其他蓝牙设备，实现点对点，或点对多点的无线交互功能。

通过使用蓝牙 API，一个 Android 应用程序能够实现如下所示的功能。

- 扫描其他蓝牙设备。
- 查询本地蓝牙适配器（local Bluetooth adapter）用于配对蓝牙设备。
- 建立 RFCOMM 信道（channels）。
- 通过服务发现（service discovery）连接其他设备。
- 数据通信。
- 管理多个连接。

Android 平台的蓝牙系统是基于 BlueZ 实现的，是通过 Linux 中一套完整的蓝牙协议栈开源实现的。当前 BlueZ 被广泛应用于各种 Linux 版本中，并被芯片公司移植到各种芯片平台上所使用。在 Linux 2.6 内核中已经包含了完整的 BlueZ 协议栈，在 Android 系统中已经移植并嵌入进了 BlueZ 的用户空间实现，并且随着硬件技术的发展而不断更新。

蓝牙技术实际上是一种短距离无线电技术。在 Android 系统的蓝牙模块中，除了使用 Kernel 支持外，还需要用户空间的 BlueZ 的支持。

Android 平台中蓝牙模块的基本层次结构如图 19-1 所示。

Android 平台中蓝牙系统从上到下主要包括 Java 框架中的 BlueTooth 类、Android 适配库、BlueZ 库、驱动程序和协议，这几部分的系统结构如图 19-2 所示。

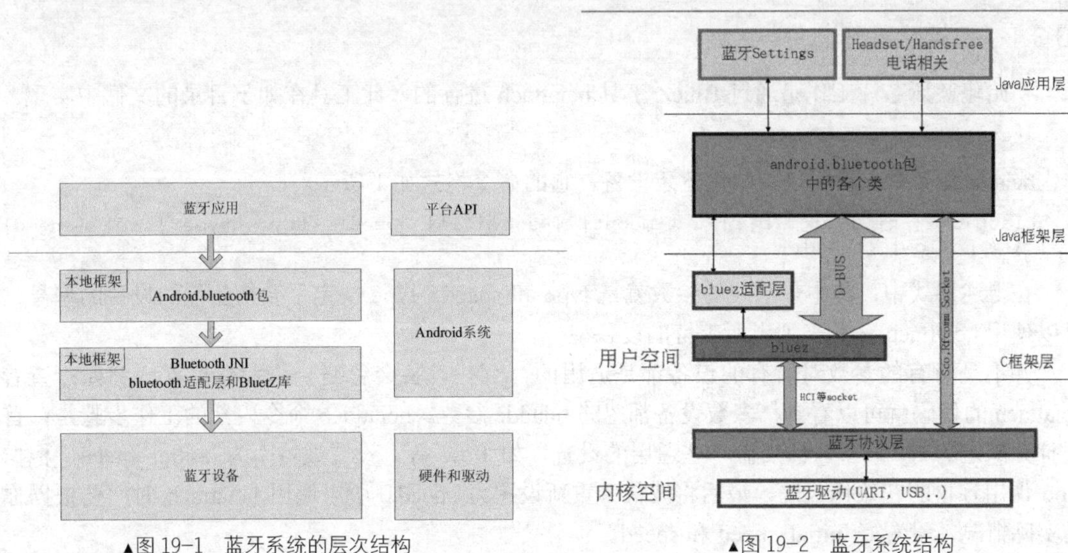

▲图 19-1 蓝牙系统的层次结构 ▲图 19-2 蓝牙系统结构

对图 19-2 所示的各个层次结构具体说明如下。

（1）blueZ 库

Android 蓝牙设备管理的库的路径如下所示。

```
external/bluez/
```

可以分别生成库 libbluetooth.so、libbluedroid.so 和 hcidump 等众多相关工具和库。blueZ 库提供了对用户空间蓝牙的支持，在里面包含了主机控制协议 HCI 以及其他众多内核实现协议的接口，并且实现了所有蓝牙应用模式 Profile。

（2）蓝牙的 JNI 部分

此部分的代码路径如下所示。

```
frameworks/base/core/jni/
```

（3）Java 框架层

Java 框架层的实现代码保存在如下路径。

```
frameworks/base/core/java/android/bluetooth          //蓝牙部分对应应用程序的API
frameworks/base/core/java/android/Server             //蓝牙的服务部分
```

蓝牙的服务部分负责管理并使用底层本地服务，并封装成系统服务。而在 android.bluetooth 部分中包含了各个蓝牙平台的 API 部分，以供应用程序层所使用。

（4）BlueTooth 的适配库

BlueTooth 适配库的代码路径如下所示。

```
system/bluetooth/
```

在此层用于生成库 libbluedroid.so 以及相关工具和库，能够实现对蓝牙设备的管理，例如蓝牙设备的电源管理。

19.3 分析蓝牙模块的源码

要想掌握蓝牙系统的开发原理，需要首先分析 Android 中的蓝牙源码并了解其核心构造，只有这样才能对蓝牙应用开发做到游刃有余。在本节的内容中，简要介绍开源 Android 中蓝牙模块相关的代码，为读者步入本书后面知识的学习打下基础。

19.3.1　初始化蓝牙芯片

初始化蓝牙芯片工作是通过 BlueZ 工具 hciattach 进行的，此工具在如下目录的文件中实现。

```
external/bluetooth/tools
```

hciattach 命令主要用来初始化蓝牙设备，它的命令格式如下所示。

```
hciattach [-n] [-p] [-b] [-t timeout] [-s initial_speed] <tty> <type | id> [speed]
[flow|noflow] [bdaddr]
```

在上述格式中，其中最重要的参数就是 type 和 speed，type 决定了要初始化的设备的型号，可以使用 hciattach -l 来列出所支持的设备型号。

并不是所有的参数对所有的设备都是适用的，有些设备会忽略一些参数设置，例如：查看 hciattach 的代码就可以看到，多数设备都忽略 bdaddr 参数。hciattach 命令内部的工作步骤是：首先打开制定的 tty 设备，然后做一些通用的设置，如 flow 等，设置波特率为 initial_speed，根据 type 调用各自的初始化代码，最后将波特率重新设置为 speed。所以调用 hciattach 时，要根据你的实际情况，设置好 initial_speed 和 speed。

对于 type BCSP 来说，它的初始化代码只做了一件事，就是完成 BCSP 协议的同步操作，它并不对蓝牙芯片做任何的 pskey 的设置。

19.3.2　蓝牙服务

在蓝牙服务方面一般不要我们自己定义，只需要使用初始化脚本文件 init.rc 中的默认内容即可。例如下面的代码。

```
service bluetoothd /system/bin/logwrapper /system/bin/bluetoothd -d -n
    socket bluetooth stream 660 bluetooth bluetooth
    socket dbus_bluetooth stream 660 bluetooth bluetooth
    # init.rc does not yet support applying capabilities, so run as root and
    # let bluetoothd drop uid to bluetooth with the right linux capabilities
    group bluetooth net_bt_admin misc
    disabled
```

```
# baudrate change 115200 to 1152000(Bluetooth)
service changebaudrate /system/bin/logwrapper /system/xbin/bccmd_115200 -t bcsp -d
/dev/s3c2410_serial1 psset -r 0x1be 0x126e
    user bluetooth
    group bluetooth net_bt_admin
    disabled
    oneshot

#service hciattach /system/bin/logwrapper /system/bin/hciattach -n -s 1152000
/dev/s3c2410_serial1 bcsp 1152000
service hciattach /system/bin/logwrapper /system/bin/hciattach -n -s 115200
/dev/s3c2410_serial1 bcsp 115200
    user bluetooth
    group bluetooth net_bt_admin misc
    disabled

service hfag /system/bin/sdptool add --channel=10 HFAG
    user bluetooth
    group bluetooth net_bt_admin
    disabled
    oneshot

service hsag /system/bin/sdptool add --channel=11 HSAG
    user bluetooth
    group bluetooth net_bt_admin
    disabled
    oneshot

service opush /system/bin/sdptool add --channel=12 OPUSH
    user bluetooth
    group bluetooth net_bt_admin
    disabled
    oneshot

service pbap /system/bin/sdptool add --channel=19 PBAP
    user bluetooth
    group bluetooth net_bt_admin
    disabled
    oneshot
```

在上述代码中，每一个“service”后面列出了一种 Android 服务。

19.3.3　管理蓝牙电源

在 Android 系统的软下目录中实现了 libbluedroid。

```
system/bluetooth/
```

我们可以调用 rfkill 接口来控制电源管理，如果已经实现了 rfkill 接口，则无需再进行配置。如果在文件 init.rc 中已经实现了 hciattach 服务，则说明在 libbluedroid 中已经实现对其调用以操作蓝牙的初始化。

19.4　和蓝牙相关的类

经过本章前面内容的学习，已经了解了 Android 系统中蓝牙的基本知识。根据对上述从底层到应用的学习，了解了蓝牙的工作原理和机制。在本节的内容中，将详细讲解在 Android 系统中和蓝牙相关的类，为读者步入本书后面知识的学习打好基础。

19.4.1　BluetoothSocket 类

1. BluetoothSocket 类基础

类 BluetoothSocket 的定义格式如下所示。

```
public static class Gallery.LayoutParams extends ViewGroup.LayoutParams
```

类 BluetoothSocket 的定义结构如下所示。

```
java.lang.Object
android.view. ViewGroup.LayoutParams
android.widget.Gallery.LayoutParams
```

Android 的蓝牙系统和 Socket 套接字密切相关，蓝牙端的监听接口和 TCP 的端口类似，都是使用了 Socket 和 ServerSocket 类。在服务器端，使用 BluetoothServerSocket 类来创建一个监听服务端口。当一个连接被 BluetoothServerSocket 所接受，它会返回一个新的 BluetoothSocket 来管理该连接。在客户端，使用一个单独的 BluetoothSocket 类去初始化一个外接连接和管理该连接。

最通常使用的蓝牙端口是 RFCOMM，它是被 Android API 支持的类型。RFCOMM 是一个面向连接，通过蓝牙模块进行的数据流传输方式，它也被称为串行端口规范（Serial Port Profile，SPP）。

为了创建一个 BluetoothSocket 去连接到一个已知设备，使用方法 BluetoothDevice.createRfcommSocketToServiceRecord()。然后调用 connect()方法去尝试一个面向远程设备的连接。这个调用将被阻塞直到一个连接已经建立或者该链接失效。

为了创建一个 BluetoothSocket 作为服务端（或者“主机”），每当该端口连接成功后，无论它初始化为客户端，或者被接受作为服务器端，都通过方法 getInputStream()和 getOutputStream()来打开 IO 流，从而获得各自的 InputStream 和 OutputStream 对象

BluetoothSocket 类的线程是安全的，因为 close()方法总会马上放弃外界操作并关闭服务器端口。

2. BluetoothSocket 类的公共方法

（1）public void close ()

功能：马上关闭该端口并且释放所有相关的资源。在其他线程的该端口中引起阻塞，从而使系统马上抛出一个 IO 异常。

异常：IOException。

（2）public void connect ()

功能：尝试连接到远程设备。该方法将阻塞，指导一个连接建立或者失效。如果该方法没有返回异常值，则该端口现在已经建立。当设备查找正在进行的时候，创建对远程蓝牙设备的新连接不可被尝试。设备查找在蓝牙适配器上是一个重量级过程，并且肯定会降低一个设备的连接。使用 cancelDiscovery()方法会取消一个外界的查询，因为这个查询并不由活动所管理，而是作为一个系统服务来运行，所以即使它不能直接请求一个查询，应用程序也总会调用 cancelDiscovery()方法。使用方法 close()可以用来放弃从另一线程而来的调用。

异常：IOException，表示一个错误，例如连接失败。

（3）public InputStream getInputStream ()

功能：通过连接的端口获得输入数据流。即使该端口未连接，该输入数据流也会返回。不过在该数据流上的操作将抛出异常，直到相关的连接已经建立。

返回值：输入流。

异常：IOException。

（4）public OutputStream getOutputStream ()

功能：通过连接的端口获得输出数据流。即使该端口未连接，该输出数据流也会返回。不过在该数据流上的操作将抛出异常，直到相关的连接已经建立。

返回值：输出流。

异常：IOException。

（5）public BluetoothDevice getRemoteDevice ()

功能：获得该端口正在连接或者已经连接的远程设备。

返回值：远程设备。

19.4.2 BluetoothServerSocket 类

1. BluetoothServerSocket 类基础

类 BluetoothServerSocket 的格式如下所示。

```
public final class BluetoothServerSocket extends Object implements Closeable
```

类 BluetoothServerSocket 的结构如下所示。

```
java.lang.Object
android.bluetooth.BluetoothServerSocket
```

2. BluetoothServerSocket 类的公共方法

（1）public BluetoothSocketaccept (int timeout)

功能：阻塞直到超时时间内的连接建立。在一个成功建立的连接上返回一个已连接的 BluetoothSocket 类。每当该调用返回的时候，它可以在此调用去接收以后新来的连接。close()方法可以用来放弃从另一线程来的调用。

参数 timeout：表示阻塞超时时间。

返回值：已连接的 BluetoothSocket。

异常：IOException，表示出现错误，比如该调用被放弃或超时。

（2）public BluetoothSocket accept ()

功能：阻塞直到一个连接已经建立。在一个成功建立的连接上返回一个已连接的 BluetoothSocket 类。每当该调用返回的时候，它可以在此调用去接收以后新来的连接。使用 close() 方法可以用来放弃从另一线程来的调用。

返回值：已连接的 BluetoothSocket。

异常：IOException，表示出现错误，比如该调用被放弃或者超时。

（3）public void close ()

功能：马上关闭端口，并释放所有相关的资源。在其他线程的该端口中引起阻塞，从而使系统马上抛出一个 IO 异常。关闭 BluetoothServerSocket 不会关闭接受自 accept()的任意 BluetoothSocket。

异常：IOException。

19.4.3 BluetoothAdapter 类

1. BluetoothAdapter 类基础

类 BluetoothAdapter 的格式如下所示。

```
public final class BluetoothAdapter extends Object
```

类 BluetoothAdapter 的结构如下所示。

```
java.lang.Object
android.bluetooth.BluetoothAdapter
```

BluetoothAdapter 代表本地的蓝牙适配器设备，通过此类可以让用户能执行基本的蓝牙任务。例如初始化设备的搜索，查询可匹配的设备集，使用一个已知的 MAC 地址来初始化一个 BluetoothDevice 类，创建一个 BluetoothServerSocket 类以监听其他设备对本机的连接请求等。

为了得到这个代表本地蓝牙适配器的 BluetoothAdapter 类，需要调用静态方法 getDefaultAdapter()，这是所有蓝牙动作使用的第一步。当拥有本地适配器以后，用户可以获得一系列的 BluetoothDevice 对象，这些对象代表所有拥有 getBondedDevice()方法的已经匹配的设备；用 startDiscovery()方法来开始设备的搜寻；或者创建一个 BluetoothServerSocket 类，通过 listenUsingRfcommWithServiceRecord(String, UUID)方法来监听新来的连接请求。

注意　大部分方法需要 BLUETOOTH 权限，一些方法同时需要 BLUETOOTH_ADMIN 权限。

2. BluetoothAdapter 类的常量

（1）String ACTION_DISCOVERY_FINISHED

广播事件：本地蓝牙适配器已经完成设备的搜寻过程。需要 BLUETOOTH 权限接收。

常量值：android.bluetooth.adapter.action.DISCOVERY_FINISHED。

（2）String ACTION_DISCOVERY_STARTED

广播事件：本地蓝牙适配器已经开始对远程设备的搜寻过程。它通常牵涉到一个大概需时 12 秒的查询扫描过程，紧跟着是一个对每个获取到自身蓝牙名称的新设备的页面扫描。用户会发现一个把 ACTION_FOUND 常量通知为远程蓝牙设备的注册。设备查找是一个重量级过程。当查找正在进行的时候，用户不能尝试对新的远程蓝牙设备进行连接，同时存在的连接将获得有限制的带宽以及高等待时间。用户可用 cancelDiscovery()类来取消正在执行的查找进程。需要 BLUETOOTH 权限接收。

常量值：android.bluetooth.adapter.action.DISCOVERY_STARTED。

（3）String ACTION_LOCAL_NAME_CHANGED

广播活动：本地蓝牙适配器已经更改了它的蓝牙名称。该名称对远程蓝牙设备是可见的，它总是包含了一个带有名称的 EXTRA_LOCAL_NAME 附加域。需要 BLUETOOTH 权限接收。

常量值：android.bluetooth.adapter.action.LOCAL_NAME_CHANGED

（4）String ACTION_REQUEST_DISCOVERABLE

Activity 活动：显示一个请求被搜寻模式的系统活动。如果蓝牙模块当前未打开，该活动也将请求用户打开蓝牙模块。被搜寻模式和 SCAN_MODE_CONNECTABLE_DISCOVERABLE 等价。当远程设备执行查找进程的时候，它允许其发现该蓝牙适配器。从隐私安全考虑，Android 不会将被搜寻模式设置为默认状态。该意图的发送者可以选择性地运用 EXTRA_DISCOVERABLE_DURATION 这个附加域去请求发现设备的持续时间。普遍来说，对于每一请求，默认的持续时间为 120 秒，最大值则可达到 300 秒。

Android 运用 onActivityResult(int, int, Intent)回收方法来传递该活动结果的通知。被搜寻的时间（以秒为单位）将通过 resultCode 值来显示，如果用户拒绝被搜寻，或者设备产生了错误，则通过 RESULT_CANCELED 值来显示。

每当扫描模式变化的时候，应用程序可以为通过 ACTION_SCAN_MODE_CHANGED 值来监听全局的消息通知。比如，当设备停止被搜寻以后，该消息可以被系统通知給应用程序。需要 BLUETOOTH 权限。

常量值：android.bluetooth.adapter.action.REQUEST_DISCOVERABLE

（5）String ACTION_REQUEST_ENABLE

Activity 活动：显示一个允许用户打开蓝牙模块的系统活动。当蓝牙模块完成打开工作，或者当用户决定不打开蓝牙模块时，系统活动将返回该值。Android 运用 onActivityResult(int, int,

Intent)回收方法来传递该活动结果的通知。如果蓝牙模块被打开，将通过resultCode值RESULT_OK来显示；如果用户拒绝该请求，或者设备产生了错误，则通过RESULT_CANCELED值来显示。每当蓝牙模块被打开或者关闭，应用程序可以为通过ACTION_STATE_CHANGED值来监听全局的消息通知。需要BLUETOOTH权限

常量值：android.bluetooth.adapter.action.REQUEST_ENABLE。

（6）String ACTION_SCAN_MODE_CHANGED

广播活动：指明蓝牙扫描模块或者本地适配器已经发生变化。它总是包含 EXTRA_SCAN_MODE 和 EXTRA_PREVIOUS_SCAN_MODE。这两个附加域各自包含了新的和旧的扫描模式。需要BLUETOOTH权限

常量值：android.bluetooth.adapter.action.SCAN_MODE_CHANGED。

（7）String ACTION_STATE_CHANGED

广播活动：本来的蓝牙适配器的状态已经改变，例如蓝牙模块已经被打开或者关闭。它总是包含EXTRA_STATE和EXTRA_PREVIOUS_STATE。这两个附加域各自包含了新的和旧的状态。需要BLUETOOTH权限接收。

常量值：android.bluetooth.adapter.action.STATE_CHANGED。

（8）int ERROR

功能：标记该类的错误值。确保和该类中的任意其他整数常量不相等。它为需要一个标记错误值的函数提供了便利。例如：

```
Intent.getIntExtra(BluetoothAdapter.EXTRA_STATE, BluetoothAdapter.ERROR)
```

常量值：-2147483648 (0x80000000)

（9）String EXTRA_DISCOVERABLE_DURATION

功能：试图在ACTION_REQUEST_DISCOVERABLE常量中作为一个可选的整型附加域，来为短时间内的设备发现请求一个特定的持续时间。默认值为120秒，超过300秒的请求将被限制。这些值是可以变化的。

常量值：android.bluetooth.adapter.extra.DISCOVERABLE_DURATION。

（10）String EXTRA_LOCAL_NAME

功能：试图在ACTION_LOCAL_NAME_CHANGED常量中作为一个字符串附加域，来请求本地蓝牙的名称。

常量值：android.bluetooth.adapter.extra.LOCAL_NAME。

（11）String EXTRA_PREVIOUS_SCAN_MODE

功能：试图在ACTION_SCAN_MODE_CHANGED常量中作为一个整型附加域，来请求以前的扫描模式。可能值如下所示。

- SCAN_MODE_NONE。
- SCAN_MODE_CONNECTABLE。
- SCAN_MODE_CONNECTABLE_DISCOVERABLE。

常量值：android.bluetooth.adapter.extra.PREVIOUS_SCAN_MODE。

（12）String EXTRA_PREVIOUS_STATE

功能：试图在ACTION_STATE_CHANGED常量中作为一个整型附加域，来请求以前的供电状态。可以取的值如下所示。

- STATE_OFF。
- STATE_TURNING_ON。

- STATE_ON。
- STATE_TURNING_OFF。

常量值：android.bluetooth.adapter.extra.PREVIOUS_STATE。

（13）String EXTRA_SCAN_MODE

功能：试图在 ACTION_SCAN_MODE_CHANGED 常量中作为一个整型附加域，来请求当前的扫描模式，可以取的值如下所示。

- SCAN_MODE_NONE。
- SCAN_MODE_CONNECTABLE。
- SCAN_MODE_CONNECTABLE_DISCOVERABLE。

常量值：android.bluetooth.adapter.extra.SCAN_MODE。

（14）String EXTRA_STATE

功能：试图在 ACTION_STATE_CHANGED 常量中作为一个整型附加域，来请求当前的供电状态。可以取的值如下所示。

- STATE_OFF。
- STATE_TURNING_ON。
- STATE_ON。
- STATE_TURNING_OFF。

常量值：android.bluetooth.adapter.extra.STATE。

（15）int　SCAN_MODE_CONNECTABLE

功能：指明在本地蓝牙适配器中，查询扫描功能失效，但页面扫描功能有效。因此该设备不能被远程蓝牙设备发现，但如果以前曾经发现过该设备，则远程设备可以对其进行连接。

常量值：21 (0x00000015)。

（16）int SCAN_MODE_CONNECTABLE_DISCOVERABLE

功能：指明在本地蓝牙适配器中，查询扫描功能和页面扫描功能都有效。因此该设备既可以被远程蓝牙设备发现，也可以被其连接。

常量值：23 (0x00000017)。

（17）int SCAN_MODE_NONE

功能：指明在本地蓝牙适配器中，查询扫描功能和页面扫描功能都失效. 因此该设备既不可以被远程蓝牙设备发现，也不可以被其连接。

常量值：20 (0x00000014)。

（18）int STATE_OFF

功能：指明本地蓝牙适配器模块已经关闭。

常量值：10 (0x0000000a)。

（19）int STATE_ON

功能：指明本地蓝牙适配器模块已经打开，并且准备被使用。

（20）int STATE_TURNING_OFF

功能：指明本地蓝牙适配器模块正在关闭。本地客户端可以立刻尝试友好地断开任意外部连接。

常量值：13 (0x0000000d)。

（21）int　STATE_TURNING_ON

功能：指明本地蓝牙适配器模块正在打开. 然而本地客户在尝试使用这个适配器之前需要为 STATE_ON 状态而等待。

常量值：11 (0x0000000b)。

3. BluetoothAdapter 类的公共方法

（1）public boolean cancelDiscovery ()

功能：取消当前的设备发现查找进程，需要 BLUETOOTH_ADMIN 权限。因为对蓝牙适配器而言，查找是一个重量级的过程，因此这个方法必须在尝试连接到远程设备前使用用 connect()方法进行调用。发现的过程不会由活动来进行管理，但是它会作为一个系统服务来运行，因此即使它不能直接请求这样的一个查询动作，也必须取消该搜索进程。如果蓝牙状态不是 STATE_ON，这个 API 将返回 false。蓝牙打开后，等待 ACTION_STATE_CHANGED 更新成 STATE_ON。

返回值：成功则返回 true，有错误则返回 false。

（2）public static boolean checkBluetoothAddress (String address)

功能：验证如"00:43:A8:23:10:F0"之类的蓝牙地址，字母必须为大写才有效。

参数 address：字符串形式的蓝牙模块地址。

返回值：地址正确则返回 true，否则返回 false。

（3）public boolean disable ()

功能：关闭本地蓝牙适配器——不能在没有明确关闭蓝牙的用户动作中使用。这个方法友好地停止所有的蓝牙连接，停止蓝牙系统服务，以及对所有基础蓝牙硬件进行断电。没有用户的直接同意，蓝牙永远不能被禁止。这个 disable()方法只提供了一个应用，该应用包含了一个改变系统设置的用户界面（例如“电源控制”应用）。

这是一个异步调用方法：该方法将马上获得返回值，用户要通过监听 ACTION_STATE_CHANGED 值来获取随后的适配器状态改变的通知。如果该调用返回 true 值，则该适配器状态会立刻从 STATE_ON 转向 STATE_TURNING_OFF，稍后则会转为 STATE_OFF 或者 STATE_ON。如果该调用返回 false，那么系统已经有一个保护蓝牙适配器被关闭的问题——比如该适配器已经被关闭了。

需要 BLUETOOTH_ADMIN 权限。

返回值：如果蓝牙适配器的停止进程已经开启则返回 true，如果产生错误则返回 false。

（4）public boolean enable ()

功能：打开本地蓝牙适配器——不能在没有明确打开蓝牙的用户动作中使用。该方法将为基础的蓝牙硬件供电，并且启动所有的蓝牙系统服务。没有用户的直接同意，蓝牙永远不能被禁止。如果用户为了创建无线连接而打开了蓝牙模块，则其需要 ACTION_REQUEST_ENABLE 值，该值将提出一个请求用户允许以打开蓝牙模块的会话。这个 enable()值只提供了一个应用，该应用包含了一个改变系统设置的用户界面（例如“电源控制”应用）。

这是一个异步调用方法：该方法将马上获得返回值，用户要通过监听 ACTION_STATE_CHANGED 值来获取随后的适配器状态改变的通知。如果该调用 返回 true 值，则该适配器状态会立刻从 STATE_OFF 转向 STATE_TURNING_ON，稍后则会转为 STATE_OFF 或者 STATE_ON。如果该调用返回 false，那么说明系统已经有一个保护蓝牙适配器被打开的问题——比如飞行模式，或者该适配器已经被打开。

需要 BLUETOOTH_ADMIN 权限。

返回值：如果蓝牙适配器的打开进程已经开启则返回 true，如果产生错误则返回 false。

（5）public String getAddress ()

功能：返回本地蓝牙适配器的硬件地址，例如：

```
00:11:22:AA:BB:CC
```

需要 BLUETOOTH 权限。

返回值：字符串形式的蓝牙模块地址。

（6）public Set<BluetoothDevice> getBondedDevices ()

功能：返回已经匹配到本地适配器的 BluetoothDevice 类的对象集合。如果蓝牙状态不是 STATE_ON，这个 API 将返回 false。蓝牙打开后，等待 ACTION_STATE_CHANGED 更新成 STATE_ON。需要 BLUETOOTH 权限。

返回值：未被修改的 BluetoothDevice 类的对象集合，如果有错误则返回 null。

（7）public static synchronized BluetoothAdapter getDefaultAdapter ()

功能：获取对默认本地蓝牙适配器的的操作权限。目前 Andoird 只支持一个蓝牙适配器，但是 API 可以被扩展为支持多个适配器。该方法总是返回默认的适配器。

返回值：返回默认的本地适配器，如果蓝牙适配器在该硬件平台上不能被支持，则返回 null。

（8）public String getName ()

功能：获取本地蓝牙适配器的蓝牙名称，这个名称对于外界蓝牙设备而言是可见的。需要 BLUETOOTH 权限。

返回值：该蓝牙适配器名称，如果有错误则返回 null。

（9）public BluetoothDevice getRemoteDevice (String address)

功能：为给予的蓝牙硬件地址获取一个 BluetoothDevice 对象。合法的蓝牙硬件地址必须为大写，格式类似于"00:11:22:33:AA:BB"。checkBluetoothAddress(String)方法可以用来验证蓝牙地址的正确性。BluetoothDevice 类对于合法的硬件地址总会产生返回值，即使这个适配器从未见过该设备。

参数：address　合法的蓝牙 MAC 地址。

异常：IllegalArgumentException，如果地址不合法。

（10）public int getScanMode ()

功能：获取本地蓝牙适配器的当前蓝牙扫描模式，蓝牙扫描模式决定本地适配器可连接并且/或者可被远程蓝牙设备所连接。需要 BLUETOOTH 权限，可能的取值如下所示。

- SCAN_MODE_NONE。
- SCAN_MODE_CONNECTABLE。
- SCAN_MODE_CONNECTABLE_DISCOVERABLE。

如果蓝牙状态不是 STATE_ON，则这个 API 将返回 false。蓝牙打开后，等待 ACTION_STATE_CHANGED 更新成 STATE_ON。

返回值：扫描模式。

（11）public int getState ()

功能：获取本地蓝牙适配器的当前状态，需要 BLUETOOTH 类。可能的取值如下所示。

- STATE_OFF。
- STATE_TURNING_ON。
- STATE_ON。
- STATE_TURNING_OFF。

返回值：蓝牙适配器的当前状态。

（12）public boolean isDiscovering ()

功能：如果当前蓝牙适配器正处于设备发现查找进程中，则返回真值。设备查找是一个重量级过程。当查找正在进行的时候，用户不能尝试对新的远程蓝牙设备进行连接，同时存在的连接将获得有限制的带宽以及高等待时间。用户可用 cencelDiscovery()类来取消正在执行的查找进程。

应用程序也可以为 ACTION_DISCOVERY_STARTED 或者 ACTION_DISCOVERY_FINISHED 进行注册，从而当查找开始或者完成的时候，可以获得通知。

如果蓝牙状态不是 STATE_ON，这个 API 将返回 false。蓝牙打开后，等待 ACTION_STATE_CHANGED 更新成 STATE_ON。需要 BLUETOOTH 权限。

返回值：如果正在查找，则返回 true。

（13）public boolean isEnabled ()

功能：如果蓝牙正处于打开状态并可用，则返回真值，与 getBluetoothState()==STATE_ON 等价，需要 BLUETOOTH 权限。

返回值：如果本地适配器已经打开，则返回 true。

（14）public BluetoothServerSocket listenUsingRfcommWithServiceRecord (String name, UUID uuid)

功能：创建一个正在监听的安全的带有服务记录的无线射频通信（RFCOMM）蓝牙端口。一个对该端口进行连接的远程设备将被认证，对该端口的通讯将被加密。使用 accpet()方法可以获取从监听 BluetoothServerSocket 处新来的连接。该系统分配一个未被使用的无线射频通信通道来进行监听。

该系统也将注册一个服务探索协议（SDP）记录，该记录带有一个包含了特定 的通用唯一识别码（Universally Unique Identifier，UUID），服务器名称和自动分配通道的本地 SDP 服务。远程蓝牙设备可以用相同的 UUID 来查询自己的 SDP 服务器，并搜寻连接 到了哪个通道上。如果该端口已经关闭，或者如果该应用程序异常退出，则这个 SDP 记录会被移除。使用 createRfcommSocketToServiceRecord(UUID)可以从另一使用相同 UUID 的设备来连接到这个端口。需要 BLUETOOTH 权限。

参数。

- name：SDP 记录下的服务器名。
- uuid：SDP 记录下的 UUID。

返回值：一个正在监听的无线射频通信蓝牙服务端口。

异常：IOException，表示产生错误，比如蓝牙设备不可用，或者许可无效，或者通道被占用。

（15）public boolean setName (String name)

功能：设置蓝牙或者本地蓝牙适配器的昵称，这个名字对于外界蓝牙设备而言是可见的。合法的蓝牙名称最多拥有 248 位 UTF-8 字符，但是很多外界设备只能显示前 40 个字符，有些可能只限制前 20 个字符。

如果蓝牙状态不是 STATE_ON，这个 API 将返回 false。蓝牙打开后，等待 ACTION_STATE_CHANGED 更新成 STATE_ON。需要 BLUETOOTH_ADMIN 权限。

参数 name：一个合法的蓝牙名称。

返回值：如果该名称已被设定，则返回 true，否则返回 false。

（16）public boolean startDiscovery ()

功能：开始对远程设备进行查找的进程，它通常牵涉到一个大概需时 12 秒的查询扫描过程，紧跟着是一个对每个获取到自身蓝牙名称的新设备的页面扫描。这是一个异步调用方法：该方法将马上获得返回值，注册 ACTION_DISCOVERY_STARTED 和 ACTION_DISCOVERY_FINISHED 意图准确地确定该探索是处于开始阶段或者完成阶段。注册 ACTION_FOUND 以活动远程蓝牙设备已找到的通知。

设备查找是一个重量级过程。当查找正在进行的时候，用户不能尝试对新的远程蓝牙设备进行连接，同时存在的连接将获得有限制的带宽以及高等待时间。用户可用 cencelDiscovery()类来取消正在执行的查找进程。发现的过程不会由活动来进行管理，但是它会作为一个系统服务来运行，因此即使它不能直接请求这样的一个查询动作，也必须取消该搜索进程。设备搜寻只寻找已经被连接的远程设备。许多蓝牙设备默认不会被搜寻到，并且需要进入到一个特殊的模式当中。

如果蓝牙状态不是 STATE_ON，这个 API 将返回 false。蓝牙打开后，等待 ACTION_STATE_CHANGED 更新成 STATE_ON。需要 BLUETOOTH_ADMIN 权限。

返回值：成功返回 true，错误返回 false。

19.4.4　BluetoothClass.Service 类

类 BluetoothClass.Service 的格式如下所示。

```
public static final class BluetoothClass.Service extends Object
```

类 BluetoothClass.Service 的结构如下所示。

```
java.lang.Object
android.bluetooth.BluetoothClass.Service
```

类 BluetoothClass.Service 用于定义所有的服务类常量，任意 BluetoothClass 由 0 或多个服务类编码组成。在类 BluetoothClass.Service 中包含如下所示的常量。

- int AUDIO。
- int CAPTURE。
- int INFORMATION。
- int LIMITED_DISCOVERABILITY。
- int NETWORKING。
- int OBJECT_TRANSFER。
- int POSITIONING。
- int RENDER。
- int TELEPHONY。

19.4.5　BluetoothClass.Device 类

类 BluetoothClass.Device 的格式如下所示。

```
public final class BluetoothClass.Device extends Object
```

类 BluetoothClass.Device 的结构如下所示。

```
java.lang.Object
android.bluetooth.BluetoothClass.Device
```

类 BluetoothClass.Device 用于定义所有的设备类的常量，每个 BluetoothClass 有一个带有主要和较小部分的设备类进行编码。里面的常量代表主要和较小的设备类部分（完整的设备类）的组合。BluetoothClass.Device.Major 的常量只能代表主要设备类，各个常量如下所示。

BluetoothClass.Device 有一个内部类，此内部类定义了所有的主要设备类常量。内部类的定义格式如下所示。

```
class BluetoothClass.Device.Major
```

注意：到此为止，Android 中的蓝牙类介绍完毕。我们在调用这些类时除了首先确保 API Level 至少为版本 5 以上，还需注意添加相应的权限，比如使用通信需要在文件 androidmanifest.xml 中加入<uses-permission android:name="android.permission.BLUETOOTH" />权限，而在开关蓝牙时需要加入 android.permission.BLUETOOTH_ADMIN 权限。

19.5 在 Android 平台开发蓝牙应用程序

经过前面的学习，了解了在 Android 系统中和蓝牙模块相关的 API 类。其实从查找蓝牙设备到能够相互通信需要经过几个基本步骤，这几个步骤缺一不可。本节将详细讲解在 Android 平台中开发蓝牙应用程序的基本方法。

19.5.1 开发 Android 蓝牙应用程序的基本步骤

在 Android 系统中，开发蓝牙应用程序的基本步骤如下所示。

（1）设置权限

在文件 AndroidManifest.xml 中声明使用蓝牙的权限，例如下面的代码。

```
<uses-permission android:name="android.permission.BLUETOOTH"/>
<uses-permission android:name="android.permission.BLUETOOTH_ADMIN"/>
```

（2）启动蓝牙

首先要查看本机是否支持蓝牙，然后获取 BluetoothAdapter 蓝牙适配器对象。例如下面的代码。

```
BluetoothAdapter mBluetoothAdapter = BluetoothAdapter.getDefaultAdapter();
if(mBluetoothAdapter == null){
      //表明此手机不支持蓝牙
      return;
}
if(!mBluetoothAdapter.isEnabled()){    //蓝牙未开启，则开启蓝牙
          Intent enableIntent = new Intent(BluetoothAdapter.ACTION_REQUEST_ENABLE);
          startActivityForResult(enableIntent, REQUEST_ENABLE_BT);
}
//......
public void onActivityResult(int requestCode, int resultCode, Intent data){
     if(requestCode == REQUEST_ENABLE_BT){
          if(requestCode == RESULT_OK){
               //蓝牙已经开启
          }
     }
}
```

（3）发现蓝牙设备

● 使本机蓝牙处于可见（即处于易被搜索到状态）状态，便于其他设备发现本机蓝牙。演示代码如下所示。

```
//使本机蓝牙在300秒内可被搜索
private void ensureDiscoverable() {
       if (mBluetoothAdapter.getScanMode() !=
          BluetoothAdapter.SCAN_MODE_CONNECTABLE_DISCOVERABLE) {
          Intent discoverableIntent = new Intent(BluetoothAdapter.ACTION_REQUEST_
          DISCOVERABLE);
          discoverableIntent.putExtra(BluetoothAdapter.EXTRA_DISCOVERABLE_DURATION,300);
          startActivity(discoverableIntent);
       }
}
```

● 查找已经配对的蓝牙设备，即以前已经配对过的设备。演示代码如下所示。

```
Set<BluetoothDevice> pairedDevices = mBluetoothAdapter.getBondedDevices();
if (pairedDevices.size() > 0) {
   findViewById(R.id.title_paired_devices).setVisibility(View.VISIBLE);
   for (BluetoothDevice device : pairedDevices) {
      //device.getName() +" "+ device.getAddress());
   }
} else {
   mPairedDevicesArrayAdapter.add("没有找到已匹对的设备");
}
```

● 通过 mBluetoothAdapter.startDiscovery()方法来搜索设备，在此需要注册一个 BroadcastReceiver 来获得这个搜索结果。即先注册再获取信息，然后进行处理。演示代码如下所示。

```
//注册，当一个设备被发现时调用 onReceive
IntentFilter filter = new IntentFilter(BluetoothDevice.ACTION_FOUND);
    this.registerReceiver(mReceiver, filter);
//当搜索结束后调用 onReceive
filter = new IntentFilter(BluetoothAdapter.ACTION_DISCOVERY_FINISHED);
    this.registerReceiver(mReceiver, filter);
//.......
private BroadcastReceiver mReceiver = new BroadcastReceiver() {
    @Override
    public void onReceive(Context context, Intent intent) {
        String action = intent.getAction();
        if(BluetoothDevice.ACTION_FOUND.equals(action)){
             BluetoothDevice device = intent.getParcelableExtra(BluetoothDevice.
             EXTRA_DEVICE);
              // 已经配对的则跳过
             if (device.getBondState() != BluetoothDevice.BOND_BONDED) {
                  mNewDevicesArrayAdapter.add(device.getName() + "\n" + device.
                  getAddress());  //保存设备地址与名字
              }
        }else if (BluetoothAdapter.ACTION_DISCOVERY_FINISHED.equals(action)) {
        //搜索结束
            if (mNewDevicesArrayAdapter.getCount() == 0) {
                mNewDevicesArrayAdapter.add("没有搜索到设备");
            }
        }
    }
};
```

（4）建立连接

当查找到蓝牙设备后，接下来需要建立本机与其他设备之间的连接。一般在使用本机搜索其他蓝牙设备时，本机可以作为一个服务端来接收其他设备的连接。启动一个服务器端的线程，死循环等待客户端的连接，这与 ServerSocket 极为相似，此线程在准备连接之前启动。演示代码如下所示。

```
//UUID 可以看做一个端口号
private static final UUID MY_UUID =
UUID.fromString("fa87c0d0-afac-11de-8a39-0800200c9a66");
  //像一个服务器一样时刻监听是否有连接建立
  private class AcceptThread extends Thread{
      private BluetoothServerSocket serverSocket;

      public AcceptThread(boolean secure){
          BluetoothServerSocket temp = null;
          try {
              temp = mBluetoothAdapter.listenUsingRfcommWithServiceRecord(
                         NAME_INSECURE, MY_UUID);
          } catch (IOException e) {
                Log.e("app", "listen() failed", e);
          }
          serverSocket = temp;
      }

      public void run(){
          BluetoothSocket socket=null;
          while(true){
              try {
                  socket = serverSocket.accept();
              } catch (IOException e) {
                   Log.e("app", "accept() failed", e);
                   break;
              }
          }
          if(socket!=null){
```

```
                //此时可以新建一个数据交换线程，把此 socket 传进去
            }
        }

        //取消监听
        public void cancel(){
            try {
                serverSocket.close();
            } catch (IOException e) {
                Log.e("app", "Socket Type" + socketType + "close() of server failed", e);
            }
        }
}
```

（5）交换数据

当搜索到蓝牙设备后，接下来可以获取设备的地址，通过此地址获取一个 BluetoothDeviced 对象，可以将其看作是一个客户端，通过对象 device.createRfcommSocketToServiceRecord(MY_UUID) 同一个 UUID 可与服务器建立连接获取另一个 socket 对象。因为此服务端与客户端各有一个 socket 对象，所以此时它们可以互相交换数据了。演示代码如下所示。

```
//另一个设备去连接本机,相当于客户端
private class ConnectThread extends Thread{
    private BluetoothSocket socket;
    private BluetoothDevice device;
    public ConnectThread(BluetoothDevice device,boolean secure){
        this.device = device;
        BluetoothSocket tmp = null;
        try {
            tmp = device.createRfcommSocketToServiceRecord(MY_UUID_SECURE);
        } catch (IOException e) {
             Log.e("app", "create() failed", e);
        }
    }

    public void run(){
        mBluetoothAdapter.cancelDiscovery();    //取消设备查找
        try {
            socket.connect();
        } catch (IOException e) {
            try {
                socket.close();
            } catch (IOException e1) {
                 Log.e("app", "unable to close() "+
                         " socket during connection failure", e1);
            }
            connetionFailed();    //连接失败
            return;
        }
        //此时可以新创建一个数据交换线程，把此 socket 传进去
    }

      public void cancel() {
        try {
            socket.close();
        } catch (IOException e) {
            Log.e("app", "close() of connect  socket failed", e);
        }
      }
}
```

（6）建立数据通信线程

这一阶段的任务读取通信数据，演示代码如下所示。

```
//建立连接后，进行数据通信的线程
private class ConnectedThread extends Thread{
    private BluetoothSocket socket;
    private InputStream inStream;
```

```
    private OutputStream outStream;

    public ConnectedThread(BluetoothSocket socket){

        this.socket = socket;
        try {
            //获得输入输出流
            inStream = socket.getInputStream();
            outStream = socket.getOutputStream();
        } catch (IOException e) {
            Log.e("app", "temp sockets not created", e);
        }
    }

    public void run(){
        byte[] buff = new byte[1024];
        int len=0;
        //读数据需不断监听，写不需要
        while(true){
            try {
                len = inStream.read(buff);
                //把读取到的数据发送给 UI 进行显示
                Message msg = handler.obtainMessage(BluetoothChat.MESSAGE_READ,
                        len, -1, buff);
                msg.sendToTarget();
            } catch (IOException e) {
                Log.e("app", "disconnected", e);
                connectionLost();    //失去连接
                start();    //重新启动服务器
                break;
            }
        }
    }


    public void write(byte[] buffer) {
        try {
            outStream.write(buffer);
            handler.obtainMessage(BluetoothChat.MESSAGE_WRITE, -1, -1, buffer)
                    .sendToTarget();
        } catch (IOException e) {
            Log.e("app", "Exception during write", e);
        }
    }
    public void cancel() {
        try {
            socket.close();
        } catch (IOException e) {
            Log.e("app", "close() of connect socket failed", e);
        }
    }
}
```

到此为止，一个基本的蓝牙通信的基本操作已经全部完成。读者在开发此类项目时，只需按照上述流程进行即可。

19.6　实战演练——开发一个控制玩具车的蓝牙遥控器

本实例的功能是，在 Android 设备中开发一个蓝牙遥控器，通过这个遥控器可以控制玩具小车的运动轨迹。笔者为了节省成本，在淘宝网上网购了一个蓝牙模块，开始了我们这个实例之旅。

题目	目的	源码路径
实例 19-1	通过蓝牙遥控指挥玩具车	\daima\19\lanya

本实例项目的具体实现流程如下所示。

（1）将购买的蓝牙模块放在一辆玩具车上，并为其接通电源。

（2）打开 Eclipse 新建一个 Android 工程文件，命名为“lanya”。

（3）编写布局文件 main.xml，在里面插入了 5 个控制按钮，分别实现对玩具车的向前、左转、右转、后退和停止的控制。具体代码如下所示。

```
<?xml version="1.0" encoding="utf-8"?>
<AbsoluteLayout
android:id="@+id/widget0"
android:layout_width="fill_parent"
android:layout_height="fill_parent"
xmlns:android="http://schemas.android.com/apk/res/android"
>
<Button
android:id="@+id/btnF"
android:layout_width="100px"
android:layout_height="60px"
android:text="向前"
android:layout_x="130px"
android:layout_y="62px"
>
</Button>
<Button
android:id="@+id/btnL"
android:layout_width="100px"
android:layout_height="60px"
android:text="左转"
android:layout_x="20px"
android:layout_y="152px"
>
</Button>
<Button
android:id="@+id/btnR"
android:layout_width="100px"
android:layout_height="60px"
android:text="右转"
android:layout_x="240px"
android:layout_y="152px"
>
</Button>
<Button
android:id="@+id/btnB"
android:layout_width="100px"
android:layout_height="60px"
android:text="后退"
android:layout_x="130px"
android:layout_y="242px"
>
</Button>
<Button
android:id="@+id/btnS"
android:layout_width="100px"
android:layout_height="60px"
android:text="停止"
android:layout_x="130px"
android:layout_y="152px"
>
</Button>
</AbsoluteLayout>
```

（4）编写蓝牙程序控制文件 lanya.java，具体实现流程如下所示。

● 定义类 lanya，然后设置 5 个按钮对象，具体代码如下所示。

```
public class lanya extends Activity {
   private static final String TAG = "THINBTCLIENT";
   private static final boolean D = true;
```

```
    private BluetoothAdapter mBluetoothAdapter = null;
    private BluetoothSocket btSocket = null;

    private OutputStream outStream = null;
    Button mButtonF;
    Button mButtonB;
    Button mButtonL;
    Button mButtonR;
    Button mButtonS;
```

- 赋值蓝牙设备上的标准串行和要连接的蓝牙设备 MAC 地址，具体代码如下所示。

```
private   static   final   UUID   MY_UUID   =   UUID.fromString("00011101-0000-1000-8019-00805F9B34FB");//蓝牙设备上的标准串行
private static String address = "00:11:03:21:00:42"; // <==要连接的蓝牙设备 MAC 地址
```

- 编写单击“向前”按钮的处理事件，具体代码如下所示。

```
@Override
public void onCreate(Bundle savedInstanceState) {
      super.onCreate(savedInstanceState);
      setContentView(R.layout.main);
      //向前
      mButtonF=(Button)findViewById(R.id.btnF);
      mButtonF.setOnTouchListener(new Button.OnTouchListener(){

         @Override
         public boolean onTouch(View v, MotionEvent event) {
            // TODO Auto-generated method stub
            String message;
            byte[] msgBuffer;
            int action = event.getAction();
            switch(action)
            {
            case MotionEvent.ACTION_DOWN:
            try {
                 outStream = btSocket.getOutputStream();
              } catch (IOException e) {
                 Log.e(TAG, "ON RESUME: Output stream creation failed.", e);
              }
              message = "1";
             msgBuffer = message.getBytes();
              try {
                 outStream.write(msgBuffer);
              } catch (IOException e) {
                      Log.e(TAG, "ON RESUME: Exception during write.", e);
              }
            break;
            case MotionEvent.ACTION_UP:
               try {
                      outStream = btSocket.getOutputStream();
                 } catch (IOException e) {
                      Log.e(TAG, "ON RESUME: Output stream creation failed.", e);
                 }
                 message = "0";
                 msgBuffer = message.getBytes();
                 try {
                    outStream.write(msgBuffer);
                 } catch (IOException e) {
                         Log.e(TAG, "ON RESUME: Exception during write.", e);
                 }
               break;
            }
            return false;
         }
      });
```

- 编写单击“后退”按钮的处理事件，具体代码如下所示。

```
mButtonB=(Button)findViewById(R.id.btnB);
mButtonB.setOnTouchListener(new Button.OnTouchListener(){
```

```
    @Override
    public boolean onTouch(View v, MotionEvent event) {
        // TODO Auto-generated method stub
        String message;
        byte[] msgBuffer;
        int action = event.getAction();
        switch(action)
        {
        case MotionEvent.ACTION_DOWN:
        try {
            outStream = btSocket.getOutputStream();
          } catch (IOException e) {
            Log.e(TAG, "ON RESUME: Output stream creation failed.", e);
          }
         message = "3";
        msgBuffer = message.getBytes();
         try {
            outStream.write(msgBuffer);
          } catch (IOException e) {
                Log.e(TAG, "ON RESUME: Exception during write.", e);
          }
        break;

        case MotionEvent.ACTION_UP:
            try {
                outStream = btSocket.getOutputStream();
            } catch (IOException e) {
                Log.e(TAG, "ON RESUME: Output stream creation failed.", e);
            }
            message = "0";
            msgBuffer = message.getBytes();
            try {
                outStream.write(msgBuffer);
            } catch (IOException e) {
                    Log.e(TAG, "ON RESUME: Exception during write.", e);
            }
          break;
        }

        return false;
    }
});
```

- 编写单击“左转”按钮的处理事件，具体代码如下所示。

```
mButtonL=(Button)findViewById(R.id.btnL);
mButtonL.setOnTouchListener(new Button.OnTouchListener(){
    @Override
    public boolean onTouch(View v, MotionEvent event) {
        // TODO Auto-generated method stub
        String message;
        byte[] msgBuffer;
        int action = event.getAction();
        switch(action)
        {
        case MotionEvent.ACTION_DOWN:
        try {
            outStream = btSocket.getOutputStream();
          } catch (IOException e) {
            Log.e(TAG, "ON RESUME: Output stream creation failed.", e);
          }
         message = "2";
        msgBuffer = message.getBytes();
         try {
            outStream.write(msgBuffer);
          } catch (IOException e) {
                Log.e(TAG, "ON RESUME: Exception during write.", e);
          }
        break;
```

```
        case MotionEvent.ACTION_UP:
            try {
                outStream = btSocket.getOutputStream();
              } catch (IOException e) {
                Log.e(TAG, "ON RESUME: Output stream creation failed.", e);
              }
              message = "0";
              msgBuffer = message.getBytes();
              try {
                outStream.write(msgBuffer);
              } catch (IOException e) {
                    Log.e(TAG, "ON RESUME: Exception during write.", e);
              }
            break;
        }

        return false;

    }
});
```

- 编写单击“右转”按钮的处理事件，具体代码如下所示。

```
mButtonR=(Button)findViewById(R.id.btnR);
mButtonR.setOnTouchListener(new Button.OnTouchListener(){
    @Override
    public boolean onTouch(View v, MotionEvent event) {
        // TODO Auto-generated method stub
        String message;
        byte[] msgBuffer;
        int action = event.getAction();
        switch(action)
        {
        case MotionEvent.ACTION_DOWN:
        try {
              outStream = btSocket.getOutputStream();
          } catch (IOException e) {
              Log.e(TAG, "ON RESUME: Output stream creation failed.", e);
          }
          message = "4";
         msgBuffer = message.getBytes();
          try {
              outStream.write(msgBuffer);
          } catch (IOException e) {
                  Log.e(TAG, "ON RESUME: Exception during write.", e);
          }
        break;

        case MotionEvent.ACTION_UP:
            try {
                outStream = btSocket.getOutputStream();
              } catch (IOException e) {
                outStream = Log.e(TAG, "ON RESUME: Output stream creation failed.", e);
              }
              message = "0";
              msgBuffer = message.getBytes();
              try {
                outStream.write(msgBuffer);
              } catch (IOException e) {
                    Log.e(TAG, "ON RESUME: Exception during write.", e);
              }
            break;
        }

        return false;
    }
});
```

- 编写单击“停止”按钮的处理事件，具体代码如下所示。

```
mButtonS=(Button)findViewById(R.id.btnS);
```

```
        mButtonS.setOnTouchListener(new Button.OnTouchListener(){
            @Override
            public boolean onTouch(View v, MotionEvent event) {
                // TODO Auto-generated method stub
                if(event.getAction()==MotionEvent.ACTION_DOWN)
                 try {
                        outStream = btSocket.getOutputStream();
                    } catch (IOException e) {
                        Log.e(TAG, "ON RESUME: Output stream creation failed.", e);
                    }
                    String message = "0";
                    byte[] msgBuffer = message.getBytes();
                    try {
                        outStream.write(msgBuffer);
                    } catch (IOException e) {
                            Log.e(TAG, "ON RESUME: Exception during write.", e);
                    }
                return false;
            }
        });
        if (D)
            Log.e(TAG, "+++ ON CREATE +++");
            mBluetoothAdapter = BluetoothAdapter.getDefaultAdapter();
        if (mBluetoothAdapter == null) {
            Toast.makeText(this, "蓝牙设备不可用，请打开蓝牙！", Toast.LENGTH_LONG).show();
                finish();
                return;
        }
        if (!mBluetoothAdapter.isEnabled()) {
                Toast.makeText(this,"请打开蓝牙并重新运行程序！", Toast.LENGTH_LONG).show();
                finish();
                return;
        }
        if (D)
        Log.e(TAG, "+++ DONE IN ON CREATE, GOT LOCAL BT ADAPTER +++");
}
```

- 通过套接字建立蓝牙连接，如果连接失败则输出对应的失败提示。主要代码如下所示。

```
    @Override
    public void onStart() {
        super.onStart();
        if (D) Log.e(TAG, "++ ON START ++");
    }
    @Override
    public void onResume() {
        super.onResume();
        if (D) {
            Log.e(TAG, "+ ON RESUME +");
         Log.e(TAG, "+ ABOUT TO ATTEMPT CLIENT CONNECT +");
        }
        DisplayToast("正在尝试连接智能小车，请稍侯····");
        BluetoothDevice device = mBluetoothAdapter.getRemoteDevice(address);
        try {
           btSocket = device.createRfcommSocketToServiceRecord(MY_UUID);
        } catch (IOException e) {
            DisplayToast("套接字创建失败！");
        }
        DisplayToast("成功连接智能小车！可以开始操控了~~~");
        mBluetoothAdapter.cancelDiscovery();
        try {
                btSocket.connect();
                DisplayToast("连接成功建立，数据连接打开！");

        } catch (IOException e) {
                try {
                    btSocket.close();
                } catch (IOException e2) {

                        DisplayToast("连接没有建立，无法关闭套接字！");
```

```
                }
            }
            if (D)
                Log.e(TAG, "+ ABOUT TO SAY SOMETHING TO SERVER +");
    }
    @Override
    public void onPause() {
            super.onPause();
            if (D)
                Log.e(TAG, "- ON PAUSE -");
            if (outStream != null) {
                    try {
                            outStream.flush();
                    } catch (IOException e) {
                            Log.e(TAG, "ON PAUSE: Couldn't flush output stream.", e);
                    }
            }
            try {
                    btSocket.close();
            } catch (IOException e2) {

                    DisplayToast("套接字关闭失败! ");
            }
    }
    @Override
    public void onStop() {
            super.onStop();
            if (D)Log.e(TAG, "-- ON STOP --");
    }
    @Override
    public void onDestroy() {
            super.onDestroy();
            if (D) Log.e(TAG, "--- ON DESTROY ---");
    }
    public void DisplayToast(String str)
    {
        Toast toast=Toast.makeText(this, str, Toast.LENGTH_LONG);
        toast.setGravity(Gravity.TOP, 0, 220);
        toast.show();
      }
}
```

（5）在文件 AndroidManifest.xml 中声明蓝牙权限，对应代码如下所示。

```
<uses-permission android:name="android.permission.BLUETOOTH_ADMIN" />
 <uses-permission android:name="android.permission.BLUETOOTH" />
    <application android:icon="@drawable/icon" android:label="@string/app_name">
        <activity android:name=".lanya"
                  android:label="@string/app_name">
            <intent-filter>
                <action android:name="android.intent.action.MAIN" />
                <category android:name="android.intent.category.LAUNCHER" />
            </intent-filter>
        </activity>
```

到此为止，我们的蓝牙控制玩具车的实例就介绍完毕了。本实例的实现比较简单，难度大的是双向控制，即实现每个设备都可以操控另外一个设备的功能，此时就需要有蓝颜功能的电脑或第二部 Andorid 手机来完成测试了。在模拟器中因为不具备蓝牙设备，程序执行后会显示“蓝牙设备不可用，请打开蓝牙！”的提示，如图 19-3 所示。

蓝牙设备不可用，请打开蓝牙！

▲图 19-3　模拟器的运行效果

19.7 实战演练——开发一个蓝牙控制器

本实例的功能是，在 Android 设备中开发一个蓝牙控制器，通过这个控制器可以实现如下所示的功能。

- 打开蓝牙。
- 关闭蓝牙。
- 允许搜索。
- 开始搜索。
- 客户端。
- 服务器端。
- OBEX 服务器。

题目	目的	源码路径
实例 19-2	开发一个 Android 蓝牙控制器	\daima\19\Activity01

19.7.1 界面布局

本实例主界面的布局文件 main.xml，主要实现代码如下所示。

```
<?xml version="1.0" encoding="utf-8"?>
<LinearLayout xmlns:android="http://schemas.android.com/apk/res/android"
   android:orientation="vertical" android:layout_width="fill_parent"
   android:layout_height="fill_parent" android:padding="10dip">
   <Button android:layout_width="fill_parent"
      android:layout_height="wrap_content" android:text="打开蓝牙"
      android:onClick="onEnableButtonClicked" />
   <Button android:layout_width="fill_parent"
      android:layout_height="wrap_content" android:text="关闭蓝牙"
      android:onClick="onDisableButtonClicked" />
   <Button android:layout_width="fill_parent"
      android:layout_height="wrap_content" android:text="允许搜索"
      android:onClick="onMakeDiscoverableButtonClicked" />
   <Button android:layout_width="fill_parent"
      android:layout_height="wrap_content" android:text="开始搜索"
      android:onClick="onStartDiscoveryButtonClicked" />
   <Button android:layout_width="fill_parent"
      android:layout_height="wrap_content" android:text="客户端"
      android:onClick="onOpenClientSocketButtonClicked" />
   <Button android:layout_width="fill_parent"
      android:layout_height="wrap_content" android:text="服务器端"
      android:onClick="onOpenServerSocketButtonClicked" />
   <Button android:layout_width="fill_parent"
      android:layout_height="wrap_content" android:text="OBEX 服务器"
      android:onClick="onOpenOBEXServerSocketButtonClicked" />
</LinearLayout>
```

执行之后的界面效果如图 19-4 所示。

▲图 19-4　执行效果

服务器端的界面布局文件是 server_socket.xml，主要实现代码如下所示。

```
<?xml version="1.0" encoding="utf-8"?>
<LinearLayout xmlns:android="http://schemas.android.com/apk/res/android"
```

```
    android:orientation="vertical" android:layout_width="fill_parent"
    android:layout_height="fill_parent" android:padding="10dip">
    <Button android:layout_width="fill_parent"
        android:layout_height="wrap_content" android:text="Stop server"
        android:onClick="onButtonClicked" />

    <ListView android:id="@+id/android:list" android:layout_width="fill_parent"
        android:layout_height="fill_parent" />
</LinearLayout>
```

其他几个界面的布局文件和上述文件类似，为节省本书篇幅，将不再一一列出。

19.7.2　响应单击按钮

编写主界面的程序文件 Activity01.java，功能是根据用户单击屏幕中的按钮来调用对应的处理函数，例如单击“服务器端”按钮会执行函数 onOpenServerSocketButtonClicked(View view)。文件 Activity01.java 的主要实现代码如下所示。

```
public class Activity01 extends Activity
{
    /* 取得默认的蓝牙适配器 */
    private BluetoothAdapter _bluetooth = BluetoothAdapter.getDefaultAdapter();
    /* 请求打开蓝牙 */
    private static final int    REQUEST_ENABLE          = 0x1;
    /* 请求能够被搜索 */
    private static final int    REQUEST_DISCOVERABLE    = 0x2;
    /** Called when the activity is first created. */
    @Override
    public void onCreate(Bundle savedInstanceState)
    {
        super.onCreate(savedInstanceState);
        setContentView(R.layout.main);
    }
    /* 开启蓝牙 */
    public void onEnableButtonClicked(View view)
    {
        // 用户请求打开蓝牙
        //Intent enabler = new Intent(BluetoothAdapter.ACTION_REQUEST_ENABLE);
        //startActivityForResult(enabler, REQUEST_ENABLE);
        //打开蓝牙
        _bluetooth.enable();
    }
    /* 关闭蓝牙 */
    public void onDisableButtonClicked(View view)
    {
        _bluetooth.disable();
    }
    /* 使设备能够被搜索 */
    public void onMakeDiscoverableButtonClicked(View view)
    {
        Intent enabler = new Intent(BluetoothAdapter.ACTION_REQUEST_DISCOVERABLE);
        startActivityForResult(enabler, REQUEST_DISCOVERABLE);
    }
    /* 开始搜索 */
    public void onStartDiscoveryButtonClicked(View view)
    {
        Intent enabler = new Intent(this, DiscoveryActivity.class);
        startActivity(enabler);
    }
    /* 客户端 */
    public void onOpenClientSocketButtonClicked(View view)
    {
        Intent enabler = new Intent(this, ClientSocketActivity.class);
        startActivity(enabler);
    }
    /* 服务端 */
    public void onOpenServerSocketButtonClicked(View view)
```

```
    {
        Intent enabler = new Intent(this, ServerSocketActivity.class);
        startActivity(enabler);
    }
    /* OBEX 服务器 */
    public void onOpenOBEXServerSocketButtonClicked(View view)
    {
        Intent enabler = new Intent(this, OBEXActivity.class);
        startActivity(enabler);
    }
}
```

19.7.3 和指定的服务器建立连接

编写程序文件 ClientSocketActivity.java，功能是创建一个 Socket 连接，和指定的服务器建立连接。文件 ClientSocketActivity.java 的主要实现代码如下所示。

```
public class ClientSocketActivity  extends Activity
{
    private static final String TAG = ClientSocketActivity.class.getSimpleName();
    private static final int REQUEST_DISCOVERY = 0x1;;
    private Handler _handler = new Handler();
    private BluetoothAdapter  _bluetooth = BluetoothAdapter.getDefaultAdapter();
    protected void onCreate(Bundle savedInstanceState) {
        super.onCreate(savedInstanceState);
        getWindow().setFlags(WindowManager.LayoutParams.FLAG_BLUR_BEHIND,
        WindowManager.LayoutParams.FLAG_BLUR_BEHIND);
        setContentView(R.layout.client_socket);
        if (!_bluetooth.isEnabled()) {
            finish();
            return;
        }
        Intent intent = new Intent(this, DiscoveryActivity.class);
        /* 提示选择一个要连接的服务器 */
        Toast.makeText(this, "select device to connect", Toast.LENGTH_SHORT).show();
        /* 跳转到搜索的蓝牙设备列表区，进行选择 */
        startActivityForResult(intent, REQUEST_DISCOVERY);
    }
    /* 选择了服务器之后进行连接 */
    protected void onActivityResult(int requestCode, int resultCode, Intent data) {
        if (requestCode != REQUEST_DISCOVERY) {
            return;
        }
        if (resultCode != RESULT_OK) {
            return;
        }
        final BluetoothDevice device = data.getParcelableExtra(BluetoothDevice.EXTRA_
        DEVICE);
        new Thread() {
            public void run() {
                /* 连接 */
                connect(device);
            };
        }.start();
    }
    protected void connect(BluetoothDevice device) {
        BluetoothSocket socket = null;
        try {
            //创建一个 Socket 连接：只需要服务器在注册时的 UUID 号
            // socket = device.createRfcommSocketToServiceRecord(BluetoothProtocols.
            // OBEX_OBJECT_PUSH_PROTOCOL_UUID);
            socket = device.createRfcommSocketToServiceRecord(UUID.fromString
            ("a60f35f0-b93a-11de-8a39-08002009c666"));
            //连接
            socket.connect();
        } catch (IOException e) {
            Log.e(TAG, "", e);
        } finally {
```

```
                if (socket != null) {
                    try {
                        socket.close();
                    } catch (IOException e) {
                        Log.e(TAG, "", e);
                    }
                }
            }
        }
    }
```

19.7.4　搜索附近的蓝牙设备

编写程序文件 DiscoveryActivity.java，功能是搜索设备附近的蓝牙设备，并在列表中显示搜索到的蓝牙设备。文件 DiscoveryActivity.java 的主要实现代码如下所示。

```
public class DiscoveryActivity  extends ListActivity
{
    private Handler _handler = new Handler();
    /* 取得默认的蓝牙适配器 */
    private BluetoothAdapter _bluetooth = BluetoothAdapter.getDefaultAdapter();
    /* 用来存储搜索到的蓝牙设备 */
    private List<BluetoothDevice> _devices = new ArrayList<BluetoothDevice>();
    /* 是否完成搜索 */
    private volatile boolean _discoveryFinished;
    private Runnable _discoveryWorkder = new Runnable() {
        public void run()
        {
            /* 开始搜索 */
            _bluetooth.startDiscovery();
            for (;;)
            {
                if (_discoveryFinished)
                {
                    break;
                }
                try
                {
                    Thread.sleep(100);
                }
                catch (InterruptedException e){}
            }
        }
    };
    /**
     * 接收器
     * 当搜索蓝牙设备完成时调用
     */
    private BroadcastReceiver _foundReceiver = new BroadcastReceiver() {
        public void onReceive(Context context, Intent intent) {
            /* 从 intent 中取得搜索结果数据 */
            BluetoothDevice device = intent
                    .getParcelableExtra(BluetoothDevice.EXTRA_DEVICE);
            /* 将结果添加到列表中 */
            _devices.add(device);
            /* 显示列表 */
            showDevices();
        }
    };
    private BroadcastReceiver _discoveryReceiver = new BroadcastReceiver() {

        @Override
        public void onReceive(Context context, Intent intent)
        {
            /* 卸载注册的接收器 */
            unregisterReceiver(_foundReceiver);
            unregisterReceiver(this);
            _discoveryFinished = true;
        }
```

```
    };
    protected void onCreate(Bundle savedInstanceState)
    {
        super.onCreate(savedInstanceState);
        getWindow().setFlags(WindowManager.LayoutParams.FLAG_BLUR_BEHIND,
        WindowManager.LayoutParams.FLAG_BLUR_BEHIND);
        setContentView(R.layout.discovery);
        /* 如果蓝牙适配器没有打开，则结束 */
        if (!_bluetooth.isEnabled())
        {

            finish();
            return;
        }
        /* 注册接收器 */
        IntentFilter discoveryFilter = new IntentFilter(BluetoothAdapter.ACTION_
        DISCOVERY_FINISHED);
        registerReceiver(_discoveryReceiver, discoveryFilter);
        IntentFilter foundFilter = new IntentFilter(BluetoothDevice.ACTION_FOUND);
        registerReceiver(_foundReceiver, foundFilter);
        /* 显示一个对话框,正在搜索蓝牙设备 */
        SamplesUtils.indeterminate(DiscoveryActivity.this, _handler, "Scanning...",
        _discoveryWorkder, new OnDismissListener() {
            public void onDismiss(DialogInterface dialog)
            {

                for (; _bluetooth.isDiscovering();)
                {
                    _bluetooth.cancelDiscovery();
                }
                _discoveryFinished = true;
            }
        }, true);
    }
    /* 显示列表 */
    protected void showDevices()
    {
        List<String> list = new ArrayList<String>();
        for (int i = 0, size = _devices.size(); i < size; ++i)
        {
            StringBuilder b = new StringBuilder();
            BluetoothDevice d = _devices.get(i);
            b.append(d.getAddress());
            b.append('\n');
            b.append(d.getName());
            String s = b.toString();
            list.add(s);
        }
        final ArrayAdapter<String> adapter = new ArrayAdapter<String>(this, android.R.
        layout.simple_list_item_1, list);
        _handler.post(new Runnable() {
            public void run()
            {
                setListAdapter(adapter);
            }
        });
    }
    protected void onListItemClick(ListView l, View v, int position, long id)
    {
        Intent result = new Intent();
        result.putExtra(BluetoothDevice.EXTRA_DEVICE, _devices.get(position));
        setResult(RESULT_OK, result);
        finish();
    }
}
```

19.7.5 建立和 OBEX 服务器的数据传输

编写程序文件 OBEXActivity.java，功能是建立和 OBEX 服务器的数据传输。OBEX 全称为

Object Exchange，中文为对象交换，所以称之为对象交换协议。OBEX 协议通过简单地使用“PUT”和“GET”命令实现在不同的设备、不同的平台之间方便、高效地交换信息。支持的设备广泛，例如 PC、PDA、电话、摄像头、自动答录机、计算器、数据采集器和手表等。文件 OBEXActivity.java 的主要实现代码如下所示。

```
public class OBEXActivity extends Activity
{
    private static final String TAG = "@MainActivity";
    private Handler _handler = new Handler();
    private BluetoothServerSocket _server;
    private BluetoothSocket _socket;
    private static final int OBEX_CONNECT = 0x80;
    private static final int OBEX_DISCONNECT = 0x81;
    private static final int OBEX_PUT = 0x02;
    private static final int OBEX_PUT_END = 0x82;
    private static final int OBEX_RESPONSE_OK = 0xa0;
    private static final int OBEX_RESPONSE_CONTINUE = 0x90;
    private static final int BIT_MASK = 0x000000ff;
    Thread t = new Thread()
    {
        public void run()
        {
            try
            {
                _server = BluetoothAdapter.getDefaultAdapter().listenUsingRfcommWith
                ServiceRecord("OBEX", null);
                new Thread()
                {
                    public void run()
                    {
                        Log.d("@Rfcom", "begin close");
                        try
                        {
                            _socket.close();
                        }
                        catch (IOException e)
                        {
                            Log.e(TAG, "", e);
                        }
                        Log.d("@Rfcom", "end close");
                    };
                }.start();
                _socket = _server.accept();
                reader.start();
                Log.d(TAG, "shutdown thread");
            }
            catch (IOException e)
            {
                e.printStackTrace();
            }
        };
    };

    Thread reader = new Thread()
    {
        public void run()
        {
            try
            {
                Log.d(TAG, "getting inputstream");
                InputStream inputStream = _socket.getInputStream();
                OutputStream outputStream = _socket.getOutputStream();
                Log.d(TAG, "got inputstream");
                int read = -1;
                byte[] bytes = new byte[2048];
                ByteArrayOutputStream baos = new ByteArrayOutputStream(bytes.length);
                while ((read = inputStream.read(bytes)) != -1)
                {
```

```
                baos.write(bytes, 0, read);
                byte[] req = baos.toByteArray();
                int op = req[0] & BIT_MASK;
                Log.d(TAG, "read:" + Arrays.toString(req));
                Log.d(TAG, "op:" + Integer.toHexString(op));
                switch (op)
                {
                    case OBEX_CONNECT:
                        outputStream.write(new byte[] { (byte) OBEX_RESPONSE_OK, 0, 7,
                        16, 0, 4, 0 });
                        break;
                    case OBEX_DISCONNECT:
                        outputStream.write(new byte[] { (byte) OBEX_RESPONSE_OK, 0, 3, 0 });
                        break;

                    case OBEX_PUT:
                        outputStream.write(new byte[] { (byte) OBEX_RESPONSE_CONTINUE,
                        0, 3, 0 });
                        break;
                    case OBEX_PUT_END:

                        outputStream.write(new byte[] { (byte) OBEX_RESPONSE_OK, 0, 3, 0 });
                        break;

                    default:
                        outputStream.write(new byte[] { (byte) OBEX_RESPONSE_OK, 0, 3, 0 });
                }
                Log.d(TAG, new String(baos.toByteArray(), "utf-8"));
                baos = new ByteArrayOutputStream(bytes.length);
            }
            Log.d(TAG, new String(baos.toByteArray(), "utf-8"));
        }
        catch (IOException e)
        {
            e.printStackTrace();
        }
    };
};
private Thread put = new Thread() {
    public void run()
    {
    };
};
public void onCreate(Bundle savedInstanceState)
{
    super.onCreate(savedInstanceState);
    setContentView(R.layout.obex_server_socket);
    t.start();
}
protected void onActivityResult(int requestCode, int resultCode, Intent data)
{
    Log.d(TAG, data.getData().toString());
    switch (requestCode)
    {
        case (1):
            if (resultCode == Activity.RESULT_OK)
            {
                Uri contactData = data.getData();
                @SuppressWarnings("deprecation")
                Cursor c = managedQuery(contactData, null, null, null, null);
                for (; c.moveToNext();)
                {
                    Log.d(TAG, "c1---------------------------------------");
                    dump(c);
                    Uri uri = Uri.withAppendedPath(data.getData(), ContactsContract.
                    Contacts.Photo.CONTENT_DIRECTORY);
                    @SuppressWarnings("deprecation")
                    Cursor c2 = managedQuery(uri, null, null, null, null);
                    for (; c2.moveToNext();)
                    {
```

```
                        Log.d(TAG, "c2---------------------------------------");
                        dump(c2);
                    }
                }
            }
            break;
        }
    }
```

19.7.6　实现蓝牙服务器端的数据处理

编写文件 ServerSocketActivity.java，功能是实现蓝牙服务器端的数据处理，建立服务器端和客户端的连接和监听工作，其中的监听工作和停止服务器工作由独立的函数实现。文件 ServerSocketActivity.java 的主要实现代码如下所示。

```
public class ServerSocketActivity extends ListActivity
{
    /* 一些常量，代表服务器的名称 */
    public static final String PROTOCOL_SCHEME_L2CAP = "btl2cap";
    public static final String PROTOCOL_SCHEME_RFCOMM = "btspp";
    public static final String PROTOCOL_SCHEME_BT_OBEX = "btgoep";
    public static final String PROTOCOL_SCHEME_TCP_OBEX = "tcpobex";
    private static final String TAG = ServerSocketActivity.class.getSimpleName();
    private Handler _handler = new Handler();
    /* 取得默认的蓝牙适配器 */
    private BluetoothAdapter _bluetooth = BluetoothAdapter.getDefaultAdapter();
    /* 蓝牙服务器 */
    private BluetoothServerSocket _serverSocket;
    /* 线程-监听客户端的链接 */
    private Thread _serverWorker = new Thread() {
        public void run() {
            listen();
        };
    };
    protected void onCreate(Bundle savedInstanceState) {
        super.onCreate(savedInstanceState);
        getWindow().setFlags(WindowManager.LayoutParams.FLAG_BLUR_BEHIND,
                WindowManager.LayoutParams.FLAG_BLUR_BEHIND);
        setContentView(R.layout.server_socket);
        if (!_bluetooth.isEnabled()) {
            finish();
            return;
        }
        /* 开始监听 */
        _serverWorker.start();
    }
    protected void onDestroy() {
        super.onDestroy();
        shutdownServer();
    }
    protected void finalize() throws Throwable {
        super.finalize();
        shutdownServer();
    }
    /* 停止服务器 */
    private void shutdownServer() {
        new Thread() {
            public void run() {
                _serverWorker.interrupt();
                if (_serverSocket != null) {
                    try {
                        /* 关闭服务器 */
                        _serverSocket.close();
                    } catch (IOException e) {
                        Log.e(TAG, "", e);
                    }
                    _serverSocket = null;
                }
```

```
            };
        }.start();
    }
    public void onButtonClicked(View view) {
        shutdownServer();
    }
    protected void listen() {
        try {
            /* 创建一个蓝牙服务器
             * 参数分别：服务器名称、UUID
             */
            _serverSocket = _bluetooth.listenUsingRfcommWithServiceRecord(PROTOCOL_
            SCHEME_RFCOMM,UUID.fromString("a60f35f0-b93a-11de-8a39-08002009c666"));
            /* 客户端连线列表 */
            final List<String> lines = new ArrayList<String>();
            _handler.post(new Runnable() {
                public void run() {
                    lines.add("Rfcomm server started...");
                    ArrayAdapter<String> adapter = new ArrayAdapter<String>(
                            ServerSocketActivity.this,
                            android.R.layout.simple_list_item_1, lines);
                    setListAdapter(adapter);
                }
            });
            /* 接受客户端的连接请求 */
            BluetoothSocket socket = _serverSocket.accept();
            /* 处理请求内容 */
            if (socket != null) {
                InputStream inputStream = socket.getInputStream();
                int read = -1;
                final byte[] bytes = new byte[2048];
                for (; (read = inputStream.read(bytes)) > -1;) {
                    final int count = read;
                    _handler.post(new Runnable() {
                        public void run() {
                            StringBuilder b = new StringBuilder();
                            for (int i = 0; i < count; ++i) {
                                if (i > 0) {
                                    b.append(' ');
                                }
                                String s = Integer.toHexString(bytes[i] & 0xFF);
                                if (s.length() < 2) {
                                    b.append('0');
                                }
                                b.append(s);
                            }
                            String s = b.toString();
                            lines.add(s);
                            ArrayAdapter<String> adapter = new ArrayAdapter<String>(
                                    ServerSocketActivity.this,
                                    android.R.layout.simple_list_item_1, lines);
                            setListAdapter(adapter);
                        }
                    });
                }
            }
        } catch (IOException e) {
            Log.e(TAG, "", e);
        } finally {
        }
    }
}
```

在文件 AndroidManifest.xml 中声明对蓝牙设备的使用权限，具体代码如下所示。

```
<uses-permission android:name="android.permission.BLUETOOTH" />
<uses-permission android:name="android.permission.BLUETOOTH_ADMIN" />
<uses-permission android:name="android.permission.READ_CONTACTS"/>
```

到此为止，整个实例全部实现完毕，本实例需要在真实机器上进行测试。

第 20 章　低功耗蓝牙技术详解

用“蓝牙”技术可以有效地简化移动通信终端设备之间的通信，也能够成功地简化设备与因特网 Internet 之间的通信，从而使数据传输变得更加迅速高效，为无线通信拓宽道路。蓝牙采用分散式网络结构以及快跳频和短包技术，支持点对点及点对多点通信，工作在全球通用的 2.4GHz ISM（即工业、科学、医学）频段。本章将详细讲解低功耗蓝牙协议栈的基本知识，为读者步入本书后面知识的学习打下基础。

20.1 短距离无线通信技术概览

在物联网中物与网相连的最后数米，发挥关键作用的是短距离无线传输技术。目前有多种短距离无线传输技术可以应用在物联网中，在我国，除已经得到大规模应用的 RFID 之外，还有 Wi-Fi、ZigBee、蓝牙等比较成熟的技术，以及基于这些技术发展而来的新技术。这些技术各具特点，因对其传输速度、距离、耗电量等方面的要求不同，形成了各自不同的物联网应用场景。本节将简要介绍当今实现短距离无线通信的常用技术。

20.1.1 ZigBee——低功耗、自组网

ZigBee 以其鲜明的技术特点在物联网中受到了高度关注，该技术使用的频段分别为 2.4GHz、868MHz（欧洲）及 915MHz（美国）。其主要的技术特点：一是数据传输速率低，只有 10～250Kbps；二是功耗低，低传输速率带来了仅为 1 毫瓦的低发射功率。据估算，ZigBee 设备仅靠两节 5 号电池就可以维持长达 6 个月到两年左右的使用时间，这是 ZigBee 的一个独特优势；三是成本低，因为 ZigBee 传输速率低、协议简单；四是网络容量大，每个 ZigBee 网络最多可以支持 255 个设备，一个区域内可以同时存在最多 100 个 ZigBee 网络，网络组成灵活。ZigBee 芯片主要企业有德州仪器、飞思卡尔等。市场调研机构 ABI Reserch 的一份数据显示，2005 年到 2012 年，ZigBee 市场的年均复合增长率为 63%。

“ZigBee 是从家庭自动化开始的，在瑞典哥德堡就是从智能电表开始，然后进一步用到燃气表、水表、热力表等家庭各种计量表。”在 2011 中国无线世界暨物联网大会上 ZigBee 联盟大中华区代表黄家瑞说，“ZigBee 在智能电表里不仅仅是远程抄表工具，它是一个终端，也是一个网关，这些网关结合在一起，整个小区就变成了智能电网小区，智能电表可以搜集家里所有家电的用电信息。”

目前，ZigBee 正在完善其网关标准，2011 年 7 月底发布了第十个标准 ZigBee Gateway（ZigBee 网关）。ZigBee Gateway 提供了一种简单、高成本效益的互联网连接方式，使服务提供商、企业和个人消费者有机会运行这些设备并将 ZigBee 网络连接至互联网。ZigBee Gateway 是 ZigBee Network Devicesp（ZigBee 网络设备）这一新类别范畴的首个标准，这将使 ZigBee 发展进一步提速。

20.1.2 Wi-Fi——大带宽支持家庭互联

Wi-Fi 是以太网的一种无线扩展技术，如果有多个用户同时通过一个热点接入，带宽将被这些用户共享，Wi-Fi 的速率会降低。处于 2.4GHz 频段的 Wi-Fi 信号受墙壁阻隔的影响较小。Wi-Fi 的传输速率随着技术的演进还在不断提高，我国电信运营商在构建无线城市中采用的 Wi-Fi 技术部分已经升级到 802.11n，最高速率从 802.11g 标准的 11Mbps 提高到 50Mbps 以上。在 Wi-Fi 产业链中，最大的芯片企业是博通。

"在过去几年里整个 Wi-Fi 技术和产品发货量达到 20 亿个，整个 Wi-Fi 产品销售每年都是以两位数的速度持续增长。"Wi-Fi 联盟董事 Myron Hattig 说："在 2011 年我们还会销售 10 亿个产品。"

在笔记本电脑和手机上已经得到广泛应用的 Wi-Fi 正在向消费电子产品渗透，Myron Hattig 说："除了手机外，已经有 25%的消费类电子设备使用 Wi-Fi，在打印机、洗衣机上都在使用 Wi-Fi，家用电器生产商协会将 Wi-Fi 作为一个更高级别的智能电器沟通技术。Wi-Fi 可以将设备与设备相连，从而使整个家庭的家用电器、电子设备相连。"

最大 Wi-Fi 芯片制造商博通正在推动 Wi-Fi Direct 标准的商用，以支持这种设备到设备的直连。特别是在家庭互联中，相片、视频等大数据量的业务在手机、平板电脑、电视等设备中的直连应用前景广阔。Myron Hattig 告诉记者："直接技术可将平板电脑的内容展示在电视上，相关产品会在 2012 年发布。"

基于 Wi-Fi 上发展起来的 WIGIG 也是未来家庭互联市场有力的竞争技术。该技术可工作在 40～60GHz 的超高频段，其传输速度可以达到 1Gbps 以上，不能穿过墙壁。目前英特尔、高通等芯片企业在支持 WIGIG 发展，目前该技术还在完善中，如需要进一步降低功耗等。

20.1.3 蓝牙——4.0 进入低功耗时代

蓝牙能在包括移动电话、PDA、无线耳机、笔记本电脑、相关外设等众多设备之间进行无线信息交换。蓝牙采用分散式网络结构以及快跳频和短包技术，支持点对点及点对多点通信，工作在全球通用的 2.4GHz 频段，其数据速率为 1Mbps。

2010 年 7 月，以低功耗为特点的蓝牙 4.0 标准推出，蓝牙大中华区技术市场经理吕荣良将其看作蓝牙第二波发展高潮的标志，他表示："蓝牙可以跨领域应用，主要有 4 个生态系统，分别是智能手机与笔记本电脑等终端市场、消费电子市场、汽车前装市场和健身运动器材市场。"

NFC 和 UWB 曾经是十分受关注的短距离无线接入技术，但其发展已经日渐势微。业内专家认为，无线频谱的规划和利用在短距离通信中日益重要。短距离通信技术目前主要采用 2.4GHz 的开放频谱，但随着物联网的发展和大量短距离通信技术的应用，频谱需求会快速增长，视频、图像等大数据量的通信正在寻求更高频段的解决方案。

20.1.4 NFC——近场通信

NFC 是近场通信（Near Field Communication）的缩写，此技术由非接触式射频识别（RFID）演变而来，由飞利浦半导体（现恩智浦半导体）、诺基亚和索尼共同研制开发，其基础是 RFID 及互连技术。NFC 是一种短距高频的无线电技术，在 13.56MHz 频率运行于 20 厘米距离内。其传输速度有 106 Kbit/秒、212 Kbit/秒或者 424 Kbit/秒三种。目前近场通信已通过成为 ISO/IEC IS 18092 国际标准、ECMA-340 标准与 ETSI TS 102 190 标准。NFC 采用主动和被动两种读取模式。

NFC 近场通信技术是由非接触式射频识别（RFID）及互联互通技术整合演变而来，在单一芯片上结合感应式读卡器、感应式卡片和点对点的功能，能在短距离内与兼容设备进行识别和数据交换。工作频率为 13.56MHz，但是使用这种手机支付方案的用户必须更换特制的手机。目前这项技术在日韩被广泛应用。手机用户凭着配置了支付功能的手机就可以行遍全国：他们的手机可以用作机场登机验证、大厦的门禁钥匙、交通一卡通、信用卡、支付卡等。

NFC 和蓝牙（Bluetooth）都是短程通信技术，而且都被集成到移动电话。但 NFC 不需要复杂的设置程序。NFC 也可以简化蓝牙连接。NFC 略胜蓝牙的地方在于设置程序较短，但无法达到低功率蓝牙（Bluetooth Low Energy）的速度。在两台 NFC 设备相互连接的设备识别过程中，使用 NFC 来替代人工设置会使创建连接的速度大大加快，会少于十分之一秒。

注意　20.1.3 的内容引用自“ZOL 网的科技频道：http://news.zol.com.cn/article/179109.html”。

20.2 蓝牙 4.0 BLE 基础

蓝牙 4.0 是 2012 年发布的最新蓝牙版本，是 3.0 的升级版本。和传统的 3.0 版本相比，4.0 具有更加省电、成本低、3 毫秒低延迟、超长有效连接距离、AES-128 加密等特点，通常被用在蓝牙耳机、蓝牙音箱等设备上。蓝牙 4.0 最重要的特性是省电，极低的运行和待机功耗可以使一粒纽扣电池连续工作数年之久。此外，低成本和跨厂商互操作性，3 毫秒低延迟、AES-128 加密等诸多特色，可以用于计步器、心律监视器、智能仪表、传感器物联网等众多领域，大大扩展了蓝牙技术的应用范围。

蓝牙技术联盟（Bluetooth SIG）2010 年 7 月 7 日宣布，正式采纳蓝牙 4.0 核心规范（Bluetooth Core Specification Version 4.0），并启动对应的认证计划。会员厂商可以提交其产品进行测试，通过后将获得蓝牙 4.0 标准认证。

蓝牙 4.0 的主要特性如下所示。

- 超低的峰值、平均和待机模式功耗。
- 使用标准纽扣电池可运行一年乃至数年。
- 低成本。
- 不同厂商设备交互性。
- 无线覆盖范围增强。
- 完全向下兼容。
- 低延迟（APT-X）。

20.2.1　蓝牙 4.0 的优势

蓝牙 4.0 将传统蓝牙技术、高速技术和低耗能技术这三种规格集一体，与 3.0 版本相比最大的不同就是低功耗。4.0 版本的功耗较老版本降低了 90%，这样更省电。蓝牙技术联盟大中华区技术市务经理吕荣良表示：“随着蓝牙技术由手机、游戏、耳机、便携电脑和汽车等传统应用领域向物联网、医疗等新领域的扩展，对低功耗的要求会越来越高。4.0 版本强化了蓝牙在数据传输上的低功耗性能。”

低功耗版本使蓝牙技术得以延伸到采用钮扣电池供电的一些新兴市场。蓝牙低耗能技术是基于蓝牙低耗能无线技术核心规格的升级版，为开拓钟表、远程控制、医疗保健及运动感应器等广大新兴市场的应用奠定基础。

这项技术将应用于每年出售的数亿台蓝牙手机、个人电脑及掌上电脑。以最低耗能提供持久的无线连接，有效扩大相关应用产品的覆盖距离，开辟全新的网络服务。低耗能无线技术的特点在于超低的峰期、平均值及待机耗能；使装置配件和人机介面装置（HIDs）具备超低成本和轻巧的特性；更能使手机及个人电脑相关配件的成本降至最低、体积更小；全球适用之外，更具使用直觉，且能确保多种设备连接的互操作性。

蓝牙 4.0 在个人健身和健康市场的影响很大，无论是在跑步机上，或在办公室的小工具。Fitbit 无线师，耐克公司的新 Fuelband，摩托罗拉 MOTACTV，和时尚的基带是都是很好的例子。而且健身手表也承诺使用蓝牙跟踪体力活动和心率。

另外蓝牙 4.0 依旧向下兼容，包含经典蓝牙技术规范和最高速度 24Mbps 的蓝牙高速技术规范。3 种技术规范可单独使用，也可同时运行。

20.2.2　Bluetooth 4.0 BLE 推动了智能设备的兴起

到目前为止，当大家谈到可穿戴设备时都要提到一个参数：支持蓝牙还是用无线网络与智能手机相连。这是衡量可穿戴设备是否能与智能手机上的软件顺利“对话”的主要依据。其实在过去的一段时间内，大家已经习惯了“Bluetooth X.0 版”的说法，其实从 Bluetooth 4.0 开始，这项技术被 Bluetooth SIG（Special Interest Group，负责推动蓝牙技术标准的开发和将其授权给制造商的非营利组织）改名为 Bluetooth Smart 或 Bluetooth Smart Ready。Bluetooth SIG 首席营销官 Suke Jawanda 对 PingWest 说：“未来 Bluctooth SIG 也将继续淡化 X.0 的概念，将更加强调 Bluetooth Smart，原因是 X.0 是说给极客听的，而 Bluetooth SIG 希望普通消费者也能听懂。”

可穿戴设备与智能手机之间的数据传输方式对蓝牙技术的要求也与以往不同。Suke Jawanda 用自己手腕上的 Fitbit Flex 举例，“过去当我们谈到蓝牙技术和数据传输，主要考虑的是类似 Spotify 这种在一个较长的时间段里输送数据的需求，现在像 Fitbit Flex 是先收集数据，再断续的在某些‘时刻’里将数据传送到用户的手机上，两种数据传输方式不同。Bluetooth Smart 的低耗能技术就可以满足这一需求了。”

Suke Jawanda 向 PingWest 解释说：“现在不少设备制造商都在强调自己支持蓝牙低耗能技术（Bluetooth Low Energy），其实它只是 Bluetooth Smart 其中的一个功能。支持 Bluetooth Smart 的设备都支持蓝牙低耗能技术”。

虽然特意强调低耗能技术是给消费者造成一种“省电省流量”的印象，但是换个角度来看，每个产品说明自己支持 Bluetooth Smart 的背后就是可穿戴设备为什么在此时流行的重要原因之一——Bluetooth Smart 对操作系统和硬件设备的支持情况，决定了可穿戴设备能否以较低的成本与软件进行数据传输，接下来才是解决软件获得数据之后怎么处理的问题。

根据 Suke Jawanda 的介绍，Bluetooth Smart 对可穿戴设备的支持分成硬件和软件。以 Fitbit 为例，如果 Fitbit 开发一款新的产品，支持 Bluetooth Smart，同时要求软件，也就是从 iOS 或者 Android——操作系统层面要支持 Bluetooth Smart，苹果是从 iOS 5（iPhone 4S 以及以上版本的手机）开始支持 Bluetooth Smart，而 Google 直到 Android 4.3 才开始支持。

当然 Bluetooth Smart 并不是唯一推动穿戴设备发展的原因，有些可穿戴设备可以用其他方式传输数据，例如无线网络，但是人们不可能一直在无线网络环境下生活。

那除了可穿戴设备外，汽车除了用蓝牙接打电话，还能做些什么？Suke Jawanda 说：“现在我们知道汽车能做到的是通过蓝牙进行语音操作、接打电话，未来我们想象蓝牙技术可以不再用钥匙，你的手机就可以作为车钥匙；另一个是利用更多的传感器收集数据，让车与车之间‘对话’，例如你的车可以知道前后三辆车的时速，当他们减速时你的车能提醒你前方的车在减速时可能遇到什么情况等。但这里最大的问题是汽车行业技术滞后，比如你现在看到的一个汽车领域的新技术，真正应用到生产、被推广恐怕是 2～3 年后的事情，而且人买一辆车的期待是要用 10～15 年的，也就是你买了一辆车之后 10 年内可能都体验不到汽车领域的新技术了，这个问题现在还没有很好的解决方案”。

20.3 低功耗蓝牙基础

BLE（Bluetooth Low Energy，低功耗蓝牙）是对传统蓝牙 BR/EDR 技术的补充。尽管 BLE 和传统蓝牙都称之为蓝牙标准，且共享射频，但是 BLE 是一个完全不一样的技术。BLE 不具备和传统蓝牙 BR/EDR 的兼容性，是专为小数据率、离散传输的应用而设计的。在本节的内容中，将详细讲解低功耗蓝牙技术的基本知识。

20.3.1 低功耗蓝牙的架构

BLE 协议架构总体上分成 3 层，从下到上分别是：控制器（Controller）、主机（Host）和应用端（Apps）。三者可以在同一芯片类实现，也可以分不同芯片内实现，控制器（Controller）处理射频数据解析、接收和发送，主机（Host）控制不同设备之间如何进行数据交换；应用端（Apps）实现具体应用。

（1）控制器 Controller

Controller 实现射频相关的模拟和数字部分，完成最基本的数据发送和接收，Controller 对外接口是天线，对内接口是主机控制器接口 HCI（Host Controller Interface）；控制器包含物理层 PHY（Physical Layer），链路层 LL（Linker Layer），直接测试模式 DTM（Direct Test Mode）以及主机控制器接口 HCI。

- 物理层 PHY

GFSK 信号调制，2402MHz～2480MHz，40 个 channel，每两个 channel 间隔 2MHz（经典蓝牙协议是 1MHz），数据传输速率是 1Mbps。

- 直接测试模式 DTM

为射频物理层测试接口，射频数据分析之用。

- 链路层 LL

基于物理层 PHY 之上，实现数据通道分发，状态切换，数据包校验，加密等；链路层 LL 分 2 种通道：广播通道（advertising channels）和数据通道（data channels）；广播通道有 3 个，37ch（2402MHz），38ch（2426MHz），39ch（2480MHz），每次广播都会往这 3 个通道同时发送（并不会在这 3 个通道之间跳频），为防止某个通道被其他设备阻塞，以至于设备无法配对或广播数据，之所以定 3 个广播通道是一种权衡，少了可能会被阻塞，多了加大功耗。还有一个有意思的事情是，3 个广播通道刚好避开了 Wi-Fi 的 1ch、6ch、11ch，所以在 BLE 广播的时候，不至于被 Wi-Fi 影响（冒出一个很邪恶想法，如果要干扰 BLE 广播数据，一个最最简单的办法，就是同时阻塞 3 个广播通道，哈哈）；当 BLE 匹配之后，链路层 LL 由广播通道切换到数据通道，数据通道 37 个，数据传输的时候会在这 37 个通道间切换，切换规则在设备间匹配时候约定。

（2）主机 Host/控制器 controller 接口 HCI

HCI 作为一种接口，存在于主机 Host 和控制器 controller 当中，控制器 Host 通过 HCI 发送数据和事件给主机，主机 Host 通过 HCI 发送命令和数据给控制器 controller. HCI 逻辑上定义一系列的命令，事件；物理上有 UART、SDIO、USB，实际可能包含里面的任意 1 种或几种。

20.3.2 低功耗蓝牙分类

BLE 通常应用在传感器和智能手机或者平板的通信中。到目前为止，只有很少的智能机和平板支持 BLE，如：iPhone 4S 以后的苹果手机，Motorola Razr 和 the new iPad 及其以后的 iPad。安卓手机也逐渐支持 BLE，安卓的 BLE 标准在 2013 年 7 月 24 日刚发布。智能机和平板会带双模蓝牙的基带和协议栈，协议栈中包括 GATT 及以下的所有部分，但是没有 GATT 之上的具体协议。所以，这些具体的协议需要在应用程序中实现，实现时需要基于各个GATT API集。这样有利于在智能机端简单地实现具体协议，也可以在智能机端简单地开发出一套基于GATT的私有协议。

在现实应用中，低功耗蓝牙分为单模（Bluetooth Smart）和双模（Bluetooth Smart Ready）两种设备。BLE和蓝牙BR/EDR有所区分，这样可以让我们用三种方式将蓝牙技术集成到具体设备中。因为不再是所有现有的蓝牙设备都可以和另一个蓝牙设备进行互联，所以准确描述产品中蓝牙的版本是非常重要的。在接下来的内容中，将详细讲解单模蓝牙和双模蓝牙的基本知识。

（1）单模蓝牙

单模蓝牙设备被称为Bluetooth Smart设备，并且有专用的Logo，如图20-1所示。

在现实应用中，手表、运动传感器等小型设备通常是基于低功耗单模蓝牙的。为了实现极低的功耗效果，在硬件和软件上都进行了优化，这样的设备只能支持BLE。单模蓝牙芯片往往是一个带有单模蓝牙协议栈的产品，这个协议栈通常是芯片商免费提供的。

（2）双模蓝牙

双模蓝牙设备被称为Bluetooth Smart Ready设备，并且有专用的Logo，如图20-2所示。

▲图20-1　Bluetooth Smart设备

▲图20-2　Bluetooth Smart Ready设备

双模设备支持蓝牙BR/EDR和BLE。在双模设备中，BR/EDR和BLE技术使用同一个射频前端和天线。典型的双模设备有智能手机、平板电脑、PC和Gateway。这些设备可以接收到通过BLE或者蓝牙BR/EDR设备发送过来的数据，这些设备往往都有足够的供电能力。双模设备和BLE设备通信的功耗低于双模设备和蓝牙BR/EDR设备通信的功耗。在使用双模解决方案时，需要用一个外部处理器才可以实现蓝牙协议栈。

20.3.3　集成方式

尽管有单模和双模方案的区别，但是在设备中集成蓝牙技术的方式有多种，其中最为常用的方式有模块和芯片。

（1）模块

在现实应用中，最简单和快速的方式是使用一个嵌入式模块。此类模块包含了天线、嵌入了协议栈并提供多种不同的接口：UART、USB、SPI和I²C，可以通过这些接口和您的处理器连接。模块会提供一种简单的接口来控制蓝牙的功能。很多的模块公司都会提供带CE、FCC和IC认证的产品。这样的模块可以只是蓝牙BR/EDR的，双模式的或者单模式的。

如果是蓝牙BR/EDR和双模的方案，还可以采用HCI模块。HCI模块只是不带蓝牙协议栈，其他的和上述的模块是一样的。所以，这样的模块会更便宜。HCI模块只是提供了硬件接口，在这样的方案中，蓝牙协议栈需要第三方提供。这样的第三方协议栈需要能在主设备的处理器中运行，如斯图曼提供的BlueCode+SR。使用HCI模块需要将软件移植到最终的硬件中。

从理论上讲，提供单模的HCI模块也是可以的。然而，所有的芯片公司都已经将GATT集成到他们的芯片中，所以市面上不会有HCI单模模块出现。

（2）芯片

通过芯片来集成BLE是从物料角度最低成本的方式，但是，这需要很多的前期工作和花费大量的时间。虽然在软件上只需要将协议栈移植到目标平台之中即可，但是，硬件方面则需要对RF的layout和天线的设计非常有经验。这些公司提供的BLE芯片有：Broadcom、CSR、EM Microelectronic、Nordic和TI。

20.3.4　低功耗蓝牙的特点

在实际应用过程中，BLE 的低功耗并不是通过优化空中的无线射频传输实现的，而是通过改变协议的设计来实现的。为了实现极低的功耗效果，通常 BLE 协议设计为：在不必要射频的时候，彻底将空中射频关断。

与传统蓝牙 BR\EDR 相比，BLE 通过如下三大特性实现低功耗效果。

- 缩短无线开启时间。
- 快速建立连接。
- 降低收发峰值功耗（具体由芯片决定）。

缩短无线开启时间的第一个技巧是只用 3 个“广告”信道，第二个技巧是通过优化协议栈来降低工作周期。一个在广播的设备可以自动和一个在搜索的设备快速建立连接，所以可以在 3 毫秒内完成连接的建立和数据的传输。

在现实应用中，低功耗设计可能会带来一些牺牲，例如音频数据无法通过 BLE 来进行传输。尽管如此，BLE 仍然是一种非常出色的技术，依然会支持跳频（37 个数据信道），并且采用了一种改进的 GFSK 调制方法来提高链路的稳定性。BLE 也仍是非常安全的技术，因为在芯片级提供了 128 bit AES 加密。

单模设备可以作为 Master 或者 Slave，但是不能同时充当两种角色。这意味着 BLE 只能建立简单的星状拓扑，不能实现散射网。在 BLE 的无线电规范中，定义了低功耗蓝牙的最高数据率为 305kbps，但是，这只是理论数据。在实际应用中，数据的吞吐量取决于上层协议栈。而 UART 的速度、处理器的能力和主设备都会影响数据吞吐能力。

高数据吞吐能力的 BLE 只有通过私有方案或者基于 ATT notification 才能实现。事实上，如果是高数据率或高数据量的应用，蓝牙 BR/EDR 通常显得更加省电。

20.3.5　BLE 和传统蓝牙 BR/EDR 技术的对比

BLE 和传统蓝牙 BR/EDR 技术相比的细节如表 20-1 所示。

表 20-1　BLE 和传统蓝牙 BR/EDR 技术的对比

	Bluetooth BR/EDR	Bluetooth low energy
Frequency	2400～2483.5 MHz	2400～2483.5 MHz
Deep Sleep	～80 μA	<5 μA
Idle	～8 mA	～1 mA
Peak Current	20～40 mA	10～30 mA
Range	500m (Class 1) / 50m (Class 2)	100m
Min. Output Power	0 dBm (Class 1) /～6 dBm (Class 2)	-20 dBm
Max. Output Power	+20 dBm (Class 1) / +4 dBm (Class 2)	+10 dBm
Receiver Sensitivity	≥ -70 dBm	≥ -70 dBm
Encryption	64 bit / 128 bit	AES-128 bit
Connection Time	100 ms	3 ms
Frequency Hopping	Yes	Yes
Advertising Channel	32	3
Data Channel	79	37
Voice capable	Yes	No

20.4 蓝牙规范

蓝牙规范即 Bluetooth Profile，Bluetooth SIG 定义了许多 Profile。Profile 目的是要确保 Bluetooth 设备间的互通性（interoperability），但是 Bluetooth 产品无需实现所有的 Bluetooth 规范 Profile。本节将详细讲解蓝牙规范的基本知识。

20.4.1 Bluetooth 系统中的常用规范

在 Bluetooth 系统中，定义了如下常用的规范。

（1）蓝牙立体声音讯传输协议 A2DP

蓝牙立体声音讯传输协议（Advance Audio Distribution Profile），功能是播放立体声。

（2）基本图像规范

基本图像规范（Basic Imaging Profile）的功能是在装置之间传送图像，可以将其再细分为如下所示的类别。

- Image Push。
- Image Pull。
- Advanced Image Printing。
- Automatic Archive。
- Remote Camera。
- Remote Display。

（3）基本打印规范

基本打印规范（Basic Printing Profile）可以将文件、电子邮件传至打印机打印，主要包含如下所示的分类。

- 无线电话规范（Cordless Telephony Profile），设置了蓝牙无线电话之间沟通的规范。
- 内通讯规范（Intercom Profile）：是另类的 TCS（Telephone Control protocol Specification）基底规范，两个 Bluetooth 通讯设备间沟通的规范。
- 拨号网络规范：Baseband、LMP、L2CAP、SDP、RFCOMM 协定所需要的传输需求。
- 传真规范（Fax Profile）：能传输传真的资料。
- 人机界面规范（Human Interface Device Profile）：可以支援鼠标和键盘功能。
- 头戴式通话器规范（Headset Profile）：能够将声音传送到蓝牙耳机设备。
- 序列埠规范（Serial Port Profile）：用来取代有线的 RS-232 Cable。
- SIM 卡存取规范（SIM Access Profile）：用于存取手机内的 SIM 卡。
- 同步规范（Synchronization Profile）：建立在 Serial Port Profile、Generic Access Profile 与 Generic Access Profile 之上。
- 档案传输规范（File Transfer Profile）：Bluetooth 可以利用 OBEX 通讯协定来传送档案。
- 泛用存取规范（Generic Access Profile）：用来建立连线。
- 泛用物件交换规范（Generic Object Exchange Profile）：使用 OBEX 进行物件交换。
- 物件交换规范（Object Push Profile）：Bluetooth 利用 OBEX 通讯协定在两个设备间交换资料。
- 个人局域网路规范（Personal Area Networking Profile）：可以支持蓝牙网络第三层协定。
- 电话簿存取规范（Phone Book Access Profile）：可以在装置之间互换电话簿。
- 影像分享规范（Video Distribution Profile）：可以使用 H.263 编码算法来分享影像信息。

20.4.2　蓝牙协议体系结构

整个蓝牙协议体系结构可分为底层硬件模块、中间协议层和高端应用层三大部分。链路管理层（LMP）、基带层（BBP）和蓝牙无线电信道构成蓝牙的底层模块。BBP 层负责跳频和蓝牙数据及信息帧的传输。LMP 层负责连接的建立和拆除以及链路的安全和控制，它们为上层软件模块提供了不同的访问入口，但是两个模块接口之间的消息和数据传递必须通过蓝牙主机控制器接口的解释才能进行。也就是说，中间协议层包括逻辑链路控制与适配协议（L2CAP）、服务发现协议（SDP）、串口仿真协议（RFCOMM）和电话控制协议规范（TCS）。L2CAP 完成数据拆装、服务质量控制、协议复用和组提取等功能，是其他上层协议实现的基础，因此也是蓝牙协议栈的核心部分。SDP 为上层应用程序提供一种机制来发现网络中可用的服务及其特性。在蓝牙协议栈的最上部是高端应用层，它对应于各种应用模型的剖面，是剖面的一部分，目前定义了 13 种剖面。

（1）蓝牙低层模块

蓝牙的低层模块是蓝牙技术的核心，是任何蓝牙设备都必须包括的部分。蓝牙工作在 2.4GHZ 的 ISM 频段。采用了蓝牙结束的设备讲能够提供高达 720kbit/s 的数据交换速率。

蓝牙支持电路交换和分组交换两种技术，分别定义了两种链路类型，即面向连接的同步链路（SCO）和面向无连接的异步链路（ACL）。为了在很低的功率状态下也能使蓝牙设备处于连接状态，蓝牙规定了三种节能状态，即停等（Park）状态、保持（Hold）状态和呼吸（Sniff）状态。这几种工作模式按照节能效率以升序排依次是：Sniff 模式、Hold 模式、Park 模式。

蓝牙采用三种纠错方案，分别是 1/3 前向纠错（FEC）、2/3 前向纠错和自动重发（ARQ）。前向纠错的目的是减少重发的可能性，但同时也增加了额外开销。然而在一个合理的无错误率环境中，多余的投标会减少输出，故分组定义的本身也保持灵活的方式，因此，在软件中可定义是否采用 FEC。一般而言，在信道的噪声干扰比较大时蓝牙系统会使用前向纠错方案，以保证通信质量：对于 SCO 链路，使用 1/3 前向纠错；对于 ACL 链路，使用 2/3 前向纠错。在无编号的自动请求重发方案中，一个时隙传送的数据必须在下一个时隙得到收到的确认。只有数据在收端通过了报头错误检测和循环冗余校验（CRC）后认为无错时，才向发端发回确认消息，否则返回一个错误消息。

蓝牙系统的移动性和开放性使得安全问题变得极其重要。虽然蓝牙系统所采用的调频技术就已经提供了一定的安全保障，但是蓝牙系统仍然需要链路层和应用层的安全管理。在链路层中，蓝牙系统提供了认证、加密和密钥管理等功能。每个用户都有一个个人标识码（PIN），它会被译成 128bit 的链路密钥（Link Key）来进行单双向认证。一旦认证完毕，链路就会以不同长度的密码（Encryphon Key）来加密（此密码以 bit 为单位增减，最大的长度为 128bit）链路层安全机制提供了大量的认证方案和一个灵活的加密方案（即允许协商密码的长度）。当来自不同国家的设备互相通信时，这种机制是极其重要的，因为某些国家会指定最大密码长度。蓝牙系统会选取微微网中各个设备的最小的最大允许密码长度。例如，美国允许 128bit 的密码长度，而西班牙仅允许 48bit，这样当两国的设备互通时，将选择 48bit 来加密。蓝牙系统也支持高层协议栈的不同应用体内的特殊的安全机制。例如两台计算机在进行商业卡信息交流时，一台计算机就只能访问另一台计算机的该项业务，而无权访问其他业务。蓝牙安全机制依赖 PIN 在设备间建立信任关系，一旦这种关系建立起来了，这些 PIN 就可以存储在设备中以便将来更快捷地连接。

（2）软件模块

L2CAP 是数据链路层的一部分，位于基带协议之上。L2CAP 向上层提供面向连接的和无连接的数据服务，它的功能包括：协议的复用能力、分组的分割和重新组装（Segmentation and Reassembly）以及提取（Group Abstraction）。L2CAP 允许高层协议和应用发送和接受高达 64K Byte

的数据分组。

SDP 为应用提供了一个发现可用协议和决定这些可用协议的特性的方法。蓝牙环境下的服务发现与传统的网络环境下的服务发现有很大的不同，在蓝牙环境下，移动的 RF 环境变化很大，因此业务的参数也是不断变换的。SDP 将强调蓝牙环境的独特的特性。蓝牙使用基于客户/服务器机制定义了根据蓝牙服务类型和属性发现服务的方法，还提供了服务浏览的方法。

RFCOMM 是射频通信协议，它可以仿真串行电缆接口协议，符合 ETSI0710 串口仿真协议。通过 RFCOMM，蓝牙可以在无线环境下实现对高层协议，如 PPP、TCP/IP、WAP 等的支持。另外，RFCOMM 可以支持 AT 命令集，从而可以实现移动电话机和传真机及调制解调器之间的无线连接。

蓝牙对语音的支持是它与 WLAN 相区别的一个重要的标志。蓝牙电话控制规范是一个基于 ITU-T 建议 Q.931 的采用面向比特的协议，它定义了用于蓝牙设备间建立语音和数据呼叫的呼叫控制信令以及用于处理蓝牙 TCS 设备的移动性管理过程。

20.4.3 低功耗（BLE）蓝牙协议

BLE 不再支持传统蓝牙 BR/EDR 的协议，例如传统蓝牙中的 SPP 协议在 BLE 中就不复存在。在 BLE 应用中，所有的协议或者服务都是基于 GATT（Generic Attribute Profile）的。尽管有些传统蓝牙中的协议，如 HID 被移植到了 BLE 中，但是在 BLE 的应用中必须区分协议和服务。其中服务描述了特点（及它们的 UUID），服务描述自身有什么特点和形式，并且描述清楚如何应用这些特点以及需要什么安全机制。而应用协议定义了其使用的服务，说明是传感器端还是接收端，定义 GATT 的角色（Server/Client）和 GAP 的角色（Peripheral/Central）。

和蓝牙 BR/EDR 协议相比，因为所有的功能都是集成在 GATT 终端，这些基于其上的应用协议只是对 GATT 提供的功能的使用，所以基于 GATT 的应用协议非常简单。

20.4.4 现有的基于 GATT 的协议/服务

截止到 2013 年 7 月，现有的基于 GATT 的协议/服务如表 20-2 所示。

表 20-2 基于 GATT 的协议/服务

GATT-Based Specifications（Qualifiable）		Adopted Version
ANP	Alert Notification Profile	1.0
ANS	Alert Notification Service	1.0
BAS	Battery Service	1.0
BLP	Blood Pressure Profile	1.0
BLS	Blood Pressure Service	1.0
CPP	Cycling Power Profile	1.0
CPS	Cycling Power Service	1.0
CSCP	Cycling Speed and Cadence Profile	1.0
CSCS	Cycling Speed and Cadence Service	1.0
CTS	Current Time Service	1.0
DIS	Device Information Service	1.1
FMP	Find Me Profile	1.0
GLP	Glucose Profile	1.0
HIDS	HID Service	1.0

续表

GATT-Based Specifications（Qualifiable）		Adopted Version
HOGP	HID over GATT Profile	1.0
HTP	Health Thermometer Profile	1.0
HTS	Health Thermometer Service	1.0
HRP	Heart Rate Profile	1.0
HRS	Heart Rate Service	1.0
IAS	Immediate Alert Service	1.0
LLS	Link Loss Service	1.0
LNP	Location and Navigation Profile	1.0
LNS	Location and Navigation Service	1.0
NDCS	Next DST Change Service	1.0
PASP	Phone Alert Status Profile	1.0
PASS	Phone Alert Status Service	1.0
PXP	Proximity Profile	1.0
RSCP	Running Speed and Cadence Profile	1.0
RSCS	Running Speed and Cadence Service	1.0
RTUS	Reference Time Update Service	1.0
ScPP	Scan Parameters Profile	1.0
ScPS	Scan Parameters Service	1.0
TIP	Time Profile	1.0
TPS	Tx Power Service	1.0

20.4.5 双模协议栈

图 20-3 展示了斯图曼双模协议栈 BlueCode+SR 的具体架构，在此架构图中包含了 SPP、HDP 和 GATT 所需要的所有部分。

20.4.6 单模协议栈

图 20-4 展示了单模协议栈的一种典型协议栈设计。

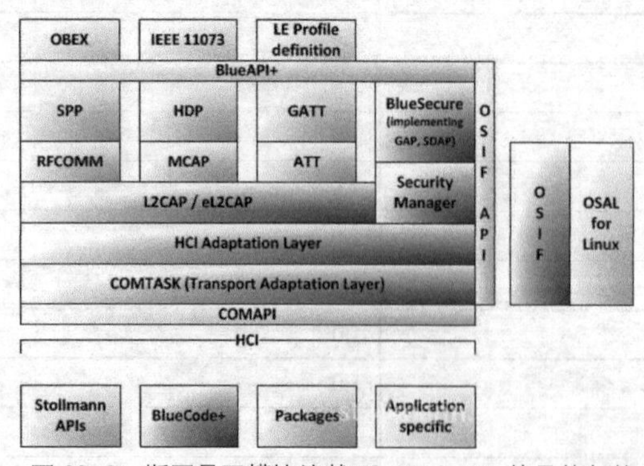

▲图 20-3 斯图曼双模协议栈 BlueCode+SR 的具体架构

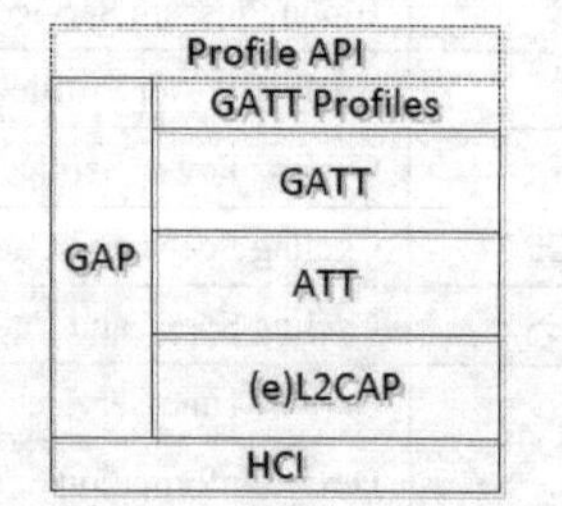

▲图 20-4 单模协议栈的一种设计

在单模协议栈中一般不会包含具体协议，所以需要在具体的应用程序中实现每一个具体应用对应的协议。这和传统蓝牙有非常大的区别，传统蓝牙会在协议栈中实现每个具体应用相关的协议，如 SPP、HDP 等。和双模协议栈相比，BLE 无需一个主处理器来实现它的协议栈，所以极低功耗的集成成为可能。大多数的单模芯片或者模块都是自带协议栈的。

因为 BLE 单模产品（芯片或者模块）中的协议栈只是实现了 GATT 层，所以通常需要将具体应用对应的协议集成到该单模产品之中。甚至芯片商都开始提供带有具体协议和 sample code 的 SDK。但是，仍然没有真正能拿到手的解决方案。

20.5 低功耗蓝牙协议栈详解

在大家的印象中，提到协议栈时都会想到开放式系统互联（OSI）协议栈，OSI 协议栈定义了厂商们如何才能生产可以与其他厂商的产品一起工作的产品。协议栈是指一组协议的集合，例如把大象装到冰箱里需要 3 步，每步就是一个协议，3 步组成一个协议栈。本节将详细讲解低功耗蓝牙协议栈的基本知识，为读者步入本书后面知识的学习打下基础。

20.5.1 蓝牙协议栈基础

蓝牙协议栈就是 SIG（Special Intersted Group）定义的一组协议的规范，目标是允许遵循规范的蓝牙应用应用能够进行相互间操作，图 20-5 展示了完整蓝牙协议栈和部分 Profile。

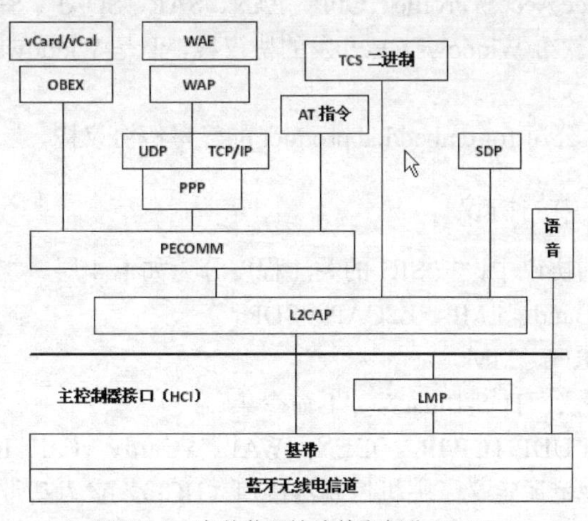

▲图 20-5 完整蓝牙协议栈和部分 Profile

在蓝牙系统中，Profile 是配置文件，配置文件定义了可能的应用，蓝牙配置文件表达了一般行为，蓝牙设备可以通过这些行为与其他设备进行通信。在蓝牙系统中定义了广泛的配置文件，描述了许多不同类型的使用案例。按照蓝牙规格中提供的指导，开发商可以创建应用程序以与其他符合蓝牙规格的设备协同工作。到目前为止，在蓝牙系统中一共有 20 多个 Profile，在 www.bluetooth.com 中有各个 Profile 的详细说明文档。在这些众多的协议栈中，r 协议栈的具体构成如下所示。

- Widcomm：第一个 Windows 上的协议栈，由 Widcomm 公司开发，也就是现在的 Broadcom。
- Microsoft Windows stack：Windows XP SP2 中包括了这个内建的协议栈，开发者也可以调用其 API 开发第三方软件。
- Toshiba stack：这也是基于 Windows 的，不支持第三方开发，但它把协议栈授权给一些 laptop 商，例如 Sony。它支持的 Profile 有：SPP、DUN、FAX、LAP、OPP、FTP、HID、HCRP、PAN、BIP、HSP、HFP、A2DP、AVRCP、GAVDP。

- BlueSoleil：著名的 IVT 公司的产品，该产品可以用于桌面和嵌入式，也支持第三方开发，例如 DUN、FAX、HFP、HSP、LAP、OBEX、OPP、PAN SPP、AV、BIP、FTP、GAP、HID、SDAP、SYNC。
- Blues：是 Linux 官方协议栈，该协议栈的上层用 Socket 封装，便于开发者使用，通过 DBUS 与其他应用程序通信。
- Affix：是 NOKIA 公司的协议栈，在 Symbian 系统上运行。
- BlueDragon：是东软公司产品，在 2002 年 6 月就通过了蓝牙的认证，支持的 Profile 有：SDP、Serial-DevB、AVCTP、AVRCP-Controller、AVRCP-Target、Headset-AG、Headset-HS、OPP-Client、OPP-Server、CT-GW、CT-Term、Intercom、FT-Server、FT-Client、GAP、SDAP、Serial-DevA、AVDTP、GAVDP、A2DP-Source、A2DP-Sink。
- BlueMagic：这是美国 Open Interface 公司 for portable embedded divce 的协议栈，iPhone（apple）、nav-u（sony）等很多电子产品都用该商业的协议栈，BlueMagic 3.0 是第一个通过 bluetooth 协议栈 1.1 认证的协议栈。
- BCHS-Bluecore Host Software：这是蓝牙芯片 CSR 的协议栈，同时也提供了一些上层应用的 Profile 的库，当然了也是为嵌入式产品提供的服务，支持的 Profile 有：A2DP、AVRCP、PBAP、BIP,BPP、CTP、DUN、FAX、FM API、FTP GAP、GAVDP、GOEP、HCRP、Headset、HF1.5、HID、ICP、JSR82、LAP Message Access Profile、OPP、PAN、SAP、SDAP、SPP、SYNC、SYNC ML。
- Windows CE：微软给 Windows CE 开发的协议栈，但是 Windows CE 本身也支持其他的协议栈。
- BlueLet：是 IVT 公司 for embedded product 的轻量级协议栈。

20.5.2　蓝牙协议体系中的协议

在蓝牙协议体系中的协议中，按 SIG 的关注程度分为如下 4 层。

- 核心协议：BaseBand、LMP、L2CAP、SDP。
- 电缆替代协议：RFCOMM。
- 电话传送控制协议：TCS-Binary、AT 命令集。
- 选用协议：PPP、UDP/TCP/IP、OBEX、WAP、vCard、vCal、IrMC、WAE。

除上述协议层外，规范还定义了主机控制器接口（HCI），它为基带控制器、连接管理器、硬件状态和控制寄存器提供命令接口。在图 1 中，HCI 位于 L2CAP 的下层，但 HCI 也可位于 L2CAP 上层。

蓝牙核心协议由 SIG 制定的蓝牙专用协议组成，绝大部分蓝牙设备都需要核心协议（加上无线部分），而其他协议则根据应用的需要而定。总之，电缆替代协议、电话控制协议和被采用的协议在核心协议基础上构成了面向应用的协议。

在现实应用中，常用蓝牙核心协议类型如下所示。

（1）基带协议

基带和链路控制层确保微微网内各蓝牙设备单元之间由射频构成的物理连接。蓝牙的射频系统是一个跳频系统，其任一分组在指定时隙、指定频率上发送。它使用查询和分页进程同步不同设备间的发送频率和时钟，为基带数据分组提供了两种物理连接方式，即面向连接（SCO）和无连接（ACL），而且，在同一射频上可实现多路数据传送。ACL 适用于数据分组，SCO 适用于话音以及话音与数据的组合，所有的话音和数据分组都附有不同级别的前向纠错（FEC）或循环冗余校验（CRC），而且可进行加密。此外，对于不同数据类型（包括连接管理信息和控制信息）都分配一个特殊通道。

可使用各种用户模式在蓝牙设备间传送话音，面向连接的话音分组只需经过基带传输，而不到达L2CAP。话音模式在蓝牙系统内相对简单，只需开通话音连接就可传送话音。

（2）连接管理协议（LMP）

该协议负责各蓝牙设备间连接的建立。它通过连接的发起、交换、核实，进行身份认证和加密，通过协商确定基带数据分组大小。它还控制无线设备的电源模式和工作周期，以及微微网内设备单元的连接状态。

（3）逻辑链路控制和适配协议（L2CAP）

该协议是基带的上层协议，可以认为它与 LMP 并行工作，它们的区别在于，当业务数据不经过LMP时，L2CAP为上层提供服务。L2CAP向上层提供面向连接的和无连接的数据服务，它采用了多路技术、分割和重组技术、群提取技术。L2CAP允许高层协议以64k字节长度收发数据分组。虽然基带协议提供了SCO和ACL两种连接类型，但L2CAP只支持ACL。

（4）服务发现协议（SDP）

发现服务在蓝牙技术框架中起着至关紧要的作用，它是所有用户模式的基础。使用SDP可以查询到设备信息和服务类型，从而在蓝牙设备间建立相应的连接。

（5）电缆替代协议（RFCOMM）

RFCOMM是基于ETSI-07.10规范的串行线仿真协议。它在蓝牙基带协议上仿真RS-232控制和数据信号，为使用串行线传送机制的上层协议（如OBEX）提供服务。

（6）电话控制协议

- 二元电话控制协议（TCS-Binary 或 TCSBIN）：是面向比特的协议，它定义了蓝牙设备间建立语音和数据呼叫的控制信令，定义了处理蓝牙 TCS 设备群的移动管理进程。基于ITU TQ.931建议的TCSBinary被指定为蓝牙的二元电话控制协议规范
- AT 命令集电话控制协议： SIG 定义了控制多用户模式下移动电话和调制解调器的 AT 命令集，该AT命令集基于ITU TV.250建议和GSM07.07，它还可以用于传真业务

（7）选用协议

- 点对点协议（PPP）：在蓝牙技术中，PPP位于RFCOMM上层，完成点对点的连接
- TCP/UDP/IP：该协议是由互联网工程任务组制定，广泛应用于互联网通信的协议。在蓝牙设备中，使用这些协议是为了与互联网相连接的设备进行通信
- 对象交换协议（OBEX）：IrOBEX（简写为OBEX）是由红外数据协会（IrDA）制定的会话层协议，它采用简单的和自发的方式交换目标。OBEX是一种类似于HTTP的协议，它假设传输层是可靠的，采用客户机/服务器模式，独立于传输机制和传输应用程序接口（API）

例如电子名片交换格式（vCard）、电子日历及日程交换格式（vCal）都是开放性规范，它们都没有定义传输机制，而只是定义了数据传输格式。SIG采用vCard/vCal规范，是为了进一步促进个人信息交换

- 无线应用协议（WAP）：该协议是由无线应用协议论坛制定的，它融合了各种广域无线网络技术，其目的是将互联网内容和电话传送的业务传送到数字蜂窝电话和其他无线终端上

20.5.3 Android的低功耗蓝牙协议栈

为了确保 Android 系统可以更好的支持蓝牙 4.0 BLE，Broadcom 公司特意推出了适应于Android平台的开源低功耗蓝牙协议栈BlueDroid，其开发文档和API是开源代码，在如下所示的地址保存。

```
https://github.com/briandbl/framework
```

在上述开源代码中，低功耗蓝牙 API 支持 Android 平台上的低功耗蓝牙通信功能。通过使用 BlueDroid 协议栈，Android 应用程序可以枚举、发现并访问低功耗蓝牙的外部设备，并且实现了低功耗蓝牙规范。

从 Android 4.2 版本开始，低功耗蓝牙模块的整体结构如图 20-6 所示。

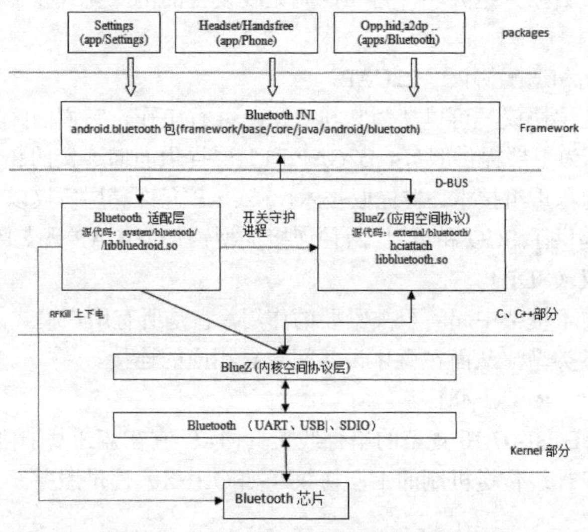

▲图 20-6　低功耗蓝牙模块的整体结构

> **注意**　虽然从 Android 4.2 版本开始，JNI 部分的代码在 packages 层中实现，但是为了便于读者从视觉上更加容易接受，特将 JNI 部分绘制在了 Framework 层中。

20.6 TI 公司的低功耗蓝牙

BLE 低功耗蓝牙协议有很多个版本，不同的厂商提供的低功耗蓝牙协议会有所区别。本节将详细讲解 TI（德州仪器）公司提供的 BLE 低功耗蓝牙协议的基本知识，为读者步入本书后面知识的学习打下基础。

20.6.1　获取 TI 公司的低功耗蓝牙协议栈

TI（德州仪器）公司提供了多个版本的 BLE 低功耗蓝牙协议栈，读者可以登录其官方网站来下载，如图 20-7 所示。

▲图 20-7　TI 公司的官方站点

笔者下载的版本是 BLE-CC254x-1.3.exe，双击此文件后可以进行安装工作，具体安装过程如下所示。

（1）首先弹出“解压缩”界面，在此单击“Next”按钮，如图 20-8 所示。

（2）弹出同意安装协议界面，在此选择“I accept...”选项，单击“Next”按钮，如图 20-9 所示。

（3）来到选择安装路径界面，通过“Browse...”按钮可以选择安装路径，如图 20-10 所示。

（4）来到准备安装界面，单击“Install”按钮开始安装，如图 20-11 所示。

（5）弹出安装进度界面，此过程需要耐心等待。如图 20-12 所示。

（6）最后弹出安装完成界面，整个安装过程结束。如图 20-13 所示。

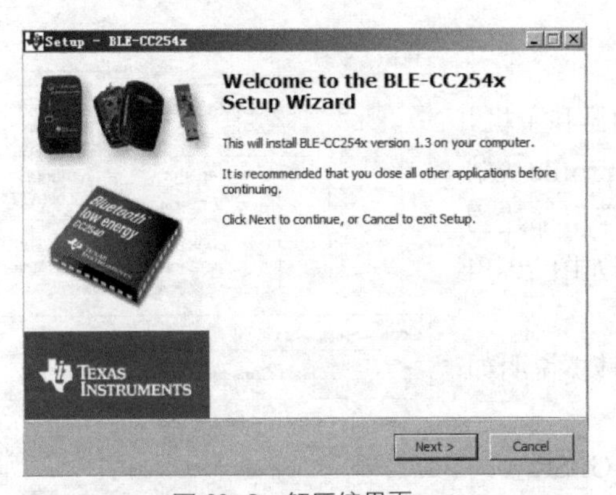

▲图 20-8 解压缩界面

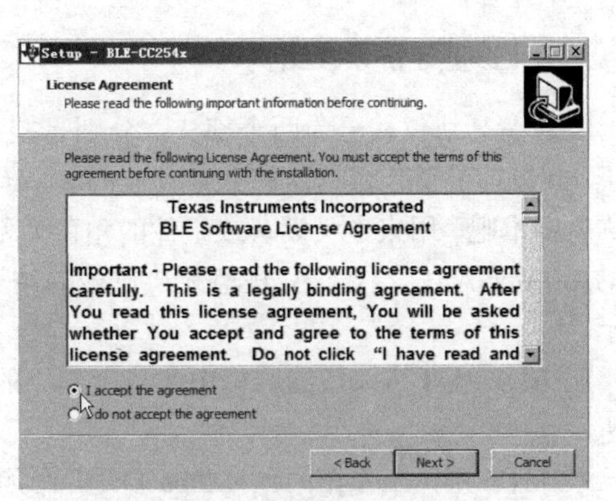

▲图 20-9 同意安装协议界面

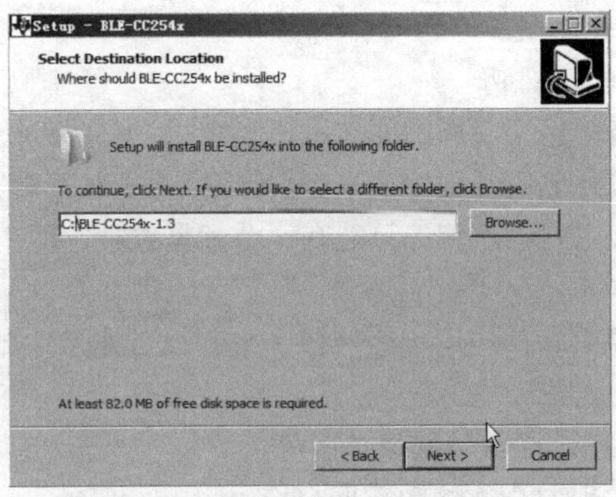

▲图 20-10 选择安装路径界面

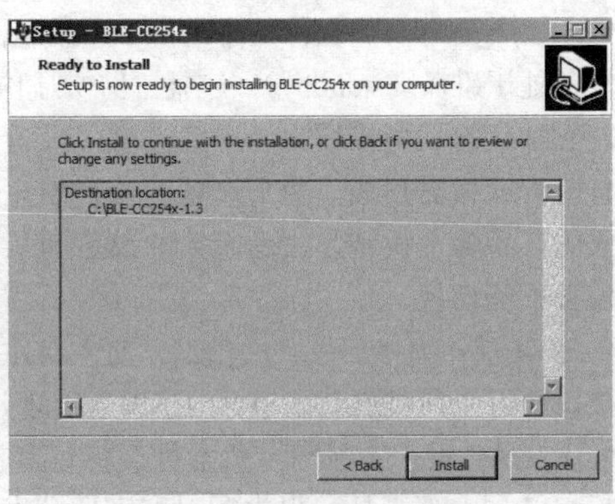

▲图 20-11 准备安装界面

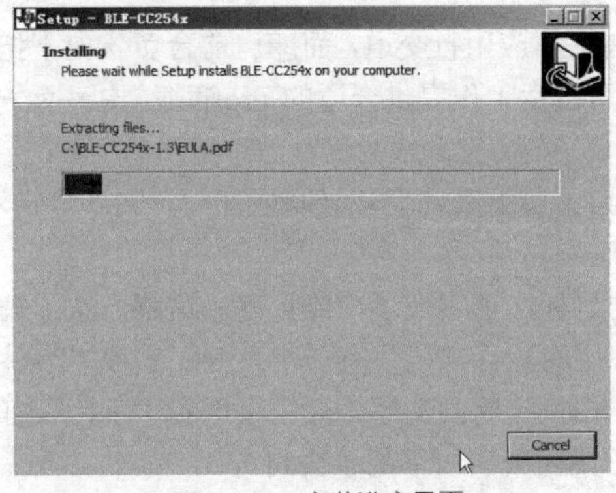

▲图 20-12 安装进度界面

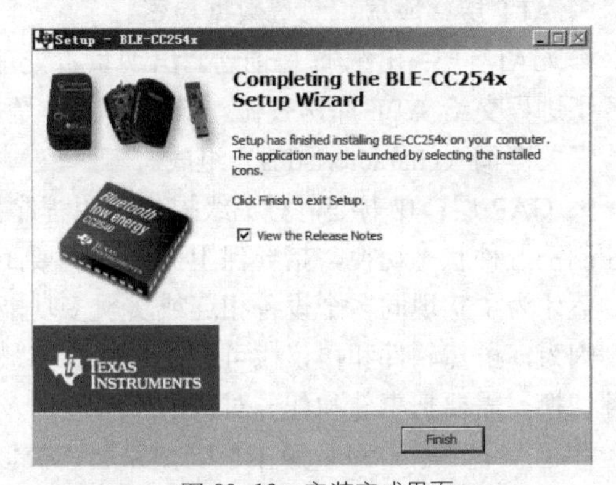

▲图 20-13 安装完成界面

安装完成后，需要使用 IAR 集成开发环境打开工程文件。例如 TI 公司在“Projects\ble\SimpleBLEPeripheral\”目录下提供了实例工程，通过使用 IAR 工具打开“.eww”文件的方式可以浏览整个工程。

注意　读者可以自行下载并安装 IAR 集成开发环境。

20.6.2 分析 TI 公司的低功耗蓝牙协议栈

注意　下面的内容参考自 TI 公司的官方资料 *CC2540Bluetooth Low Energy Software Developer's Guide* (Rev. B)，部分图片也是直接引用自上述参考文档。

1. BLE 蓝牙协议栈结构

BLE 蓝牙协议栈分为两个部分，分别是控制器和主机。对于 4.0 以前的蓝牙，这两部分是分开的。所有 profile（姑且称为剧本吧，用来定义设备或组件的角色）和应用都建构在 GAP 或 GATT 之上。BLE 蓝牙协议栈的结构如图 20-14 所示。

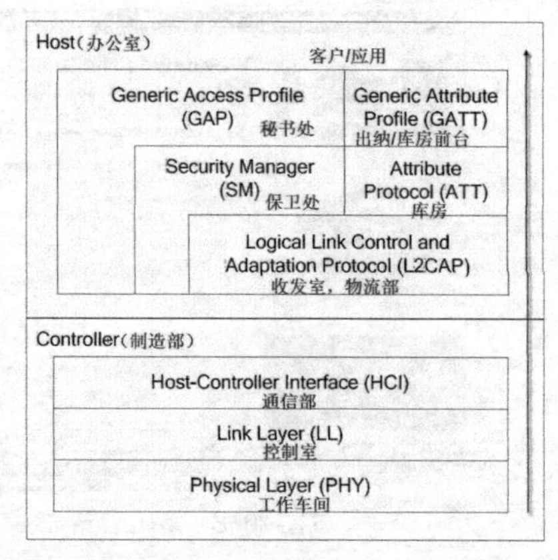

▲图 20-14　BLE 蓝牙协议栈的结构

在 BLE 蓝牙协议栈的结构中，从上到下的具体说明如下所示。

- PHY 层，工作车间：1Mbps 自适应跳频 GFSK（高斯频移键控），运行在免许可证使用的 2.4GHz 频段中。
- LL 层：为以下五种状态之一 RF 控制器、控制室、控制设备处于准备（standby）；广播；监听/扫描（scan）；初始化。五种状态切换描述为：未连接时，设备广播信息（向周围邻居讲“我来了”），另外一个设备一直监听或按需扫描，两个设备连接初始化（搬几把椅子到院子），设备连接上了（开聊）。发起聊天的设备为主设备，接受聊天的设备为从设备，同一次聊天只能有一个意见领袖，即主设备和从设备不能切换。
- HCI 层：为接口层，通信部，向上为主机提供软件应用程序接口（API），对外为外部硬件控制接口，可以通过串口、SPI、USB 来实现设备控制。
- L2CAP 层：物流部，负责行李打包盒拆封处，提供数据封装服务。
- SM 层：保卫处，提供配对和密匙分发，实现安全连接和数据交换。
- ATT 层：库房，负责数据检索。
- GATT 层：出纳/库房前台，出纳负责处理向上与应用打交道，而库房前台负责向下把检索任务子进程交给 ATT 库房去做，其关键工作是把为检索工作提供合适的 profile 结构，而 profile 由检索关键词（characteristics）组成。
- GAP 层：秘书处，对上级提供应用程序接口，对下级管理各级职能部门，尤其是指示 LL 层控制室五种状态切换，指导保卫处做好机要工作。

蓝牙为了实现同多个设备相连或实现多功能的目标，也实现了功能扩充，这就产生了调度问题。因为，虽然软件和协议栈可扩充，但终究最底层的执行部门只有一个。为了实现多事件和多任务切换，需要把事件和任务对应的应用，以及其相关的提供支撑“办公室”和“工厂”打包起来，并起一个名字 OSAL 操作系统抽象层，类似于集团公司以下的子公司。

如果实现软件和硬件的低耦合，使软件不经改动或很少改动即可应用在另外的硬件上，这样就方便硬件改造、升级、迁移后软件的移植。HAL 硬件抽象层正是用来抽象各种硬件的资源，告知给软件。其作用类似于嵌入式系统设备驱动的定义硬件资源的“.h”头文件，其角色类似于现代工厂的设备管理部。

2. BLE 低功耗蓝牙系统架构

BLE 低功耗蓝牙系统架构如图 20-15 所示。

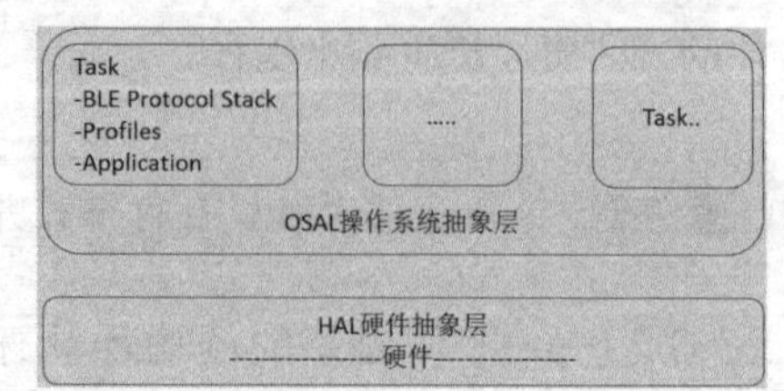

▲图 20-15　BLE 低功耗蓝牙系统架构图

由此可见，BLE 低功耗蓝牙软件有两个主要组成，分别是 OSAL 操作系统抽象层和 HAL 硬件抽象层，多个 Task 任务和事件在 OSAL 管理下工作，而每个任务和事件又包括 3 个组成部分，分别是 BLE 协议栈、profiles 和应用程序。

（1）OSAL 操作系统抽象层

OSAL 作为调度核心，BLE 协议栈、profile 定义、所有的应用都围绕它来实现。OSAL 不是传统大家使用的操作系统，而是一个允许软件建立和执行事件的循环。软件功能是由任务事件来实现的，创建的任务事件需要完成如下所示的工作。

- 创建 task identifier 任务 ID。
- 编写任务初始化（task initialization routine）进程，并需要添加到 OSAL 初始化进程中，这就是说系统启动后不能动态添加功能。
- 编写任务处理程序。
- 提供消息服务。

BLE 协议栈的各层都是以 OSAL 任务方式实现，由于 LL 控制室的时间要求最为迫切，所以其任务优先级最高。为了实现任务管理，OSAL 通过消息处理（messageprocess），存储管理，计时器定时等附加服务实现。

（2）系统启动流程

为了使用 OSAL，在 main 函数的最后要启动一个名叫 osal_start_system 的进程，该进程会调用由特定应用决定的启动函数 osalInitTasks（来启动系统）。osalInitTasks 逐个调用 BLE 协议栈各层的启动进程来初始化协议栈。随后，设置一个任务的 8bit 任务 ID（task ID），跳入循环等待执行任务，系统启动完成。

（3）任务事件与事件处理

- 进程优先级和任务 ID

➢ 任务优先级决定于任务 ID，任务 ID 越小，优先级越高。

➢ BLE 协议栈各层的任务优先级比应用程序的高。

➢ 初始化协议栈后，越早调入的任务，任务 ID 越高，优先级越低，即系统倾向于处理新到的任务。

- 事件变量和旗语

每个事件任务由对应的 16bit 事件变量来标示，事件状态由旗号（taskflag）来标示。如果事件处理程序已经完成，但其旗号并没有移除，OSAL 会认为事情还没有完成而继续在该程序中不返回。比如，在 SimpleBLEPeripheral 实例工程中，当事件 START_DEVICE_EVT 发生，其处理函数 SimpleBLEPeripheral_ProcessEvent 就运行，结束后返回 16bit 事件变量，并清除旗语 SBP_START_DEVICE_EVT。

- 事件处理表单

每当 OSAL 事件检测到了有任务事件，其相应的处理进程将被添加到由处理进程指针构成的事件处理表单中，该表单名叫 taskArr（taskarray）。taskArr 中各个事件进程的顺序和 osalInitTasks 初始化函数中任务 ID 的顺序是对应的。

- 事件调度的方法

有两种事件调度的方法，最简单的方法是使用 osal_set_event 函数（函数原型在 OSAL.h 文件中），在这个函数中，用户可以像定义函数参数一样设置任务 ID 和事件旗语。第二种方法是使用 osal_start_timerEx 函数（函数原型在 OSAL_Timers.h 文件中），使用方法同 osal_set_event 函数，而第三个以毫秒为单位的参数 osal_start_timerEx 则指示该事件处理必须要在这个限定时间内，通过定时器来为事件处理计时。

（4）存储管理

存储管理类似于 Linux 嵌入式系统内存分配 C 函数 mem_alloc，OSAL 利用 osal_mem_alloc 提供基本的存储管理，但 osal_mem_alloc 只有一个用于定义 byte 数的参数。对应的内存释放函数为 osal_mem_free。

（5）进程间通信——通过消息机制实现

不同的子系统通过 OSAL 的消息机制通信。消息即为数据，数据种类和长度都不限定。消息收发的过程描述如下所示。

在接收信息时调用函数 osal_msg_allocate 创建消息占用内存空间(已经包含了 osal_mem_alloc 函数功能)，需要为该函数指定空间大小，该函数返回内存空间地址指针，利用该指针就可把所需数据拷贝到该空间。

在发送数据时调用函数 osal_msg_send，需为该函数指定发送目标任务，OSAL 通过旗语 SYS_EVENT_MSG 告知目标任务，目标任务的处理函数调用 osal_msg_receive 来接收发来的数据。建议每个 OSAL 任务都有一个消息处理函数，每当任务收到一个消息后，通过消息的种类来确定需要本任务做相应处理。消息接收并处理完成，调用函数 osal_msg_deallocate 来释放内存（已经包含了 osal_mem_free 函数功能）。

3. 硬件抽象层 HAL

当新的硬件平台做好后，只需修改 HAL，而不需修改 HAL 之上的协议栈的其他组件和应用程序。

4. BLE 低功耗蓝牙协议栈

（1）BLE 库文件

TI 蓝牙协议栈是以单独一个库文件提供的，并没有提供源代码，因此不做深入说明。对于 TI 的 BLE 实例应用来说，这个单独库文件完全够用，因为已经列出了所有的库文件。

（2）GAP 秘书处

- 角色（即服务/功能）

在 TI 实例中，GAP 运行在如下四种角色的一种。

➢ Broadcaster：广播员——我在，但只可远观，不可连接。

➢ Observer：观察员——看看谁在，但我只远观，不连接。

➢ Peripheral：外设（从机）——我在，谁要我就跟谁走，协议栈单层连接。

➢ Central：核心（主机）——看看谁在，并且愿意跟我走我就带她/他走，协议栈单层或多层连接，目前最多支持 3 个同时连接。

虽然指标显示 BLE 可以同时扮演多个角色，但是在 TI 提供的 BLE 实例应用中默认只支持外设角色。每一种角色都由一个剧本（roleprofile）来定义。

- 连接

在主从机连接过程中，一个典型的低功耗蓝牙系统同时包含外设和核心（主机），两者的连接过程是：外设角色向外发送自己的信息（设备地址、名字等），主机收到外设广播信息后，发送扫描请求（scanrequest）给外设，外设响应主机的请求，连接建立完成。

连接参数主要有通信间隙（connectioninterval）、外设鄙视（slavelatency）、最大耐心等待时间（supervisiontimeout）等，具体说明如下所示。

➢ 通信间隙——蓝牙通信是间断的、跳频的，每次连接都可能选择不同的子频带。跳频的好处是避免频道拥塞，间断连接的好处是节省功耗，通信间隙就是指两次连接之间的时间间隔。这个间隔以 1.25ms 为基本单位，最小 6 单位最大 3200 单位，间隙越小通信越及时，间隙越大功耗越低。

➢ 外设鄙视——外设与主机建立连接以后，没事的时候主机总会定期发送问候信息到外设，外设懒得搭理，这些主机发送的信息就浮云般飘过。可以忽略的连接事件个数从 0 到 499 个，最多不超过 32 秒。有效连接间隙= 连接间隙×（1+ 外设鄙视）。

➢ 最大耐心等待时间——指的是为了创建一个连接，主机允许的最大等候时间，在这个时间内，不停地尝试连接。范围是 10～3200 个通信间隙基本单位（1.25ms）。

以上 3 参数大小设置优劣是显而易见的，连接参数的设置请请阅读本小节后面的内容。

假如主机采用从机并不舒坦的参数来请求连接，有如主从机已经连接了，但从机有想法了，要改参数条约。通过“连接参数更新请求（ConnectionParameter Update Request）”来解决问题，交由 L2CAP“收发室物流处”处理。

在实现加密处理时可以利用配对实现，利用密匙来加密授权连接。典型的过程是：外设向主机请求口令一个（passkey）以便进行配对，待主机发送了正确的口令之后，连接通信通过主从机互换密码来校验。由于蓝牙通信是间断通信，如果一个应用需要经常通信，而每次通信都要重新申请连接，那将是劳神费力的，为此 GAP 安全卫士（Security Profile，SM）提供了一种长期签证（long-termset of keys），叫做绑定（bonding），使得每次建立连接通关流程简便并快捷。

（3）出纳 GATT

GATT 负责两个设备间通信的数据交互。共有两种角色：出纳员（GATTClient）和银行（GATTServer），银行提供资金，出纳从银行存取款。银行可以同时面对多个出纳员。这两种角色和主从机等角色是无关的。

GATT 把工作拆分成几部分来实现：读关键词（CharacteristicValue）和描述符（CharacteristicDescriptor），用来去库房查找提取数据，并写/读关键词和描述符。

GATT 银行（GATTServer）的业务部门（API）主要提供两个主要的功能：一是服务功能，注册或销毁服务（serviceattribute），并作为回调函数（callbackfunction）；二是管理功能，添加或删除 GATT 银行业务。

一个角色定义的剧本可以同时定义多个角色，每个角色的服务（service）、关键词（characteristic）、关键值（characteristic value）、描述符（descriptor）都以句柄（attribute）形式保存在角色提供的服务上。所有的服务都是一个 gattAttribute_t 类型的 array，在文件 gatt.h.中定义。

（4）调用 GAP 和 GATT 的一般过程

调用 GAP 和 GATT 的一般过程如下所示。

- API 调用。
- 协议栈响应并返回。
- 协议栈发送一个 OSAL 消息（数据）去调用相应任务事件。
- 调用任务去接收和处理消息。
- 消息清除。

以设备初始化为 GAP 外设角色来举例说明，外设角色由其剧本（GAPperipheral role profile）来决定，实例程序在文件 peripheral.c 内。

- 调用 API 函数 GAP_DeviceInit。
- GAP 检查了一下说，好，可以初始化，返回值为 SUCCESS (0x00)，并通知 BLE 干活。
- BLE 协议栈发送 OSAL 消息给外设角色剧本（peripheral roleprofile），消息内容包括要干什么（eventvalue）GAP_MSG_EVENT 和指标是什么（opcodevalue，参数）。
- 角色剧本的服务任务就收到了事件请求 SYS_EVENT_MSG，表示有消息来了。
- 角色剧本接收消息，并拆看到底是什么事，接着把消息数据转换（cast）成具体要干事情，并完成相应的工作（这里为 gapDeviceInitDoneEvent_t）。
- 角色剧本清除消息并返回。例如 GATT 客户端设备想从 GATT 服务器端读取数据，即 GATT 出纳想从 GATT 银行那边取点钱出来。

- 应用程序调用 GATT 子进程 API 函数 GATT_ReadCharValue，传递的参数为连接句柄、关键词句柄和自身任务的 ID。
- GATT 答应了这个请求，返回值为 SUCCESS (0x00)，向下告知 BLE 有活干了。
- BLE 协议栈在下次建立蓝牙连接时，发送取钱的指令给银行，当银行说好，我们正好有柜员没事在干剪指甲，于是把钱取出来交给了 BLE。
- BLE 接着就把取到的钱包成消息（OSAL message），通过出纳 GATT 返回给了应用程序。消息内包含 GATT_MSG_EVENT 和修改了的 ATT_READ_RSP。
- 应用程序接收到了从 OSAL 来的 SYS_EVENT_MSG 事件，表示钱可能到了。
- 应用程序接收消息，拆包检查，并拿走需要的钱。
- 最后应用程序把包装袋销毁。

（5）GAP 角色剧本 Profiles

在 TI 的 BLE 实例应用中提供了 3 中 GAP 角色剧本，分别是保卫处角色和几种 GATT 出纳/库管示例程序服务角色。

- GAP 外设剧本

其 API 函数在 peripheral.h 中定义，包括如下所示的信息。

- GAPROLE_ADVERT_ENABLED——广播使能。
- GAPROLE_ADVERT_DATA——包含在广播里的信息。
- GAPROLE_SCAN_RSP_DATA——外设用于回复主机扫描请求的信息。
- GAPROLE_ADVERT_OFF_TIME——表示外设关闭广播持续时间，该值为零表示无限期关闭广播直到下一次广播使能信号到来。
- GAPROLE_PARAM_UPDATE_ENABLE——使能自动更新连接参数，可以让外设连接失败时自动调整连接参数以便重新连接。
- GAPROLE_MIN_CONN_INTERVAL——设置最小连接间隙，默认值为 80 个单位（每单位 1.25ms）。
- GAPROLE_MIN_CONN_INTERVAL——设置最大连接间隙，默认值为 3200 个单位。
- GAPROLE_SLAVE_LATENCY——外设鄙视参数，默认为零。
- GAPROLE_TIMEOUT_MULTIPLIER——最大耐心等待时间，默认为 1000 个单位。

函数 GAPRole_StartDevice 用来初始化 GAP 外设角色，其唯一的参数是 gapRolesCBs_t，这个参数是一个包含两个函数指针的结构体，这两函数是 pfnStateChange 和 pfnRssiRead，前者标示状态，后者标示 RSSI 已经被读走了。

- 多角色同时扮演

在此以设备同时为外设和广播员两种角色，方法是去除前文外设的定义剧本 peripheral.c 和 peripheral.h，添加新的剧本 peripheralBroadcaster.c 和 peripheralBroadcaster.h；定义处理器值（preprocessorvalue）PLUS_BROADCASTER。

- GAP 主机剧本

与外设剧本相似，主机剧本的 API 函数在 central.h 中定义，包括 GAPCentralRole_GetParameter 和 GAPCentralRole_SetParameter 以及其他。如 GAPROLE_PARAM_UPDATE_ENABLE 连接参数自动更新使能的功能，跟外设角色的一样。

GAPCentralRole_StartDevice 函数用来初始化 GAP 主机角色，其唯一的参数是 gapCentralRolesCBs_t,，这个参数是一个包含两个函数指针的结构体，这两函数是 eventCB 和 rssiCB，每次 GAP 时间发生，前者都会被调用，后者标示 RSSI 已经被读走。

- GAP 绑定管理器剧本

GAP 绑定管理器剧本用于保持长期的连接。同时支持外设配置和主机配置。当建立了配对连接后，如果绑定使能，绑定管理器就维护这个连接。主要参数有。

- GAPBOND_PAIRING_MODE。
- GAPBOND_MITM_PROTECTION。
- GAPBOND_IO_CAPABILITIES。
- GAPBOND_IO_CAP_DISPLAY_ONLY。
- GAPBOND_BONDING_ENABLED。

函数 GAPBondMgr_Register 用来初始化 GAP 主机角色，其唯一的参数是 gapBondCBs_t，这个参数是一个包含两个函数指针的结构体，这两函数是 pairStateCB 和 passcodeCB，前者返回状态，后者用于配对时产生 6 为数字口令（passcode）。

- 编写一个剧本来创建（定义）新的角色（功能、服务）

以 SimpleGATT Profile 为剧本名称，包含两个文件 simpleGATTProfile.c 和 simpleGATTProfile.h。包含如下所示的主要 API 函数。

SimpleProfile_AddService——用于初始化的进程，作用是添加服务句柄（serviceattributes）到句柄组（attributetable）内，寄存器读取和回写。

- SimpleProfile_SetParameter——设置剧本（profile）关键词（characteristics）。
- SimpleProfile_GetParameter——获取获取设置剧本关键词。
- SimpleProfile_RegisterAppCBs。
- simpleProfile_ReadAttrCB。
- simpleProfile_WriteAttrCB。
- simpleProfile_HandleConnStatusCB。

此实例剧本共有如下所示 id5 个关键词。

- SIMPLEPROFILE_CHAR1。
- ……
- SIMPLEPROFILE_CHAR5。

为节省本书的篇幅，TI 公司的低功耗蓝牙协议栈的基本知识介绍完毕。有关此协议栈的具体知识，请读者登录其官方站点查看帮助文档。

20.7 使用蓝牙控制电风扇

在本节的内容中，将通过一个具体实例的实现过程，详细讲解使用蓝牙技术传输数据的过程。本实例的功能是，使用蓝牙将 Android 应用程序和风扇建立连接，在 Android 应用程序中可以设置一个湿度值，并根据设置的湿度值来控制风扇的转速。

实例	功能	源码路径
实例 20-1	使用蓝牙控制电风扇的转动	\daima\20\lan

本实例的功能是使用蓝牙控制电风扇的转动，在开始之前使用蓝牙将 Android 应用程序和电风扇建立了连接，Android 应用程序是建立在 Arduino 开发板之上的，当然也可以建立在任何 Android 开发板中，也包括 Android 可穿戴设备。蓝牙、开发板、风扇的连接如图 20-16 所示。

20.7.1 准备 DHT 传感器

为了降低开发成本，在本实例中使用的是 DHT 公司的低端传感器产品。在使用之前，需要将文件 DHT.cpp 和文件 DHT.h 包含 Ardunio 开发板的库文件夹的工程中。其中文件 DHT.h 是一个头文件，功能是设置需要多少时间进行转换跟踪，具体实现代码如下所示。

```
#ifndef DHT_H
#define DHT_H
#if ARDUINO >= 100
 #include "Arduino.h"
#else
 #include "WProgram.h"
#endif

/* DHT library

MIT license
written by Adafruit Industries
*/

// how many timing transitions we need to keep track of. 2 * number bits + extra
#define MAXTIMINGS 85

#define DHT11 11
#define DHT22 22
#define DHT21 21
#define AM2301 21

class DHT {
 private:
  uint8_t data[6];
  uint8_t _pin, _type, _count;
  boolean read(void);
  unsigned long _lastreadtime;
  boolean firstreading;

 public:
  DHT(uint8_t pin, uint8_t type, uint8_t count=6);
  void begin(void);
  float readTemperature(bool S=false);
  float convertCtoF(float);
  float readHumidity(void);

};
#endif
```

▲图 20-16　硬件设备的连接

文件 DHT.cpp 的功能是提供 DHT 的温度传感器和湿度传感器测试，具体实现代码如下所示。

```
#include "DHT.h"

DHT::DHT(uint8_t pin, uint8_t type, uint8_t count) {
  _pin = pin;
  _type = type;
  _count = count;
  firstreading = true;
}

void DHT::begin(void) {
  // 建立指针!
  pinMode(_pin, INPUT);
  digitalWrite(_pin, HIGH);
  _lastreadtime = 0;
}

//boolean S == Scale.  True == Farenheit; False == Celcius
float DHT::readTemperature(bool S) {
  float f;

  if (read()) {
    switch (_type) {
    case DHT11:
      f = data[2];
      if(S)
          f = convertCtoF(f);

      return f;
```

```
    case DHT22:
    case DHT21:
      f = data[2] & 0x7F;
      f *= 256;
      f += data[3];
      f /= 10;
      if (data[2] & 0x80)
    f *= -1;
      if(S)
    f = convertCtoF(f);

      return f;
    }
  }
  Serial.print("Read fail");
  return NAN;
}

float DHT::convertCtoF(float c) {
    return c * 9 / 5 + 32;
}

float DHT::readHumidity(void) {
  float f;
  if (read()) {
    switch (_type) {
    case DHT11:
      f = data[0];
      return f;
    case DHT22:
    case DHT21:
      f = data[0];
      f *= 256;
      f += data[1];
      f /= 10;
      return f;
    }
  }
  Serial.print("Read fail");
  return NAN;
}


boolean DHT::read(void) {
  uint8_t laststate = HIGH;
  uint8_t counter = 0;
  uint8_t j = 0, i;
  unsigned long currenttime;

  // 提升指针，并设置等待 250 毫秒
  digitalWrite(_pin, HIGH);
  delay(250);

  currenttime = millis();
  if (currenttime < _lastreadtime) {
    // 此时翻转
    _lastreadtime = 0;
  }
  if (!firstreading && ((currenttime - _lastreadtime) < 2000)) {
    return true; // return last correct measurement
    //delay(2000 - (currenttime - _lastreadtime));
  }
  firstreading = false;
  /*
    Serial.print("Currtime: "); Serial.print(currenttime);
    Serial.print(" Lasttime: "); Serial.print(_lastreadtime);
  */
  _lastreadtime = millis();
```

```
  data[0] = data[1] = data[2] = data[3] = data[4] = 0;

  //降低 20 毫秒
  pinMode(_pin, OUTPUT);
  digitalWrite(_pin, LOW);
  delay(20);
  cli();
  digitalWrite(_pin, HIGH);
  delayMicroseconds(40);
  pinMode(_pin, INPUT);

  // 读取时间
  for ( i=0; i< MAXTIMINGS; i++) {
    counter = 0;
    while (digitalRead(_pin) == laststate) {
      counter++;
      delayMicroseconds(1);
      if (counter == 255) {
        break;
      }
    }
    laststate = digitalRead(_pin);

    if (counter == 255) break;

    // 忽略前三个转换
    if ((i >= 4) && (i%2 == 0)) {
      // shove each bit into the storage bytes
      data[j/8] <<= 1;
      if (counter > _count)
        data[j/8] |= 1;
      j++;
    }

  }

  sei();

  /*
  Serial.println(j, DEC);
  Serial.print(data[0], HEX); Serial.print(", ");
  Serial.print(data[1], HEX); Serial.print(", ");
  Serial.print(data[2], HEX); Serial.print(", ");
  Serial.print(data[3], HEX); Serial.print(", ");
  Serial.print(data[4], HEX); Serial.print(" =? ");
  Serial.println(data[0] + data[1] + data[2] + data[3], HEX);
  */

  // 读取到 40 位便进行校验并比较
  if ((j >= 40) &&
      (data[4] == ((data[0] + data[1] + data[2] + data[3]) & 0xFF)) ) {
    return true;
  }


  return false;

}
```

编写测试程序 DHTtester.pde 来获取 DHT 传感器温度和湿度值，具体实现代码如下所示。

```
DHT dht(DHTPIN, DHTTYPE);

void setup() {
  Serial.begin(9600);
  Serial.println("DHTxx test!");

  dht.begin();
}
```

```
void loop() {
  // Reading temperature or humidity takes about 250 milliseconds!
  // Sensor readings may also be up to 2 seconds 'old' (its a very slow sensor)
  float h = dht.readHumidity();
  float t = dht.readTemperature();

  // check if returns are valid, if they are NaN (not a number) then something went wrong!
  if (isnan(t) || isnan(h)) {
    Serial.println("Failed to read from DHT");
  } else {
    Serial.print("Humidity: ");
    Serial.print(h);
    Serial.print(" %\t");
    Serial.print("Temperature: ");
    Serial.print(t);
    Serial.println(" *C");
  }
}
```

20.7.2 实现 Android 测试 APP

开始编写 Android 应用程序，首先编写界面布局文件 activity_control.xml，功能是设置蓝牙启动测试按钮，具体实现代码如下所示。

```
<RelativeLayout
      xmlns:android="http://schemas.android.com/apk/res/android"
      xmlns:tools="http://schemas.android.com/tools"
        android:layout_width="match_parent"
        android:layout_height="match_parent"
        >

   <LinearLayout
      android:layout_width="match_parent"
      android:layout_height="match_parent"
      android:orientation="vertical"
      android:paddingBottom="@dimen/activity_vertical_margin"
      android:paddingLeft="@dimen/activity_horizontal_margin"
      android:paddingRight="@dimen/activity_horizontal_margin"
      android:paddingTop="@dimen/activity_vertical_margin"
      tools:context=".ControlActivity" >

      <TextView
         android:id="@+id/textView1"
         android:layout_width="wrap_content"
         android:layout_height="wrap_content"
         android:layout_alignParentLeft="true"
         android:layout_alignParentTop="true"
         android:layout_marginLeft="17dp"
         android:layout_marginTop="15dp"
         android:text="" />

      <Button
         android:id="@+id/connect"
         android:layout_width="wrap_content"
         android:layout_height="wrap_content"
         android:layout_alignLeft="@+id/textView1"
         android:layout_below="@+id/textView1"
         android:layout_marginTop="10dp"
         android:text="Connect"
         android:layout_gravity="center" />

      <ImageView
         android:id="@+id/power_button"
         android:layout_width="130dp"
         android:layout_height="130dp"
         android:contentDescription="@string/power_control"
         android:src="@drawable/humidigator_off"
         android:layout_gravity="center_horizontal"
```

```
        android:layout_weight="7" />

    <!-- In this TextView, the specific height of 36dp is given because I wanted to
make the temperature
    invisible for my project. I couldn't delete it out of the Arduino, I couldn't figure
out why. I kept getting
    errors. Make this "wrap_content" and it will go back to normal. -->
    <TextView
        android:id="@+id/current_data"
        android:layout_width="wrap_content"
        android:layout_height="36dp"
        android:layout_marginTop="69dp"
        android:layout_gravity="center_horizontal"
        android:gravity="center_horizontal"
        android:textSize="16sp"
        android:textColor="@color/white"
        android:text="Press 'Connect' to connect to Humidigator" />


    <LinearLayout
        android:orientation="vertical"
        android:layout_width="match_parent"
        android:layout_height="wrap_content"
        android:gravity="center"
        android:layout_weight="12"
        android:layout_marginBottom="20dp"
            >

        <LinearLayout
            android:layout_width="match_parent"
            android:layout_height="wrap_content"
            android:orientation="horizontal"
            android:gravity="center"
            android:id="@+id/setHumLayout"
            >


            <TextView
                android:layout_width="wrap_content"
                android:layout_height="wrap_content"
                android:gravity="center"
                android:text="Set Humidity:"
                android:textSize="30dp"
                android:textColor="@color/white"
                android:textStyle="normal"
                android:layout_marginRight="15dp"
            />

            <TextView
                android:id="@+id/humidity"
                    android:layout_width="wrap_content"
                android:layout_height="wrap_content"
                android:text="50"
                android:textSize="30dp"
                android:layout_marginTop="3dp"
                android:textColor="@color/white"
                android:layout_gravity="center_vertical" />
            <TextView
                android:gravity="center"
                android:layout_width="wrap_content"
                android:layout_height="match_parent"
                android:text="%"
                android:textSize="30dp"
                android:layout_marginTop="3dp"
                android:textColor="@color/white"
                android:layout_gravity="center_vertical"
            />
```

```
        </LinearLayout>

            <SeekBar
                android:id="@+id/humidity_seek"
            android:layout_width="300dp"
            android:layout_height="wrap_content"
            android:layout_gravity="center"
            android:progress="50"
                />
        </LinearLayout>
    </LinearLayout>
</RelativeLayout>
```

然后编写主 Android 应用程序文件 ControlActivity.java，功能根据用户触摸屏幕操作来开启关闭蓝牙测试开关，将用户设置的值通过蓝牙传递给 DHT 传感器程序以控制电风扇的转动。文件 ControlActivity.java 的具体实现代码如下所示。

```
public class ControlActivity extends Activity implements OnClickListener,
OnSeekBarChangeListener{

    Button Connect;
    ToggleButton OnOff;
    TextView Result;
    private String dataToSend;

    boolean powerOn = false;

    private static final String TAG = "Ian";

    private BluetoothAdapter mBluetoothAdapter = null;
    private BluetoothSocket btSocket = null;
    private OutputStream outStream = null;
    private static String address = "20:13:08:28:03:16";
    private static final UUID MY_UUID = UUID
            .fromString("00001101-0000-1000-8000-00805F9B34FB");
    private InputStream inStream = null;
    Handler handler = new Handler();
    byte delimiter = 10;
    boolean stopWorker = false;
    int readBufferPosition = 0;
    byte[] readBuffer = new byte[1024];

    private TextView setHum;
    private SeekBar humSeek;
    private LinearLayout setHumLayout;

    ImageView powerButton;

    @Override
    protected void onCreate(Bundle savedInstanceState) {
        super.onCreate(savedInstanceState);
        setContentView(R.layout.activity_control);

        CheckBt();
        BluetoothDevice device = mBluetoothAdapter.getRemoteDevice(address);
        Log.e("Ian", device.toString());

        Connect = (Button) findViewById(R.id.connect);
        powerButton = (ImageView) findViewById(R.id.power_button);
        Result = (TextView) findViewById(R.id.current_data);
        setHum = (TextView) findViewById(R.id.humidity);
        humSeek = (SeekBar) findViewById(R.id.humidity_seek);
        setHumLayout = (LinearLayout) findViewById(R.id.setHumLayout);


        Connect.setOnClickListener(this);
        powerButton.setOnClickListener(this);
        humSeek.setOnSeekBarChangeListener(this);
```

```
        powerButton.setVisibility(View.INVISIBLE);
        setHumLayout.setVisibility(View.INVISIBLE);
        humSeek.setVisibility(View.INVISIBLE);


    }


    private void CheckBt() {
        mBluetoothAdapter = BluetoothAdapter.getDefaultAdapter();

        if (!mBluetoothAdapter.isEnabled()) {
            Toast.makeText(getApplicationContext(), "Bluetooth Disabled !",
                    Toast.LENGTH_SHORT).show();
        }

        if (mBluetoothAdapter == null) {
            Toast.makeText(getApplicationContext(),
                    "Bluetooth null !", Toast.LENGTH_SHORT)
                    .show();
        }
    }

        public void Connect() {
            Log.d(TAG, address);
            BluetoothDevice device = mBluetoothAdapter.getRemoteDevice(address);
            Log.d(TAG, "Connecting to ... " + device);
            mBluetoothAdapter.cancelDiscovery();
            try {
                btSocket = device.createRfcommSocketToServiceRecord(MY_UUID);
                btSocket.connect();
                Log.d(TAG, "Connection made.");
            } catch (IOException e) {
                try {
                    btSocket.close();
                } catch (IOException e2) {
                    Log.d(TAG, "Unable to end the connection");
                }
                Log.d(TAG, "Socket creation failed");
            }

            beginListenForData();
        }

    private void writeData(String data) {
        try {
            outStream = btSocket.getOutputStream();
        } catch (IOException e) {
            Log.d(TAG, "Bug BEFORE Sending stuff", e);
        }

        String message = data;
        byte[] msgBuffer = message.getBytes();

        try {
            outStream.write(msgBuffer);
        } catch (IOException e) {
            Log.d(TAG, "Bug while sending stuff", e);
        }
    }

    @Override
    protected void onDestroy() {
        super.onDestroy();

            try {
                btSocket.close();
            } catch (IOException e) {
```

```
        }
    }

    public void beginListenForData()    {
        try {
            inStream = btSocket.getInputStream();
        } catch (IOException e) {
        }

        Thread workerThread = new Thread(new Runnable()
        {
            public void run()
            {
               while(!Thread.currentThread().isInterrupted() && !stopWorker)
               {
                    try
                    {
                        int bytesAvailable = inStream.available();
                        if(bytesAvailable > 0)
                        {
                            byte[] packetBytes = new byte[bytesAvailable];
                            inStream.read(packetBytes);
                            for(int i=0;i<bytesAvailable;i++)
                            {
                                byte b = packetBytes[i];
                                if(b == delimiter)
                                {
                                    byte[] encodedBytes = new byte[readBufferPosition];
                                    System.arraycopy(readBuffer, 0, encodedBytes, 0,
                                    encodedBytes.length);
                                    final String data = new String(encodedBytes,
                                    "US-ASCII");
                                    readBufferPosition = 0;
                                    handler.post(new Runnable()
                                    {
                                        public void run()
                                        {

                                            Result.setText(data);
                                            Result.setTextSize(25);

                                            /* You also can use Result.setText(data); it
                                            won't display multilines
                                            */

                                        }
                                    });
                                }
                                else
                                {
                                    readBuffer[readBufferPosition++] = b;
                                }
                            }
                        }
                    }
                    catch (IOException ex)
                    {
                        stopWorker = true;
                    }
               }
            }
        });

        workerThread.start();
    }
```

```
    @Override
    public boolean onCreateOptionsMenu(Menu menu) {
        // Inflate the menu; this adds items to the action bar if it is present.
        getMenuInflater().inflate(R.menu.control, menu);
        return true;

    }


    @Override
    public void onClick(View control) {
        // TODO Auto-generated method stub
        switch (control.getId()) {
        case R.id.connect:
                Connect();
                powerButton.setVisibility(View.VISIBLE);
                setHumLayout.setVisibility(View.VISIBLE);
                humSeek.setVisibility(View.VISIBLE);

                Connect.setVisibility(View.INVISIBLE);
            break;
        case R.id.power_button:
            if (!powerOn){
                powerOn = true;
                powerButton.setImageResource(R.drawable.humidigator_on);
                dataToSend = "1";
                writeData(dataToSend);
            }
            else if (powerOn){
                powerOn = false;
                powerButton.setImageResource(R.drawable.humidigator_off);
                dataToSend = "0";
                writeData(dataToSend);
            }

            break;
        }
    }

    @Override
    public void onProgressChanged(SeekBar seekBar, int progress,
            boolean fromUser) {
        // TODO Auto-generated method stub
        setHum.setText(Integer.toString(humSeek.getProgress()));
    }
    @Override
    public void onStartTrackingTouch(SeekBar seekBar) {
        // TODO Auto-generated method stub

    }
    @Override
    public void onStopTrackingTouch(SeekBar seekBar) {
        // TODO Auto-generated method stub

    }
}
```

到此为止，整个实例介绍完毕，执行后的效果如图 20-17 所示。

本实例的目的是演示通过蓝牙和传感器设备控制硬件的过程，说明蓝牙技术在近距离传输中的作用。从本实例就可以感受到可穿戴设备必然会得到广泛接受和支持。

▲图 20-17　执行效果

第21章　语音识别技术详解

语音识别技术是 Android SDK 中比较重要且比较新颖的一项技术，在 Android 穿戴设备应用中可以通过语音来控制设备。本章将详细讲解在 Android 设备中使用语音识别技术的基本知识，为步入本书后面知识的学习打下基础。

21.1 语音识别技术基础

语音识别是一门交叉学科，经过多年的发展，语音识别技术取得了显著进步，逐渐从实验室走向市场。据专家预计，在未来 10 年内，语音识别技术将进入工业、家电、通信、汽车电子、医疗、家庭服务、消费电子产品等各个领域。很多专家都认为语音识别技术是 2000 年至 2010 年间信息技术领域十大重要的科技发展技术之一。语音识别技术所涉及的领域包括：信号处理、模式识别、概率论和信息论、发声机理和听觉机理、人工智能等。本节将详细讲解语音识别技术的基本知识。

21.1.1　语音识别的发展历史

1952 年，贝尔研究所 Davis 等研究员成功研制了世界上第一个能识别 10 个英文数字发音的实验系统。

1960 年，英国的 Denes 等研究成功了世界上第一个计算机语音识别系统。

20 世纪 70 年代以后，大规模的语音识别得到了好的发展契机，在小词汇量、孤立词的识别方面取得了实质性的进展。DARPA（Defense Advanced Research Projects Agency）是在 70 年代由美国国防部高级研究计划署资助的一项 10 年计划，其旨在支持语言理解系统的研究开发工作。

80 年代以后，研究的重点逐渐转向大词汇量、非特定人连续语音识别。在研究思路上也发生了重大变化，即由传统的基于标准模板匹配的技术思路开始转向基于统计模型（HMM）的技术思路。此外，再次提出了将神经网络技术引入语音识别问题的技术思路。

80 年代，美国国防部高级研究计划署又资助了一项为期 10 年的 DARPA 战略计划，其中包括噪声下的语音识别和会话（口语）识别系统，识别任务设定为“(1000 单词) 连续语音数据库管理”。

到了 90 年代，这一 DARPA 计划仍在持续进行中。其研究重点已转向识别装置中的自然语言处理部分，识别任务设定为“航空旅行信息检索”。

90 年代以后，在语音识别的系统框架方面并没有什么重大突破。但是，在语音识别技术的应用及产品化方面出现了很大进展。

1981 年，日本在其第五代计算机计划中提出了有关语音识别“输入-输出”自然语言的宏伟目标，虽然没能实现预期目标，但是有关语音识别技术的研究有了大幅度的加强和进展。

1987 年起，日本又拟出新的国家项目：高级人机口语接口和自动电话翻译系统。

21.1.2　技术发展历程

目前 IBM 语音研究小组在大词汇语音识别方面处于领先地位，在 70 年代开始了它的大词汇语音识别研究工作的。AT&T 的贝尔研究所也开始了一系列有关非特定人语音识别的实验。这一研究历经 10 年，其成果是确立了如何制作用于非特定人语音识别的标准模板的方法。在这一时期所取得的重大进展有如下 3 点。

（1）隐式马尔科夫模型（HMM）技术的成熟和不断完善成为语音识别的主流方法。

（2）以知识为基础的语音识别的研究日益受到重视。在进行连续语音识别的时候，除了识别声学信息外，更多地利用各种语言知识，诸如构词、句法、语义、对话背景方面等的知识来帮助进一步对语音做出识别和理解。同时在语音识别研究领域，还产生了基于统计概率的语言模型。

（3）人工神经网络在语音识别中的应用研究的兴起。在这些研究中，大部分采用基于反向传播算法（BP 算法）的多层感知网络。人工神经网络具有区分复杂的分类边界的能力，显然它十分有助于模式划分。特别是在电话语音识别方面，由于其有着广泛的应用前景，成了当前语音识别应用的一个热点。

另外，面向个人用途的连续语音听写机技术也日趋完善。这方面，最具代表性的是 IBM 的 ViaVoice 和 Dragon 公司的 Dragon Dictate 系统。这些系统具有说话人自适应能力，新用户不需要对全部词汇进行训练，便可在使用中不断提高识别率。

21.2　Text-To-Speech 技术详解

Text-To-Speech 简称 TTS，是 Android 1.6 版本中比较重要的新功能，能够将所指定的文本转成不同语言音频输出。TTS 可以方便地嵌入到游戏或者应用程序中，增强用户体验。本节将详细讲解在 Android 系统中使用 Text-To-Speech 技术的基本知识，为读者步入本书后面知识的学习打下基础。

21.2.1　Text-To-Speech 基础

TTS engine 依托于当前 Android Platform 所支持的几种主要的语言：English、French、German、Italian 和 Spanish 五大语言（注意：暂时没有中文，目前至少 Google 的科学家们还没有把中文玩到炉火纯青的地步）。TTS 可以将文本随意地转换成以上任意 5 种语言的语音输出。与此同时，对于个别的语言版本将取决于不同的时区，例如，对于 English，在 TTS 中可以分别输出美式和英式两种不同的版本。

既然能支持如此庞大的数据量，TTS 引擎对于资源的优化采取预加载的方法。根据一系列的参数信息从库中提取相应的资源，并加载到当前系统中。尽管当前大部分加载有 Android 操作系统的设备都通过这套引擎来提供 TTS 功能，但由于一些设备的存储空间非常有限，而影响到 TTS 无法最大限度的发挥功能，算是当前的一个瓶颈。为此开发小组引入了检测模块，让利用这项技术的应用程序或者游戏针对不同的设备可以有相应的优化调整，从而避免由于此项功能的限制，影响到整个应用程序的使用。比较稳妥的做法是让用户自行选择是否有足够的空间或者需求来加载此项资源，下边给出了一个标准的检测方法。

```
Intent checkIntent = new Intent();
checkIntent.setAction(TextToSpeech.Engine.ACTION_CHECK_TTS_DATA);
startActivityForResult(checkIntent, MY_DATA_CHECK_CODE);
```

如果当前系统允许创建一个“android.speech.tts.TextToSpeech”的 Object 对象，则说明已经提供 TTS 功能的支持，将检测返回结果中给出“CHECK_VOICE_DATA_PASS”的标记。如果系统不支持这项功能，那么用户可以选择是否加载这项功能，从而让设备支持输出多国语言的语音功能“Multi-lingual Talking”。“ACTION_INSTALL_TTS_DATA”intent 将用户引入 Android market 中的 TTS 下载界面。下载完成后将自动完成安装，下边是实现上述过程的完整代码（androidres.com）。

```
private TextToSpeech mTts;
protected void onActivityResult(
        int requestCode, int resultCode, Intent data) {
    if (requestCode == MY_DATA_CHECK_CODE) {
        if (resultCode == TextToSpeech.Engine.CHECK_VOICE_DATA_PASS) {
            // success, create the TTS instance
            mTts = new TextToSpeech(this, this);
        } else {
            // missing data, install it
            Intent installIntent = new Intent();
           installIntent.setAction(
                TextToSpeech.Engine.ACTION_INSTALL_TTS_DATA);
            startActivity(installIntent);
        }
    }
}
```

TextToSpeech 实体和 OnInitListener 都需要引用当前 Activity 的 Context 作为构造参数。OnInitListener()的用处是通知系统当前 TTS Engine 已经加载完成，并处于可用状态。

21.2.2 Text-To-Speech 的实现流程

（1）首先检查 TTS 数据是否可用，如下面的代码。

```
view plaincopy to clipboardprint?
//检查 TTS 数据是否已经安装并且可用
    Intent checkIntent = new Intent();
    checkIntent.setAction(TextToSpeech.Engine.ACTION_CHECK_TTS_DATA);
    startActivityForResult(checkIntent, REQ_TTS_STATUS_CHECK);
protected  void onActivityResult(int requestCode, int resultCode, Intent data) {
    if(requestCode == REQ_TTS_STATUS_CHECK)
    {
        switch (resultCode) {
        case TextToSpeech.Engine.CHECK_VOICE_DATA_PASS:
            //这个返回结果表明 TTS Engine 可以用
        {
            mTts = new TextToSpeech(this, this);
            Log.v(TAG, "TTS Engine is installed!");
        }
            break;
        case TextToSpeech.Engine.CHECK_VOICE_DATA_BAD_DATA:
            //需要的语音数据已损坏
        case TextToSpeech.Engine.CHECK_VOICE_DATA_MISSING_DATA:
            //缺少需要语言的语音数据
        case TextToSpeech.Engine.CHECK_VOICE_DATA_MISSING_VOLUME:
            //缺少需要语言的发音数据
        {
            //这二种情况都表明数据有错,重新下载安装需要的数据
            Log.v(TAG, "Need language stuff:"+resultCode);
            Intent dataIntent = new Intent();
            dataIntent.setAction(TextToSpeech.Engine.ACTION_INSTALL_TTS_DATA);
            startActivity(dataIntent);

        }
            break;
        case TextToSpeech.Engine.CHECK_VOICE_DATA_FAIL:
            //检查失败
        default:
            Log.v(TAG, "Got a failure. TTS apparently not available");
            break;
        }
    }
    else
```

```
    {
        //其他 Intent 返回的结果
    }
    }
```

（2）然后初始化 TTS，如下面的代码。

```
view plaincopy to clipboardprint?
//实现 TTS 初始化接口
    @Override
    public void onInit(int status) {
        // TODO Auto-generated method stub
        //TTS Engine 初始化完成
        if(status == TextToSpeech.SUCCESS)
        {
            int result = mTts.setLanguage(Locale.US);
            //设置发音语言
            if(result == TextToSpeech.LANG_MISSING_DATA || result == TextToSpeech.
            LANG_NOT_SUPPORTED)
            //判断语言是否可用
            {
                Log.v(TAG, "Language is not available");
                speakBtn.setEnabled(false);
            }
            else
            {
    mTts.speak("This is an example of speech synthesis.", TextToSpeech.QUEUE_ADD, null);
                speakBtn.setEnabled(true);
            }
        }
    }
```

（3）接下来需要设置发音语言，如下面的代码。

```
view plaincopy to clipboardprint?
public void onItemSelected(AdapterView<?> parent, View view,
        int position, long id) {
    // TODO Auto-generated method stub
    int pos = langSelect.getSelectedItemPosition();
    int result = -1;
    switch (pos) {
    case 0:
    {
        inputText.setText("I love you");
        result = mTts.setLanguage(Locale.US);
    }
        break;
    case 1:
    {
        inputText.setText("Je t'aime");
        result = mTts.setLanguage(Locale.FRENCH);
    }
        break;
    case 2:
    {
        inputText.setText("Ich liebe dich");
        result = mTts.setLanguage(Locale.GERMAN);
    }
        break;
    case 3:
    {
        inputText.setText("Ti amo");
        result = mTts.setLanguage(Locale.ITALIAN);
    }
        break;
    case 4:
    {
        inputText.setText("Te quiero");
        result = mTts.setLanguage(new Locale("spa", "ESP"));
    }
```

```
        break;
    default:
        break;
    }
    //设置发音语言
    if(result == TextToSpeech.LANG_MISSING_DATA || result == TextToSpeech.LANG_NOT_
    SUPPORTED)
    //判断语言是否可用
    {
        Log.v(TAG, "Language is not available");
        speakBtn.setEnabled(false);
    }
    else
    {
        speakBtn.setEnabled(true);
    }
}
```

（4）最近设置点击 Button 按钮发出声音，如下面的代码。

```
view plaincopy to clipboardprint?
public void onClick(View v) {
    //朗读输入框里的内容
    mTts.speak(inputText.getText().toString(), TextToSpeech.QUEUE_ADD, null);
}
```

21.2.3 实战演练——使用 Text-To-Speech 技术实现语音识别

在接下来的内容中，将通过一个具体实例的实现过程，详细讲解在 Android 设备中使用 Text-To-Speech 技术实现语音识别的过程。

实例	功能	源码路径
实例 21-1	使用 Text-To-Speech 实现语音识别	\daima\21\TextSpeech

本实例的功能是在屏幕中显示一个文本框供用户输入需要读的内容，点击按钮后开始读取文本框中输入的内容，本实例支持中文。本实例的具体实现流程如下所示。

（1）编写布局文件 main.xml，功能是在屏幕中显示一个可输入内容的文本框，具体实现代码如下所示。

```
<LinearLayout xmlns:android="http://schemas.android.com/apk/res/android"
    android:orientation="vertical" android:layout_width="fill_parent"
    android:layout_height="fill_parent">
    <EditText android:id="@+id/EditText01" android:text="hello guan"
        android:layout_width="fill_parent" android:layout_height="wrap_content">
    </EditText>
    <Button android:text="朗读" android:id="@+id/Button01"
        android:layout_width="wrap_content" android:layout_height="wrap_content">
    </Button>
</LinearLayout>
```

（2）编写程序文件 speechActivity.java，功能是根据用户在文本框中输入的内容实现语音识别，具体实现代码如下所示。

```
public class speechActivity extends Activity  implements TextToSpeech.OnInitListener{
    private TextToSpeech mSpeech;
    private Button btn;
    static final int TTS_CHECK_CODE = 0;
    private EditText mEditText;
    private TTS myTts;

    @Override
    public void onCreate(Bundle savedInstanceState) {
        super.onCreate(savedInstanceState);
        setContentView(R.layout.main);
        btn = (Button) findViewById(R.id.Button01);
```

```
        mEditText = (EditText) findViewById(R.id.EditText01);
        //btn.setEnabled(false);
        myTts = new TTS(this, ttsInitListener, true);
        mSpeech = new TextToSpeech(this, this);
        btn.setOnClickListener(new OnClickListener() {
            public void onClick(View v) {
                myTts.setLanguage("CHINESE");
                  myTts.speak(mEditText.getText().toString(), 0, null);
                mSpeech.speak(mEditText.getText().toString(),
                        TextToSpeech.QUEUE_FLUSH, null);
            }
        });
    }

    private TTS.InitListener ttsInitListener = new TTS.InitListener() {
            public void onInit(int version) {
              myTts.setLanguage("CHINESE");
              myTts.speak("爸爸", 0, null);
            }
    };

    public void onInit(int status) {
        if (status == TextToSpeech.SUCCESS) {
            int result = mSpeech.setLanguage(Locale.CHINA);
            if (result == TextToSpeech.LANG_MISSING_DATA
                    || result == TextToSpeech.LANG_NOT_SUPPORTED) {
                Log.e("lanageTag", "not use");
            } else {
                btn.setEnabled(true);
                mSpeech.speak("我喜欢你", TextToSpeech.QUEUE_FLUSH,
                        null);
            }
        }
    }

    @Override
    protected void onDestroy() {
        if (mSpeech != null) {
            mSpeech.stop();
            mSpeech.shutdown();
        }
        super.onDestroy();
    }
}
```

▲图 21-1　执行效果

执行后的效果如图 21-1 所示。

21.3 Voice Recognition 技术详解

大家应该知道，苹果手机 iPhone 语音识别用的是 Google 的技术，作为 Google 力推的 Android 自然会将其核心技术往 Android 系统里面植入，并结合 Google 的云端技术将其发扬光大。本节将详细讲解 Voice Recognition 技术的基本知识。

21.3.1　Voice Recognition 技术基础

在 Android 中使用 Google 的 Voice Recognition 的方法极其简单。Android 自带的 API 例子中，是通过一个 Intent 的 Action 动作来完成的，主要有以下两种模式。

- ACTION_RECOGNIZE_SPEECH：一般语音识别，在这种模式下人们可以捕捉到语音处理后的文字列。
- ACTION_WEB_SEARCH：网络搜索。

我们看在 API Demo 源码中为我们提供的语音识别实例，具体实现代码如下所示。

```
package com.example.android.apis.app;

import com.example.android.apis.R;

import android.app.Activity;
import android.content.Intent;
import android.content.pm.PackageManager;
import android.content.pm.ResolveInfo;
import android.os.Bundle;
import android.speech.RecognizerIntent;
import android.view.View;
import android.view.View.OnClickListener;
import android.widget.ArrayAdapter;
import android.widget.Button;
import android.widget.ListView;

import java.util.ArrayList;
import java.util.List;

/**
 *用 API 开发的抽象语音识别代码
 */
public class VoiceRecognition extends Activity implements OnClickListener {

    private static final int VOICE_RECOGNITION_REQUEST_CODE = 1234;

    private ListView mList;

    /**
     *呼叫与活动首先被创造
     */
    @Override
    public void onCreate(Bundle savedInstanceState) {
        super.onCreate(savedInstanceState);
        //从它的 XML 布局描述的 UI
        setContentView(R.layout.voice_recognition);

        //得到最新互作用的显示项目
        Button speakButton = (Button) findViewById(R.id.btn_speak);
        mList = (ListView) findViewById(R.id.list);

        //检查看公认活动是否存在
        PackageManager pm = getPackageManager();
        List<ResolveInfo> activities = pm.queryIntentActivities(
                new Intent(RecognizerIntent.ACTION_RECOGNIZE_SPEECH), 0);
        if (activities.size() != 0) {
            speakButton.setOnClickListener(this);
        } else {
            speakButton.setEnabled(false);
            speakButton.setText("Recognizer not present");
        }
    }

    /**
     *点击“开始识别按钮”后的处理事件
     */
    public void onClick(View v) {
        if (v.getId() == R.id.btn_speak) {
            startVoiceRecognitionActivity();
        }
    }
    /**
     *发送开始语音识别信号
     */
    private void startVoiceRecognitionActivity() {
        Intent intent = new Intent(RecognizerIntent.ACTION_RECOGNIZE_SPEECH);
        intent.putExtra(RecognizerIntent.EXTRA_LANGUAGE_MODEL,
```

```
                RecognizerIntent.LANGUAGE_MODEL_FREE_FORM);
        intent.putExtra(RecognizerIntent.EXTRA_PROMPT, "Speech recognition demo");
        startActivityForResult(intent, VOICE_RECOGNITION_REQUEST_CODE);
    }

    /**
     *处理识别结果
     */
    @Override
    protected void onActivityResult(int requestCode, int resultCode, Intent data) {
        if (requestCode == VOICE_RECOGNITION_REQUEST_CODE && resultCode == RESULT_OK) {
            // Fill the list view with the strings the recognizer thought it could have heard
            ArrayList<String> matches = data.getStringArrayListExtra(
                    RecognizerIntent.EXTRA_RESULTS);
            mList.setAdapter(new ArrayAdapter<String>(this, android.R.layout.simple_
            list_item_1, matches));
        }
        super.onActivityResult(requestCode, resultCode, data);
    }
}
```

上述代码保存在 Google 的 API 开源文件中，原理和实现代码十分简单，感兴趣的读者可以学习一下。上述源码执行后，用户通过点击【Speak!】按钮显示界面如图 21-2 所示；用户说完话后将提交到云端搜索，如图 21-3 所示；在云端搜索完成后将返回打印数据，如图 21-4 所示。

▲图 21-2　单击按钮后

▲图 21-3　说完后

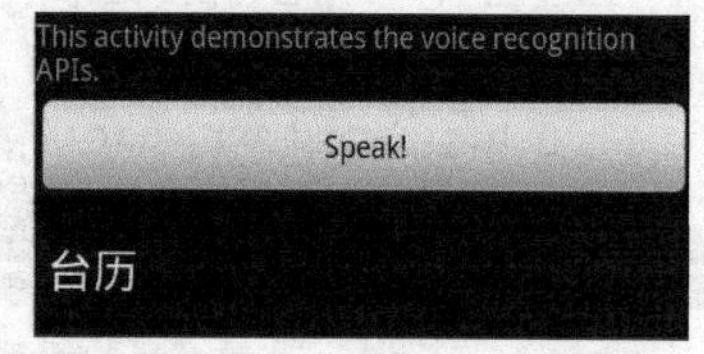

▲图 21-4　返回识别结果

21.3.2　实战演练——使用 Voice Recognition 技术实现语音识别

在接下来的内容中，将通过一个具体实例的实现过程，详细讲解在 Android 设备中使用 Voice Recognition 技术实现语音识别的过程。本实例的功能是在屏幕中显示一个文本框供用户输入需要读的内容，点击按钮后开始读取文本框中输入的内容，本实例支持中文。

实例	功能	源码路径
实例 21-2	使用 Voice Recognition 实现语音识别	\daima\21\yuyin

本实例的具体实现流程如下所示。

（1）编写布局文件 activity_main.xml，功能是在屏幕中设置一个图标按钮，单击按钮后可以说话，并在下方列表中显示语音识别结果。文件 activity_main.xml 的具体实现代码如下所示。

```
<LinearLayout xmlns:android="http://schemas.android.com/apk/res/android"
    xmlns:tools="http://schemas.android.com/tools"
    android:layout_width="match_parent"
    android:layout_height="match_parent"
    android:orientation="vertical"
    android:paddingBottom="@dimen/activity_vertical_margin"
    android:paddingLeft="@dimen/activity_horizontal_margin"
    android:paddingRight="@dimen/activity_horizontal_margin"
    android:paddingTop="@dimen/activity_vertical_margin"
    tools:context=".MainActivity" >
```

```
    <TextView
        android:layout_width="match_parent"
        android:layout_height="wrap_content"
        android:gravity="center_horizontal"
        android:layout_margin="20dp"
        android:text="@string/start_voice"
        android:textSize="16sp" />

    <Button
        android:id="@+id/mic_button"
        android:layout_width="wrap_content"
        android:layout_height="wrap_content"
        android:layout_gravity="center_horizontal"
        android:background="@drawable/ic_mic" />

    <ListView
        android:id="@+id/result_list"
        android:layout_width="fill_parent"
        android:layout_height="0dip"
        android:layout_weight="1" />

</LinearLayout>
```

（2）编写主 Activity 文件 MainActivity.java，功能是监听用户单击屏幕的图标按钮，调用识别函数进行语音识别，并调用函数 onActivityResult 显示识别结果。文件 MainActivity.java 的具体实现代码如下所示。

```
public class MainActivity extends Activity {

    private static final int REQUEST_CODE = 1;
    private ListView mResultList;
    private Button mMicButton;

    @Override
    protected void onCreate(Bundle savedInstanceState) {
        super.onCreate(savedInstanceState);
        setContentView(R.layout.activity_main);

        mMicButton = (Button) findViewById(R.id.mic_button);
        mResultList = (ListView) findViewById(R.id.result_list);

        PackageManager pm = getPackageManager();
        List<ResolveInfo> activities = pm.queryIntentActivities(new Intent
        (RecognizerIntent.ACTION_RECOGNIZE_SPEECH), 0);
        if (activities == null || activities.size() == 0) {
            mMicButton.setEnabled(false);
            Toast.makeText(getApplicationContext(), "Not Supported", Toast.LENGTH_LONG).
            show();
        }
        mMicButton.setOnClickListener(new OnClickListener() {
            @Override
            public void onClick(View v) {
                runVoiceRecognition();
            }
        });
    }

    private void runVoiceRecognition() {
        Intent intent = VoiceRecognitionIntentFactory.getFreeFormRecognizeIntent("Speak
        Now...");
        startActivityForResult(intent, REQUEST_CODE);
    }

    @Override
    protected void onActivityResult(int requestCode, int resultCode, Intent data) {
        if (requestCode == REQUEST_CODE && resultCode == RESULT_OK) {
            ArrayList<String> matches = data.getStringArrayListExtra(RecognizerIntent.
            EXTRA_RESULTS);
```

```
            mResultList.setAdapter(new ArrayAdapter<String>(this, android.R.layout.
            simple_list_item_1, matches));
        }

        super.onActivityResult(requestCode, resultCode, data);
    }

    @Override
    public boolean onCreateOptionsMenu(Menu menu) {
        // Inflate the menu; this adds items to the action bar if it is present.
        getMenuInflater().inflate(R.menu.voice, menu);
        return true;
    }

}
```

（3）编写文件 VoiceRecognitionIntentFactory.java，功能是调用 Android 内置的 Voice Recognition 技术实现语音识别功能，具体实现代码如下所示。

```
public class VoiceRecognitionIntentFactory {
    public static final int ACTION_GET_LANGUAGE_DETAILS_REQUEST_CODE = 88811;
    private static final int MAX_RESULTS = 100;

    // Suppress default constructor for noninstantiability
    private VoiceRecognitionIntentFactory() { }

    public static Intent getSimpleRecognizerIntent(String prompt)
    {
        Intent intent = getBlankRecognizeIntent();
        intent.putExtra(RecognizerIntent.EXTRA_LANGUAGE_MODEL,
        RecognizerIntent.LANGUAGE_MODEL_WEB_SEARCH);
        intent.putExtra(RecognizerIntent.EXTRA_PROMPT, prompt);
        return intent;
    }

    public static Intent getBlankRecognizeIntent()
    {
        Intent intent = new Intent(RecognizerIntent.ACTION_RECOGNIZE_SPEECH);
        return intent;
    }

    public static Intent getFreeFormRecognizeIntent(String prompt){
        Intent intent = getBlankRecognizeIntent();
        intent.putExtra(RecognizerIntent.EXTRA_LANGUAGE_MODEL,
        RecognizerIntent.LANGUAGE_MODEL_FREE_FORM);
        intent.putExtra(RecognizerIntent.EXTRA_PROMPT, prompt);
        return intent;
    }

    public static Intent getWebSearchRecognizeIntent()
    {
        Intent intent = new Intent(RecognizerIntent.ACTION_WEB_SEARCH);
        return intent;
    }

    public static Intent getHandsFreeRecognizeIntent()
    {
        Intent intent = new Intent(RecognizerIntent.ACTION_VOICE_SEARCH_HANDS_FREE);
        return intent;
    }

    public static Intent getPossilbeWebSearchRecognizeIntent(String prompt)
    {
        Intent intent = getWebSearchRecognizeIntent();
        intent.putExtra(RecognizerIntent.EXTRA_LANGUAGE_MODEL,
        RecognizerIntent.LANGUAGE_MODEL_WEB_SEARCH);
        intent.putExtra(RecognizerIntent.EXTRA_PROMPT, prompt);
        intent.putExtra(RecognizerIntent.EXTRA_MAX_RESULTS, MAX_RESULTS);
        intent.putExtra(RecognizerIntent.EXTRA_WEB_SEARCH_ONLY, false);
```

```
        intent.putExtra(RecognizerIntent.EXTRA_PARTIAL_RESULTS, true);
        return intent;
    }

    public static Intent getLanguageDetailsIntent()
    {
        Intent intent = new Intent(RecognizerIntent.ACTION_GET_LANGUAGE_DETAILS);
        return intent;
    }
}
```

到此为止，整个实例介绍完毕，执行后的效果如图 21-5 所示。

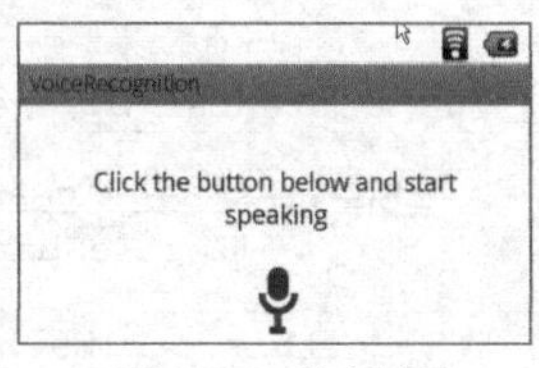

▲图 21-5　执行效果

21.4 实战演练——开发一个语音识别系统

本节将通过一个具体实例的实现过程，详细讲解在 Android 设备中开发一个综合性语音识别系统的过程。本实例的功能是，使用 TTS 技术实现一个功能强大的语音识别系统。

实例	功能	源码路径
实例 21-3	开发一个综合性语音识别系统	https://github.com/gast-lib/gast-lib

本实例的功能是查看设备中传感器的基本信息，本实例源码来源于：

```
https://github.com/gast-lib/gast-lib
```

21.4.1 验证是否支持所需要的语言

编写文件 TextToSpeechStartupListener.java 用于初始化 TextToSpeech 对象，功能是验证是否支持所需要的语言，在必要时需要下载额外的数据。文件 TextToSpeechStartupListener.java 的主要实现代码如下所示。

```
public interface TextToSpeechStartupListener
{
    public void onSuccessfulInit(TextToSpeech tts);

    /**
     * language data is required, to install call
     * {@link TextToSpeechInitializer#installLanguageData()}
     */
    public void onRequireLanguageData();

    /**
     * The app has already requested language data, and is waiting for it.
     */
    public void onWaitingForLanguageData();

    /**
     * initialization failed and can never complete.
     */
    public void onFailedToInit();
}
```

21.4.2 实现 TTS 的初始化工作

编写文件 TextToSpeechInitializer.java，通过方法 createTextToSpeech 实现 TTS 的初始化工作，并通过方法 setTextToSpeechSettings 使用 Locale 来检查对不同语言类型的支持，并且此方法还可以实现 TextToSpeech 的设置或根据语言数据的可用性来回调处理意外的事件。文件 TextToSpeech Initializer.java 的具体实现代码如下所示。

```
public class TextToSpeechInitializer
{
```

```
    private static final String TAG = "TextToSpeechInitializer";

    private TextToSpeech tts;

    private TextToSpeechStartupListener callback;

    private Context context;

    /**
     * creates by checking {@link TextToSpeech#isLanguageAvailable(Locale)}
     */
    public TextToSpeechInitializer(Context context, final Locale loc,
            TextToSpeechStartupListener callback)
    {
        this.callback = callback;
        this.context = context;
        createTextToSpeech(loc);
    }
    /**
     * creating a TTS
     * @param locale {@link Locale} for the desired language
     */
    private void createTextToSpeech(final Locale locale)
    {
        tts = new TextToSpeech(context, new OnInitListener()
        {
            @Override
            public void onInit(int status)
            {
                if (status == TextToSpeech.SUCCESS)
                {
                    setTextToSpeechSettings(locale);
                } else
                {
                    Log.e(TAG, "error creating text to speech");
                    callback.onFailedToInit();
                }
            }
        });
    }

    /**
     * perform the initialization checks after the
     * TextToSpeech is done
     */
    private void setTextToSpeechSettings(final Locale locale)
    {
        Locale defaultOrPassedIn = locale;
        if (locale == null)
        {
            defaultOrPassedIn = Locale.getDefault();
        }
        // check if language is available
        switch (tts.isLanguageAvailable(defaultOrPassedIn))
        {
            case TextToSpeech.LANG_AVAILABLE:
            case TextToSpeech.LANG_COUNTRY_AVAILABLE:
            case TextToSpeech.LANG_COUNTRY_VAR_AVAILABLE:
                Log.d(TAG, "SUPPORTED");
                tts.setLanguage(locale);
                callback.onSuccessfulInit(tts);
                break;
            case TextToSpeech.LANG_MISSING_DATA:
                Log.d(TAG, "MISSING_DATA");
                // check if waiting, by checking
                // a shared preference
                if (LanguageDataInstallBroadcastReceiver
                        .isWaiting(context))
                {
```

```
                    Log.d(TAG, "waiting for data...");
                    callback.onWaitingForLanguageData();
                } else
                {
                    Log.d(TAG, "require data...");
                    callback.onRequireLanguageData();
                }
                break;
            case TextToSpeech.LANG_NOT_SUPPORTED:
                Log.d(TAG, "NOT SUPPORTED");
                callback.onFailedToInit();
                break;
        }
    }

    public void setOnUtteranceCompleted(OnUtteranceCompletedListener whenDoneSpeaking)
    {
        int result = tts.setOnUtteranceCompletedListener(whenDoneSpeaking);
        if (result == TextToSpeech.ERROR)
        {
            Log.e(TAG, "failed to add utterance listener");
        }
    }

    /**
     * helper method to display a dialog if waiting for a download
     */
    public void installLanguageData(Dialog ifAlreadyWaiting)
    {
        boolean alreadyDidThis = LanguageDataInstallBroadcastReceiver
                .isWaiting(context);
        if (alreadyDidThis)
        {
            // give the user the ability to try again..
            // throw exception here?
            ifAlreadyWaiting.show();
        } else
        {
            installLanguageData();
        }
    }

    public void installLanguageData()
    {
        // set waiting for the download
        LanguageDataInstallBroadcastReceiver.setWaiting(context, true);

        Intent installIntent = new Intent();
        installIntent.setAction(TextToSpeech.Engine.ACTION_INSTALL_TTS_DATA);
        context.startActivity(installIntent);
    }
}
```

21.4.3 开启语言检查功能

编写文件 CommonTtsMethods.java，功能是通过方法 getLanguageAvailableDescription 来描述我们设备所支持的 Locale 所需的代码。通过方法 installLanguageData 来启动安装语言数据的操作，并通过方法 startDataCheck 开启语言检查的操作。文件 CommonTtsMethods.java 的具体实现代码如下所示。

```
public class CommonTtsMethods
{
    private static final String TAG = "CommonTtsMethods";

    private static final String NEW_LINE = "\n";

    public static final int SPEECH_DATA_CHECK_CODE = 19327;
```

```
    /**
     * start the language data install
     */
    public static void installLanguageData(final Context context)
    {
        //waiting for the download
        LanguageDataInstallBroadcastReceiver.setWaiting(context, true);

        //don't actually do it in test mode, just register a receiver
        Intent installIntent = new Intent();
        installIntent.setAction(
            TextToSpeech.Engine.ACTION_INSTALL_TTS_DATA);
        context.startActivity(installIntent);

    }

    /**
     * get a descriptions of all the languages available as determined by
     * {@link TextToSpeech#isLanguageAvailable(Locale)}
     */
    public static String getLanguageAvailableDescription(TextToSpeech tts)
    {
        StringBuilder sb = new StringBuilder();
        for (Locale loc : Locale.getAvailableLocales())
        {
            int availableCheck = tts.isLanguageAvailable(loc);
            sb.append(loc.toString()).append(" ");
            switch (availableCheck)
            {
                case TextToSpeech.LANG_AVAILABLE:
                    break;
                case TextToSpeech.LANG_COUNTRY_AVAILABLE:
                    sb.append("COUNTRY_AVAILABLE");
                    break;
                case TextToSpeech.LANG_COUNTRY_VAR_AVAILABLE:
                    sb.append("COUNTRY_VAR_AVAILABLE");
                    break;
                case TextToSpeech.LANG_MISSING_DATA:
                    sb.append("MISSING_DATA");
                    break;
                case TextToSpeech.LANG_NOT_SUPPORTED:
                    sb.append("NOT_SUPPORTED");
                    break;
            }
            sb.append(NEW_LINE);
        }
        return sb.toString();
    }

    public static void startDataCheck(Activity callingActivity)
    {
        Log.d(TAG, "launching speech check");
        Intent checkIntent = new Intent();
        checkIntent.setAction(TextToSpeech.Engine.ACTION_CHECK_TTS_DATA);
        callingActivity.startActivityForResult(checkIntent,
                SPEECH_DATA_CHECK_CODE);
    }
}
```

21.4.4　跟踪语言数据的安装状况

编写文件 LanguageDataInstallBroadcastReceiver.java，功能是通过广播侦听器来跟踪语言数据在什么时候完成安装。文件 LanguageDataInstallBroadcastReceiver.java 的具体实现代码如下所示。

```
public class LanguageDataInstallBroadcastReceiver extends BroadcastReceiver
{
    private static final String TAG = "LanguageDataInstallBroadcastReceiver";
```

```
    private static final String PREFERENCES_NAME = "installedLanguageData";

    private static final String WAITING_PREFERENCE_NAME =
            "WAITING_PREFERENCE_NAME";

    private static final Boolean WAITING_DEFAULT = false;

    public LanguageDataInstallBroadcastReceiver()
    {
    }

    @Override
    public void onReceive(Context context, Intent intent)
    {
        if (intent.getAction().equals(
                TextToSpeech.Engine.ACTION_TTS_DATA_INSTALLED))
        {
            Log.d(TAG, "language data preference: " + intent.getAction());
            // clear waiting state
            setWaiting(context, false);
        }
    }

    /**
     * check if the receiver is waiting for a language data install
     */
    public static boolean isWaiting(Context context)
    {
        SharedPreferences preferences;
        preferences =
                context.getSharedPreferences(PREFERENCES_NAME,
                        Context.MODE_WORLD_READABLE);
        boolean waiting =
                preferences
                        .getBoolean(WAITING_PREFERENCE_NAME, WAITING_DEFAULT);
        return waiting;
    }

    /**
     * start waiting by setting a flag
     */
    public static void setWaiting(Context context, boolean waitingStatus)
    {
        SharedPreferences preferences;
        preferences =
                context.getSharedPreferences(PREFERENCES_NAME,
                        Context.MODE_WORLD_WRITEABLE);
        Editor editor = preferences.edit();
        editor.putBoolean(WAITING_PREFERENCE_NAME, waitingStatus);
        editor.commit();
    }
}
```

21.4.5 转换语言并处埋结果

编写文件 TextToSpeechInitializerByAction.java，功能是通过方法 convertLocaleToVoice 将 Locale 转换为语音，通过方法 handleOnActivityResult 处理来自 ACTION_CHECK_TTS_DATA 的结果。文件 TextToSpeechInitializerByAction.java 的具体实现代码如下所示。

```
public class TextToSpeechInitializerByAction
{
    private static final String TAG = "TextToSpeechInitializerByAction";

    private TextToSpeech tts;

    private TextToSpeechStartupListener callback;
```

```
    private Activity activity;

    private Locale targetLocale;

    /**
     * creates by checking {@link TextToSpeech#isLanguageAvailable(Locale)}
     */
    public TextToSpeechInitializerByAction(Activity activity, String voiceToCheck,
          TextToSpeechStartupListener callback, Locale targetLocale)
    {
        this.callback = callback;
        this.activity = activity;
        this.targetLocale = targetLocale;
        startDataCheck(activity, voiceToCheck);
    }

    /**
     * version of the constructor that converts the {@link Locale}
     * to the proper voice to check
     */
    public TextToSpeechInitializerByAction(Activity activity,
          TextToSpeechStartupListener callback, Locale targetLocale)
    {
        this(activity, convertLocaleToVoice(targetLocale),
              callback, targetLocale);
    }

    /**
     * voice name as defined by
     * {@link TextToSpeech.Engine#EXTRA_CHECK_VOICE_DATA_FOR}
     */
    public void startDataCheck(Activity activity, String voiceToCheck)
    {
        Intent check = new Intent();
        check.setAction(TextToSpeech.Engine.ACTION_CHECK_TTS_DATA);
        Log.d(TAG, "launching speech check");
        if (voiceToCheck != null && voiceToCheck.length() > 0)
        {
            Log.d(TAG, "adding voice check for: " + voiceToCheck);
            // needs to be in an ArrayList
            ArrayList<String> voicesToCheck = new ArrayList<String>();
            voicesToCheck.add(voiceToCheck);
            check.putStringArrayListExtra(
                    TextToSpeech.Engine.EXTRA_CHECK_VOICE_DATA_FOR,
                    voicesToCheck);
        }
        activity.startActivityForResult(check,
              CommonTtsMethods.SPEECH_DATA_CHECK_CODE);
    }

    /**
     * handle onActivityResult from call to
     * {@link #startDataCheck(Activity, String)}
     */
    public void handleOnActivityResult(Context launchFrom,
          int requestCode, int resultCode, Intent data)
    {
        if (requestCode == CommonTtsMethods.SPEECH_DATA_CHECK_CODE)
        {
            switch (resultCode)
            {
                case TextToSpeech.Engine.CHECK_VOICE_DATA_PASS:
                    // success, create the TTS instance
                    Log.d(TAG, "has language data");
                    tts = new TextToSpeech(launchFrom, new OnInitListener()
                    {
                        @Override
                        public void onInit(int status)
                        {
```

```
                    if (targetLocale != null)
                    {
                        tts.setLanguage(targetLocale);
                    }
                    if (status == TextToSpeech.SUCCESS)
                    {
                        callback.onSuccessfulInit(tts);
                    } else
                    {
                        callback.onFailedToInit();
                    }
                }
            });
            break;
        case TextToSpeech.Engine.CHECK_VOICE_DATA_MISSING_VOLUME:
        case TextToSpeech.Engine.CHECK_VOICE_DATA_FAIL:
        case TextToSpeech.Engine.CHECK_VOICE_DATA_BAD_DATA:
        case TextToSpeech.Engine.CHECK_VOICE_DATA_MISSING_DATA:
            Log.d(TAG, "no language data");
            callback.onRequireLanguageData();
        }
    }
}

public void installLanguageData()
{
    // waiting for the download
    LanguageDataInstallBroadcastReceiver.setWaiting(activity, true);

    // don't actually do it in test mode, just register a receiver
    Intent installIntent = new Intent();
    installIntent.setAction(TextToSpeech.Engine.ACTION_INSTALL_TTS_DATA);
    activity.startActivity(installIntent);
}

public static String convertLocaleToVoice(Locale loc)
{
    // The format of each voice is: lang-COUNTRY-variant where COUNTRY and
    // variant are optional (ie, "eng" or "eng-USA" or "eng-USA-FEMALE").
    String country = loc.getISO3Country();
    String language = loc.getISO3Language();
    StringBuilder sb = new StringBuilder();
    sb.append(language);
    if (country.length() > 0)
    {
        sb.append("-");
        sb.append(country);
    }
    return sb.toString();
}
}
```

21.4.6 实现语音阅读测试

编写文件 TextToSpeechDemo.java 实现语音阅读测试，需要通过方法 playScript 来阅读脚本，这包括声音信号、静音、合成声音和预录制语音等，并且通过方法 onDone 重新启用视图。另外，还实现了如下所示的功能。

- 使用不同格式的对话框来通知用户在初始化过程中发生的事情。
- 根据具体应用情况来确定是否启用和关闭 Speak 按钮。
- 在初始化 TextToSpeech 时停止使用 Speak 按钮。

文件 TextToSpeechDemo.java 的具体实现代码如下所示。

```
public class TextToSpeechDemo extends Activity implements
    TextToSpeechStartupListener
{
```

```
    private static final String TAG = "TextToSpeechDemo";
    private Button speak;
    private Button stopSpeak;

    private static final String LAST_SPOKEN = "lastSpoken";

    private TextToSpeechInitializer ttsInit;
    private TextToSpeech tts;

    @Override
    public void onCreate(Bundle savedInstanceState)
    {
        super.onCreate(savedInstanceState);
        setContentView(R.layout.ttsdemo);
        hookButtons();
        init();
    }

    private void hookButtons()
    {
        speak = (Button) findViewById(R.id.btn_speak);
        speak.setOnClickListener(new View.OnClickListener()
        {
            @Override
            public void onClick(View v)
            {
                setViewToWhileSpeaking();
                playScript();
            }
        });

        stopSpeak = (Button) findViewById(R.id.btn_stop_speak);
        stopSpeak.setOnClickListener(new View.OnClickListener()
        {
            @Override
            public void onClick(View v)
            {
                setViewToDoneSpeaking();
                tts.stop();
            }
        });
    }

    private void init()
    {
        deactivateUi();
        ttsInit = new TextToSpeechInitializer(this, Locale.getDefault(), this);
    }

    @Override
    public void onSuccessfulInit(TextToSpeech tts)
    {
        Log.d(TAG, "successful init");
        this.tts = tts;
        activateUi();
        setTtsListener();
    }

    /**
     * set the TTS listener to call {@link #onDone(String)} depending on the
     * Build.Version.SDK_INT
     */
    private void setTtsListener()
    {
        if (Build.VERSION.SDK_INT >= 15)
        {
            int listenerResult =
                    tts.setOnUtteranceProgressListener(new UtteranceProgressListener()
                    {
```

```
                    @Override
                    public void onDone(String utteranceId)
                    {
                        TextToSpeechDemo.this.onDone(utteranceId);
                    }

                    @Override
                    public void onError(String utteranceId)
                    {
                        Log.e(TAG, "TTS error");
                    }

                    @Override
                    public void onStart(String utteranceId)
                    {
                        Log.d(TAG, "TTS start");
                    }
                });
        if (listenerResult != TextToSpeech.SUCCESS)
        {
            Log.e(TAG, "failed to add utterance progress listener");
        }
    }
    else
    {
        int listenerResult =
                tts.setOnUtteranceCompletedListener(
                    new OnUtteranceCompletedListener()
                {
                    @Override
                    public void onUtteranceCompleted(String utteranceId)
                    {
                        TextToSpeechDemo.this.onDone(utteranceId);
                    }
                });
        if (listenerResult != TextToSpeech.SUCCESS)
        {
            Log.e(TAG, "failed to add utterance completed listener");
        }
    }
}

private void onDone(final String utteranceId)
{
    Log.d(TAG, "utterance completed: " + utteranceId);
    runOnUiThread(new Runnable()
    {
        @Override
        public void run()
        {
            if (utteranceId.equals(LAST_SPOKEN))
            {
                setViewToDoneSpeaking();
            }
        }
    });
}

@Override
public void onFailedToInit()
{
    DialogInterface.OnClickListener onClickOk = makeOnFailedToInitHandler();
    AlertDialog a =
            new AlertDialog.Builder(this).setTitle("Error")
                    .setMessage("Unable to create text to speech")
                    .setNeutralButton("Ok", onClickOk).create();
    a.show();
}
```

```
@Override
public void onRequireLanguageData()
{
    DialogInterface.OnClickListener onClickOk =
            makeOnClickInstallDialogListener();
    DialogInterface.OnClickListener onClickCancel =
            makeOnFailedToInitHandler();
    AlertDialog a =
            new AlertDialog.Builder(this)
                    .setTitle("Error")
                    .setMessage(
                            "Requires Language data to proceed, " +
                            "would you like to install?")
                    .setPositiveButton("Ok", onClickOk)
                    .setNegativeButton("Cancel", onClickCancel).create();
    a.show();
}

@Override
public void onWaitingForLanguageData()
{
    // either wait for install
    DialogInterface.OnClickListener onClickWait =
            makeOnFailedToInitHandler();
    DialogInterface.OnClickListener onClickInstall =
            makeOnClickInstallDialogListener();

    AlertDialog a =
            new AlertDialog.Builder(this)
                    .setTitle("Info")
                    .setMessage(
                            "Please wait for the language data to finish" +
                            " installing and try again.")
                    .setNegativeButton("Wait", onClickWait)
                    .setPositiveButton("Retry", onClickInstall).create();
    a.show();
}

private DialogInterface.OnClickListener makeOnClickInstallDialogListener()
{
    return new DialogInterface.OnClickListener()
    {
        @Override
        public void onClick(DialogInterface dialog, int which)
        {
            ttsInit.installLanguageData();
        }
    };
}

private DialogInterface.OnClickListener makeOnFailedToInitHandler()
{
    return new DialogInterface.OnClickListener()
    {
        @Override
        public void onClick(DialogInterface dialog, int which)
        {
            finish();
        }
    };
}

private void playScript()
{
    Log.d(TAG, "started script");
    // setup

    // id to send back when saying the last phrase
    // so the app can re-enable the "speak" button
```

```
    HashMap<String, String> lastSpokenWord = new HashMap<String, String>();
    lastSpokenWord.put(TextToSpeech.Engine.KEY_PARAM_UTTERANCE_ID,
            LAST_SPOKEN);

    // add earcon
    final String EARCON_NAME = "[tone]";
    tts.addEarcon(EARCON_NAME, "root.gast.playground", R.raw.tone);

    // add prerecorded speech
    final String CLOSING = "[Thank you]";
    tts.addSpeech(CLOSING, "root.gast.playground",
            R.raw.enjoytestapplication);

    // pass in null to most of these because we do not want a callback to
    // onDone
    tts.playEarcon(EARCON_NAME, TextToSpeech.QUEUE_ADD, null);
    tts.playSilence(1000, TextToSpeech.QUEUE_ADD, null);
    tts.speak("Attention readers: Use the try button to experiment with"
            + " Text to Speech. Use the diagnostics button to see "
            + "detailed Text to Speech engine information.",
            TextToSpeech.QUEUE_ADD, null);
    tts.playSilence(500, TextToSpeech.QUEUE_ADD, null);
    tts.speak(CLOSING, TextToSpeech.QUEUE_ADD, lastSpokenWord);

}

// activate and deactivate the UI based on various states

private void deactivateUi()
{
    Log.d(TAG, "deactivate ui");
    // don't enable until the initialization is complete
    speak.setEnabled(false);
}

private void activateUi()
{
    Log.d(TAG, "activate ui");
    speak.setEnabled(true);
}

public void setViewToWhileSpeaking()
{
    stopSpeak.setVisibility(View.VISIBLE);
    speak.setVisibility(View.GONE);
}

public void setViewToDoneSpeaking()
{
    stopSpeak.setVisibility(View.GONE);
    speak.setVisibility(View.VISIBLE);
}

@Override
protected void onDestroy()
{
    if (tts != null)
    {
        tts.shutdown();
    }
    super.onDestroy();
}
}
```

21.4.7 保证系统可以实现正确的语音识别

编写文件 LanguageDetailsChecker.java，功能是确保系统可以实现正确的语音识别，此过程使用了异步调用和 BroadcastIntent 方式实现。类 LanguageDetailsChecker 在接收到检查结果后，会回调 onLanguageDetailsListener 来处理异步调用。文件 LanguageDetailsChecker.java 的具体实现代码如下所示。

```
public class LanguageDetailsChecker extends BroadcastReceiver
{
    private static final String TAG = "LanguageDetailsChecker";

    private List<String> supportedLanguages;

    private String languagePreference;

    private OnLanguageDetailsListener doAfterReceive;

    public LanguageDetailsChecker(OnLanguageDetailsListener doAfterReceive)
    {
        supportedLanguages = new ArrayList<String>();
        this.doAfterReceive = doAfterReceive;
    }

    @Override
    public void onReceive(Context context, Intent intent)
    {
        Bundle results = getResultExtras(true);
        if (results.containsKey(RecognizerIntent.EXTRA_LANGUAGE_PREFERENCE))
        {
            languagePreference =
                    results.getString(RecognizerIntent.EXTRA_LANGUAGE_PREFERENCE);
        }
        if (results.containsKey(RecognizerIntent.EXTRA_SUPPORTED_LANGUAGES))
        {
            supportedLanguages =
                    results.getStringArrayList(
                            RecognizerIntent.EXTRA_SUPPORTED_LANGUAGES);
        }

        if (doAfterReceive != null)
        {
            doAfterReceive.onLanguageDetailsReceived(this);
        }
    }

    public String matchLanguage(Locale toCheck)
    {
        String matchedLanguage = null;
        // modify the returned languages to look like the output from
        // Locale.toString()
        String targetLanguage = toCheck.toString().replace('_', '-');
        for (String supportedLanguage : supportedLanguages)
        {
            // use contains, so that partial matches are possible
            // for example, if the Locale is
            // en-US-POSIX, it will still match en-US
            // and that if the target language is en, it will match something
            Log.d(TAG, targetLanguage + " contains " + supportedLanguage);
            if ((targetLanguage.contains(supportedLanguage))
                    || supportedLanguage.contains(targetLanguage))
            {
                matchedLanguage = supportedLanguage;
            }
        }
        return matchedLanguage;
    }

    /**
     * @return the supportedLanguages
     */
    public List<String> getSupportedLanguages()
    {
        return supportedLanguages;
```

```
    }

    /**
     * @return the languagePreference
     */
    public String getLanguagePreference()
    {
        return languagePreference;
    }

    public String toString()
    {
        StringBuilder sb = new StringBuilder();
        sb.append("Language Preference: ").append(getLanguagePreference())
              .append("\n");
        sb.append("languages supported: ").append("\n");
        for (String lang : getSupportedLanguages())
        {
            sb.append(" ").append(lang).append("\n");
        }
        return sb.toString();
    }
}
```

21.4.8 显示语音识别的结果

编写文件SpeechRecognitionResultsActivity.java显示语音识别的结果，具体实现代码如下所示。

```
public class SpeechRecognitionResultsActivity extends Activity
{
    private static final String TAG = "SpeechRecognitionResultsActivity";

    /**
     * for passing in the input
     */
    public static String WHAT_YOU_ARE_TRYING_TO_SAY_INTENT_INPUT =
          "WHAT_YOU_ARE_TRYING_TO_SAY_INPUT";

    private ListView log;

    private TextView resultsSummary;

    @Override
    protected void onCreate(Bundle savedInstanceState)
    {
        super.onCreate(savedInstanceState);
        setContentView(R.layout.speechrecognition_result);
        Log.d(TAG, "SpeechRecognition Pending intent received");
        hookButtons();
        init();
    }

    private void hookButtons()
    {
        log = (ListView) findViewById(R.id.lv_resultlog);
        resultsSummary = (TextView) findViewById(R.id.tv_speechResultsSummary);
    }

    private void init()
    {
        if (getIntent() != null)
        {
            if (getIntent().hasExtra(WHAT_YOU_ARE_TRYING_TO_SAY_INTENT_INPUT))
            {
                String whatSayFromIntent =
                      getIntent().getStringExtra(
                            WHAT_YOU_ARE_TRYING_TO_SAY_INTENT_INPUT);
                resultsSummary.setText(whatSayFromIntent);
            }
```

```
            String whatSayFromIntent =
                    getIntent().getStringExtra(
                            WHAT_YOU_ARE_TRYING_TO_SAY_INTENT_INPUT);
            resultsSummary.setText(whatSayFromIntent);

            if (getIntent().hasExtra(RecognizerIntent.EXTRA_RESULTS))
            {
                List<String> results =
                        getIntent().getStringArrayListExtra(
                                RecognizerIntent.EXTRA_RESULTS);
                ArrayAdapter<String> adapter =
                        new ArrayAdapter<String>(this,
                                R.layout.speechresultactivity_listitem,
                                R.id.tv_speech_activity_result, results);
                log.setAdapter(adapter);
            }
            else
            {
                // if RESULT_EXTRA is not present, the recognition had an
                // error
                DialogInterface.OnClickListener onClickFinish =
                        new DialogInterface.OnClickListener()
                        {
                            @Override
                            public void onClick(DialogInterface dialog,
                                    int which)
                            {
                                finish();
                            }
                        };
                AlertDialog a =
                        new AlertDialog.Builder(this)
                                .setTitle(
                                        getResources().getString(
                                                R.string.d_info))
                                .setMessage(
                                        getResources()
                                                .getString(
                                                R.string.
                                                speechRecognitionFailed))
                                .setPositiveButton(
                                        getResources().getString(R.string.d_ok),
                                        onClickFinish).create();
                a.show();
            }
        }
    }
}
```

21.4.9 处理回调

编写文件SpeechRecognizingActivity.java，此文件实现了文件SpeechRecognitionResultsActivity.java 的扩展类，功能是处理各个来自初始化和接收识别结果的回调。文件 SpeechRecognizingActivity.java 执行之初便会执行语音可用性的检查工作，具体实现代码如下所示。

```
public abstract class SpeechRecognizingActivity extends Activity implements
        RecognitionListener
{
    private static final String TAG = "SpeechRecognizingActivity";

    /**
     * code to identify return recognition results
     */
    public static final int VOICE_RECOGNITION_REQUEST_CODE = 1234;

    public static final int UNKNOWN_ERROR = -1;
```

```
private SpeechRecognizer recognizer;

@Override
protected void onCreate(Bundle savedInstanceState)
{
    super.onCreate(savedInstanceState);

    boolean recognizerIntent =
        SpeechRecognitionUtil.isSpeechAvailable(this);
    if (!recognizerIntent)
    {
        speechNotAvailable();
    }
    boolean direct = SpeechRecognizer.isRecognitionAvailable(this);
    if (!direct)
    {
        directSpeechNotAvailable();
    }
}

protected void checkForLanguage(final Locale language)
{
    OnLanguageDetailsListener andThen = new OnLanguageDetailsListener()
    {
        @Override
        public void onLanguageDetailsReceived(LanguageDetailsChecker data)
        {
            // do a best match
            String languageToUse = data.matchLanguage(language);
            languageCheckResult(languageToUse);
        }
    };
    SpeechRecognitionUtil.getLanguageDetails(this, andThen);
}

/**
 * execute the RecognizerIntent, then call
 * {@link #receiveWhatWasHeard(List, List)} when done
 */
public void recognize(Intent recognizerIntent)
{
    startActivityForResult(recognizerIntent,
        VOICE_RECOGNITION_REQUEST_CODE);
}

/**
 * Handle the results from the RecognizerIntent.
 */
@Override
protected void
    onActivityResult(int requestCode, int resultCode, Intent data)
{
    if (requestCode == VOICE_RECOGNITION_REQUEST_CODE)
    {
        if (resultCode == RESULT_OK)
        {
            List<String> heard =
                    data.
                    getStringArrayListExtra
                        (RecognizerIntent.EXTRA_RESULTS);
            float[] scores =
                    data.
                    getFloatArrayExtra
                        (RecognizerIntent.ACTION_GET_LANGUAGE_DETAILS);
            if (scores == null)
            {
                for (int i = 0; i < heard.size(); i++)
                {
                    Log.d(TAG, i + ": " + heard.get(i));
```

```
                }
            }
            else
            {
                for (int i = 0; i < heard.size(); i++)
                {
                    Log.d(TAG, i + ": " + heard.get(i) + " score: "
                            + scores[i]);
                }
            }

            receiveWhatWasHeard(heard, scores);
        }
        else
        {
            Log.d(TAG, "error code: " + resultCode);
            recognitionFailure(UNKNOWN_ERROR);
        }
    }
    super.onActivityResult(requestCode, resultCode, data);
}

/**
 * called when speech is not available on this device, and when
 * {@link #recognize(Intent)} will not work
 */
abstract protected void speechNotAvailable();

/**
 * called when {@link SpeechRecognizer} cannot be used on this device and
 * {@link #recognizeDirectly(Intent)} will not work
 */
abstract protected void directSpeechNotAvailable();

/**
 * call back the result from {@link #checkForLanguage(Locale)}
 *
 * @param languageToUse
 *          the language string to use or null if failure
 */
abstract protected void languageCheckResult(String languageToUse);

/**
 * result of speech recognition
 *
 * @param heard
 *          possible speech to text conversions
 * @param confidenceScores
 *          the confidence for the strings in heard
 */
abstract protected void receiveWhatWasHeard(List<String> heard,
        float[] confidenceScores);

/**
 * @param code
 *          If using {@link #recognizeDirectly(Intent) it will be
 *          the error code from {@link SpeechRecognizer}
 *          if using {@link #recognize(Intent)}
 *          it will be {@link #UNKNOWN_ERROR}.
 */
abstract protected void recognitionFailure(int errorCode);

//direct speech recognition methods follow

/**
 * Uses {@link SpeechRecognizer} to perform recognition and then calls
 * {@link #receiveWhatWasHeard(List, float[])} with the results <br>
 * check {@link SpeechRecognizer.isRecognitionAvailable(context)} before
 * calling this method otherwise if it isn't available the code will report
```

```
 * an error
 */
public void recognizeDirectly(Intent recognizerIntent)
{
    // SpeechRecognizer requires EXTRA_CALLING_PACKAGE, so add if it's not
    // here
    if (!recognizerIntent.hasExtra(RecognizerIntent.EXTRA_CALLING_PACKAGE))
    {
        recognizerIntent.putExtra(RecognizerIntent.EXTRA_CALLING_PACKAGE,
                "com.dummy");
    }
    SpeechRecognizer recognizer = getSpeechRecognizer();
    recognizer.startListening(recognizerIntent);
}

@Override
public void onResults(Bundle results)
{
    Log.d(TAG, "full results");
    receiveResults(results);
}

@Override
public void onPartialResults(Bundle partialResults)
{
    Log.d(TAG, "partial results");
    receiveResults(partialResults);
}

/**
 * common method to process any results bundle from {@link SpeechRecognizer}
 */
private void receiveResults(Bundle results)
{
    if ((results != null)
            && results.containsKey(SpeechRecognizer.RESULTS_RECOGNITION))
    {
        List<String> heard =
                results.getStringArrayList(SpeechRecognizer.RESULTS_RECOGNITION);
        float[] scores =
                results.getFloatArray(SpeechRecognizer.RESULTS_RECOGNITION);
        receiveWhatWasHeard(heard, scores);
    }
}

@Override
public void onError(int errorCode)
{
    recognitionFailure(errorCode);
}

/**
 * stop the speech recognizer
 */
@Override
protected void onPause()
{
    if (getSpeechRecognizer() != null)
    {
        getSpeechRecognizer().stopListening();
        getSpeechRecognizer().cancel();
        getSpeechRecognizer().destroy();
    }
    super.onPause();
}

/**
 * lazy initialize the speech recognizer
 */
```

```
    private SpeechRecognizer getSpeechRecognizer()
    {
        if (recognizer == null)
        {
            recognizer = SpeechRecognizer.createSpeechRecognizer(this);
            recognizer.setRecognitionListener(this);
        }
        return recognizer;
    }

    // other unused methods from RecognitionListener...

    @Override
    public void onReadyForSpeech(Bundle params)
    {
        Log.d(TAG, "ready for speech " + params);
    }

    @Override
    public void onEndOfSpeech()
    {
    }

    /**
     * @see android.speech.RecognitionListener#onBeginningOfSpeech()
     */
    @Override
    public void onBeginningOfSpeech()
    {
    }

    @Override
    public void onBufferReceived(byte[] buffer)
    {
    }

    @Override
    public void onRmsChanged(float rmsdB)
    {
    }

    @Override
    public void onEvent(int eventType, Bundle params)
    {
    }

    public void onPartialResultsUnsupported(Bundle partialResults)
    {
        Log.d(TAG, "partial results");
        if (partialResults
                .containsKey(SpeechRecognitionUtil.UNSUPPORTED_GOOGLE_RESULTS))
        {
            String[] heard =
                    partialResults
                            .getStringArray(SpeechRecognitionUtil.UNSUPPORTED_GOOGLE_RESULTS);
            float[] scores =
                    partialResults
                            .getFloatArray(SpeechRecognitionUtil.UNSUPPORTED_GOOGLE_RESUL
                            TS_CONFIDENCE);
            receiveWhatWasHeard(Arrays.asList(heard), scores);
        }
        else
        {
            receiveResults(partialResults);
        }
    }
}
```

到此为止，整个实例介绍完毕。为了节省篇幅，在书中没有讲解布局文件的实现过程，这部分内容请读者参阅开源代码。执行后的效果如图 21-6 所示。

开始说话界面　　　　语音识别设置界面

▲图 21-6　执行效果

第22章　手势识别技术详解

手势识别技术是 Android SDK 中比较重要且比较新颖的一项技术，在 Android 设备应用中可以通过手势来灵活地操控设备的运行。本章将详细讲解在 Android 设备中使用手势识别技术的基本知识，为步入本书后面知识的学习打下基础。

22.1 手势识别技术基础

对于触摸屏设备来说，其消息传递机制包括按下、抬起和移动这几种，用户只需要简单地实现重载 onTouch 或者设置触摸侦听器 setOnTouchListener 即可处理触摸事件。但是有时为了提高应用程序的用户体验，需要识别用户的手势。本节将详细讲解在 Android 设备中实现手势识别的基本知识。

22.1.1 类 GestureDetector 基础

在 Android 系统中，专门提供了手势识别类 GestureDetector。在 Android 设备中，通过类 GestureDetector 可以识别很多的手势，通过其 nTouchEvent(event)方法可以完成不同手势的识别。类 GestureDetector 对外提供了两个接口：OnGestureListener 和 OnDoubleTapListener，另外还提供了一个内部类 SimpleOnGestureListener。

（1）GestureDetector.OnDoubleTapListener 接口：用来通知 DoubleTap 事件，类似于鼠标的双击事件。此接口中各个成员的具体说明如下所示。

- onDoubleTap(MotionEvent e)：在二次双击 Touch down 时触发。
- onDoubleTapEvent(MotionEvent e)：通知 DoubleTap 手势中的事件，包含 down、up 和 move 事件（这里指的是在双击之间发生的事件，例如，在同一个地方双击会产生 DoubleTap 手势，而在 DoubleTap 手势里面还会发生 down 和 up 事件，这两个事件由该函数通知）；双击的第二下 Touch down 和 up 都会触发，可用 e.getAction()区分。
- onSingleTapConfirmed(MotionEvent e)：用来判定该次点击是 SingleTap 而不是 DoubleTap，如果连续点击两次就是 DoubleTap 手势，如果只点击一次，系统等待一段时间后没有收到第二次点击，则判定该次点击为 SingleTap 而不是 DoubleTap，然后触发 SingleTapConfirmed 事件。这个方法不同于 onSingleTapUp，它是在 GestureDetector 确信用户在第一次触摸屏幕后，没有紧跟着第二次触摸屏幕，也就是不是“双击”的时候触发。

（2）GestureDetector.OnGestureListener 接口：用来通知普通的手势事件，该接口有如下所示的 6 个回调函数。

- onDown(MotionEvent e)：down 事件。
- onSingleTapUp(MotionEvent e)：一次点击 up 事件，在 Touch down 后没有滑动。
- onLongPress：用户长按触摸屏，由多个 MotionEvent ACTION_DOWN 触发。
- onShowPress(MotionEvent e)：down 事件发生而 move 或 up 还没发生前触发该事件。

● onFling(MotionEvent e1, MotionEvent e2, float velocityX, float velocityY)：滑动手势事件，Touch 了滑动一点距离后，在 ACTION_UP 时才会触发。各个参数的具体说明如下所示。

➢ e1：第 1 个 ACTION_DOWN MotionEvent 并且只有一个。

➢ e2：最后一个 ACTION_MOVE MotionEvent。

➢ velocityX：*X* 轴上的移动速度，像素/秒。

➢ velocityY：*Y* 轴上的移动速度，像素/秒，触发条件：*X* 轴的坐标位移大于 FLING_MIN_DISTANCE，且移动速度大于 FLING_MIN_VELOCITY 个像素/秒。

● onScroll(MotionEvent e1, MotionEvent e2, float distanceX, float distanceY)：在屏幕上拖动事件。无论是用手拖动 view，或者是以抛的动作滚动，都会多次触发。这个方法在 ACTION_MOVE 动作发生时就会触发。

22.1.2 使用类 GestureDetector

Android 系统的事件处理机制是基于 Listener（监听器）实现的，触摸屏相关的事件就是通过 onTouchListener 实现的。另外，在 Android 系统中，所有 View 的子类都可以通过 setOnTouchListener()、setOnKeyListener()等方法来添加对某一类事件的监听器。并且 Listener 一般会以 Interface（接口）的方式来提供，其中包含一个或多个 abstract（抽象）方法，我们需要实现这些方法来完成 onTouch()、onKey()等操作。这样，当我们给某个 view 设置了事件 Listener，并实现了其中的抽象方法以后，程序便可以在特定的事件被 dispatch 到该 view 的时候，通过 callback 函数给予适当的响应。

在 Android 开发应用中，有多种使用类 GestureDetector 的方法。

1. 第一种

（1）首先通过 GestureDetector 的构造方法将 SimpleOnGestureListener 对象传递进去，这样 GestureDetector 能处理不同的手势了。

```
public GestureDetector(Context context, GestureDetector.OnGestureListener listener)
```

（2）然后在 onTouch 方法中实现 OnTouchListener 监听。

```
private OnTouchListener gestureTouchListener = new OnTouchListener() {
            public boolean onTouch(View v, MotionEvent event) {
          return gDetector.onTouchEvent(event);
      }
  };
```

2. 第二种

（1）首先使用如下所示的方法构建场景。

```
private GestureDetector mGestureDetector;
mGestureListener = new BookOnGestureListener();
```

（2）然后使用 new 新建构造出来的 GestureDetector 对象。

```
mGestureDetector = new GestureDetector(mGestureListener);
class BookOnGestureListener implements OnGestureListener {
```

（3）最后实现事件处理。

```
public boolean onTouchEvent(MotionEvent event) {
                mGestureListener.onTouchEvent(event);
}
```

3. 第三种

（1）首先在当前类中创建一个 GestureDetector 实例。

```
private GestureDetector mGestureDetector;
```

（2）然后创建一个 Listener 来实时监听当前面板操作手势。

```
class LearnGestureListener extends GestureDetector.SimpleOnGestureListener
```

（3）在初始化时，将 Listener 实例关联当前的 GestureDetector 实例。

```
mGestureDetector = new GestureDetector(this, new LearnGestureListener());
```

（4）使用方法 onTouchEvent 作为入口检测，通过传递 MotionEvent 参数来监听操作手势。

```
mGestureDetector.onTouchEvent(event)
```

例如，如下所示的演示代码。

```
private GestureDetector mGestureDetector;
@Override
public void onCreate(Bundle savedInstanceState) {
   super.onCreate(savedInstanceState);
   mGestureDetector = new GestureDetector(this, new LearnGestureListener());
}
@Override
public boolean onTouchEvent(MotionEvent event) {
   if (mGestureDetector.onTouchEvent(event))
      return true;
   else
      return false;
}
class LearnGestureListener extends GestureDetector.SimpleOnGestureListener{
   @Override
   public boolean onSingleTapUp(MotionEvent ev) {
      Log.d("onSingleTapUp",ev.toString());
      return true;
   }
   @Override
   public void onShowPress(MotionEvent ev) {
      Log.d("onShowPress",ev.toString());
   }
   @Override
   public void onLongPress(MotionEvent ev) {
      Log.d("onLongPress",ev.toString());
   }
   @Override
   public boolean onScroll(MotionEvent e1, MotionEvent e2, float distanceX, float
distanceY) {
      Log.d("onScroll",e1.toString());
      return true;
   }
   @Override
   public boolean onDown(MotionEvent ev) {
      Log.d("onDownd",ev.toString());
      return true;
   }
   @Override
   public boolean onFling(MotionEvent e1, MotionEvent e2, float velocityX, float
velocityY) {
      Log.d("d",e1.toString());
      Log.d("e2",e2.toString());
      return true;
   }
}
```

4．第四种

（1）首先创建一个 GestureDetector 的对象，传入 listener 对象，在接收到的 onTouchEvent 中将 event 传给 GestureDetector 进行分析，listener 会回调给我们相应的动作。

（2）通过 GestureDetector.SimpleOnGestureListener（Framework 帮我们简化了）实现了 OnGestureListener 和 OnDoubleTapListener 两个接口类，只需要继承它并重写其中的回调即可。

（3）设置在第一次单击 down 时，给 Hanlder 发送了一个延时的消息，例如，延时 300ms。如果在 300ms 里发生了第二次单击的 down 事件，那么就认为是双击事件，并移除之前发送的延时消息。如果 300ms 后仍没有第二次的 down 消息，那么就判定为 SingleTapConfirmed 事件（当然，此时用户的手指应已完成第一次点击的 up 过程）。第三次单击的判定和双击的判定类似，只是多了一次发送延时消息的过程。

例如，如下所示的演示代码。

```
private GestureDetector mGestureDetector;
@Override
public void onCreate(Bundle savedInstanceState) {
  super.onCreate(savedInstanceState);
  mGestureDetector = new GestureDetector(this, new MyGestureListener());
}
@Override
public boolean onTouchEvent(MotionEvent event) {
 return mGestureDetector.onTouchEvent(event);
}
class MyGestureListener extends GestureDetector.SimpleOnGestureListener{
  @Override
  public boolean onSingleTapUp(MotionEvent ev) {
    Log.d("onSingleTapUp",ev.toString());
    return true;
  }
  @Override
  public void onShowPress(MotionEvent ev) {
    Log.d("onShowPress",ev.toString());
  }
  @Override
  public void onLongPress(MotionEvent ev) {
    Log.d("onLongPress",ev.toString());
  }
…
}
```

22.1.3 手势识别处理事件和方法

在 Android 系统中实现手势识别功能时，通常通过如下所示 id 事件和方法实现。

（1）boolean onDoubleTap(MotionEvent e)：双击的第二下 Touch down 时触发。

（2）boolean onDoubleTapEvent(MotionEvent e)：双击的第二下 Touch down 和 up 都会触发，可用 e.getAction()区分。

（3）boolean onDown(MotionEvent e)：Touch down 时触发。

（4）boolean onFling(MotionEvent e1, MotionEvent e2, float velocityX, float velocityY)：Touch 了滑动一点距离后，up 时触发。

（5）void onLongPress(MotionEvent e)：Touch 了不移动一直 Touch down 时触发。

（6）boolean onScroll(MotionEvent e1, MotionEvent e2, float distanceX, float distanceY)：Touch 了滑动时触发。

（7）void onShowPress(MotionEvent e)：Touch 了还没有滑动时触发。

注意

onDown 和 onLongPress 的具体对比如下所示。

- onDown 只要 Touch down 一定立刻触发。
- 而 Touch down 后过一会儿没有滑动，先触发 onShowPress 再是 onLongPress。

由此可见，Touch down 后一直不滑动，会按照 onDown→onShowPress→onLongPress 顺序进行触发。

（8）boolean onSingleTapConfirmed(MotionEvent e)和 boolean onSingleTapUp(MotionEvent e)：这两个函数都是在 Touch down 后又没有滑动（onScroll），又没有长按（onLongPress），然后 Touch up 时触发。

（9）onDown→onSingleTapUp→onSingleTapConfirmed：点击一下非常快地（不滑动）Touch up。

（10）onDown→onShowPress→onSingleTapUp→onSingleTapConfirmed：点击一下稍微慢点地（不滑动）Touch up。

22.2 实战演练——通过触摸方式移动图片

本节将通过一个具体实例的实现过程来讲解在 Android 屏幕中通过触摸点击的方式移动图片的方法和具体实现流程。

实例	功能	源码路径
实例 22-1	在屏幕中通过触摸方式移动图片	\daima\22\move

22.2.1　实例说明

在触摸屏手机中，点击移动照片的功能十分常见。在本实例中，用 ImageView 控件来显示 Drawable 中的照片，在程序运行后将照片放在屏幕中央。通过 onTouchEvent 来处理点击、拖动、放开等事件来完成拖动图片的功能。并且设置了 ImageView 的单击监听事件，让用户在单击图片的同时恢复到图片的初始位置。

22.2.2　具体实现

编写主程序文件，具体实现流程如下所示。

（1）通过 DisplayMetrics 获取屏幕对象，分别用 intScreenX 和 intScreenY 取得屏幕解析像素，并分别设置图片的宽高。具体代码如下所示。

```
public void onCreate(Bundle savedInstanceState)
{
   super.onCreate(savedInstanceState);
   setContentView(R.layout.main);

   /* 取得屏幕对象 */
   DisplayMetrics dm = new DisplayMetrics();
   getWindowManager().getDefaultDisplay().getMetrics(dm);

   /* 取得屏幕解析像素 */
   intScreenX = dm.widthPixels;
   intScreenY = dm.heightPixels;

   /* 设置图片的宽高 */
   intWidth = 100;
   intHeight = 100;
```

（2）将图片从 Drawable 中赋值给 ImageView 控件来呈现在屏幕中，并通过方法 RestoreButton()初始化按钮使其位置居中。具体代码如下所示。

```
/*通过 findViewById 构造器创建 ImageView 对象*/
mImageView01 =(ImageView) findViewById(R.id.myImageView1);
/*将图片从 Drawable 赋值给 ImageView 来呈现*/
mImageView01.setImageResource(R.drawable.baby);

/* 初始化按钮位置居中 */
RestoreButton();
```

（3）定义点击监听事件 setOnClickListener，当用户点击 ImageView 图片时，将图片还原到初始位置显示。具体代码如下所示。

```
    /* 当点击 ImageView，还原初始位置 */
    mImageView01.setOnClickListener(new Button.OnClickListener()
    {
      @Override
      public void onClick(View v)
      {
        RestoreButton();
      }
    });
}
```

（4）定义 onTouchEvent(MotionEvent event)覆盖触控事件。首先取得手指触控屏幕的位置，然后实现触控事件的处理，分别实现点击屏幕、移动位置和离开屏幕这三个动作处理。具体代码如下所示。

```
/*覆盖触控事件*/
public boolean onTouchEvent(MotionEvent event)
{
  /*取得手指触控屏幕的位置*/
  float x = event.getX();
  float y = event.getY();

  try
  {
    /*触控事件的处理*/
    switch (event.getAction())
    {
      /*点击屏幕*/
      case MotionEvent.ACTION_DOWN:
        picMove(x, y);
          break;
      /*移动位置*/
      case MotionEvent.ACTION_MOVE:
        picMove(x, y);
          break;
      /*离开屏幕*/
      case MotionEvent.ACTION_UP:
        picMove(x, y);
          break;
    }
  }catch(Exception e)
    {
      e.printStackTrace();
    }
  return true;
}
```

（5）定义方法 picMove(float x, float y)来移动屏幕中的图片，具体代码如下所示。

```
/*移动图片的方法*/
private void picMove(float x, float y)
{
  /*默认微调图片与指针的相对位置*/
  mX=x-(intWidth/2);
  mY=y-(intHeight/2);

  /*防图片超过屏幕的相关处理*/
  /*防止屏幕向右超过屏幕*/
  if((mX+intWidth)>intScreenX)
  {
    mX = intScreenX-intWidth;
  }
  /*防止屏幕向左超过屏幕*/
  else if(mX<0)
  {
```

```
    mX = 0;
  }
  /*防止屏幕向下超过屏幕*/
  else if ((mY+intHeight)>intScreenY)
  {
    mY=intScreenY-intHeight;
  }
  /*防止屏幕向上超过屏幕*/
  else if (mY<0)
  {
    mY = 0;
  }
  /*通过 log 来查看图片位置*/
  Log.i("jay", Float.toString(mX)+","+Float.toString(mY));
  /* 以 setLayoutParams 方法，重新安排 Layout 上的位置 */
  mImageView01.setLayoutParams
  (
    new AbsoluteLayout.LayoutParams
    (intWidth,intHeight,(int) mX,(int)mY)
  );
}
```

（6）定义方法 RestoreButton()来还原 ImageView 图片到初始位置，具体代码如下所示。

```
/* 还原 ImageView 位置的事件处理 */
public void RestoreButton()
{
  intDefaultX = ((intScreenX-intWidth)/2);
  intDefaultY = ((intScreenY-intHeight)/2);
  /*Toast 还原位置坐标*/
  mMakeTextToast
  (
    "("+
    Integer.toString(intDefaultX)+
    ","+
    Integer.toString(intDefaultY)+")",true
  );

  /* 以 setLayoutParams 方法，重新安排 Layout 上的位置 */
  mImageView01.setLayoutParams
  (
    new AbsoluteLayout.LayoutParams
    (intWidth,intHeight,intDefaultX,intDefaultY)
  );
}
```

执行后效果如图 22-1 所示，可以通过鼠标点击的方式移动图片的位置，如图 22-2 所示。

▲图 22-1　执行效果

▲图 22-2　移动图片

22.3 实战演练——实现各种手势识别

在本节的内容中，将通过一个具体实例的实现过程来讲解在 Android 系统中实现各种常见手势识别方法和具体实现流程。

实例	功能	源码路径
实例 22-2	在屏幕中实现各种常见的手势识别	\daima\22\Gesture

本实例的具体实现流程如下所示。

（1）布局文件 main.xml 非常简单，具体实现代码如下所示。

```
<LinearLayout xmlns:android="http://schemas.android.com/apk/res/android"
    android:orientation="vertical"
    android:layout_width="fill_parent"
    android:layout_height="fill_parent"
    >
<TextView
    android:layout_width="fill_parent"
    android:layout_height="wrap_content"
    android:text="@string/hello"
    />
</LinearLayout>
```

（2）主 Activity 的实现文件是 mainActivity.java，功能是为了更好地演示手势识别效果，将屏幕顶部的电池等图标和标题内容隐藏。文件 mainActivity.java 的具体实现代码如下所示。

```
public class mainActivity extends Activity{
    private GestureDetector mGestureDetector;;
    @Override
    public void onCreate(Bundle savedInstanceState) {
        super.onCreate(savedInstanceState);
        mGestureDetector = new GestureDetector(this, new MyGestureListener());

        if(getRequestedOrientation()!=ActivityInfo.SCREEN_ORIENTATION_LANDSCAPE){
            setRequestedOrientation(ActivityInfo.SCREEN_ORIENTATION_LANDSCAPE);
        }

        this.getWindow().setFlags(WindowManager.LayoutParams.FLAG_FULLSCREEN,
        WindowManager.LayoutParams.FLAG_FULLSCREEN);
        //隐去电池等图标和一切修饰部分（状态栏部分）
        this.requestWindowFeature(Window.FEATURE_NO_TITLE);
        // 隐去标题栏（程序的名字）
        setContentView(new MyView(this));
    }

    @Override
    public boolean onTouchEvent(MotionEvent event) {
       if (mGestureDetector.onTouchEvent(event))
            return true;
       else
            return false;
    }
}
```

（3）编写文件 MyGestureListener.java，功能是监听触摸屏幕中的各种常用的手势，具体实现代码如下所示。

```
public class MyGestureListener extends SimpleOnGestureListener implements
        OnGestureListener {
    @Override
    public boolean onDoubleTap(MotionEvent e) {
        // TODO Auto-generated method stub
        MyView.x=e.getX();
```

```
        MyView.y=e.getY();
        return super.onDoubleTap(e);
    }
    @Override
    public boolean onDoubleTapEvent(MotionEvent e) {
        // TODO Auto-generated method stub
        MyView.x=e.getX();
        MyView.y=e.getY();
        return super.onDoubleTapEvent(e);
    }
    @Override
    public boolean onDown(MotionEvent e) {
        // TODO Auto-generated method stub
        MyView.x=e.getX();
        MyView.y=e.getY();
        return super.onDown(e);
    }
    @Override
    public boolean onFling(MotionEvent e1, MotionEvent e2, float velocityX,
            float velocityY) {
        // TODO Auto-generated method stub
        MyView.x=e2.getX();
        MyView.y=e2.getY();
        return super.onFling(e1, e2, velocityX, velocityY);
    }
    @Override
    public void onLongPress(MotionEvent e) {
        // TODO Auto-generated method stub
        MyView.x=e.getX();
        MyView.y=e.getY();
        super.onLongPress(e);
    }
    @Override
    public boolean onScroll(MotionEvent e1, MotionEvent e2, float distanceX,
            float distanceY) {
        // TODO Auto-generated method stub
        MyView.x=e2.getX();
        MyView.y=e2.getY();
        return super.onScroll(e1, e2, distanceX, distanceY);
    }
    @Override
    public void onShowPress(MotionEvent e) {
        // TODO Auto-generated method stub
        super.onShowPress(e);
    }
    @Override
    public boolean onSingleTapConfirmed(MotionEvent e) {
        // TODO Auto-generated method stub
        MyView.x=e.getX();
        MyView.y=e.getY();
        return super.onSingleTapConfirmed(e);
    }
    @Override
    public boolean onSingleTapUp(MotionEvent e) {
        // TODO Auto-generated method stub
        return super.onSingleTapUp(e);
    }
}
```

（4）编写文件 MyView.java，功能是根据监听到的用户手势创建不同的视图。文件 MyView.java 的具体实现代码如下所示。

```
public class MyView extends SurfaceView implements SurfaceHolder.Callback {
    SurfaceHolder holder;
    static float x;
    static float y;
    public MyView(Context context) {
        super(context);
        holder = this.getHolder();//获取holder
```

```
        holder.addCallback(this);
    }
    @Override
    public void surfaceChanged(SurfaceHolder holder, int format, int width,
            int height) {
        // TODO Auto-generated method stub

    }

    @Override
    public void surfaceCreated(SurfaceHolder holder) {
        // TODO Auto-generated method stub
        new Thread(new MyThread()).start();
    }

    @Override
    public void surfaceDestroyed(SurfaceHolder holder) {
        // TODO Auto-generated method stub

    }
    public class MyThread implements Runnable {
        Paint paint=new Paint();
        @Override
        public void run() {
            // TODO Auto-generated method stub
            while(true){
                Canvas canvas = holder.lockCanvas(null);//获取画布
                paint.setColor(Color.BLACK);
//                canvas.drawRect(0, 0, 320, 480, paint);竖屏
                canvas.drawRect(0, 0, 480, 320, paint);
                paint.setColor(Color.GREEN);
                canvas.drawRect(x-5, y-5, x+5, y+5, paint);

                holder.unlockCanvasAndPost(canvas);// 解锁画布，提交画好的图像
                try {
                    Thread.sleep(100);
                } catch (InterruptedException e) {
                    // TODO Auto-generated catch block
                    e.printStackTrace();
                }
            }
        }
    }
}
```

到此为止，整个实例介绍完毕，执行后的效果如图 22-3 所示。在真机中运行后，会实现手势识别功能。

▲图 22-3　执行效果

22.4　实战演练——实现手势拖动和缩放图片效果

在本节的内容中，将通过一个具体实例的实现过程，来讲解在 Android 系统中实现手势拖动和缩放图片效果的方法和具体实现流程。

实例	功能	源码路径
实例 22-3	在屏幕中实现手势拖动和缩放图片效果	\daima\22\TutorialZoomActivity1

22.4.1　实现布局文件

布局文件 main.xml 非常简单，具体实现代码如下所示。

```
<com.sonyericsson.zoom.ImageZoomView
    xmlns:android="http://schemas.android.com/apk/res/android"
    android:id="@+id/zoomview"
    android:layout_width="fill_parent"
    android:layout_height="fill_parent"
    />
```

22.4.2　监听用户选择的设置选项

编写主 Activity 文件 TutorialZoomActivity1.java，功能是监听用户选择的设置选项，具体实现代码如下所示。

```
public class TutorialZoomActivity1 extends Activity {

    /** Constant used as menu item id for setting zoom control type */
    private static final int MENU_ID_ZOOM = 0;

    /** Constant used as menu item id for setting pan control type */
    private static final int MENU_ID_PAN = 1;

    /** Constant used as menu item id for resetting zoom state */
    private static final int MENU_ID_RESET = 2;

    /** Image zoom view */
    private ImageZoomView mZoomView;

    /** Zoom state */
    private ZoomState mZoomState;

    /** Decoded bitmap image */
    private Bitmap mBitmap;

    /** On touch listener for zoom view */
    private SimpleZoomListener mZoomListener;

    @Override
    public void onCreate(Bundle savedInstanceState) {
        super.onCreate(savedInstanceState);
        setContentView(R.layout.main);
        mZoomState = new ZoomState();
        mBitmap = BitmapFactory.decodeResource(getResources(), R.drawable.image800x600);
        mZoomListener = new SimpleZoomListener();
        mZoomListener.setZoomState(mZoomState);
        mZoomView = (ImageZoomView)findViewById(R.id.zoomview);
        mZoomView.setZoomState(mZoomState);
        mZoomView.setImage(mBitmap);
        mZoomView.setOnTouchListener(mZoomListener);
        resetZoomState();
    }
    @Override
    protected void onDestroy() {
        super.onDestroy();
        mBitmap.recycle();
        mZoomView.setOnTouchListener(null);
        mZoomState.deleteObservers();
    }
    @Override
    public boolean onCreateOptionsMenu(Menu menu) {
        menu.add(Menu.NONE, MENU_ID_ZOOM, 0, R.string.menu_zoom);
```

```
        menu.add(Menu.NONE, MENU_ID_PAN, 1, R.string.menu_pan);
        menu.add(Menu.NONE, MENU_ID_RESET, 2, R.string.menu_reset);
        return super.onCreateOptionsMenu(menu);
    }
    @Override
    public boolean onOptionsItemSelected(MenuItem item) {
        switch (item.getItemId()) {
            case MENU_ID_ZOOM:
                mZoomListener.setControlType(ControlType.ZOOM);
                break;
            case MENU_ID_PAN:
                mZoomListener.setControlType(ControlType.PAN);
                break;
            case MENU_ID_RESET:
                resetZoomState();
                break;
        }
        return super.onOptionsItemSelected(item);
    }
    /**
     *缩放状态复位并通知观察者
     */
    private void resetZoomState() {
        mZoomState.setPanX(0.5f);
        mZoomState.setPanY(0.5f);
        mZoomState.setZoom(1f);
        mZoomState.notifyObservers();
    }
}
```

这样单击模拟器中的“MENU”按钮后，将会在屏幕底部弹出 3 个选项，如图 22-4 所示。

▲图 22-4　在屏幕底部弹出 3 个选项

22.4.3　获取并设置移动位置和缩放值

编写文件 ZoomState.java，分别获取并设置移动位置和缩放值，具体实现代码如下所示。

```
public class ZoomState extends Observable {
    private float mZoom;
    private float mPanX;
    private float mPanY;

    /**
     *获取当前 X 轴的位置
     *
     * @return current x-pan
     */
    public float getPanX() {
        return mPanX;
    }
```

```
    /**
     * 获取当前 Y 轴的位置
     */
    public float getPanY() {
        return mPanY;
    }

    /**
     *获取当前的缩放值
     */
    public float getZoom() {
        return mZoom;
    }

    /**
     *帮助在 X 轴计算缩放值
     *
     * @param aspectQuotient (Aspect ratio content) / (Aspect ratio view)
     * @return Current zoom value in x-dimension
     */
    public float getZoomX(float aspectQuotient) {
        return Math.min(mZoom, mZoom * aspectQuotient);
    }

    /**
     *帮助在 Y 轴计算缩放值
     *
     * @param aspectQuotient (Aspect ratio content) / (Aspect ratio view)
     * @return Current zoom value in y-dimension
     */
    public float getZoomY(float aspectQuotient) {
        return Math.min(mZoom, mZoom / aspectQuotient);
    }

    /**
     *设置 X 轴移动位置
     */
    public void setPanX(float panX) {
        if (panX != mPanX) {
            mPanX = panX;
            setChanged();
        }
    }
    /**
     * 设置 Y 轴移动位置
     */
    public void setPanY(float panY) {
        if (panY != mPanY) {
            mPanY = panY;
            setChanged();
        }
    }
    /**
     * 设置缩放
     */
    public void setZoom(float zoom) {
        if (zoom != mZoom) {
            mZoom = zoom;
            setChanged();
        }
    }
}
```

22.4.4　在不同的缩放状态下绘制图像视图

编写文件 ImageZoomView.java，功能是在不同的缩放状态水平绘制图像视图，具体实现代码如下所示。

```
public class ImageZoomView extends View implements Observer {
    /**使用的对象绘制位图 */
    private final Paint mPaint = new Paint(Paint.FILTER_BITMAP_FLAG);
    /**绘制矩形 */
    private final Rect mRectSrc = new Rect();
    /** Rectangle used (and re-used) for specifying drawing area on canvas. */
    private final Rect mRectDst = new Rect();
    /** 放大绘图 */
    private Bitmap mBitmap;
    private float mAspectQuotient;
    /** 缩放状态 */
    private ZoomState mState;
    /**
     *构造函数
     */
    public ImageZoomView(Context context, AttributeSet attrs) {
        super(context, attrs);
    }
    /**
     *设置图像的位图
     *
     * @param bitmap The bitmap to view and zoom into
     */
    public void setImage(Bitmap bitmap) {
        mBitmap = bitmap;
        calculateAspectQuotient();
        invalidate();
    }
    /**
     *设置对象保持放大状态
     *
     * @param state The zoom state
     */
    public void setZoomState(ZoomState state) {
        if (mState != null) {
            mState.deleteObserver(this);
        }
        mState = state;
        mState.addObserver(this);
        invalidate();
    }
    // 私有方法
    private void calculateAspectQuotient() {
        if (mBitmap != null) {
            mAspectQuotient = (((float)mBitmap.getWidth()) / mBitmap.getHeight())
                    / (((float)getWidth()) / getHeight());
        }
    }
    // Superclass overrides
    @Override
    protected void onDraw(Canvas canvas) {
        if (mBitmap != null && mState != null) {
            final int viewWidth = getWidth();
            final int viewHeight = getHeight();
            final int bitmapWidth = mBitmap.getWidth();
            final int bitmapHeight = mBitmap.getHeight();
            final float panX = mState.getPanX();
            final float panY = mState.getPanY();
            final float zoomX = mState.getZoomX(mAspectQuotient) * viewWidth / bitmapWidth;
            final float zoomY = mState.getZoomY(mAspectQuotient) * viewHeight /
            bitmapHeight;

            //
            mRectSrc.left = (int)(panX * bitmapWidth - viewWidth / (zoomX * 2));
            mRectSrc.top = (int)(panY * bitmapHeight - viewHeight / (zoomY * 2));
            mRectSrc.right = (int)(mRectSrc.left + viewWidth / zoomX);
            mRectSrc.bottom = (int)(mRectSrc.top + viewHeight / zoomY);
            mRectDst.left = getLeft();
            mRectDst.top = getTop();
```

```
            mRectDst.right = getRight();
            mRectDst.bottom = getBottom();

            // Adjust source rectangle so that it fits within the source image.
            if (mRectSrc.left < 0) {
                mRectDst.left += -mRectSrc.left * zoomX;
                mRectSrc.left = 0;
            }
            if (mRectSrc.right > bitmapWidth) {
                mRectDst.right -= (mRectSrc.right - bitmapWidth) * zoomX;
                mRectSrc.right = bitmapWidth;
            }
            if (mRectSrc.top < 0) {
                mRectDst.top += -mRectSrc.top * zoomY;
                mRectSrc.top = 0;
            }
            if (mRectSrc.bottom > bitmapHeight) {
                mRectDst.bottom -= (mRectSrc.bottom - bitmapHeight) * zoomY;
                mRectSrc.bottom = bitmapHeight;
            }

            canvas.drawBitmap(mBitmap, mRectSrc, mRectDst, mPaint);
        }
    }

    @Override
    protected void onLayout(boolean changed, int left, int top, int right, int bottom)
{
        super.onLayout(changed, left, top, right, bottom);

        calculateAspectQuotient();
    }
    //实现观察者
    public void update(Observable observable, Object data) {
        invalidate();
    }
}
```

22.4.5 根据监听到的手势实现图片缩放

编写文件 SimpleZoomListener.java，功能是监听用户设置选项，并根据监听到的手势实现图片缩放功能。文件 SimpleZoomListener.java 的具体实现代码如下所示。

```
public class SimpleZoomListener implements View.OnTouchListener {
    /**
     * 使用哪种类型的控制方法
     */
    public enum ControlType {
        PAN, ZOOM
    }
    /**状态被触摸事件控制*/
    private ZoomState mState;
    /** 当前设置的控制类型*/
    private ControlType mControlType = ControlType.ZOOM;
    /** X-coordinate of previously handled touch event */
    private float mX;
    /** Y-coordinate of previously handled touch event */
    private float mY;
    /**
     * 设置缩放状态
     * @param state Zoom state
     */
    public void setZoomState(ZoomState state) {
        mState = state;
    }

    /**
     * Sets the control type to use
```

```
     *
     * @param controlType Control type
     */
    public void setControlType(ControlType controlType) {
        mControlType = controlType;
    }

    // implements View.OnTouchListener
    public boolean onTouch(View v, MotionEvent event) {
        final int action = event.getAction();
        final float x = event.getX();
        final float y = event.getY();

        switch (action) {
            case MotionEvent.ACTION_DOWN:
                mX = x;
                mY = y;
                break;

            case MotionEvent.ACTION_MOVE: {
                final float dx = (x - mX) / v.getWidth();
                final float dy = (y - mY) / v.getHeight();

                if (mControlType == ControlType.ZOOM) {
                    mState.setZoom(mState.getZoom() * (float)Math.pow(20, -dy));
                    mState.notifyObservers();
                } else {
                    mState.setPanX(mState.getPanX() - dx);
                    mState.setPanY(mState.getPanY() - dy);
                    mState.notifyObservers();
                }
                mX = x;
                mY = y;
                break;
            }
        }
        return true;
    }
}
```

到此为止，执行后将可以通过手势来移动图片或缩放图片，如图 22-5 所示。

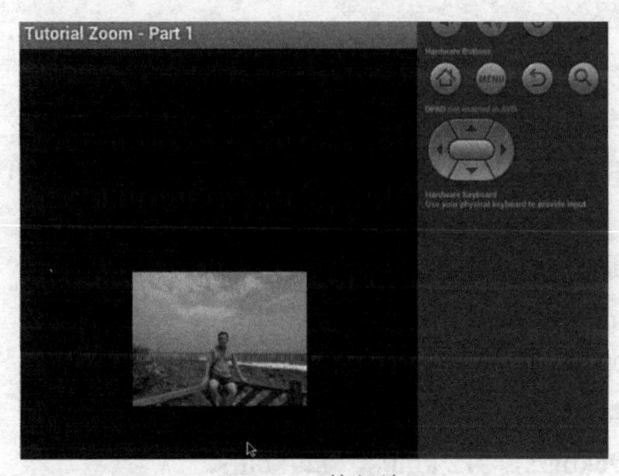

▲图 22-5　执行效果

第 23 章 NFC 近场通信技术详解

NFC 是近场通信（Near Field Communication）的缩写，此技术由非接触式射频识别（RFID）演变而来，由飞利浦半导体（现恩智浦半导体）、诺基亚和索尼共同研制开发，其基础是 RFID 及互联技术。NFC 是一种短距高频的无线电技术，在 13.56MHz 频率运行于 20cm 距离内。其传输速度有 106kbit/s、212kbit/s 或 424kbit/s 3 种。目前近场通信已通过成为 ISO/IEC IS 18092 国际标准、ECMA-340 标准与 ETSI TS 102 190 标准。NFC 采用主动和被动两种读取模式。本章将详细讲解在 Android 设备中使用近场通信技术的基本知识，为步入本书后面知识的学习打下基础。

23.1 近场通信技术基础

NFC 近场通信技术是由非接触式射频识别（RFID）及互联互通技术整合演变而来，在单一芯片上结合感应式读卡器、感应式卡片和点对点的功能，能在短距离内与兼容设备进行识别和数据交换。工作频率为 13.56MHz，但是使用这种手机支付方案的用户必须更换特制的手机。目前这项技术在日韩被广泛应用。手机用户凭着配置了支付功能的手机就可以行遍全国：他们的手机可以用作机场登机验证、大厦的门禁钥匙、交通一卡通、信用卡、支付卡等。本节将简要讲解 NFC 技术的基本知识。

23.1.1 NFC 技术的特点

近场通信是基于 RFID 技术发展起来的一种近距离无线通信技术。与 RFID 一样，近场通信信息也是通过频谱中无线频率部分的电磁感应耦合方式传递，但两者之间还是存在很大的区别。近场通信的传输范围比 RFID 小，RFID 的传输范围可以达到 0~1m，但由于近场通信采取了独特的信号衰减技术，相对于 RFID 来说，近场通信具有成本低、带宽高、能耗低等特点。

在现实应用中，近场通信技术主要特征如下所示。

- 用于近距离（10cm 以内）安全通信的无线通信技术。
- 射频频率：13.56MHz。
- 射频兼容：ISO 14443，ISO 15693，Felica 标准。
- 数据传输速度：106kbit/s，212 kbit/s，424kbit/s。

23.1.2 NFC 的工作模式

在现实应用中，NFC 技术有如下所示的的 3 种工作模式。

- 卡模式（Card emulation）：此模式其实就是相当于一张采用 RFID 技术的 IC 卡。可以替代现在大量的 IC 卡（包括信用卡）场合商场刷卡、公交卡、门禁管制，车票，门票等。此种方式下，有一个极大的优点，那就是卡片通过非接触读卡器的 RF 域来供电，即便是寄主设备（如手机）没电也可以工作。

- 点对点模式（P2P mode）：此模式和红外线差不多，可用于数据交换，只是传输距离较短，传输创建速度较快，传输速度也快些，功耗低（蓝牙也类似）。将两个具备NFC功能的设备链接，能实现数据点对点传输，如下载音乐、交换图片或者同步设备地址薄。因此通过NFC，多个设备如数位相机、PDA、计算机和手机之间都可以交换资料或者服务。
- 读卡器模式（Reader/Writer mode）：作为非接触读卡器使用，如从海报或者展览信息电子标签上读取相关信息。

23.1.3 NFC和蓝牙的对比

在现实应用中，NFC和蓝牙（Bluetooth）都是短程通信技术，而且都被集成到移动电话。但NFC不需要复杂的设置程序，并且也可以简化蓝牙连接。NFC略胜蓝牙的地方在于设置程序较短，但无法达到低功率蓝牙（Bluetooth Low Energy）的速度。在两台NFC设备相互链接的设备识别过程中，使用NFC来替代人工设置会使创建链接的速度大大加快，会少于1/10s。

NFC的最大数据传输量是424 kbit/s，这远小于Bluetooth V2.1 (2.1Mbit/s)。虽然NFC在传输速度与距离比不上Bluetooth，但是NFC技术不需要电源，对于移动电话或是移动消费性电子产品来说，NFC的使用比较方便。NFC的短距离通信特性正是其优点，由于耗电量低，一次只和一台机器链接，拥有较高的保密性与安全性，NFC有利于信用卡交易时避免被盗用。NFC的目标并非是取代蓝牙等其他无线技术，而是在不同的场合、不同的领域起到相互补充的作用。

NFC技术和蓝牙技术相比，主要支持的功能参数支持见表23-1。

表23-1　NFC技术和蓝牙技术的参数对比

说　明	NFC	Bluetooth	Bluetooth Low Energy
RFID 兼容	ISO 18000-3	active	active
标准化机构	ISO/IEC	Bluetooth SIG	Bluetooth SIG
网络标准	ISO 13157 etc.	IEEE 802.15.1	IEEE 802.15.1
网络类型	Point-to-Point	WPAN	WPAN
加密	not with RFID	available	available
范围	< 0.2m	～10m (class 2)	～1m (class 3)
频率	13.56MHz	2.4～2.5GHz	2.4～2.5GHz
Bit rate	424kbit/s	2.1Mbit/s	～1.0Mbit/s
设置程序	< 0.1s	< 6s	< 1s
功耗	< 15mA (read)	varies with class	< 15mA (xmit)

23.2 射频识别技术详解

射频识别即RFID（Radio Frequency Identification）技术，又称无线射频识别，是NFC技术的一个子集。RFID是一种通信技术，可以通过无线电信号识别特定目标并读写相关数据，而无需识别系统与特定目标之间建立机械或光学接触。在现实中，常用的RFID技术有低频（125～134.2kHz）、高频（13.56MHz）、超高频、微波等技术。RFID读写器也分移动式的和固定式的，目前RFID技术应用很广，例如，图书馆、门禁系统和食品安全溯源等。在本节的内容中，将详细讲解射频识别技术RFID的基本知识。

23.2.1　RFID 技术简介

RFID 是一种无线通信技术，可以通过无线电信号识别特定目标并读写相关数据，而无需识别系统与特定目标之间建立机械或者光学接触。从概念上来讲，RFID 类似于条码扫描，对于条码技术而言，它是将已编码的条形码附着于目标物，并使用专用的扫描读写器利用光信号将信息由条形磁传送到扫描读写器；而 RFID 则使用专用的 RFID 读写器及专门的可附着于目标物的 RFID 标签，利用频率信号将信息由 RFID 标签传送至 RFID 读写器。

无线电的信号是通过调成无线电频率的电磁场，把数据从附着在物品上的标签上传送出去，以自动辨识与追踪该物品。某些标签在识别时从识别器发出的电磁场中就可以得到能量，并不需要电池；也有标签本身拥有电源，并可以主动发出无线电波（调成无线电频率的电磁场）。标签包含了电子存储的信息，数米之内都可以识别。与条形码不同的是，射频标签不需要处在识别器视线之内，也可以嵌入被追踪物体之内。

在现实应用中，有许多行业都运用了射频识别技术。将标签附着在一辆正在生产中的汽车，厂方便可以追踪此车在生产线上的进度。仓库可以追踪药品的所在。射频标签也可以附于牲畜与宠物上，方便对牲畜与宠物的积极识别（积极识别意思是防止数只牲畜使用同一个身份）。射频识别的身份识别卡可以使员工得以进入锁住的建筑部分，汽车上的射频应答器也可以用来征收收费路段与停车场的费用。

某些射频标签附在衣物、个人财物上，甚至于植入人体之内。由于这项技术可能会在未经本人许可的情况下读取个人信息，这项技术也会有侵犯个人隐私的忧患。

23.2.2　RFID 技术的组成

从结构上讲 RFID 是一种简单的无线系统，只有两个基本器件，该系统用于控制、检测和跟踪物体。系统由一个询问器和很多应答器组成。在最初的技术领域中，应答器是指能够传输信息回复信息的电子模块。近年来，由于射频技术发展迅猛，应答器有了新的说法和含义，又被叫作智能标签或标签。RFID 电子标签的阅读器通过天线与 RFID 电子标签进行无线通信，可以实现对标签识别码和内存数据的读出或写入操作。RFID 技术可识别高速运动物体并可同时识别多个标签，操作快捷方便。

伴随着 RFID 技术的不断发展，其具体组成如下所示。

- 应答器：由天线、耦合元件及芯片组成，一般来说都是用标签作为应答器，每个标签具有唯一的电子编码，附着在物体上标识目标对象。
- 阅读器：由天线、耦合元件，芯片组成，读取（有时还可以写入）标签信息的设备，可设计为手持式 RFID 读写器（如 C5000W）或固定式读写器。
- 应用软件系统：是应用层软件，主要是把收集的数据进一步处理，并为人们所使用。

23.2.3　RFID 技术的特点

射频识别系统最重要的优点是非接触识别，它能穿透雪、雾、冰、涂料、尘垢和条形码无法使用的恶劣环境阅读标签，并且阅读速度极快，大多数情况下不到 100ms。有源式射频识别系统的速写能力也是重要的优点，可用于流程跟踪和维修跟踪等交互式业务。

制约射频识别系统发展的主要问题是不兼容的标准。射频识别系统的主要厂商提供的都是专用系统，导致不同的应用和不同的行业采用不同厂商的频率和协议标准，这种混乱和割据的状况已经制约了整个射频识别行业的增长。许多欧美组织正在着手解决这个问题，并已经取得了一些成绩。标准化必将刺激射频识别技术的大幅度发展和广泛应用。

RFID 技术的主要特点如下所示。

- 快速扫描：RFID 辨识器可以同时辨识读取数个 RFID 标签。
- 体积小型化、形状多样化：RFID 在读取上并不受尺寸大小与形状限制，不需为了读取精确度而配合纸张的固定尺寸和印刷品质。此外，RFID 标签更可往小型化与多样形态发展，以应用于不同产品。
- 抗污染能力和耐久性：传统条形码的载体是纸张，因此容易受到污染，但 RFID 对水、油和化学药品等物质具有很强抵抗性。此外，由于条形码是附于塑料袋或外包装纸箱上，所以特别容易受到折损；RFID 卷标是将数据存在芯片中，因此可以免受污损。
- 可重复使用：当今的条形码印刷上去之后就无法更改，RFID 标签则可以重复地新增、修改、删除 RFID 卷标内储存的数据，方便信息的更新。
- 穿透性和无屏障阅读：在被覆盖的情况下，RFID 能够穿透纸张、木材和塑料等非金属或非透明的材质，并能够进行穿透性通信。而条形码扫描机必须在近距离而且没有物体阻挡的情况下，才可以辨读条形码。
- 数据的记忆容量大：一维条形码的容量是 50Bytes，二维条形码最大的容量可储存 2 至 3000 字符，RFID 最大的容量则有数 MegaBytes。随着记忆载体的发展，数据容量也有不断扩大的趋势。未来物品所需携带的资料量会越来越大，对卷标所能扩充容量的需求也相应增加。
- 安全性：由于 RFID 承载的是电子式信息，其数据内容可经由密码保护，使其内容不易被伪造及变造。

RFID 因其所具备的远距离读取、高储存量等特性而备受瞩目。它不仅可以帮助一个企业大幅提高货物、信息管理的效率，还可以让销售企业和制造企业互联，从而更加准确地接收反馈信息，控制需求信息，优化整个供应链。

23.2.4 RFID 技术的工作原理

RFID 技术的基本工作原理是：当标签进入磁场后，接收解读器发出的射频信号，凭借感应电流所获得的能量发送出存储在芯片中的产品信息（Passive Tag，无源标签或被动标签），或者由标签主动发送某一频率的信号（Active Tag，有源标签或主动标签），解读器读取信息并解码后，送至中央信息系统进行有关数据处理。

一套完整的 RFID 系统是由阅读器（Reader）与电子标签（TAG）也就是所谓的应答器（Transponder）及应用软件系统 3 个部分所组成，其工作原理是 Reader 发射一特定频率的无线电波能量给 Transponder，用以驱动 Transponder 电路将内部的数据送出，此时 Reader 便依序接收解读数据，送给应用程序做相应处理。

以 RFID 卡片阅读器及电子标签之间的通信及能量感应方式来看，可以大致上将 RFID 分成感应耦合（Inductive Coupling）及后向散射耦合（Backscatter Coupling）两种。通常低频的 RFID 大都采用第一种方式，而较高频大多采用第二种方式。

阅读器根据使用的结构和技术不同可以是读或读/写装置，是 RFID 系统信息控制和处理中心。阅读器通常由耦合模块、收发模块、控制模块和接口单元组成。阅读器和应答器之间一般采用半双工通信方式进行信息交换，同时阅读器通过耦合给无源应答器提供能量和时序。在实际应用中，可进一步通过 Ethernet 或 WLAN 等实现对物体识别信息的采集、处理及远程传送等管理功能。应答器是 RFID 系统的信息载体，应答器大多是由耦合元件（线圈、微带天线等）和微芯片组成无源单元。

23.3 Android 系统中的 NFC

NFC 通信总是由一个发起者（initiator）和一个接受者（target）组成。通常 initiator 主动发送电磁场（RF），可以为被动式接受者（passive target）提供电源。其工作的基本原理和收音机类似。正是由于被动式接受者可以通过发起者提供电源，因此 target 可以有非常简单的形式，如标签、卡和 Sticker 的形式。另外，NFC 也支持点到点的通信（peer to peer），此时参与通信的双方都有电源支持。在 Android 系统的 NFC 模块应用中，Android 手机通常是作为通信中的发起者，也就是作为 NFC 的读写器。Android 手机也可以模拟作为 NFC 通信的接受者，并且从 Android 2.3.3 起也支持 P2P 通信。Android 系统支持如下所示的的 NFC 标准。

- NfcANFC-A (ISO 14443-3A)。
- NfcBNFC-B (ISO 14443-3B)。
- NfcFNFC-F (JIS 6319-4)。
- NfcVNFC-V (ISO 15693)。
- IsoDepISO-DEP (ISO 14443-4)。
- MifareClassic。
- MifareUltralight。

在 Android 系统中，NFC 模块从上到下的结构如图 23-1 所示。

```
--------------------- /system/framework/framework.jar----------------------

android.nfc              标准接口 ( NFCAdapter/NfcManager)
android.nfc.tech          标签技术

----------------------------- /system/Nfc.apk-----------------------------

com.android.nfc          NFC 服务相关
      .DeviceHost           底层设备接口原型
      .NfcService           NFC 服务 实现 DeviceHostListener 接口
com.android.nfc.dhimpl   NFC 功能底层实现-com.android.nfc.DeviceHost (NXP)
      .NativeNfcManager       implements DeviceHost
      JNI-> com_android_nfc_NativeNfcManager.cpp (libnfc_jni.so)
      .NativeNfcSecureElement
      JNI-> com_android_nfc_NativeNfcSecureElement.cpp (libnfc_jni.so)

----------------------- /system/lib/libnfc___.so--------------------------

libnfc-nxp => libnfc.so, libnfc_ndef.so
libnfc-nci => libnfc-nci.so
```

▲图 23-1　NFC 模块从上到下的结构

在本节的内容中，将详细讲解 Android 系统中 NFC 模块源码的基本知识。

23.3.1　分析 Java 层

在 Android 系统中，NFC 模块的 Java 层代码位于如下所示的目录中。

```
\frameworks\base\core\java\android\nfc\
```

在上述目录中，包含了用来与本地 NFC 适配器进行交互的顶层类，这些类可以表示被检测到的 tags 和用 NDEF 数据格式。在 NFC 的 Java 层中，常用的顶层类如下所示。

（1）NfcManager

类 NfcManager 在文件\frameworks\base\core\java\android\nfc\NfcManager.java 中定义，这是一个 NFC Adapter 的管理器，可以列出所有此 Android 设备支持的 NFC Adapter，只不过大部分

Android 设备只有一个 NFC Adapter，所以在大部分情况下可以直接用静态方法 getDefaultAdapter (context)来取适配器。文件\frameworks\base\core\java\android\nfc\NfcManager.java 的具体实现代码如下所示。

```
public final class NfcManager {
   private final NfcAdapter mAdapter;

   /**
    * @hide
    */
   public NfcManager(Context context) {
      NfcAdapter adapter;
      context = context.getApplicationContext();
      if (context == null) {
         throw new IllegalArgumentException(
                 "context not associated with any application (using a mock context?)");
      }
      try {
         adapter = NfcAdapter.getNfcAdapter(context);
      } catch (UnsupportedOperationException e) {
         adapter = null;
      }
      mAdapter = adapter;
   }

   /**
    * Get the default NFC Adapter for this device.
    *
    * @return the default NFC Adapter
    */
   public NfcAdapter getDefaultAdapter() {
      return mAdapter;
   }
}
```

（2）NfcAdapter

类在文件\frameworks\base\core\java\android\nfc\NfcAdapter.java 中定义，此类表示本设备的 NFC Adapter，可以定义 Intent 来请求将系统检测到 tags 的提醒发送到我们的 Activity，并提供方法去注册前台 tag 提醒发布和前台 NDEF 推送。前台 NDEF 推送是当前 Android 版本唯一支持的 P2P NFC 通信方式。文件\frameworks\base\core\java\android\nfc\NfcAdapter.java 的具体实现代码如下所示。

```
public final class NfcAdapter {
   static final String TAG = "NFC";
   public  static  final  String  ACTION_NDEF_DISCOVERED  =  "android.nfc.action.NDEF_
DISCOVERED";
   @SdkConstant(SdkConstantType.ACTIVITY_INTENT_ACTION)
   public  static  final  String  ACTION_TECH_DISCOVERED  =  "android.nfc.action.TECH_
DISCOVERED";
   @SdkConstant(SdkConstantType.ACTIVITY_INTENT_ACTION)
   public  static  final  String  ACTION_TAG_DISCOVERED  =  "android.nfc.action.TAG_
DISCOVERED";
   public static final String ACTION_TAG_LEFT_FIELD = "android.nfc.action.TAG_LOST";
   public static final String EXTRA_TAG = "android.nfc.extra.TAG";
   public static final String EXTRA_NDEF_MESSAGES = "android.nfc.extra.NDEF_MESSAGES";
   public static final String EXTRA_ID = "android.nfc.extra.ID";
   @SdkConstant(SdkConstantType.BROADCAST_INTENT_ACTION)
   public static final String ACTION_ADAPTER_STATE_CHANGED =
         "android.nfc.action.ADAPTER_STATE_CHANGED";
   public static final String EXTRA_ADAPTER_STATE = "android.nfc.extra.ADAPTER_STATE";

   public static final int STATE_OFF = 1;
   public static final int STATE_TURNING_ON = 2;
   public static final int STATE_ON = 3;
   public static final int STATE_TURNING_OFF = 4;
```

```
/** @hide */
public static final int FLAG_NDEF_PUSH_NO_CONFIRM = 0x1;

/** @hide */
public static final String ACTION_HANDOVER_TRANSFER_STARTED =
        "android.nfc.action.HANDOVER_TRANSFER_STARTED";

/** @hide */
public static final String ACTION_HANDOVER_TRANSFER_DONE =
        "android.nfc.action.HANDOVER_TRANSFER_DONE";

/** @hide */
public static final String EXTRA_HANDOVER_TRANSFER_STATUS =
        "android.nfc.extra.HANDOVER_TRANSFER_STATUS";

/** @hide */
public static final int HANDOVER_TRANSFER_STATUS_SUCCESS = 0;
/** @hide */
public static final int HANDOVER_TRANSFER_STATUS_FAILURE = 1;

/** @hide */
public static final String EXTRA_HANDOVER_TRANSFER_URI =
        "android.nfc.extra.HANDOVER_TRANSFER_URI";

// Guarded by NfcAdapter.class
static boolean sIsInitialized = false;

// Final after first constructor, except for
// attemptDeadServiceRecovery() when NFC crashes - we accept a best effort
// recovery
static INfcAdapter sService;
static INfcTag sTagService;
static HashMap<Context, NfcAdapter> sNfcAdapters = new HashMap();
//guard by NfcAdapter.class
static NfcAdapter sNullContextNfcAdapter;  // protected by NfcAdapter.class

final NfcActivityManager mNfcActivityManager;
final Context mContext;
public interface OnNdefPushCompleteCallback {
    public void onNdefPushComplete(NfcEvent event);
}
public interface CreateNdefMessageCallback {
    public NdefMessage createNdefMessage(NfcEvent event);
}
// TODO javadoc
public interface CreateBeamUrisCallback {
    public Uri[] createBeamUris(NfcEvent event);
}
private static boolean hasNfcFeature() {
    IPackageManager pm = ActivityThread.getPackageManager();
    if (pm == null) {
        Log.e(TAG, "Cannot get package manager, assuming no NFC feature");
        return false;
    }
    try {
        return pm.hasSystemFeature(PackageManager.FEATURE_NFC);
    } catch (RemoteException e) {
        Log.e(TAG, "Package manager query failed, assuming no NFC feature", e);
        return false;
    }
}
public static synchronized NfcAdapter getNfcAdapter(Context context) {
    if (!sIsInitialized) {
        /* is this device meant to have NFC */
        if (!hasNfcFeature()) {
            Log.v(TAG, "this device does not have NFC support");
            throw new UnsupportedOperationException();
        }
```

```
            sService = getServiceInterface();
            if (sService == null) {
                Log.e(TAG, "could not retrieve NFC service");
                throw new UnsupportedOperationException();
            }
            try {
                sTagService = sService.getNfcTagInterface();
            } catch (RemoteException e) {
                Log.e(TAG, "could not retrieve NFC Tag service");
                throw new UnsupportedOperationException();
            }

            sIsInitialized = true;
        }
        if (context == null) {
            if (sNullContextNfcAdapter == null) {
                sNullContextNfcAdapter = new NfcAdapter(null);
            }
            return sNullContextNfcAdapter;
        }
        NfcAdapter adapter = sNfcAdapters.get(context);
        if (adapter == null) {
            adapter = new NfcAdapter(context);
            sNfcAdapters.put(context, adapter);
        }
        return adapter;
    }
    /** get handle to NFC service interface */
    private static INfcAdapter getServiceInterface() {
        /* get a handle to NFC service */
        IBinder b = ServiceManager.getService("nfc");
        if (b == null) {
            return null;
        }
        return INfcAdapter.Stub.asInterface(b);
    }
    public static NfcAdapter getDefaultAdapter(Context context) {
        if (context == null) {
            throw new IllegalArgumentException("context cannot be null");
        }
        context = context.getApplicationContext();
        if (context == null) {
            throw new IllegalArgumentException(
                    "context not associated with any application (using a mock context?)");
        }
        /* use getSystemService() for consistency */
        NfcManager manager = (NfcManager) context.getSystemService(Context.NFC_SERVICE);
        if (manager == null) {
            // NFC not available
            return null;
        }
        return manager.getDefaultAdapter();
    }
    @Deprecated
    public static NfcAdapter getDefaultAdapter() {
        // introduced in API version 9 (GB 2.3)
        // deprecated in API version 10 (GB 2.3.3)
        // removed from public API in version 16 (ICS MR2)
        // should maintain as a hidden API for binary compatibility for a little longer
        Log.w(TAG, "WARNING: NfcAdapter.getDefaultAdapter() is deprecated, use " +
                "NfcAdapter.getDefaultAdapter(Context) instead", new Exception());

        return NfcAdapter.getNfcAdapter(null);
    }

    NfcAdapter(Context context) {
        mContext = context;
```

```
        mNfcActivityManager = new NfcActivityManager(this);
    }
    public Context getContext() {
        return mContext;
    }
    public INfcAdapter getService() {
        isEnabled();  // NOP call to recover sService if it is stale
        return sService;
    }
    public INfcTag getTagService() {
        isEnabled();  // NOP call to recover sTagService if it is stale
        return sTagService;
    }
    public void attemptDeadServiceRecovery(Exception e) {
        Log.e(TAG, "NFC service dead - attempting to recover", e);
        INfcAdapter service = getServiceInterface();
        if (service == null) {
            Log.e(TAG, "could not retrieve NFC service during service recovery");
            // nothing more can be done now, sService is still stale, we'll hit
            // this recovery path again later
            return;
        }
        // assigning to sService is not thread-safe, but this is best-effort code
        // and on a well-behaved system should never happen
        sService = service;
        try {
            sTagService = service.getNfcTagInterface();
        } catch (RemoteException ee) {
            Log.e(TAG, "could not retrieve NFC tag service during service recovery");
            // nothing more can be done now, sService is still stale, we'll hit
            // this recovery path again later
        }

        return;
    }
    public boolean isEnabled() {
        try {
            return sService.getState() == STATE_ON;
        } catch (RemoteException e) {
            attemptDeadServiceRecovery(e);
            return false;
        }
    }
    public int getAdapterState() {
        try {
            return sService.getState();
        } catch (RemoteException e) {
            attemptDeadServiceRecovery(e);
            return NfcAdapter.STATE_OFF;
        }
    }
    public boolean enable() {
        try {
            return sService.enable();
        } catch (RemoteException e) {
            attemptDeadServiceRecovery(e);
            return false;
        }
    }
    public boolean disable() {
        try {
            return sService.disable(true);
        } catch (RemoteException e) {
            attemptDeadServiceRecovery(e);
            return false;
        }
    }

    public void setBeamPushUris(Uri[] uris, Activity activity) {
```

```
        if (activity == null) {
            throw new NullPointerException("activity cannot be null");
        }
        if (uris != null) {
            for (Uri uri : uris) {
                if (uri == null) throw new NullPointerException("Uri not " +
                        "allowed to be null");
                String scheme = uri.getScheme();
                if (scheme == null || (!scheme.equalsIgnoreCase("file") &&
                        !scheme.equalsIgnoreCase("content"))) {
                    throw new IllegalArgumentException("URI needs to have " +
                            "either scheme file or scheme content");
                }
            }
        }
        mNfcActivityManager.setNdefPushContentUri(activity, uris);
    }
    public void setBeamPushUrisCallback(CreateBeamUrisCallback callback, Activity
activity) {
        if (activity == null) {
            throw new NullPointerException("activity cannot be null");
        }
        mNfcActivityManager.setNdefPushContentUriCallback(activity, callback);
    }
    public void setNdefPushMessage(NdefMessage message, Activity activity,
            Activity ... activities) {
        int targetSdkVersion = getSdkVersion();
        try {
            if (activity == null) {
                throw new NullPointerException("activity cannot be null");
            }
            mNfcActivityManager.setNdefPushMessage(activity, message, 0);
            for (Activity a : activities) {
                if (a == null) {
                    throw new NullPointerException("activities cannot contain null");
                }
                mNfcActivityManager.setNdefPushMessage(a, message, 0);
            }
        } catch (IllegalStateException e) {
            if (targetSdkVersion < android.os.Build.VERSION_CODES.JELLY_BEAN) {
                // Less strict on old applications - just log the error
                Log.e(TAG, "Cannot call API with Activity that has already " +
                        "been destroyed", e);
            } else {
                // Prevent new applications from making this mistake, re-throw
                throw(e);
            }
        }
    }

    /**
     * @hide
     */
    public void setNdefPushMessage(NdefMessage message, Activity activity, int flags) {
        if (activity == null) {
            throw new NullPointerException("activity cannot be null");
        }
        mNfcActivityManager.setNdefPushMessage(activity, message, flags);
    }
    public void setNdefPushMessageCallback(CreateNdefMessageCallback callback, Activity
    activity,Activity ... activities) {
        int targetSdkVersion = getSdkVersion();
        try {
            if (activity == null) {
                throw new NullPointerException("activity cannot be null");
            }
            mNfcActivityManager.setNdefPushMessageCallback(activity, callback, 0);
            for (Activity a : activities) {
                if (a == null) {
```

```
                throw new NullPointerException("activities cannot contain null");
            }
            mNfcActivityManager.setNdefPushMessageCallback(a, callback, 0);
        }
    } catch (IllegalStateException e) {
        if (targetSdkVersion < android.os.Build.VERSION_CODES.JELLY_BEAN) {
            // Less strict on old applications - just log the error
            Log.e(TAG, "Cannot call API with Activity that has already " +
                    "been destroyed", e);
        } else {
            // Prevent new applications from making this mistake, re-throw
            throw(e);
        }
    }
}
public void setNdefPushMessageCallback(CreateNdefMessageCallback callback, Activity
activity, int flags) {
    if (activity == null) {
        throw new NullPointerException("activity cannot be null");
    }
    mNfcActivityManager.setNdefPushMessageCallback(activity, callback, flags);
}
public void setOnNdefPushCompleteCallback(OnNdefPushCompleteCallback callback,
        Activity activity, Activity ... activities) {
    int targetSdkVersion = getSdkVersion();
    try {
        if (activity == null) {
            throw new NullPointerException("activity cannot be null");
        }
        mNfcActivityManager.setOnNdefPushCompleteCallback(activity, callback);
        for (Activity a : activities) {
            if (a == null) {
                throw new NullPointerException("activities cannot contain null");
            }
            mNfcActivityManager.setOnNdefPushCompleteCallback(a, callback);
        }
    } catch (IllegalStateException e) {
        if (targetSdkVersion < android.os.Build.VERSION_CODES.JELLY_BEAN) {
            // Less strict on old applications - just log the error
            Log.e(TAG, "Cannot call API with Activity that has already " +
                    "been destroyed", e);
        } else {
            // Prevent new applications from making this mistake, re-throw
            throw(e);
        }
    }
}
public void enableForegroundDispatch(Activity activity, PendingIntent intent,
        IntentFilter[] filters, String[][] techLists) {
    if (activity == null || intent == null) {
        throw new NullPointerException();
    }
    if (!activity.isResumed()) {
        throw new IllegalStateException("Foreground dispatch can only be enabled " +
                "when your activity is resumed");
    }
    try {
        TechListParcel parcel = null;
        if (techLists != null && techLists.length > 0) {
            parcel = new TechListParcel(techLists);
        }
        ActivityThread.currentActivityThread().registerOnActivityPausedListener
        (activity, mForegroundDispatchListener);
        sService.setForegroundDispatch(intent, filters, parcel);
    } catch (RemoteException e) {
        attemptDeadServiceRecovery(e);
    }
}
public void disableForegroundDispatch(Activity activity) {
```

```
ActivityThread.currentActivityThread().unregisterOnActivityPausedListener(activity,
            mForegroundDispatchListener);
        disableForegroundDispatchInternal(activity, false);
    }

    OnActivityPausedListener mForegroundDispatchListener = new OnActivityPaused
Listener() {
        @Override
        public void onPaused(Activity activity) {
            disableForegroundDispatchInternal(activity, true);
        }
    };

    void disableForegroundDispatchInternal(Activity activity, boolean force) {
        try {
            sService.setForegroundDispatch(null, null, null);
            if (!force && !activity.isResumed()) {
                throw new IllegalStateException("You must disable foreground dispatching " +
                        "while your activity is still resumed");
            }
        } catch (RemoteException e) {
            attemptDeadServiceRecovery(e);
        }
    }
    @Deprecated
    public void enableForegroundNdefPush(Activity activity, NdefMessage message) {
        if (activity == null || message == null) {
            throw new NullPointerException();
        }
        enforceResumed(activity);
        mNfcActivityManager.setNdefPushMessage(activity, message, 0);
    }
    @Deprecated
    public void disableForegroundNdefPush(Activity activity) {
        if (activity == null) {
            throw new NullPointerException();
        }
        enforceResumed(activity);
        mNfcActivityManager.setNdefPushMessage(activity, null, 0);
        mNfcActivityManager.setNdefPushMessageCallback(activity, null, 0);
        mNfcActivityManager.setOnNdefPushCompleteCallback(activity, null);
    }
    public boolean enableNdefPush() {
        try {
            return sService.enableNdefPush();
        } catch (RemoteException e) {
            attemptDeadServiceRecovery(e);
            return false;
        }
    }
    public boolean disableNdefPush() {
        try {
            return sService.disableNdefPush();
        } catch (RemoteException e) {
            attemptDeadServiceRecovery(e);
            return false;
        }
    }
    public boolean isNdefPushEnabled() {
        try {
            return sService.isNdefPushEnabled();
        } catch (RemoteException e) {
            attemptDeadServiceRecovery(e);
            return false;
        }
    }
    public void dispatch(Tag tag) {
        if (tag == null) {
```

```
            throw new NullPointerException("tag cannot be null");
        }
        try {
            sService.dispatch(tag);
        } catch (RemoteException e) {
            attemptDeadServiceRecovery(e);
        }
    }

    /**
     * @hide
     */
    public void setP2pModes(int initiatorModes, int targetModes) {
        try {
            sService.setP2pModes(initiatorModes, targetModes);
        } catch (RemoteException e) {
            attemptDeadServiceRecovery(e);
        }
    }

    /**
     * @hide
     */
    public INfcAdapterExtras getNfcAdapterExtrasInterface() {
        if (mContext == null) {
            throw new UnsupportedOperationException("You need a context on NfcAdapter to
            use the "+ " NFC extras APIs");
        }
        try {
            return sService.getNfcAdapterExtrasInterface(mContext.getPackageName());
        } catch (RemoteException e) {
            attemptDeadServiceRecovery(e);
            return null;
        }
    }

    void enforceResumed(Activity activity) {
        if (!activity.isResumed()) {
            throw new IllegalStateException("API cannot be called while activity is paused");
        }
    }

    int getSdkVersion() {
        if (mContext == null) {
            return android.os.Build.VERSION_CODES.GINGERBREAD; // best guess
        } else {
            return mContext.getApplicationInfo().targetSdkVersion;
        }
    }
}
```

（3）NdefMessage 和 NdefRecord

NDEF 是 NFC 论坛定义的数据结构，用来将有效的数据存储到 NFC tags 中，如文本、URL 和其他 MIME 类型。一个 NdefMessage 扮演一个容器，这个容器存储那些发送和读到的数据。一个 NdefMessage 对象包含 0 或多个 NdefRecord，每个 NDEF record 有一个类型，如文本、URL、智慧型海报/广告，或其他 MIME 数据。在 NdefMessage 中的第一个 NfcRecord 的类型，功能是发送 tag 到一个 Android 设备上的 Activity。其中类 NdefMessage 在文件\frameworks\base\core\java\android\nfc\NdefMessage.java 中定义，具体实现代码如下所示。

```
public final class NdefMessage implements Parcelable {
    private final NdefRecord[] mRecords;
    public NdefMessage(byte[] data) throws FormatException {
        if (data == null) throw new NullPointerException("data is null");
        ByteBuffer buffer = ByteBuffer.wrap(data);

        mRecords = NdefRecord.parse(buffer, false);
```

```
        if (buffer.remaining() > 0) {
            throw new FormatException("trailing data");
        }
    }
    public NdefMessage(NdefRecord record, NdefRecord ... records) {
        // validate
        if (record == null) throw new NullPointerException("record cannot be null");

        for (NdefRecord r : records) {
            if (r == null) {
                throw new NullPointerException("record cannot be null");
            }
        }

        mRecords = new NdefRecord[1 + records.length];
        mRecords[0] = record;
        System.arraycopy(records, 0, mRecords, 1, records.length);
    }
    public NdefMessage(NdefRecord[] records) {
        // validate
        if (records.length < 1) {
            throw new IllegalArgumentException("must have at least one record");
        }
        for (NdefRecord r : records) {
            if (r == null) {
                throw new NullPointerException("records cannot contain null");
            }
        }

        mRecords = records;
    }
    public NdefRecord[] getRecords() {
        return mRecords;
    }
    public int getByteArrayLength() {
        int length =0;
        for (NdefRecord r : mRecords) {
            length += r.getByteLength();
        }
        return length;
    }
    public byte[] toByteArray() {
        int length = getByteArrayLength();
        ByteBuffer buffer = ByteBuffer.allocate(length);

        for (int i=0; i<mRecords.length; i++) {
            boolean mb = (i == 0);  // first record
            boolean me = (i == mRecords.length - 1);  // last record
            mRecords[i].writeToByteBuffer(buffer, mb, me);
        }

        return buffer.array();
    }

    @Override
    public int describeContents() {
        return 0;
    }

    @Override
    public void writeToParcel(Parcel dest, int flags) {
        dest.writeInt(mRecords.length);
        dest.writeTypedArray(mRecords, flags);
    }

    public static final Parcelable.Creator<NdefMessage> CREATOR =
            new Parcelable.Creator<NdefMessage>() {
        @Override
```

```
        public NdefMessage createFromParcel(Parcel in) {
            int recordsLength = in.readInt();
            NdefRecord[] records = new NdefRecord[recordsLength];
            in.readTypedArray(records, NdefRecord.CREATOR);
            return new NdefMessage(records);
        }
        @Override
        public NdefMessage[] newArray(int size) {
            return new NdefMessage[size];
        }
    };

    @Override
    public int hashCode() {
        return Arrays.hashCode(mRecords);
    }
    @Override
    public boolean equals(Object obj) {
        if (this == obj) return true;
        if (obj == null) return false;
        if (getClass() != obj.getClass()) return false;
        NdefMessage other = (NdefMessage) obj;
        return Arrays.equals(mRecords, other.mRecords);
    }

    @Override
    public String toString() {
        return "NdefMessage " + Arrays.toString(mRecords);
    }
}
```

（4）Tag

类Tag在文件\frameworks\base\core\java\android\nfc\Tag.java中定义，此类用于标识一个被动的NFC目标，如tag、card和钥匙挂扣，甚至是一个电话模拟的NFC卡。当一个tag被检测到时，一个tag对象将被创建并且封装到一个Intent里，然后NFC发布系统将这个Intent用startActivity发送到注册了接受这种Intent的Activity里。我们可以用方法getTechList()来得到这个tag支持的技术细节，并创建一个android.nfc.tech提供的相应的TagTechnology对象。文件Tag.java的具体实现代码如下所示。

```
public final class Tag implements Parcelable {
    final byte[] mId;
    final int[] mTechList;
    final String[] mTechStringList;
    final Bundle[] mTechExtras;
    final int mServiceHandle;  // for use by NFC service, 0 indicates a mock
    final INfcTag mTagService; // interface to NFC service, will be null if mock tag

    int mConnectedTechnology;
    public Tag(byte[] id, int[] techList, Bundle[] techListExtras, int serviceHandle,
            INfcTag tagService) {
        if (techList == null) {
            throw new IllegalArgumentException("rawTargets cannot be null");
        }
        mId = id;
        mTechList = Arrays.copyOf(techList, techList.length);
        mTechStringList = generateTechStringList(techList);
        // Ensure mTechExtras is as long as mTechList
        mTechExtras = Arrays.copyOf(techListExtras, techList.length);
        mServiceHandle = serviceHandle;
        mTagService = tagService;

        mConnectedTechnology = -1;
    }
    public static Tag createMockTag(byte[] id, int[] techList, Bundle[] techListExtras) {
        // set serviceHandle to 0 and tagService to null to indicate mock tag
        return new Tag(id, techList, techListExtras, 0, null);
```

```
}

private String[] generateTechStringList(int[] techList) {
    final int size = techList.length;
    String[] strings = new String[size];
    for (int i = 0; i < size; i++) {
        switch (techList[i]) {
            case TagTechnology.ISO_DEP:
                strings[i] = IsoDep.class.getName();
                break;
            case TagTechnology.MIFARE_CLASSIC:
                strings[i] = MifareClassic.class.getName();
                break;
            case TagTechnology.MIFARE_ULTRALIGHT:
                strings[i] = MifareUltralight.class.getName();
                break;
            case TagTechnology.NDEF:
                strings[i] = Ndef.class.getName();
                break;
            case TagTechnology.NDEF_FORMATABLE:
                strings[i] = NdefFormatable.class.getName();
                break;
            case TagTechnology.NFC_A:
                strings[i] = NfcA.class.getName();
                break;
            case TagTechnology.NFC_B:
                strings[i] = NfcB.class.getName();
                break;
            case TagTechnology.NFC_F:
                strings[i] = NfcF.class.getName();
                break;
            case TagTechnology.NFC_V:
                strings[i] = NfcV.class.getName();
                break;
            case TagTechnology.NFC_BARCODE:
                strings[i] = NfcBarcode.class.getName();
                break;
            default:
                throw new IllegalArgumentException("Unknown tech type " + techList[i]);
        }
    }
    return strings;
}
public Tag rediscover() throws IOException {
    if (getConnectedTechnology() != -1) {
        throw new IllegalStateException("Close connection to the technology first!");
    }

    if (mTagService == null) {
        throw new IOException("Mock tags don't support this operation.");
    }
    try {
        Tag newTag = mTagService.rediscover(getServiceHandle());
        if (newTag != null) {
            return newTag;
        } else {
            throw new IOException("Failed to rediscover tag");
        }
    } catch (RemoteException e) {
        throw new IOException("NFC service dead");
    }
}
public boolean hasTech(int techType) {
    for (int tech : mTechList) {
        if (tech == techType) return true;
    }
    return false;
}
public Bundle getTechExtras(int tech) {
```

```
        int pos = -1;
        for (int idx = 0; idx < mTechList.length; idx++) {
          if (mTechList[idx] == tech) {
              pos = idx;
              break;
          }
        }
        if (pos < 0) {
            return null;
        }

        return mTechExtras[pos];
    }
    @Override
    public String toString() {
        StringBuilder sb = new StringBuilder("TAG: Tech [");
        String[] techList = getTechList();
        int length = techList.length;
        for (int i = 0; i < length; i++) {
            sb.append(techList[i]);
            if (i < length - 1) {
                sb.append(", ");
            }
        }
        sb.append("]");
        return sb.toString();
    }

    /*package*/ static byte[] readBytesWithNull(Parcel in) {
        int len = in.readInt();
        byte[] result = null;
        if (len >= 0) {
            result = new byte[len];
            in.readByteArray(result);
        }
        return result;
    }
    /*package*/ static void writeBytesWithNull(Parcel out, byte[] b) {
        if (b == null) {
            out.writeInt(-1);
            return;
        }
        out.writeInt(b.length);
        out.writeByteArray(b);
    }
    @Override
    public void writeToParcel(Parcel dest, int flags) {
        // Null mTagService means this is a mock tag
        int isMock = (mTagService == null)?1:0;

        writeBytesWithNull(dest, mId);
        dest.writeInt(mTechList.length);
        dest.writeIntArray(mTechList);
        dest.writeTypedArray(mTechExtras, 0);
        dest.writeInt(mServiceHandle);
        dest.writeInt(isMock);
        if (isMock == 0) {
            dest.writeStrongBinder(mTagService.asBinder());
        }
    }

    public static final Parcelable.Creator<Tag> CREATOR =
            new Parcelable.Creator<Tag>() {
        @Override
        public Tag createFromParcel(Parcel in) {
            INfcTag tagService;

            // Tag fields
            byte[] id = Tag.readBytesWithNull(in);
```

```
            int[] techList = new int[in.readInt()];
            in.readIntArray(techList);
            Bundle[] techExtras = in.createTypedArray(Bundle.CREATOR);
            int serviceHandle = in.readInt();
            int isMock = in.readInt();
            if (isMock == 0) {
                tagService = INfcTag.Stub.asInterface(in.readStrongBinder());
            }
            else {
                tagService = null;
            }

            return new Tag(id, techList, techExtras, serviceHandle, tagService);
        }

        @Override
        public Tag[] newArray(int size) {
            return new Tag[size];
        }
    };
    public synchronized void setConnectedTechnology(int technology) {
        if (mConnectedTechnology == -1) {
            mConnectedTechnology = technology;
        } else {
            throw new IllegalStateException("Close other technology first!");
        }
    }
    public int getConnectedTechnology() {
        return mConnectedTechnology;
    }
    public void setTechnologyDisconnected() {
        mConnectedTechnology = -1;
    }
}
```

（5）“tech\”子目录

除此之外，在“frameworks\base\core\java\android\nfc\tech\”目录中包含了查询 tag 属性和进行 I/O 操作的类。这些类分别标示一个 tag 支持的不同的 NFC 技术标准，如图 23-2 所示。

▲图 23-2 “frameworks\base\core\java\android\nfc\tech\”目录

在如图 23-2 所示的目录中，类 TagTechnology 是表 23-2 中所有有 Tag Technology 类必须实现的。

表 23-2 必须实现 TagTechnology 的类

类	说　明
NfcA	支持 ISO 14443-3A 标准的操作。Provides access to NFC-A (ISO 14443-3A) properties and I/O operations
NfcB	Provides access to NFC-B (ISO 14443-3B) properties and I/O operations
NfcF	Provides access to NFC-F (JIS 6319-4) properties and I/O operations
NfcV	Provides access to NFC-V (ISO 15693) properties and I/O operations
IsoDep	Provides access to ISO-DEP (ISO 14443-4) properties and I/O operations
Ndef	提供对那些被格式化为 NDEF 的 tag 的数据的访问和其他操作

而类 NdefFormatable 对那些可以被格式化成 NDEF 格式的 tag 提供一个格式化的操作。文件 frameworks\base\core\java\android\nfc\tech\NdefFormatable.java 的具体实现代码如下所示。

```
public final class NdefFormatable extends BasicTagTechnology {
    private static final String TAG = "NFC";
    public static NdefFormatable get(Tag tag) {
        if (!tag.hasTech(TagTechnology.NDEF_FORMATABLE)) return null;
        try {
            return new NdefFormatable(tag);
        } catch (RemoteException e) {
            return null;
```

```
        }
    }
    /*package*/ void format(NdefMessage firstMessage, boolean makeReadOnly) throws
    IOException,FormatException {
        checkConnected();

        try {
            int serviceHandle = mTag.getServiceHandle();
            INfcTag tagService = mTag.getTagService();
            int errorCode = tagService.formatNdef(serviceHandle, MifareClassic.KEY_ DEFAULT);
            switch (errorCode) {
                case ErrorCodes.SUCCESS:
                    break;
                case ErrorCodes.ERROR_IO:
                    throw new IOException();
                case ErrorCodes.ERROR_INVALID_PARAM:
                    throw new FormatException();
                default:
                    // Should not happen
                    throw new IOException();
            }
            // Now check and see if the format worked
            if (!tagService.isNdef(serviceHandle)) {
                throw new IOException();
            }

            // Write a message, if one was provided
            if (firstMessage != null) {
                errorCode = tagService.ndefWrite(serviceHandle, firstMessage);
                switch (errorCode) {
                    case ErrorCodes.SUCCESS:
                        break;
                    case ErrorCodes.ERROR_IO:
                        throw new IOException();
                    case ErrorCodes.ERROR_INVALID_PARAM:
                        throw new FormatException();
                    default:
                        // Should not happen
                        throw new IOException();
                }
            }

            // optionally make read-only
            if (makeReadOnly) {
                errorCode = tagService.ndefMakeReadOnly(serviceHandle);
                switch (errorCode) {
                    case ErrorCodes.SUCCESS:
                        break;
                    case ErrorCodes.ERROR_IO:
                        throw new IOException();
                    case ErrorCodes.ERROR_INVALID_PARAM:
                        throw new IOException();
                    default:
                        // Should not happen
                        throw new IOException();
                }
            }
        } catch (RemoteException e) {
            Log.e(TAG, "NFC service dead", e);
        }
    }
}
```

类 MifareClassic 在文件 frameworks\base\core\java\android\nfc\tech\MifareClassic.java 中定义，如果 Android 设备支持 MIFARE，则提供对 MIFARE Classic 目标的属性和 I/O 操作。文件 MifareClassic.java 的具体实现代码如下所示。

```
public final class MifareClassic extends BasicTagTechnology {
```

```
private static final String TAG = "NFC";
public static final byte[] KEY_DEFAULT =
        {(byte)0xFF,(byte)0xFF,(byte)0xFF,(byte)0xFF,(byte)0xFF,(byte)0xFF};
public static final byte[] KEY_MIFARE_APPLICATION_DIRECTORY =
        {(byte)0xA0,(byte)0xA1,(byte)0xA2,(byte)0xA3,(byte)0xA4,(byte)0xA5};
public static final byte[] KEY_NFC_FORUM =
        {(byte)0xD3,(byte)0xF7,(byte)0xD3,(byte)0xF7,(byte)0xD3,(byte)0xF7};

/** A MIFARE Classic compatible card of unknown type */
public static final int TYPE_UNKNOWN = -1;
/** A MIFARE Classic tag */
public static final int TYPE_CLASSIC = 0;
/** A MIFARE Plus tag */
public static final int TYPE_PLUS = 1;
/** A MIFARE Pro tag */
public static final int TYPE_PRO = 2;

/** Tag contains 16 sectors, each with 4 blocks. */
public static final int SIZE_1K = 1024;
/** Tag contains 32 sectors, each with 4 blocks. */
public static final int SIZE_2K = 2048;
/**
 * Tag contains 40 sectors. The first 32 sectors contain 4 blocks and the last 8 sectors
 * contain 16 blocks.
 */
public static final int SIZE_4K = 4096;
/** Tag contains 5 sectors, each with 4 blocks. */
public static final int SIZE_MINI = 320;

/** Size of a MIFARE Classic block (in bytes) */
public static final int BLOCK_SIZE = 16;

private static final int MAX_BLOCK_COUNT = 256;
private static final int MAX_SECTOR_COUNT = 40;

private boolean mIsEmulated;
private int mType;
private int mSize;
public static MifareClassic get(Tag tag) {
    if (!tag.hasTech(TagTechnology.MIFARE_CLASSIC)) return null;
    try {
        return new MifareClassic(tag);
    } catch (RemoteException e) {
        return null;
    }
}
/** @hide */
public MifareClassic(Tag tag) throws RemoteException {
    super(tag, TagTechnology.MIFARE_CLASSIC);
    NfcA a = NfcA.get(tag);  // MIFARE Classic is always based on NFC a
    mIsEmulated = false;
    switch (a.getSak()) {
    case 0x01:
    case 0x08:
        mType = TYPE_CLASSIC;
        mSize = SIZE_1K;
        break;
    case 0x09:
        mType = TYPE_CLASSIC;
        mSize = SIZE_MINI;
        break;
    case 0x10:
        mType = TYPE_PLUS;
        mSize = SIZE_2K;
        // SecLevel = SL2
        break;
    case 0x11:
        mType = TYPE_PLUS;
        mSize = SIZE_4K;
```

```
            // Seclevel = SL2
            break;
        case 0x18:
            mType = TYPE_CLASSIC;
            mSize = SIZE_4K;
            break;
        case 0x28:
            mType = TYPE_CLASSIC;
            mSize = SIZE_1K;
            mIsEmulated = true;
            break;
        case 0x38:
            mType = TYPE_CLASSIC;
            mSize = SIZE_4K;
            mIsEmulated = true;
            break;
        case 0x88:
            mType = TYPE_CLASSIC;
            mSize = SIZE_1K;
            // NXP-tag: false
            break;
        case 0x98:
        case 0xB8:
            mType = TYPE_PRO;
            mSize = SIZE_4K;
            break;
        default:
            // Stack incorrectly reported a MifareClassic. We cannot handle this
            // gracefully - we have no idea of the memory layout. Bail.
            throw new RuntimeException(
                    "Tag incorrectly enumerated as MIFARE Classic, SAK = " + a.getSak());
        }
    }
    public int getSectorCount() {
        switch (mSize) {
        case SIZE_1K:
            return 16;
        case SIZE_2K:
            return 32;
        case SIZE_4K:
            return 40;
        case SIZE_MINI:
            return 5;
        default:
            return 0;
        }
    }
    public int getBlockCountInSector(int sectorIndex) {
        validateSector(sectorIndex);

        if (sectorIndex < 32) {
            return 4;
        } else {
            return 16;
        }
    }
    public int blockToSector(int blockIndex) {
        validateBlock(blockIndex);
        if (blockIndex < 32 * 4) {
            return blockIndex / 4;
        } else {
            return 32 + (blockIndex - 32 * 4) / 16;
        }
    }
    public int sectorToBlock(int sectorIndex) {
        if (sectorIndex < 32) {
            return sectorIndex * 4;
        } else {
            return 32 * 4 + (sectorIndex - 32) * 16;
```

```
        }
    }
    private boolean authenticate(int sector, byte[] key, boolean keyA) throws IOException {
        validateSector(sector);
        checkConnected();
        byte[] cmd = new byte[12];
        // First byte is the command
        if (keyA) {
            cmd[0] = 0x60; // phHal_eMifareAuthentA
        } else {
            cmd[0] = 0x61; // phHal_eMifareAuthentB
        }
        // Second byte is block address
        // Authenticate command takes a block address. Authenticating a block
        // of a sector will authenticate the entire sector.
        cmd[1] = (byte) sectorToBlock(sector);
        // Next 4 bytes are last 4 bytes of UID
        byte[] uid = getTag().getId();
        System.arraycopy(uid, uid.length - 4, cmd, 2, 4);
        // Next 6 bytes are key
        System.arraycopy(key, 0, cmd, 6, 6);
        try {
            if (transceive(cmd, false) != null) {
                return true;
            }
        } catch (TagLostException e) {
            throw e;
        } catch (IOException e) {
            // No need to deal with, will return false anyway
        }
        return false;
    }
    public byte[] readBlock(int blockIndex) throws IOException {
        validateBlock(blockIndex);
        checkConnected();

        byte[] cmd = { 0x30, (byte) blockIndex };
        return transceive(cmd, false);
    }
    public void writeBlock(int blockIndex, byte[] data) throws IOException {
        validateBlock(blockIndex);
        checkConnected();
        if (data.length != 16) {
            throw new IllegalArgumentException("must write 16-bytes");
        }

        byte[] cmd = new byte[data.length + 2];
        cmd[0] = (byte) 0xA0; // MF write command
        cmd[1] = (byte) blockIndex;
        System.arraycopy(data, 0, cmd, 2, data.length);

        transceive(cmd, false);
    }
    public void increment(int blockIndex, int value) throws IOException {
        validateBlock(blockIndex);
        validateValueOperand(value);
        checkConnected();

        ByteBuffer cmd = ByteBuffer.allocate(6);
        cmd.order(ByteOrder.LITTLE_ENDIAN);
        cmd.put( (byte) 0xC1 );
        cmd.put( (byte) blockIndex );
        cmd.putInt(value);

        transceive(cmd.array(), false);
    }
    public void decrement(int blockIndex, int value) throws IOException {
        validateBlock(blockIndex);
        validateValueOperand(value);
```

```
        checkConnected();

        ByteBuffer cmd = ByteBuffer.allocate(6);
        cmd.order(ByteOrder.LITTLE_ENDIAN);
        cmd.put( (byte) 0xC0 );
        cmd.put( (byte) blockIndex );
        cmd.putInt(value);

        transceive(cmd.array(), false);
    }
    public void transfer(int blockIndex) throws IOException {
        validateBlock(blockIndex);
        checkConnected();

        byte[] cmd = { (byte) 0xB0, (byte) blockIndex };

        transceive(cmd, false);
    }
    public void restore(int blockIndex) throws IOException {
        validateBlock(blockIndex);
        checkConnected();

        byte[] cmd = { (byte) 0xC2, (byte) blockIndex };

        transceive(cmd, false);
    }
    public void setTimeout(int timeout) {
        try {
            int err = mTag.getTagService().setTimeout(TagTechnology.MIFARE_CLASSIC, timeout);
            if (err != ErrorCodes.SUCCESS) {
                throw new IllegalArgumentException("The supplied timeout is not valid");
            }
        } catch (RemoteException e) {
            Log.e(TAG, "NFC service dead", e);
        }
    }
    public int getTimeout() {
        try {
            return mTag.getTagService().getTimeout(TagTechnology.MIFARE_CLASSIC);
        } catch (RemoteException e) {
            Log.e(TAG, "NFC service dead", e);
            return 0;
        }
    }

    private static void validateSector(int sector) {
        if (sector < 0 || sector >= MAX_SECTOR_COUNT) {
            throw new IndexOutOfBoundsException("sector out of bounds: " + sector);
        }
    }

    private static void validateBlock(int block) {
        // Just looking for obvious out of bounds...
        if (block < 0 || block >= MAX_BLOCK_COUNT) {
            throw new IndexOutOfBoundsException("block out of bounds: " + block);
        }
    }

    private static void validateValueOperand(int value) {
        if (value < 0) {
            throw new IllegalArgumentException("value operand negative");
        }
    }
}
```

类MifareUltralight在文件frameworks\base\core\java\android\nfc\tech\MifareUltralight.java中定义，如果Android设备支持MIFARE，则提供对MIFARE Ultralight目标的属性和I/O操作，具体实现代码如下所示。

```
public final class MifareUltralight extends BasicTagTechnology {
    private static final String TAG = "NFC";
    /** A MIFARE Ultralight compatible tag of unknown type */
    public static final int TYPE_UNKNOWN = -1;
    /** A MIFARE Ultralight tag */
    public static final int TYPE_ULTRALIGHT = 1;
    /** A MIFARE Ultralight C tag */
    public static final int TYPE_ULTRALIGHT_C = 2;
    /** Size of a MIFARE Ultralight page in bytes */
    public static final int PAGE_SIZE = 4;
    private static final int NXP_MANUFACTURER_ID = 0x04;
    private static final int MAX_PAGE_COUNT = 256;
    /** @hide */
    public static final String EXTRA_IS_UL_C = "isulc";
    private int mType;
    public static MifareUltralight get(Tag tag) {
        if (!tag.hasTech(TagTechnology.MIFARE_ULTRALIGHT)) return null;
        try {
            return new MifareUltralight(tag);
        } catch (RemoteException e) {
            return null;
        }
    }
    /** @hide */
    public MifareUltralight(Tag tag) throws RemoteException {
        super(tag, TagTechnology.MIFARE_ULTRALIGHT);
        // Check if this could actually be a MIFARE
        NfcA a = NfcA.get(tag);
        mType = TYPE_UNKNOWN;
        if (a.getSak() == 0x00 && tag.getId()[0] == NXP_MANUFACTURER_ID) {
            Bundle extras = tag.getTechExtras(TagTechnology.MIFARE_ULTRALIGHT);
            if (extras.getBoolean(EXTRA_IS_UL_C)) {
                mType = TYPE_ULTRALIGHT_C;
            } else {
                mType = TYPE_ULTRALIGHT;
            }
        }
    }
    public byte[] readPages(int pageOffset) throws IOException {
        validatePageIndex(pageOffset);
        checkConnected();

        byte[] cmd = { 0x30, (byte) pageOffset};
        return transceive(cmd, false);
    }
    public void writePage(int pageOffset, byte[] data) throws IOException {
        validatePageIndex(pageOffset);
        checkConnected();
        byte[] cmd = new byte[data.length + 2];
        cmd[0] = (byte) 0xA2;
        cmd[1] = (byte) pageOffset;
        System.arraycopy(data, 0, cmd, 2, data.length);
        transceive(cmd, false);
    }
    public void setTimeout(int timeout) {
        try {
            int err = mTag.getTagService().setTimeout(
                    TagTechnology.MIFARE_ULTRALIGHT, timeout);
            if (err != ErrorCodes.SUCCESS) {
                throw new IllegalArgumentException("The supplied timeout is not valid");
            }
        } catch (RemoteException e) {
            Log.e(TAG, "NFC service dead", e);
        }
    }
    public int getTimeout() {
        try {
            return mTag.getTagService().getTimeout(TagTechnology.MIFARE_ULTRALIGHT);
        } catch (RemoteException e) {
            Log.e(TAG, "NFC service dead", e);
```

```
            return 0;
        }
    }

    private static void validatePageIndex(int pageIndex) {
        // Do not be too strict on upper bounds checking, since some cards
        // may have more addressable memory than they report.
        // Note that issuing a command to an out-of-bounds block is safe - the
        // tag will wrap the read to an addressable area. This validation is a
        // helper to guard against obvious programming mistakes.
        if (pageIndex < 0 || pageIndex >= MAX_PAGE_COUNT) {
            throw new IndexOutOfBoundsException("page out of bounds: " + pageIndex);
        }
    }
}
```

23.3.2　分析 JNI 部分

在 Android 系统中，NFC 模块的 JNI 部分代码在如下所示的目录中实现。

```
\packages\apps\Nfc\nxp\jni
```

JNI 部分代码向上会跟 Framework 层的 Java 代码进行交互，向下会跟 libnfc 层进行交互。JNI 层的核心文件是 com_android_nfc_NativeNfcManager.cpp，功能是初始化并启动 NFC 服务，并扫描和读取 tag。文件 com_android_nfc_NativeNfcManager.cpp 的主要实现代码如下所示。

```
static void client_kill_deferred_call(void* arg)
{
   struct nfc_jni_native_data *nat = (struct nfc_jni_native_data *)arg;

   nat->running = FALSE;
}

static void kill_client(nfc_jni_native_data *nat)
{
   phDal4Nfc_Message_Wrapper_t  wrapper;
   phLibNfc_DeferredCall_t     *pMsg;

   usleep(50000);

   ALOGD("Terminating client thread...");

   pMsg = (phLibNfc_DeferredCall_t*)malloc(sizeof(phLibNfc_DeferredCall_t));
   pMsg->pCallback = client_kill_deferred_call;
   pMsg->pParameter = (void*)nat;

   wrapper.msg.eMsgType = PH_LIBNFC_DEFERREDCALL_MSG;
   wrapper.msg.pMsgData = pMsg;
   wrapper.msg.Size     = sizeof(phLibNfc_DeferredCall_t);

   phDal4Nfc_msgsnd(gDrvCfg.nClientId, (struct msgbuf *)&wrapper, sizeof(phLibNfc_
Message_t), 0);
}

static void nfc_jni_ioctl_callback(void *pContext, phNfc_sData_t *pOutput, NFCSTATUS
status) {
   struct nfc_jni_callback_data * pCallbackData = (struct nfc_jni_callback_data *)
pContext;
   LOG_CALLBACK("nfc_jni_ioctl_callback", status);

   /* Report the callback status and wake up the caller */
   pCallbackData->status = status;
   sem_post(&pCallbackData->sem);
}

static void nfc_jni_deinit_download_callback(void *pContext, NFCSTATUS status)
{
   struct nfc_jni_callback_data * pCallbackData = (struct nfc_jni_callback_data *)
```

```
pContext;
   LOG_CALLBACK("nfc_jni_deinit_download_callback", status);

   /* Report the callback status and wake up the caller */
   pCallbackData->status = status;
   sem_post(&pCallbackData->sem);
}

static int nfc_jni_download_locked(struct nfc_jni_native_data *nat, uint8_t update)
{
    uint8_t OutputBuffer[1];
    uint8_t InputBuffer[1];
    struct timespec ts;
    NFCSTATUS status = NFCSTATUS_FAILED;
    phLibNfc_StackCapabilities_t caps;
    struct nfc_jni_callback_data cb_data;
    bool result;

    /* Create the local semaphore */
    if (!nfc_cb_data_init(&cb_data, NULL))
    {
       goto clean_and_return;
    }

    if(update)
    {
        //deinit
        TRACE("phLibNfc_Mgt_DeInitialize() (download)");
        REENTRANCE_LOCK();
        status = phLibNfc_Mgt_DeInitialize(gHWRef, nfc_jni_deinit_download_callback,
        (void *)&cb_data);
        REENTRANCE_UNLOCK();
        if (status != NFCSTATUS_PENDING)
        {
            ALOGE("phLibNfc_Mgt_DeInitialize() (download) returned 0x%04x[%s]", status,
            nfc_jni_get_status_name(status));
        }

        clock_gettime(CLOCK_REALTIME, &ts);
        ts.tv_sec += 5;

        /* Wait for callback response */
        if(sem_timedwait(&cb_data.sem, &ts))
        {
            ALOGW("Deinitialization timed out (download)");
        }

        if(cb_data.status != NFCSTATUS_SUCCESS)
        {
            ALOGW("Deinitialization FAILED (download)");
        }
        TRACE("Deinitialization SUCCESS (download)");
    }

    result = performDownload(nat, false);

    if (!result) {
        status = NFCSTATUS_FAILED;
        goto clean_and_return;
    }

    TRACE("phLibNfc_Mgt_Initialize()");
    REENTRANCE_LOCK();
    status = phLibNfc_Mgt_Initialize(gHWRef, nfc_jni_init_callback, (void *)&cb_data);
    REENTRANCE_UNLOCK();
    if(status != NFCSTATUS_PENDING)
    {
        ALOGE("phLibNfc_Mgt_Initialize() (download) returned 0x%04x[%s]", status,
        nfc_jni_get_status_name(status));
```

```
        goto clean_and_return;
    }
    TRACE("phLibNfc_Mgt_Initialize() returned 0x%04x[%s]", status, nfc_jni_get_status_
    name(status));

    if(sem_wait(&cb_data.sem))
    {
       ALOGE("Failed to wait for semaphore (errno=0x%08x)", errno);
       status = NFCSTATUS_FAILED;
       goto clean_and_return;
    }

    /* Initialization Status */
    if(cb_data.status != NFCSTATUS_SUCCESS)
    {
        status = cb_data.status;
        goto clean_and_return;
    }

    /* ====== CAPABILITIES ======= */
    REENTRANCE_LOCK();
    status = phLibNfc_Mgt_GetstackCapabilities(&caps, (void*)nat);
    REENTRANCE_UNLOCK();
    if (status != NFCSTATUS_SUCCESS)
    {
       ALOGW("phLibNfc_Mgt_GetstackCapabilities returned 0x%04x[%s]", status,
       nfc_jni_get_status_name(status));
    }
    else
    {
        ALOGD("NFC capabilities: HAL = %x, FW = %x, HW = %x, Model = %x, HCI = %x, Full_FW
= %d, Rev = %d, FW Update Info = %d",
              caps.psDevCapabilities.hal_version,
              caps.psDevCapabilities.fw_version,
              caps.psDevCapabilities.hw_version,
              caps.psDevCapabilities.model_id,
              caps.psDevCapabilities.hci_version,
              caps.psDevCapabilities.full_version[NXP_FULL_VERSION_LEN-1],
              caps.psDevCapabilities.full_version[NXP_FULL_VERSION_LEN-2],
              caps.psDevCapabilities.firmware_update_info);
    }

    /*Download is successful*/
    status = NFCSTATUS_SUCCESS;

clean_and_return:
   nfc_cb_data_deinit(&cb_data);
   return status;
}

static int nfc_jni_configure_driver(struct nfc_jni_native_data *nat)
{
    char value[PROPERTY_VALUE_MAX];
    int result = FALSE;
    NFCSTATUS status;

    /* ====== CONFIGURE DRIVER ======= */
    /* Configure hardware link */
    gDrvCfg.nClientId = phDal4Nfc_msgget(0, 0600);

    TRACE("phLibNfc_Mgt_ConfigureDriver(0x%08x)", gDrvCfg.nClientId);
    REENTRANCE_LOCK();
    status = phLibNfc_Mgt_ConfigureDriver(&gDrvCfg, &gHWRef);
    REENTRANCE_UNLOCK();
    if(status == NFCSTATUS_ALREADY_INITIALISED) {
          ALOGW("phLibNfc_Mgt_ConfigureDriver() returned 0x%04x[%s]", status,
          nfc_jni_get_status_name(status));
    }
    else if(status != NFCSTATUS_SUCCESS)
```

```
    {
        ALOGE("phLibNfc_Mgt_ConfigureDriver() returned 0x%04x[%s]", status, nfc_jni_get_
        status_name(status));
        goto clean_and_return;
    }
    TRACE("phLibNfc_Mgt_ConfigureDriver() returned 0x%04x[%s]", status, nfc_jni_get_
    status_name(status));

    if(pthread_create(&(nat->thread), NULL, nfc_jni_client_thread, nat) != 0)
    {
        ALOGE("pthread_create failed");
        goto clean_and_return;
    }

    driverConfigured = TRUE;

clean_and_return:
    return result;
}

static int nfc_jni_unconfigure_driver(struct nfc_jni_native_data *nat)
{
    int result = FALSE;
    NFCSTATUS status;

    /* Unconfigure driver */
    TRACE("phLibNfc_Mgt_UnConfigureDriver()");
    REENTRANCE_LOCK();
    status = phLibNfc_Mgt_UnConfigureDriver(gHWRef);
    REENTRANCE_UNLOCK();
    if(status != NFCSTATUS_SUCCESS)
    {
        ALOGE("phLibNfc_Mgt_UnConfigureDriver() returned error 0x%04x[%s] -- this should
        never happen", status, nfc_jni_get_status_name( status));
    }
    else
    {
        ALOGD("phLibNfc_Mgt_UnConfigureDriver() returned 0x%04x[%s]", status,
        nfc_jni_get_status_name(status));
        result = TRUE;
    }

    driverConfigured = FALSE;

    return result;
}

/* Initialization function */
static int nfc_jni_initialize(struct nfc_jni_native_data *nat) {
    struct timespec ts;
    uint8_t resp[16];
    NFCSTATUS status;
    phLibNfc_StackCapabilities_t caps;
    phLibNfc_SE_List_t SE_List[PHLIBNFC_MAXNO_OF_SE];
    uint8_t i, No_SE = PHLIBNFC_MAXNO_OF_SE, SmartMX_index = 0, SmartMX_detected = 0;
    phLibNfc_Llcp_sLinkParameters_t LlcpConfigInfo;
    struct nfc_jni_callback_data cb_data;
    uint8_t firmware_status;
    uint8_t update = TRUE;
    int result = JNI_FALSE;
    const hw_module_t* hw_module;
    nfc_pn544_device_t* pn544_dev = NULL;
    int ret = 0;
    ALOGD("Start Initialization\n");

    /* Create the local semaphore */
    if (!nfc_cb_data_init(&cb_data, NULL))
    {
        goto clean_and_return;
```

```
}
/* Get EEPROM values and device port from product-specific settings */
ret = hw_get_module(NFC_HARDWARE_MODULE_ID, &hw_module);
if (ret) {
   ALOGE("hw_get_module() failed.");
   goto clean_and_return;
}
ret = nfc_pn544_open(hw_module, &pn544_dev);
if (ret) {
   ALOGE("Could not open pn544 hw_module.");
   goto clean_and_return;
}
if (pn544_dev->num_eeprom_settings == 0 || pn544_dev->eeprom_settings == NULL) {
    ALOGE("Could not load EEPROM settings");
    goto clean_and_return;
}

/* Reset device connected handle */
device_connected_flag = 0;

/* Reset stored handle */
storedHandle = 0;

/* Initialize Driver */
if(!driverConfigured)
{
    nfc_jni_configure_driver(nat);
}

/* ====== INITIALIZE ======= */

TRACE("phLibNfc_Mgt_Initialize()");
REENTRANCE_LOCK();
status = phLibNfc_Mgt_Initialize(gHWRef, nfc_jni_init_callback, (void *)&cb_data);
REENTRANCE_UNLOCK();
if(status != NFCSTATUS_PENDING)
{
   ALOGE("phLibNfc_Mgt_Initialize() returned 0x%04x[%s]", status, nfc_jni_get_
   status_name(status));
   update = FALSE;
   goto force_download;
}
TRACE("phLibNfc_Mgt_Initialize returned 0x%04x[%s]", status, nfc_jni_get_
status_name(status));

/* Wait for callback response */
if(sem_wait(&cb_data.sem))
{
   ALOGE("Failed to wait for semaphore (errno=0x%08x)", errno);
   goto clean_and_return;
}

/* Initialization Status */
if(cb_data.status != NFCSTATUS_SUCCESS)
{
   update = FALSE;
   goto force_download;
}

/* ====== CAPABILITIES ======= */

REENTRANCE_LOCK();
status = phLibNfc_Mgt_GetstackCapabilities(&caps, (void*)nat);
REENTRANCE_UNLOCK();
if (status != NFCSTATUS_SUCCESS)
{
   ALOGW("phLibNfc_Mgt_GetstackCapabilities returned 0x%04x[%s]", status,
   nfc_jni_get_status_name(status));
}
```

```
    else
    {
        ALOGD("NFC capabilities: HAL = %x, FW = %x, HW = %x, Model = %x, HCI = %x, Full_FW
= %d, Rev = %d, FW Update Info = %d",
              caps.psDevCapabilities.hal_version,
              caps.psDevCapabilities.fw_version,
              caps.psDevCapabilities.hw_version,
              caps.psDevCapabilities.model_id,
              caps.psDevCapabilities.hci_version,
              caps.psDevCapabilities.full_version[NXP_FULL_VERSION_LEN-1],
              caps.psDevCapabilities.full_version[NXP_FULL_VERSION_LEN-2],
              caps.psDevCapabilities.firmware_update_info);
    }

    /* ====== FIRMWARE VERSION ======= */
    if(caps.psDevCapabilities.firmware_update_info)
    {
force_download:
        for (i=0; i<3; i++)
        {
            TRACE("Firmware version not UpToDate");
            status = nfc_jni_download_locked(nat, update);
            if(status == NFCSTATUS_SUCCESS)
            {
                ALOGI("Firmware update SUCCESS");
                break;
            }
            ALOGW("Firmware update FAILED");
            update = FALSE;
        }
        if(i>=3)
        {
            ALOGE("Unable to update firmware, giving up");
            goto clean_and_return;
        }
    }
    else
    {
        TRACE("Firmware version UpToDate");
    }
    /* ====== EEPROM SETTINGS ======= */

    // Update EEPROM settings
    TRACE("******  START EEPROM SETTINGS UPDATE ******");
    for (i = 0; i < pn544_dev->num_eeprom_settings; i++)
    {
       char eeprom_property[PROPERTY_KEY_MAX];
       char eeprom_value[PROPERTY_VALUE_MAX];
       uint8_t* eeprom_base = &(pn544_dev->eeprom_settings[i*4]);
       TRACE("> EEPROM SETTING: %d", i);

       // Check for override of this EEPROM value in properties
       snprintf(eeprom_property, sizeof(eeprom_property), "debug.nfc.eeprom.%02X%02X",
               eeprom_base[1], eeprom_base[2]);
       TRACE(">> Checking property: %s", eeprom_property);
       if (property_get(eeprom_property, eeprom_value, "") == 2) {
           int eeprom_value_num = (int)strtol(eeprom_value, (char**)NULL, 16);
           ALOGD(">> Override EEPROM addr 0x%02X%02X with value %02X",
                   eeprom_base[1], eeprom_base[2], eeprom_value_num);
           eeprom_base[3] = eeprom_value_num;
       }

       TRACE(">> Addr: 0x%02X%02X set to: 0x%02X", eeprom_base[1], eeprom_base[2],
               eeprom_base[3]);
       gInputParam.buffer = eeprom_base;
       gInputParam.length = 0x04;
       gOutputParam.buffer = resp;

       REENTRANCE_LOCK();
```

```
    status = phLibNfc_Mgt_IoCtl(gHWRef, NFC_MEM_WRITE, &gInputParam, &gOutputParam,
    nfc_jni_ioctl_callback, (void *)&cb_data);
    REENTRANCE_UNLOCK();
    if (status != NFCSTATUS_PENDING) {
      ALOGE("phLibNfc_Mgt_IoCtl() returned 0x%04x[%s]", status, nfc_jni_get_status_
      name(status));
      goto clean_and_return;
    }
    /* Wait for callback response */
    if(sem_wait(&cb_data.sem))
    {
      ALOGE("Failed to wait for semaphore (errno=0x%08x)", errno);
      goto clean_and_return;
    }

    /* Initialization Status */
    if (cb_data.status != NFCSTATUS_SUCCESS)
    {
      goto clean_and_return;
    }
  }
  TRACE("******  ALL EEPROM SETTINGS UPDATED  ******");

  /* ====== SECURE ELEMENTS ======= */

  REENTRANCE_LOCK();
  ALOGD("phLibNfc_SE_GetSecureElementList()");
  status = phLibNfc_SE_GetSecureElementList(SE_List, &No_SE);
  REENTRANCE_UNLOCK();
  if (status != NFCSTATUS_SUCCESS)
  {
    ALOGD("phLibNfc_SE_GetSecureElementList(): Error");
    goto clean_and_return;
  }

  ALOGD("\n> Number of Secure Element(s) : %d\n", No_SE);
  /* Display Secure Element information */
  for (i = 0; i < No_SE; i++)
  {
    if (SE_List[i].eSE_Type == phLibNfc_SE_Type_SmartMX) {
      ALOGD("phLibNfc_SE_GetSecureElementList(): SMX detected, handle=%p",
      (void*)SE_List[i].hSecureElement);
    } else if (SE_List[i].eSE_Type == phLibNfc_SE_Type_UICC) {
      ALOGD("phLibNfc_SE_GetSecureElementList(): UICC detected, handle=%p",
      (void*)SE_List[i].hSecureElement);
    }

    /* Set SE mode - Off */
    REENTRANCE_LOCK();
    status = phLibNfc_SE_SetMode(SE_List[i].hSecureElement,
        phLibNfc_SE_ActModeOff, nfc_jni_se_set_mode_callback,
        (void *)&cb_data);
    REENTRANCE_UNLOCK();
    if (status != NFCSTATUS_PENDING)
    {
      ALOGE("phLibNfc_SE_SetMode() returned 0x%04x[%s]", status,
          nfc_jni_get_status_name(status));
      goto clean_and_return;
    }
    ALOGD("phLibNfc_SE_SetMode() returned 0x%04x[%s]", status,
        nfc_jni_get_status_name(status));

    /* Wait for callback response */
    if(sem_wait(&cb_data.sem))
    {
      ALOGE("Failed to wait for semaphore (errno=0x%08x)", errno);
      goto clean_and_return;
    }
  }
```

```
    /* ====== LLCP ======= */

    /* LLCP Params */
    TRACE("****** NFC Config Mode NFCIP1 - LLCP ******");
    LlcpConfigInfo.miu    = nat->miu;
    LlcpConfigInfo.lto    = nat->lto;
    LlcpConfigInfo.wks    = nat->wks;
    LlcpConfigInfo.option = nat->opt;

    REENTRANCE_LOCK();
    status = phLibNfc_Mgt_SetLlcp_ConfigParams(&LlcpConfigInfo,
                                              nfc_jni_llcpcfg_callback,
                                              (void *)&cb_data);
    REENTRANCE_UNLOCK();
    if(status != NFCSTATUS_PENDING)
    {
       ALOGE("phLibNfc_Mgt_SetLlcp_ConfigParams returned 0x%04x[%s]", status,
           nfc_jni_get_status_name(status));
       goto clean_and_return;
    }
    TRACE("phLibNfc_Mgt_SetLlcp_ConfigParams returned 0x%04x[%s]", status,
        nfc_jni_get_status_name(status));

    /* Wait for callback response */
    if(sem_wait(&cb_data.sem))
    {
       ALOGE("Failed to wait for semaphore (errno=0x%08x)", errno);
       goto clean_and_return;
    }

    /* ===== DISCOVERY ==== */
    nat->discovery_cfg.NfcIP_Mode = nat->p2p_initiator_modes;  //initiator
    nat->discovery_cfg.NfcIP_Target_Mode = nat->p2p_target_modes;  //target
    nat->discovery_cfg.Duration = 300000; /* in ms */
    nat->discovery_cfg.NfcIP_Tgt_Disable = FALSE;

    /* Register for the card emulation mode */
    REENTRANCE_LOCK();
    ret = phLibNfc_SE_NtfRegister(nfc_jni_transaction_callback,(void *)nat);
    REENTRANCE_UNLOCK();
    if(ret != NFCSTATUS_SUCCESS)
    {
        ALOGD("phLibNfc_SE_NtfRegister returned 0x%02x",ret);
        goto clean_and_return;
    }
    TRACE("phLibNfc_SE_NtfRegister returned 0x%x\n", ret);


    /* ====== END ======= */

    ALOGI("NFC Initialized");

    result = TRUE;

 clean_and_return:
    if (result != TRUE)
    {
       if(nat)
       {
          kill_client(nat);
       }
    }
    if (pn544_dev != NULL) {
       nfc_pn544_close(pn544_dev);
    }
    nfc_cb_data_deinit(&cb_data);

    return result;
 }
```

23.3.3　分析底层

在 Android 系统中，NFC 模块的底层部分有驱动部分和 libnfc 库部分两大类。驱动部分在"device"目录中实现，例如，"device\samsung\tuna\nfc"目录中保存了设备厂家三星电子提供的 hardware lib。而"libnfc-nci"和"libnfc-nxp"目录中的底层文件则负责 NFC 数据的读取和解析工作。

23.4 编写 NFC 程序

当 Android 手机开启了 NFC 程序，并且检测到一个 TAG 后，TAG 分发系统会自动创建一个封装了 NFC TAG 信息的 Intent。如果多于一个应用程序能够处理这个 Intent，那么手机就会弹出一个对话框，让用户选择处理该 TAG 的 Activity。在 TAG 分发系统中定义了 3 种 Intent，按优先级从高到低排列顺序依次是。

- NDEF_DISCOVERED。
- TECH_DISCOVERED。
- TAG_DISCOVERED。

当 Android 设备检测到有 NFC Tag 靠近时，会根据 Action 申明的顺序给对应的 Activity 发送含 NFC 消息的 Intent。此处我们使用的 intent-filter 的 Action 类型为 TECH_DISCOVERED，从而可以处理所有类型为 ACTION_TECH_DISCOVERED 并且使用的技术为 nfc_tech_filter.xml 文件中定义的类型的 TAG。

当 Android 手机检测到一个 TAG 时，启用 Activity 的匹配过程如图 23-3 所示。

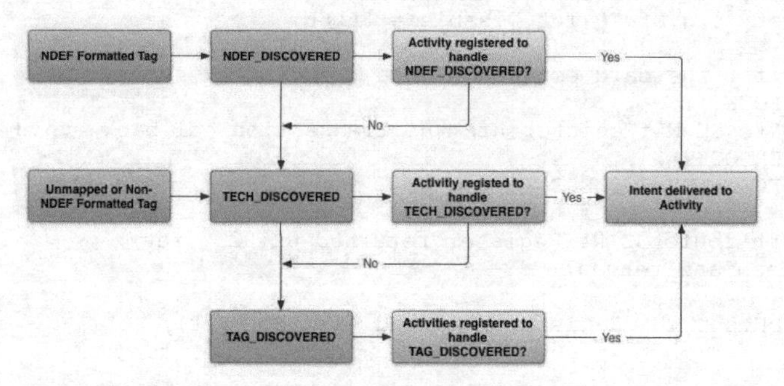

▲图 23-3　启用 Activity 的匹配过程

23.4.1　在 Android 系统编写 NFC APP 的方法

在 Android 系统中，编写 NFC APP 的基本流程如下所示。

（1）在相关的 androidManifest 文件中设置 NFC 权限，具体代码如下所示。

```
<uses-permission android:name="android.permission.NFC" />
```

（2）设置 SDK 的级别限制，例如，设置为 API 10。

```
<uses-sdk android:minSdkVersion="10"/>
```

（3）声明特殊功能的限制权限，通过如下声明可以让应用程序在 Google Play 上声明使用者必须拥有 NFC 功能。

```
<uses-feature android:name="android.hardware.nfc" android:required="true" />
```

（4）实现 NFC 标签过滤。在 Activity 的 Intent 过滤 XML 声明中，可以同时声明过滤如下 3

种 action（动作），但是需要提前知道系统在发送 Intent 时拥有的优先级。

- 动作一：过滤 ACTION_TAG_DISCOVERED

```
<intent-filter>
   <action android:name="android.nfc.action.TAG_DISCOVERED"/>
   <category android:name="android.intent.category.DEFAULT"/>
</intent-filter>
```

这个最简单，也是最后一个被尝试接受 intent 的选项。

- 动作二：过滤 ACTION_NDEF_DISCOVERED

```
<intent-filter>
<action android:name="android.nfc.action.NDEF_DISCOVERED"/>
<category android:name="android.intent.category.DEFAULT"/>
<data android:mimeType="text/plain" />
 </intent-filter>
```

其中最重要的是 data 的 mimeType 类型，此定义越准确，Intent 指向这个 Activity 的成功率就越高，否则系统可能不会发出你想要的 NDEF intent 了。

- 动作三：过滤 ACTION_TECH_DISCOVERED

首先需要在<project-path>/res/xml 下面创建一个过滤规则文件，可以任意命名，例如，可以叫作 nfc_tech_filter.xml。这个里面定义的是 nfc 实现的各种标准，每一个 NFC 卡都会符合多个不同的标准。可以在检测到 NFC 标签后，使用 getTechList()方法来查看所检测的 tag 到底支持哪些 NFC 标准。在一个 nfc_tech_filter.xml 文件中可以定义多个<tech-list>结构组，每一组代表声明只接受同时满足这些标准的 NFC 标签。如 A 组表示，只有同时满足 IsoDep、NfcA、NfcB、NfcF 这 4 个标准的 NFC 标签的 Intent 才能进入。A 与 B 组之间的关系就是只要满足其中一个就可以了。换句话说，我们的 NFC 标签技术满足 A 的声明也可以，满足 B 的声明也可以。

```
<resources xmlns:xliff="urn:oasis:names:tc:xliff:document:1.2">
<tech-list> --------------------------------A 组
<tech>android.nfc.tech.IsoDep</tech>               <tech>android.nfc.tech.NfcA</tech>
<tech>android.nfc.tech.NfcB</tech> <tech>android.nfc.tech.NfcF</tech>
</tech-list>
<tech-list>------------------------------------------B 组
<tech>android.nfc.tech.NfcV</tech>                 <tech>android.nfc.tech.Ndef</tech>
<tech>android.nfc.tech.NdefFormatable</tech>
<tech>android.nfc.tech.MifareClassic</tech>
<tech>android.nfc.tech.MifareUltralight</tech>
 </tech-list>
</resources>
```

在 androidManifest 文件中，声明 xml 过滤的举例代码如下所示。

```
<activity>
 <intent-filter>
<action android:name="android.nfc.action.TECH_DISCOVERED"/>
</intent-filter>
<meta-data android:name="android.nfc.action.TECH_DISCOVERED"        android:resource=
"@xml/nfc_tech_filter" />-------------这个就是你的资源文件名
 </activity>
```

（5）创建 NFC 标签的前台分发系统。什么是 NFC 的前台发布系统？就是说当已经打开我们的应用的时候，那么通过这个前台发布系统的设置，可以让已经启动的 Activity 拥有更高的优先级来依据在代码中定义的标准来过滤和处理 Intent，而不是让别的声明了 Intent Filter 的 Activity 来干扰，甚至连自己声明在文件 androidManifest 中的 Intent Filter 都不会来干扰。也就是说，Foreground Dispatch 的优先级大于 Intent Filter。此时有如下所示的两种情况。

- 第一种情况：当 Activity 没有启动的时候去扫描 Tag，那么系统中所有的 Intent Filter 都将一起参与过滤。

● 第二种情况：当 Actiity 启动去扫描 Tag 时，那么将直接使用在 Foreground Dispatch 中代码写入的过滤标准。如果这个标准没有命中任何 Intent，那么系统将使用所有 Activity 声明的 Intent Filter xml 来过滤。

例如，在 OnCreate 中你可以添加如下所示的代码。

```
// Create a generic PendingIntent that will be deliver to this activity. The NFC stack
will fill in the intent with the details of the discovered tag before delivering to this
activity.
   mPendingIntent = PendingIntent.getActivity(this, 0,
             new Intent(this, getClass()).addFlags(Intent.FLAG_ACTIVITY_SINGLE_TOP), 0);
        // 做一个 Intent Filter 过滤你想要的 action，这里过滤的是 ndef
        // IntentFilter ndef = new IntentFilter(NfcAdapter.ACTION_NDEF_DISCOVERED);
//如果对 action 的定义有更高的要求，如 data 的要求，可以使用如下的代码来定义 IntentFilter
//       try {
//            ndef.addDataType("*/*");
//       } catch (MalformedMimeTypeException e) {
//            // TODO Auto-generated catch block
//            e.printStackTrace();
//       }
      //生成 IntentFilter
        mFilters = new IntentFilter[] {
             ndef,
        };
        // 做一个 tech-list，可以看到是二维数据，每一个一维数组之间的关系是或，但是一个一维数组之内
        // 的各个项就是与的关系了
        mTechLists = new String[][] {
             new String[] { NfcF.class.getName()},
             new String[]{NfcA.class.getName()},
             new String[]{NfcB.class.getName()},
             new String[]{NfcV.class.getName()}
             };
```

在 onPause 和 onResume 中需要加入如下相应的代码。

```
public void onPause() {
 super.onPause();
 //反注册 mAdapter.disableForegroundDispatch(this);
}
public void onResume() {
 super.onResume();
 //设定 intentfilter 和 tech-list。如果两个都为 null，就代表优先接收任何形式的 TAG action。也就是
 说系统会主动发 TAG intent。
 mAdapter.enableForegroundDispatch(this, mPendingIntent, mFilters, mTechLists);
}
```

23.4.2　实战演练——使用 NFC 发送消息

当 Android 设备检测到有 NFC Tag 时，预期的行为是触发最合适的 Activity 来处理检测到的 Tag，这是因为 NFC 通常是在非常近的距离才起作用（<4m）。如果在此时需要用户来选择合适的应用来处理 Tag，则很容易断开与 Tag 之间的通信，因此我们需要选择合适的 Intent Filter 只处理你想读写的 Tag 类型。Android 系统支持两种 NFC 消息发送机制，分别是 Intent 发送机制和前台 Activity 消息发送机制。

● Intent 发送机制：当系统检测到 Tag 时，Android 系统提供 Manifest 中定义的 Intent Filter 来选择合适的 Activity 来处理对应的 Tag，当有多个 Activity 可以处理对应的 Tag 类型时，则会显示 Activity 选择窗口由用户选择，如图 23-4 所示。

▲图 23-4　选择窗口

● 前台 Activity 消息发送机制：允许一个在前台运行的 Activity 在读写 NFC Tag 具有优先权，

此时如果 Android 检测到有 NFC Tag，如果前台允许的 Activity 可以处理该种类型的 Tag，则该 Activity 具有优先权，而不出现 Activity 选择窗口。

上述两种方法基本上都是使用 Intent Filter 来指明 Activity 可以处理的 Tag 类型，一个是使用 Android 的 Manifest 来说明，一个是通过代码来声明。

在接下来的内容中，将通过一个具体实例的实现过程，来讲解在 Android 系统中使用 NFC 消息发送机制的基本方法。

实例	功能	源码路径
实例 23-1	演示 NFC 消息发送机制的基本用法	\daima\23\NFCDemo

本实例的具体实现流程如下所示。

（1）在文件 AndroidManifest.xml 中声明 NFC 权限，具体实现代码如下所示。

```
<manifest xmlns:android="http://schemas.android.com/apk/res/android"
     package="com.pstreets.nfc"
     android:versionCode="1"
     android:versionName="1.0">
   <uses-sdk android:minSdkVersion="10" />
    <uses-permission android:name="android.permission.NFC" />
    <uses-feature android:name="android.hardware.nfc" android:required="true" />
   <application android:icon="@drawable/icon" android:label="@string/app_name">
       <activity android:name=".NFCDemoActivity"
               android:label="@string/app_name"
               android:launchMode="singleTop">
           <intent-filter>
               <action android:name="android.intent.action.MAIN" />
               <category android:name="android.intent.category.LAUNCHER" />
           </intent-filter>
           <intent-filter>
             <action android:name="android.nfc.action.NDEF_DISCOVERED"/>
             <data android:mimeType="text/plain" />
           </intent-filter>
           <intent-filter
               >
               <action
                   android:name="android.nfc.action.TAG_DISCOVERED"
                   >
               </action>
               <category
                   android:name="android.intent.category.DEFAULT"
                   >
               </category>
           </intent-filter>
            <!-- Add a technology filter -->
           <intent-filter>
               <action android:name="android.nfc.action.TECH_DISCOVERED" />
           </intent-filter>
           <meta-data android:name="android.nfc.action.TECH_DISCOVERED"
               android:resource="@xml/filter_nfc"
           />
       </activity>
       <activity android:name=".MainActivity"
                android:label="@string/app_name">
           <intent-filter>
               <action android:name="android.intent.action.MAIN" />
               <category android:name="android.intent.category.LAUNCHER" />
           </intent-filter>
       </activity>
    </application>
</manifest>
```

这样通过上述声明代码，当 Android 检测到有 Tag 时，会显示 Activity 选择窗口，就会显示前面图 23-4 的 Reading Example 效果。

（2）编写布局文件 main.xml，功能是通过文本控件显示当前的扫描状态，具体实现代码如下所示。

```
<RelativeLayout
  xmlns:android="http://schemas.android.com/apk/res/android"
  android:layout_width="match_parent"
  android:layout_height="match_parent"
  android:text="@string/title">
  <TableLayout
      android:id="@+id/purchScanTable1"
      android:layout_width="wrap_content"
      android:layout_height="wrap_content">
      <TableRow
          android:id="@+id/table1Row1"
          android:layout_width="wrap_content"
          android:layout_height="wrap_content">
          <TextView
              android:id="@+id/status_label"
              android:layout_width="wrap_content"
              android:layout_height="wrap_content"
              android:text="Current Status:  " />
          <TextView
              android:id="@+id/status_data"
              android:layout_width="wrap_content"
              android:layout_height="wrap_content"
              android:text=" Scan a Tag" />
      </TableRow>
      <View
           android:id="@+id/purchScanES1"
           android:layout_width="match_parent"
          android:layout_height="55px"
          android:layout_below="@id/status_label"
          android:background="#000000" />
      <TableRow
          android:id="@+id/table1Row2"
          android:layout_width="wrap_content"
          android:layout_height="wrap_content">
          <TextView
              android:id="@+id/block_0_label"
              android:layout_width="wrap_content"
              android:layout_height="wrap_content"
              android:text="BLOCK 0:  " />
          <TextView
              android:id="@+id/block_0_data"
              android:layout_width="wrap_content"
              android:layout_height="wrap_content"
              android:text=" " />
      </TableRow>
      <TableRow
          android:id="@+id/table1Row3"
          android:layout_width="wrap_content"
          android:layout_height="wrap_content">
          <TextView
              android:id="@+id/block_1_label"
              android:layout_width="wrap_content"
              android:layout_height="wrap_content"
              android:text="BLOCK 1:  " />
          <TextView
              android:id="@+id/block_1_data"
              android:layout_width="wrap_content"
              android:layout_height="wrap_content"
              android:text=" " />
      </TableRow>
  </TableLayout>
  <View
           android:id="@+id/purchScanES1"
           android:layout_width="match_parent"
          android:layout_height="75px"
          android:layout_below="@id/purchScanTable1"
```

```
        android:background="#000000" />
    <Button
            android:id="@+id/clear_but"
            android:layout_width="fill_parent"
            android:layout_height="wrap_content"
            android:layout_below="@id/purchScanES1"
            android:gravity="center_horizontal"
            android:text="Clear" />
</RelativeLayout>
```

（3）编写程序文件 NFCDemoActiviy.java，当在前台运行 NFCDemoActiviy 时，如果希望只有它来处理 Mifare 类型的 Tag，此时可以使用前台消息发送机制。文件 NFCDemoActiviy.java 的具体实现代码如下所示。

```
public class NFCDemoActivity extends Activity {
    private NfcAdapter mAdapter;
    private PendingIntent mPendingIntent;
    private IntentFilter[] mFilters;
    private String[][] mTechLists;
    private TextView mText;
    private int mCount = 0;

    @Override
    public void onCreate(Bundle savedState) {
        super.onCreate(savedState);

        setContentView(R.layout.foreground_dispatch);
        mText = (TextView) findViewById(R.id.text);
        mText.setText("Scan a tag");

        mAdapter = NfcAdapter.getDefaultAdapter(this);

        // Create a generic PendingIntent that will be deliver to this activity. The NFC stack
        // will fill in the intent with the details of the discovered tag before delivering to
        // this activity.
        mPendingIntent = PendingIntent.getActivity(this, 0,
                new Intent(this, getClass()).addFlags(Intent.FLAG_ACTIVITY_SINGLE_TOP), 0);

        // Setup an intent filter for all MIME based dispatches
        IntentFilter ndef = new IntentFilter(NfcAdapter.ACTION_TECH_DISCOVERED);
        try {
            ndef.addDataType("*/*");
        } catch (MalformedMimeTypeException e) {
            throw new RuntimeException("fail", e);
        }
        mFilters = new IntentFilter[] {
                ndef,
        };

        // Setup a tech list for all MifareClassic tags
        mTechLists = new String[][] { new String[] { MifareClassic.class.getName() } };
    }

    @Override
    public void onResume() {
        super.onResume();
        mAdapter.enableForegroundDispatch(this, mPendingIntent, mFilters, mTechLists);
    }

    @Override
    public void onNewIntent(Intent intent) {
        Log.i("Foreground dispatch", "Discovered tag with intent: " + intent);
        mText.setText("Discovered tag " + ++mCount + " with intent: " + intent);
    }

    @Override
    public void onPause() {
        super.onPause();
```

```
        mAdapter.disableForegroundDispatch(this);
    }
}
```

这样在执行本实例后，每当靠近一次 Tag，计数就会增加 1。执行效果如图 23-5 所示。

23.4.3　实战演练——使用 NFC 读写 Mifare Tag

Mifare Tag 可以有 1KB、2KB、4KB，其内存分区大同小异，下面的图 23-6 给出了 1KB 容量的 Tag 的内存分布。

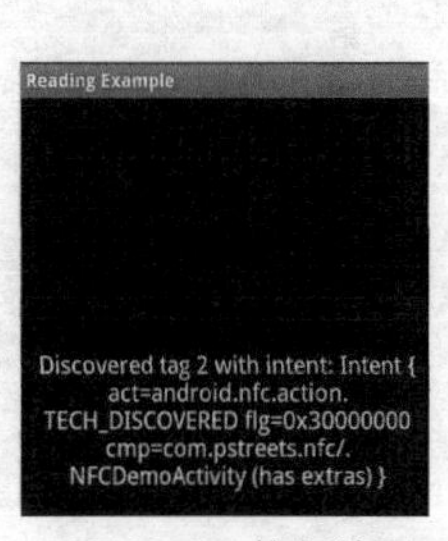

▲图 23-5　执行效果

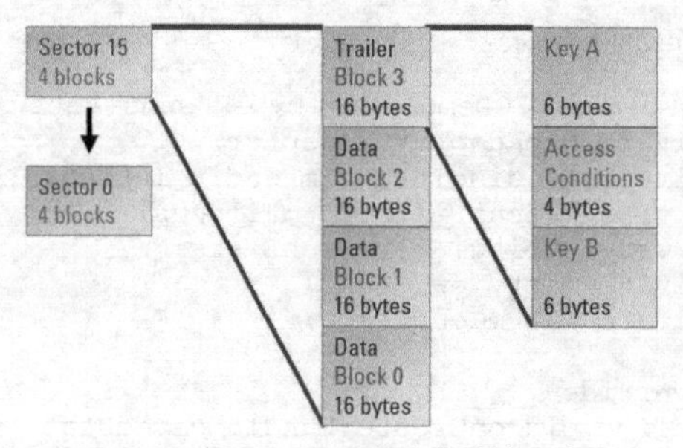

▲图 23-6　Mifare Tag 的内存分布示意图

在上图中，数据被分为 16 个区（Sector），每个区有 4 个块（Block），每个块可以存放 16 字节的数据，其大小为 16×4×16 =1024B。每个区最后一个块称为 Trailer，主要用来存放读写该区 Block 数据的 Key，可以有 A 和 B 两个 Key，每个 Key 长度为 6 个字节，缺省的 Key 值一般为全 FF 或是 0，由 MifareClassic.KEY_DEFAULT 定义。因此在读写 Mifare Tag 时，首先需要有正确的 Key 值（起到保护的作用），只有在鉴权成功后才可以读写该区数据。

在接下来的内容中，将通过一个具体实例的实现过程，来讲解在 Android 系统中使用 NFC 读写 Mifare Tag 的基本方法。

实例	功能	源码路径
实例 23-2	使用 NFC 读写 Mifare Tag	\daima\23\MifareDemo

本实例的具体实现流程如下所示。

（1）在文件 AndroidManifest.xml 中声明 NFC 权限，然后编写布局文件 main.xml，通过文本控件显示当前的扫描状态。

（2）为了方便读写 Mifare Tag，在本实例中定义了 4 个 Mifare 相关类，分别是 MifareClassCard、MifareSector、MifareBlock 和 MifareKey。并将检测到的 NFC Tag 封装成 Tag 类，存放到 Intent 的 NfcAdapter.EXTRA_TAG Extra 数据包中，我们可以使用 MifareClassic.get(Tag)获取对象的 MifareClassic 类。最终会通过文件 MainActivity.java 在屏幕中显示读取的结果，具体实现代码如下所示。

```
public class MainActivity extends ListActivity {
    // NFC parts
    private static NfcAdapter mAdapter;
    private static PendingIntent mPendingIntent;
    private static IntentFilter[] mFilters;
    private static String[][] mTechLists;
    private static final int AUTH = 1;
    private static final int EMPTY_BLOCK_0 = 2;
    private static final int EMPTY_BLOCK_1 = 3;
    private static final int NETWORK = 4;
    private static final String TAG = "mifare";
    /** Called when the activity is first created. */
```

```
@Override
public void onCreate(Bundle savedInstanceState) {
    super.onCreate(savedInstanceState);
    setContentView(R.layout.main);
    // Capture Purchase button from layout
    Button scanBut = (Button) findViewById(R.id.clear_but);
    // Register the onClick listener with the implementation above
    scanBut.setOnClickListener(new OnClickListener(){
        public void onClick(View view) {
            clearFields();
        }});
    mAdapter = NfcAdapter.getDefaultAdapter(this);
    // Create a generic PendingIntent that will be deliver to this activity.
    // The NFC stack
    // will fill in the intent with the details of the discovered tag before
    // delivering to
    // this activity.
    mPendingIntent = PendingIntent.getActivity(this, 0, new Intent(this,
            getClass()).addFlags(Intent.FLAG_ACTIVITY_SINGLE_TOP), 0);
    // Setup an intent filter for all MIME based dispatches
    IntentFilter ndef = new IntentFilter(NfcAdapter.ACTION_TECH_DISCOVERED);
    try {
        ndef.addDataType("*/*");
    } catch (MalformedMimeTypeException e) {
        throw new RuntimeException("fail", e);
    }
    mFilters = new IntentFilter[] { ndef, };
    // Setup a tech list for all NfcF tags
    mTechLists = new String[][] { new String[] { MifareClassic.class
            .getName() } };
    Intent intent = getIntent();
    resolveIntent(intent);
}
void resolveIntent(Intent intent) {
    // 1) Parse the intent and get the action that triggered this intent
    String action = intent.getAction();
    // 2) Check if it was triggered by a tag discovered interruption.
    if (NfcAdapter.ACTION_TECH_DISCOVERED.equals(action)) {
        // 3) Get an instance of the TAG from the NfcAdapter
        Tag tagFromIntent = intent.getParcelableExtra(NfcAdapter.EXTRA_TAG);
        // 4) Get an instance of the Mifare classic card from this TAG
        // intent
        MifareClassic mfc = MifareClassic.get(tagFromIntent);
        MifareClassCard mifareClassCard=null;
        try { // 5.1) Connect to card
            mfc.connect();
            boolean auth = false;
            // 5.2) and get the number of sectors this card has..and loop
            // thru these sectors
            int secCount = mfc.getSectorCount();
            mifareClassCard= new MifareClassCard(secCount);
            int bCount = 0;
            int bIndex = 0;
            for (int j = 0; j < secCount; j++) {
                MifareSector mifareSector = new MifareSector();
                mifareSector.sectorIndex = j;
                // 6.1) authenticate the sector
                auth = mfc.authenticateSectorWithKeyA(j,
                        MifareClassic.KEY_DEFAULT);
                mifareSector.authorized = auth;
                if (auth) {
                    // 6.2) In each sector - get the block count
                    bCount = mfc.getBlockCountInSector(j);
                    bCount =Math.min(bCount, MifareSector.BLOCKCOUNT);
                    bIndex = mfc.sectorToBlock(j);
                    for (int i = 0; i < bCount; i++) {
                        // 6.3) Read the block
                        byte []data = mfc.readBlock(bIndex);
                        MifareBlock mifareBlock = new MifareBlock(data);
```

```
                    mifareBlock.blockIndex = bIndex;
                    // 7) Convert the data into a string from Hex
                    // format.
                    bIndex++;
                    mifareSector.blocks[i] = mifareBlock;
                }
                mifareClassCard.setSector(mifareSector.sectorIndex,
                        mifareSector);
            } else { // Authentication failed - Handle it
            }
        }
        ArrayList<String> blockData=new ArrayList<String>();
        int blockIndex=0;
        for(int i=0;i<secCount;i++){

            MifareSector mifareSector=mifareClassCard.getSector(i);
            for(int j=0;j<MifareSector.BLOCKCOUNT;j++){
                MifareBlock mifareBlock=mifareSector.blocks[j];
                byte []data=mifareBlock.getData();
                blockData.add("Block "+ blockIndex++ +" : "+
                Converter.getHexString(data, data.length));
            }
        }
        String []contents=new String[blockData.size()];
        blockData.toArray(contents);
         setListAdapter(new ArrayAdapter<String>(this,
                android.R.layout.simple_list_item_1, contents));
        getListView().setTextFilterEnabled(true);

    } catch (IOException e) {
        Log.e(TAG, e.getLocalizedMessage());
        showAlert(3);
    }finally{
        if(mifareClassCard!=null){
            mifareClassCard.debugPrint();
        }
    }
  }// End of method
}
private void showAlert(int alertCase) {
    // prepare the alert box
    AlertDialog.Builder alertbox = new AlertDialog.Builder(this);
    switch (alertCase) {
    case AUTH:// Card Authentication Error
        alertbox.setMessage("Authentication Failed ");
        break;
    case EMPTY_BLOCK_0: // Block 0 Empty
        alertbox.setMessage("Failed reading ");
        break;
    case EMPTY_BLOCK_1:// Block 1 Empty
        alertbox.setMessage("Failed reading 0");
        break;
    case NETWORK: // Communication Error
        alertbox.setMessage("Tag reading error");

        break;
    }
    // set a positive/yes button and create a listener
    alertbox.setPositiveButton("OK", new DialogInterface.OnClickListener() {

        // Save the data from the UI to the database - already done
        public void onClick(DialogInterface arg0, int arg1) {
            clearFields();
        }
    });
    // display box
    alertbox.show();
}
private void clearFields() {
```

```
    }
    @Override
    public void onResume() {
        super.onResume();
        mAdapter.enableForegroundDispatch(this, mPendingIntent, mFilters,
                mTechLists);
    }
    @Override
    public void onNewIntent(Intent intent) {
        Log.i("Foreground dispatch", "Discovered tag with intent: " + intent);
        resolveIntent(intent);
        // mText.setText("Discovered tag " + ++mCount + " with intent: " +
        // intent);
    }
    @Override
    public void onPause() {
        super.onPause();
        mAdapter.disableForegroundDispatch(this);
    }
}
```

本实例执行后将在屏幕中显示读取结果，这样便完成了 NFC 通信功能。执行效果如图 23-7 所示。

▲图 23-7　执行效果

第 24 章　拍照解析条形码技术详解

在当前手机系统中，无论是智能手机还是普通的手机，基本上都支持手机拍照和录制视频功能，这些功能都是通过手机上的摄像头实现的。在 Android 系统中，上述照相机功能是通过 Camera 系统实现的。另外，通过摄像头可以实现条形码解析技术。本章将详细讲解在 Android 设备中使用摄像头条形码技术的基本知识，为步入本书后面知识的学习打下基础。

24.1 Android 拍照系统介绍

在 Android 系统中，Camera（照相机/拍照）系统提供了取景器、视频录制和拍摄相片等功能，并且还具有各种控制类的接口。另外，在 Camera 系统中还提供了 Java 层的接口和本地接口。其中 Java 框架中的 Camera 类实现了 Java 层相机接口，为照相机类和扫描类使用。而 Camera 的本地接口可以给本地程序调用，作为视频输入环节应用于摄像机和视频电话领域。

Android 照相机系统的基本层次结构如图 24-1 所示。

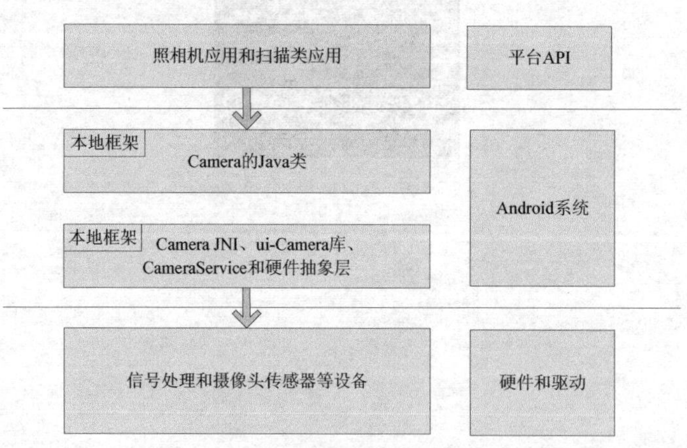

▲图 24-1　照相机系统的层次结构

Android 系统中的 Camera 系统包括了 Camera 驱动程序层、Camera 硬件抽象层、AudioService、Camera 本地库、Camera 的 Java 框架类和 Java 应用层对 Camera 系统的调用。Camera 系统的具体结构如图 24-2 所示。

在如图 24-2 所示的结构中，各个构成层次的具体说明如下所示。

（1）Camera 系统的 Java 层，代码路径如下：

```
frameworks/base/core/java/android/hardware/
```

其中文件 Camera.java 是主要实现的文件，对应的 Java 层次的类是 android.hardware.Camera，这个类是和 JNI 中定义的类是一个，有些方法通过 JNI 的方式调用本地代码得到，有些方法自己实现。

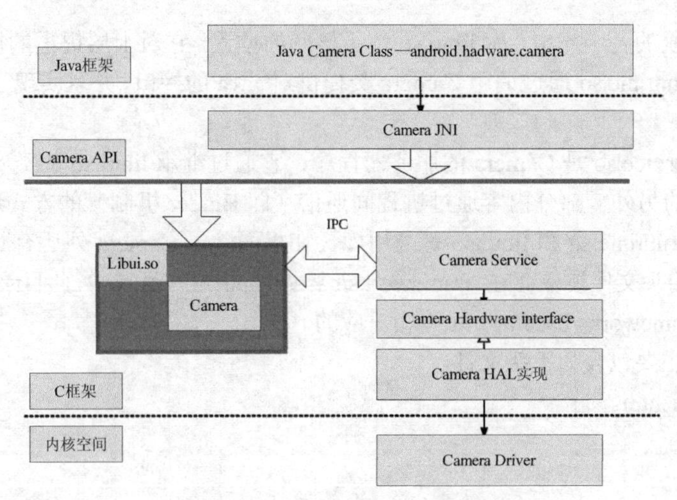

▲图 24-2 Camera 的系统结构

（2）Camera 系统的 Java 本地调用部分（JNI），代码路径如下：

```
frameworks/base/core/jni/android_hardware_Camera.cpp
```

这部分内容编译成为目标文件 libandroid_runtime.so，主要的头文件在以下的目录中：

```
frameworks/base/include/ui/
```

（3）Camera 本地框架，其中头文件路径如下：

```
frameworks/native/include/ui
```

或：

```
frameworks/av/include/camera/
```

源代码路径如下：

```
frameworks/native/libs/ui
```

或：

```
frameworks/av/camera/
```

这部分的内容被编译成库 libui.so 或 libcamera_client.so。

（4）Camera 服务部分，代码路径如下：

```
frameworks/av/services/camera/libcameraservice/
```

这部分内容被编译成库 libcameraservice.so。

为了实现一个具体功能的 Camera 驱动程序，在最底层还需要一个硬件相关的 Camera 库（例如，通过调用 video for linux 驱动程序和 Jpeg 编码程序实现）。这个库将被 Camera 的服务库 libcameraservice.so 调用。

（5）摄像头驱动程序

此部分是基于 Linux 的 Video for Linux 视频驱动框架。

（6）硬件抽象层

硬件抽象层中的接口代码路径如下：

```
frameworks/base/include/ui/
```

或：

```
frameworks/av/include/camera/
```

其中的核心文件是 CameraHardwareInterface.h。

在 Camera 系统的各个库中，库 libui.so 位于核心的位置，它对上层提供的接口主要是 Camera 类，类 libandroid_runtime.so 通过调用 Camera 类提供对 Java 的接口，并且实现了 android.hardware.camera 类。

库 libcameraservice.so 是 Camera 的服务器程序，它通过继承 libui.so 的类实现服务器的功能，并且与 libui.so 中的另外一部分内容通过进程间通信（即 Binder 机制）的方式进行通信。

库 libandroid_runtime.so 和 libui.so 是公用库，在里面除了 Camera 外还有其他方面的功能。

Camera 部分的头文件被保存在"frameworks/base/include/ui/"目录中，此目录是和库 libmedia.so 的源文件目录"frameworks/base/libs/ui/"相对应的。

在 Camera 中主要包含如下头文件。

- ICameraClient.h。
- Camera.h。
- ICamera.h。
- ICameraService.h。
- CameraHardwareInterface.h。

文件 Camera.h 提供了对上层的接口，而其他的几个头文件都是提供一些接口类（即包含了纯虚函数的类），这些接口类必须被实现类继承才能够使用。

当整个 Camera 在运行的时候，可以大致上分成 Client 和 Server 两个部分，它们分别在两个进程中运行，它们之间使用 Binder 机制实现进程间通信。这样在客户端调用接口，功能则在服务器中实现，但是在客户端中调用就好像直接调用服务器中的功能，进程间通信的部分对上层程序不可见。

从框架结构上来看，文件 ICameraService.h、ICameraClient.h 和 ICamera.h 中的三个类定义了 Camera 的接口和架构，ICameraService.cpp 和 Camera.cpp 两个文件用于实现 Camera 架构，Camera 的具体功能在下层调用硬件相关的接口来实现。

24.1.1　分析拍照系统的底层程序

在 Linux 系统中，Camera 驱动程序使用了 Linux 标准的 Video for Linux 2（V4L2）驱动程序。无论是内核空间还是用户空间，都使用 V4L2 驱动程序框架来定义数据类和控制类。所以在移植 Android 中的 Camera 系统时，也是用标准的 V4L2 驱动程序作为 Camera 的驱动程序。在 Camera 系统中，V4L2 驱动程序的任务是获得 Video 数据。

1. V4L2 API

V4L2 是 V4L 的升级版本，为 Linux 下视频设备程序提供了一套接口规范，包括一套数据结构和底层 V4L2 驱动接口。V4L2 驱动程序向用户空间提供字符设备，主设备号是 81。对于视频设备来说，次设备号是 0～63。如果次设备号在 64～127 之间的是 Radio 设备，次设备号在 192～223 之间的是 Teletext 设备，次设备号在 224～255 之间的是 VBI 设备，

V4L2 中常用的结构体在内核文件"include/linux/videodev2.h"中定义，代码如下所示。

```
struct v4l2_requestbuffers                  //申请帧缓冲，对应命令 VIDIOC_REQBUFS
       struct v4l2_capability               //视频设备的功能，对应命令 VIDIOC_QUERYCAP
       struct v4l2_input                    //视频输入信息，对应命令 VIDIOC_ENUMINPUT
       struct v4l2_standard                 //视频的制式，如 PAL，NTSC，对应命令 VIDIOC_ENUMSTD
       struct v4l2_format                   //帧的格式，对应命令 VIDIOC_G_FMT、VIDIOC_S_FMT 等
       struct v4l2_buffer                   //驱动中的一帧图像缓存，对应命令 VIDIOC_QUERYBUF
       struct v4l2_crop                     //视频信号矩形边框
       v4l2_std_id                          //视频制式
```

常用的 ioctl 接口命令也在文件“include/linux/videodev2.h”中定义，代码如下所示。

```
VIDIOC_REQBUFS //分配内存
       VIDIOC_QUERYBUF         //把 VIDIOC_REQBUFS 中分配的数据缓存转换成物理地址
       VIDIOC_QUERYCAP         //查询驱动功能
       VIDIOC_ENUM_FMT         //获取当前驱动支持的视频格式
       VIDIOC_S_FMT           //设置当前驱动的频捕获格式
       VIDIOC_G_FMT           //读取当前驱动的频捕获格式
       VIDIOC_TRY_FMT          //验证当前驱动的显示格式
       VIDIOC_CROPCAP           //查询驱动的修剪能力
       VIDIOC_S_CROP           //设置视频信号的矩形边框
       VIDIOC_G_CROP          //读取视频信号的矩形边框
       VIDIOC_QBUF            //把数据从缓存中读取出来
       VIDIOC_DQBUF           //把数据放回缓存队列
       VIDIOC_STREAMON         //开始视频显示函数
       VIDIOC_STREAMOFF          //结束视频显示函数
       VIDIOC_QUERYSTD           //检查当前视频设备支持的标准，如 PAL 或 NTSC
```

2. 操作 V4L2 的流程

在 V4L2 中提供了很多访问接口，我们可以根据具体需要选择操作方法。不过需要注意的是，很少有驱动完全实现了所有的接口功能。所以在使用时需要参考驱动源码，或仔细阅读驱动提供者的使用说明。接下来就简单列举出一种 V4L2 的操作流程供读者朋友们参考。

（1）打开设备文件，具体代码如下所示。

```
int fd = open(Devicename,mode);
Devicename: /dev/video0、/dev/video1 ……
Mode: O_RDWR [| O_NONBLOCK]
```

如果需要使用非阻塞模式调用视频设备，当没有可用的视频数据时不会阻塞，而会立刻返回。

（2）获取设备的 capability，具体代码如下所示。

```
struct v4l2_capability capability;
int ret = ioctl(fd, VIDIOC_QUERYCAP, &capability);
```

在此需要查看设备具有什么功能，如是否具有视频输入特性。

（3）选择视频输入，代码如下。

```
struct v4l2_input input;
//……开始初始化 input
int ret = ioctl(fd, VIDIOC_QUERYCAP, &input);
```

每一个视频设备可以有多个视频输入，如果只有一路输入，则可以没有这个功能。

（4）检测视频支持的制式，具体代码如下所示。

```
v4l2_std_id std;
do {
            ret = ioctl(fd, VIDIOC_QUERYSTD, &std);
 } while (ret == -1 && errno == EAGAIN);
              switch (std) {
              case V4L2_STD_NTSC:
            //……
              case V4L2_STD_PAL:
            //……
              }
```

（5）设置视频捕获格式，具体代码如下所示。

```
struct v4l2_format fmt;
fmt.type = V4L2_BUF_TYPE_VIDEO_OUTPUT;
fmt.fmt.pix.pixelformat = V4L2_PIX_FMT_UYVY;
fmt.fmt.pix.height = height;
fmt.fmt.pix.width = width;
fmt.fmt.pix.field = V4L2_FIELD_INTERLACED;
ret = ioctl(fd, VIDIOC_S_FMT, &fmt);
if(ret) {
```

```
        perror("VIDIOC_S_FMT\n");
        close(fd);
        return -1;
}
```

（6）向驱动申请帧缓存，具体代码如下所示。

```
struct v4l2_requestbuffers req;
 if (ioctl(fd, VIDIOC_REQBUFS, &req) == -1) {
        return -1;
 }
```

在结构 v4l2_requestbuffers 中定义了缓存的数量，驱动会根据这个申请对应数量的视频缓存。通过多个缓存可以建立 FIFO，这样可以提高视频采集的效率。

（7）获取每个缓存的信息，并 mmap 到用户空间。具体代码如下所示。

```
typedef struct VideoBuffer {
        void *start;
        size_t length;
} VideoBuffer;
          VideoBuffer* buffers = calloc( req.count, sizeof(*buffers) );
struct v4l2_buffer buf;
          for (numBufs = 0; numBufs < req.count; numBufs++) {//映射所有的缓存
        memset( &buf, 0, sizeof(buf) );
        buf.type = V4L2_BUF_TYPE_VIDEO_CAPTURE;
        buf.memory = V4L2_MEMORY_MMAP;
        buf.index = numBufs;
        if (ioctl(fd, VIDIOC_QUERYBUF, &buf) == -1) {//获取到对应 index 的缓存信息，此处主要
利用 length 信息及 offset 信息来完成后面的 mmap 操作
                return -1;
        }
buffers[numBufs].length = buf.length;
        // 转换成相对地址
        buffers[numBufs].start = mmap(NULL, buf.length,
                PROT_READ | PROT_WRITE,
                MAP_SHARED,
                fd, buf.m.offset);
 if (buffers[numBufs].start == MAP_FAILED) {
                return -1;
        }
```

（8）开始采集视频，具体代码如下所示。

```
int buf_type= V4L2_BUF_TYPE_VIDEO_CAPTURE;
int ret = ioctl(fd, VIDIOC_STREAMON, &buf_type);
```

（9）取出 FIFO 缓存中已经采样的帧缓存，具体代码如下所示。

```
struct v4l2_buffer buf;
memset(&buf,0,sizeof(buf));
buf.type=V4L2_BUF_TYPE_VIDEO_CAPTURE;
buf.memory=V4L2_MEMORY_MMAP;
buf.index=0;//此值由下面的 ioctl 返回
if (ioctl(fd, VIDIOC_DQBUF, &buf) == -1)
{
        return -1;
}
```

通过上述代码，可以根据返回的 buf.index 找到对应的 mmap 映射好的缓存，实现取出视频数据的功能。

（10）将刚刚处理完的缓冲重新入队列尾，这样可以循环采集，具体代码如下所示。

```
if (ioctl(fd, VIDIOC_QBUF, &buf) == -1) {
        return -1;
}
```

（11）停止视频的采集，具体代码如下所示。

```
int ret = ioctl(fd, VIDIOC_STREAMOFF, &buf_type);
```

（12）关闭视频设备，具体代码如下所示。

```
close(fd);
```

3. V4L2 驱动框架

在上述使用 V4L2 的流程中，各个操作都需要有底层 V4L2 驱动的支持。在内核中有一些非常完善的例子。例如，在 Linux-2.6.26 内核目录“/drivers/media/video//zc301/”中，文件 zc301_core.c 实现了 ZC301 视频驱动代码。

（1）V4L2 驱动注册、注销函数

在 Video 核心层文件“drivers/media/video/videodev.c”中提供了注册函数，具体代码如下所示。

```
int video_register_device(struct video_device *vfd, int type, int nr)
```

- video_device：要构建的核心数据结构。
- Type：表示设备类型，此设备号的基地址受此变量的影响。
- Nr：如果 end-base>nr>0：次设备号=base（基准值，受 type 影响）+nr；否则将系统自动分配合适的次设备号。

我们具体需要的驱动只需构建 video_device 结构，然后调用注册函数既可。例如，在文件 zc301_core.c 中的实现代码如下所示。

```
err = video_register_device(cam->v4ldev, VFL_TYPE_GRABBER,
                            video_nr[dev_nr]);
```

在 Video 核心层文件“drivers/media/video/videodev.c”中提供了如下注销函数。

```
void video_unregister_device(struct video_device *vfd)
```

（2）构建 struct video_device

在结构 video_device 中包含了视频设备的属性和操作方法，具体可以参考文件 zc301_core.c，代码如下所示。

```
strcpy(cam->v4ldev->name, "ZC0301[P] PC Camera");
 cam->v4ldev->owner = THIS_MODULE;
 cam->v4ldev->type = VID_TYPE_CAPTURE | VID_TYPE_SCALES;
 cam->v4ldev->fops = &zc0301_fops;
 cam->v4ldev->minor = video_nr[dev_nr];
 cam->v4ldev->release = video_device_release;
 video_set_drvdata(cam->v4ldev, cam);
```

在上述 zc301 的驱动中并没有实现 struct video_device 中的很多操作函数，例如，vidioc_querycap、vidioc_g_fmt_cap，这是因为在 struct file_operations zc0301_fops 中的 zc0301_ioctl 实现了前面的所有 ioctl 操作，所以无需在 struct video_device 再次实现 struct video_device 中的操作。

另外也可以使用下面的代码来构建 struct video_device。

```
static struct video_device camif_dev =
 {
   .name = "s3c2440 camif",
   .type = VID_TYPE_CAPTURE|VID_TYPE_SCALES|VID_TYPE_SUBCAPTURE,
   .fops = &camif_fops,
   .minor = -1,
   .release = camif_dev_release,
   .vidioc_querycap = vidioc_querycap,
   .vidioc_enum_fmt_cap = vidioc_enum_fmt_cap,
   .vidioc_g_fmt_cap = vidioc_g_fmt_cap,
   .vidioc_s_fmt_cap = vidioc_s_fmt_cap,
   .vidioc_queryctrl = vidioc_queryctrl,
   .vidioc_g_ctrl = vidioc_g_ctrl,
   .vidioc_s_ctrl = vidioc_s_ctrl,
 };
 static struct file_operations camif_fops =
```

```
{
  .owner = THIS_MODULE,
  .open = camif_open,
  .release = camif_release,
  .read = camif_read,
  .poll = camif_poll,
  .ioctl = video_ioctl2, /* V4L2 ioctl handler */
  .mmap = camif_mmap,
  .llseek = no_llseek,
};
```

结构 video_ioctl2 在文件 videodev.c 中是实现的，video_ioctl2 中会根据 ioctl 不同的 cmd 来调用 video_device 中的操作方法。

4. 实现 Video 核心层

具体实现代码请参考内核文件“/drivers/media/video/videodev.c”，实现流程如下所示。

（1）注册 256 个视频设备，代码如下所示。

```
static int __init videodev_init(void)
        {
                int ret;
                if (register_chrdev(VIDEO_MAJOR, VIDEO_NAME, &video_fops)) {
                        return -EIO;
                }
                ret = class_register(&video_class);
                ......
        }
```

在上述代码中注册了 256 个视频设备和 video_class 类，video_fops 为这 256 个设备共同的操作方法。

（2）实现 V4L2 驱动的注册函数，具体代码如下所示。

```
int video_register_device(struct video_device *vfd, int type, int nr)
        {
                int i=0;
                int base;
                int end;
                int ret;
                char *name_base;
        switch(type) //根据不同的type确定设备名称、次设备号
                {
                        case VFL_TYPE_GRABBER:
                                base=MINOR_VFL_TYPE_GRABBER_MIN;
                                end=MINOR_VFL_TYPE_GRABBER_MAX+1;
                                name_base = "video";
                                break;
                        case VFL_TYPE_VTX:
                                base=MINOR_VFL_TYPE_VTX_MIN;
                                end=MINOR_VFL_TYPE_VTX_MAX+1;
                                name_base = "vtx";
                                break;
                        case VFL_TYPE_VBI:
                                base=MINOR_VFL_TYPE_VBI_MIN;
                                end=MINOR_VFL_TYPE_VBI_MAX+1;
                                name_base = "vbi";
                                break;
                        case VFL_TYPE_RADIO:
                                base=MINOR_VFL_TYPE_RADIO_MIN;
                                end=MINOR_VFL_TYPE_RADIO_MAX+1;
                                name_base = "radio";
                                break;
                        default:
                                printk(KERN_ERR "%s called with unknown type: %d\n",
                                        __func__, type);
                                return -1;
```

```
        }
    /* 计算出次设备号 */
        mutex_lock(&videodev_lock);
        if (nr >= 0 && nr < end-base) {
            /* use the one the driver asked for */
            i = base+nr;
            if (NULL != video_device[i]) {
                mutex_unlock(&videodev_lock);
                return -ENFILE;
            }
        } else {
            /* use first free */
            for(i=base;i<end;i++)
                if (NULL == video_device[i])
                    break;
            if (i == end) {
                mutex_unlock(&videodev_lock);
                return -ENFILE;
            }
        }
       video_device[i]=vfd; //保存 video_device 结构指针到系统的结构数组中，最终的次设备
                            //号和 i 相关
       vfd->minor=i;
       mutex_unlock(&videodev_lock);
       mutex_init(&vfd->lock);
    /* sysfs class */
        memset(&vfd->class_dev, 0x00, sizeof(vfd->class_dev));
        if (vfd->dev)
            vfd->class_dev.parent = vfd->dev;
      vfd->class_dev.class = &video_class;
      vfd->class_dev.devt = MKDEV(VIDEO_MAJOR, vfd->minor);
      sprintf(vfd->class_dev.bus_id, "%s%d", name_base, i - base);
      //最后在/dev 目录下的名称
  ret = device_register(&vfd->class_dev);//结合 udev 或 mdev 可以实现自动在/dev 下创建设备节点
    ……
    }
```

从上面的注册函数代码中可以看出，注册 V4L2 驱动的过程只是创建了设备节点，例如，“/dev/video0”，并且保存了 video_device 结构指针。

（3）打开视频驱动。当使用下面的代码在用户空间调用 open()函数打开对应的视频文件时：

```
int fd = open(/dev/video0, O_RDWR);
```

对应“/dev/video0”目录的文件操作结构是在文件“/drivers/media/video/videodev.c”中定义的 video_fops。代码如下所示。

```
static const struct file_operations video_fops=
    {
        .owner = THIS_MODULE,
        .llseek = no_llseek,
        .open = video_open,
    };
```

上述代码只是实现了 open 操作，后面的其他操作需要使用 video_open()来实现，具体代码如下所示。

```
static int video_open(struct inode *inode, struct file *file)
    {
        unsigned int minor = iminor(inode);
        int err = 0;
        struct video_device *vfl;
        const struct file_operations *old_fops;
    if(minor>=VIDEO_NUM_DEVICES)
            return -ENODEV;
        mutex_lock(&videodev_lock);
        vfl=video_device[minor];
        if(vfl==NULL) {
            mutex_unlock(&videodev_lock);
```

```
                request_module("char-major-%d-%d", VIDEO_MAJOR, minor);
                mutex_lock(&videodev_lock);
                vfl=video_device[minor]; //根据次设备号取出 video_device 结构
                if (vfl==NULL) {
                        mutex_unlock(&videodev_lock);
                        return -ENODEV;
                }
        }
        old_fops = file->f_op;
        file->f_op = fops_get(vfl->fops);//替换此打开文件的 file_operation 结构。后面
        //的其他针对此文件的操作都由新的结构来负责。也就是由每个具体的 video_device 的 fops 负责
        if(file->f_op->open)
                err = file->f_op->open(inode,file);
        if (err) {
                fops_put(file->f_op);
                file->f_op = fops_get(old_fops);
        }
        ......
}
```

24.1.2　分析拍照系统的硬件抽象层

在 Android 2.1 及其以前的版本中，Camera 系统的硬件抽象层的头文件保存在如下目录中。

```
frameworks/base/include/ui/
```

在 Android 2.2 及其以后的版本中，Camera 系统的硬件抽象层的头文件保存在如下目录中。

```
frameworks/av/include/camera/
```

在上述目录中主要包含了如下头文件。

- CameraHardwareInterface.h：在里面定义了 C++接口类，此类需要根据系统的情况来实现继承。
- CameraParameters.h：在里面定义了 Camera 系统的参数，可以在本地系统的各个层次中使用这些参数。
- Camera.h：在里面提供了 Camera 系统本地对上层的接口。

1. Android 2.1 及其以前的版本

在 Android 2.1 及其以前的版本中，在文件 CameraHardwareInterface.h 中首先定义了硬件抽象层接口的回调函数类型，对应代码如下所示。

```
/** startPreview()使用的回调函数*/
typedef void (*preview_callback)(const sp<IMemory>& mem, void* user);

/** startRecord()使用的回调函数*/
typedef void (*recording_callback)(const sp<IMemory>& mem, void* user);

/** takePicture()使用的回调函数*/
typedef void (*shutter_callback)(void* user);

/** takePicture()使用的回调函数*/
typedef void (*raw_callback)(const sp<IMemory>& mem, void* user);

/** takePicture()使用的回调函数*/
typedef void (*jpeg_callback)(const sp<IMemory>& mem, void* user);

/** autoFocus()使用的回调函数*/
typedef void (*autofocus_callback)(bool focused, void* user);
```

然后定义类 CameraHardwareInterface，并在类中定义了各个接口函数。具体代码如下所示。

```
class CameraHardwareInterface : public virtual RefBase {
public:
    virtual ~CameraHardwareInterface() { }
    virtual sp<IMemoryHeap>         getPreviewHeap() const = 0;
    virtual sp<IMemoryHeap>         getRawHeap() const = 0;
```

```
    virtual status_t    startPreview(preview_callback cb, void* user) = 0;
    virtual bool useOverlay() {return false;}
    virtual status_t setOverlay(const sp<Overlay> &overlay) {return BAD_VALUE;}
    virtual void        stopPreview() = 0;
    virtual bool        previewEnabled() = 0;
    virtual status_t    startRecording(recording_callback cb, void* user) = 0;
    virtual void        stopRecording() = 0;
    virtual bool        recordingEnabled() = 0;
    virtual void        releaseRecordingFrame(const sp<IMemory>& mem) = 0;
    virtual status_t    autoFocus(autofocus_callback,
                               void* user) = 0;
    virtual status_t    takePicture(shutter_callback,
                                 raw_callback,
                                 jpeg_callback,
                                 void* user) = 0;
    virtual status_t    cancelPicture(bool cancel_shutter,
                                   bool cancel_raw,
                                   bool cancel_jpeg) = 0;
    /** Return the camera parameters. */
    virtual CameraParameters  getParameters() const = 0;
    virtual void release() = 0;
    virtual status_t dump(int fd, const Vector<String16>& args) const = 0;
};
extern "C" sp<CameraHardwareInterface> openCameraHardware();
};
```

可以将上述代码中的接口分为如下几类。

- 取景预览：startPreview、stopPreview、useOverlay 和 setOverlay。
- 录制视频：startRecording、stopRecording、recordingEnabled 和 releaseRecordingFrame。
- 拍摄照片：takePicture 和 cancelPicture。
- 辅助功能：autoFocus（自动对焦）、setParameters 和 getParameters。

2. Android 2.2 及其以后的版本

在 Android 2.2 及其以后的版本中，在文件 Camera.h 中首先定义了通知信息的枚举值，对应代码如下所示。

```
enum {
    CAMERA_MSG_ERROR            = 0x001,   //错误信息
    CAMERA_MSG_SHUTTER           = 0x002,  //快门信息
    CAMERA_MSG_FOCUS            = 0x004,  //聚焦信息
    CAMERA_MSG_ZOOM            = 0x008,  //缩放信息
    CAMERA_MSG_PREVIEW_FRAME    = 0x010,  //帧预览信息
    CAMERA_MSG_VIDEO_FRAME     = 0x020,  //视频帧信息
    CAMERA_MSG_POSTVIEW_FRAME   = 0x040,  //拍照后停止帧信息
    CAMERA_MSG_RAW_IMAGE       = 0x080,  //原始数据格式照片信息
    CAMERA_MSG_COMPRESSED_IMAGE = 0x100,  //压缩格式照片信息
    CAMERA_MSG_ALL_MSGS        = 0x1FF  //所有信息
};
```

然后在文件 CameraHardwareInterface.h 中定义如下三个回调函数。

```
//通知回调
typedef void (*notify_callback)(int32_t msgType,
                           int32_t ext1,
                           int32_t ext2,
                           void* user);
//数据回调
typedef void (*data_callback)(int32_t msgType,
                         const sp<IMemory>& dataPtr,
                         void* user);
//带有时间戳的数据回调
typedef void (*data_callback_timestamp)(nsecs_t timestamp,
                                   int32_t msgType,
                                   const sp<IMemory>& dataPtr,
                                   void* user);
```

然后定义类 CameraHardwareInterface，在类中的各个函数的具体实现和其他 Android 版本中的相同。区别是回调函数不再由各个函数分别设置，所以在 startPreview 和 startRecording 缺少了回调函数的指针和 void*类型的附加参数。主要实现代码如下所示。

```
class CameraHardwareInterface : public virtual RefBase {
public:
    virtual ~CameraHardwareInterface() { }
    virtual sp<IMemoryHeap>         getPreviewHeap() const = 0;
    virtual sp<IMemoryHeap>         getRawHeap() const = 0;
    virtual void setCallbacks(notify_callback notify_cb,
                        data_callback data_cb,
                        data_callback_timestamp data_cb_timestamp,
    virtual void        enableMsgType(int32_t msgType) = 0;
    virtual void        disableMsgType(int32_t msgType) = 0;
    virtual bool        msgTypeEnabled(int32_t msgType) = 0;
    virtual status_t     startPreview() = 0;
    virtual status_t     getBufferInfo(sp<IMemory>& Frame, size_t *alignedSize) = 0;
    virtual bool         useOverlay() {return false;}
    virtual status_t      setOverlay(const sp<Overlay> &overlay) {return BAD_VALUE;}
    virtual void        stopPreview() = 0;
    virtual bool        previewEnabled() = 0;
    virtual status_t     startRecording() = 0;
    virtual void        stopRecording() = 0;
    virtual bool        recordingEnabled() = 0;
    virtual void        releaseRecordingFrame(const sp<IMemory>& mem) = 0;
    virtual status_t     autoFocus() = 0;
    virtual status_t     cancelAutoFocus() = 0;
    virtual status_t     takePicture() = 0;
    virtual status_t     cancelPicture() = 0;
    virtual CameraParameters  getParameters() const = 0;
    virtual status_t sendCommand(int32_t cmd, int32_t arg1, int32_t arg2) = 0;
    virtual void release() = 0;
    virtual status_t d
}
```

因为在新版本的 Camera 系统中增加了 sendCommand()，所以需要在文件 Camera.h 中增加新命令和返回值。具体实现代码如下所示。

```
// 函数 sendCommand()使用的命令类型
enum {
    CAMERA_CMD_START_SMOOTH_ZOOM     = 1,
    CAMERA_CMD_STOP_SMOOTH_ZOOM      = 2,
    CAMERA_CMD_SET_DISPLAY_ORIENTATION = 3,
};

// 错误类型
enum {
    CAMERA_ERROR_UKNOWN  = 1,
    CAMERA_ERROR_SERVER_DIED = 100
};
```

3. 实现 Camera 硬件抽象层

在函数 startPreview 的实现过程中，保存预览回调函数并建立预览线程。在预览线程的循环中，等待视频数据的到达；视频帧到达后调用预览回调函数，将视频帧送出。

取景器预览的主要步骤如下所示。

（1）在初始化的过程中，建立预览数据的内存队列（多种方式）。

（2）在函数 startPreview 中建立预览线程。

（3）在预览线程的循环中，等待视频数据到达。

（4）视频到达后使用预览回调机制将视频向上传送。

在此过程不需要使用预览回调函数，可以直接将视频数据输入到 Overlay 上。 如果使用 Overlay 实现取景器，则需要有以下两个变化：

- 在函数 setOverlay()中，从 ISurface 接口中取得 Overlay 类；
- 在预览线程的循环中，不是用预览回调函数直接将数据输入到 Overlay 上。

录制视频的主要步骤如下所示。

（1）在函数 startRecording 的实现（或者在 setCallbacks ）中保存录制视频回调函数。

（2）录制视频可以使用自己的线程，也可以使用预览线程。

（3）通过调用录制回调函数的方式将视频帧送出。

当调用函数 releaseRecordingFrame 后，表示上层通知 Camera 硬件抽象层，这一帧的内存已经用完，可以进行下一次的处理。如果在 V4L2 驱动程序中使用原始数据（RAW），则视频录制的数据和取景器预览的数据为同一数据。当调用 releaseRecordingFrame()时，通常表示编码器已经完成了对当前视频帧的编码，对这块内存进行释放。在这个函数的实现中，可以设置标志位，标记帧内存可以再次使用。

由此可见，对于 Linux 系统来说，摄像头驱动部分大多使用 Video for Linux 2（V4L2）驱动程序，在此处主要的处理流程可以如下所示。

（1）如果使用映射内核内存的方式（V4L2_MEMORY_MMAP），构建预览的内存 MemoryHeapBase 需要从 V4L2 驱动程序中得到内存指针。

（2）如果使用用户空间内存的方式（V4L2_MEMORY_USERPTR），MemoryHeapBase 中开辟的内存是在用户空间建立的。

（3）在预览的线程中，使用 VIDIOC_DQBUF 调用阻塞等待视频帧的到来，处理完成后使用 VIDIOC_QBUF 调用将帧内存再次压入队列，然后等待下一帧的到来。

24.1.3 分析拍照系统的 Java 部分

在文件“packages/apps/Camera/src/com/android/camera/Camera.java”中，已经包含了对 Camera 的调用。在文件 Camera.java 中包含的对包的引用代码如下所示。

```
import android.hardware.Camera.PictureCallback;
import android.hardware.Camera.Size;
```

然后定义类 Camera，此类继承了活动 Activity 类，在它的内部包含了一个 android.hardware. Camera。对应代码如下所示。

```
public class Camera extends Activity implements View.OnClickListener, SurfaceHolder.
Callback{
android.hardware.Camera mCameraDevice;
}
```

调用 Camera 功能的代码如下所示。

```
mCameraDevice.takePicture(mShutterCallback, mRawPictureCallback, mJpegPicture Callback);
mCameraDevice.startPreview();
mCameraDevice.stopPreview();
startPreview、stopPreview 和 takePicture 等接口就是通过 JAVA 本地调用（JNI）来实现的。
frameworks/base/core/java/android/hardware/目录中的 Camera.java 文件提供了一个 JAVA 类：
Camera。
public class Camera {
}
```

在类 Camera 中，大部分代码使用 JNI 调用下层得到，如下面的代码。

```
public void setParameters(Parameters params) {
      Log.e(TAG, "setParameters()");
      //params.dump();
      native_setParameters(params.flatten());
}
```

还有下面的代码。

```
public final void setPreviewDisplay(SurfaceHolder holder) {
    setPreviewDisplay(holder.getSurface());
}
private native final void setPreviewDisplay(Surface surface);
```

在上面代码中，两个 setPreviewDisplay 参数不同，后一个是本地方法，参数为 Surface 类型，前一个通过调用后一个实现，但自己的参数以 SurfaceHolder 为类型。

（1）Camera 的 Java 本地调用部分

在 Android 系统中，Camera 驱动的 Java 本地调用（JNI）部分在如下文件中实现。

```
frameworks/base/core/jni/android_hardware_Camera.cpp
```

在文件 android_hardware_Camera.cpp 中定义了一个 JNINativeMethod（Java 本地调用方法）类型的数组 gMethods，具体代码如下所示。

```
static JNINativeMethod camMethods[] = {
{"native_setup","(Ljava/lang/Object;)V",(void*)android_hardware_Camera_native_setup
},
  {"native_release","()V",(void*)android_hardware_Camera_release },
  {"setPreviewDisplay","(Landroid/view/Surface;)V",(void
*)android_hardware_Camera_setPreviewDisplay },
  {"startPreview","()V",(void *)android_hardware_Camera_startPreview },
  {"stopPreview", "()V", (void *)android_hardware_Camera_stopPreview },
  {"setHasPreviewCallback","(Z)V",(void *)android_hardware_Camera_setHasPreviewCallback },
  {"native_autoFocus","()V",(void *)android_hardware_Camera_autoFocus },
  {"native_takePicture", "()V", (void *)android_hardware_Camera_takePicture },
  {"native_setParameters","(Ljava/lang/String;)V",(void
*)android_hardware_Camera_setParameters },
  {"native_getParameters",  "()Ljava/lang/String;",(void  *)android_hardware_Camera_
getParameters }
};
```

JNINativeMethod 的第一个成员是一个字符串，表示 Java 本地调用方法的名称，此名称是在 Java 程序中调用的名称；第二个成员也是一个字符串，表示 Java 本地调用方法的参数和返回值；第三个成员是 Java 本地调用方法对应的 C 语言函数。

通过函数 register_android_hardware_Camera()将 gMethods 注册为类“android/media/Camera”，其主要实现如下所示。

```
int register_android_hardware_Camera(JNIEnv *env)
{
// Register native functions
return      AndroidRuntime::registerNativeMethods(env,       "android/hardware/Camera",
camMethods, NELEM(camMethods));
}
```

其中类“android/hardware/Camera”和 Java 类 android.hardware.Camera 相对应。

（2）Camera 的本地库 libui.so

文件“frameworks/base/libs/ui/Camera.cpp”用于实现文件 Camera.h 中提供的接口，其中最重要的代码片段如下所示。

```
sp<Camera> Camera::create(const sp<ICamera>& camera)
{
    ALOGV("create");
    if (camera == 0) {
        ALOGE("camera remote is a NULL pointer");
        return 0;
    }

   sp<Camera> c = new Camera(-1);
   if (camera->connect(c) == NO_ERROR) {
       c->mStatus = NO_ERROR;
       c->mCamera = camera;
       camera->asBinder()->linkToDeath(c);
```

```
        return c;
    }
    return 0;
}
```

函数 connect()的实现代码如下所示。

```
sp<Camera> Camera::connect(int cameraId, const String16& clientPackageName,
        int clientUid)
{
    return CameraBaseT::connect(cameraId, clientPackageName, clientUid);
}
```

函数 connect()通过调用 getCameraService 得到一个 ICameraService，再通过 ICameraService 的 cs->connect(c)得到一个 ICamera 类型的指针。调用 connect()函数会得到一个 Camera 类型的指针。在正常情况下，已经初始化完成了 Camera 的成员 mCamera。

函数 startPreview()的实现代码如下所示：

```
status_t Camera::startPreview()
{
    ALOGV("startPreview");
    sp <ICamera> c = mCamera;
    if (c == 0) return NO_INIT;
    return c->startPreview();
}
```

其他函数的实现过程也与函数 setDataSource 类似。在库 libmedia.so 中的其他一些文件与头文件的名称相同，分别是：

```
frameworks/base/libs/ui/ICameraClient.cpp
frameworks/base/libs/ui/ICamera.cpp
frameworks/base/libs/ui/ICameraService.cpp
```

此处的类 BnCameraClient 和 BnCameraService 虽然实现了 onTransact()函数，但是由于还有纯虚函数没有实现，所以都是不能实例化这个类。

（3）Camera 服务 libcameraservice.so

目录“frameworks/av/services/camera/libcameraservice/”实现一个 Camera 的服务，此服务是继承 ICameraService 的具体实现。在此目录下和硬件抽象层“桩”实现相关的文件说明如下所示。

- CameraHardwareStub.cpp：Camera 硬件抽象层“桩”实现。
- CameraHardwareStub.h：Camera 硬件抽象层“桩”实现的接口。
- CannedJpeg.h：包含一块 JPEG 数据，在拍照片时作为 JPEG 数据。
- FakeCamera.h 和 FakeCamera.cpp：实现假的 Camera 黑白格取景器效果。

在文件 Android.mk 中，使用宏 USE_CAMERA_STUB 决定是否使用真的 Camera，如果宏为真，则使用 CameraHardwareStub.cpp 和 FakeCamera.cpp 构造一个假的 Camera，如果为假则使用 CameraService.cpp 构造一个实际上的 Camera 服务。文件 Android.mk 的主要代码如下所示。

```
LOCAL_MODULE:= libcamerastub
LOCAL_SHARED_LIBRARIES:= libui
include $(BUILD_STATIC_LIBRARY)
endif # USE_CAMERA_STUB
#
# libcameraservice
#

include $(CLEAR_VARS)
LOCAL_SRC_FILES:=               \
    CameraService.cpp
LOCAL_SHARED_LIBRARIES:= \
    libui \
    libutils \
    libcutils \
```

```
   libmedia
LOCAL_MODULE:= libcameraservice
LOCAL_CFLAGS+=-DLOG_TAG=\"CameraService\"
ifeq ($(USE_CAMERA_STUB), true)
LOCAL_STATIC_LIBRARIES += libcamerastub
LOCAL_CFLAGS += -include CameraHardwareStub.h
else
LOCAL_SHARED_LIBRARIES += libcamera
endif
include $(BUILD_SHARED_LIBRARY)
```

文件 CameraService.cpp 继承了 BnCameraService 的实现，在此类内部又定义了类 Client，CameraService::Client 继承了 BnCamera。在运作的过程中，函数 CameraService::connect()用于得到一个 CameraService::Client。在使用过程中，主要是通过调用这个类的接口来实现完成 Camera 的功能。因为 CameraService::Client 本身继承了 BnCamera 类，而 BnCamera 类继承了 ICamera，所以可以将此类当成 ICamera 来使用。

类 CameraService 和 CameraService::Client 的结果如下所示。

```
class CameraService : public BnCameraService
{
class Client : public BnCamera {};
wp<Client>              mClient;
}
```

在 CameraService 中，静态函数 instantiate()用于初始化一个 Camera 服务，此函数的代码如下所示。

```
void CameraService::instantiate() {
defaultServiceManager()->addService( String16("media.camera"), new CameraService());
}
```

其实函数 CameraService::instantiate()注册了一个名称为“media.camera”的服务，此服务和文件 Camera.cpp 中调用的名称相对应。

Camera 整个运作机制是：在文件 Camera.cpp 中调用 ICameraService 的接口，此时实际上调用的是 BpCameraService。而 BpCameraService 通过 Binder 机制和 BnCameraService 实现两个进程的通信。因为 BpCameraService 的实现就是此处的 CameraService，所以 Camera.cpp 虽然是在另外一个进程中运行，但是调用 ICameraService 的接口就像直接调用一样，从函数 connect()中可以得到一个 ICamera 类型的指针，整个指针的实现实际上是 CameraService::Client。

上述 Camera 功能的具体实现就是 CameraService::Client 所实现的，其构造函数如下所示。

```
CameraService::Client::Client(const sp<CameraService>& cameraService,
      const sp<ICameraClient>& cameraClient) :
mCameraService(cameraService), mCameraClient(cameraClient), mHardware(0)
{
mHardware = openCameraHardware();
mHasFrameCallback = false;
}
```

在构造函数中，通过调用 openCameraHardware()得到一个 CameraHardwareInterface 类型的指针，并作为其成员 mHardware。以后对实际的 Camera 的操作都通过对这个指针进行，这是一个简单的直接调用关系。

其实真正的 Camera 功能已经通过实现 CameraHardwareInterface 类来完成。在这个库中，文件 CameraHardwareStub.h 和 CameraHardwareStub.cpp 定义了一个“桩”模块的接口，可以在没有 Camera 硬件的情况下使用。例如，在仿真器的情况下使用的文件就是文件 CameraHardwareStub.cpp 和它依赖的文件 FakeCamera.cpp。

类 CameraHardwareStub 的结构如下所示。

```
class CameraHardwareStub : public CameraHardwareInterface {
class PreviewThread : public Thread {
};
};
```

在类 CameraHardwareStub 中包含了线程类 PreviewThread，此线程可以处理 Preview，即负责刷新取景器的内容。实际的 Camera 硬件接口通常可以通过对 V4L2 捕获驱动的调用来实现，同时还需要一个 JPEG 编码程序将从驱动中取出的数据编码成 JPEG 文件。

在文件 FakeCamera.h 和 FakeCamera.cpp 中实现了类 FakeCamera，用于实现一个假的摄像头输入数据的内存。定义代码如下所示。

```
class FakeCamera {
public:
    FakeCamera(int width, int height);
    ~FakeCamera();

    void setSize(int width, int height);
    void getNextFrameAsRgb565(uint16_t *buffer);//获取 RGB565 格式的预览帧
    void getNextFrameAsYuv422(uint8_t *buffer);//获取 Yuv422 格式的预览帧
    status_t dump(int fd, const Vector<String16>& args);

private:
    void drawSquare(uint16_t *buffer, int x, int y, int size, int color, int shadow);
    void drawCheckerboard(uint16_t *buffer, int size);

    static const int kRed = 0xf800;
    static const int kGreen = 0x07c0;
    static const int kBlue = 0x003e;

    int         mWidth, mHeight;
    int         mCounter;
    int         mCheckX, mCheckY;
    uint16_t    *mTmpRgb16Buffer;
};
```

当在 CameraHardwareStub 中设置参数后会调用函数 initHeapLocked()，此函数的实现代码如下所示。

```
void CameraHardwareStub::initHeapLocked()
{
     int picture_width, picture_height;
    mParameters.getPictureSize(&picture_width, &picture_height);
    //建立内存堆栈，创建两块内存
    mRawHeap = new MemoryHeapBase(picture_width * 2 * picture_height);

    int preview_width, preview_height;
    mParameters.getPreviewSize(&preview_width, &preview_height);
    LOGD("initHeapLocked: preview size=%dx%d", preview_width, preview_height);

    // 从参数中获取信息
    int how_big = preview_width * preview_height * 2;

    // If we are being reinitialized to the same size as before, no
    // work needs to be done.
    if (how_big == mPreviewFrameSize)
        return;

    mPreviewFrameSize = how_big;

    // Make a new mmap'ed heap that can be shared across processes.
    // use code below to test with pmem
    mPreviewHeap = new MemoryHeapBase(mPreviewFrameSize * kBufferCount);
    // 建立内存队列
    for (int i = 0; i < kBufferCount; i++) {
        mBuffers[i] = new MemoryBase(mPreviewHeap, i * mPreviewFrameSize,
mPreviewFrameSize);
    }
```

```
    // Recreate the fake camera to reflect the current size.
    delete mFakeCamera;
    mFakeCamera = new FakeCamera(preview_width, preview_height);
}
```

定义函数 startPreview()来创建一个线程，此函数的实现代码如下所示。

```
status_t CameraHardwareStub::startPreview(preview_callback cb, void* user)
{
    Mutex::Autolock lock(mLock);
    if (mPreviewThread != 0) {
        // already running
        return INVALID_OPERATION;
    }
    mPreviewCallback = cb;
    mPreviewCallbackCookie = user;
    mPreviewThread = new PreviewThread(this);//建立视频预览线程
    return NO_ERROR;
}
```

通过上面建立的线程可以调用预览回调机制，将预览的数据传递给上层的 CameraService。

创建预览线程函数 previewThread，建立一个循环以得到假的摄像头输入数据的来源，并通过预览回调函数将输出传到上层中去。函数 previewThread 的主要实现代码如下所示。

```
int CameraHardwareStub::previewThread()
{
    mLock.lock();
       int previewFrameRate = mParameters.getPreviewFrameRate();
        //发现在当前缓冲的堆的之内垂距
        ssize_t offset = mCurrentPreviewFrame * mPreviewFrameSize;
        sp<MemoryHeapBase> heap = mPreviewHeap;
            // 假设假照相机内部状态没有变化
        // (or is thread safe)
        FakeCamera* fakeCamera = mFakeCamera;
        sp<MemoryBase> buffer = mBuffers[mCurrentPreviewFrame];
            mLock.unlock();
    if (buffer != 0) {
        //多久计算等待在框架之间
        int delay = (int)(1000000.0f / float(previewFrameRate));
            //这是总是合法的，即使内存消亡仍然在我们的过程中被映射
           void *base = heap->base();
           //用假照相机填装当前框架
        uint8_t *frame = ((uint8_t *)base) + offset;
        fakeCamera->getNextFrameAsYuv422(frame);

                // Notify the client of a new frame.
        mPreviewCallback(buffer, mPreviewCallbackCookie);
            //推进缓冲头
        mCurrentPreviewFrame = (mCurrentPreviewFrame + 1) % kBufferCount;
        //等待它...
        usleep(delay);
    }
    return NO_ERROR;
}
```

在上述文件中还定义了其他的函数，函数的功能一看便知，在此为节省本书篇幅将不再一一进行详细讲解，请读者参考开源的代码文件。

24.2 开发拍照应用程序

在开发 Android 应用程序的过程中，有如下两种方式可以调用系统的摄像头实现拍照功能。

- 通过 Intent 调用系统的照相机 Activity
- 通过编码调用 Camera API

本节将详细讲解上述两种方式的具体实现流程。

24.2.1 通过 Intent 调用系统的照相机 Activity

在现实中的很多应用场景中，都需要使用摄像头去拍摄照片或视频，然后在照片或视频的基础之上进行处理。但是因为 Android 系统源码是开源的，所以很多设备厂商均可使用，造成定制比较混乱的状况发生。一般来说，在需要用到摄像头拍照或摄像的时候，均会直接调用系统现有的相机应用去进行拍照或摄像，我们只取它拍摄的结果进行处理，这样就避免了不同设备的摄像头的一些细节问题。

当通过 Intent 直接调用系统提供的照相机功能时，复用其拍照 Activity 是最简单和最方便的办法，此时不需要考虑手机的兼容性问题，如预览拍照图片大小等。如下面的演示代码。

```
Intent i = new Intent("android.media.action.IMAGE_CAPTURE");
startActivityForResult(i, C.REQUEST_CODE_CAMERA);
```

然后通过瑞安演示代码在 onActivityResult 中获取返回的数据。

```
if(resultCode==RESULT_OK){
    Bundle extras = data.getExtras();
    if(extras!=null){
      Bitmap bitmap = (Bitmap) extras.get(C.CODE_PHOTO_BITMAP_DATA);
      … …
    }
}
```

但是通过上述过程这样获取的图片是缩小过的，如果要获取原始的相片，则需要在调用的时候就指定相片生成的路径，如下面的演示代码。

```
i.putExtra(MediaStore.EXTRA_OUTPUT, Uri.fromFile(new File(tempPath)));
```

然后指定相片的类型，如下面的演示代码。

```
i.putExtra("outputFormat", Bitmap.CompressFormat.PNG.name());
```

也可以设置拍照的横竖屏，如下面的演示代码。

```
i.putExtra(MediaStore.EXTRA_SCREEN_ORIENTATION,Configuration.ORIENTATION_LANDSCAPE);
```

24.2.2 调用 Camera API 拍照

除了调用内置的拍照系统之外，还可以直接调用摄像头的 API 来编写类似取景框的拍照 Activity，这种方式更加灵活。但是，需要考虑底层的具体结构，需要了解 Android 的版本、摄像头的分辨率组合等信息，这样才能够编写出各种屏幕分辨率组合的布局等。

另外，Android 提供的 SDK（android.hardware.Camera）不能正常的使用竖屏（Portrait Layout）加载照相机，当用竖屏模式加载照相机时会产生如下所示的问题。

- 照相机成像左倾 90 度。
- 照相机成像长宽比例不对（失比）。

造成上述问题的原因是摄像头对照物的映射是 Android 底层固定的，以 landscape 方式为正，并且产生大小为 320×480 的像。此时可以通过编写自定义代码来校正上述问题，具体编码思路如下所示。

（1）设置 Camera 的参数来实现转变图片预览角度，但是这种方式并不是对所有的 Camera 都有效果的。如下面的演示代码。

```
Camera.Parameters parameters = camera.getParameters();
parameters.set("orientation", "portrait");
parameters.set("rotation", 90);
camera.setParameters(parameters);
```

（2）在获取拍摄相片之后进行旋转校正，但是如果图片太大会造成内存溢出的问题。如下面的演示代码。

```
Matrix m = new Matrix();
m.postRotate(90);
m.postScale(balance, balance);
Bitmap.createBitmap(tempBitmap, 0, 0, tempBitmap.getWidth(), tempBitmap.getHeight(),m,
true);
```

在 Android 系统中，当通过 Camera API 方式实现拍照功能时，需要用到如下所示的类。

（1）Camera 类：最主要的类，用于管理 Camera 设备，常用的方法如下所示。

- open()：通过 open 方法获取 Camera 实例。
- setPreviewDisplay(SurfaceHolder)：设置预览拍照。
- startPreview()：开始预览。
- stopPreview()：停止预览。
- release()：释放 Camera 实例。
- takePicture(Camera.ShutterCallback shutter, Camera.PictureCallback raw, Camera.PictureCallback jpeg)：这是拍照要执行的方法，包含了 3 个回调参数。其中参数 shutter 是快门按下时的回调，参数 raw 是获取拍照原始数据的回调，参数 jpeg 是获取经过压缩成 jpg 格式的图像数据。
- Camera.PictureCallback 接口：该回调接口包含了一个 onPictureTaken(byte[]data, Camera camera)方法。在这个方法中可以保存图像数据。

（2）SurfaceView 类：用于控制预览界面。其中 SurfaceHolder.Callback 接口用于处理预览的事件，需要实现如下所示的 3 个方法。

- surfaceCreated(SurfaceHolderholder)：预览界面创建时调用，每次界面改变后都会重新创建，需要获取相机资源并设置 SurfaceHolder。
- surfaceChanged(SurfaceHolderholder, int format, int width, int height)：在预览界面发生变化时调用，每次界面发生变化之后需要重新启动预览。
- surfaceDestroyed(SurfaceHolderholder)：预览销毁时调用，停止预览，释放相应资源。

24.2.3　总结 Camera 拍照的流程

经过本节前面内容的介绍，相比大家对 Camera 拍照的基本知识有了一个全新的了解。在此根据笔者的开发经验，总结出使用 Camera 实现拍照应用的基本流程，希望对广大读者起到一个抛砖引玉的作用。

（1）如果你想在自己的应用中使用摄像头，需要在 AndroidManifest.xml 中增加以下权限声明代码。

```
<uses-permissionandroid:name="android.permission.CAMERA"/>
```

（2）设定摄像头布局。这一步开发工作的基础，也就是说我们希望在应用程序中增加多少辅助性元素，如摄像头各种功能按钮等。在本文中我们采取最简方式，除了拍照外，没有多余摄像头功能。下面我们一起看一下本文示例将要用到的布局文件“camera_surface.xml”。

```
<LinearLayoutxmlns:android="http://schemas.android.com/apk/res/android"
    android:layout_width="fill_parent"android:layout_height="fill_parent"
    androidrientation="vertical">
    <SurfaceViewandroid:id="@+id/surface_camera"
    android:layout_width="fill_parent"android:layout_height="10dip"
    android:layout_weight="1">
    </SurfaceView>
</LinearLayout>
```

在此不要在资源文件名称中使用大写字母，如果把该文件命名为“CameraSurface.xml”，会给你带来不必要的麻烦。

上述该布局非常简单，只有一个LinearLayout视图组，在它下面只有一个SurfaceView视图，也就是我们的摄像头屏幕。

（3）编写摄像头实现代码。在此创建一个名为“CameraView”的Activity类，实现SurfaceHolder.Callback接口。

```
publicclassCamaraViewextendsActivityimplementsSurfaceHolder.Callback
```

接口SurfaceHolder.Callback被用来接收摄像头预览界面变化的信息，它实现了如下3个方法。

- surfaceChanged：当预览界面的格式和大小发生改变时，该方法被调用。
- surfaceCreated：在初次实例化、预览界面被创建时，该方法被调用。
- surfaceDestroyed：当预览界面被关闭时，该方法被调用。

接下来看一下在摄像头应用中如何使用这个接口，首先看一下在Activity类中的onCreate方法，其中通过下面的代码设置摄像头预览界面将通过全屏显示，并且没有“标题（title）”，并设置屏幕格式为“半透明”。

```
getWindow().setFormat(PixelFormat.TRANSLUCENT);
requestWindowFeature(Window.FEATURE_NO_TITLE);
getWindow().setFlags(WindowManager.LayoutParams.FLAG_FULLSCREEN,
WindowManager.LayoutParams.FLAG_FULLSCREEN);
```

然后通过setContentView来设定Activity的布局为前面我们创建的camera_surface，并创建一个SurfaceView对象，从xml文件中获得它。对应代码如下所示。

```
setContentView(R.layout.camera_surface);
mSurfaceView=(SurfaceView)findViewById(R.id.surface_camera);
mSurfaceHolder=mSurfaceView.getHolder();
mSurfaceHolder.addCallback(this);
mSurfaceHolder.setType(SurfaceHolder.SURFACE_TYPE_PUSH_BUFFERS);
```

通过以上代码从SurfaceView中获得了holder，并增加callback功能到“this”。这意味着我们的操作（Activity）将可以管理这个SurfaceView。

再看callback功能的实现代码。

```
publicvoidsurfaceCreated(SurfaceHolderholder){
mCamera=Camera.open();
```

上面的mCamera是“Camera”类的一个对象。在surfaceCreated方法中我们“打开”摄像头。这就是启动它的方式。

接下来定义方法surfaceChanged()让摄像头做好拍照准备，设定它的参数，并开始在Android屏幕中启动预览画面。在此使用了“semaphore”参数来防止冲突：当mPreviewRunning为true时，意味着摄像头处于激活状态，并未被关闭，因此我们可以使用它。

```
public void surfaceChanged(SurfaceHolderholder,intformat,intw,inth){
  if(mPreviewRunning){
  mCamera.stopPreview();
}
Camera.Parametersp=mCamera.getParameters();
  p.setPreviewSize(w,h);
  mCamera.setParameters(p);
  try{
  mCamera.setPreviewDisplay(holder);
  }catch(IOExceptione){
  e.printStackTrace();
}
mCamera.startPreview();
mPreviewRunning=true;
  }
```

```
publicvoidsurfaceDestroyed(SurfaceHolderholder){
  mCamera.stopPreview();
  mPreviewRunning=false;
  mCamera.release();
}
```

通过上述方法代码停止了摄像头，并释放相关的资源。正如大家所看到的，在此设置mPreviewRunning 为 false 以防止在 surfaceChanged 方法中的冲突。这是因为我们已经关闭了摄像头，而且我们不能再设置其参数或在摄像头中启动图像预览。

最后我们看下面的最重要的方法：

```
Camera.PictureCallbackmPictureCallback=newCamera.PictureCallback(){
  publicvoidonPictureTaken(byte[]imageData,Camerac){
}
};
```

在拍照时该方法被调用。例如，我们可以在界面上创建一个 OnClickListener，当点击屏幕时调用 PictureCallBack 方法，此方法会向你提供图像的字节数组，然后你可以使用 Android 提供的 Bitmap 和 BitmapFactory 类，将其从字节数组转换成你想要的图像格式。

24.2.4　实战演练——获取系统现有相机拍摄的图片

在接下来的内容中，将通过一个具体实例来讲解调用 Android 系统拍照模块功能的方法。

题目	功能	源码路径
实例 24-1	获取系统现有相机拍摄的图片	\daima\24\PaiZhao

本实例的功能是调用 Android 系统的拍照模块获取系统现有相机拍摄的图片。本实例的具体实现流程如下所示。

（1）编写布局文件 activity_main.xml，功能是在屏幕中设置两个调用系统内置拍照模块的按钮。文件 activity_main.xml 的具体实现代码如下所示。

```
<LinearLayout xmlns:android="http://schemas.android.com/apk/res/android"
    xmlns:tools="http://schemas.android.com/tools"
    android:layout_width="match_parent"
    android:layout_height="match_parent"
    android:orientation="vertical"
    android:paddingBottom="@dimen/activity_vertical_margin"
    android:paddingLeft="@dimen/activity_horizontal_margin"
    android:paddingRight="@dimen/activity_horizontal_margin"
    android:paddingTop="@dimen/activity_vertical_margin"
    tools:context=".MainActivity" >
    <Button
        android:id="@+id/btn_bySysCamera"
        android:layout_width="wrap_content"
        android:layout_height="wrap_content"
        android:text="调用系统照相机拍照" />
    <Button
        android:id="@+id/btn_bySysVideoCamera"
        android:layout_width="wrap_content"
        android:layout_height="wrap_content"
        android:text="调用系统摄像头摄像" />
</LinearLayout>
```

（2）编写文件 MainActivity.java，功能是根据用户单击的按钮来执行调用摄像头拍照功能和录制视频功能。文件 MainActivity.java 的具体实现代码如下所示。

```
public class MainActivity extends Activity {
    private Button btn_bySysCamera, btn_bySysVideoCamera;

    @Override
    protected void onCreate(Bundle savedInstanceState) {
```

```
        super.onCreate(savedInstanceState);
        setContentView(R.layout.activity_main);

        btn_bySysCamera = (Button) findViewById(R.id.btn_bySysCamera);
        btn_bySysVideoCamera = (Button) findViewById(R.id.btn_bySysVideoCamera);

        btn_bySysCamera.setOnClickListener(click);
        btn_bySysVideoCamera.setOnClickListener(click);
    }

    private View.OnClickListener click = new View.OnClickListener() {

        @Override
        public void onClick(View v) {
            Intent intent=null;
            switch (v.getId()) {
            case R.id.btn_bySysCamera:
                intent = new Intent(MainActivity.this, SysCameraActivity.class);
                startActivity(intent);
                break;
            case R.id.btn_bySysVideoCamera:
                intent = new Intent(MainActivity.this, SysVideoCameraActivity.class);
                startActivity(intent);
                break;
            default:
                break;
            }
        }
    };
}
```

（3）编写布局文件 activity_syscamera.xml，功能是在屏幕中实现调用系统拍照功能的两个按钮。文件 activity_syscamera.xml 的具体实现代码如下所示。

```
<LinearLayout xmlns:android="http://schemas.android.com/apk/res/android"
    android:layout_width="match_parent"
    android:layout_height="match_parent"
    android:orientation="vertical" >
    <Button
        android:id="@+id/btn_StartCamera"
        android:layout_width="wrap_content"
        android:layout_height="wrap_content"
        android:text="系统相机拍照--指定路径到 SD 卡" />
    <Button
        android:id="@+id/btn_StartCameraInGallery"
        android:layout_width="wrap_content"
        android:layout_height="wrap_content"
        android:text="系统相机拍照--默认图库" />
    <ImageView
        android:id="@+id/iv_CameraImg"
        android:layout_width="match_parent"
        android:layout_height="match_parent" />
</LinearLayout>
```

（4）编写程序文件 SysCameraActivity.java，功能是根据用户的单击按钮执行拍照功能和获取照片信息功能。文件 SysCameraActivity.java 的具体实现代码如下所示。

```
public class SysCameraActivity extends Activity {
    private Button btn_StartCamera, btn_StartCameraInGallery;
    private ImageView iv_CameraImg;

    private static final String TAG = "main";
    private static final String FILE_PATH = "/sdcard/syscamera.jpg";

    @Override
    protected void onCreate(Bundle savedInstanceState) {
        super.onCreate(savedInstanceState);
        setContentView(R.layout.activity_syscamera);
```

```
        btn_StartCamera = (Button) findViewById(R.id.btn_StartCamera);
        btn_StartCameraInGallery = (Button) findViewById(R.id.btn_StartCameraInGallery);
        iv_CameraImg = (ImageView) findViewById(R.id.iv_CameraImg);

        btn_StartCamera.setOnClickListener(click);
        btn_StartCameraInGallery.setOnClickListener(click);
    }

    private View.OnClickListener click = new View.OnClickListener() {

        @Override
        public void onClick(View v) {

            Intent intent = null;
            switch (v.getId()) {
            // 指定相机拍摄照片保存地址
            case R.id.btn_StartCamera:
                intent = new Intent();
                // 指定开启系统相机的 Action
                intent.setAction(MediaStore.ACTION_IMAGE_CAPTURE);
                intent.addCategory(Intent.CATEGORY_DEFAULT);
                // 根据文件地址创建文件
                File file = new File(FILE_PATH);
                if (file.exists()) {
                    file.delete();
                }
                // 把文件地址转换成 Uri 格式
                Uri uri = Uri.fromFile(file);
                // 设置系统相机拍摄照片完成后图片文件的存放地址
                intent.putExtra(MediaStore.EXTRA_OUTPUT, uri);
                startActivityForResult(intent, 0);
                break;
            // 不指定相机拍摄照片保存地址
            case R.id.btn_StartCameraInGallery:
                intent = new Intent();
                // 指定开启系统相机的 Action
                intent.setAction(MediaStore.ACTION_IMAGE_CAPTURE);
                intent.addCategory(Intent.CATEGORY_DEFAULT);
                startActivityForResult(intent, 1);
                break;
            default:
                break;
            }

        }
    };

    @Override
    protected void onActivityResult(int requestCode, int resultCode, Intent data) {
        Log.i(TAG, "系统相机拍照完成，resultCode="+resultCode);

        if (requestCode == 0) {
            File file = new File(FILE_PATH);
            Uri uri = Uri.fromFile(file);
            iv_CameraImg.setImageURI(uri);
        } else if (requestCode == 1) {
            Log.i(TAG, "默认 content 地址："+data.getData());
            iv_CameraImg.setImageURI(data.getData());
        }
    }
}
```

（5）编写布局文件 activity_sysvideocamera.xml，功能是在屏幕中实现调用系统录制视频功能的按钮。文件 activity_sysvideocamera.xml 的具体实现代码如下所示。

```
<LinearLayout xmlns:android="http://schemas.android.com/apk/res/android"
    android:layout_width="match_parent"
    android:layout_height="match_parent"
    android:orientation="vertical" >
```

```
    <Button
        android:id="@+id/btn_StartVideoCamera"
        android:layout_width="wrap_content"
        android:layout_height="wrap_content"
        android:text="开启系统摄像头录像" />
</LinearLayout>
```

（6）编写程序文件 SysVideoCameraActivity.java，功能是根据用户的单击按钮执行拍视频录制功能。文件 SysVideoCameraActivity.java 的具体实现代码如下所示。

```
public class SysVideoCameraActivity extends Activity {
    private Button btn_StartVideoCamera;
    private static final String FILE_PATH = "/sdcard/sysvideocamera.3gp";
    private static final String TAG="main";
    @Override
    protected void onCreate(Bundle savedInstanceState) {
        super.onCreate(savedInstanceState);
        setContentView(R.layout.activity_sysvideocamera);

        btn_StartVideoCamera = (Button) findViewById(R.id.btn_StartVideoCamera);
        btn_StartVideoCamera.setOnClickListener(click);
    }

    private View.OnClickListener click = new View.OnClickListener() {

        @Override
        public void onClick(View v) {
            Intent intent = new Intent();
            intent.setAction("android.media.action.VIDEO_CAPTURE");
            intent.addCategory("android.intent.category.DEFAULT");
            File file = new File(FILE_PATH);
            if(file.exists()){
                file.delete();
            }
            Uri uri = Uri.fromFile(file);
            intent.putExtra(MediaStore.EXTRA_OUTPUT, uri);
            startActivityForResult(intent, 0);
        }
    };

    @Override
    protected void onActivityResult(int requestCode, int resultCode, Intent data) {
        Log.i(TAG, "拍摄完成，resultCode="+requestCode);
    }

}
```

到此为止，整个实例介绍完毕。执行后的效果如图 24-3 所示。

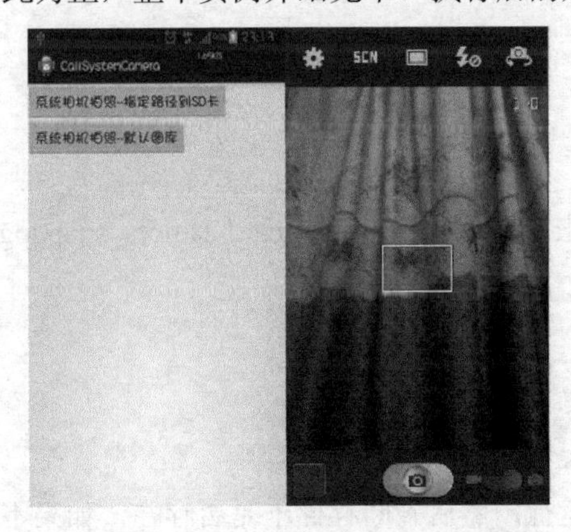

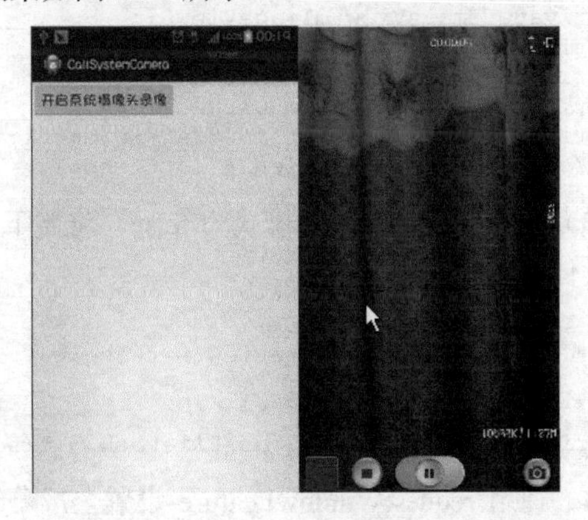

▲图 24-3　执行效果

24.2.5　实战演练——使用 Camera 预览并拍照

在接下来的内容中，将通过一个具体实例来讲解使用 Camera 实现预览和拍照功能的方法。

题目	目的	源码路径
实例 24-2	使用 Camera 预览并拍照	\daima\24\PaiZhao

本实例实现了一个简单的拍照功能，在实例中以 Activity 为基础，在 Layout 中配置了 3 个按钮，分别实现预览、关闭相机和拍照处理功能。当单击“拍照”按钮后会将屏幕中拍的画面截取下来并存储到 SD 卡中，然后将拍下来的图片显示在 Activity 中的 ImageView 控件中。为避免拍照相片造成的存储卡垃圾暂存堆栈，在离开程序前删除临时文件。本实例的主程序文件是 example.java，其具体实现流程如下所示。

（1）引用 PictureCallback 作为取得拍照后的事件，具体实现代码如下所示。

```
/* 引用 Camera 类 */
import android.hardware.Camera;

/* 引用 PictureCallback 作为取得拍照后的事件 */
import android.hardware.Camera.PictureCallback;
import android.hardware.Camera.ShutterCallback;
import android.os.Bundle;
import android.util.DisplayMetrics;
import android.util.Log;
import android.view.SurfaceHolder;
import android.view.SurfaceView;
import android.view.View;
import android.view.Window;
import android.widget.Button;
import android.widget.ImageView;
import android.widget.TextView;
import android.widget.Toast;
```

（2）创建私有 Camera 对象，然后分别创建 mImageView01、mTextView01、TAG、mSurfaceView01、mSurfaceHolder01 和 intScreenY 作为预览相片之用。具体实现代码如下所示。

```
/* 使 Activity 实现 SurfaceHolder.Callback */
public class example10 extends Activity
implements SurfaceHolder.Callback
{
  /* 创建私有 Camera 对象 */
  private Camera mCamera01;
  private Button mButton01, mButton02, mButton03;

  /* 作为 review 照下来的相片之用 */
  private ImageView mImageView01;
  private TextView mTextView01;
  private String TAG = "HIPPO";
  private SurfaceView mSurfaceView01;
  private SurfaceHolder mSurfaceHolder01;
  //private int intScreenX, intScreenY;
```

（3）设置默认相机预览模式为 false，将照下来的图片存储在“/sdcard/camera_snap.jpg”目录下。具体实现代码如下所示。

```
  /* 默认相机预览模式为 false */
  private boolean bIfPreview = false;

  /* 将照下来的图片存储在此 */
  private String strCaptureFilePath = "/sdcard/camera_snap.jpg";
```

（4）使用 requestWindowFeature 设置全屏幕运行，然后判断存储卡是否存在，如果不存在则提醒用户未安装存储卡。具体实现代码如下所示。

```
public void onCreate(Bundle savedInstanceState)
{
  super.onCreate(savedInstanceState);

  /* 使应用程序全屏幕运行，不使用 title bar */
  requestWindowFeature(Window.FEATURE_NO_TITLE);
  setContentView(R.layout.main);

  /* 判断存储卡是否存在 */
  if(!checkSDCard())
  {
    /* 提醒 User 未安装存储卡 */
    mMakeTextToast
    (
      getResources().getText(R.string.str_err_nosd).toString(),
      true
    );
  }
```

(5) 通过 DisplayMetrics 对象 dm 获取屏幕解析像素，然后以 SurfaceView 作为相机预览之用，绑定 SurfaceView 后获取 SurfaceHolder 对象，通过 setFixedSize 可以设置预览大小。具体实现代码如下所示。

```
/* 取得屏幕解析像素 */
DisplayMetrics dm = new DisplayMetrics();
getWindowManager().getDefaultDisplay().getMetrics(dm);
mTextView01 = (TextView) findViewById(R.id.myTextView1);
mImageView01 = (ImageView) findViewById(R.id.myImageView1);
/* 以 SurfaceView 作为相机 Preview 之用 */
mSurfaceView01 = (SurfaceView) findViewById(R.id.mSurfaceView1);

/* 绑定 SurfaceView，取得 SurfaceHolder 对象 */
mSurfaceHolder01 = mSurfaceView01.getHolder();
/* Activity 必须实现 SurfaceHolder.Callback */
mSurfaceHolder01.addCallback(example10.this);

/* 额外的设置预览大小设置，在此不使用 */
/*以 SURFACE_TYPE_PUSH_BUFFERS(3)作为 SurfaceHolder 显示类型 */
mSurfaceHolder01.setType
(SurfaceHolder.SURFACE_TYPE_PUSH_BUFFERS);

mButton01 = (Button)findViewById(R.id.myButton1);
mButton02 = (Button)findViewById(R.id.myButton2);
mButton03 = (Button)findViewById(R.id.myButton3);
```

(6) 编写打开相机和预览按钮事件，自定义初始化打开相机函数。具体实现代码如下所示。

```
/* 打开相机及 Preview */
mButton01.setOnClickListener(new Button.OnClickListener()
{
  @Override
  public void onClick(View arg0)
  {
    // TODO Auto-generated method stub

    /* 自定义初始化打开相机函数 */
    initCamera();
  }
});
```

(7) 设置停止预览按钮事件，自定义重置相机并关闭相机预览函数。具体实现代码如下所示。

```
/* 停止 Preview 及相机 */
mButton02.setOnClickListener(new Button.OnClickListener()
{
  @Override
  public void onClick(View arg0)
  {
    /* 自定义重置相机，并关闭相机预览函数 */
```

```
    resetCamera();
  }
});
```

（8）设置停止拍照按钮事件，当存储卡存在才允许拍照，自定义函数 takePicture()实现拍照功能。具体实现代码如下所示。

```
    /* 拍照 */
    mButton03.setOnClickListener(new Button.OnClickListener()
    {
      @Override
      public void onClick(View arg0)
      {
        // TODO Auto-generated method stub

        /* 当存储卡存在才允许拍照，存储暂存图像文件 */
        if(checkSDCard())
        {
          /* 自定义拍照函数 */
          takePicture();
        }
        else
        {
          /* 存储卡不存在显示提示 */
          mTextView01.setText
          (
            getResources().getText(R.string.str_err_nosd).toString()
          );
        }
      }
    });
}
```

（9）定义方法 initCamera()，如果相机处于非预览模式则打开相机。具体实现代码如下所示。

```
  /* 自定义初始相机函数 */
  private void initCamera()
  {
    if(!bIfPreview)
    {
      /* 若相机非在预览模式，则打开相机 */
      mCamera01 = Camera.open();
    }

    if (mCamera01 != null && !bIfPreview)
    {
      Log.i(TAG, "inside the camera");

      /* 创建 Camera.Parameters 对象 */
      Camera.Parameters parameters = mCamera01.getParameters();

      /* 设置相片格式为 JPEG */
      parameters.setPictureFormat(PixelFormat.JPEG);

      /* 指定 preview 的屏幕大小 */
      parameters.setPreviewSize(320, 240);

      /* 设置图片分辨率大小 */
      parameters.setPictureSize(320, 240);

      /* 将 Camera.Parameters 设置予 Camera */
      mCamera01.setParameters(parameters);

      /* setPreviewDisplay 唯一的参数为 SurfaceHolder */
      mCamera01.setPreviewDisplay(mSurfaceHolder01);

      /* 立即运行 Preview */
      mCamera01.startPreview();
      bIfPreview = true;
```

```
    }
}
```

（10）定义方法 takePicture()来调用 takePicture()实现拍照并撷取图像。具体实现代码如下所示。

```
/* 拍照撷取图像 */
private void takePicture()
{
  if (mCamera01 != null && bIfPreview)
  {
    /* 调用 takePicture()方法拍照 */
    mCamera01.takePicture
    (shutterCallback, rawCallback, jpegCallback);
  }
}
```

（11）定义方法 resetCamera()来实现相机重置，具体实现代码如下所示。

```
/* 相机重置 */
private void resetCamera()
{
  if (mCamera01 != null && bIfPreview)
  {
    mCamera01.stopPreview();
    /* 扩展学习，释放 Camera 对象 */
    //mCamera01.release();
    mCamera01 = null;
    bIfPreview = false;
  }
}

private ShutterCallback shutterCallback = new ShutterCallback()
{
  public void onShutter()
  {
    // Shutter has closed
  }
};

private PictureCallback rawCallback = new PictureCallback()
{
  public void onPictureTaken(byte[] _data, Camera _camera)
  {
    // TODO Handle RAW image data
  }
};
```

（12）定义方法 delFile(String strFileName)来自定义删除临时文件，具体实现代码如下所示。

```
/* 自定义删除文件函数 */
private void delFile(String strFileName)
{
  try
  {
    File myFile = new File(strFileName);
    if(myFile.exists())
    {
      myFile.delete();
    }
  }
  catch (Exception e)
  {
    Log.e(TAG, e.toString());
    e.printStackTrace();
  }
}
```

（13）定义方法 mMakeTextToast(String str, boolean isLong)来输出提示语句，具体实现代码如下所示。

```
public void mMakeTextToast(String str, boolean isLong)
{
  if(isLong==true)
  {
    Toast.makeText(example10.this, str, Toast.LENGTH_LONG).show();
  }
  else
  {
    Toast.makeText(example10.this, str, Toast.LENGTH_SHORT).show();
  }
}
```

（14）定义方法 checkSDCard()来检查是否有存储卡，具体实现代码如下所示。

```
private boolean checkSDCard()
{
  /* 判断存储卡是否存在 */
  if(android.os.Environment.getExternalStorageState().equals
  (android.os.Environment.MEDIA_MOUNTED))
  {
    return true;
  }
  else
  {
    return false;
  }
}
public void surfaceChanged
(SurfaceHolder surfaceholder, int format, int w, int h)
{
  Log.i(TAG, "Surface Changed");
}

public void surfaceCreated(SurfaceHolder surfaceholder)
{
  // TODO Auto-generated method stub
  Log.i(TAG, "Surface Changed");
}
```

在文件 AndroidManifest.xml 中声明 android.permission.CAMERA 权限，具体实现代码如下所示。

```
<uses-permission android:name="android.permission.CAMERA">
```

执行后的效果如图 24-4 所示，分别单击“点击预览”“拍照”和“关闭相机”按钮后可以实现对应的功能。

▲图 24-4　执行效果

24.2.6　实战演练——使用 Camera API 方式拍照

在接下来的内容中，将通过一个具体实例来讲解使用 Camera API 方式拍照的方法。

题目	目的	源码路径
实例 24-3	使用 Camera API 方式拍照	\daima\24\AndroidCamera

本实例的具体实现流程如下所示。

（1）在布局文件 main.xml 中插入一个 Capture 按钮，主要实现代码如下所示。

```
<FrameLayout
  android:id="@+id/camera_preview"
  android:layout_width="fill_parent"
  android:layout_height="fill_parent"
  android:layout_weight="1"
  />

<Button
```

```
    android:id="@+id/button_capture"
    android:text="Capture"
    android:layout_width="wrap_content"
    android:layout_height="wrap_content"
    android:layout_gravity="center"
    />
```

（2）在文件 AndroidManifest.xml 中添加使用 Camera 相关的声明，具体代码如下所示。

```
<uses-permission android:name="android.permission.CAMERA" />
<uses-feature android:name="android.hardware.camera" />
<uses-feature android:name="android.hardware.camera.autofocus" />
<uses-permission android:name="android.permission.WRITE_EXTERNAL_STORAGE" />
```

（3）编写 AndroidCameraActivity 类，具体实现代码如下所示。

```
public class AndroidCameraActivity extends Activity implements OnClickListener,
PictureCallback {
    private CameraSurfacePreview mCameraSurPreview = null;
    private Button mCaptureButton = null;
    private String TAG = "Dennis";

    @Override
    protected void onCreate(Bundle savedInstanceState) {
        super.onCreate(savedInstanceState);
        setContentView(R.layout.main);

        // Create our Preview view and set it as the content of our activity.
        FrameLayout preview = (FrameLayout) findViewById(R.id.camera_preview);
        mCameraSurPreview = new CameraSurfacePreview(this);
        preview.addView(mCameraSurPreview);

      // Add a listener to the Capture button
        mCaptureButton = (Button) findViewById(R.id.button_capture);
        mCaptureButton.setOnClickListener(this);
    }

    @Override
    public void onPictureTaken(byte[] data, Camera camera) {

        //save the picture to sdcard
        File pictureFile = getOutputMediaFile();
        if (pictureFile == null){
            Log.d(TAG, "Error creating media file, check storage permissions: ");
            return;
        }

        try {
            FileOutputStream fos = new FileOutputStream(pictureFile);
            fos.write(data);
            fos.close();

            Toast.makeText(this, "Image has been saved to "+pictureFile.getAbsolute
            Path(),Toast.LENGTH_LONG).show();
        } catch (FileNotFoundException e) {
            Log.d(TAG, "File not found: " + e.getMessage());
        } catch (IOException e) {
            Log.d(TAG, "Error accessing file: " + e.getMessage());
        }

        // Restart the preview and re-enable the shutter button so that we can take another picture
        camera.startPreview();

        //See if need to enable or not
        mCaptureButton.setEnabled(true);
    }

    @Override
    public void onClick(View v) {
        mCaptureButton.setEnabled(false);
```

```
        // get an image from the camera
        mCameraSurPreview.takePicture(this);
    }

    private File getOutputMediaFile(){
        //get the mobile Pictures directory
        File   picDir   =   Environment.getExternalStoragePublicDirectory(Environment.
DIRECTORY_PICTURES);

        //get the current time
        String timeStamp = new SimpleDateFormat("yyyyMMdd_HHmmss").format(new Date());

        return new File(picDir.getPath() + File.separator + "IMAGE_"+ timeStamp + ".jpg");
    }
}
```

通过上述实现代码，可以看出通过 Camera 方式实现拍照的如下基本流程。

（1）通过 Camera.open()来获取 Camera 实例。

（2）创建 Preview 类，需要继承 SurfaceView 类并实现 SurfaceHolder.Callback 接口。

（3）为相机设置 Preview。

（4）构建一个 Preview 的 Layout 来预览相机。

（5）为拍照建立 Listener 以获取拍照后的回调。

（6）拍照并保存文件。

（7）释放 Camera。

本实例需要在有摄像头的真机上运行，拍完照之后可以在 SD 卡中的“Pictures”目录下找到保存的照片。

24.3 使用拍照方式解析二维码

QR Code 码是由日本的一个公司于 1994 年 9 月研制的一种矩阵二维码符号，它具有一维条码及其他二维条码所具有的信息容量大、可靠性高、可表示汉字及图像多种信息、保密防伪性强等优点。在本节的内容中，将详细讲解使用相机解析 QR Code 码的方法。

24.3.1　QR Code 码的特点

从 QR Code 码的英文名称 Quick Response Code 可以看出，超高速识读特点是 QR Code 码区别于四一七条码、Data Matrix 等二维码的主要特性。由于在用 CCD 识读 QR Code 码时，整个 QR Code 码符号中信息的读取是通过 QR Code 码符号的位置探测图形，用硬件来实现，因此，信息识读过程所需时间很短，它具有超高速识读特点。用 CCD 二维条码识读设备，每秒可识读 30 个含有 100 个字符的 QR Code 码符号；对于含有相同数据信息的四一七条码符号，每秒仅能识读 3 个符号；对于 Data Martix 矩阵码，每秒仅能识读 2～3 个符号。QRCode 码的超高速识读特性，使它能够广泛应用于工业自动化生产线管理等领域。

QR Code 码具有全方位（360°）识读特点,这是 QR Code 码优于行排式二维条码如四一七条码的另一主要特点,由于四一七条码是将一维条码符号在行排高度上的截短来实现的,因此,它很难实现全方位识读，其识读方位角仅为±10°，能够有效地表示中国汉字和日本汉字。由于 QR Code 码用特定的数据压缩模式表示中国汉字和日本汉字，它仅用 13bit 可表示一个汉字，而四一七条码、Data Martix 等二维码没有特定的汉字表示模式，因此仅用字节表示模式来表示汉字，在用字节模式表示汉字时，需用 16bit（二个字节）表示一个汉字，因此 QR Code 码比其它的二维条码表示汉字的效率提高了 20%。

24.3.2 实战演练——使用 Android 相机解析二维码

在接下来的内容中，将通过一个具体实例来讲解使用 Android 相机解析 QR Code 二维码的方法。

题目	目的	源码路径
实例 24-4	使用 Camera API 方式拍照	\daima\24\qr

本实例的具体实现流程如下所示。

（1）分别创建私有 Camera 对象 mCamera01、mButton01、mButton02 和 mButton03，然后设置默认相机预览模式为 false，具体代码如下所示。

```
/* 创建私有 Camera 对象 */
private Camera mCamera01;
private Button mButton01, mButton02, mButton03;
/* 作为 review 照下来的相片之用 */
private ImageView mImageView01;
private String TAG = "HIPPO";
private SurfaceView mSurfaceView01;
private SurfaceHolder mSurfaceHolder01;

/* 默认相机预览模式为 false */
private boolean bIfPreview = false;

/** Called when the activity is first created. */
```

（2）设置应用程序全屏幕运行，并添加红色正方形红框 View 供用户对准条形码，然后将创建的红色方框添加至此 Activity 中，具体代码如下所示。

```
public void onCreate(Bundle savedInstanceState)
{
  super.onCreate(savedInstanceState);
  /* 使应用程序全屏幕运行，不使用 title bar */
  requestWindowFeature(Window.FEATURE_NO_TITLE);
  setContentView(R.layout.main);
  /* 添加红色正方形红框 View，供 User 对准条形码 */
  DrawCaptureRect mDraw = new DrawCaptureRect
  (
    example203.this,
    110, 10, 100, 100,
    getResources().getColor(R.drawable.lightred)
  );

  /* 将创建的红色方框添加至此 Activity 中 */
  addContentView
  (
    mDraw,
    new LayoutParams
    (
      LayoutParams.WRAP_CONTENT, LayoutParams.WRAP_CONTENT
    )
  );
```

（3）分别取得屏幕解析像素，绑定 SurfaceView 并设置预览大小，具体代码如下所示。

```
/* 取得屏幕解析像素 */
DisplayMetrics dm = new DisplayMetrics();
getWindowManager().getDefaultDisplay().getMetrics(dm);

mImageView01 = (ImageView) findViewById(R.id.myImageView1);

/* 以 SurfaceView 作为相机 Preview 之用 */
mSurfaceView01 = (SurfaceView) findViewById(R.id.mSurfaceView1);

/* 绑定 SurfaceView，取得 SurfaceHolder 对象 */
mSurfaceHolder01 = mSurfaceView01.getHolder();
```

```
/* Activity 必须实现 SurfaceHolder.Callback */
mSurfaceHolder01.addCallback(example203.this);

/* 额外的设置预览大小设置，在此不使用 */
//mSurfaceHolder01.setFixedSize(320, 240);

/*
 * 以 SURFACE_TYPE_PUSH_BUFFERS(3)
 * 作为 SurfaceHolder 显示类型
 * */
mSurfaceHolder01.setType
(SurfaceHolder.SURFACE_TYPE_PUSH_BUFFERS);

mButton01 = (Button)findViewById(R.id.myButton1);
mButton02 = (Button)findViewById(R.id.myButton2);
mButton03 = (Button)findViewById(R.id.myButton3);
```

（4）编写单击方法 mButton01 按钮的响应程序，单击后打开相机及预览二维条形码，具体代码如下所示。

```
/* 打开相机及预览二维条形码 */
mButton01.setOnClickListener(new Button.OnClickListener()
{
  @Override
  public void onClick(View arg0)
  {
    // TODO Auto-generated method stub

    /* 自定义初始化打开相机函数 */
    initCamera();
  }
});
```

（5）编写方法单击 mButton02 按钮的响应程序，单击后停止预览，具体代码如下所示。

```
/* 停止预览 */
mButton02.setOnClickListener(new Button.OnClickListener()
{
  @Override
  public void onClick(View arg0)
  {
    // TODO Auto-generated method stub

    /* 自定义重置相机，并关闭相机预览函数 */
    resetCamera();
  }
});
```

（6）编写单击 mButton03 按钮后的响应程序，单击后拍照处理并生成二维条形码，具体代码如下所示。

```
    /* 拍照 QR Code 二维条形码 */
    mButton03.setOnClickListener(new Button.OnClickListener()
    {
      @Override
      public void onClick(View arg0)
      {
        // TODO Auto-generated method stub
        /* 自定义拍照函数 */
        takePicture();
      }
    });
  }
```

（7）定义方法 initCamera()用于自定义初始相机函数，具体代码如下所示。

```
  /* 自定义初始相机函数 */
  private void initCamera()
  {
```

```
    if(!bIfPreview)
    {
      /* 若相机非在预览模式，则打开相机 */
      mCamera01 = Camera.open();
    }

    if (mCamera01 != null && !bIfPreview)
    {
      Log.i(TAG, "inside the camera");

      /* 创建 Camera.Parameters 对象 */
      Camera.Parameters parameters = mCamera01.getParameters();

      /* 设置相片格式为 JPEG */
      parameters.setPictureFormat(PixelFormat.JPEG);

      /* 指定 preview 的屏幕大小 */
      parameters.setPreviewSize(160, 120);

      /* 设置图片分辨率大小 */
      parameters.setPictureSize(160, 120);

      /* 将 Camera.Parameters 设置予 Camera */
      mCamera01.setParameters(parameters);

      /* setPreviewDisplay 唯一的参数为 SurfaceHolder */
      mCamera01.setPreviewDisplay(mSurfaceHolder01);

      /* 立即运行 Preview */
      mCamera01.startPreview();
      bIfPreview = true;
    }
  }
```

（8）定义方法 takePicture()用于拍照并获取图像，具体代码如下所示。

```
/* 拍照撷取图像 */
private void takePicture()
{
  if (mCamera01 != null && bIfPreview)
  {
    /* 调用 takePicture()方法拍照 */
    mCamera01.takePicture
    (shutterCallback, rawCallback, jpegCallback);
  }
}
```

（9）定义方法 resetCamera()来实现相机重置，然后释放 Camera 对象，具体代码如下所示。

```
/* 相机重置 */
private void resetCamera()
{
  if (mCamera01 != null && bIfPreview)
  {
    mCamera01.stopPreview();
    /*释放 Camera 对象 */
    //mCamera01.release();
    mCamera01 = null;
    bIfPreview = false;
  }
}
private ShutterCallback shutterCallback = new ShutterCallback()
{
  public void onShutter()
  {
    // Shutter has closed
  }
};
private PictureCallback rawCallback = new PictureCallback()
```

```
  {
    public void onPictureTaken(byte[] _data, Camera _camera)
    {
      // TODO Handle RAW image data
    }
  };
```

（10）定义方法 onPictureTaken，对传入的图片进行处理。首先设置 onPictureTaken 传入的第一个参数即为相片的 byte；然后使用 Matrix.postScale 方法缩小图像大小；接下来创建新的 Bitmap 对象；然后获取 4:3 图片的居中红色框部分 100 像素×100 像素，并将拍照的图文件以 ImageView 显示出来；最后将传入的图文件译码成字符串，并定义方法 mMakeTextToast 输出提示。具体代码如下所示。

```
  private PictureCallback jpegCallback = new PictureCallback()
  {
    public void onPictureTaken(byte[] _data, Camera _camera)
    {
      try
      {
        /* onPictureTaken 传入的第一个参数即为相片的 byte */
        Bitmap bm =
        BitmapFactory.decodeByteArray(_data, 0, _data.length);

        int resizeWidth = 160;
        int resizeHeight = 120;
        float scaleWidth = ((float) resizeWidth) / bm.getWidth();
        float scaleHeight = ((float) resizeHeight) / bm.getHeight();

        Matrix matrix = new Matrix();
        /* 使用 Matrix.postScale 方法缩小 Bitmap Size*/
        matrix.postScale(scaleWidth, scaleHeight);

        /* 创建新的 Bitmap 对象 */
        Bitmap resizedBitmap = Bitmap.createBitmap
        (bm, 0, 0, bm.getWidth(), bm.getHeight(), matrix, true);

        /* 撷取 4:3 图片的居中红色框部分 100 像素×100 像素 */
        Bitmap resizedBitmapSquare = Bitmap.createBitmap
        (resizedBitmap, 30, 10, 100, 100);

        /* 将拍照的图文件以 ImageView 显示出来 */
        mImageView01.setImageBitmap(resizedBitmapSquare);

        /* 将传入的图文件译码成字符串 */
        String strQR2 = decodeQRImage(resizedBitmapSquare);
        if(strQR2!="")
        {
          if (URLUtil.isNetworkUrl(strQR2))
          {
            /* OMIA 规范，网址条形码，打开浏览器上网 */
            mMakeTextToast(strQR2, true);
            Uri mUri = Uri.parse(strQR2);
            Intent intent = new Intent(Intent.ACTION_VIEW, mUri);
            startActivity(intent);
          }
          else if(eregi("wtai://",strQR2))
          {
            /* OMIA 规范，手机拨打电话格式 */
            String[] aryTemp01 = strQR2.split("wtai://");
            Intent myIntentDial = new Intent
            (
              "android.intent.action.CALL",
              Uri.parse("tel:"+aryTemp01[1])
            );
            startActivity(myIntentDial);
          }
          else if(eregi("TEL:",strQR2))
```

```
          {
            /* OMIA 规范，手机拨打电话格式 */
            String[] aryTemp01 = strQR2.split("TEL:");
            Intent myIntentDial = new Intent
            (
              "android.intent.action.CALL",
              Uri.parse("tel:"+aryTemp01[1])
            );
            startActivity(myIntentDial);
          }
          else
          {
            /* 若仅是文字，则以 Toast 显示出来 */
            mMakeTextToast(strQR2, true);
          }
        }

        /* 显示完图文件，立即重置相机，并关闭预览 */
        resetCamera();

        /* 再重新启动相机继续预览 */
        initCamera();
      }
      catch (Exception e)
      {
        Log.e(TAG, e.getMessage());
      }
    }
  };

  public void mMakeTextToast(String str, boolean isLong)
  {
    if(isLong==true)
    {
      Toast.makeText(example203.this, str, Toast.LENGTH_LONG).show();
    }
    else
    {
      Toast.makeText(example203.this, str, Toast.LENGTH_SHORT).show();
    }
  }
```

（11）定义方法 checkSDCard()来判断记忆卡是否存在，具体代码如下所示。

```
  private boolean checkSDCard()
  {
    /* 判断记忆卡是否存在 */
    if(android.os.Environment.getExternalStorageState().equals
    (android.os.Environment.MEDIA_MOUNTED))
    {
      return true;
    }
    else
    {
      return false;
    }
  }
```

（12）定义方法 decodeQRImage(Bitmap myBmp)来解码传入的 Bitmap 图片，主要代码如下所示。

```
  /* 解码传入的 Bitmap 图片 */
  public String decodeQRImage(Bitmap myBmp)
  {
    String strDecodedData = "";
    try
    {
      QRCodeDecoder decoder = new QRCodeDecoder();
      strDecodedData  = new String
      (decoder.decode(new AndroidQRCodeImage(myBmp)));
    }
```

```
  catch(Exception e)
  {
    e.printStackTrace();
  }
  return strDecodedData;
}
```

（13）定义类 DrawCaptureRect 来绘制相机预览画面里的正方形方框，具体代码如下所示。

```
/* 绘制相机预览画面里的正方形方框 */
class DrawCaptureRect extends View
{
  private int colorFill;
  private int intLeft,intTop,intWidth,intHeight;

  public DrawCaptureRect
  (
    Context context, int intX, int intY, int intWidth,
    int intHeight, int colorFill
  )
  {
    super(context);
    this.colorFill = colorFill;
    this.intLeft = intX;
    this.intTop = intY;
    this.intWidth = intWidth;
    this.intHeight = intHeight;
  }

  @Override
  protected void onDraw(Canvas canvas)
  {
    Paint mPaint01 = new Paint();
    mPaint01.setStyle(Paint.Style.FILL);
    mPaint01.setColor(colorFill);
    mPaint01.setStrokeWidth(1.0F);
    /* 在画布上绘制红色的四条方边框作为瞄准器 */
    canvas.drawLine
    (
      this.intLeft, this.intTop,
      this.intLeft+intWidth, this.intTop, mPaint01
    );
    canvas.drawLine
    (
      this.intLeft, this.intTop,
      this.intLeft, this.intTop+intHeight, mPaint01
    );
    canvas.drawLine
    (
      this.intLeft+intWidth, this.intTop,
      this.intLeft+intWidth, this.intTop+intHeight, mPaint01
    );
    canvas.drawLine
    (
      this.intLeft, this.intTop+intHeight,
      this.intLeft+intWidth, this.intTop+intHeight, mPaint01
    );
    super.onDraw(canvas);
  }
}
```

（14）定义方法 eregi 实现自定义比较字符串处理，具体代码如下所示。

```
/* 自定义比较字符串函数 */
public static boolean eregi(String strPat, String strUnknow)
{
  String strPattern = "(?i)"+strPat;
  Pattern p = Pattern.compile(strPattern);
  Matcher m = p.matcher(strUnknow);
  return m.find();
```

```
    }

    @Override
    public void surfaceChanged
    (SurfaceHolder surfaceholder, int format, int w, int h)
    {
      // TODO Auto-generated method stub
      Log.i(TAG, "Surface Changed");
    }

    @Override
    public void surfaceCreated(SurfaceHolder surfaceholder)
    {
      // TODO Auto-generated method stub
      Log.i(TAG, "Surface Changed");
    }

    @Override
    public void surfaceDestroyed(SurfaceHolder surfaceholder)
    {
      // TODO Auto-generated method stub
      Log.i(TAG, "Surface Destroyed");
    }

    @Override
    protected void onPause()
    {
      // TODO Auto-generated method stub
      super.onPause();
    }
}
```

执行后能够通过手机拍照的方式实现二维码解析，如图 24-5 所示。

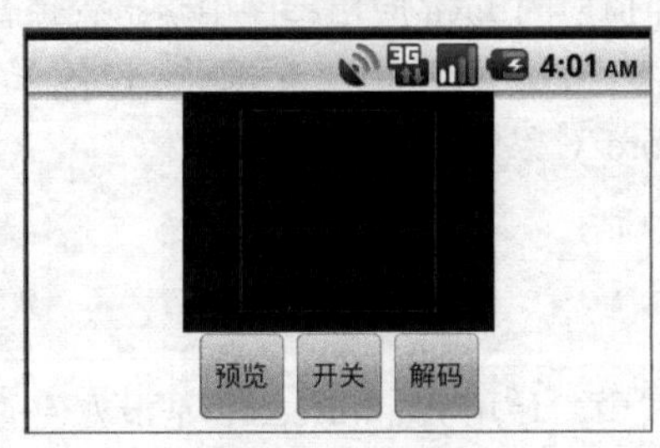

▲图 24-5　执行效果

第 25 章　麦克风音频录制技术详解

在当今的智能手机中，几乎每一款手机都具备录音功能。在 Android 系统中，同样也可以实现录音处理。本章将详细讲解在 Android 设备中使用麦克风（传感器）来录制音频技术的基本知识，为步入本书后面知识的学习打下基础。

25.1 使用 MediaRecorder 接口录制音频

在 Android 系统中，通常采用 MediaRecorder 接口实现录制音频和视频功能。在录制音频文件之前，需要设置音频源、输出格式、录制时间和编码格式等。本节将详细讲解在 Android 设备中使用 MediaRecorder 接口录制音频的方法。

25.1.1　类 MediaRecorder 详解

类 AudioRecord 在 Android 顶层的 Java 应用程序中负责管理音频资源，通过 AudioRecord 对象来完成“pulling”（读取）数据，以记录从平台音频输入设备产生的数据。在 Android 应用中，可以通过以下方法从 AudioRecord 对象中读取音频数据。

- read(byte[], int, int)。
- read(short[], int, int)。
- read(ByteBuffer, int)。

在上述读取音频数据的方式中，使用 AudioRecord 是最方便的。

在 Android 系统中，当创建 AudioRecord 对象时会初始化 AudioRecord，并和音频缓冲区连接以缓冲新的音频数据。根据构造时指定的缓冲区大小，可以决定 AudioRecord 能够记录多长的数据。一般来说，从硬件设备读取的数据应小于整个记录缓冲区。

MediaRecorder 的内部类是 AudioRecord.OnRecordPositionUpdateListener，当 AudioRecord 收到一个由 setNotificationMarkerPosition(int)设置的通知标志，或由 setPositionNotificationPeriod(int)设置的周期更新记录的进度状态时，需要回调这个接口。

在类 MediaRecorder 中，包含了表 25-1 中列出的常用方法。

表 25-1　类 MediaRecorder 中的常用方法

方法名称	描　述
public void setAudioEncoder (int audio_encoder)	设置刻录的音频编码，其值可以通过 MediaRecoder 内部类的 MediaRecorder.AudioEncoder 的几个常量：AAC、AMR_NB、AMR_WB、DEFAULT 来设置
public void setAudioEncodingBitRate (int bitRate)	设置音频编码比特率
public void setAudioSource (int audio_source)	设置音频的来源，其值可以通过 MediaRecoder 内部类的 MediaRecorder.AudioSource 的几个常量来设置，通常设置的值 MIC：来源于麦克风

续表

方法名称	描 述
public void setCamera (Camera c)	设置摄像头用于刻录
public void setOutputFormat (int output_format)	设置输出文件的格式，其值可以通过 MediaRecoder 内部类 MediaRecorder. OutputFormat 的一些常量字段来设置，如一些 3gp(THREE_GPP)、MP4(MPEG4)等
setOutputFile(String path)	设置输出文件的路径
setVideoEncoder(int video_encoder)	设置视频的编码格式，其值可以通过 MediaRecoder 内部类的 MediaRecorder. VideoEncoder 的几个常量：H263、H264、MPEG_4_SP 来设置
setVideoSource(int video_source)	设置刻录视频来源，其值可以通过 MediaRecorder 的内部类 MediaRecorder. VideoSource 来设置，如可以设置刻录视频来源为摄像头：CAMERA
setVideoEncodingBitRate(int bitRate)	设置编码的比特率
setVideoSize(int width, int height)	设置视频的大尺寸
public void start()	开始刻录
public void prepare()	预期做准备
public void stop()	停止
public void release()	释放该对象资源

在 AudioRecord 中，有一个受保护的方法 protected void finalize()，用于通知 VM 回收此对象内存。方法 finalize ()只能用在运行的应用程序没有任何线程再使用此对象，来告诉垃圾回收器回收此对象。方法 finalize ()用于释放系统资源，由垃圾回收器清除此对象。在执行期间，调用方法 finalize ()可能会立即抛出未定义异常，但是可以忽略。

注意　VM 保证对象可以一次或多次调用 finalize()，但并不保证 finalize()会马上执行。例如，对象 B 的 finalize()可能延迟执行，等待对象 A 的 finalize()延迟回收 A 的内存。为了安全起见，请看 ReferenceQueue，它提供了更多地控制 VM 的垃圾回收。

25.1.2 实战演练——使用 MediaRecorder 录制音频

在接下来将通过一个具体实例的实现过程，来讲解使用 MediaRecorder 实现音频录制的方法。在本实例中插入了 4 个按钮，分别实现录音、停止录音、播放录音和删除录音这 4 种操作。为了能够不限制录音的长度，现将录音暂时保存到存储卡，当录音完毕后，再将录音文件显示在 ListView 列表中。单击文件后，可以播放或删除录音文件。

实例	功能	源码路径
实例 25-1	使用 MediaRecorder 录制音频	\daima\25\luzhi

本实例的实现文件是 example.java，具体实现流程如下所示。

（1）分别构造 4 个按钮对象和两个文本对象，然后设置按钮状态不可选，具体实现代码如下所示。

```
/** Called when the activity is first created. */
@Override
public void onCreate(Bundle savedInstanceState)
{
  super.onCreate(savedInstanceState);
  setContentView(R.layout.main);
  /* 设置 4 个按钮和 2 个文本 */
  myButton1 = (ImageButton) findViewById(R.id.ImageButton01);
  myButton2 = (ImageButton) findViewById(R.id.ImageButton02);
```

```
    myButton3 = (ImageButton) findViewById(R.id.ImageButton03);
    myButton4 = (ImageButton) findViewById(R.id.ImageButton04);
    myListView1 = (ListView) findViewById(R.id.ListView01);
    myTextView1 = (TextView) findViewById(R.id.TextView01);
    /* 设置按钮状态不可选 */
    myButton2.setEnabled(false);
    myButton3.setEnabled(false);
    myButton4.setEnabled(false);
```

（2）通过方法 sdCardExit 判断是否插入 SD 卡，然后将获取的 SD 卡路径作为录音的文件位置，并取得 SD 卡目录里的所有“.amr”格式的文件，最后将 ArrayAdapter 添加 ListView 对象中以列表显示录音文件，具体实现代码如下所示。

```
/* 判断是否插入 SD 卡 */
sdCardExit = Environment.getExternalStorageState().equals(
    android.os.Environment.MEDIA_MOUNTED);
/* 取得 SD Card 路径作为录音的文件位置 */
if (sdCardExit)
  myRecAudioDir = Environment.getExternalStorageDirectory();
/* 取得 SD Card 目录里的所有.amr 文件 */
getRecordFiles();
adapter = new ArrayAdapter<String>(this,
    R.layout.my_simple_list_item, recordFiles);
/* 将 ArrayAdapter 添加 ListView 对象中 */
myListView1.setAdapter(adapter);
```

（3）编写单击录音按钮后的录音处理事件，先创建录音频文件，然后设置录音来源为麦克风，最后通过 myTextView1.setText("录音中")设置录音过程显示的提示文本，具体实现代码如下所示。

```
/* 单击录音按钮的处理事件 */
myButton1.setOnClickListener(new ImageButton.OnClickListener()
{
  @Override
  public void onClick(View arg0)
  {
    try
    {
      if (!sdCardExit)
      {
        Toast.makeText(example.this, "请插入 SD Card",
            Toast.LENGTH_LONG).show();
        return;
      }
      /* 创建录音频文件 */
      myRecAudioFile = File.createTempFile(strTempFile, ".amr",
          myRecAudioDir);
      mMediaRecorder01 = new MediaRecorder();
      /* 设置录音来源为麦克风 */
      mMediaRecorder01
          .setAudioSource(MediaRecorder.AudioSource.MIC);
      mMediaRecorder01
          .setOutputFormat(MediaRecorder.OutputFormat.DEFAULT);
      mMediaRecorder01
          .setAudioEncoder(MediaRecorder.AudioEncoder.DEFAULT);
      mMediaRecorder01.setOutputFile(myRecAudioFile
          .getAbsolutePath());
      mMediaRecorder01.prepare();
      mMediaRecorder01.start();
      myTextView1.setText("录音中");
      myButton2.setEnabled(true);
      myButton3.setEnabled(false);
      myButton4.setEnabled(false);
      isStopRecord = false;
    }
    catch (IOException e)
    {
      // TODO Auto-generated catch block
      e.printStackTrace();
```

```
        }
    }
});
```

（4）编写单击停止按钮的处理事件，首先使用方法 mMediaRecorder01.stop()停止录音，然后将录音文件名传递给 Adapter，具体实现代码如下所示。

```
/* 停止 */
myButton2.setOnClickListener(new ImageButton.OnClickListener()
{
  @Override
  public void onClick(View arg0)
  {
    // TODO Auto-generated method stub
    if (myRecAudioFile != null)
    {
      /* 停止录音 */
      mMediaRecorder01.stop();
      mMediaRecorder01.release();
      mMediaRecorder01 = null;
      /* 将录音频文件名给 Adapter */
      adapter.add(myRecAudioFile.getName());
      myTextView1.setText("停止: " + myRecAudioFile.getName());
      myButton2.setEnabled(false);
      isStopRecord = true;
    }
  }
});
```

（5）编写单击播放按钮的处理事件，单击后将打开播放的程序，主要代码如下所示。

```
/* 播放 */
myButton3.setOnClickListener(new ImageButton.OnClickListener()
{
  @Override
  public void onClick(View arg0)
  {
    // TODO Auto-generated method stub
    if (myPlayFile != null && myPlayFile.exists())
    {
      /* 打开播放的程序 */
      openFile(myPlayFile);
    }
  }
});
```

（6）编写单击删除按钮的处理事件，首先在 Adapter 删除录音文件名，然后删除录制的文件，具体实现代码如下所示。

```
/* 删除事件 */
myButton4.setOnClickListener(new ImageButton.OnClickListener()
{
  @Override
  public void onClick(View arg0)
  {
    if (myPlayFile != null)
    {
      /* 先将 Adapter 删除文件名 */
      adapter.remove(myPlayFile.getName());
      /* 删除文件 */
      if (myPlayFile.exists())
        myPlayFile.delete();
      myTextView1.setText("完成删除");
    }
  }
});
```

（7）编写单击 Adapter 列表中某个录制文件的处理事件，如果有文件，单击后将删除及播放按钮设置为 Enable 不可用，然后输出选择提示语句，具体实现代码如下所示。

```
    myListView1.setOnItemClickListener
    (new AdapterView.OnItemClickListener()
    {
      @Override
      public void onItemClick(AdapterView<?> arg0, View arg1,
          int arg2, long arg3)
      {
        /* 当有点击文档名时将删除及播放按钮 Enable */
        myButton3.setEnabled(true);
        myButton4.setEnabled(true);
        myPlayFile = new File(myRecAudioDir.getAbsolutePath()
            + File.separator
            + ((CheckedTextView) arg1).getText());
        myTextView1.setText("你选的是："
            + ((CheckedTextView) arg1).getText());
      }
    });
  }
```

（8）定义方法 onStop()实现停止录音操作，具体实现代码如下所示。

```
  protected void onStop()
  {
    if (mMediaRecorder01 != null && !isStopRecord)
    {
      /* 停止录音 */
      mMediaRecorder01.stop();
      mMediaRecorder01.release();
      mMediaRecorder01 = null;
    }
    super.onStop();
  }
```

（9）定义方法 getRecordFiles()来获取文件的长度，在此设置只能获取“.amr”格式的文件，具体实现代码如下所示。

```
  private void getRecordFiles()
  {
    recordFiles = new ArrayList<String>();
    if (sdCardExit)
    {
      File files[] = myRecAudioDir.listFiles();
      if (files != null)
      {
        for (int i = 0; i < files.length; i++)
        {
          if (files[i].getName().indexOf(".") >= 0)
          {
            /* 只取.amr 文件 */
            String fileS = files[i].getName().substring(
                files[i].getName().indexOf("."));
            if (fileS.toLowerCase().equals(".amr"))
              recordFiles.add(files[i].getName());
          }
        }
      }
    }
  }
```

（10）定义方法 openFile(File f)来打开播放指定的录音文件，主要代码如下所示。

```
  /* 打开播放录音文件的程序 */
  private void openFile(File f)
  {
    Intent intent = new Intent();
    intent.addFlags(Intent.FLAG_ACTIVITY_NEW_TASK);
    intent.setAction(android.content.Intent.ACTION_VIEW);
    String type = getMIMEType(f);
    intent.setDataAndType(Uri.fromFile(f), type);
    startActivity(intent);
  }
```

（11）定义方法 getMIMEType(File f)设置系统可接受的文件类型，在此设置了 audio 类型、image 类型和其他类型，具体实现代码如下所示。

```
private String getMIMEType(File f)
{
  String end = f.getName().substring(
     f.getName().lastIndexOf(".") + 1, f.getName().length())
     .toLowerCase();
  String type = "";
  if (end.equals("mp3") || end.equals("aac") || end.equals("aac")
     || end.equals("amr") || end.equals("mpeg")
     || end.equals("mp4"))
  {
   type = "audio";
  }
  else if (end.equals("jpg") || end.equals("gif")
     || end.equals("png") || end.equals("jpeg"))
  {
   type = "image";
  }
  else
  {
   type = "*";
  }
  type += "/*";
  return type;
 }
}
```

还需在文件 AndroidManifest.xml 中声明录音权限，具体实现代码如下所示。

```
<uses-permission android:name="android.permission.RECORD_AUDIO">
```

至此，整个实例介绍完毕，执行后的效果如图 25-1 所示。当单击“录音”按钮时开始录音处理，如图 25-2 所示。当单击“停止”按钮后停止录音处理，并在列表中显示录制的音频文件。当选中音频文件并单击“删除”按钮后会删除选中音频文件，单击“播放”按钮后会播放选中的音频文件。

▲图 25-1　初始效果

▲图 25-2　正在录音

25.2 使用 AudioRecord 接口录制音频

类 AudioRecord 可以在 Java 应用程序中管理音频资源，通过 AudioRecord 对象来完成“pulling”（读取）数据的方法，以记录从平台音频输入设备产生的数据。本节将详细讲解在 Android 设备中使用 AudioRccord 接口录制音频的知识。

25.2.1 AudioRecord 的常量

AudioRecord 中包含的常量如下所示。

- public static final int ERROR：表示操作失败，常量值：–1 (0xffffffff)。
- public static final int ERROR_BAD_VALUE：表示使用了一个不合理的值导致的失败，常量值：–2 (0xfffffffe)。
- public static final int ERROR_INVALID_OPERATION：表示不恰当的方法导致的失败，常量值：–3 (0xfffffffd)。

- public static final int RECORDSTATE_RECORDING：指示 AudioRecord 录制状态为“正在录制”，常量值：3 (0x00000003)。
- public static final int RECORDSTATE_STOPPED：指示 AudioRecord 录制状态为“不在录制”，常量值：1 (0x00000001)。
- public static final int STATE_INITIALIZED：指示 AudioRecord 准备就绪，常量值：1 (0x00000001)。
- public static final int STATE_UNINITIALIZED：指示 AudioRecord 状态没有初始化成功，常量值：0 (0x00000000)。
- public static final int SUCCESS：表示操作成功，常量值：0 (0x00000000)。

25.2.2　AudioRecord 的构造函数

AudioRecord 的构造函数是 AudioRecord，具体定义格式如下所示。

```
public AudioRecord (int audioSource, int sampleRateInHz, int channelConfig, int
audioFormat, int bufferSizeInBytes)
```

各个参数的具体说明如下所示。

- audioSource：录制源。
- sampleRateInHz：默认采样率，单位 Hz。 44100Hz 是当前唯一能保证在所有设备上工作的采样率，在一些设备上还有 22050，16000 或 11025。
- channelConfig：描述音频通道设置。
- audioFormat：音频数据保证支持此格式。
- bufferSizeInBytes：在录制过程中，音频数据写入缓冲区的总数（字节）。从缓冲区读取的新音频数据总会小于此值。用 getMinBufferSize(int, int, int)返回 AudioRecord 实例创建成功后的最小缓冲区，如果其设置的值比 getMinBufferSize()还小，则会导致初始化失败。

25.2.3　AudioRecord 的公共方法

AudioRecord 中的公共方法如下所示。

（1）public int getAudioFormat ()：返回设置的音频数据格式。

（2）public int getAudioSource ()：返回音频录制源。

（3）public int getChannelConfiguration ()：返回设置的频道设置。

（4）public int getChannelCount ()：返回设置的频道数目。

（5）public static int getMinBufferSize (int sampleRateInHz, int channelConfig, int audioFormat)：返回成功创建 AudioRecord 对象所需要的最小缓冲区大小。 注意：这个大小并不保证在负荷下的流畅录制，应根据预期的频率来选择更高的值，AudioRecord 实例在推送新数据时使用此值。

各个参数的具体说明如下。

- sampleRateInHz：默认采样率，单位 Hz。
- channelConfig：描述音频通道设置。
- audioFormat：音频数据保证支持此格式，参见 ENCODING_PCM_16BIT。

如果硬件不支持录制参数，或输入了一个无效的参数，则返回 ERROR_BAD_VALUE，如果硬件查询到输出属性没有实现，或最小缓冲区用 byte 表示，则返回 ERROR。

（6）public int getNotificationMarkerPosition ()：返回通知，标记框架中的位置。

（7）public int getPositionNotificationPeriod ()：返回通知，更新框架中的时间位置。

（8）public int getRecordingState ()：返回 AudioRecord 实例的录制状态。

（9）public int getSampleRate ()：返回设置的音频数据样本采样率，单位 Hz。

（10）public int getState ()：返回 AudioRecord 实例的状态。

（11）public int read (short[] audioData, int offsetInShorts, int sizeInShorts)：从音频硬件录制缓冲区读取数据。

各个参数的具体说明如下所示。

- audioData：写入的音频录制数据。
- offsetInShorts：目标数组 audioData 的起始偏移量。
- sizeInShorts：请求读取的数据大小。

返回值是返回 short 型数据，表示读取到的数据，如果对象属性没有初始化，则返回 ERROR_INVALID_OPERATION，如果参数不能解析成有效的数据或索引，则返回 ERROR_BAD_VALUE。返回数值不会超过 sizeInShorts。

（12）public int read (byte[] audioData, int offsetInBytes, int sizeInBytes)：从音频硬件录制缓冲区读取数据，读入缓冲区的总 byte 数，如果对象属性没有初始化，则返回 ERROR_INVALID_OPERATION，如果参数不能解析成有效的数据或索引，则返回 ERROR_BAD_VALUE。读取的总 byte 数不会超过 sizeInBytes。各个参数的具体说明如下所示。

- audioData：写入的音频录制数据。
- offsetInBytes：audioData 的起始偏移值，单位 byte。
- sizeInBytes：读取的最大字节数。

（13）public int read (ByteBuffer audioBuffer, int sizeInBytes)：从音频硬件录制缓冲区读取数据，直接复制到指定缓冲区。如果 audioBuffer 不是直接的缓冲区，此方法总是返回 0。读入缓冲区的总 byte 数，如果对象属性没有初始化，则返回 ERROR_INVALID_OPERATION，如果参数不能解析成有效的数据或索引，则返回 ERROR_BAD_VALUE。读取的总 byte 数不会超过 sizeInBytes。各个参数的具体说明如下所示。

- audioBuffer：存储写入音频录制数据的缓冲区。
- sizeInBytes：请求的最大字节数。

（14）public void release ()：释放本地 AudioRecord 资源。对象不能经常使用此方法，而且在调用 release()后，必须设置引用为 null。

（15）public int setNotificationMarkerPosition (int markerInFrames)：如果设置了 setRecordPositionUpdateListener(OnRecordPositionUpdateListener) 或 setRecordPositionUpdateListener(OnRecordPositionUpdateListener, Handler)，则通知监听者设置位置标记。

参数 markerInFrames 表示在框架中快速标记位置。

（16）public int setPositionNotificationPeriod (int periodInFrames)：如果设置了 setRecordPositionUpdateListener(OnRecordPositionUpdateListener) 或 setRecordPositionUpdateListener(OnRecordPositionUpdateListener, Handler)，则通知监听者设置时间标记。

参数 periodInFrames 表示在框架中快速更新时间标记。

（17）public void setRecordPositionUpdateListener (AudioRecord.OnRecordPositionUpdateListener listener, Handler handler)：当之前设置的标志已经成立，或者周期录制位置更新时，设置处理监听者。使用此方法来将 Handler 和别的线程联系起来，来接收 AudioRecord 事件，比创建 AudioTrack 实例更好一些。

参数 handler 用来接收事件通知消息。

（18）public void setRecordPositionUpdateListener (AudioRecord.OnRecordPositionUpdateListener

listener)：当之前设置的标志已经成立，或者周期录制位置更新时，设置处理监听者。

（19）public void startRecording ()：表示 AudioRecord 实例开始进行录制。

25.2.4 AudioRecord 的受保护方法

在 AudioRecord 中受保护方法是 protected void finalize ()，此方法用于通知 VM 回收此对象内存。只能被用在运行的应用程序没有任何线程再使用此对象，来告诉垃圾回收器回收此对象。

方法 finalize ()用于释放系统资源，由垃圾回收器清除此对象。默认没有实现，由 VM 来决定，但子类根据需要可重写 finalize()。在执行期间，调用此方法可能会立即抛出未定义异常，但是可以忽略。

注意

VM 保证对象可以一次或多次调用 finalize()，但并不保证 finalize()会马上执行。例如，对象 B 的 finalize()可能延迟执行，等待对象 A 的 finalize()延迟回收 A 的内存。为了安全起见，请看 ReferenceQueue，它提供了更多地控制 VM 的垃圾回收。另外，需要在 Activity 的线程里面创建 AudioRecord 对象，可以在独立的线程里面进行读取数据，否则手机录音时会出错。

25.2.5 实战演练——使用 AudioRecord 录制音频

在 Android 系统中，因为 MediaRecorder 可以直接把麦克风的数据存到文件，并且能够直接进行编码（如 AMR、MP3 等），而 AudioRecord 则是读取麦克风的音频流。本实例使用 AudioRecord 读取音频流，使用 AudioTrack 播放音频流，通过“边读边播放”及增大音量的方式实现一个简单的助听器程序。在接下来将通过一个具体实例的实现过程，来讲解使用 AudioRecord 实现音频录制的方法。

题目	目的	源码路径
实例 25-2	使用 AudioRecord 录制音频	\daima\25\testRecord

本实例的功能是使用 AudioRecord 录制音频，具体实现流程如下所示。

（1）编写布局文件 main.xml，主要代码如下所示。

```
<?xml version="1.0" encoding="utf-8"?>
<LinearLayout xmlns:android="http://schemas.android.com/apk/res/android"
    android:orientation="vertical" android:layout_width="fill_parent"
    android:layout_height="fill_parent">
    <Button android:layout_height="wrap_content" android:id="@+id/btnRecord"
        android:layout_width="fill_parent" android:text="开始边录边放"></Button>
    <Button android:layout_height="wrap_content"
        android:layout_width="fill_parent"  android:text=" 停 止 "  android:id="@+id/btn
Stop"></Button>
    <Button android:layout_height="wrap_content" android:id="@+id/btnExit"
        android:layout_width="fill_parent" android:text="退出"></Button>
    <TextView android:id="@+id/TextView01" android:layout_height="wrap_content"
        android:text="程序音量调节" android:layout_width="fill_parent"></TextView>
    <SeekBar android:layout_height="wrap_content" android:id="@+id/skbVolume"
        android:layout_width="fill_parent"></SeekBar>
</LinearLayout>
```

（2）在文件 AndroidManifest.xml 中声明权限，主要代码如下所示。

```
<uses-permission android:name="android.permission.RECORD_AUDIO"></uses-permission>
```

（3）编写文件 TestRecordActivity.java，主要代码如下所示。

```
public class TestRecordActivity extends Activity {
Button btnRecord, btnStop, btnExit;
```

```
SeekBar skbVolume;//调节音量
boolean isRecording = false;//是否录放的标记
static final int frequency = 44100;
static final int channelConfiguration = AudioFormat.CHANNEL_CONFIGURATION_MONO;
static final int audioEncoding = AudioFormat.ENCODING_PCM_16BIT;
int recBufSize,playBufSize;
AudioRecord audioRecord;
AudioTrack audioTrack;

@Override
public void onCreate(Bundle savedInstanceState) {
    super.onCreate(savedInstanceState);
    setContentView(R.layout.main);
    setTitle("助听器");
    recBufSize = AudioRecord.getMinBufferSize(frequency,
            channelConfiguration, audioEncoding);
    playBufSize=AudioTrack.getMinBufferSize(frequency,
            channelConfiguration, audioEncoding);
    audioRecord = new AudioRecord(MediaRecorder.AudioSource.MIC, frequency,
            channelConfiguration, audioEncoding, recBufSize);
    audioTrack = new AudioTrack(AudioManager.STREAM_MUSIC, frequency,
            channelConfiguration, audioEncoding,
            playBufSize, AudioTrack.MODE_STREAM);
    btnRecord = (Button) this.findViewById(R.id.btnRecord);
    btnRecord.setOnClickListener(new ClickEvent());
    btnStop = (Button) this.findViewById(R.id.btnStop);
    btnStop.setOnClickListener(new ClickEvent());
    btnExit = (Button) this.findViewById(R.id.btnExit);
    btnExit.setOnClickListener(new ClickEvent());
    skbVolume=(SeekBar)this.findViewById(R.id.skbVolume);
    skbVolume.setMax(100);//音量调节的极限
    skbVolume.setProgress(70);//设置 seekbar 的位置值
    audioTrack.setStereoVolume(0.7f, 0.7f);//设置当前音量大小
    skbVolume.setOnSeekBarChangeListener(new SeekBar.OnSeekBarChangeListener() {

        @Override
        public void onStopTrackingTouch(SeekBar seekBar) {
            float vol=(float)(seekBar.getProgress())/(float)(seekBar.getMax());
            audioTrack.setStereoVolume(vol, vol);//设置音量
        }

        @Override
        public void onStartTrackingTouch(SeekBar seekBar) {
        }

        public void onProgressChanged(SeekBar seekBar, int progress,
                boolean fromUser) {
        }
    });
}
protected void onDestroy() {
    super.onDestroy();
    android.os.Process.killProcess(android.os.Process.myPid());
}

class ClickEvent implements View.OnClickListener {

    @Override
    public void onClick(View v) {
        if (v == btnRecord) {
            isRecording = true;
            new RecordPlayThread().start();// 开一条线程边录边放
        } else if (v == btnStop) {
            isRecording = false;
        } else if (v == btnExit) {
            isRecording = false;
            TestRecordActivity.this.finish();
        }
    }
}
```

```
    class RecordPlayThread extends Thread {
        public void run() {
            try {
                byte[] buffer = new byte[recBufSize];
                audioRecord.startRecording();//开始录制
                audioTrack.play();//开始播放

                while (isRecording) {
                    //从 MIC 保存数据到缓冲区
                    int bufferReadResult = audioRecord.read(buffer, 0,
                            recBufSize);

                    byte[] tmpBuf = new byte[bufferReadResult];
                    System.arraycopy(buffer, 0, tmpBuf, 0, bufferReadResult);
                    //写入数据即播放
                    audioTrack.write(tmpBuf, 0, tmpBuf.length);
                }
                audioTrack.stop();
                audioRecord.stop();
            } catch (Throwable t) {
                Toast.makeText(TestRecordActivity.this, t.getMessage(), 1000);
            }
        }
    };
}
```

执行后可以实现录音功能，效果如图 25-3 所示。程序音量调节只是程序内部调节音量而已，要调到最大音量还需要手动设置系统音量。

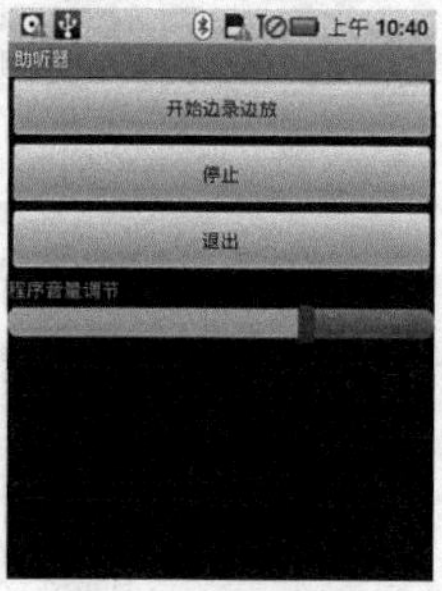

▲图 25-3　执行效果

25.3　实战演练——麦克风录音综合实例

本实例讲解将麦克风作为外部传感器设备，演示了实现综合录音处理功能的具体方法。

实例	功能	源码路径
实例 25-3	麦克风录音综合实例	https://github.com/gast-lib/gast-lib

本实例源码是开源代码，来源于如下地址，读者可以自行登录并下载。

```
https://github.com/gast-lib/gast-lib/
```

25.3.1　获取录音源的最大振幅

在录制音频的过程中，探测录音源的最大振幅十分有用。在 Android 系统中，通过 MediaRecorder 可以获取音源的最大振幅值。编写文件 MaxAmplitudeRecorder.java，功能是定期记录录制音源的最大振幅值，具体实现代码如下所示。

```
public class MaxAmplitudeRecorder
{
    private static final String TAG = "MaxAmplitudeRecorder";
    private static final long DEFAULT_CLIP_TIME = 1000;
    private long clipTime = DEFAULT_CLIP_TIME;
    private AmplitudeClipListener clipListener;
    private boolean continueRecording;
    private MediaRecorder recorder;
    private String tmpAudioFile;
    private AsyncTask task;
    public MaxAmplitudeRecorder(long clipTime, String tmpAudioFile,
            AmplitudeClipListener clipListener, AsyncTask task)
    {
        this.clipTime = clipTime;
        this.clipListener = clipListener;
```

```
        this.tmpAudioFile = tmpAudioFile;
        this.task = task;
    }
//检测设备是否有麦克风
    public boolean startRecording() throws IOException
    {
        Log.d(TAG, "recording maxAmplitude");

        recorder = AudioUtil.prepareRecorder(tmpAudioFile);

        // when an error occurs just stop recording
        recorder.setOnErrorListener(new MediaRecorder.OnErrorListener()
        {
            @Override
            public void onError(MediaRecorder mr, int what, int extra)
            {
                // log it
                new RecorderErrorLoggerListener().onError(mr, what, extra);
                // stop recording
                stopRecording();
            }
        });

        //possible RuntimeException if Audio recording channel is occupied
        recorder.start();

        continueRecording = true;
        boolean heard = false;
        recorder.getMaxAmplitude();
        while (continueRecording)
        {
            Log.d(TAG, "waiting while recording...");
            waitClipTime();
            if (task != null)
            {
                Log.d(TAG, "continue recording: " + continueRecording + " cancelled after
                waiting? " + task.isCancelled());
            }
            //in case external code stopped this while read was happening
            if ((!continueRecording) || ((task != null) && task.isCancelled()))
            {
                break;
            }

            int maxAmplitude = recorder.getMaxAmplitude();
            Log.d(TAG, "current max amplitude: " + maxAmplitude);

            heard = clipListener.heard(maxAmplitude);
            if (heard)
            {
                stopRecording();
            }
        }

        Log.d(TAG, "stopped recording max amplitude");
        done();

        return heard;
    }

    private void waitClipTime()
    {
        try
        {
            Thread.sleep(clipTime);
        } catch (InterruptedException e)
        {
            Log.d(TAG, "interrupted");
        }
    }
```

```
    /**
     * 停止录制清空资源
     */
    public void done()
    {
        Log.d(TAG, "stop recording on done");
        if (recorder != null)
        {
            try
            {
                recorder.stop();
            } catch (Exception e)
            {
                Log.d(TAG, "failed to stop");
                return;
            }
            recorder.release();
        }
    }
    public boolean isRecording()
    {
        return continueRecording;
    }
    public void stopRecording()
    {
        continueRecording = false;
    }
}
```

25.3.2 实现异步音频录制功能

编写文件 RecordAmplitudeTask.java，功能是实现异步音频录制功能，此功能可以保证在录制音频时做其他的事情。文件 RecordAmplitudeTask.java 的具体实现代码如下所示。

```
public class RecordAmplitudeTask extends
        AsyncTask<AmplitudeClipListener, Void, Boolean>
{
    private static final String TAG = "RecordAmplitudeTask";

    private TextView status;
    private TextView log;
    private Context context;
    private String taskName;

    private static final String TEMP_AUDIO_DIR_NAME = "temp_audio";

    /**
     * time between amplitude checks
     */
    private static final int CLIP_TIME = 1000;

    public RecordAmplitudeTask(Context context, TextView status, TextView log,
            String taskName)
    {
        this.context = context;
        this.status = status;
        this.log = log;
        this.taskName = taskName;
    }

    @Override
    protected void onPreExecute()
    {
        // tell UI recording is starting
        status.setText(context.getResources().getString(
                R.string.audio_status_recording)
                + " for " + taskName);
        AudioTaskUtil.appendToStartOfLog(log, "started " + taskName);
```

```
        super.onPreExecute();
    }

    /**
     * note: only uses the first listener passed in
     */
    @Override
    protected Boolean doInBackground(AmplitudeClipListener... listeners)
    {
        if (listeners.length == 0)
        {
            return false;
        }

        Log.d(TAG, "recording amplitude");
        // construct recorder, using only the first listener passed in
        AmplitudeClipListener listener = listeners[0];
        String appStorageLocation =
            context.getExternalFilesDir(TEMP_AUDIO_DIR_NAME).getAbsolutePath()
                + File.separator + "audio.3gp";
        MaxAmplitudeRecorder recorder =
              new MaxAmplitudeRecorder(CLIP_TIME, appStorageLocation,
                  listener, this);

        //set to true if the recorder successfully detected something
        //false if it was canceled or otherwise stopped
        boolean heard = false;
        try
        {
            // start recording
            heard = recorder.startRecording();
        } catch (IOException io)
        {
            Log.e(TAG, "failed to record", io);
            heard = false;
        } catch (IllegalStateException se)
        {
            Log.e(TAG, "failed to record, recorder not setup properly", se);
            heard = false;
        } catch (RuntimeException se)
        {
            Log.e(TAG, "failed to record, recorder already being used", se);
            heard = false;
        }

        return heard;
    }

    @Override
    protected void onPostExecute(Boolean result)
    {
        // update UI
        if (result)
        {
            AudioTaskUtil.appendToStartOfLog(log, "heard clap at "
                    + AudioTaskUtil.getNow());
        }
        else
        {
            AudioTaskUtil.appendToStartOfLog(log, "heard no claps");
        }
        setDoneMessage();
        super.onPostExecute(result);
    }

    @Override
    protected void onCancelled()
    {
        AudioTaskUtil.appendToStartOfLog(log, "cancelled " + taskName);
        setDoneMessage();
```

```
            super.onCancelled();
        }

        private void setDoneMessage()
        {
            status.setText(context.getResources().getString(
                    R.string.audio_status_stopped));
        }
    }
```

25.3.3　监听是否超越最大值

编写文件 SingleClapDetector.java 定义一个声控开关，功能是监听最大振幅是否超越了我们设置的最大值。文件 SingleClapDetector.java 的具体实现代码如下所示。

```
public class SingleClapDetector implements AmplitudeClipListener
{
    private static final String TAG = "SingleClapDetector";
    /**
     * required loudness to determine it is a clap
     */
    private int amplitudeThreshold;
    /**
     * requires a little of noise by the user to trigger, background noise may
     * trigger it
     */
    public static final int AMPLITUDE_DIFF_LOW = 10000;
    public static final int AMPLITUDE_DIFF_MED = 18000;
    /**
     * requires a lot of noise by the user to trigger. background noise isn't
     * likely to be this loud
     */
    public static final int AMPLITUDE_DIFF_HIGH = 25000;

    private static final int DEFAULT_AMPLITUDE_DIFF = AMPLITUDE_DIFF_MED;

    public SingleClapDetector()
    {
        this(DEFAULT_AMPLITUDE_DIFF);
    }
    public SingleClapDetector(int amplitudeThreshold)
    {
        this.amplitudeThreshold = amplitudeThreshold;
    }
    @Override
    public boolean heard(int maxAmplitude)
    {
        boolean clapDetected = false;
        if (maxAmplitude >= amplitudeThreshold)
        {
            Log.d(TAG, "heard a clap");
            clapDetected = true;
        }
        return clapDetected;
    }
}
```

25.3.4　录制音频

编写文件 AudioClipRecorder.java，功能是通过 AudioRecorder 录制音频，具体实现代码如下所示。

```
public class AudioClipRecorder
{
    private static final String TAG = "AudioClipRecorder";

    private AudioRecord recorder;
    private AudioClipListener clipListener;
```

```
private boolean continueRecording;

public static final int RECORDER_SAMPLERATE_CD = 44100;
public static final int RECORDER_SAMPLERATE_8000 = 8000;

private static final int DEFAULT_BUFFER_INCREASE_FACTOR = 3;

private AsyncTask task;

private boolean heard;

public AudioClipRecorder(AudioClipListener clipListener)
{
   this.clipListener = clipListener;
   heard = false;
   task = null;
}

public AudioClipRecorder(AudioClipListener clipListener, AsyncTask task)
{
   this(clipListener);
   this.task = task;
}

public boolean startRecording()
{
   return startRecording(RECORDER_SAMPLERATE_8000,
         AudioFormat.ENCODING_PCM_16BIT);
}
public boolean startRecordingForTime(int millisecondsPerAudioClip,
      int sampleRate, int encoding)
{
   float percentOfASecond = (float) millisecondsPerAudioClip / 1000.0f;
   int numSamplesRequired = (int) ((float) sampleRate * percentOfASecond);
   int bufferSize =
         determineCalculatedBufferSize(sampleRate, encoding,
               numSamplesRequired);

   return doRecording(sampleRate, encoding, bufferSize,
         numSamplesRequired, DEFAULT_BUFFER_INCREASE_FACTOR);
}

public boolean startRecording(final int sampleRate, int encoding)
{
   int bufferSize = determineMinimumBufferSize(sampleRate, encoding);
   return doRecording(sampleRate, encoding, bufferSize, bufferSize,
         DEFAULT_BUFFER_INCREASE_FACTOR);
}

private int determineMinimumBufferSize(final int sampleRate, int encoding)
{
   int minBufferSize =
         AudioRecord.getMinBufferSize(sampleRate,
               AudioFormat.CHANNEL_IN_MONO, encoding);
   return minBufferSize;
}

private int determineCalculatedBufferSize(final int sampleRate,
      int encoding, int numSamplesInBuffer)
{
   int minBufferSize = determineMinimumBufferSize(sampleRate, encoding);

   int bufferSize;
   // each sample takes two bytes, need a bigger buffer
   if (encoding == AudioFormat.ENCODING_PCM_16BIT)
   {
      bufferSize = numSamplesInBuffer * 2;
   }
   else
```

```
        {
            bufferSize = numSamplesInBuffer;
        }

        if (bufferSize < minBufferSize)
        {
            Log.w(TAG, "Increasing buffer to hold enough samples "
                    + minBufferSize + " was: " + bufferSize);
            bufferSize = minBufferSize;
        }

        return bufferSize;
    }
    private boolean doRecording(final int sampleRate, int encoding,
            int recordingBufferSize, int readBufferSize,
            int bufferIncreaseFactor)
    {
        if (recordingBufferSize == AudioRecord.ERROR_BAD_VALUE)
        {
            Log.e(TAG, "Bad encoding value, see logcat");
            return false;
        }
        else if (recordingBufferSize == AudioRecord.ERROR)
        {
            Log.e(TAG, "Error creating buffer size");
            return false;
        }

        // give it extra space to prevent overflow
        int increasedRecordingBufferSize =
            recordingBufferSize * bufferIncreaseFactor;

        recorder =
                new AudioRecord(AudioSource.MIC, sampleRate,
                        AudioFormat.CHANNEL_IN_MONO, encoding,
                        increasedRecordingBufferSize);

        final short[] readBuffer = new short[readBufferSize];

        continueRecording = true;
        Log.d(TAG, "start recording, " + "recording bufferSize: "
                + increasedRecordingBufferSize
                + " read buffer size: " + readBufferSize);

        //Note: possible IllegalStateException
        //if audio recording is already recording or otherwise not available
        //AudioRecord.getState() will be AudioRecord.STATE_UNINITIALIZED
        recorder.startRecording();

        while (continueRecording)
        {
            int bufferResult = recorder.read(readBuffer, 0, readBufferSize);
            //in case external code stopped this while read was happening
            if ((!continueRecording) || ((task != null) && task.isCancelled()))
            {
                break;
            }
            // check for error conditions
            if (bufferResult == AudioRecord.ERROR_INVALID_OPERATION)
            {
                Log.e(TAG, "error reading: ERROR_INVALID_OPERATION");
            }
            else if (bufferResult == AudioRecord.ERROR_BAD_VALUE)
            {
                Log.e(TAG, "error reading: ERROR_BAD_VALUE");
            }
            else
            // no errors, do processing
            {
                heard = clipListener.heard(readBuffer, sampleRate);
```

```
                if (heard)
                {
                    stopRecording();
                }
            }
        }
        done();

        return heard;
    }

    public boolean isRecording()
    {
        return continueRecording;
    }

    public void stopRecording()
    {
        continueRecording = false;
    }

    public void done()
    {
        Log.d(TAG, "shut down recorder");
        if (recorder != null)
        {
            recorder.stop();
            recorder.release();
            recorder = null;
        }
    }

    private void setOnPositionUpdate(final short[] audioData,
            final int sampleRate, int numSamplesInBuffer)
    {
        // possibly do it that way
        // setOnNotification(audioData, sampleRate, numSamplesInBuffer);

        OnRecordPositionUpdateListener positionUpdater =
                new OnRecordPositionUpdateListener()
                {
                    @Override
                    public void onPeriodicNotification(AudioRecord recorder)
                    {
                        // no need to read the audioData again since it was just
                        // read
                        heard = clipListener.heard(audioData, sampleRate);
                        if (heard)
                        {
                            Log.d(TAG, "heard audio");
                            stopRecording();
                        }
                    }

                    @Override
                    public void onMarkerReached(AudioRecord recorder)
                    {
                        Log.d(TAG, "marker reached");
                    }
                };
        // get notified after so many samples collected
        recorder.setPositionNotificationPeriod(numSamplesInBuffer);
        recorder.setRecordPositionUpdateListener(positionUpdater);
    }
}
```

25.3.5 巨响检测

编写文件 LoudNoiseDetector.java，功能是通过检测是否听到一声巨响的方式来实现声控开关功能。这样可以判断当前的录音源的音量大小，还可以判断录音源的声音是否响亮。文件 LoudNoiseDetector.java 的具体实现代码如下所示。

```
public class LoudNoiseDetector implements AudioClipListener
{
    private static final String TAG = "LoudNoiseDetector";
    private double volumeThreshold;
    public static final int DEFAULT_LOUDNESS_THRESHOLD = 2000;
    private static final boolean DEBUG = true;
    public LoudNoiseDetector()
    {
        volumeThreshold = DEFAULT_LOUDNESS_THRESHOLD;
    }
    public LoudNoiseDetector(double volumeThreshold)
    {
        this.volumeThreshold = volumeThreshold;
    }
    @Override
    public boolean heard(short[] data, int sampleRate)
    {
        boolean heard = false;
        // use rms to take the entire audio signal into account
        // and discount any one single high amplitude
        double currentVolume = rootMeanSquared(data);
        if (DEBUG)
        {
            Log.d(TAG, "current: " + currentVolume + " threshold: "
                    + volumeThreshold);
        }
        if (currentVolume > volumeThreshold)
        {
            Log.d(TAG, "heard");
            heard = true;
        }
        return heard;
    }
    private double rootMeanSquared(short[] nums)
    {
        double ms = 0;
        for (int i = 0; i < nums.length; i++)
        {
            ms += nums[i] * nums[i];
        }
        ms /= nums.length;
        return Math.sqrt(ms);
    }
}
```

25.3.6　检测一致性频率

编写文件 ConsistentFrequencyDetector.java 分析音频数据来检测一致性频率，设置用户必须发出 3 秒钟声音来触发声控开关。文件 ConsistentFrequencyDetector.java 的具体实现代码如下所示。

```
public class ConsistentFrequencyDetector implements AudioClipListener
{
    private static final String TAG = "ConsistentFrequencyDetector";
    private LinkedList<Integer> frequencyHistory;
    private int rangeThreshold;
    private int silenceThreshold;
    public static final int DEFAULT_SILENCE_THRESHOLD = 2000;
    private static final boolean DEBUG = false;
    public ConsistentFrequencyDetector(int historySize, int rangeThreshold,
            int silenceThreshold)
    {
        frequencyHistory = new LinkedList<Integer>();
        // pre-fill so modification is easy
        for (int i = 0; i < historySize; i++)
        {
            frequencyHistory.add(Integer.MAX_VALUE);
```

```
        }
        this.rangeThreshold = rangeThreshold;
        this.silenceThreshold = silenceThreshold;
    }
    @Override
    public boolean heard(short[] audioData, int sampleRate)
    {
        int frequency = ZeroCrossing.calculate(sampleRate, audioData);
        frequencyHistory.addFirst(frequency);
        // since history is always full, just remove the last
        frequencyHistory.removeLast();
        int range = calculateRange();
        if (DEBUG)
        {
            Log.d(TAG, "range: " + range + " threshold " + rangeThreshold
                    + " loud: " + AudioUtil.rootMeanSquared(audioData));
        }
        boolean heard = false;
        if (range < rangeThreshold)
        {
            // only trigger it isn't silence
            if (AudioUtil.rootMeanSquared(audioData) > silenceThreshold)
            {
                Log.d(TAG, "heard");
                heard = true;
            }
            else
            {
                Log.d(TAG, "not loud enough");
            }
        }
        return heard;
    }

    private int calculateRange()
    {
        int min = Integer.MAX_VALUE;
        int max = Integer.MIN_VALUE;
        for (Integer val : frequencyHistory)
        {
            if (val >= max)
            {
                max = val;
            }
            if (val < min)
            {
                min = val;
            }
        }
        if (DEBUG)
        {
            StringBuilder sb = new StringBuilder();
            for (Integer val : frequencyHistory)
            {
                sb.append(val).append(" ");
            }
            Log.d(TAG, sb.toString() + " [" + (max - min) + "]");
        }
        return max - min;
    }
}
```

到此为止，整个实例介绍完毕。这样便循序渐进地完成了使用 AudioRecord 来录制原始音频数据实现声控开关的信号处理算法的过程。这样读者可以以此实例为基础开发类似的音频采集系统，也可以实现其他信号处理的算法工作。

第 26 章　基于图像处理的人脸识别技术详解

从 Android 2.2 版本以后，Android 对多媒体框架进行了很大的调整，抛弃了原来的 OpenCore 框架，改用 StageFright 框架，仅仅对 OpenCore 中的 omx-component 部分做了引用。主要是为了录像和视频电话功能，另外在混音和多摄像头支持方面也做了增强。除此之外，Android 系统的全新的版本对图形和图像处理更加游刃有余，无论是色彩的美观性还是性能优化，都得到了质的提升。本章将详细讲解在 Android 设备中使用图像处理技术实现人脸识别的基本知识，为步入本书后面知识的学习打下基础。

26.1 二维图形处理详解

在多媒体应用中，离不开绚丽图像的点缀，我们美好的生活离不开这些精美的图片来修饰。无论是二维图像还是三维图像，都给我们带来了绚丽的色彩和视觉冲击。本节将详细讲解在 Android 系统中使用 Graphics 类处理二维图像的知识，为读者步入本书后面知识的学习打下基础。

26.1.1 类 Graphics 基础

在 Android 系统中，图形处理类 Graphics 的功能十分强大。类 Graphics 是一个全能的绘图类，不但可以绘制 2D 图像，而且可以轻松地为这些图像填充不同的颜色。为什么 Graphics 这么强呢？原因是它有很多子类，通过这些子类可以实现不同的功能。在类 Graphics 中有如下 10 个非常重要的子类。

- 护法 Color 类。
- 护法 Paint 类。
- 护法 Canvas 画布。
- 护法 Rect 矩形类。
- 护法 NinePatch 类。
- 护法 Matrix 类。
- 护法 Bitmap 类。
- 护法使用 BitmapFactory 类。
- 护法使用 Typeface 类。
- 护法 Shader 类。

26.1.2 实战演练——使用 Graphics 类

类 Color 的完整写法是 Android.Graphics.Color，其功能是设置颜色。虽然在 Android 平台上有很多种表示颜色的方法，但是类 Color 中提供的设置颜色方法是最好用的。在类 Color 中包含了如下 12 种最常用的颜色。

- Color.BLACK。
- Color.BLUE。
- Color.CYAN。
- Color.DKGRAY。
- Color.GRAY。
- Color.GREEN。
- Color.LTGRAY。
- Color.MAGENTA。
- Color.RED。
- Color.TRANSPARENT。
- Color.WHITE。
- Color.YELLOW。

类 Color 的功能是通过如下 3 个静态方法实现的。

（1）static int argb(int alpha, int red, int green, int blue)：构造一个包含透明对象的颜色。

（2）static int rgb(int red, int green, int blue)：构造一个标准的颜色对象。

（3）static int parseColor(String colorString)：解析一种颜色字符串的值，如传入 Color.BLACK。

类 Color 返回的都是一个整型结果，例如，返回 0xff00ff00 表示绿色，返回 0xffff0000 表示红色。我们可以将这个 DWORD 型看作 AARRGGBB，AA 代表 Aphla 透明色，后面的就不难理解，每个分成 WORD 正好为 0～255。

在接下来的内容中，将通过一个具体实例来讲解使用类 Color 更改文字颜色的方法。

实例	功能	源码路径
实例 26-1	使用类 Color 更改文字的颜色	\daima\26\yan

本实例的具体实现流程如下所示。

（1）编写布局文件 main.xml，在里面使用了两个 TextView 对象，主要代码如下所示。

```
<?xml version="1.0" encoding="utf-8"?>
<LinearLayout xmlns:android="http://schemas.android.com/apk/res/android"
 android:orientation="vertical"
 android:layout_width="fill_parent"
 android:layout_height="fill_parent"
 >
 <TextView
 android:id="@+id/myTextView01"
 android:layout_width="fill_parent"
 android:layout_height="wrap_content"
 android:text="@string/str_textview01"
 />
 <TextView
 android:id="@+id/myTextView02"
 android:layout_width="fill_parent"
 android:layout_height="wrap_content"
 android:text="@string/str_textview02"
 />
</LinearLayout>
```

（2）编写主文件 yan.java，功能是调用各个公用文件来实现具体的功能。在里面分别新建了两个类成员变量 mTextView01 和 mTextView02，这两个变量在 onCreate 之初，以 findViewById 方法使之初始化为 layout（main.xml）里的 TextView 对象。在当中使用了 Resource 类及 Drawable 类，分别创建了 resources 对象及 HippoDrawable 对象，并调用了 setBackgroundDrawable 来更改 mTextView01 的文字底纹。并且使用 setText 方法来更改 TextView 里的文字。而在 mTextView02

中，使用了类 Android.Graphics.Color 中的颜色常数，并使用 setTextColor 来更改文字的前景色。文件 yan.java 的主要代码如下所示。

```
package dfzy.yan;

import dfzy.yan.R;
import android.app.Activity;
import android.content.res.Resources;
import android.graphics.Color;
import android.graphics.drawable.Drawable;
import android.os.Bundle;
import android.widget.TextView;
public class yan extends Activity
{
  private TextView mTextView01;
  private TextView mTextView02;
  @Override
  public void onCreate(Bundle savedInstanceState)
  {
    super.onCreate(savedInstanceState);
    setContentView(R.layout.main);
    mTextView01 = (TextView) findViewById(R.id.myTextView01);
    mTextView01.setText("使用的是 Drawable 背景色文本。");
    Resources resources = getBaseContext().getResources();
    Drawable HippoDrawable = resources.getDrawable(R.drawable.white);
    mTextView01.setBackgroundDrawable(HippoDrawable);
    mTextView02 = (TextView) findViewById(R.id.myTextView02);
    mTextView02.setTextColor(Color.MAGENTA);
  }
}
```

调试运行后的效果如图 26-1 所示。

2:44
使用的是Drawable背景色文本。
使用graphics.Color设置文本颜色

▲图 26-1　运行效果

26.1.3　实战演练——使用 Color 类和 Paint 类实现绘图处理

在接下来的示例代码中，将使用 Color 类和 Paint 类实现了绘图处理功能。本实例是类 Paint 的经典应用之一，虽然是和 Color 联手完成的，但是 Paint 在此实例中发挥了更大的作用。

实例	功能	源码路径
实例 26-2	使用类 Color 更改文字的颜色	\daima\26\PaintCH

本实例的具体实现流程如下所示。

（1）编写布局文件 main.xml，具体代码如下所示。

```
<LinearLayout xmlns:Android="http://schemas.Android.com/apk/res/Android"
    Android:orientation="vertical"
    Android:layout_width="fill_parent"
    Android:layout_height="fill_parent"
    >
<TextView
    Android:layout_width="fill_parent"
    Android:layout_height="wrap_content"
    Android:text="@string/hello"
    />
</LinearLayout>
```

（2）编写文件 Activity.java，通过“mGameView = new GameView(this)”，用 Activity 类的

setContentView 方法来设置要显示的具体 View 类。文件 Activity.java 的主要代码如下所示。

```
public class Activity01 extends Activity
{
    @Override
    public void onCreate(Bundle savedInstanceState)
    {
        super.onCreate(savedInstanceState);

        mGameView = new GameView(this);

        setContentView(mGameView);
    }
}
```

（3）编写文件 PaintCH.java 来绘制一个指定的图形，主要实现代码如下所示。

```
    /* 声明 Paint 对象 */
    private Paint mPaint    = null;
    public draw(Context context)
    {
        super(context);
        /* 构建对象 */
        mPaint = new Paint();
        /* 开启线程 */
        new Thread(this).start();
    }
    public void onDraw(Canvas canvas)
    {
        super.onDraw(canvas);
        /* 设置 Paint 为无锯齿 */
        mPaint.setAntiAlias(true);
        /* 设置 Paint 的颜色 */
        mPaint.setColor(Color.WHITE);
        mPaint.setColor(Color.BLUE);
        mPaint.setColor(Color.YELLOW);
        mPaint.setColor(Color.GREEN);
        /* 同样是设置颜色 */
        mPaint.setColor(Color.rgb(255, 0, 0));
        /* 提取颜色 */
        Color.red(0xcccccc);
        Color.green(0xcccccc);
        /* 设置 Paint 的颜色和 Alpha 值(a,r,g,b) */
        mPaint.setARGB(255, 255, 0, 0);
        /* 设置 Paint 的 Alpha 值 */
        mPaint.setAlpha(220);
        /* 这里可以设置为另外一个 Paint 对象 */
        // mPaint.set(new Paint());
        /* 设置字体的尺寸 */
        mPaint.setTextSize(14);
        // 设置 Paint 的风格为“空心”.
        // 当然也可以设置为“实心”(Paint.Style.FILL)
        mPaint.setStyle(Paint.Style.STROKE);
        // 设置“空心”的外框的宽度
        mPaint.setStrokeWidth(5);
        /* 得到 Paint 的一些属性 */
        Log.i(TAG, "paint 的颜色: " + mPaint.getColor());
        Log.i(TAG, "paint 的 Alpha: " + mPaint.getAlpha());
        Log.i(TAG, "paint 的外框的宽度: " + mPaint.getStrokeWidth());
        Log.i(TAG, "paint 的字体尺寸: " + mPaint.getTextSize());
        /* 绘制一个矩形 */
        // 肯定是一个空心的矩形
        canvas.drawRect((320 - 80) / 2, 20, (320 - 80) / 2 + 80, 20 + 40, mPaint);
        /* 设置风格为实心 */
        mPaint.setStyle(Paint.Style.FILL);
        mPaint.setColor(Color.GREEN);
        /* 绘制绿色实心矩形 */
        canvas.drawRect(0, 20, 40, 20 + 40, mPaint);
    }
    // 触笔事件
```

```
    public boolean onTouchEvent(MotionEvent event)
    {
        return true;
    }
    // 按键按下事件
    public boolean onKeyDown(int keyCode, KeyEvent event)
    {
        return true;
    }
    // 按键弹起事件
    public boolean onKeyUp(int keyCode, KeyEvent event)
    {
        return false;
    }
    public boolean onKeyMultiple(int keyCode, int repeatCount, KeyEvent event)
    {
        return true;
    }
    public void run()
    {
        while (!Thread.currentThread().isInterrupted())
        {
            try
            {
                Thread.sleep(100);
            }
            catch (InterruptedException e)
            {
                Thread.currentThread().interrupt();
            }
            // 使用 postInvalidate 可以直接在线程中更新界面
            postInvalidate();
        }
    }
}
```

执行后的效果如图 26-2 所示。

▲图 26-2　执行效果

26.2 二维动画处理详解

在多媒体领域中，具有视觉冲击力的动画永远是程序员们谈论的话题之一。动画和简单的图像相比，更具有震撼性的效果。Android 系统为我们提供了一套完整的动画框架，使得开发者可以用它来开发各种动画效果。本节将详细讲解 Android 系统处理二维动画的知识。

26.2.1　类 Drawable 详解

Android SDK 为程序员提供了一个功能强大的类——Drawable。虽然 Drawable 是一个很抽象的概念，但是可以实现动画效果。为了深入了解 Drawable 的基本知识，先通过一个简单的例子来认识它。在这个例子中，使用 Drawable 的子类 ShapeDrawable 绘制了一幅图，具体实现流程如下所示。

（1）创建一个 OvalShape（椭圆）。

（2）使用刚创建的 OvalShape 构造一个 ShapeDrawable 对象 mDrawable。

（3）设置 mDrawable 的颜色。

（4）设置 mDrawable 的大小。

（5）将 mDrawable 画在 testViewCH 的画布上。

例子的具体代码如下所示。

```
public class testViewCH extends View {
private ShapeDrawable mDrawable;
public testViewCH(Context context) {
super(context);
int x = 10;
int y = 10;
int width = 300;
int height = 50;
mDrawable = new ShapeDrawable(new OvalShape());
mDrawable.getPaint().setColor(0xff74AC23);
mDrawable.setBounds(x, y, x + width, y + height);
}
protected void onDraw(Canvas canvas)
super.onDraw(canvas);
canvas.drawColor(Color.WHITE);//画白色背景
mDrawable.draw(canvas);
}
}
```

▲图 26-3　执行效果

上述代码的执行效果如图 26-3 所示。

例子虽然简单，但是却让我们明白了 Drawable 就是一个可画的对象，可能是一张位图（BitmapDrawable），也可能是一个图形（ShapeDrawable），还有可能是一个图层（LayerDrawable）。在项目中可以根据画图的需求，创建相应的可画对象，就可以将这个可画对象当作一块“画布（Canvas）”，在其上面操作可画对象，并最终将这种可画对象显示在画布上，有点类似于“内存画布”。

26.2.2　实现 Tween Animation 动画

通过本章前面内容中的学习，了解了 Drawable 可以实现动画效果。其实 Drawable 的功能何止如此，它更加强大的功能是可以显示 Animation。在 Android SDK 中提供了如下两种 Animation。

（1）Tween Animation：通过对场景里的对象不断做平移、缩放、旋转等变换来产生动画效果。

（2）Frame Animation：跟电影似的顺序播放事先做好的图像。

由此可见，Android 平台为我们提供了如下两类动画。

- Tween 动画：用于对场景里的对象不断进行图像变换来产生动画效果，可以把对象进行缩小、放大、旋转和渐变等操作。
- Frame 动画：用于顺序播放事先做好的图像。

在使用 Animation 前需要学会定义 Animation 的方法，Animation 是以 XML 格式定义的，定义好的 XML 文件存放在“res\anim”目录中。Tween Animation 与 Frame Animation 的定义、使用都有很大的差异。

Tween 动画通过对 View 的内容完成一系列的图形变换，通过平移、缩放、旋转、改变透明度来实现动画效果。在 XML 文件中，Tween 动画主要包括以下 4 种动画效果。

- Alpha：渐变透明度动画效果。
- Scale：渐变尺寸伸缩动画效果。
- Translate：画面转移位置移动动画效果。
- Rotate：画面转移旋转动画效果。

在 Java 代码中，Tween 动画对应以下 4 种动画效果。

- AlphaAnimation：渐变透明度动画效果。
- ScaleAnimation：渐变尺寸伸缩动画效果。
- TranslateAnimation：画面转换位置移动动画效果。
- RotateAnimation：画面转移旋转动画效果。

Tween 动画是通过预先定义一组指令，这些指令指定了图形变换的类型、触发时间、持续时间，程序沿着时间线执行这些指令就可以实现动画效果。我们可以首先定义 Animation 动画对象，然后设置该动画的一些属性，最后通过 startAnimation()方法开始动画效果。

26.2.3　实战演练——实现 Tween 动画效果

在接下来的实例中，将通过一个具体实例来讲解在 Android 系统中实现 Tween 动画效果的方法。

实例	功能	源码路径
实例 26-3	使用类 Color 更改文字的颜色	\daima\26\PaintCH

实例文件 TweenCH.java 的主要实现代码如下所示。

```
/* 定义 Alpha 动画 */
private Animation    mAnimationAlpha        = null;

/* 定义 Scale 动画 */
private Animation    mAnimationScale        = null;

/* 定义 Translate 动画 */
private Animation    mAnimationTranslate    = null;

/* 定义 Rotate 动画 */
private Animation    mAnimationRotate    = null;

/* 定义 Bitmap 对象 */
Bitmap              mBitQQ              = null;

public example9(Context context)
{
    super(context);

    /* 装载资源 */
    mBitQQ = ((BitmapDrawable) getResources().getDrawable(R.drawable.qq)).getBitmap();
}

public void onDraw(Canvas canvas)
{
    super.onDraw(canvas);

    /* 绘制图片 */
    canvas.drawBitmap(mBitQQ, 0, 0, null);
}

public boolean onKeyUp(int keyCode, KeyEvent event)
{
    switch ( keyCode )
    {
    case KeyEvent.KEYCODE_DPAD_UP:
        /* 创建 Alpha 动画 */
        mAnimationAlpha = new AlphaAnimation(0.1f, 1.0f);
        /* 设置动画的时间 */
        mAnimationAlpha.setDuration(3000);
        /* 开始播放动画 */
        this.startAnimation(mAnimationAlpha);
        break;
    case KeyEvent.KEYCODE_DPAD_DOWN:
        /* 创建 Scale 动画 */
        mAnimationScale =new ScaleAnimation(0.0f, 1.0f, 0.0f, 1.0f,
```

```
                                        Animation.RELATIVE_TO_SELF, 0.5f,
                                        Animation.RELATIVE_TO_SELF, 0.5f);
            /* 设置动画的时间 */
            mAnimationScale.setDuration(500);
            /* 开始播放动画 */
            this.startAnimation(mAnimationScale);
            break;
        case KeyEvent.KEYCODE_DPAD_LEFT:
            /* 创建 Translate 动画 */
            mAnimationTranslate = new TranslateAnimation(10, 100,10, 100);
            /* 设置动画的时间 */
            mAnimationTranslate.setDuration(1000);
            /* 开始播放动画 */
            this.startAnimation(mAnimationTranslate);
            break;
        case KeyEvent.KEYCODE_DPAD_RIGHT:
            /* 创建 Rotate 动画 */
            mAnimationRotate=new RotateAnimation(0.0f, +360.0f,
                                         Animation.RELATIVE_TO_SELF,0.5f,
                                         Animation.RELATIVE_TO_SELF, 0.5f);
            /* 设置动画的时间 */
            mAnimationRotate.setDuration(1000);
            /* 开始播放动画 */
            this.startAnimation(mAnimationRotate);
            break;
        }
        return true;
    }
}
```

执行后可以通过键盘的上下左右键实现动画效果，如图 26-4 所示。

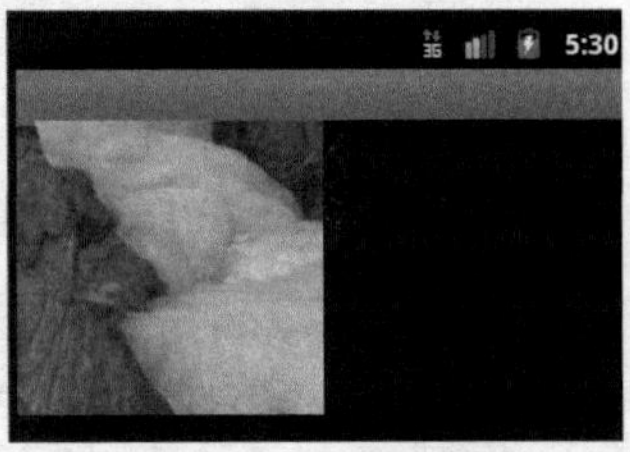

▲图 26-4 执行效果

26.2.4 实战演练——使用 Tween Animation 实现 Tween 动画效果

在 Android SDK 提供了两种使用 Tween Animation 的方法，分别是直接从 XML 资源中读取 Animation 和使用 Animation 子类的构造函数来初始化 Animation 对象。本实例是从 XML 资源中读取 Animation 的，具体实现流程如下所示。

（1）创建 Android 工程。

（2）导入一张图片资源。

（3）将“res\layout\main.xml”目录中的 TextView 取代为 ImageView。

（4）在 res 下创建新的文件夹“anim”，并在此文件夹下面定义 Animation XML 文件。

（5）修改 OnCreate()中的代码，显示动画资源。

在接下来的内容中，将通过一个具体实例来讲解使用 Tween Animation 实现 Tween 动画效果的方法。

实例	功能	源码路径
实例 26-4	使用 Tween Animation 方法实现 Tween 动画效果	\daima\26\testDrawableCH

本实例的实现文件是 testDrawableCH.java，在文件中使用 AnimationUtils 提供一个加载动画的函数，除了函数 loadAnimation()外，读者可以到 Android SDK 中去详细了解其他函数。文件 testDrawableCH.java 的主要代码如下所示。

```
public class testDrawableCH extends Activity {
    LinearLayout mLinearLayout;
    protected void onCreate(Bundle savedInstanceState) {
        super.onCreate(savedInstanceState);
        setContentView(R.layout.main);
        ImageView spaceshipImage = (ImageView) findViewById(R.id.spaceshipImage);
        Animation hyperspaceJumpAnimation = AnimationUtils.loadAnimation(this,
```

```
        R.anim.hyperspace_jump);
        spaceshipImage.startAnimation(hyperspaceJumpAnimation);
    }
}
```

执行显示一个动画效果界面，如图 26-5 所示。

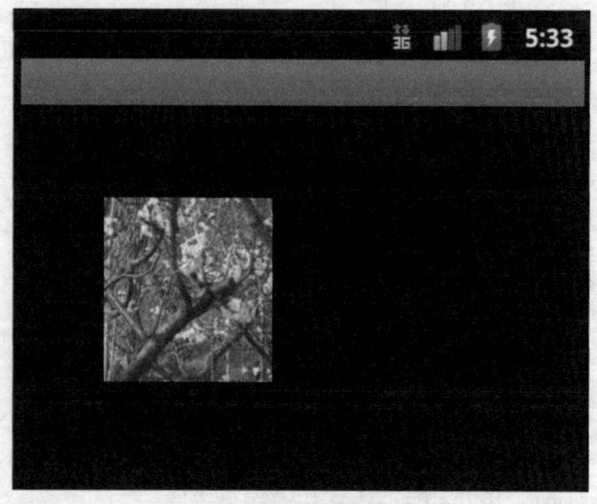

▲图 26-5　执行效果

26.2.5　实现 Frame Animation 动画效果

在我们日常生活中见到的最多的可能就是 Frame 动画了，在 Android SDK 中，使用类 AnimationDrawable 来定义并使用 Frame Animation 动画。SDK 的位置如下所示。

- Tween animation：android.view.animation 包。
- Frame animation：android.graphics.drawable.AnimationDrawable 类。

（1）AnimationDrawable 介绍

AnimationDrawable 的功能是获取、设置动画的属性，里面最为常用的方法如下所示。

- int getDuration()：获取动画的时长。
- int getNumberOfFrames()：获取动画的帧数。
- boolean isOneShot()：获取 oneshot 属性。
- Void setOneShot(boolean oneshot)：设置 oneshot 属性。
- void inflate(Resurce r,XmlPullParser p,AttributeSet attrs)：增加、获取帧动画。
- Drawable getFrame(int index)：获取某帧的 Drawable 资源。
- void addFrame(Drawable frame,int duration)：为当前动画增加帧（资源，持续时长）。
- void start()：开始动画。
- void run()：外界不能直接调用，使用 start()替代。
- boolean isRunning()：当前动画是否在运行。
- void stop()：停止当前动画。

（2）Frame Animation 格式定义

我们既可以在 XML Resource 中定义 Frame Animation，也可以使用 AnimationDrawable 中的 API 来定义。由于 Tween Animation 和 Frame Animation 有着很大的不同，所以定义 XML 的格式也很不相同。

定义 Frame Animation 的格式是：首先是 animation-list 根节点，animation-list 根节点中包含多个 item 子节点，每个 item 节点定义一帧动画，定义当前帧的 drawable 资源和当前帧持续的时间。表 26-1 中对节点中的元素进行了详细说明。

表 26-1 XML 属性元素说明

XML 属性	说 明
drawable	当前帧引用的 drawable 资源
duration	当前帧显示的时间（毫秒为单位）
oneshot	如果为 true，表示动画只播放一次停止在最后一帧上，如果设置为 false 表示动画循环播放
variablePadding	如果为真允许 drawable 根据被选择的现状而变动
visible	规定 drawable 的初始可见性，默认为 flase

（3）使用 Frame 动画

使用 Frame 动画的方法十分简单，只需要创建一个 AnimationDrawabledF 对象来表示 Frame 动画，然后通过 addFrame 方法把每一帧要显示的内容添加进去，最后通过 start 方法就可以播放这个动画了，同时还可以通过 setOneShot 方法设置是否重复播放。

Frame 动画主要是通过 AnimationDrawable 类来实现的，用 start()和 stop()这两个重要的方法分别启动和停止动画。Frame 动画一般通过 XML 文件配置，在 Android 工程的“res/anim”目录下创建一个 XML 配置文件，该配置文件有一个<animation-list>根元素和若干个<item>子元素。

26.2.6 实战演练——播放 GIF 动画

虽然 Android 很开放，但是还没有达到“来者不拒”的境界，例如，在很多情况下不能播放 GIF 动画。要想让 Android 播放 GIF 动画，需要我们做一些背后的工作。接下来的实例演示了一个很好的背后工作，即先对 GIF 图像进行解码，然后将 GIF 中的每一帧取出来保存到一个容器中，然后根据需要连续绘制每一帧，这样就可以轻松地实现 GIF 动画的播放。

实例	功能	源码路径
实例 26-5	在 Android 中实播放 GIF 动画	\daima\26\GIFCH

本实例的实现文件是 GameView.java，通过此文件解析 GIF 动画文件并设置显示效果，主要代码如下所示。

```
public class GameView extends View implements Runnable
{
    Context       mContext    = null;
    /* 声明GifFrame对象 */
    GifFrame    mGifFrame    = null;
    public GameView(Context context)
    {
        super(context);
        mContext = context;
        /* 解析GIF动画 */
    mGifFrame=GifFrame.CreateGifImage(fileConnect(this.getResources().
openRawResource(R.drawable.gif1)));
        /* 开启线程 */
        new Thread(this).start();
    }
    public void onDraw(Canvas canvas)
    {
        super.onDraw(canvas);
        /* 下一帧 */
        mGifFrame.nextFrame();
        /* 得到当前帧的图片 */
        Bitmap b=mGifFrame.getImage();

        /* 绘制当前帧的图片 */
        if(b!=null)
            canvas.drawBitmap(b,10,10,null);
    }
```

```
    /*线程处理*/
    public void run()
    {
        while (!Thread.currentThread().isInterrupted())
        {
            try
            {
                Thread.sleep(100);
            }
            catch (InterruptedException e)
            {
                Thread.currentThread().interrupt();
            }
            //使用 postInvalidate 可以直接在线程中更新界面
            postInvalidate();
        }
    }
    /* 读取文件 */
    public byte[] fileConnect(InputStream is)
    {
        try
        {ByteArrayOutputStream baos = new ByteArrayOutputStream();
            int ch = 0;
            while( (ch = is.read()) != -1)
            {
                baos.write(ch);
            }
            byte[] datas = baos.toByteArray();
            baos.close();
            baos = null;
            is.close();
            is = null;
            return datas;
        }
        catch(Exception e)
        {
            return null;
        }
    }
}
```

▲图 26-6　执行效果

执行后可以在屏幕中实现一个 GIF 动画效果，如图 26-6 所示。

26.3 Android 人脸识别技术详解

2006 年 8 月，Google 收购了 Neven Vision 公司，该公司拥有 10 多项应用于移动设备领域的图像识别的专利。这样 Google 获得了图像识别的技术，并将这项技术加入到了 Android 系统中。本节将详细讲解 Android 人脸识别技术的基本知识。

26.3.1　分析人脸识别模块的源码

（1）底层文件

在 Android 人脸识别系统中，底层库文件的实现路径是：

```
android/external/neven/
```

在上述目录中保存了实现人脸识别的算法代码。

（2）接口文件

在 Android 人脸识别系统中，Java 层接口文件的实现路径是：

```
frameworks\base\media\java\android\media\FaceDetector.java
```

在 Android 系统中，自带的人脸识别模块只能识别出人脸在画面中的位置、中点、眼间距和角度等基本特性，提供给拍照性质的应用使用，从基本功能中不能得出明显的特征数据。通过分析 Android 源码可知，自带的人脸识别系统的 Java 层接口存在如下所示的限制。

- 只能接受 Bitmap 格式的数据。
- 只能识别双眼距离大于 20 像素的人脸像（当然，这个可在 framework 层中修改）。
- 只能检测出人脸的位置（双眼的中心点及距离）。
- 不能对人脸进行匹配（查找指定的脸谱）。

在开发 Android 应用程序的过程中，通过需要开发如下两类人脸识别程序。

（1）为 Camera 添加人脸识别的功能：使得 Camera 的取景器上能标识出人脸范围，如果硬件支持则可以对人脸进行对焦。

（2）为相册程序添加按人脸索引相册的功能：按人脸索引相册和按人脸进行分组，可以快速搜索相册。

26.3.2 实战演练——使用内置模块实现人脸识别

接下来将通过一个具体实例的实现过程，来讲解使用 Android 自带的识别库实现人脸识别功能的具体方法。

题目	目的	源码路径
实例 26-6	利用 Android 自带的识别库实现人脸识别	\daima\26\mydetect

本实例的布局文件是 main.xml，功能是通过一个图片控件显示预识别的图像，具体实现代码如下所示。

```
<LinearLayout xmlns:android="http://schemas.android.com/apk/res/android"
    android:orientation="vertical"
    android:layout_width="fill_parent"
    android:layout_height="fill_parent"
    >
    <ImageView android:id="@+id/image"
        android:layout_width="wrap_content"
        android:layout_height="wrap_content"
        android:src="@drawable/baby"
        />
</LinearLayout>
```

编写程序文件 MyDetectActivity.java，功能是调用 Android 自带的识别库实现人脸识别。文件 MyDetectActivity.java 的具体实现代码如下所示。

```
public class MyDetectActivity extends Activity {
    private ImageView mImageView;
    private Bitmap mBitmap;
    private float mScale = 1F;

    @Override
    public void onCreate(Bundle savedInstanceState) {
        super.onCreate(savedInstanceState);
        setContentView(R.layout.main);
        mImageView = (ImageView) this.findViewById(R.id.image);
        detect();
    }

    private void handleFace(FaceDetector.Face f) {
        PointF midPoint = new PointF();
        int r = ((int) (f.eyesDistance() * mScale * 1.5));
        f.getMidPoint(midPoint);
        midPoint.x *= mScale;
        midPoint.y *= mScale;
```

```
        Canvas c = new Canvas(mBitmap);
        Paint p = new Paint();
        p.setAntiAlias(true);
        p.setAlpha(0x80);
        c.drawCircle(midPoint.x, midPoint.y, r, p);
        mImageView.setImageBitmap(mBitmap);
    }

    private void detect() {
        Matrix matrix = new Matrix();
        FaceDetector.Face[] mFaces = new FaceDetector.Face[3];
        int mNumFaces = 0;

        mBitmap = BitmapFactory.decodeResource(getResources(), R.drawable.baby);
        if (mBitmap == null) {
            return;
        }
        if (mBitmap.getWidth() > 256) {
            mScale = 256.0F / mBitmap.getWidth();
        }
        matrix.setScale(mScale, mScale);
        Bitmap faceBitmap = Bitmap.createBitmap(mBitmap, 0, 0, mBitmap
                .getWidth(), mBitmap.getHeight(), matrix, true);

        mScale = 1.0F / mScale;
        if (faceBitmap != null) {
            FaceDetector detector = new FaceDetector(faceBitmap.getWidth(),
                    faceBitmap.getHeight(), mFaces.length);
            mNumFaces = detector.findFaces(faceBitmap, mFaces);
            if (mNumFaces > 0) {
                for (int i = 0; i < mNumFaces; i++) {
                    handleFace(mFaces[i]);
                }
            }
        }
    }
}
```

执行后将实现人脸识别。

26.3.3　实战演练——实现人脸识别

接下来将通过一个具体实例的实现过程，来讲解在 Android 系统中实现人脸识别功能的具体方法。

题目	目的	源码路径
实例 26-7	在 Android 系统中实现人脸识别功能	\daima\26\Face

编写程序文件 MyImageView.java，功能是在屏幕中绘制我们预先准备的人物素材图像。文件 MyImageView.java 的具体实现代码如下所示。

```
class MyImageView extends ImageView {
    private Bitmap mBitmap;
    private Canvas mCanvas;
    private int mBitmapWidth = 200;
    private int mBitmapHeight = 200;
    private Paint mPaint = new Paint(Paint.ANTI_ALIAS_FLAG);
    private int mDisplayStyle = 0;
    private int [] mPX = null;
    private int [] mPY = null;

    public MyImageView(Context c) {
        super(c);
        init();
    }
```

```
public MyImageView(Context c, AttributeSet attrs) {
   super(c, attrs);
   init();
}


private void init() {
   mBitmap = Bitmap.createBitmap(mBitmapWidth, mBitmapHeight, Bitmap.Config.RGB_565);
   mCanvas = new Canvas(mBitmap);

   mPaint.setStyle(Paint.Style.STROKE);
   mPaint.setStrokeCap(Paint.Cap.ROUND);
   mPaint.setColor(0x80ff0000);
   mPaint.setStrokeWidth(3);
}

public Bitmap getBitmap() {
   return mBitmap;
}

@Override
public void setImageBitmap(Bitmap bm) {
   if (bm != null) {
      mBitmapWidth = bm.getWidth();
      mBitmapHeight = bm.getHeight();

      mBitmap = Bitmap.createBitmap(mBitmapWidth, mBitmapHeight, Bitmap.Config.
      RGB_565);
      mCanvas = new Canvas();
      mCanvas.setBitmap(mBitmap);
      mCanvas.drawBitmap(bm, 0, 0, null);
   }

   super.setImageBitmap(bm);
}

@Override
protected void onSizeChanged(int w, int h, int oldw, int oldh) {
   super.onSizeChanged(w, h, oldw, oldh);

   mBitmapWidth = (mBitmap != null) ? mBitmap.getWidth() : 0;
   mBitmapHeight = (mBitmap != null) ? mBitmap.getHeight() : 0;
   if (mBitmapWidth == w && mBitmapHeight == h) {
      return;
   }

   if (mBitmapWidth < w) mBitmapWidth = w;
   if (mBitmapHeight < h) mBitmapHeight = h;
}
// set up detected face features for display
public void setDisplayPoints(int [] xx, int [] yy, int total, int style) {
   mDisplayStyle = style;
   mPX = null;
   mPY = null;

   if (xx != null && yy != null && total > 0) {
      mPX = new int[total];
      mPY = new int[total];

      for (int i = 0; i < total; i++) {
         mPX[i] = xx[i];
         mPY[i] = yy[i];
      }
   }
}

@Override
```

```
    protected void onDraw(Canvas canvas) {
        super.onDraw(canvas);

        if (mBitmap != null) {
            canvas.drawBitmap(mBitmap, 0, 0, null);

            if (mPX != null && mPY != null) {
                for (int i = 0; i < mPX.length; i++) {
                    if (mDisplayStyle == 1) {
                        canvas.drawCircle(mPX[i], mPY[i], 10.0f, mPaint);
                    } else {
                        canvas.drawRect(mPX[i] - 20,  mPY[i] - 20, mPX[i] + 20,  mPY[i] + 20,
                        mPaint);
                    }
                }
            }
        }
    }
}
```

编写文件 TutorialOnFaceDetect.java，功能是编写人脸识别算法来实现人脸识别功能，具体实现代码如下所示。

```
public class TutorialOnFaceDetect extends Activity {
    private MyImageView mIV;
    private Bitmap mFaceBitmap;
    private int mFaceWidth = 200;
    private int mFaceHeight = 200;
    private static final int MAX_FACES = 10;
    private static String TAG = "TutorialOnFaceDetect";
    private static boolean DEBUG = false;

    protected static final int GUIUPDATE_SETFACE = 999;
    protected Handler mHandler = new Handler(){
        // @Override
        public void handleMessage(Message msg) {
            mIV.invalidate();
            super.handleMessage(msg);
        }
    };

    @Override
    public void onCreate(Bundle savedInstanceState) {
        super.onCreate(savedInstanceState);

        mIV = new MyImageView(this);
        setContentView(mIV, new LayoutParams(LayoutParams.WRAP_CONTENT, LayoutParams.
        WRAP_CONTENT));

        // load the photo
        Bitmap b = BitmapFactory.decodeResource(getResources(), R.drawable.face3);
        mFaceBitmap = b.copy(Bitmap.Config.RGB_565, true);
        b.recycle();

        mFaceWidth = mFaceBitmap.getWidth();
        mFaceHeight = mFaceBitmap.getHeight();
        mIV.setImageBitmap(mFaceBitmap);
        mIV.invalidate();

        // perform face detection in setFace() in a background thread
        doLengthyCalc();
    }

    public void setFace() {
        FaceDetector fd;
        FaceDetector.Face [] faces = new FaceDetector.Face[MAX_FACES];
        PointF eyescenter = new PointF();
        float eyesdist = 0.0f;
        int [] fpx = null;
```

```
        int [] fpy = null;
        int count = 0;

        try {
            fd = new FaceDetector(mFaceWidth, mFaceHeight, MAX_FACES);
            count = fd.findFaces(mFaceBitmap, faces);
        } catch (Exception e) {
            Log.e(TAG, "setFace(): " + e.toString());
            return;
        }

        // check if we detect any faces
        if (count > 0) {
            fpx = new int[count * 2];
            fpy = new int[count * 2];

            for (int i = 0; i < count; i++) {
                try {
                    faces[i].getMidPoint(eyescenter);
                    eyesdist = faces[i].eyesDistance();

                    // set up left eye location
                    fpx[2 * i] = (int)(eyescenter.x - eyesdist / 2);
                    fpy[2 * i] = (int)eyescenter.y;

                    // set up right eye location
                    fpx[2 * i + 1] = (int)(eyescenter.x + eyesdist / 2);
                    fpy[2 * i + 1] = (int)eyescenter.y;

                    if (DEBUG)
                        Log.e(TAG, "setFace(): face " + i + ": confidence = " +
                        faces[i].confidence()
                                + ", eyes distance = " + faces[i].eyesDistance()
                                + ", pose = ("+ faces[i].pose(FaceDetector.Face.EULER_X) + ","
                                + faces[i].pose(FaceDetector.Face.EULER_Y) + ","
                                + faces[i].pose(FaceDetector.Face.EULER_Z) + ")"
                                + ", eyes midpoint = (" + eyescenter.x + "," + eyescenter.y +")");
                } catch (Exception e) {
                    Log.e(TAG, "setFace(): face " + i + ": " + e.toString());
                }
            }
        }

        mIV.setDisplayPoints(fpx, fpy, count * 2, 1);
    }

    private void doLengthyCalc() {
        Thread t = new Thread() {
            Message m = new Message();

            public void run() {
                try {
                    setFace();
                    m.what = TutorialOnFaceDetect.GUIUPDATE_SETFACE;
                    TutorialOnFaceDetect.this.mHandler.sendMessage(m);
                } catch (Exception e) {
                    Log.e(TAG, "doLengthyCalc(): " + e.toString());
                }
            }
        };

        t.start();
    }

}
```

执行后将实现人脸识别，执行效果如图 26-7 所示。

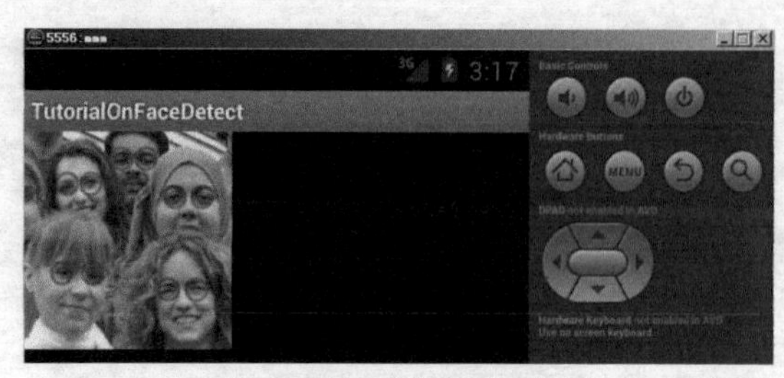

▲图 26-7　执行效果

26.3.4　实战演练——从照片中取出人脸

接下来将通过一个具体实例的实现过程，来讲解在 Android 系统中通过人脸识别技术取出人脸的具体方法。

题目	目的	源码路径
实例 26-8	从照片中取出人脸	\daima\26\FaceIdentifying

编写布局文件 main.xml，功能是在屏幕中通过两个图片控件显示两幅图像，具体实现代码如下所示。

```
<RelativeLayout xmlns:android="http://schemas.android.com/apk/res/android"
    xmlns:tools="http://schemas.android.com/tools"
    android:layout_width="match_parent"
    android:layout_height="match_parent"
    android:paddingBottom="@dimen/activity_vertical_margin"
    android:paddingLeft="@dimen/activity_horizontal_margin"
    android:paddingRight="@dimen/activity_horizontal_margin"
    android:paddingTop="@dimen/activity_vertical_margin"
    tools:context=".MainActivity" >
<ImageView
    android:id="@+id/img1"
    android:layout_width="wrap_content"
    android:layout_height="wrap_content"
    android:background="@drawable/img1"
    />
<ImageView
    android:id="@+id/img2"
    android:layout_width="100dp"
    android:layout_height="100dp"
    android:layout_below="@+id/img1"
    />
</RelativeLayout>
```

编写程序文件 MainActivity.java，功能是在屏幕上方加载显示预处理的图片，通过人脸识别算法在下方显示取出的人脸。文件 MainActivity.java 的具体实现代码如下所示。

```
public class MainActivity extends Activity {
    float eyedistance;
    private int imageWidth, imageHeight;
    private int numberOfFace = 5;
    private FaceDetector myFaceDetect;
    private FaceDetector.Face[] myFace;
    float myEyesDistance;
    int numberOfFaceDetected;
    Bitmap myBitmap;
    @Override
    protected void onCreate(Bundle savedInstanceState) {
        super.onCreate(savedInstanceState);
        setContentView(R.layout.activity_main);
        ImageView img1=(ImageView)findViewById(R.id.img1);
        ImageView img2=(ImageView)findViewById(R.id.img2);
          BitmapFactory.Options BitmapFactoryOptionsbfo = new BitmapFactory. Options();
          BitmapFactoryOptionsbfo.inPreferredConfig = Bitmap.Config.RGB_565;
```

```
            myBitmap = BitmapFactory.decodeResource(getResources(), R.drawable.img1,
            BitmapFactoryOptionsbfo);
            imageWidth = myBitmap.getWidth();
            imageHeight = myBitmap.getHeight();
            if(imageWidth%2!=0)
            {
                //如果图片的宽度是奇数无法识别人脸，这里要把奇数变为偶数
               myBitmap=zoomImage(myBitmap, imageWidth+1, imageHeight);
                imageWidth = myBitmap.getWidth();
                imageHeight = myBitmap.getHeight();
            }

            myFace = new FaceDetector.Face[numberOfFace];
            myFaceDetect = new FaceDetector(imageWidth, imageHeight, numberOfFace);
            numberOfFaceDetected = myFaceDetect.findFaces(myBitmap, myFace);
            Toast.makeText(this, "找到"+numberOfFaceDetected+"张脸", 1).show();
            for(int i=0; i < numberOfFaceDetected; i++)
            {
                Face face = myFace[i];
                PointF myMidPoint = new PointF();
                face.getMidPoint(myMidPoint);//人脸的中间点
                myEyesDistance = face.eyesDistance();//这是一只眼睛距离中心点的距离，你
                //要得到脸部大概长度，乘以 2
          if(i==0)
           {
        Bitmap      newbit=myBitmap.createBitmap(myBitmap,(int)(myMidPoint.x-myEyes
Distance),(int)(myMidPoint.y-myEyesDistance/2),(int)(myEyesDistance*2),(int)
(myEyesDistance*2));
        img2.setImageBitmap(newbit);
        }
            }

        }

        @Override
        public boolean onCreateOptionsMenu(Menu menu) {
            getMenuInflater().inflate(R.menu.main, menu);
            return true;
        }


        public  Bitmap zoomImage(Bitmap bgimage, double newWidth,
                double newHeight) {
        // 获取这个图片的宽和高
        float width = bgimage.getWidth();
        float height = bgimage.getHeight();
        // 创建操作图片用的 matrix 对象
        Matrix matrix = new Matrix();
        // 计算宽高缩放率
        float scaleWidth = ((float) newWidth) / width;
        float scaleHeight = ((float) newHeight) / height;
        // 缩放图片动作
        matrix.postScale(scaleWidth, scaleHeight);
        Bitmap bitmap = Bitmap.createBitmap(bgimage, 0, 0, (int) width,
                    (int) height, matrix, true);
        return bitmap;
    }
}
```

执行后的效果如图 26-8 所示。

▲图 26-8 执行效果

第 27 章　行走轨迹记录器

随着 2008 奥运会在北京的成功举办，全民健身热潮已经日益深入民心。长期在办公室工作的都市一族纷纷来到户外亲密接触大自然，在这种背景环境下，为传感器应用开发提供了良好的舞台。本章将通过一个综合实例的实现过程，详细讲解利用 Android 传感器技术开发行走轨迹记录系统的方法，为读者步入现实工作岗位打下基础。

27.1 系统功能模块介绍

本章轨迹记录器的功能是，通过 Android 传感器记录当前的位置、速率、海拔、记录频率和距离等信息，并且可以将轨迹信息打包上传或分享。本章行走轨迹记录器的构成模块结构如图 27-1 所示。

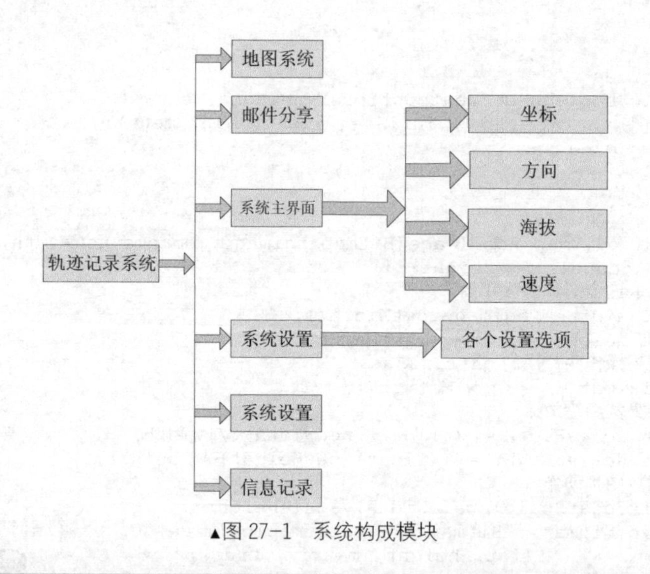

▲图 27-1　系统构成模块

27.2 系统主界面

系统主界面是运行程序后首先呈现在用户面前的界面。本节将详细讲解本章行走轨迹记录器主界面的具体实现流程。

27.2.1　布局文件

本系统主界面的布局文件是 main.xml，功能是通过文本控件显示当前的位置信息和传感器信息，具体实现代码如下所示。

```
<ScrollView xmlns:android="http://schemas.android.com/apk/res/android"
        android:id="@+id/scroll" android:layout_width="fill_parent"
        android:layout_height="fill_parent" android:background="#000000">
    <LinearLayout
            android:layout_width="fill_parent" android:layout_height="fill_parent"
            android:orientation="vertical">
        <TextView android:id="@+id/textStatus" android:layout_width="wrap_content"
            android:layout_height="wrap_content"/>
        <TableLayout android:id="@+id/TableGPS"
                android:layout_width="fill_parent"    android:layout_height="wrap_
content"
                android:stretchColumns="1" android:background="#000000">
            <TableRow android:background="#333333" android:layout_margin="1dip">
                <TextView android:id="@+id/txtDateTimeAndProvider"
                        android:gravity="left"    android:textStyle="bold"    android:
padding="2dip"
                        android:layout_span="2"/>
            </TableRow>
            <TableRow android:background="#333333" android:layout_margin="1dip">
                <TextView android:textStyle="bold" android:text="@string/txt_latitude"
                        android:padding="3dip" android:textSize="17sp"/>
                <TextView android:id="@+id/txtLatitude" android:gravity="left"
                        android:padding="3dip" android:textColor="#e8a317"
                        android:textStyle="bold" android:textSize="18sp"/>
            </TableRow>
            <TableRow android:background="#333333" android:layout_margin="1dip">
                <TextView android:textStyle="bold" android:text="@string/txt_longitude"
                        android:padding="3dip" android:textSize="17sp"/>
                <TextView android:id="@+id/txtLongitude" android:gravity="left"
                        android:padding="3dip" android:textColor="#e8a317"
                        android:textStyle="bold" android:textSize="18sp"/>
            </TableRow>
            <TableRow android:background="#333333" android:layout_margin="1dip">
                <TextView android:id="@+id/lblAltitude" android:textStyle="bold"
                        android:text="@string/txt_altitude" android:padding="3dip"/>
                <TextView android:id="@+id/txtAltitude" android:gravity="left"
                        android:padding="3dip"/>
            </TableRow>
            <TableRow android:background="#333333" android:layout_margin="1dip">
                <TextView android:id="@+id/lblSpeed" android:textStyle="bold"
                        android:text="@string/txt_speed" android:padding="3dip"/>
                <TextView android:id="@+id/txtSpeed" android:gravity="left"
                        android:padding="3dip"/>
            </TableRow>
            <TableRow android:background="#333333" android:layout_margin="1dip">
                <TextView android:id="@+id/lblDirection" android:textStyle="bold"
                        android:text="@string/txt_direction" android:padding="3dip"/>
                <TextView android:id="@+id/txtDirection" android:gravity="left"
                        android:padding="3dip"/>
            </TableRow>
            <TableRow android:background="#333333" android:layout_margin="1dip">
                <TextView android:id="@+id/lblSatellites" android:textStyle="bold"
                        android:text="@string/txt_satellites" android:padding="3dip"/>
                <TextView android:id="@+id/txtSatellites" android:gravity="left"
                        android:padding="3dip"/>
            </TableRow>
            <TableRow android:background="#333333" android:layout_margin="1dip">
                <TextView android:id="@+id/lblAccuracy" android:textStyle="bold"
                        android:text="@string/txt_accuracy" android:padding="3dip"/>
                <TextView android:id="@+id/txtAccuracy" android:gravity="left"
                        android:padding="3dip"/>
            </TableRow>
        </TableLayout>
        <LinearLayout android:orientation="vertical"
                android:layout_width="fill_parent" android:layout_height="wrap_content">
            <!--
                <Button android:id="@+id/buttonStart" android:layout_width="120px"
                android:layout_height="wrap_content" android:tag="Start"
                android:text="Start" /> <Button android:id="@+id/buttonStop"
```

```
            android:layout_width="120px" android:layout_height="wrap_content"
            android:tag="Stop" android:text="Stop" />
        -->
        <ToggleButton android:id="@+id/buttonOnOff"
                android:layout_width="fill_parent"
                android:layout_height= "wrap_content"
                android:textOn="@string/btn_stop_logging"
                android:textOff= "@string/btn_start_logging"/>
    </LinearLayout>
    <TableLayout android:id="@+id/TableSummary"
            android:layout_width="fill_parent"android:layout_height=
            "wrap_content"
            android:background="#222222">
        <TableRow android:layout_width="fill_parent" android:layout_height="fill_
parent">
            <TextView  android:id="@+id/lblLoggingTo"  android:layout_width="wrap_
content"
                android:textSize="9dip"    android:layout_height="fill_parent"
android:textStyle="italic"
                android:paddingLeft="8dip"    android:text="@string/summary_
loggingto"/>
            <TextView  android:id="@+id/txtLoggingTo"  android:layout_width="wrap_
content"
                android:paddingLeft="3dip"        android:textSize="9dip"
android:textStyle="italic"
                android:layout_height="fill_parent"/>
        </TableRow>
        <TableRow android:layout_width="fill_parent"
              android:layout_height="fill_parent">
            <TextView     android:id="@+id/lblFrequency"     android:layout_width=
"wrap_content"
                android:textSize="9dip"    android:layout_height="fill_parent"
android:textStyle="italic"
                android:paddingLeft="8dip" android:text="@string/summary_freq_
every"/>
            <TextView  android:id="@+id/txtFrequency"  android:layout_width="wrap_
content"
                android:paddingLeft="3dip"        android:textSize="9dip"
android:textStyle="italic"
                android:layout_height="fill_parent"/>
        </TableRow>
        <TableRow android:layout_width="fill_parent"
              android:layout_height="fill_parent">
            <TextView  android:id="@+id/lblDistance"  android:layout_width="wrap_
content"
                android:textSize="9dip"    android:layout_height="fill_parent"
android:textStyle="italic"
                android:paddingLeft="8dip" android:text="@string/summary_dist"/>
            <TextView  android:id="@+id/txtDistance"  android:layout_width="wrap_
content"
                android:paddingLeft="3dip"        android:textSize="9dip"
android:textStyle="italic"
                android:layout_height="fill_parent"/>
        </TableRow>
        <TableRow android:layout_width="fill_parent"
              android:layout_height="fill_parent">
            <TextView  android:id="@+id/lblFileName"  android:layout_width="wrap_
content"
                android:textSize="9dip"    android:layout_height="fill_parent"
android:textStyle="italic"
                android:paddingLeft="8dip"    android:text="@string/summary_
current_filename"/>
            <TextView  android:id="@+id/txtFileName"  android:layout_width="wrap_
content"
                android:paddingLeft="3dip" android:textSize="9dip"
                android:textStyle="italic"
                android:layout_height="fill_parent"/>
        </TableRow>
        <TableRow android:layout_width="fill_parent"
```

```
                android:layout_height="fill_parent" android:id="@+id/trAutoEmail">
            <TextView     android:id="@+id/lblAutoEmail"     android:layout_width=
"wrap_content"
                    android:textSize="9dip"    android:layout_height="fill_parent"
android:textStyle="italic"
                    android:paddingLeft="8dip"        android:text="@string/summary_
autoemail"/>
            <TextView android:id="@+id/txtAutoEmail" android:layout_width="wrap_content"
                    android:paddingLeft="3dip"              android:textSize="9dip"
android:textStyle="italic"
                   android:layout_height="fill_parent"/>
        </TableRow>
      </TableLayout>
      <!-- <TextView android:id="@+id/lblSummary" android:layout_width="fill_parent"
    android:layout_height="wrap_content" android:textStyle="italic"
    android:textSize="11dip" /> -->
   </LinearLayout>
</ScrollView>
```

27.2.2 实现主 Activity

本系统的主 Activity 是 GpsMainActivity，实现文件是 GpsMainActivity.java，具体实现流程如下所示。

（1）定义更新 UI 线程的类 GpsMainActivity，获取 ToggleButton 按钮的开关来显示位置信息，具体实现代码如下所示。

```
public class GpsMainActivity extends Activity implements OnCheckedChangeListener,
        IGpsLoggerServiceClient
{
    /**
     * 用处理器更新 UI 线程
     */
    public final Handler handler = new Handler();
    private static Intent serviceIntent;
    private GpsLoggingService loggingService;
    /**
     * 提供一个连接到 GPS 记录的服务
     */
    private ServiceConnection gpsServiceConnection = new ServiceConnection()
    {
        public void onServiceDisconnected(ComponentName name)
        {
            loggingService = null;
        }
        public void onServiceConnected(ComponentName name, IBinder service)
        {
            loggingService = ((GpsLoggingService.GpsLoggingBinder) service).getService();
            GpsLoggingService.SetServiceClient(GpsMainActivity.this);
            // 设置切换按钮，显示现有的位置信息
            ToggleButton buttonOnOff = (ToggleButton) findViewById(R.id.buttonOnOff);
            if (Session.isStarted())
            {
                buttonOnOff.setChecked(true);
                DisplayLocationInfo(Session.getCurrentLocationInfo());
            }
            buttonOnOff.setOnCheckedChangeListener(GpsMainActivity.this);
        }
    };
```

（2）定义第一次创建样式时触发的方法，具体实现代码如下所示。

```
/**
 * 第一次创建样式时触发的事件
 */
@Override
public void onCreate(Bundle savedInstanceState)
{
```

```
    SharedPreferences    prefs    =    PreferenceManager.getDefaultSharedPreferences
(getBaseContext());
    String lang = prefs.getString("locale_override", "");
    if (!lang.equalsIgnoreCase(""))
    {
        Locale locale = new Locale(lang);
        Locale.setDefault(locale);
        Configuration config = new Configuration();
        config.locale = locale;
        getBaseContext().getResources().updateConfiguration(config,
                getBaseContext().getResources().getDisplayMetrics());
    }
    super.onCreate(savedInstanceState);
    Utilities.LogInfo("GPSLogger started");
    setContentView(R.layout.main);
    GetPreferences();
    StartAndBindService();
}
```

（3）启动定位服务并绑定到当前的 Activity 界面，具体实现代码如下所示。

```
private void StartAndBindService()
{
    Utilities.LogDebug("StartAndBindService - binding now");
    serviceIntent = new Intent(this, GpsLoggingService.class);
    // Start the service in case it isn't already running
    startService(serviceIntent);
    // Now bind to service
    bindService(serviceIntent, gpsServiceConnection, Context.BIND_AUTO_CREATE);
    Session.setBoundToService(true);
}
```

（4）当按钮关闭则停止系统的监听服务，具体实现代码如下所示。

```
private void StopAndUnbindServiceIfRequired()
{
    if(Session.isBoundToService())
    {
        unbindService(gpsServiceConnection);
        Session.setBoundToService(false);
    }
    if(!Session.isStarted())
    {
        Utilities.LogDebug("StopServiceIfRequired - Stopping the service");
        //serviceIntent = new Intent(this, GpsLoggingService.class);
        stopService(serviceIntent);
    }
}

@Override
protected void onPause()
{
    StopAndUnbindServiceIfRequired();
    super.onPause();
}

@Override
protected void onDestroy()
{
    StopAndUnbindServiceIfRequired();
    super.onDestroy();

}
```

（5）当切换按钮被单击时调用方法 onCheckedChanged，具体实现代码如下所示。

```
public void onCheckedChanged(CompoundButton buttonView, boolean isChecked)
{
    if (isChecked)
    {
```

```
        GetPreferences();

        loggingService.StartLogging();
    }
    else
    {
        loggingService.StopLogging();
    }
}
```

（6）根据用户设置选项值显示一个具有良好可读性的视图界面，具体实现代码如下所示。

```
    private void ShowPreferencesSummary()
    {
        TextView txtLoggingTo = (TextView) findViewById(R.id.txtLoggingTo);
        TextView txtFrequency = (TextView) findViewById(R.id.txtFrequency);
        TextView txtDistance = (TextView) findViewById(R.id.txtDistance);
        TextView txtAutoEmail = (TextView) findViewById(R.id.txtAutoEmail);
        if (!AppSettings.shouldLogToKml() && !AppSettings.shouldLogToGpx())
        {
            txtLoggingTo.setText(R.string.summary_loggingto_screen);
        }
        else if (AppSettings.shouldLogToGpx() && AppSettings.shouldLogToKml())
        {
            txtLoggingTo.setText(R.string.summary_loggingto_both);
        }
        else
        {
            txtLoggingTo.setText((AppSettings.shouldLogToGpx() ? "GPX" : "KML"));
        }
        if (AppSettings.getMinimumSeconds() > 0)
        {
            String descriptiveTime = Utilities.GetDescriptiveTimeString(AppSettings.
            getMinimumSeconds(),getBaseContext());
            txtFrequency.setText(descriptiveTime);
        }
        else
        {
            txtFrequency.setText(R.string.summary_freq_max);
        }
        if (AppSettings.getMinimumDistance() > 0)
        {
            if (AppSettings.shouldUseImperial())
            {
                int minimumDistanceInFeet = Utilities.MetersToFeet(AppSettings.
                getMinimumDistance());
                txtDistance.setText(((minimumDistanceInFeet == 1)
                    ? getString(R.string.foot)
                    : String.valueOf(minimumDistanceInFeet) + getString(R.string.feet)));
            }
            else
            {
                txtDistance.setText(((AppSettings.getMinimumDistance() == 1)
                    ? getString(R.string.meter)
                    : String.valueOf(AppSettings.getMinimumDistance()) + getString
                    (R.string.meters)));
            }
        }
        else
        {
            txtDistance.setText(R.string.summary_dist_regardless);
        }
        if (AppSettings.isAutoEmailEnabled())
        {
            String autoEmailResx;
            if (AppSettings.getAutoEmailDelay() == 0)
            {
                autoEmailResx = "autoemail_frequency_whenistop";
            }
            else
```

```
            {

                autoEmailResx = "autoemail_frequency_"
                        + String.valueOf(AppSettings.getAutoEmailDelay()).replace(".", "");
                //.replace(".0", "")
            }
            String autoEmailDesc = getString(getResources().getIdentifier(autoEmailResx,
            "string", getPackageName()));

//          String autoEmailDesc = getString(getResources().getIdentifier(
//                  getPackageName() + ":string/" + autoEmailResx, null, null));

            txtAutoEmail.setText(autoEmailDesc);
        }
        else
        {
            TableRow trAutoEmail = (TableRow) findViewById(R.id.trAutoEmail);
            trAutoEmail.setVisibility(View.INVISIBLE);
        }
    }
```

（7）根据用户选择的菜单项调用执行不同的处理方法，具体实现代码如下所示。

```
public boolean onOptionsItemSelected(MenuItem item)
{
    int itemId = item.getItemId();
    Utilities.LogInfo("Option item selected - " + String.valueOf(item.getTitle()));
    switch (itemId)
    {
        case R.id.mnuSettings:
            Intent settingsActivity = new Intent(getBaseContext(), GpsSettingsActivity.
            class);
            startActivity(settingsActivity);
            break;
        case R.id.mnuOSM:
            UploadToOpenStreetMap();
            break;
        case R.id.mnuAnnotate:
            Annotate();
            break;
        case R.id.mnuShare:
            Share();
            break;
        case R.id.mnuEmailnow:
            EmailNow();
            break;
        case R.id.mnuExit:
            loggingService.StopLogging();
            loggingService.stopSelf();
            System.exit(0);
            break;
    }
    return false;
}
private void EmailNow()
{
    if(Utilities.IsEmailSetup(getBaseContext()))
    {
        loggingService.ForceEmailLogFile();
    }
    else
    {
        Intent emailSetup = new Intent(getBaseContext(), AutoEmailActivity.class);
        startActivity(emailSetup);
    }
}
```

（8）通过方法 Share 实现轨迹分享功能，允许用户发送带位置的 GPX/KML 文件的位置，或者只使用一个提供者，可以使用的分享方式有 Facebook、短信、电子邮件、推特和蓝牙。方法 Share

的具体实现代码如下所示。

```
private void Share()
{
    try
    {
        final String locationOnly = getString(R.string.sharing_location_only);
        final File gpxFolder = new File(Environment.getExternalStorageDirectory(),
        "GPSLogger");
        if (gpxFolder.exists())
        {
            String[] enumeratedFiles = gpxFolder.list();
            List<String>fileList=new ArrayList<String>(Arrays.asList (enumeratedFiles));
            Collections.reverse(fileList);
            fileList.add(0, locationOnly);
            final String[] files = fileList.toArray(new String[0]);
            final Dialog dialog = new Dialog(this);
            dialog.setTitle(R.string.sharing_pick_file);
            dialog.setContentView(R.layout.filelist);
            ListView thelist = (ListView) dialog.findViewById(R.id.listViewFiles);
            thelist.setAdapter(new ArrayAdapter<String>(getBaseContext(),
                    android.R.layout.simple_list_item_single_choice, files));
            thelist.setOnItemClickListener(new OnItemClickListener()
            {
                public void onItemClick(AdapterView<?> av, View v, int index, long arg)
                {
                    dialog.dismiss();
                    String chosenFileName = files[index];
                    final Intent intent = new Intent(Intent.ACTION_SEND);
                    // intent.setType("text/plain");
                    intent.setType("*/*");
                    if (chosenFileName.equalsIgnoreCase(locationOnly))
                    {
                        intent.setType("text/plain");
                    }
                    intent.putExtra(Intent.EXTRA_SUBJECT,
                    getString(R.string.sharing_mylocation));
                    if (Session.hasValidLocation())
                    {
                        String bodyText = getString(R.string.sharing_latlong_text,
                                String.valueOf(Session.getCurrentLatitude()),
                                String.valueOf(Session.getCurrentLongitude()));
                        intent.putExtra(Intent.EXTRA_TEXT, bodyText);
                        intent.putExtra("sms_body", bodyText);
                    }
                    if (chosenFileName.length() > 0
                            && !chosenFileName.equalsIgnoreCase(locationOnly))
                    {
                        intent.putExtra(Intent.EXTRA_STREAM,
                                Uri.fromFile(new File(gpxFolder, chosenFileName)));
                    }
                    startActivity(Intent.createChooser(intent,
                    getString(R.string.sharing_via)));
                }
            });
            dialog.show();
        }
        else
        {
            Utilities.MsgBox(getString(R.string.sorry),
            getString(R.string.no_files_found), this);
        }
    }
    catch (Exception ex)
    {
        Utilities.LogError("Share", ex);
    }
}
```

（9）编写方法UploadToOpenStreetMap上传一个跟踪GPS记录的对象，具体实现代码如下所示。

```
private void UploadToOpenStreetMap()
{
    if(!Utilities.IsOsmAuthorized(getBaseContext()))
    {
        startActivity(Utilities.GetOsmSettingsIntent(getBaseContext()));
        return;
    }
    final String goToOsmSettings = getString(R.string.menu_settings);

    final File gpxFolder=new File(Environment.getExternalStorageDirectory(),"GPSLogger");
    if (gpxFolder.exists())
    {
        FilenameFilter select = new FilenameFilter()
        {

            public boolean accept(File dir, String filename)
            {
                return filename.toLowerCase().contains(".gpx");
            }
        };
        String[] enumeratedFiles = gpxFolder.list(select);
        List<String> fileList = new ArrayList<String>(Arrays.asList(enumeratedFiles));
        Collections.reverse(fileList);
        fileList.add(0, goToOsmSettings);
        final String[] files = fileList.toArray(new String[0]);
        final Dialog dialog = new Dialog(this);
        dialog.setTitle(R.string.osm_pick_file);
        dialog.setContentView(R.layout.filelist);
        ListView thelist = (ListView) dialog.findViewById(R.id.listViewFiles);
        thelist.setAdapter(new ArrayAdapter<String>(getBaseContext(),
                android.R.layout.simple_list_item_single_choice, files));
        thelist.setOnItemClickListener(new OnItemClickListener()
        {
            public void onItemClick(AdapterView<?> av, View v, int index, long arg)
            {

                dialog.dismiss();
                String chosenFileName = files[index];

                if(chosenFileName.equalsIgnoreCase(goToOsmSettings))
                {
                    startActivity(Utilities.GetOsmSettingsIntent(getBaseContext()));
                }
                else
                {
                    OSMHelper osm = new OSMHelper(GpsMainActivity.this);
                    Utilities.ShowProgress(GpsMainActivity.this,
getString(R.string.osm_uploading), getString(R.string.please_wait));
                    osm.UploadGpsTrace(chosenFileName);
                }
            }
        });
        dialog.show();
    }
    else
    {
        Utilities.MsgBox(getString(R.string.sorry),getString(R.string.no_files_found),this);
    }
}
```

（10）通过方法Annotate提示用户输入，然后添加文本日志文件，具体实现代码如下所示。

```
private void Annotate()
{
    if (!AppSettings.shouldLogToGpx() && !AppSettings.shouldLogToKml())
    {
```

```
        return;
    }

    if (!Session.shoulAllowDescription())
    {
        Utilities.MsgBox(getString(R.string.not_yet),
                getString(R.string.cant_add_description_until_next_point),
                GetActivity());
        return;
    }
    AlertDialog.Builder alert = new AlertDialog.Builder(GpsMainActivity.this);
    alert.setTitle(R.string.add_description);
    alert.setMessage(R.string.letters_numbers);
    //设置一个 EditText 视图用来获取用户的输入
    final EditText input = new EditText(getBaseContext());
    alert.setView(input);
    alert.setPositiveButton(R.string.ok, new DialogInterface.OnClickListener()
    {
        public void onClick(DialogInterface dialog, int whichButton)
        {
            final String desc = Utilities.CleanDescription(input.getText().toString());
            Annotate(desc);
        }
    });
    alert.setNegativeButton(R.string.cancel, new DialogInterface.OnClickListener()
    {
        public void onClick(DialogInterface dialog, int whichButton)
        {
            // Cancelled.
        }
    });
    alert.show();
}
```

（11）编写方法 ClearForm 清理当前屏幕视图，并删除所有获取的值，具体实现代码如下所示。

```
public void ClearForm()
{
    TextView tvLatitude = (TextView) findViewById(R.id.txtLatitude);
    TextView tvLongitude = (TextView) findViewById(R.id.txtLongitude);
    TextView tvDateTime = (TextView) findViewById(R.id.txtDateTimeAndProvider);
    TextView tvAltitude = (TextView) findViewById(R.id.txtAltitude);
    TextView txtSpeed = (TextView) findViewById(R.id.txtSpeed);
    TextView txtSatellites = (TextView) findViewById(R.id.txtSatellites);
    TextView txtDirection = (TextView) findViewById(R.id.txtDirection);
    TextView txtAccuracy = (TextView) findViewById(R.id.txtAccuracy);
    tvLatitude.setText("");
    tvLongitude.setText("");
    tvDateTime.setText("");
    tvAltitude.setText("");
    txtSpeed.setText("");
    txtSatellites.setText("");
    txtDirection.setText("");
    txtAccuracy.setText("");
}
```

（12）在顶部的状态标签设置信息，具体实现代码如下所示。

```
private void SetStatus(String message)
{
    TextView tvStatus = (TextView) findViewById(R.id.textStatus);
    tvStatus.setText(message);
    Utilities.LogInfo(message);
}
```

（13）设置表中的卫星视图，具体实现代码如下所示。

```
private void SetSatelliteInfo(int number)
{
    Session.setSatelliteCount(number);
```

```
    TextView txtSatellites = (TextView) findViewById(R.id.txtSatellites);
    txtSatellites.setText(String.valueOf(number));
}
```

（14）处理指定的位置坐标，并将结果显示在视图中，具体实现代码如下所示。

```
private void DisplayLocationInfo(Location loc)
{
    try
    {
        if (loc == null)
        {
            return;
        }
        Session.setLatestTimeStamp(System.currentTimeMillis());
        TextView tvLatitude = (TextView) findViewById(R.id.txtLatitude);
        TextView tvLongitude = (TextView) findViewById(R.id.txtLongitude);
        TextView tvDateTime = (TextView) findViewById(R.id.txtDateTimeAndProvider);
        TextView tvAltitude = (TextView) findViewById(R.id.txtAltitude);
        TextView txtSpeed = (TextView) findViewById(R.id.txtSpeed);
        TextView txtSatellites = (TextView) findViewById(R.id.txtSatellites);
        TextView txtDirection = (TextView) findViewById(R.id.txtDirection);
        TextView txtAccuracy = (TextView) findViewById(R.id.txtAccuracy);
        String providerName = loc.getProvider();
        if (providerName.equalsIgnoreCase("gps"))
        {
            providerName = getString(R.string.providername_gps);
        }
        else
        {
            providerName = getString(R.string.providername_celltower);
        }
        tvDateTime.setText(new Date().toLocaleString()
                + getString(R.string.providername_using, providerName));
        tvLatitude.setText(String.valueOf(loc.getLatitude()));
        tvLongitude.setText(String.valueOf(loc.getLongitude()));
        if (loc.hasAltitude())
        {
            double altitude = loc.getAltitude();
            if (AppSettings.shouldUseImperial())
            {
                tvAltitude.setText(String.valueOf(Utilities.MetersToFeet(altitude))
                        + getString(R.string.feet));
            }
            else
            {
                tvAltitude.setText(String.valueOf(altitude) + getString(R.string.meters));
            }
        }
        else
        {
            tvAltitude.setText(R.string.not_applicable);
        }          if (loc.hasSpeed())
        {
            float speed = loc.getSpeed();
            if (AppSettings.shouldUseImperial())
            {
                txtSpeed.setText(String.valueOf(Utilities.MetersToFeet(speed))
                        + getString(R.string.feet_per_second));
            }
            else
            {
                txtSpeed.setText(String.valueOf(speed) + getString(R.string.meters_
                per_second));
            }
        }
        else
        {
            txtSpeed.setText(R.string.not_applicable);
        }
```

```
        if (loc.hasBearing())
        {
            float bearingDegrees = loc.getBearing();
            String direction;
            direction = Utilities.GetBearingDescription(bearingDegrees, getBaseContext());
            txtDirection.setText(direction + "(" + String.valueOf(Math.round(bearing
            Degrees))+ getString(R.string.degree_symbol) + ")");
        }
        else
        {
            txtDirection.setText(R.string.not_applicable);
        }
        if (!Session.isUsingGps())
        {
            txtSatellites.setText(R.string.not_applicable);
            Session.setSatelliteCount(0);
        }
        if (loc.hasAccuracy())
        {
            float accuracy = loc.getAccuracy();
            if (AppSettings.shouldUseImperial())
            {
                txtAccuracy.setText(getString(R.string.accuracy_within,
                        String.valueOf(Utilities.MetersToFeet(accuracy)),
                        getString(R.string.feet)));
            }
            else
            {
                txtAccuracy.setText(getString(R.string.accuracy_within,
                String.valueOf(accuracy),getString(R.string.meters)));
            }
        }
        else
        {
            txtAccuracy.setText(R.string.not_applicable);
        }
    }
    catch (Exception ex)
    {
        SetStatus(getString(R.string.error_displaying, ex.getMessage()));
    }
}
```

在主 Activity 的实现过程中用到了系统服务 Activity，其实现文件是 GpsLoggingService.java，功能是提供了本系统所需要的后台服务方法。文件 GpsLoggingService.java 的具体实现流程如下所示。

（1）定义可以调用的类和方法，具体实现代码如下所示。

```
class GpsLoggingBinder extends Binder
{
    public GpsLoggingService getService()
    {
        Utilities.LogDebug("GpsLoggingBinder.getService");
        return GpsLoggingService.this;
    }
}
```

（2）建立基于用户偏好设置的电邮自动计时器，具体实现代码如下所示。

```
private void SetupAutoEmailTimers()
{
    Utilities.LogDebug("GpsLoggingService.SetupAutoEmailTimers");
    Utilities.LogDebug("isAutoEmailEnabled - " + String.valueOf(AppSettings.isAuto
EmailEnabled()));
    Utilities.LogDebug("Session.getAutoEmailDelay - " + String.valueOf(Session.
getAutoEmailDelay()));
    if (AppSettings.isAutoEmailEnabled() && Session.getAutoEmailDelay() > 0)
    {
```

```
        Utilities.LogDebug("Setting up email alarm");
        long triggerTime = System.currentTimeMillis()
                + (long) (Session.getAutoEmailDelay() * 60 * 60 * 1000);
        alarmIntent = new Intent(getBaseContext(), AlarmReceiver.class);
        PendingIntent sender = PendingIntent.getBroadcast(this, 0, alarmIntent,
                PendingIntent.FLAG_UPDATE_CURRENT);
        AlarmManager am = (AlarmManager) getSystemService(ALARM_SERVICE);
        am.set(AlarmManager.RTC_WAKEUP, triggerTime, sender);
    }
    else
    {
        Utilities.LogDebug("Checking if alarmIntent is null");
        if (alarmIntent != null)
        {
            Utilities.LogDebug("alarmIntent was null, canceling alarm");
            CancelAlarm();
        }
    }
}
```

（3）如果调用用户选择自动邮件日志文件方法，则停止记录操作，具体实现代码如下所示。

```
private void AutoEmailLogFileOnStop()
{
    Utilities.LogDebug("GpsLoggingService.AutoEmailLogFileOnStop");
    Utilities.LogVerbose("isAutoEmailEnabled - " + AppSettings.isAutoEmailEnabled());
    // autoEmailDelay 0 means send it when you stop logging.
    if (AppSettings.isAutoEmailEnabled() && Session.getAutoEmailDelay() == 0)
    {
        Session.setEmailReadyToBeSent(true);
        AutoEmailLogFile();
    }
}
```

（4）调用自动电子邮件辅助处理文件，并将其发送，具体实现代码如下所示。

```
private void AutoEmailLogFile()
{
    Utilities.LogDebug("GpsLoggingService.AutoEmailLogFile");
    Utilities.LogVerbose("isEmailReadyToBeSent - " + Session.isEmailReadyToBeSent());
    if (Session.getCurrentFileName() != null && Session.getCurrentFileName().length() > 0
            && Session.isEmailReadyToBeSent())
    {
        if(IsMainFormVisible())
        {
            Utilities.ShowProgress(mainServiceClient.GetActivity(), getString(R.string.
            autoemail_sending),getString(R.string.please_wait));
        }
        Utilities.LogInfo("Emailing Log File");
        AutoEmailHelper aeh = new AutoEmailHelper(GpsLoggingService.this);
        aeh.SendLogFile(Session.getCurrentFileName(), false);
        SetupAutoEmailTimers();

        if(IsMainFormVisible())
        {
            Utilities.HideProgress();
        }
    }
}
protected void ForceEmailLogFile()
{

    Utilities.LogDebug("GpsLoggingService.ForceEmailLogFile");
    if (Session.getCurrentFileName() != null && Session.getCurrentFileName().length() > 0)
    {
        if(IsMainFormVisible())
        {
            Utilities.ShowProgress(mainServiceClient.GetActivity(), getString(R.string.
            autoemail_sending),getString(R.string.please_wait));
        }
```

```
        Utilities.LogInfo("Force emailing Log File");
        AutoEmailHelper aeh = new AutoEmailHelper(GpsLoggingService.this);
        aeh.SendLogFile(Session.getCurrentFileName(), true);

        if(IsMainFormVisible())
        {
            Utilities.HideProgress();
        }
    }
}
```

（5）设置此服务的活动形式，活动形式需要调用 IGpsLoggerServiceClient，具体实现代码如下所示。

```
protected static void SetServiceClient(IGpsLoggerServiceClient mainForm)
{
    mainServiceClient = mainForm;
}
```

（6）根据用户的偏好设置选择并填充 AppSettings 对象，并设置电子邮件的定时器，具体实现代码如下所示。

```
private void GetPreferences()
{
    Utilities.LogDebug("GpsLoggingService.GetPreferences");
    Utilities.PopulateAppSettings(getBaseContext());
    Utilities.LogDebug("Session.getAutoEmailDelay: " + Session.getAutoEmailDelay());
    Utilities.LogDebug("AppSettings.getAutoEmailDelay:"+AppSettings.getAutoEmailDelay());
    if (Session.getAutoEmailDelay() != AppSettings.getAutoEmailDelay())
    {
        Utilities.LogDebug("Old autoEmailDelay - " + String.valueOf(Session.
        getAutoEmailDelay())
                + "; New -" + String.valueOf(AppSettings.getAutoEmailDelay()));
        Session.setAutoEmailDelay(AppSettings.getAutoEmailDelay());
        SetupAutoEmailTimers();
    }
}
```

（7）实现复位处理，具体实现代码如下所示。

```
protected void StartLogging()
{
    Utilities.LogDebug("GpsLoggingService.StartLogging");
    Session.setAddNewTrackSegment(true);
    if (Session.isStarted())
    {
        return;
    }
    Utilities.LogInfo("Starting logging procedures");
    startForeground(NOTIFICATION_ID, null);
    Session.setStarted(true);
    GetPreferences();
    Notify();
    ResetCurrentFileName();
    ClearForm();
    StartGpsManager();
}
```

（8）停止记录，删除通知，停止 GPS，通过定时器停止邮件，具体实现代码如下所示。

```
protected void StopLogging()
{
    Utilities.LogDebug("GpsLoggingService.StopLogging");
    Session.setAddNewTrackSegment(true);
    Utilities.LogInfo("Stopping logging");
    Session.setStarted(false);
    // Email log file before setting location info to null
    AutoEmailLogFileOnStop();
    CancelAlarm();
```

```
    Session.setCurrentLocationInfo(null);
    stopForeground(true);
    RemoveNotification();
    StopGpsManager();
    StopMainActivity();
}
```

（9）状态栏中显示通知，具体实现代码如下所示。

```
private void Notify()
{
    Utilities.LogDebug("GpsLoggingService.Notify");
    if (AppSettings.shouldShowInNotificationBar())
    {
        gpsNotifyManager=(NotificationManager)getSystemService(NOTIFICATION_SERVICE);
        ShowNotification();
    }
    else
    {
        RemoveNotification();
    }
}
```

（10）如果图标是可见的，则隐藏状态栏中的通知，具体实现代码如下所示。

```
private void RemoveNotification()
{
    Utilities.LogDebug("GpsLoggingService.RemoveNotification");
    try
    {
        if (Session.isNotificationVisible())
        {
            gpsNotifyManager.cancelAll();
        }
    }
    catch (Exception ex)
    {
        Utilities.LogError("RemoveNotification", ex);
    }
    finally
    {
        // notificationVisible = false;
        Session.setNotificationVisible(false);
    }
}
```

（11）在状态条中显示GPS记录器的通知图标，具体实现代码如下所示。

```
private void ShowNotification()
{
    Utilities.LogDebug("GpsLoggingService.ShowNotification");
    // What happens when the notification item is clicked
    Intent contentIntent = new Intent(this, GpsMainActivity.class);
    PendingIntent pending = PendingIntent.getActivity(getBaseContext(),0,contentIntent,
          android.content.Intent.FLAG_ACTIVITY_NEW_TASK);
    Notification nfc = new Notification(R.drawable.gpsloggericon2, null,
    System.currentTimeMillis());
    nfc.flags |= Notification.FLAG_ONGOING_EVENT;
    NumberFormat nf = new DecimalFormat("###.######");
    String contentText = getString(R.string.gpslogger_still_running);
    if (Session.hasValidLocation())
    // if (currentLatitude != 0 && currentLongitude != 0)
    {
        contentText = nf.format(Session.getCurrentLatitude()) + ","
              + nf.format(Session.getCurrentLongitude());
    }
    nfc.setLatestEventInfo(getBaseContext(),
```

```
getString(R.string.gpslogger_still_running),
        contentText, pending);
    gpsNotifyManager.notify(NOTIFICATION_ID, nfc);
    Session.setNotificationVisible(true);
}
```

（12）根据用户的偏好设置选项启动GPS功能，具体实现代码如下所示。

```
private void StartGpsManager()
{
    Utilities.LogDebug("GpsLoggingService.StartGpsManager");
    GetPreferences();
    gpsLocationListener = new GeneralLocationListener(this);
    towerLocationListener = new GeneralLocationListener(this);
    gpsLocationManager = (LocationManager) getSystemService(Context.LOCATION_SERVICE);
    towerLocationManager = (LocationManager) getSystemService(Context.LOCATION_SERVICE);
    CheckTowerAndGpsStatus();
    if (Session.isGpsEnabled() && !AppSettings.shouldPreferCellTower())
    {
        Utilities.LogInfo("Requesting GPS location updates");
        // gps satellite based
        gpsLocationManager.requestLocationUpdates(LocationManager.GPS_PROVIDER,
                AppSettings.getMinimumSeconds() * 1000, AppSettings.getMinimumDistance(),
                gpsLocationListener);
        gpsLocationManager.addGpsStatusListener(gpsLocationListener);
        Session.setUsingGps(true);
    }
    else if (Session.isTowerEnabled())
    {
        Utilities.LogInfo("Requesting tower location updates");
        Session.setUsingGps(false);
        // isUsingGps = false;
        // Cell tower and wifi based
        towerLocationManager.requestLocationUpdates(LocationManager.NETWORK_PROVIDER,
                AppSettings.getMinimumSeconds() * 1000, AppSettings.getMinimumDistance(),
                towerLocationListener);
    }
    else
    {
        Utilities.LogInfo("No provider available");
        Session.setUsingGps(false);
        SetStatus(R.string.gpsprovider_unavailable);
        SetFatalMessage(R.string.gpsprovider_unavailable);
        StopLogging();
        return;
    }
    SetStatus(R.string.started);
}
```

（13）周期性检查是否已经启动GPS和信号塔，具体实现代码如下所示。

```
private void CheckTowerAndGpsStatus()
{
    Session.setTowerEnabled(towerLocationManager.isProviderEnabled
(LocationManager.NETWORK_PROVIDER));
    Session.setGpsEnabled(gpsLocationManager.isProviderEnabled
(LocationManager.GPS_PROVIDER));
}
```

（14）停止位置管理服务，具体实现代码如下所示。

```
private void StopGpsManager()
{
    Utilities.LogDebug("GpsLoggingService.StopGpsManager");
    if (towerLocationListener != null)
    {
        towerLocationManager.removeUpdates(towerLocationListener);
    }
    if (gpsLocationListener != null)
    {
```

```
        gpsLocationManager.removeUpdates(gpsLocationListener);
        gpsLocationManager.removeGpsStatusListener(gpsLocationListener);
    }
    SetStatus(getString(R.string.stopped));
}
```

（15）基于用户偏好设置当前文件名，具体实现代码如下所示。

```
private void ResetCurrentFileName()
{
    Utilities.LogDebug("GpsLoggingService.ResetCurrentFileName");
    String newFileName;
    if (AppSettings.shouldCreateNewFileOnceADay())
    {
        // 20100114.gpx
        SimpleDateFormat sdf = new SimpleDateFormat("yyyyMMdd");
        newFileName = sdf.format(new Date());
        Session.setCurrentFileName(newFileName);
    }
    else
    {
        // 20100114183329.gpx
        SimpleDateFormat sdf = new SimpleDateFormat("yyyyMMddHHmmss");
        newFileName = sdf.format(new Date());
        Session.setCurrentFileName(newFileName);
    }
    if (IsMainFormVisible())
    {
        mainServiceClient.onFileName(newFileName);
    }
}
```

（16）为客户端显示一个状态信息，具体实现代码如下所示。

```
void SetStatus(String status)
{
    if (IsMainFormVisible())
    {
        mainServiceClient.OnStatusMessage(status);
    }
}
```

到此为止，系统的主 Activity 和服务 Activity 的实现过程介绍完毕，执行后的效果如图 27-2 所示。

单击设备中的“MENU”按钮后，在屏幕下方弹出选项设置界面，如图 27-3 所示。

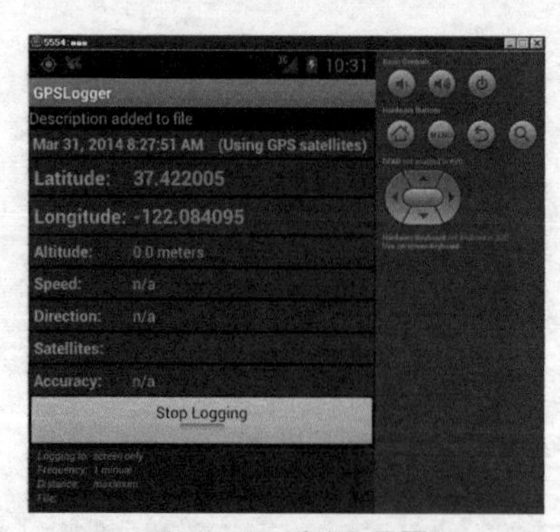

▲图 27-2　系统主界面

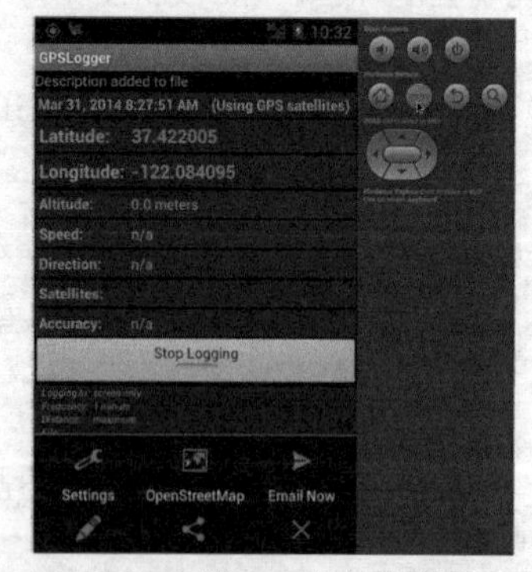

▲图 27-3　屏幕下方弹出选项设置界面

27.3 系统设置

当单击图 27-3 中的“Settings”选项后，会弹出系统设置界面，如图 27-4 所示。

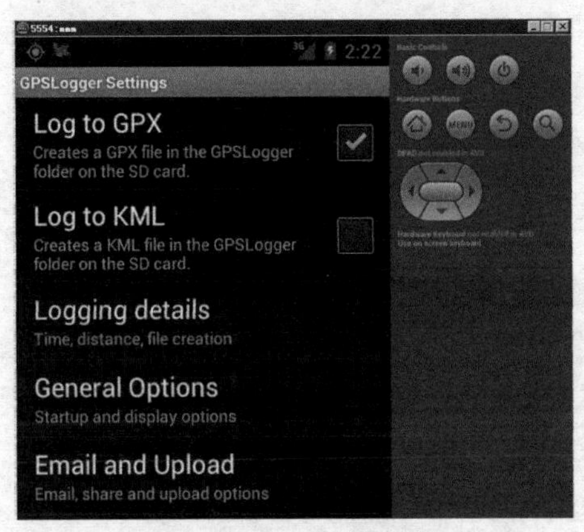

▲图 27-4 系统设置界面

在系统设置界面中，可以设置系统的常用选项参数。在本节的内容中，将详细讲解系统设置模块的实现过程。

27.3.1 选项设置

编写文件 AppSettings.java，功能是根据用户各个选项的值来设置系统，例如，设置保存为 GPX（GPS eXchange Format 的缩写，译为 GPS 交换格式，是一个 XML 格式，为应用软体设计的通用 GPS 数据格式，可以用来描述路点、轨迹、路程）格式数据文件或 KML（是一种文件格式，用于在地球浏览器中显示地理数据，如 Google 地球、Google 地图和谷歌手机地图）格式数据文件。文件 AppSettings.java 的具体实现代码如下所示。

```
public class AppSettings extends Application
{
    // -------------------------------------------------------
    //用户设置
    // -------------------------------------------------------
    private static boolean useImperial = false;
    private static boolean newFileOnceADay;
    private static boolean preferCellTower;
    private static boolean useSatelliteTime;
    private static boolean logToKml;
    private static boolean logToGpx;
    private static boolean showInNotificationBar;
    private static int minimumDistance;
    private static int minimumSeconds;
    private static String newFileCreation;
    private static Float autoEmailDelay = 0f;
    private static boolean autoEmailEnabled = false;
    private static String smtpServer;
    private static String smtpPort;
    private static String smtpUsername;
    private static String smtpPassword;
    private static String autoEmailTarget;
    private static boolean smtpSsl;
    private static boolean debugToFile;
    public static boolean shouldUseImperial()
    {
        return useImperial;
    }
    static void setUseImperial(boolean useImperial)
    {
        AppSettings.useImperial = useImperial;
    }
    /**
     * @return the 一天更新一个新文件
     */
```

```
public static boolean shouldCreateNewFileOnceADay()
{
   return newFileOnceADay;
}
static void setNewFileOnceADay(boolean newFileOnceADay)
{
   AppSettings.newFileOnceADay = newFileOnceADay;
}
public static boolean shouldPreferCellTower()
{
   return preferCellTower;
}
static void setPreferCellTower(boolean preferCellTower)
{
   AppSettings.preferCellTower = preferCellTower;
}
public static boolean shouldUseSatelliteTime()
{
   return useSatelliteTime;
}
static void setUseSatelliteTime(boolean useSatelliteTime)
{
   AppSettings.useSatelliteTime = useSatelliteTime;
}
public static boolean shouldLogToKml()
{
   return logToKml;
}
static void setLogToKml(boolean logToKml)
{
   AppSettings.logToKml = logToKml;
}
public static boolean shouldLogToGpx()
{
   return logToGpx;
}
static void setLogToGpx(boolean logToGpx)
{
   AppSettings.logToGpx = logToGpx;
}
public static boolean shouldShowInNotificationBar()
{
   return showInNotificationBar;
}
static void setShowInNotificationBar(boolean showInNotificationBar)
{
   AppSettings.showInNotificationBar = showInNotificationBar;
}
public static int getMinimumDistance()
{
   return minimumDistance;
}
static void setMinimumDistance(int minimumDistance)
{
   AppSettings.minimumDistance = minimumDistance;
}
public static int getMinimumSeconds()
{
   return minimumSeconds;
}

/**
 * @param minimumSeconds
 *          the minimumSeconds to set
 */
static void setMinimumSeconds(int minimumSeconds)
{
   AppSettings.minimumSeconds = minimumSeconds;
}
```

27.3.2 生成 GPX 文件和 KML 文件

在系统设置界面中，可以设置指定的文件来保存行走轨迹。本系统提供了两种保存轨迹的文件格式，分别是 GPX 和 KML，如图 27-5 所示。

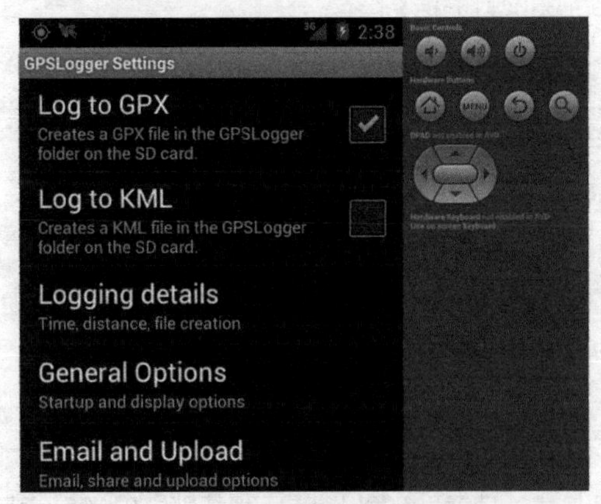

▲图 27-5 设置保存轨迹的文件格式

编写文件 Gpx10FileLogger.java 生成 GPX 格式的文件，具体实现代码如下所示。

```
class Gpx10FileLogger implements IFileLogger
{
    private final static Object lock = new Object();
    private File gpxFile = null;
    private boolean useSatelliteTime = false;
    private boolean addNewTrackSegment;
    private int satelliteCount;

    Gpx10FileLogger(File gpxFile, boolean useSatelliteTime, boolean addNewTrackSegment,
int satelliteCount)
    {
        this.gpxFile = gpxFile;
        this.useSatelliteTime = useSatelliteTime;
        this.addNewTrackSegment = addNewTrackSegment;
        this.satelliteCount = satelliteCount;
    }
    public void Write(Location loc) throws Exception
    {
        try
        {
            Date now;
            if (useSatelliteTime)
            {
                now = new Date(loc.getTime());
            }
            else
            {
                now = new Date();
            }
            String dateTimeString = Utilities.GetIsoDateTime(now);

            if (!gpxFile.exists())
            {
                gpxFile.createNewFile();
                FileOutputStream initialWriter = new FileOutputStream(gpxFile, true);
                BufferedOutputStream initialOutput = new BufferedOutputStream
                (initialWriter);
                String initialXml = "<?xml version=\"1.0\"?>"
                        + "<gpx version=\"1.0\" creator=\"GPSLogger - http://gpslogger.
                        mendhak.com/\" "+ "xmlns:xsi=\"http://www.w3.org/2001/XMLSchema-
```

```
                instance\" "+ "xmlns=\"http://www.topografix.com/GPX/1/0\" "
                + "xsi:schemaLocation=\"http://www.topografix.com/GPX/1/0 "
                + "http://www.topografix.com/GPX/1/0/gpx.xsd\">"
                + "<time>" + dateTimeString + "</time>" + "<bounds />" +
                "<trk></trk></gpx>";
        initialOutput.write(initialXml.getBytes());
        initialOutput.flush();
        initialOutput.close();
    }
    DocumentBuilderFactory factory = DocumentBuilderFactory.newInstance();
    DocumentBuilder builder = factory.newDocumentBuilder();
    Document doc = builder.parse(gpxFile);
    Node trkSegNode;
    NodeList trkSegNodeList = doc.getElementsByTagName("trkseg");
    if(addNewTrackSegment || trkSegNodeList.getLength()==0)
    {
        NodeList trkNodeList = doc.getElementsByTagName("trk");
        trkSegNode = doc.createElement("trkseg");
        trkNodeList.item(0).appendChild(trkSegNode);
    }
    else
    {
        trkSegNode = trkSegNodeList.item(trkSegNodeList.getLength()-1);
    }
    Element trkptNode = doc.createElement("trkpt");
    Attr latAttribute = doc.createAttribute("lat");
    latAttribute.setValue(String.valueOf(loc.getLatitude()));
    trkptNode.setAttributeNode(latAttribute);
    Attr lonAttribute = doc.createAttribute("lon");
    lonAttribute.setValue(String.valueOf(loc.getLongitude()));
    trkptNode.setAttributeNode(lonAttribute);
    if(loc.hasAltitude())
    {
        Node eleNode = doc.createElement("ele");
        eleNode.appendChild(doc.createTextNode(String.valueOf(loc.getAltitude())));
        trkptNode.appendChild(eleNode);
    }
    Node timeNode = doc.createElement("time");
    timeNode.appendChild(doc.createTextNode(dateTimeString));
    trkptNode.appendChild(timeNode);
    trkSegNode.appendChild(trkptNode);
    if(loc.hasBearing())
    {
        Node courseNode = doc.createElement("course");
        courseNode.appendChild(doc.createTextNode(String.valueOf(loc.getBearing())));
        trkptNode.appendChild(courseNode);
    }
    if(loc.hasSpeed())
    {
        Node speedNode = doc.createElement("speed");
        speedNode.appendChild(doc.createTextNode(String.valueOf(loc.getSpeed())));
        trkptNode.appendChild(speedNode);
    }
    Node srcNode = doc.createElement("src");
    srcNode.appendChild(doc.createTextNode(loc.getProvider()));
    trkptNode.appendChild(srcNode);
    if(Session.getSatelliteCount() > 0)
    {
        Node satNode = doc.createElement("sat");
        satNode.appendChild(doc.createTextNode(String.valueOf(satelliteCount)));
        trkptNode.appendChild(satNode);
    }
    String newFileContents = Utilities.GetStringFromNode(doc);
    synchronized(lock)
    {
        FileOutputStream fos = new FileOutputStream(gpxFile, false);
```

```
            fos.write(newFileContents.getBytes());
            fos.close();
        }
    }
    catch (Exception e)
    {
        Utilities.LogError("Gpx10FileLogger.Write", e);
        throw new Exception("Could not write to GPX file");
    }
}
public void Annotate(String description) throws Exception
{
    if (!gpxFile.exists())
    {
        return;
    }
    try
    {
        DocumentBuilderFactory factory = DocumentBuilderFactory.newInstance();
        DocumentBuilder builder = factory.newDocumentBuilder();
        Document doc = builder.parse(gpxFile);
        NodeList trkptNodeList = doc.getElementsByTagName("trkpt");
        Node lastTrkPt = trkptNodeList.item(trkptNodeList.getLength()-1);
        Node nameNode = doc.createElement("name");
        nameNode.appendChild(doc.createTextNode(description));
        lastTrkPt.appendChild(nameNode);
        Node descNode = doc.createElement("desc");
        descNode.appendChild(doc.createTextNode(description));
        lastTrkPt.appendChild(descNode);
        String newFileContents = Utilities.GetStringFromNode(doc);
        synchronized(lock)
        {
            FileOutputStream fos = new FileOutputStream(gpxFile, false);
            fos.write(newFileContents.getBytes());
            fos.close();
        }
    }
    catch(Exception e)
    {
        Utilities.LogError("Gpx10FileLogger.Annotate", e);
        throw new Exception("Could not annotate GPX file");
    }
}
}
```

编写文件Kml10FileLogger.java生成KML格式文件，具体实现代码如下所示。

```
class Kml10FileLogger implements IFileLogger
{
    private final static Object lock = new Object();
    private boolean useSatelliteTime;
    private File kmlFile;
    private FileLock kmlLock;
    Kml10FileLogger(File kmlFile, boolean useSatelliteTime)
    {
        this.useSatelliteTime = useSatelliteTime;
        this.kmlFile = kmlFile;
    }
    public void Write(Location loc) throws Exception
    {
        try
        {
            Date now;
            if(useSatelliteTime)
            {
                now = new Date(loc.getTime());
            }
            else
            {
                now = new Date();
```

```
        }
        String dateTimeString = Utilities.GetIsoDateTime(now);
        if(!kmlFile.exists())
        {
            kmlFile.createNewFile();
            FileOutputStream initialWriter = new FileOutputStream(kmlFile, true);
            BufferedOutputStream initialOutput=new BufferedOutputStream(initialWriter);
            String initialXml = "<?xml version=\"1.0\"?>"
                    + "<kml xmlns=\"http://www.opengis.net/kml/2.2\"><Document>"
                    + "<Placemark><LineString><extrude>1</extrude><tessellate>1</tessellate>"
                    + "<altitudeMode>absolute</altitudeMode>"
                    + "<coordinates></coordinates></LineString></Placemark>"
                    + "</Document></kml>";
            initialOutput.write(initialXml.getBytes());
            initialOutput.flush();
            initialOutput.close();
        }
        DocumentBuilderFactory factory = DocumentBuilderFactory.newInstance();
        DocumentBuilder builder = factory.newDocumentBuilder();
        Document doc = builder.parse(kmlFile);
        NodeList coordinatesList = doc.getElementsByTagName("coordinates");
        if(coordinatesList.item(0) != null)
        {
            Node coordinates = coordinatesList.item(0);
            Node coordTextNode = coordinates.getFirstChild();
            if(coordTextNode == null)
            {
                coordTextNode = doc.createTextNode("");
                coordinates.appendChild(coordTextNode);
            }
            String coordText = coordinates.getFirstChild().getNodeValue();
            coordText = coordText + "\n" + String.valueOf(loc.getLongitude()) + ","
                    + String.valueOf(loc.getLatitude()) + "," + String.valueOf
                    (loc.getAltitude());
            coordinates.removeChild(coordinates.getFirstChild());
            coordinates.appendChild(doc.createTextNode(coordText));
        }
        Node documentNode = doc.getElementsByTagName("Document").item(0);
        Node newPlacemark = doc.createElement("Placemark");
        Node timeStamp = doc.createElement("TimeStamp");
        Node whenNode = doc.createElement("when");
        Node whenNodeText = doc.createTextNode(dateTimeString);
        whenNode.appendChild(whenNodeText);
        timeStamp.appendChild(whenNode);
        newPlacemark.appendChild(timeStamp);
        Node newPoint = doc.createElement("Point");
        Node newCoords = doc.createElement("coordinates");
        newCoords.appendChild(doc.createTextNode(String.valueOf(loc.getLongitude())
        + ","+ String.valueOf(loc.getLatitude()) + "," + String.valueOf(loc.
        getAltitude())));
        newPoint.appendChild(newCoords);
        newPlacemark.appendChild(newPoint);
        documentNode.appendChild(newPlacemark);
        String newFileContents = Utilities.GetStringFromNode(doc);
        synchronized(lock)
        {
            FileOutputStream fos = new FileOutputStream(kmlFile, false);
            fos.write(newFileContents.getBytes());
            fos.close();
        }
    }
    catch(Exception e)
    {
        Utilities.LogError("Kml10FileLogger.Write", e);
        throw new Exception("Could not write to KML file");
    }
}
```

```
    public void Annotate(String description) throws Exception
    {
        if(!kmlFile.exists())
        {
            return;
        }
        try
        {
            DocumentBuilderFactory factory = DocumentBuilderFactory.newInstance();
            DocumentBuilder builder = factory.newDocumentBuilder();
            Document doc = builder.parse(kmlFile);
            NodeList placemarkList = doc.getElementsByTagName("Placemark");
            Node lastPlacemark = placemarkList.item(placemarkList.getLength() - 1);
            Node annotation = doc.createElement("name");
            annotation.appendChild(doc.createTextNode(description));
            lastPlacemark.appendChild(annotation);
            String newFileContents = Utilities.GetStringFromNode(doc);
            synchronized(lock)
            {
                FileOutputStream fos = new FileOutputStream(kmlFile, false);
                fos.write(newFileContents.getBytes());
                fos.close();
            }
        }
        catch(Exception e)
        {
            Utilities.LogError("Kml10FileLogger.Annotate", e);
            throw new Exception("Could not annotate KML file");
        }
    }
}
```

27.4 邮件分享提醒

在系统设置模块中,可以设置系统邮件来分享行走轨迹信息。其中邮件分享提醒界面如图 27-6 所示。

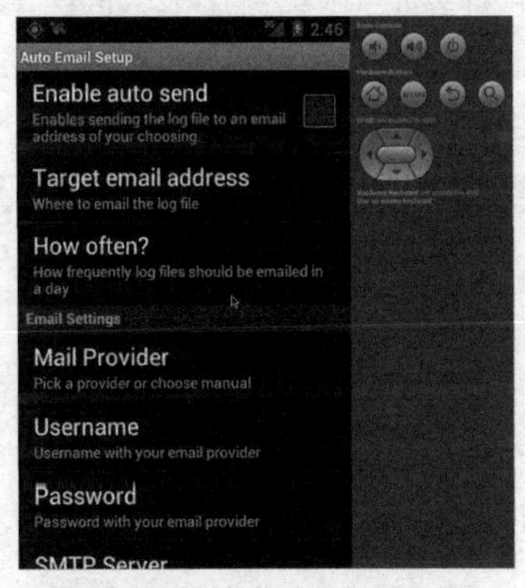

▲图 27-6　邮件分享提醒界面

27.4.1　基本邮箱设置

编写文件 AutoEmailActivity.java，功能是设置发送邮件的邮箱地址、密码、邮件服务器等信息，这样可以在使用时实现邮件自动发送功能。文件 AutoEmailActivity.java 的具体实现代码如下所示。

```
public class AutoEmailActivity extends PreferenceActivity implements
        OnPreferenceChangeListener, IMessageBoxCallback, IAutoSendHelper,
```

```
    OnPreferenceClickListener
{
    private final Handler    handler    = new Handler();
    @Override
    public void onCreate(Bundle savedInstanceState)
    {
        super.onCreate(savedInstanceState);
        addPreferencesFromResource(R.xml.autoemailsettings);
        CheckBoxPreference chkEnabled = (CheckBoxPreference) findPreference("autoemail_
        enabled");
        chkEnabled.setOnPreferenceChangeListener(this);
        ListPreference lstPresets = (ListPreference) findPreference("autoemail_preset");
        lstPresets.setOnPreferenceChangeListener(this);
        EditTextPreference txtSmtpServer = (EditTextPreference) findPreference("smtp_
        server");
        EditTextPreference txtSmtpPort = (EditTextPreference) findPreference("smtp_
        port");
        txtSmtpServer.setOnPreferenceChangeListener(this);
        txtSmtpPort.setOnPreferenceChangeListener(this);
        Preference testEmailPref = (Preference) findPreference("smtp_testemail");
        testEmailPref.setOnPreferenceClickListener(this);
    }
    public boolean onPreferenceClick(Preference preference)
    {
        if (!IsFormValid())
        {
            Utilities.MsgBox(getString(R.string.autoemail_invalid_form),
                    getString(R.string.autoemail_invalid_form_message),
                    AutoEmailActivity.this);
            return false;
        }
        Utilities.ShowProgress(this, getString(R.string.autoemail_sendingtest),
                getString(R.string.please_wait));
        CheckBoxPreference chkUseSsl = (CheckBoxPreference) findPreference("smtp_ssl");
        EditTextPreference txtSmtpServer = (EditTextPreference) findPreference("smtp_
        server");
        EditTextPreference txtSmtpPort = (EditTextPreference) findPreference("smtp_
        port");
        EditTextPreference txtUsername = (EditTextPreference) findPreference("smtp_
        username");
        EditTextPreference txtPassword = (EditTextPreference) findPreference("smtp_
        password");
        EditTextPreference txtTarget = (EditTextPreference) findPreference("autoemail_
        target");
        AutoEmailHelper aeh = new AutoEmailHelper(null);
        aeh.SendTestEmail(txtSmtpServer.getText(), txtSmtpPort.getText(),
                txtUsername.getText(), txtPassword.getText(),
                chkUseSsl.isChecked(), txtTarget.getText(),
                AutoEmailActivity.this, AutoEmailActivity.this);
                return true;
    }
    private boolean IsFormValid()
    {
        CheckBoxPreference chkEnabled = (CheckBoxPreference) findPreference("autoemail_
        enabled");
        EditTextPreference txtSmtpServer = (EditTextPreference) findPreference("smtp_
        server");
        EditTextPreference txtSmtpPort = (EditTextPreference) findPreference("smtp_
        port");
        EditTextPreference txtUsername = (EditTextPreference) findPreference("smtp_
        username");
        EditTextPreference txtPassword = (EditTextPreference) findPreference("smtp_
        password");
        EditTextPreference txtTarget = (EditTextPreference) findPreference("autoemail_
        target");
        if (chkEnabled.isChecked())
        {
            if (txtSmtpServer.getText() != null
                    && txtSmtpServer.getText().length() > 0
```

```
                && txtSmtpPort.getText() != null
                && txtSmtpPort.getText().length() > 0
                && txtUsername.getText() != null
                && txtUsername.getText().length() > 0
                && txtPassword.getText() != null
                && txtPassword.getText().length() > 0
                && txtTarget.getText() != null
                && txtTarget.getText().length() > 0)
        {
            return true;
        }
        else
        {
            return false;
        }
    }
    return true;
}
public boolean onKeyDown(int keyCode, KeyEvent event)
{
    if (keyCode == KeyEvent.KEYCODE_BACK)
    {
        if (!IsFormValid())
        {
            Utilities.MsgBox(getString(R.string.autoemail_invalid_form),
                    getString(R.string.autoemail_invalid_form_message),
                    this);
            return false;
        }
        else
        {
            return super.onKeyDown(keyCode, event);
        }
    }
    else
    {
        return super.onKeyDown(keyCode, event);
    }
}
public void MessageBoxResult(int which)
{
    finish();
}
public boolean onPreferenceChange(Preference preference, Object newValue)
{
    if (preference.getKey().equals("autoemail_preset"))
    {
        int newPreset = Integer.valueOf(newValue.toString());
        switch (newPreset)
        {
            case 0:
                // Gmail
                SetSmtpValues("smtp.gmail.com", "465", true);
                break;
            case 1:
                // Windows live mail
                SetSmtpValues("smtp.live.com", "587", false);
                break;
            case 2:
                // Yahoo
                SetSmtpValues("smtp.mail.yahoo.com", "465", true);
                break;
            case 99:
                // manual
                break;
        }
    }
    return true;
}
```

```
    private void SetSmtpValues(String server, String port, boolean useSsl)
    {
        SharedPreferences prefs = PreferenceManager
            .getDefaultSharedPreferences(getBaseContext());
        SharedPreferences.Editor editor = prefs.edit();
        EditTextPreference txtSmtpServer = (EditTextPreference) findPreference("smtp_
        server");
        EditTextPreference txtSmtpPort = (EditTextPreference) findPreference("smtp_
        port");
        CheckBoxPreference chkUseSsl = (CheckBoxPreference) findPreference("smtp_ssl");
        // Yahoo
        txtSmtpServer.setText(server);
        editor.putString("smtp_server", server);
        txtSmtpPort.setText(port);
        editor.putString("smtp_port", port);
        chkUseSsl.setChecked(useSsl);
        editor.putBoolean("smtp_ssl", useSsl);
        editor.commit();
    }
    String   testResults;
    public void OnRelay(boolean connectionSuccess, String message)
    {
        testResults = message;
        handler.post(showTestResults);
    }
    private final Runnable   showTestResults   = new Runnable()
                                       {
                                           public void run()
                                           {
                                               TestEmailResults();
                                           }
                                       };
    private void TestEmailResults()
    {
        Utilities.HideProgress();
        Utilities.MsgBox(getString(R.string.autoemail_testresult_title),
            testResults, this);
    }
}
```

27.4.2　实现邮件发送功能

编写文件 AutoEmailHelper.java，功能是使用邮件设置模块的邮箱来发送邮件信息，具体实现代码如下所示。

```
public class AutoEmailHelper implements IAutoSendHelper
{
    private GpsLoggingService   mainActivity;
    private boolean             forcedSend    = false;
    public AutoEmailHelper(GpsLoggingService activity)
    {
        this.mainActivity = activity;
    }
    public void SendLogFile(String currentFileName, boolean forcedSend)
    {
        this.forcedSend = forcedSend;
        Thread t = new Thread(new AutoSendHandler(currentFileName, this));
        t.start();
    }
    void SendTestEmail(String smtpServer, String smtpPort,
            String smtpUsername, String smtpPassword, boolean smtpUseSsl,
            String emailTarget, Activity callingActivity, IAutoSendHelper helper)
    {
        Thread t = new Thread(new TestEmailHandler(helper, smtpServer,
            smtpPort, smtpUsername, smtpPassword, smtpUseSsl, emailTarget));
        t.start();
    }
    public void OnRelay(boolean connectionSuccess, String errorMessage)
```

```
        {
            if (!connectionSuccess)
            {
                mainActivity.handler.post(mainActivity.updateResultsEmailSendError);
            }
            else
            {
                // This was a success
                Utilities.LogInfo("Email sent");
                if (!forcedSend)
                {
                    Utilities.LogDebug("setEmailReadyToBeSent = false");
                    Session.setEmailReadyToBeSent(false);
                }
            }
        }
    }
    class AutoSendHandler implements Runnable
    {
        private String          currentFileName;
        private IAutoSendHelper   helper;
        public AutoSendHandler(String currentFileName,    IAutoSendHelper helper)
        {
            this.currentFileName = currentFileName;
            this.helper = helper;
        }
        public void run()
        {
            File gpxFolder = new File(Environment.getExternalStorageDirectory(),
                    "GPSLogger");
            if (!gpxFolder.exists())
            {
                helper.OnRelay(true, null);
                return;
            }
            File gpxFile = new File(gpxFolder.getPath(), currentFileName + ".gpx");
            File kmlFile = new File(gpxFolder.getPath(), currentFileName + ".kml");
            File foundFile = null;
            if (kmlFile.exists())
            {
                foundFile = kmlFile;
            }
            if (gpxFile.exists())
            {
                foundFile = gpxFile;
            }
            if (foundFile == null)
            {
                helper.OnRelay(true, null);
                return;
            }
            String[] files = new String[]
            { foundFile.getAbsolutePath() };
            File zipFile = new File(gpxFolder.getPath(), currentFileName + ".zip");
            try
            {
                Utilities.LogInfo("Zipping file");
                ZipHelper zh = new ZipHelper(files, zipFile.getAbsolutePath());
                zh.Zip();
                Mail m = new Mail(AppSettings.getSmtpUsername(),
                        AppSettings.getSmtpPassword());
                String[] toArr =
                { AppSettings.getAutoEmailTarget() };
                m.setTo(toArr);
                m.setFrom(AppSettings.getSmtpUsername());
                m.setSubject("GPS Log file generated at "
                        + Utilities.GetReadableDateTime(new Date()) + " - "
                        + zipFile.getName());
                m.setBody(zipFile.getName());
```

```
            m.setPort(AppSettings.getSmtpPort());
            m.setSecurePort(AppSettings.getSmtpPort());
            m.setSmtpHost(AppSettings.getSmtpServer());
            m.setSsl(AppSettings.isSmtpSsl());
            m.addAttachment(zipFile.getAbsolutePath());
            Utilities.LogInfo("Sending email...");
            if (m.send())
            {
                helper.OnRelay(true, "Email was sent successfully.");
            }
            else
            {
                helper.OnRelay(false, "Email was not sent.");
            }
        }
        catch (Exception e)
        {
            helper.OnRelay(false, e.getMessage());
            Utilities.LogError("AutoSendHandler.run", e);
        }
    }
}

class TestEmailHandler implements Runnable
{
    String          smtpServer;
    String          smtpPort;
    String          smtpUsername;
    String          smtpPassword;
    boolean          smtpUseSsl;
    String          emailTarget;
    IAutoSendHelper   helper;
    public TestEmailHandler(IAutoSendHelper helper, String smtpServer,
            String smtpPort, String smtpUsername, String smtpPassword,
            boolean smtpUseSsl, String emailTarget)
    {
        this.smtpServer = smtpServer;
        this.smtpPort = smtpPort;
        this.smtpPassword = smtpPassword;
        this.smtpUsername = smtpUsername;
        this.smtpUseSsl = smtpUseSsl;
        this.emailTarget = emailTarget;
        this.helper = helper;
    }
    public void run()
    {
        try
        {
            Mail m = new Mail(smtpUsername, smtpPassword);
            String[] toArr =
            { emailTarget };
            m.setTo(toArr);
            m.setFrom(smtpUsername);
            m.setSubject("Test Email from GPSLogger at "
                    + Utilities.GetReadableDateTime(new Date()));
            m.setBody("Test Email from GPSLogger at "
                    + Utilities.GetReadableDateTime(new Date()));
            m.setPort(smtpPort);
            m.setSecurePort(smtpPort);
            m.setSmtpHost(smtpServer);
            m.setSsl(smtpUseSsl);
            m.setDebuggable(true);
            Utilities.LogInfo("Sending email...");
            if (m.send())
            {
                helper.OnRelay(true, "Email was sent successfully.");
            }
            else
            {
```

```
            helper.OnRelay(false, "Email was not sent.");
        }
    }
    catch (Exception e)
    {
        helper.OnRelay(false, e.getMessage());
        Utilities.LogError("AutoSendHandler.run", e);
    }
  }
}
```

27.5 上传 OSM 地图

OSM 是 Open Street Map 的缩写，这是一个网上地图协作计划，目标是创造一个内容自由且能让所有人编辑的世界地图。OSM 的地图由用户根据手提 GPS 装置、航空摄影照片、其他自由内容，甚至单靠地方智慧绘制。网站里的地图图像及向量数据皆以共享创意姓名标示，用相同方式分享 2.0 授权。OSM 网站的灵感来自维基百科等网站，经注册的用户可上载 GPS 路径及使用内置的编辑程式编辑数据。在上传地图信息之前，需要先获得授权标识。当单击“OpenStreetMap”选项后会来到授权提示界面，在此界面会显示授权提示信息，如图 27-7 所示。

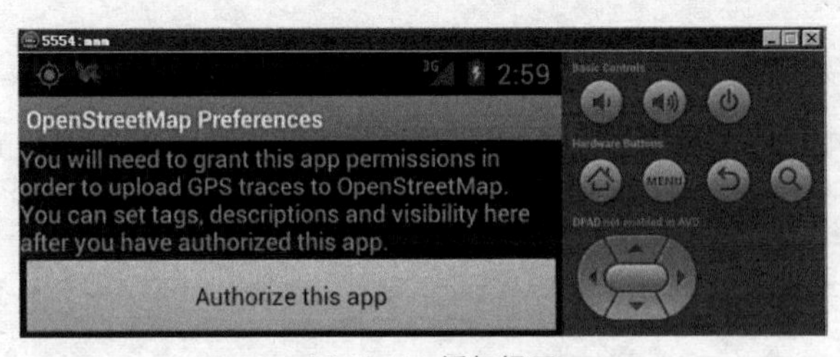

▲图 27-7　授权提示

27.5.1　授权提示布局文件

授权提示界面的布局文件是 osmauth.xml，具体实现代码如下所示。

```
<LinearLayout
  xmlns:android="http://schemas.android.com/apk/res/android"
  android:layout_width="fill_parent"
  android:layout_height="fill_parent">
    <TableLayout android:id="@+id/TableOSM"
        android:layout_width="fill_parent" android:layout_height="wrap_content"
        android:stretchColumns="1" android:background="#000000">
        <TableRow></TableRow>
        <TableRow>
          <TextView  android:id="@+id/lblAuthorizeDescription"  android:layout_height=
"wrap_content"
          android:text="@string/osm_lbl_authorize_description"
android:layout_width="wrap_content"></TextView>
      </TableRow>
      <TableRow>
           <Button android:id="@+id/btnAuthorizeOSM"
                android:text="@string/osm_lbl_authorize"   android:layout_height="wrap_
content" android:layout_width="wrap_content"/>
      </TableRow>
    </TableLayout>
</LinearLayout>
```

通过上述代码在屏幕中显示授权提示信息，并在屏幕下方显示了一个激活按钮。当单击激活按钮后，会触发文件 OSMAuthorizationActivity.java，具体实现代码如下所示。

```
public class OSMAuthorizationActivity extends Activity implements
        OnClickListener
```

```
{
    private static OAuthProvider    provider;
    private static OAuthConsumer    consumer;
    @Override
    public void onCreate(Bundle savedInstanceState)
    {
        super.onCreate(savedInstanceState);
        setContentView(R.layout.osmauth);
        final Intent intent = getIntent();
        final Uri myURI = intent.getData();
        if (myURI != null && myURI.getQuery() != null
            && myURI.getQuery().length() > 0)
        {
            //User has returned! Read the verifier info from querystring
            String oAuthVerifier = myURI.getQueryParameter("oauth_verifier");
            try
            {
                SharedPreferences prefs = PreferenceManager.getDefaultSharedPreferences
                (getBaseContext());

                if (provider == null)
                {
                    provider = Utilities.GetOSMAuthProvider(getBaseContext());
                }
                if (consumer == null)
                {
                    //In case consumer is null, re-initialize from stored values.
                    consumer = Utilities.GetOSMAuthConsumer(getBaseContext());
                }
                //Ask OpenStreetMap for the access token. This is the main event.
                provider.retrieveAccessToken(consumer, oAuthVerifier);

                String osmAccessToken = consumer.getToken();
                String osmAccessTokenSecret = consumer.getTokenSecret();

                //Save for use later.
                SharedPreferences.Editor editor = prefs.edit();
                editor.putString("osm_accesstoken", osmAccessToken);
                editor.putString("osm_accesstokensecret", osmAccessTokenSecret);
                editor.commit();

                //Now go away
                startActivity(new Intent(getBaseContext(), GpsMainActivity.class));
                finish();
            }
            catch (Exception e)
            {
                Utilities.LogError("OSMAuthorizationActivity.onCreate-user has returned",e);
                Utilities.MsgBox(getString(R.string.sorry), getString(R.string.osm_
                auth_error), this);
            }
        }
        Button authButton = (Button) findViewById(R.id.btnAuthorizeOSM);
        authButton.setOnClickListener(this);
    }
    public void onClick(View v)
    {
        try
        {
            //User clicks. Set the consumer and provider up.
            consumer = Utilities.GetOSMAuthConsumer(getBaseContext());
            provider = Utilities.GetOSMAuthProvider(getBaseContext());
            String authUrl;
            //Get the request token and request token secret
            authUrl = provider.retrieveRequestToken(consumer, OAuth.OUT_OF_BAND);
            //Save for later
            SharedPreferences prefs = PreferenceManager.getDefaultSharedPreferences
            (getBaseContext());
```

```
            SharedPreferences.Editor editor = prefs.edit();
            editor.putString("osm_requesttoken", consumer.getToken());
            editor.putString("osm_requesttokensecret",consumer.getTokenSecret());
            editor.commit();
            //Open browser, send user to OpenStreetMap.org
            Uri uri = Uri.parse(authUrl);
            Intent intent = new Intent(Intent.ACTION_VIEW, uri);
            startActivity(intent);
        }
        catch (Exception e)
        {
            Utilities.LogError("OSMAuthorizationActivity.onClick", e);
            Utilities.MsgBox(getString(R.string.sorry), getString(R.string.osm_auth_
            error), this);
        }
    }
}
```

27.5.2 实现文件上传

编写文件 OSMHelper.java，功能是在获取权限后上传 OpenStreetMap 轨迹文件，具体实现代码如下所示。

```
public class OSMHelper implements IOsmHelper
{
    private GpsMainActivity mainActivity;
    public OSMHelper(GpsMainActivity activity)
    {
        this.mainActivity = activity;
    }

    public void UploadGpsTrace(String fileName)
    {
        File gpxFolder=new File(Environment.getExternalStorageDirectory(),"GPSLogger");
        File chosenFile = new File(gpxFolder, fileName);
        OAuthConsumer consumer = Utilities.GetOSMAuthConsumer(mainActivity.
        getBaseContext());
        String gpsTraceUrl = mainActivity.getString(R.string.osm_gpstrace_url);

        SharedPreferences prefs = PreferenceManager.getDefaultSharedPreferences
        (mainActivity.getBaseContext());
        String description = prefs.getString("osm_description", "");
        String tags = prefs.getString("osm_tags", "");
        String visibility = prefs.getString("osm_visibility", "private");

        Thread t = new Thread(new OsmUploadHandler(this, consumer, gpsTraceUrl, chosenFile,
        description, tags, visibility));
        t.start();
    }
    public void OnComplete()
    {
        mainActivity.handler.post(mainActivity.updateOsmUpload);
    }
    private class OsmUploadHandler implements Runnable
    {
        OAuthConsumer consumer;
        String gpsTraceUrl;
        File chosenFile;
        String description;
        String tags;
        String visibility;
        IOsmHelper helper;

        public OsmUploadHandler(IOsmHelper helper, OAuthConsumer consumer, String
        gpsTraceUrl, File chosenFile, String description, String tags, String visibility)
        {
            this.consumer = consumer;
```

```
        this.gpsTraceUrl = gpsTraceUrl;
        this.chosenFile = chosenFile;
        this.description = description;
        this.tags = tags;
        this.visibility = visibility;
        this.helper = helper;
    }

    public void run()
    {
        try
        {
            HttpPost request = new HttpPost(gpsTraceUrl);

            consumer.sign(request);
            MultipartEntity entity = new MultipartEntity(HttpMultipartMode.BROWSER_
            COMPATIBLE);

            FileBody gpxBody = new FileBody(chosenFile);
            entity.addPart("file", gpxBody);
            if(description == null || description.length() <= 0)
            {
                description = "GPSLogger for Android";
            }

            entity.addPart("description", new StringBody(description));
            entity.addPart("tags", new StringBody(tags));
            entity.addPart("visibility", new StringBody(visibility));

            request.setEntity(entity);
            DefaultHttpClient httpClient = new DefaultHttpClient();
            HttpResponse response = httpClient.execute(request);
            int statusCode = response.getStatusLine().getStatusCode();
            Utilities.LogDebug("OSM Upload - " + String.valueOf(statusCode));
            helper.OnComplete();

        }
        catch(Exception e)
        {
                Utilities.LogError("OsmUploadHelper.run", e);
        }
    }
  }
}
interface IOsmHelper
{
    public void OnComplete();
}
```

到此为止，本章实例的主要模块介绍完毕。为了节省本书的篇幅，有关本实例其余模块的具体实现过程，请读者参阅本书附带程序的内容。

第 28 章　手势音乐播放器

超级女声，中国好声音，我是歌手……近年来选秀节目深受观众的欢迎，这些精彩的音乐节目为忙碌了一天的用户带来了轻松一刻。在智能手机的众多应用中，音乐播放一直是最受消费者欢迎的功能之一。本章将通过一个综合实例的实现过程，详细讲解在 Android 系统中通过手势识别技术开发一个音乐播放器的方法。

28.1 系统功能模块介绍

本章手势音乐播放器实例来源于 Apache Licence 项目，本项目的地址是。

```
https://github.com/telecapoland/jamendo-android
```

本项目源码的下载地址是。

```
git clone https://github.com/telecapoland/jamendo-android.git
```

本章手势音乐播放器的功能是，在 Android 系统中开发一个音乐播放器，可以通过手势来控制音乐的播放。本章手势音乐播放器的构成模块结构如图 28-1 所示。

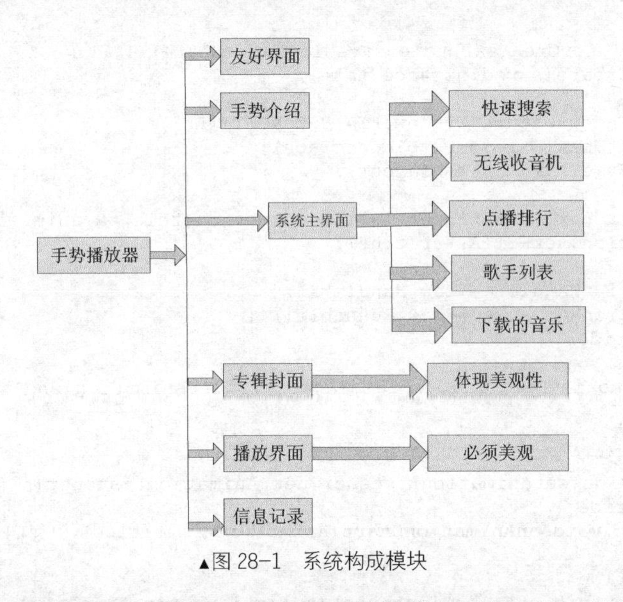

▲图 28-1　系统构成模块

28.2 系统主界面

系统主界面是运行程序后首先呈现在用户面前的界面。本节将详细讲解实现本章手势音乐播放器主界面的具体流程。

通过分析 AndroidManifest.xml 清单文件可知，本系统主界面的布局文件是 splashscreen.xml，功能是通过动画技术显示一个手势操控说明提示背景，然后过渡到系统标志界面。文件 splashscreen.xml 的具体实现代码如下所示。

```
<LinearLayout xmlns:android="http://schemas.android.com/apk/res/android"
    android:layout_width="match_parent" android:layout_height="match_parent"
    android:orientation="vertical" android:id="@+id/splashlayout"
    android:background="#fff">
    <ImageView android:layout_height="wrap_content" android:id="@+id/imageView1"
        android:layout_width="fill_parent" android:src="@drawable/splash_jamendo"
        android:layout_weight="1" android:padding="10dp"></ImageView>
    <LinearLayout android:layout_height="wrap_content"
        android:id="@+id/linearLayout1" android:layout_width="fill_parent"
        android:padding="10dp">
        <TextView android:id="@+id/textView1" android:layout_height="wrap_content"
            android:layout_weight="1" android:layout_width="fill_parent"></TextView>
        <ImageView android:layout_height="wrap_content"
            android:src="@drawable/teleca_logo_splash" android:id="@+id/imageView2"
            android:layout_width="fill_parent" android:layout_weight="1"></ImageView>
    </LinearLayout>
</LinearLayout>
```

本系统的主 Activity 是 SplashscreenActivity，即本系统的入口文件为 SplashscreenActivity.java，功能是通过调用动画模块显示动画加载特效，具体实现流程如下所示。

```
public class SplashscreenActivity extends Activity {
    public final static String FIRST_RUN_PREFERENCE = "first_run";

    private Animation endAnimation;

    private Handler endAnimationHandler;
    private Runnable endAnimationRunnable;

    /* (non-Javadoc)
     * @see android.app.Activity#onCreate(android.os.Bundle)
     */
    @Override
    protected void onCreate(Bundle savedInstanceState) {
        super.onCreate(savedInstanceState);

        requestWindowFeature(Window.FEATURE_NO_TITLE);
        setContentView(R.layout.splashscreen);
        findViewById(R.id.splashlayout);

        endAnimation = AnimationUtils.loadAnimation(this, R.anim.fade_out);
        endAnimation.setFillAfter(true);

        endAnimationHandler = new Handler();
        endAnimationRunnable = new Runnable() {
            @Override
            public void run() {
                findViewById(R.id.splashlayout).startAnimation(endAnimation);
            }
        };

        endAnimation.setAnimationListener(new AnimationListener() {
            @Override
            public void onAnimationStart(Animation animation) {    }

            @Override
            public void onAnimationRepeat(Animation animation) { }

            @Override
            public void onAnimationEnd(Animation animation) {
                HomeActivity.launch(SplashscreenActivity.this);
                SplashscreenActivity.this.finish();
            }
        });
```

```
        showTutorial();
    }

    final void showTutorial() {
        boolean showTutorial = PreferenceManager.getDefaultSharedPreferences(this).
        getBoolean(FIRST_RUN_PREFERENCE, true);
        if (showTutorial) {
            final TutorialDialog dlg = new TutorialDialog(this);
            dlg.setOnDismissListener(new DialogInterface.OnDismissListener() {
                @Override
                public void onDismiss(DialogInterface dialog) {
                    CheckBox cb = (CheckBox) dlg.findViewById(R.id.toggleFirstRun);
                    if (cb != null && cb.isChecked()) {
                        SharedPreferences prefs = PreferenceManager.getDefaultShared
                        Preferences(SplashscreenActivity.this);
                        prefs.edit().putBoolean(FIRST_RUN_PREFERENCE, false).commit();
                    }
                    endAnimationHandler.removeCallbacks(endAnimationRunnable);
                    endAnimationHandler.postDelayed(endAnimationRunnable, 2000);
                }
            });
            dlg.show();

        } else {
            endAnimationHandler.removeCallbacks(endAnimationRunnable);
            endAnimationHandler.postDelayed(endAnimationRunnable, 1500);
        }
    }
}
```

在上述代码中加载了动画特效样式的手势使用说明信息，首先通过静态方法 AnimationUtils.loadAnimation 加载了 anim 下的 fade_out 淡出动画效果，然后调用 setFillAfter 方法设置终止填充工作。布局文件 fade_out.xml 实现了动画特效，具体实现代码如下所示。

```
<set xmlns:android="http://schemas.android.com/apk/res/android"
    android:zAdjustment="bottom" android:fillAfter="false">
    <alpha android:fromAlpha="1.0" android:toAlpha="0"
        android:duration="400" />
</set>
```

由此可见，本系统的动画特效是通过透明度变化来生成淡出动画（Alpha 从完全不透名 fromAlpha=“1.0”到完全透明 toAlpha=“0”）效果的，并设置了动画的持续时间 duration 为 400ms。

到此为止，系统主界面的实现过程介绍完毕，执行后的效果如图 28-2 所示。

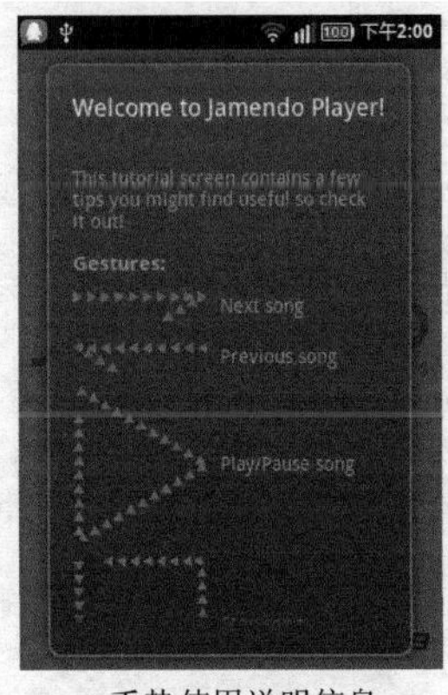

手势使用说明信息

标志美观界面

▲图 28-2　执行效果

28.3 系统列表界面

当加载并完动画特效后，系统会进入系统列表界面。本节将详细讲解实现本章实例系统列表界面的具体流程，为读者步入本书后面知识的学习打下基础。

28.3.1 布局文件

系统列表界面的布局文件是 main.xml，功能是在屏幕顶部通过 Gallery 控件显示专辑的封面，在屏幕下方通过 ListView 控件列表系统的主要功能模块。文件 main.xml 的具体实现代码如下所示。

```
<LinearLayout xmlns:android="http://schemas.android.com/apk/res/android"
    android:orientation="vertical" android:layout_width="fill_parent"
    android:layout_height="fill_parent" android:background="#ffffff">

    <com.teleca.jamendo.util.FixedViewFlipper
        android:orientation="vertical" android:id="@+id/ViewFlipper"
        android:layout_width="fill_parent" android:layout_height="75dip"
        android:background="@drawable/gradient_dark_purple">

        <!-- (0) Loading -->
        <LinearLayout android:orientation="vertical"
            android:layout_width="fill_parent" android:layout_height="fill_parent"
            android:layout_marginLeft="15dip" android:gravity="left|center_vertical">
            <com.teleca.jamendo.widget.ProgressBar
                android:id="@+id/ProgressBar" android:layout_width="wrap_content"
                android:layout_height="wrap_content">
            </com.teleca.jamendo.widget.ProgressBar>
        </LinearLayout>

        <!-- (1) Gallery -->
        <LinearLayout android:orientation="vertical"
            android:layout_width="fill_parent" android:layout_height="fill_parent"
            android:gravity="center">
            <Gallery android:id="@+id/Gallery" android:layout_width="fill_parent"
                android:layout_height="wrap_content" android:spacing="0px" />
        </LinearLayout>

        <!-- (2) Failure -->
        <LinearLayout android:orientation="vertical"
            android:layout_width="fill_parent" android:layout_height="fill_parent"
            android:layout_marginLeft="15dip" android:gravity="left|center_vertical">
            <com.teleca.jamendo.widget.FailureBar
                android:id="@+id/FailureBar" android:layout_width="wrap_content"
                android:layout_height="wrap_content">
            </com.teleca.jamendo.widget.FailureBar>
        </LinearLayout>
    </com.teleca.jamendo.util.FixedViewFlipper>

    <android.gesture.GestureOverlayView
        android:id="@+id/gestures"
        android:layout_width="fill_parent" android:layout_height="fill_parent"

        android:gestureStrokeType="multiple"
        android:eventsInterceptionEnabled="false" android:orientation="vertical">

        <ListView android:id="@+id/HomeListView"
            android:layout_width="fill_parent" android:layout_height="fill_parent"
            android:divider="#000" />
    </android.gesture.GestureOverlayView>

</LinearLayout>
```

根据上述代码可知，页面是在主 LinearLayout 布局内嵌套了如下所示的两个布局。

（1）com.teleca.jamendo.util.FixedViewFlipper：页面顶部用于展示专辑图片。自定义控件FixedViewFlipper继承于ViewFilpper，ViewAnimator用于切换加载到其中的两个或更多的View，但是每次只能显示一个。如果需要，也可以在特定时间间隔自动切换。com.teleca.jamendo.util.FixedViewFlipper布局的实现代码如下所示。

```
public class FixedViewFlipper extends ViewFlipper {

    public FixedViewFlipper(Context context) {
        super(context);
    }

    public FixedViewFlipper(Context context, AttributeSet attrs) {
        super(context, attrs);
    }

    @Override
    protected void onDetachedFromWindow() {

        int apiLevel = Build.VERSION.SDK_INT;

        if (apiLevel >= 7) {
            try {
                super.onDetachedFromWindow();
            } catch (IllegalArgumentException e) {
                Log.w("Jamendo", "Android project  issue 6191  workaround.");
                /* Quick catch and continue on api level 7, the Eclair 2.1 */
            } finally {
                super.stopFlipping();
            }
        } else {
            super.onDetachedFromWindow();
        }
    }
}
```

通过上述代码，在FixedViewFlipper中分别包含了Loading、Gallery和Failure三个View。其中Loading 是用来显示加载中页面，Gallery是显示加载完成后正常显示的图片，Failure是显示因为网络或其他原因引起的加载失败的界面。运行过程的执行效果如图29-3所示。

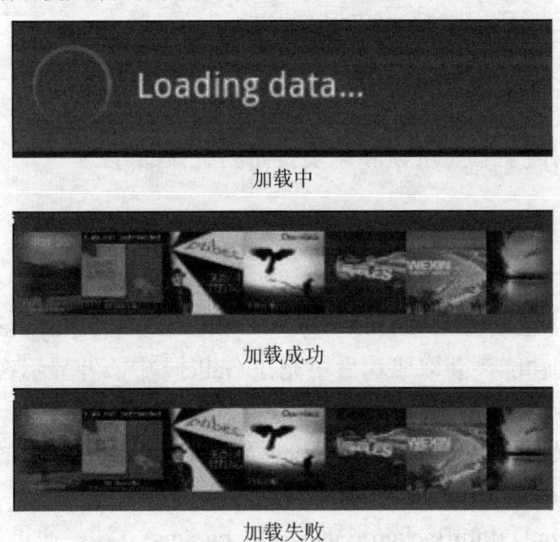

▲图29-3 执行效果

其中Loading使用了自定义布局com.teleca.jamendo.widget.ProgressBar，具体实现代码如下所示。

```
public class ProgressBar extends LinearLayout {
    protected TextView mTextView;
```

```
    public ProgressBar(Context context, AttributeSet attrs) {
        super(context, attrs);
        init();
    }
    public ProgressBar(Context context) {
        super(context);
        init();
    }
    private void init(){
        LayoutInflater.from(getContext()).inflate(R.layout.progress_bar, this);

        mTextView = (TextView)findViewById(R.id.ProgressTextView);
    }
    public void setText(int resid){
        mTextView.setText(resid);
    }
}
```

Failure 使用了自定义布局 com.teleca.jamendo.widget.FailureBar，具体实现代码如下所示。

```
public class FailureBar extends LinearLayout{
    protected TextView mTextView;
    protected Button mRetryButton;

    public FailureBar(Context context, AttributeSet attrs) {
        super(context, attrs);
        init();
    }
    public FailureBar(Context context) {
        super(context);
        init();
    }
    private void init(){
        LayoutInflater.from(getContext()).inflate(R.layout.failure_bar, this);

        mTextView = (TextView)findViewById(R.id.ProgressTextView);
        mRetryButton = (Button)findViewById(R.id.RetryButton);
    }
    public void setText(int resid){
        mTextView.setText(resid);
    }
    public void setOnRetryListener(OnClickListener l){
        mRetryButton.setOnClickListener(l);
    }
}
```

（2）android.gesture.GestureOverlayView：手势视图，采用垂直列表方式显示。这部分比较简单，在一个手势视图中内嵌了 ListView 和现实页面的主体部分。

28.3.2　程序文件

系统列表界面的程序文件是 HomeActivity，具体实现流程如下所示。

（1）通过 onCreate 方法实现了页面的手势监听 mGestureOverlayView 和通过执行 NewsTask 来异步加载 album 列表的功能，获取成功后填充在 Gallery 中，并显示在页面最上面。

（2）定义内部类 NewsTask，此类继承于抽象类 AsyncTask。

（3）在 onPreExecute 方法中设置 mViewFlipper 显示第一个 view 即上面介绍的 FixedViewFlipper 内的 Loading，设置 ProgressBar 控件中 Text 的显示文字。

（4）定义 public Album[] doInBackground(Void... params)方法，通过封装好的方法从服务器端获取流行的专辑列表，并通过 publishProgress(e)和 onProgressUpdate 将捕获的异常信息在 UI 线程中显示即更新 UI。

（5）定义 onPostExecute 方法，功能是在 UI 线程中调用通过 doInBackground 异步得到的数据

（Album[] album），并将数据显示在页面上。

文件 HomeActivity 的具体实现代码如下所示。

```
public class HomeActivity extends Activity implements OnAlbumClickListener {
   private static final String TAG = "HomeActivity";
   private ViewFlipper mViewFlipper;
   private Gallery mGallery;
   private ProgressBar mProgressBar;
   private FailureBar mFailureBar;
   private ListView mHomeListView;
   private PurpleAdapter mBrowseJamendoPurpleAdapter;
   private PurpleAdapter mMyLibraryPurpleAdapter;
   private GestureOverlayView mGestureOverlayView;

   /**
    * Launch Home activity helper
    * @param c context where launch home from (used by SplashscreenActivity)
    */
   public static void launch(Context c){
      Intent intent = new Intent(c, HomeActivity.class);
      intent.setFlags(Intent.FLAG_ACTIVITY_CLEAR_TOP );
      c.startActivity(intent);
   }

   /** Called when the activity is first created. */
   @Override
   public void onCreate(Bundle savedInstanceState) {
      super.onCreate(savedInstanceState);
      requestWindowFeature(Window.FEATURE_NO_TITLE);
      setContentView(R.layout.main);

      mHomeListView = (ListView)findViewById(R.id.HomeListView);
      mGallery = (Gallery)findViewById(R.id.Gallery);
      mProgressBar = (ProgressBar)findViewById(R.id.ProgressBar);
      mFailureBar = (FailureBar)findViewById(R.id.FailureBar);
      mViewFlipper = (ViewFlipper)findViewById(R.id.ViewFlipper);

      mGestureOverlayView = (GestureOverlayView) findViewById(R.id.gestures);
      mGestureOverlayView.addOnGesturePerformedListener(JamendoApplication
            .getInstance().getPlayerGestureHandler());
      new NewsTask().execute((Void)null);
   }
   @Override
   protected void onRestoreInstanceState(Bundle savedInstanceState) {
      // commented out, was causing "Wrong state class -- expecting View State" on view
rotation
      // super.onRestoreInstanceState(savedInstanceState);
   }
   @Override
   public void onAlbumClicked(Album album) {
      PlayerActivity.launch(this, album);
   }
   @Override
   public boolean onCreateOptionsMenu(Menu menu) {
      MenuInflater inflater = getMenuInflater();
      inflater.inflate(R.menu.home, menu);
      return super.onCreateOptionsMenu(menu);
   }

   @Override
   public boolean onPrepareOptionsMenu(Menu menu) {
      if(JamendoApplication.getInstance().getPlayerEngineInterface()  ==  null  ||
JamendoApplication.getInstance().getPlayerEngineInterface().getPlaylist() == null){
         menu.findItem(R.id.player_menu_item).setVisible(false);
      } else {
         menu.findItem(R.id.player_menu_item).setVisible(true);
      }
      return super.onPrepareOptionsMenu(menu);
   }
```

```
    @Override
    public boolean onOptionsItemSelected(MenuItem item) {
        switch(item.getItemId()){
        case R.id.player_menu_item:
            PlayerActivity.launch(this, (Playlist)null);
            break;
        case R.id.about_menu_item:
            new AboutDialog(this).show();
            break;
        case R.id.settings_menu_item:
            SettingsActivity.launch(this);
            break;
        default:
        }
        return super.onOptionsItemSelected(item);
    }
    @Override
    public boolean onKeyDown(int keyCode, KeyEvent event) {
        // support for search key
        // TODO remove hardcoded '1' value
        if(keyCode == KeyEvent.KEYCODE_SEARCH){
            mHomeListView.setSelection(1);
            PurpleEntry entry = (PurpleEntry) mHomeListView.getAdapter().getItem(1);
            entry.getListener().performAction();
        }
        return super.onKeyDown(keyCode, event);
    }

    @Override
    protected void onPause() {
        super.onPause();
    }

    @Override
    protected void onResume() {
        fillHomeListView();
        boolean gesturesEnabled = PreferenceManager.getDefaultSharedPreferences(this).
        getBoolean("gestures", true);
        mGestureOverlayView.setEnabled(gesturesEnabled);
        super.onResume();
    }

    /**
     * Fills ListView with clickable menu items
     */
    private void fillHomeListView(){
        mBrowseJamendoPurpleAdapter = new PurpleAdapter(this);
        mMyLibraryPurpleAdapter = new PurpleAdapter(this);
        ArrayList<PurpleEntry> browseListEntry = new ArrayList<PurpleEntry>();
        ArrayList<PurpleEntry> libraryListEntry = new ArrayList<PurpleEntry>();

        // BROWSE JAMENDO

        browseListEntry.add(new PurpleEntry(R.drawable.list_search, R.string.search,
        new PurpleListener(){
            @Override
            public void performAction() {
                SearchActivity.launch(HomeActivity.this);
            }
        }));
        browseListEntry.add(new PurpleEntry(R.drawable.list_radio, R.string.radio, new
        PurpleListener(){
            @Override
            public void performAction() {
                RadioActivity.launch(HomeActivity.this);
            }
        }));

        browseListEntry.add(new PurpleEntry(R.drawable.list_top, R.string.most_
```

```
        listened, new PurpleListener(){
            @Override
            public void performAction() {
                new Top100Task(HomeActivity.this, R.string.loading_top100, R.string.
                top100_fail).execute();
            }
        }));

        // MY LIBRARY

        libraryListEntry.add(new PurpleEntry(R.drawable.list_playlist, R.string.
        playlists, new PurpleListener(){
            @Override
            public void performAction() {
                BrowsePlaylistActivity.launch(HomeActivity.this, Mode.Normal);
            }
        }));

        // check if we have personalized client then add starred albums
        final String userName = PreferenceManager.getDefaultSharedPreferences(this).
        getString("user_name", null);
        if(userName != null && userName.length() > 0){
            libraryListEntry.add(new PurpleEntry(R.drawable.list_cd, R.string.albums,
            new PurpleListener(){
                @Override
                public void performAction() {
                    StarredAlbumsActivity.launch(HomeActivity.this, userName);
                }
            }));
        }

        /* following needs jamendo authorization (not documented yet on the wiki)
         * listEntry.add(new PurpleEntry(R.drawable.list_mail, "Inbox"));
         */

        // show this list item only if the SD Card is present
        if(Environment.getExternalStorageState().equals(Environment.MEDIA_MOUNTED)){
            libraryListEntry.add(new PurpleEntry(R.drawable.list_download, R.string.
            download, new PurpleListener(){
                @Override
                public void performAction() {
                    DownloadActivity.launch(HomeActivity.this);
                }
            }));
        }

//        listEntry.add(new PurpleEntry(R.drawable.list_star, R.string.favorites, new
          PurpleListener(){
//
//            @Override
//            public void performAction() {
//                Playlist playlist = new DatabaseImpl(HomeActivity.this).getFavorites();
//                JamendroidApplication.getInstance().getPlayerEngine().openPlaylist
                  (playlist);
//                PlaylistActivity.launch(HomeActivity.this, true);
//            }
//
//        }));

        // attach list data to adapters
        mBrowseJamendoPurpleAdapter.setList(browseListEntry);
        mMyLibraryPurpleAdapter.setList(libraryListEntry);

        // separate adapters on one list
        SeparatedListAdapter separatedAdapter = new SeparatedListAdapter(this);
        separatedAdapter.addSection(getString(R.string.browse_jamendo),
        mBrowseJamendoPurpleAdapter);
        separatedAdapter.addSection(getString(R.string.my_library),
        mMyLibraryPurpleAdapter);
```

```
    mHomeListView.setAdapter(separatedAdapter);
    mHomeListView.setOnItemClickListener(mHomeItemClickListener);
}

/**
 * Launches menu actions
 */
private OnItemClickListener mHomeItemClickListener = new OnItemClickListener(){

    @Override
    public void onItemClick(AdapterView<?> adapterView, View view, int index,
            long time) {
        try{
            PurpleListener listener = ((PurpleEntry)adapterView.getAdapter().getItem
            (index)).getListener();
            if(listener != null){
                listener.performAction();
            }
        }catch (ClassCastException e) {
            Log.w(TAG, "Unexpected position number was occurred");
        }
    }
};

/**
 * Executes news download, JamendoGet2Api.getPopularAlbumsWeek
 */
private class NewsTask extends AsyncTask<Void, WSError, Album[]> {

    @Override
    public void onPreExecute() {
        mViewFlipper.setDisplayedChild(0);
        mProgressBar.setText(R.string.loading_news);
        super.onPreExecute();
    }

    @Override
    public Album[] doInBackground(Void... params) {
        JamendoGet2Api server = new JamendoGet2ApiImpl();
        Album[] albums = null;
        try {
            albums = server.getPopularAlbumsWeek();
        } catch (JSONException e) {
            e.printStackTrace();
        } catch (WSError e){
            publishProgress(e);
        }
        return albums;
    }

    @Override
    public void onPostExecute(Album[] albums) {

        if(albums != null && albums.length > 0){
            mViewFlipper.setDisplayedChild(1);
            ImageAdapter albumsAdapter = new ImageAdapter(HomeActivity.this);
            albumsAdapter.setList(albums);
            mGallery.setAdapter(albumsAdapter);
            mGallery.setOnItemClickListener(mGalleryListener);
            mGallery.setSelection(albums.length/2, true); // animate to center

        } else {
            mViewFlipper.setDisplayedChild(2);
            mFailureBar.setOnRetryListener(new OnClickListener(){

                @Override
                public void onClick(View v) {
                    new NewsTask().execute((Void)null);
```

```
                }
            });
            mFailureBar.setText(R.string.connection_fail);
        }
        super.onPostExecute(albums);
    }
    @Override
    protected void onProgressUpdate(WSError... values) {
        Toast.makeText(HomeActivity.this, values[0].getMessage(), Toast.LENGTH_
        LONG).show();
        super.onProgressUpdate(values);
    }
}
/**
 * Background task fetching top 100 tracks from the remote server
 * It's sort of ugly circumvention of Get2 API limitations<br>
 * 1 rss + 2 json requests
 */
private class Top100Task extends LoadingDialog<Void, Playlist>{
    public Top100Task(Activity activity, int loadingMsg, int failMsg) {
        super(activity, loadingMsg, failMsg);
    }
    @Override
    public Playlist doInBackground(Void... params) {
        JamendoGet2Api server = new JamendoGet2ApiImpl();
        int[] id = null;
        try {
            id = server.getTop100Listened();
            // if loading rss failed and no tracks are there - report an error
            if (id == null) {
                publishProgress(new WSError((String) getResources().getText
                (R.string.top100_fail)));
                return null;
            }
            Album[] albums = server.getAlbumsByTracksId(id);
            Track[] tracks = server.getTracksByTracksId(id, JamendoApplication.
            getInstance().getStreamEncoding());
            if(albums == null || tracks == null)
                return null;
            Hashtable<Integer, PlaylistEntry> hashtable = new Hashtable<Integer,
            PlaylistEntry>();
            for(int i = 0; i < tracks.length; i++){
                PlaylistEntry playlistEntry = new PlaylistEntry();
                playlistEntry.setAlbum(albums[i]);
                playlistEntry.setTrack(tracks[i]);
                hashtable.put(tracks[i].getId(), playlistEntry);
            }
            // creating playlist in the correct order
            Playlist playlist = new Playlist();
            for(int i =0; i < id.length; i++){
                playlist.addPlaylistEntry(hashtable.get(id[i]));
            }
            return playlist;
        } catch (JSONException e) {
            e.printStackTrace();
        } catch (WSError e) {
            publishProgress(e);
        }
        return null;
    }
    @Override
    public void doStuffWithResult(Playlist playlist) {
        if(playlist.size() <= 0){
            failMsg();
            return;
        }

        PlayerActivity.launch(HomeActivity.this, playlist);
    }
```

```
    }

    private OnItemClickListener mGalleryListener = new OnItemClickListener(){
        @Override
        public void onItemClick(AdapterView<?> adapterView, View view, int position,
                long id) {
            Album album = (Album) adapterView.getItemAtPosition(position);
            PlayerActivity.launch(HomeActivity.this, album);
        }
    };
}
```

在上述代码中提到了 Album 获取到数据的情况，具体实现过程是首先设置 mViewFlipper 显示 Gallery，然后实例化 ImageAdapter albumsAdapter，并将 album 列表传递进去。接下来 Gallery 设置适配器为 albumsAdapter，并设置 gallery 的 item 点击事件。最后设置 Gallery 选中 Album 中间的 item。上述功能是通过文件 ImageAdapter.java 实现的，具体实现代码如下所示。

```
public class ImageAdapter extends AlbumAdapter {

    int mIconSize;


    public ImageAdapter(Activity context) {
        super(context);
        mIconSize = (int)context.getResources().getDimension(R.dimen.icon_size);
    }

    @Override
    public View getView(int position, View convertView, ViewGroup parent) {
        RemoteImageView i;

        if (convertView == null) {
            i = new RemoteImageView(mContext);
            i.setScaleType(RemoteImageView.ScaleType.FIT_CENTER);
            i.setLayoutParams(new Gallery.LayoutParams(mIconSize, mIconSize));
        } else {
            i = (RemoteImageView) convertView;
        }
        i.setDefaultImage(R.drawable.no_cd);
        i.setImageUrl(mList.get(position).getImage());
        return i;
    }
    static class ViewHolder {
        RemoteImageView image;
    }
}
```

通过上述实现代码可知，ImageAdapter 继承了 AlbumAdapter，而 AlbumAdapter 继承了 ArrayListAdapter，ArrayListAdapter 继承了 BaseAdapter。通过查看 ImageAdapter 和 AlbumAdapter 代码，可以看出两者都是强制实现了抽象类 ArrayListAdapter 中的抽象方法 abstract public View getView(int position, View convertView, ViewGroup parent)，也就是说 ImageAdapter 可以看作是直接继承 ArrayListAdapter，而继承 AlbumAdapter 的作用是确定了 ArrayListAdapter<T>中 T 的具体类型为 Album。文件 AlbumAdapter.java 的具体实现代码如下所示。

```
public class AlbumAdapter extends ArrayListAdapter<Album> {
    public AlbumAdapter(Activity context) {
        super(context);
    }

    @Override
    public View getView(int position, View convertView, ViewGroup parent) {
        View row=convertView;

        ViewHolder holder;
```

```
        if (row==null) {
            LayoutInflater inflater = mContext.getLayoutInflater();
            row=inflater.inflate(R.layout.album_row, null);

            holder = new ViewHolder();
            holder.image = (RemoteImageView)row.findViewById(R.id.AlbumRowImageView);
            holder.albumText = (TextView)row.findViewById(R.id.AlbumRowAlbumTextView);
            holder.artistText = (TextView)row.findViewById(R.id.AlbumRowArtistTextView);
            holder.progressBar = (ProgressBar)row.findViewById(R.id.AlbumRowRatingBar);

            row.setTag(holder);
        }
        else{
            holder = (ViewHolder) row.getTag();
        }

        holder.image.setDefaultImage(R.drawable.no_cd);
        holder.image.setImageUrl(mList.get(position).getImage(),position,
        getListView());
        holder.albumText.setText(mList.get(position).getName());
        holder.artistText.setText(mList.get(position).getArtistName());
        holder.progressBar.setMax(10);
        holder.progressBar.setProgress((int) (mList.get(position).getRating()*10));

        return row;
    }
    static class ViewHolder {
        RemoteImageView image;
        TextView albumText;
        TextView artistText;
        ProgressBar progressBar;
    }
}
```

在类 ArrayListAdapter 中不但实现了 BaseAdapter 中的几个方法，而且还实现了向 Adapter 中传入了和 List 操作相关的函数，在 ImageAdapter 中实现了 Adapter 中最重要的 getView 方法。在 getView 中将 Album 中每个 item 显示出来。在 ImageAdapter 的 getView 中使用了 RemoteImageView，而不是 Android 提供的 ImageView。文件 ArrayListAdapter.java 的具体实现代码如下所示。

```
public abstract class ArrayListAdapter<T> extends BaseAdapter{

    protected ArrayList<T> mList;
    protected Activity mContext;
    protected ListView mListView;

    public ArrayListAdapter(Activity context){
        this.mContext = context;
    }

    @Override
    public int getCount() {
        if(mList != null)
            return mList.size();
        else
            return 0;
    }

    @Override
    public Object getItem(int position) {
        return mList == null ? null : mList.get(position);
    }

    @Override
    public long getItemId(int position) {
        return position;
    }

    @Override
```

```
    abstract public View getView(int position, View convertView, ViewGroup parent);

    public void setList(ArrayList<T> list){
        this.mList = list;
        notifyDataSetChanged();
    }

    public ArrayList<T> getList(){
        return mList;
    }

    public void setList(T[] list){
        ArrayList<T> arrayList = new ArrayList<T>(list.length);
        for (T t : list) {
            arrayList.add(t);
        }
        setList(arrayList);
    }

    public ListView getListView(){
        return mListView;
    }

    public void setListView(ListView listView){
        mListView = listView;
    }

}
```

除此之外，在文件 HomeActivity.java 中还调用了类 SeparatedListAdapter 中的功能。ListView 是由 Browse Jamendo 和 My Library 两个分栏组成的，在通过 Listview 显示多分栏的简单实现过程中，首先定义一个 ListView 的适配器 Adapter——SeparatedListAdapter，SeparatedListAdapter 可以合并接受如下所示的其他的分栏 Adapter。

```
separatedAdapter.addSection(getString(R.string.browse_jamendo)
mBrowseJamendoPurpleAdapter);separatedAdapter.addSection(getString(R.string.my_libra
ry)
mMyLibraryPurpleAdapter)
```

其中 mBrowseJamendoPurpleAdapter 和 mMyLibraryPurpleAdapter 分别是看到的两个分栏所显示出来的内容。文件 SeparatedListAdapter.java 的具体实现代码如下所示。

```
public class SeparatedListAdapter extends BaseAdapter {

    public final Map<String,Adapter> sections = new LinkedHashMap<String,Adapter>();
    public final ArrayAdapter<String> headers;
    public final static int TYPE_SECTION_HEADER = 0;

    public SeparatedListAdapter(Context context) {
        headers = new ArrayAdapter<String>(context, R.layout.list_header);
    }

    public void addSection(String section, Adapter adapter) {
        this.headers.add(section);
        this.sections.put(section, adapter);
    }

    public Object getItem(int position) {
        for(Object section : this.sections.keySet()) {
            Adapter adapter = sections.get(section);
            int size = adapter.getCount() + 1;

            // check if position inside this section
            if(position == 0) return section;
            if(position < size) return adapter.getItem(position - 1);

            // otherwise jump into next section
```

```
            position -= size;
        }
        return null;
    }

    public int getCount() {
        // total together all sections, plus one for each section header
        int total = 0;
        for(Adapter adapter : this.sections.values())
            total += adapter.getCount() + 1;
        return total;
    }

    public int getViewTypeCount() {
        // assume that headers count as one, then total all sections
        int total = 1;
        for(Adapter adapter : this.sections.values())
            total += adapter.getViewTypeCount();
        return total;
    }

    public int getItemViewType(int position) {
        int type = 1;
        for(Object section : this.sections.keySet()) {
            Adapter adapter = sections.get(section);
            int size = adapter.getCount() + 1;

            // check if position inside this section
            if(position == 0) return TYPE_SECTION_HEADER;
            if(position < size) return type + adapter.getItemViewType(position - 1);

            // otherwise jump into next section
            position -= size;
            type += adapter.getViewTypeCount();
        }
        return -1;
    }

    public boolean areAllItemsSelectable() {
        return false;
    }

    public boolean isEnabled(int position) {
        return (getItemViewType(position) != TYPE_SECTION_HEADER);
    }

    @Override
    public View getView(int position, View convertView, ViewGroup parent) {
        int sectionnum = 0;
        for(Object section : this.sections.keySet()) {
            Adapter adapter = sections.get(section);
            int size = adapter.getCount() + 1;

            // check if position inside this section
            if(position == 0) return headers.getView(sectionnum, convertView, parent);
            if(position < size) return adapter.getView(position - 1, convertView, parent);

            // otherwise jump into next section
            position -= size;
            sectionnum++;
        }
        return null;
    }

    @Override
    public long getItemId(int position) {
        return position;
    }

}
```

在上述代码中，涉及了如下所示的 5 个方法。

- addSection(String section, Adapter adapter)：用来添加分栏 Adapter 两个参数分别是分栏名和对应的 Adapter。
- getItem(int position)：用来获取 position 对应 Item，根据 position 和每个分栏的 size 关系来确定点击是哪个分栏中的对应的 Item 并返回相应的 Item。
- getViewTypeCount：以 int 数值类型返回列表拥有的 itemview 的个数，不是 Item 个数，而是不同布局的 Itemview 个数。
- isEnabled(int position)：返回 true，如果项目在指定的位置不是一个分隔符。
- getView(int position, View convertView, ViewGroup parent)：这是 Adapter 中的核心方法，功能是将数据显示到 Listview 中。因为有两个分栏，所以根据 position 值来确定显示哪个分栏的内容及要显示的标题。

另外，在文件 HomeActivity.java 中还定义了类 Top100Task，此类用于在点击 Most listened 时从服务器加载数据。Top100Task 不是通过像 NewsTask 那样直接继承 AsyncTask 来实现的，而是通过继承一个可以被复用的 LoadingDialog<Void, Playlist>类来实现的。类 Top100Task 继承于 LoadingDialog，LoadingDialog 继承于 AsyncTask，其实实现方法和 NewsTask 类似，都是通过 AsyncTask 来实现异步加载数据功能的。文件 LoadingDialog.java 的具体实现代码如下所示。

```
public abstract class LoadingDialog<Input, Result> extends AsyncTask<Input, WSError, 
Result>{
    private ProgressDialog mProgressDialog;
    protected Activity mActivity;
    private int mLoadingMsg;
    private int mFailMsg;
    public LoadingDialog(Activity activity, int loadingMsg, int failMsg){
        this.mActivity = activity;
        this.mLoadingMsg = loadingMsg;
        this.mFailMsg = failMsg;
    }
    @Override
    public void onCancelled() {

        if( mActivity instanceof PlayerActivity)
            {
            PlayerActivity pa = (PlayerActivity)mActivity;
            pa.doCloseActivity();
            }

        failMsg();
        super.onCancelled();
    }
    @Override
    public void onPreExecute() {
        String title = "";
        String message = mActivity.getString(mLoadingMsg);
        mProgressDialog = ProgressDialog.show(mActivity, title, message, true, true, new 
OnCancelListener(){

            @Override
            public void onCancel(DialogInterface dialogInterface) {
                LoadingDialog.this.cancel(true);
            }
        });
        super.onPreExecute();
    }
    @Override
    public abstract Result doInBackground(Input... params);
    @Override
    public void onPostExecute(Result result) {
        super.onPostExecute(result);
```

```
            mProgressDialog.dismiss();
            if(result != null){
                doStuffWithResult(result);
            } else {

                if( mActivity instanceof PlayerActivity)
                    {
                    PlayerActivity pa = (PlayerActivity)mActivity;
                    pa.doCloseActivity();
                    }
                failMsg();

            }
        }

        protected void failMsg(){
            Toast.makeText(mActivity, mFailMsg, 2000).show();
        }
        /**
         * Very abstract function hopefully very meaningful name,
         * executed when result is other than null
         *
         * @param result
         * @return
         */
        public abstract void doStuffWithResult(Result result);

        @Override
        protected void onProgressUpdate(WSError... values) {
            Toast.makeText(mActivity, values[0].getMessage(), Toast.LENGTH_LONG).show();
            this.cancel(true);
            mProgressDialog.dismiss();
            super.onProgressUpdate(values);
        }

        public void doCancel()
        {
            mProgressDialog.dismiss();
        }

    }
```

到此为止，系统列表界面的主要功能全部介绍完毕。执行后的效果如图 28-4 所示。

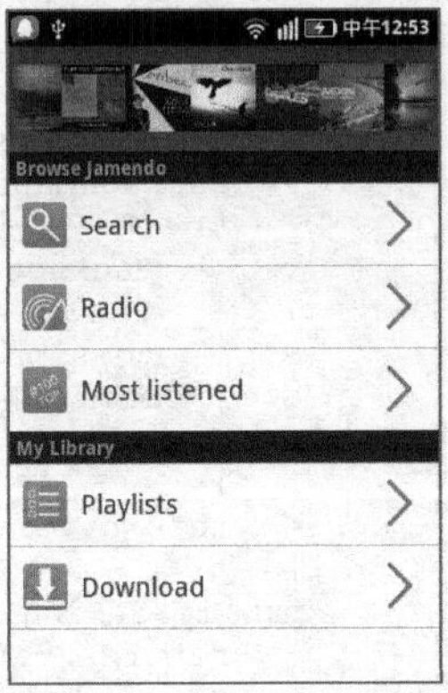

▲图 28-4　执行效果

28.4 实现公共类

本实例的核心公共类是 JamendoApplication，此类继承自类 Application，代表本程序的唯一实例，主要功能是维护整个程序会用到的公共的资源实例。本节将详细讲解本实例公共类的具体实现流程。

28.4.1 核心公共类 JamendoApplication

本实例的公共类用于维护整个程序会用到的公共的资源实例，例如，图片缓存 ImageCache、Web 请求缓存 RequestCache、播放引擎 PlayerEngine、当前正在使用的 MediaPlayer、均衡器 Equalizer、播放列表 PlayList、下载管理器 DownloadManager 及手势处理器等。其中文件 JamendoApplication.java 的具体实现代码如下所示。

```
public void onCreate() {
    super.onCreate();
    mImageCache = new ImageCache();
    mRequestCache = new RequestCache();
```

```
        Caller.setRequestCache(mRequestCache);
        instance = this;
        mDownloadManager = new DownloadManagerImpl(this);
    }
    public ImageCache getImageCache() {
        return mImageCache;
    }
    public void setConcretePlayerEngine(PlayerEngine playerEngine) {
        this.mServicePlayerEngine = playerEngine;
    }
    public PlayerEngine getPlayerEngineInterface() {
        // request service bind
        if (mIntentPlayerEngine == null) {
            mIntentPlayerEngine = new IntentPlayerEngine();
        }
        return mIntentPlayerEngine;
    }
    public GesturesHandler getPlayerGestureHandler(){
        if(mPlayerGestureHandler == null){
            mPlayerGestureHandler = new GesturesHandler(this, new PlayerGestureCommand
            Regiser(getPlayerEngineInterface()));
        }
        return mPlayerGestureHandler;
    }
    public void setPlayerEngineListener(PlayerEngineListener l) {
        getPlayerEngineInterface().setListener(l);
    }
    public PlayerEngineListener fetchPlayerEngineListener() {
        return mPlayerEngineListener;
    }
    public Playlist fetchPlaylist() {
        return mPlaylist;
    }
    public String getVersion() {
        String version = "0.0.0";
        PackageManager packageManager = getPackageManager();
        try {
            PackageInfo packageInfo = packageManager.getPackageInfo(
                    getPackageName(), 0);
            version = packageInfo.versionName;
        } catch (NameNotFoundException e) {
            e.printStackTrace();
        }
        return version;
    }
    public String getDownloadFormat() {
        return PreferenceManager.getDefaultSharedPreferences(this).getString(
                "download_format", JamendoGet2Api.ENCODING_MP3);
    }
    public String getStreamEncoding() {
        // http://groups.google.com/group/android-developers/msg/c546760177b22197
        // According to JBQ: ogg files are supported but not streamable
        return JamendoGet2Api.ENCODING_MP3;
    }
    public DownloadManager getDownloadManager() {
        return mDownloadManager;
    }
    private class IntentPlayerEngine implements PlayerEngine {
        @Override
        public Playlist getPlaylist() {
            return mPlaylist;
        }
        @Override
        public boolean isPlaying() {
            if (mServicePlayerEngine == null) {
                // service does not exist thus no playback possible
                return false;
            } else {
                return mServicePlayerEngine.isPlaying();
```

```
        }
    }
    @Override
    public void next() {
        if (mServicePlayerEngine != null) {
            playlistCheck();
            mServicePlayerEngine.next();
        } else {
            startAction(PlayerService.ACTION_NEXT);
        }
    }
    @Override
    public void openPlaylist(Playlist playlist) {
        mPlaylist = playlist;
        if(mServicePlayerEngine != null){
            mServicePlayerEngine.openPlaylist(playlist);
        }
    }
    @Override
    public void pause() {
        if (mServicePlayerEngine != null) {
            mServicePlayerEngine.pause();
        }
    }
    @Override
    public void play() {
        if (mServicePlayerEngine != null) {
            playlistCheck();
            mServicePlayerEngine.play();
        } else {
            startAction(PlayerService.ACTION_PLAY);
        }
    }
    @Override
    public void prev() {
        if (mServicePlayerEngine != null) {
            playlistCheck();
            mServicePlayerEngine.prev();
        } else {
            startAction(PlayerService.ACTION_PREV);
        }
    }
    @Override
    public void setListener(PlayerEngineListener playerEngineListener) {
        mPlayerEngineListener = playerEngineListener;
        // we do not want to set this listener if Service
        // is not up and a new listener is null
        if (mServicePlayerEngine != null || mPlayerEngineListener != null) {
            startAction(PlayerService.ACTION_BIND_LISTENER);
        }
    }
    @Override
    public void skipTo(int index) {
        if (mServicePlayerEngine != null) {
            mServicePlayerEngine.skipTo(index);
        }
    }
    @Override
    public void stop() {
        startAction(PlayerService.ACTION_STOP);
        // stopService(new Intent(JamendoApplication.this,
        // PlayerService.class));
    }
    private void startAction(String action) {
        Intent intent = new Intent(JamendoApplication.this,
                PlayerService.class);
        intent.setAction(action);
        startService(intent);
    }
```

```
        private void playlistCheck() {
            if (mServicePlayerEngine != null) {
                if (mServicePlayerEngine.getPlaylist() == null
                        && mPlaylist != null) {
                    mServicePlayerEngine.openPlaylist(mPlaylist);
                }
            }
        }
        @Override
        public void setPlaybackMode(PlaylistPlaybackMode aMode) {
            mPlaylist.setPlaylistPlaybackMode(aMode);
        }
        @Override
        public PlaylistPlaybackMode getPlaybackMode() {
            return mPlaylist.getPlaylistPlaybackMode();
        }

        @Override
        public void forward(int time) {
            if(mServicePlayerEngine != null){
                mServicePlayerEngine.forward( time );
            }

        }
        @Override
        public void rewind(int time) {
            if(mServicePlayerEngine != null){
                mServicePlayerEngine.rewind( time );
            }

        }

    }
}
```

28.4.2　缓存图片资源

类 ImageCache 用于缓存从网络上下载的图片资源到内存中，这样可以节省手机流量。类 ImageCache 继承于类 WeakHashMap，WeakHashMap 是基于 key-value 对实现，其特点是当除了自己对某个 key 的引用外，不存在其他指向这个 key 的引用时，这种 Map 会丢弃该 key 对应的 value 的值。文件 ImageCache.java 的具体实现代码如下所示。

```
public class ImageCache extends WeakHashMap<String, Bitmap> {

    private static final long serialVersionUID = 1L;

    public boolean isCached(String url){
        return containsKey(url) && get(url) != null;
    }
}
```

28.4.3　类 RequestCache

类 RequestCache 可以结合 LinkedList 和 Hashtable，当缓存空间不足时能够删除最旧的 url 的策略。其中 LinkedList 用于将最新的 url 插入链表头部，并在缓存空间不足时，移除链表尾部的 url；Hashtabel 用于存储 url 请求和对应的响应数据之间的映射关系。文件 RequestCache.java 的具体实现代码如下所示。

```
public class RequestCache {
    private static int CACHE_LIMIT = 10;
    @SuppressWarnings("unchecked")
    private LinkedList history;
    private Hashtable<String, String> cache;
    @SuppressWarnings("unchecked")
```

```
    public RequestCache(){
        history = new LinkedList();
        cache = new Hashtable<String, String>();
    }

    @SuppressWarnings("unchecked")
    public void put(String url, String data){
        history.add(url);
        // too much in the cache, we need to clear something
        if(history.size() > CACHE_LIMIT){
            String old_url = (String) history.poll();
            cache.remove(old_url);
        }
        cache.put(url, data);
    }
    public String get(String url){
        return cache.get(url);
    }
}
```

28.5 手势操作

手势操作是本实例的核心，我们可以通过手势来控制对当前音乐的播放控制。本节将详细讲解手势操作功能的具体实现过程，为读者步入本书后面知识的学习打下基础。

28.5.1 Android 提供的手势操作 API

在 Android 系统中通过 GestureLibrary 来表示手势库，同时通过工具类 GestureLibraries 可以从不同的数据源来创建手势库。其中 GestureLibrary 是一个抽象类，有 FileGestureLibrary 和 ResourceGestureLibrary 两个子类，从名字可以看出这两个手势库的数据来源，分别是文件和 Android raw 资源。这两个类作为 GestureLibraries 类的私有内部类实现。

类 GestureLibrary 实际上是对另一个类 GestureStore 的一层封装，下面来看下 GestureLibrary 的代码。

```
public abstract class GestureLibrary {
    protected final GestureStore mStore;

    protected GestureLibrary() {
        mStore = new GestureStore();
    }

    // 保存手势库
    public abstract boolean save();

    // 加载手势库
    public abstract boolean load();

    public boolean isReadOnly() {
        return false;
    }

    /** @hide */
    public Learner getLearner() {
        return mStore.getLearner();
    }

    public void setOrientationStyle(int style) {
        mStore.setOrientationStyle(style);
    }

    public int getOrientationStyle() {
        return mStore.getOrientationStyle();
```

```
    }

    public void setSequenceType(int type) {
        mStore.setSequenceType(type);
    }

    public int getSequenceType() {
        return mStore.getSequenceType();
    }

    public Set<String> getGestureEntries() {
        return mStore.getGestureEntries();
    }

    public ArrayList<Prediction> recognize(Gesture gesture) {
        return mStore.recognize(gesture);
    }

    public void addGesture(String entryName, Gesture gesture) {
        mStore.addGesture(entryName, gesture);
    }

    public void removeGesture(String entryName, Gesture gesture) {
        mStore.removeGesture(entryName, gesture);
    }

    public void removeEntry(String entryName) {
        mStore.removeEntry(entryName);
    }

    public ArrayList<Gesture> getGestures(String entryName) {
        return mStore.getGestures(entryName);
    }
}
```

GestureLibrary 的两个子类主要实现了父类的两个抽象函数，并且重写了 isReadOnly 函数，分别对应不同的数据来源。尽管如此，但是最终都是将数据转换成流的形式传给 GestureStore 实例的函数，由它完成最终的操作。例如，如下所示的演示代码。

```
    // 从 File 数据源加载手势库的类
    private static class FileGestureLibrary extends GestureLibrary {
        private final File mPath;

        public FileGestureLibrary(File path) {
            mPath = path;
        }

        @Override
        public boolean isReadOnly() {
            return !mPath.canWrite();
        }

        public boolean save() {
// 手势库没有发生变化，就不需要重新保存
            if (!mStore.hasChanged()) return true;

            final File file = mPath;

            final File parentFile = file.getParentFile();
// 第一次保存时，保证文件路径的存在
            if (!parentFile.exists()) {
                if (!parentFile.mkdirs()) {
                    return false;
                }
            }

            boolean result = false;
            try {
```

```
            file.createNewFile();
            mStore.save(new FileOutputStream(file), true); // 实际的保存操作
            result = true;
        } catch (FileNotFoundException e) {
            Log.d(LOG_TAG, "Could not save the gesture library in " + mPath, e);
        } catch (IOException e) {
            Log.d(LOG_TAG, "Could not save the gesture library in " + mPath, e);
        }

        return result;
    }

    public boolean load() {
        boolean result = false;
        final File file = mPath;
        if (file.exists() && file.canRead()) {
            try {
                // 调用 GestureStore 实例的函数加载手势库文件
                mStore.load(new FileInputStream(file), true);
                result = true;
            } catch (FileNotFoundException e) {
                Log.d(LOG_TAG, "Could not load the gesture library from " + mPath, e);
            } catch (IOException e) {
                Log.d(LOG_TAG, "Could not load the gesture library from " + mPath, e);
            }
        }

        return result;
    }
}

// 从 Android 的 res/raw 目录加载对应的手势库文件
 private static class ResourceGestureLibrary extends GestureLibrary {
    private final WeakReference<Context> mContext;
    private final int mResourceId;

    public ResourceGestureLibrary(Context context, int resourceId) {
        mContext = new WeakReference<Context>(context);
        mResourceId = resourceId;
    }

    @Override
    public boolean isReadOnly() { // raw 目录下的手势库文件是只读的
        return true;
    }

    // 这种数据源只能加载不能保存
    public boolean save() {
        return false;
    }

    public boolean load() {
        boolean result = false;
        final Context context = mContext.get();
        if (context != null) {
            final InputStream in = context.getResources().openRawResource(mResourceId);
            try {
                mStore.load(in, true);
                result = true;
            } catch (IOException e) {
                Log.d(LOG_TAG, "Could not load the gesture library from raw resource " +
                        context.getResources().getResourceName(mResourceId), e);
            }
        }
        return result;
    }
}
```

28.5.2　使用命令模式构建手势识别系统

命令模式是指将一个请求封装为一个对象，从而使我们可以用不同的请求对客户进行参数化，对请求排队或记录请求日志，以及支持可撤销的操作。命令模式的典型结构图如图 28-5 所示。

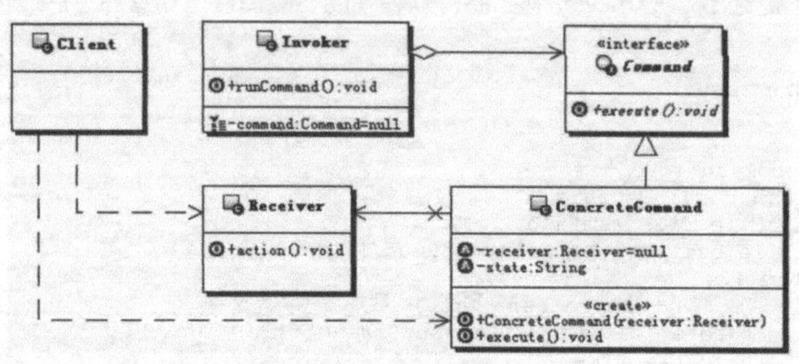

▲图 28-5　命令模式的典型结构图

在命令模式编程模型中主要涉及如下所示的 4 种角色。

- 接收者角色 Receiver：真正执行命令的对象，任何一个类都可以成为接收者，只要它能完成命令要求实现的相应功能。
- 命令角色 Command：实际包括一个 Command 接口和一个或者多个实现 Command 接口的具体命令 ConcreteCommand 类。实际命令对象通常持有 Receiver 的一个引用，并调用 Receiver 提供的功能来完成命令要执行的操作。
- 调用者角色 Invoker：通常持有命令对象，并要求命令对象执行相应的请求，Invoker 相当于是使用命令对象的入口。
- 装配者角色 Client：用于创建具体的命令对象，并设置命令对象的接收者。

28.5.3　实现抽象命令角色 Command

在本实例中，实现抽象命令角色 Command 的是文件 GestureCommand.java，此文件十分简单，整个接口只有一个 execute 函数，具体代码如下所示。

```
package com.teleca.jamendo.gestures;
public interface GestureCommand {
    void execute();
}
```

28.5.4　实现具体命令角色 ConcreteCommand

在本实例中，实现具体命令角色 ConcreteCommand 的是文件 PlayerGestureNextCommand.java、PlayerGesturePlayCommand.java、PlayerGesturePrevCommand.java 和 PlayerGestureStopCommand.java，在这 4 个具体实现类中，都持有接收者角色 PlayerEngine 的实例，由它来完成具体的控制操作。

（1）文件 PlayerGestureNextCommand.java 用于实现播放下一首命令功能，具体实现代码如下所示。

```
public class PlayerGestureNextCommand implements GestureCommand {

    PlayerEngine mPlayerEngine;

    public PlayerGestureNextCommand( PlayerEngine engine ){
        mPlayerEngine = engine;
    }

    @Override
    public void execute() {
```

```
        Log.v(JamendoApplication.TAG, "PlayerGestureNextCommand");
        mPlayerEngine.next();
    }

}
```

（2）文件 PlayerGesturePlayCommand.java 用于实现播放命令功能，具体实现代码如下所示。

```
public class PlayerGesturePlayCommand implements GestureCommand {
    PlayerEngine mPlayerEngine;
    public PlayerGesturePlayCommand(PlayerEngine engine) {
        mPlayerEngine = engine;
    }
    @Override
    public void execute() {
        Log.v(JamendoApplication.TAG, "PlayerGesturePlayCommand");
        if (mPlayerEngine.isPlaying()) {
            mPlayerEngine.pause();
        } else {
            mPlayerEngine.play();
        }
    }
}
```

（3）文件 PlayerGesturePrevCommand.java 用于实现播放上一首音乐命令功能，具体实现代码如下所示。

```
public class PlayerGesturePrevCommand implements GestureCommand {
    PlayerEngine mPlayerEngine;

    public PlayerGesturePrevCommand( PlayerEngine engine ){
        mPlayerEngine = engine;
    }
    @Override
    public void execute() {
        Log.v(JamendoApplication.TAG, "PlayerGesturePrevCommand");
        mPlayerEngine.prev();
    }
}
```

（4）文件 PlayerGestureStopCommand.java 用于实现停止当前的播放音乐命令功能，具体实现代码如下所示。

```
public class PlayerGestureStopCommand implements GestureCommand {
    PlayerEngine mPlayerEngine;

    public PlayerGestureStopCommand( PlayerEngine engine ){
        mPlayerEngine = engine;
    }
    @Override
    public void execute() {
        Log.v(JamendoApplication.TAG, "PlayerGestureStopCommand");
        mPlayerEngine.stop();
    }
}
```

28.5.5 实现命令接收者角色 Receiver

在本实例中，实现命令接收者角色 Receiver 功能的是文件 PlayerEngine.java，此类本身是一个接口，在里面定义了音频播放操作的一些通用函数。文件 PlayerEngine.java 的具体实现代码如下所示。

```
public interface PlayerEngine {
    public void openPlaylist(Playlist playlist);
    public Playlist getPlaylist();
    public void play();
    public boolean isPlaying();
    public void stop();
```

```
    public void pause();
    public void next();
    public void prev();
    public void skipTo(int index);
    public void setListener(PlayerEngineListener playerEngineListener);
    public void setPlaybackMode(PlaylistPlaybackMode aMode);
    public PlaylistPlaybackMode getPlaybackMode();
    public void forward(int time);
    public void rewind(int time);
}
```

PlayerEngine 有两个实现类，分别是 IntentPlayerEngine 和 PlayerEngineImpl，这一部分放在后面分析，因为不属于本节的手势主题。

28.5.6　实现调用者角色 Invoker

在本实例中，实现调用者角色 Invoker 功能的是文件 GesturesHandler.java，功能是实现 OnGesturePerformedListener 监听器，重写 onGesturePerformed 接口函数。在函数 onGesturePerformed 中，首先判断手势库是否已经加载，在确保手势库正确加载的情况下调用 GestureLibrary.recognize 进行手势识别，并在匹配度大于 2.0 时才确认识别到了预先定义的手势。当识别到手势之后，调用命令接口定义的 execute 函数进行音乐播放的控制。文件 GesturesHandler.java 的具体实现代码如下所示。

```
public class GesturesHandler implements OnGesturePerformedListener {
    private GestureLibrary mLibrary;
    private boolean mLoaded = false;
    private GestureCommandRegister mRegister;
    public GesturesHandler(Context context, GestureCommandRegister register) {
        // mLibrary = GestureLibraries.fromRawResource(context, R.raw.gestures);
        mLibrary = GestureLibraries.fromRawResource(context, R.raw.gestures);
        load();
        setRegister(register);
    }
    private boolean load() {
        mLoaded = mLibrary.load();
        return mLoaded;
    }
    @Override
    public void onGesturePerformed(GestureOverlayView overlay, Gesture gesture) {
        if (!mLoaded) {
            if (!load()) {
                return;
            }
        }
        ArrayList<Prediction> predictions = mLibrary.recognize(gesture);
        if (predictions.size() > 0) {
            Prediction prediction = predictions.get(0);
            Log.v(JamendoApplication.TAG, "Gesture " + prediction.name
                    + " recognized with score " + prediction.score);
            if (prediction.score > 2.0) {
                GestureCommand command = getRegister().getCommand(
                        prediction.name);
                if (command != null) {
                    command.execute();
                }
            }
        }
    }
    public void setRegister(GestureCommandRegister mRegister) {
        this.mRegister = mRegister;
    }
    public GestureCommandRegister getRegister() {
        return mRegister;
    }
}
```

28.5.7 实现装配者角色 Client

在本实例中，通过类 GestureCommandRegister 和类 PlayGestureCommandRegister 共同作为 Client 角色，类 GestureCommandRegister 通过维护 HashMap 来充当各种具体命令角色的容器，具体实现代码如下所示。

```
public class GestureCommandRegister {
    private HashMap<String, GestureCommand> mGestures;
    public GestureCommandRegister() {
        mGestures = new HashMap<String, GestureCommand>();
    }
    public void registerCommand(String name, GestureCommand gestureCommand) {
        mGestures.put(name, gestureCommand);
    }
    public GestureCommand getCommand(String name) {
        return mGestures.get(name);
    }
}
```

类 PlayGestureCommandRegister 是 GestureCommandRegister 的子类，主要功能是实现注册 next、prev、play 和 stop 这 4 种具体播发控制命令角色的功能。类 PlayGestureCommandRegister 的具体实现代码如下所示。

```
public class PlayerGestureCommandRegiser extends GestureCommandRegister {
    public PlayerGestureCommandRegiser(PlayerEngine playerEngine) {
        super();
        registerCommand("next", new PlayerGestureNextCommand(playerEngine));
        registerCommand("prev", new PlayerGesturePrevCommand(playerEngine));
        registerCommand("play", new PlayerGesturePlayCommand(playerEngine));
        registerCommand("stop", new PlayerGestureStopCommand(playerEngine));
    }
}
```

到此为止，整个手势识别播放功能全部介绍完毕。

28.6 播放处理

在本实例中，播放一个音乐的基本流程如图 28-6 所示。

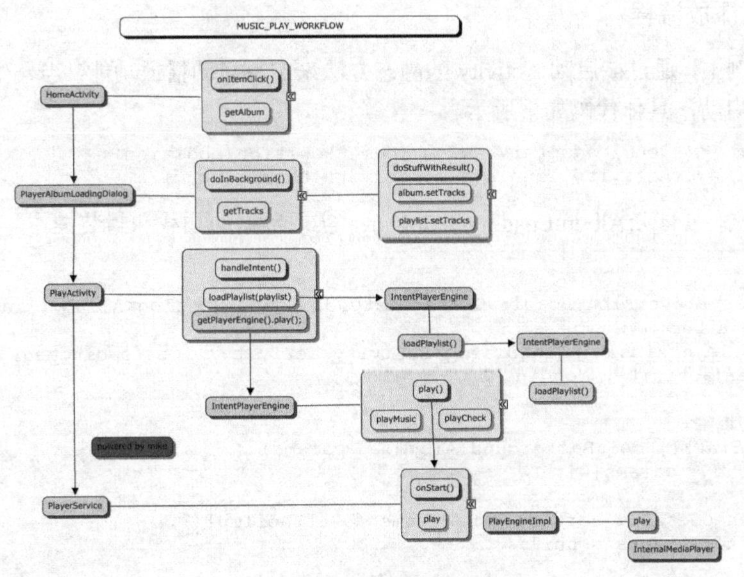

▲图 28-6 播放流程图

本节将详细讲解播放流程的具体实现过程。

28.6.1　设计播放界面

编写文件 playlist.xml 构建一个播放列表界面，可以通过手势播放列表中的音乐。文件 playlist.xml 的具体实现代码如下所示。

```
<LinearLayout xmlns:android="http://schemas.android.com/apk/res/android"
    android:orientation="vertical" android:layout_width="fill_parent"
    android:layout_height="fill_parent" android:background="#ffffff">

    <LinearLayout android:layout_width="fill_parent"
        android:layout_height="wrap_content"  android:background="@drawable/gradient_
dark_purple">

        <com.teleca.jamendo.widget.AlbumBar
            android:id="@+id/AlbumBar" android:layout_width="fill_parent"
            android:layout_height="fill_parent"></com.teleca.jamendo.widget.AlbumBar>
    </LinearLayout>

    <android.gesture.GestureOverlayView
        xmlns:android="http://schemas.android.com/apk/res/android"
android:id="@+id/gestures"
        android:layout_width="fill_parent" android:layout_height="fill_parent"
        android:layout_weight="1" android:gestureStrokeType="multiple"
        android:eventsInterceptionEnabled="false" android:orientation="vertical">

        <com.teleca.jamendo.util.FixedViewFlipper
            android:id="@+id/PlaylistViewFlipper" android:background="#fff"
            android:layout_weight="1" android:layout_width="fill_parent"
            android:layout_height="fill_parent">
            <ListView android:id="@+id/PlaylistTracksListView"
                android:layout_width="fill_parent" android:layout_height="fill_parent"
                android:divider="#000">
            </ListView>
            <TextView android:layout_width="wrap_content"
                android:layout_height="wrap_content" android:text="@string/no_tracks"
                android:layout_gravity="center"></TextView>
        </com.teleca.jamendo.util.FixedViewFlipper>
    </android.gesture.GestureOverlayView>
</LinearLayout>
```

28.6.2　分析播放流程

在本章实例中，通过被启动 Activity 的静态方法来实现启动播放功能，当发生点击 Gallery 的 Item 事件之后启动，具体代码如下所示。

```
Album album = (Album) adapterView.getItemAtPosition(position);
            PlayerActivity.launch(HomeActivity.this, album);
```

然后通过文件 PlayerAlbumLoadingDialog.java 实现数据的加载和封装操作，具体实现代码如下所示。

```
public class PlayerAlbumLoadingDialog extends LoadingDialog<Album, Track[]>{
    private Album mAlbum;
    public PlayerAlbumLoadingDialog(Activity activity, int loadingMsg, int failMsg) {
        super(activity, loadingMsg, failMsg);
    }
    @Override
    public Track[] doInBackground(Album... params) {
        mAlbum = params[0];

        JamendoGet2Api service = new JamendoGet2ApiImpl();
        Track[] tracks = null;

        try {
```

```
        tracks = service.getAlbumTracks(mAlbum, JamendoApplication.getInstance().
        getStreamEncoding());
    } catch (JSONException e) {
        e.printStackTrace();
        return null;
    } catch (WSError e) {
        publishProgress(e);
        cancel(true);
    }
    return tracks;

  }
  @Override
  public void doStuffWithResult(Track[] tracks) {

    Intent intent = new Intent(mActivity, PlayerActivity.class);
    Playlist playlist = new Playlist();
    mAlbum.setTracks(tracks);
    playlist.addTracks(mAlbum);
    intent.putExtra("playlist", playlist);
    mActivity.startActivity(intent);
  }
}
```

当数据封装工作完成以后，接下来开始进行各种播放器的准备。其中在handleIntent()方法中处理已经封装好的PlayList类，具体实现代码如下所示。

```
private void handleIntent(){
    Log.i(JamendoApplication.TAG, "PlayerActivity.handleIntent");

    // This will be result of this intent handling
    Playlist playlist = null;

    Log.d("play_workflow", "data is:" + getIntent().getData());//本身并未给data赋值，
所以执行的都是data==null的情形
    // We need to handle Uri
    if(getIntent().getData() != null){//不执行此条判断

        // Check if this intent was already once parsed
        // we don't need to do that again
        if(!getIntent().getBooleanExtra("handled", false)){
            Log.d("play_workflow", "handled is true" );
            mUriLoadingDialog = (LoadingDialog) new UriLoadingDialog(this,
            R.string.loading, R.string.loading_fail).execute();
        }

    } else {
        Log.d("play_workflow", "handled is false" );
        playlist = (Playlist) getIntent().getSerializableExtra("playlist");
        Log.d("play_workflow", "handle_intent playlist is:" + playlist);
        loadPlaylist(playlist);
    }
}
```

方法loadPlaylist()的具体实现代码如下所示。

```
private void loadPlaylist(Playlist playlist){
    Log.i(JamendoApplication.TAG, "PlayerActivity.loadPlaylist");
    if(playlist == null)
        return;

    mPlaylist = playlist;
    if(mPlaylist != getPlayerEngine().getPlaylist()){//播放列表不相等，就要重新加载
        //getPlayerEngine().stop();
        getPlayerEngine().openPlaylist(mPlaylist);
        getPlayerEngine().play();
    }
}
```

播放前的最后准备工作在文件 PlayerEngineImpl.java 中实现，具体实现代码如下所示。

```
public PlayerEngineImpl() {
    mLastFailTime = 0;
    mTimesFailed = 0;
    mHandler = new Handler();
}

@Override
public void next() {
    if(mPlaylist != null){
        mPlaylist.selectNext();
        play();
    }
}

@Override
public void openPlaylist(Playlist playlist) {
    if(!playlist.isEmpty())
        mPlaylist = playlist;
    else
        mPlaylist = null;
}

@Override
public void pause() {
    if(mCurrentMediaPlayer != null){
        // still preparing
        if(mCurrentMediaPlayer.preparing){
            mCurrentMediaPlayer.playAfterPrepare = false;
            return;
        }

        // check if we play, then pause
        if(mCurrentMediaPlayer.isPlaying()){
            mCurrentMediaPlayer.pause();
            if(mPlayerEngineListener != null)
                mPlayerEngineListener.onTrackPause();
            return;
        }
    }
}

@Override
public void play() {

    if( mPlayerEngineListener.onTrackStart() == false ){
        return; // apparently sth prevents us from playing tracks
    }

    // check if there is anything to play
    if(mPlaylist != null){

        // check if media player is initialized
        if(mCurrentMediaPlayer == null){
            mCurrentMediaPlayer = build(mPlaylist.getSelectedTrack());
        }

        // check if current media player is set to our song
        if(mCurrentMediaPlayer != null && mCurrentMediaPlayer.playlistEntry !=
        mPlaylist.getSelectedTrack()){
            cleanUp(); // this will do the cleanup job
            mCurrentMediaPlayer = build(mPlaylist.getSelectedTrack());
        }

        // check if there is any player instance, if not, abort further execution
        if(mCurrentMediaPlayer == null)
            return;

        // check if current media player is not still buffering
        if(!mCurrentMediaPlayer.preparing){
```

```
            // prevent double-press
            if(!mCurrentMediaPlayer.isPlaying()){
               // i guess this mean we can play the song
               Log.i(JamendoApplication.TAG, "Player [playing] "+mCurrentMediaPlayer.
               playlistEntry.getTrack().getName());

               // starting timer
               mHandler.removeCallbacks(mUpdateTimeTask);
               mHandler.postDelayed(mUpdateTimeTask, 1000);

               mCurrentMediaPlayer.start();
            }
         } else {
            // tell the mediaplayer to play the song as soon as it ends preparing
            mCurrentMediaPlayer.playAfterPrepare = true;
         }
      }
   }

   @Override
   public void prev() {
      if(mPlaylist != null){
         mPlaylist.selectPrev();
         play();
      }
   }

   @Override
   public void skipTo(int index) {
      mPlaylist.select(index);
      play();
   }

   @Override
   public void stop() {
      cleanUp();

      if(mPlayerEngineListener != null){
         mPlayerEngineListener.onTrackStop();
      }
   }
   private void cleanUp(){
      // nice clean-up job
      if(mCurrentMediaPlayer != null)
         try{
            mCurrentMediaPlayer.stop();
         } catch (IllegalStateException e){
            // this may happen sometimes
         } finally {
            mCurrentMediaPlayer.release();
            mCurrentMediaPlayer = null;
         }
   }

   private InternalMediaPlayer build(PlaylistEntry playlistEntry){
      final InternalMediaPlayer mediaPlayer = new InternalMediaPlayer();

      // try to setup local path
      String path = JamendoApplication.getInstance().getDownloadManager().
      getTrackPath(playlistEntry);
      if(path == null)
         // fallback to remote one
         path = playlistEntry.getTrack().getStream();

      // some albums happen to contain empty stream url, notify of error, abort playback
      if(path.length() == 0){
         if(mPlayerEngineListener != null){
            mPlayerEngineListener.onTrackStreamError();
            mPlayerEngineListener.onTrackChanged(mPlaylist.getSelectedTrack());
         }
```

```
            stop();
            return null;
        }

        try {
            mediaPlayer.setDataSource(path);
            mediaPlayer.playlistEntry = playlistEntry;
            //mediaPlayer.setScreenOnWhilePlaying(true);

            mediaPlayer.setOnCompletionListener(new OnCompletionListener(){

                @Override
                public void onCompletion(MediaPlayer mp) {
                    if(!mPlaylist.isLastTrackOnList() || mPlaylist.getPlaylist Playback
                    Mode() == PlaylistPlaybackMode.REPEAT || mPlaylist.getPlaylistPlaybackMode()
                    == PlaylistPlaybackMode.SHUFFLE_AND_REPEAT ){
                        next();
                    }else{
                        stop();
                    }
                }

            });

            mediaPlayer.setOnPreparedListener(new OnPreparedListener(){

                @Override
                public void onPrepared(MediaPlayer mp) {
                    mediaPlayer.preparing = false;

                    // we may start playing
                    if(mPlaylist.getSelectedTrack() == mediaPlayer.playlistEntry
                            && mediaPlayer.playAfterPrepare){
                        mediaPlayer.playAfterPrepare = false;
                        play();
                    }

                }

            });

            mediaPlayer.setOnBufferingUpdateListener(new OnBufferingUpdateListener(){

                @Override
                public void onBufferingUpdate(MediaPlayer mp, int percent) {
                    if(mPlayerEngineListener != null){
                        mPlayerEngineListener.onTrackBuffering(percent);
                    }
                }

            });

            mediaPlayer.setOnErrorListener(new OnErrorListener() {

                @Override
                public boolean onError(MediaPlayer mp, int what, int extra) {
                    Log.w(JamendoApplication.TAG, "PlayerEngineImpl fail, what ("+what+")
                    extra ("+extra+")");

                    if(what == MediaPlayer.MEDIA_ERROR_UNKNOWN){
                        // we probably lack network
                        if(mPlayerEngineListener != null){
                            mPlayerEngineListener.onTrackStreamError();
                        }
                        stop();
                        return true;
                    }

                    // not sure what error code -1 exactly stands for but it causes player
                    // to start to jump songs
                    // if there are more than 5 jumps without playback during 1 second then
```

```
we abort
                // further playback
                if(what == -1){
                    long failTime = System.currentTimeMillis();
                    if(failTime - mLastFailTime > FAIL_TIME_FRAME){
                        // outside time frame
                        mTimesFailed = 1;
                        mLastFailTime = failTime;
                        Log.w(JamendoApplication.TAG, "PlayerEngineImpl "+mTimesFailed+"
                        fail within FAIL_TIME_FRAME");
                    } else {
                        // inside time frame
                        mTimesFailed++;
                        if(mTimesFailed > ACCEPTABLE_FAIL_NUMBER){
                            Log.w(JamendoApplication.TAG, "PlayerEngineImpl too many
                            fails, aborting playback");
                            if(mPlayerEngineListener != null){
                                mPlayerEngineListener.onTrackStreamError();
                            }
                            stop();
                            return true;
                        }
                    }
                }
                return false;
            }
        });

        // start preparing
        Log.i(JamendoApplication.TAG, "Player [buffering] "+mediaPlayer.
        playlistEntry.getTrack().getName());
        mediaPlayer.preparing = true;
        mediaPlayer.prepareAsync();

        // this is a new track, so notify the listener
        if(mPlayerEngineListener != null){
            mPlayerEngineListener.onTrackChanged(mPlaylist.getSelectedTrack());
        }

        return mediaPlayer;
    } catch (IllegalArgumentException e) {
        e.printStackTrace();
    } catch (IllegalStateException e) {
        e.printStackTrace();
    } catch (IOException e) {
        e.printStackTrace();
    }
    return null;
}

@Override
public Playlist getPlaylist() {
    return mPlaylist;
}

@Override
public boolean isPlaying() {

    // no media player instance
    if(mCurrentMediaPlayer == null)
        return false;

    // so there is one, let's see if it's not preparing
    if(mCurrentMediaPlayer.preparing)
        return false;

    // finally
    return mCurrentMediaPlayer.isPlaying();
}

@Override
```

```
    public void setListener(PlayerEngineListener playerEngineListener) {
        mPlayerEngineListener = playerEngineListener;
    }

    @Override
    public void setPlaybackMode(PlaylistPlaybackMode aMode) {
        mPlaylist.setPlaylistPlaybackMode(aMode);
    }

    @Override
    public PlaylistPlaybackMode getPlaybackMode() {
        return mPlaylist.getPlaylistPlaybackMode();
    }

    public void forward(int time) {
        mCurrentMediaPlayer.seekTo( mCurrentMediaPlayer.getCurrentPosition()+time );

    }

    @Override
    public void rewind(int time) {
        mCurrentMediaPlayer.seekTo( mCurrentMediaPlayer.getCurrentPosition()-time );
    }
}
```

由此可见，方法 play 最终调用到的也是类 PlayEngine 中的相关方法。

接下来开始分析播放过程中的控制工作，首先通过如下代码来监听按键。

```
private OnClickListener mPlayOnClickListener = new OnClickListener(){

    @Override
    public void onClick(View v) {
        if(getPlayerEngine().isPlaying()){
            getPlayerEngine().pause();
        } else {
            getPlayerEngine().play();
        }
    }

};
```

然后通过方法 play()调用接口启动 intent，具体实现代码如下所示。

```
@Override
        public void play() {
            if (mServicePlayerEngine != null) {
                Log.d("play_workflow", "mServicePlayerEngine is not null:" );
                playlistCheck();
                mServicePlayerEngine.play();
            } else {
                Log.d("play_workflow", "mServicePlayerEngine is null");
                startAction(PlayerService.ACTION_PLAY);
            }
        }
```

接下来执行 Service 指定的操作，具体代码如下所示。

```
if(action.equals(ACTION_PLAY)){
            mPlayerEngine.play();
            return;
        }
```

最终执行的方法是 play()，具体实现代码如下所示。

```
public void play() {

    if( mPlayerEngineListener.onTrackStart() == false ){
        return; // apparently sth prevents us from playing tracks
    }

    // check if there is anything to play
```

```
    if(mPlaylist != null){

        // check if media player is initialized
        if(mCurrentMediaPlayer == null){
            mCurrentMediaPlayer = build(mPlaylist.getSelectedTrack());
        }

        // check if current media player is set to our song
        if(mCurrentMediaPlayer != null && mCurrentMediaPlayer.playlistEntry !=
        mPlaylist.getSelectedTrack()){
            cleanUp(); // this will do the cleanup job
            mCurrentMediaPlayer = build(mPlaylist.getSelectedTrack());
        }

        // check if there is any player instance, if not, abort further execution
        if(mCurrentMediaPlayer == null)
            return;

        // check if current media player is not still buffering
        if(!mCurrentMediaPlayer.preparing){

            // prevent double-press
            if(!mCurrentMediaPlayer.isPlaying()){
                // i guess this mean we can play the song
                Log.i(JamendoApplication.TAG, "Player [playing] "+mCurrentMediaPlayer.
                playlistEntry.getTrack().getName());

                // starting timer
                mHandler.removeCallbacks(mUpdateTimeTask);
                mHandler.postDelayed(mUpdateTimeTask, 1000);

                mCurrentMediaPlayer.start();
            }
        } else {
            // tell the mediaplayer to play the song as soon as it ends preparing
            mCurrentMediaPlayer.playAfterPrepare = true;
        }
    }
}
```

到此为止，音乐播放功能的核心内容介绍完毕，执行效果如图 28-7 所示。

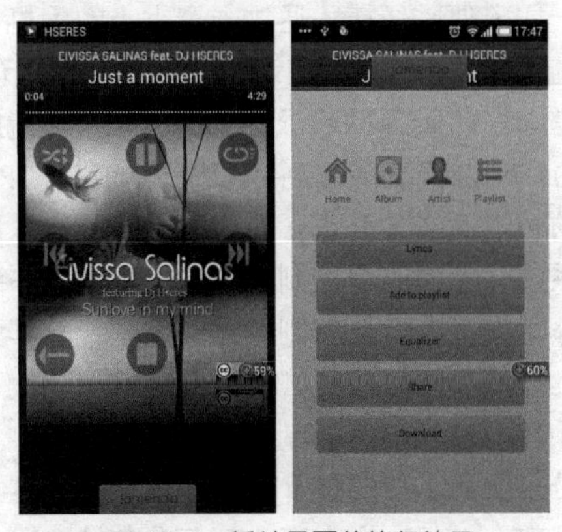

▲图 28-7　播放界面的执行效果

到此为止，本章手势音乐播放器的主要功能全部介绍完毕。本书只是重点讲解了手势识别和音乐播放功能，至于公共类的构建、数据库操作、下载操作、音乐搜索等功能，请读者参阅附带程序中的源码。

第 29 章　智能家居系统

智能家居是一个居住环境，又称智能住宅。通俗地说，它是融合了自动化控制系统、计算机网络系统、无限传感网络系统和网络通信技术于一体的网络化智能化的家居控制系统。本章通过一个综合实例的实现过程，详细讲解在 Android 系统中开发一个智能家居系统的方法，为读者步入现实工作岗位打下基础。

29.1　需求分析

智能家居将让用户有更方便的手段来管理家庭设备，例如，通过加触摸屏、无线遥控器、电话、互联网或者语音识别控制家用设备，更可以执行场景操作，使多个设备形成联动；另外，智能家居内的各种设备相互间可以通信，不需要用户指挥也能根据不同的状态互动运行，从而给用户带来最大程度的高效、便利、舒适与安全。本节将详细讲解本系统的需求分析。

29.1.1　背景介绍

家居智能化在中国的应用已经有一段时间了，但是大多数的智能家居系统仍然适用于别墅、洋房、公寓等高级住房。在世界上人口最多的国家，移动电话的应用也是非常普及，所以手机智能家居系统软件将最终成为智能家居系统中的主流产品。

对于智能家居产品，第一印象便是便捷，通过一个小小的手机，便可随时掌握、控制家里的所有常用家电设备，包括灯光、窗帘、电器、空调、地暖等，甚至天气预报，室内温湿度显示等，成为未来理想智能家居的必需品。随着各种基于 3G 和 Wi-Fi 功能的智能产品逐步应用于人们的生活中，方便直观触摸操作的移动触摸智能控制终端如 Android、iPhone、iPad 等，必将成为智能家居未来的发展趋势。传统的智能家居系统包含的主要子系统有家居布线系统、家庭网络系统、智能家居（中央）控制管理系统、家居照明控制系统、家庭安防系统、背景音乐系统、家庭影院与智能家居控制系统、多媒体系统家庭环境控制系统八大系统。其中，智能家居（中央）控制管理系统、家居照明控制系统、家庭安防系统是必备系统，家居布线系统、家庭网络系统、背景音乐系统、家庭影院与多媒体系统、家庭环境控制系统为可选系统。

通常会认为智能家居会带来生活品质的提升，其实物联网智慧家居正在改变这些观点，最显著的变化就是实用、方便、易整合。每一个家庭中都存在各种电器，不管是号称智能的冰箱、空调还是传统的电灯、电视一直以来由于标准不一都是独立工作的，从系统的角度来看，它们都是零碎的、混乱的、无序的，并不是一个有机的、可组织的整体，这些杂乱无章的电器其消耗的时间成本、管理成本、控制成本通常都是很高的，作为家庭的主人对其进行智能化管理是非常必要的。

29.1.2　传感技术的推动

传感器技术是近年来在科技界兴起的一大热点，这项技术是全球第一个利用物联网来控制灯饰及电子电器产品（ZigBee 标准产品），并将其作为智能家居主流产品走向了商业化。ZigBee 最初预计的应用领域主要包括消费电子、能源管理、卫生保健、家庭自动化、建筑自动化和工业自动化。随着物联网的兴起，ZigBee 又获得了新的应用机会。物联网的网络边缘应用最多的就是传感器或控制单元，这些是构成物联网的最基础最核心最广泛的单元细胞，而 ZigBee 能够在数千个微小的传感传动单元之间相互协调实现通信，并且这些单元只需要很少的能量，以接力的方式通过无线电波将数据从一个网络节点传到另一个节点，所以它的通信效率非常高。这种技术低功耗、抗干扰、高可靠、易组网、易扩容、易使用、易维护、便于快速大规模部署等特点顺应了物联网发展的要求和趋势。目前来看，物联网和 ZigBee 技术在智能家居、工业监测和健康保健等方面的应用有很大的融合性。

值得注意的是，随着物联网的兴起将会给 ZigBee 带来广阔的市场空间。因为物联网的目的是要将各种信息传感传动单元与互联网结合起来，从而形成一个巨大的网络，在这个巨大网络中，传感传动单元与通信网络之间需要数据的传输，而相对其他无线技术而言，ZigBee 以其在投资、建设、维护等方面的优势，必将在物联网型智能家居领域获得更广泛的应用。物联传感控制规格遂成为当今家庭智能家居自动化控制规格的主要领导者。

29.1.3 Android 与智能家居的紧密联系

2011 年 5 月 10 日，Google 在 Moscone West 会议中心举行的 Google I/O 大会上首次宣布了 Android@Home 计划。究竟什么是 Android@Home？就是通过 Android 设备来完全控制如台灯、闹钟、洗碗机等家电产品，实现居住环境的完全 Android 化。

在 2011 年举办的 Google I/O 大会展示当中，Google 为我们演示了通过摩托罗拉 XOOM 平板上对 4 台台灯进行控制。用户只需对平板上的 4 个开关进行虚拟化操作，对应的台灯就会立马做出相应反应。下一个例子，Google 向我们展示了唤醒懒人的好方法通过 Android@Home 连接在一起的闹铃系统。每当闹铃响起之际，家中音响的响度，灯光的亮度也会随之递增，面对如此先进的闹铃系统，相信再能睡的人士也会准时被其唤醒。

现在，Android@Home 计划又有了新的进展，正在有计划建立一个连接家中所有 Android 设备的云端自动化平台。因为后 PC 时代的一部分组成是云，所以 Google 正在努力地成为一个云服务提供商。随着 Google 对 Android@Home 计划的投入，越来越多的新技术也必将应用到智能家居中，这也将引爆新一轮的智能家居创新热潮。

29.2 系统功能模块介绍

本章智能家居系统的功能是，在 Android 系统中开发一个智能家居系统客户端 APP，通过这个 APP 可以实现照明控制、温度控制、查看天气和系统设置功能。本章智能家居系统的构成模块结构如图 29-1 所示。

29.3 系统主界面

系统主界面是运行程序后首先呈现在用户面前的界面。在本节的内容中，将详细讲解实现本章智能家居系统主界面的具体流程。

29.3.1 实现布局文件

通过分析 AndroidManifest.xml 清单文件可知，本系统主界面的布局文件是 main.xml，功能是通过 GridView 控件显示系统控制选项。文件 main.xml 的具体实现代码如下所示。

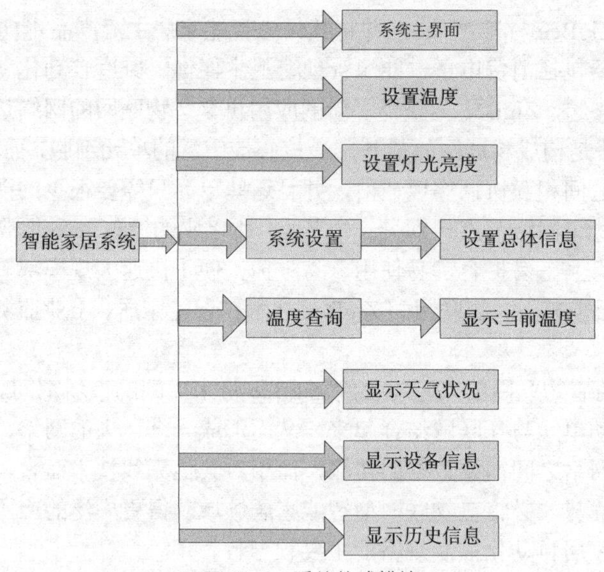

▲图 29-1　系统构成模块

```
<LinearLayout xmlns:android="http://schemas.android.com/apk/res/android"
   android:orientation="vertical"
   android:layout_width="fill_parent"
   android:layout_height="fill_parent"
   >
<TextView
   android:layout_width="fill_parent"
   android:layout_height="wrap_content"
   android:text="@string/hello"
   />

   <GridView xmlns:android="http://schemas.android.com/apk/res/android"
   android:id="@+id/gridview"
   android:layout_width="fill_parent"
   android:layout_height="fill_parent"
   android:numColumns="auto_fit"
   android:verticalSpacing="10dp"
   android:horizontalSpacing="10.dp"
   android:columnWidth="90dp"
   android:stretchMode="columnWidth"
   android:gravity="center"
   />

</LinearLayout>
```

29.3.2　实现程序文件

系统主界面的程序文件是 ClientStart.java，功能是加载布局文件 main.xml，然后在 GridView 控件中显示系统控制选项：温度、电器控制、预案管理。文件 ClientStart.java 的具体实现代码如下所示。

```
public class ClientStart extends Activity {
   /** Called when the activity is first created. */
   @Override
   public void onCreate(Bundle savedInstanceState) {
      super.onCreate(savedInstanceState);
      setContentView(R.layout.main);
      GridView gridView=(GridView)findViewById(R.id.gridview);
      ArrayList<HashMap<String, Object>> lstImageItem=
         new ArrayList<HashMap<String, Object>>();

      HashMap<String, Object> map0=new HashMap<String,Object>();
```

```
        map0.put("Image", R.drawable.app_icon);
        map0.put("Text", "温度");
        lstImageItem.add(map0);

        HashMap<String, Object> map1=new HashMap<String,Object>();
        map1.put("Image", R.drawable.sln);
        map1.put("Text", "电器控制");
        lstImageItem.add(map1);

        HashMap<String, Object> map2=new HashMap<String,Object>();
        map2.put("Image", R.drawable.open);
        map2.put("Text", "预案管理");
        lstImageItem.add(map2);

        HashMap<String, Object> map3=new HashMap<String,Object>();
        map3.put("Image", R.drawable.close);
        map3.put("Text", "4");
        lstImageItem.add(map2);

        SimpleAdapter    saImageItems=new    SimpleAdapter(this,lstImageItem,R.layout.
function,new String[]{"Image","Text"},new int[]{R.id.ItemImage,R.id.ItemText});

        gridView.setAdapter(saImageItems);

        gridView.setOnItemClickListener(new ItemClickListener());

        }
    class ItemClickListener implements OnItemClickListener{
        public void onItemClick(AdapterView<?> arg0,View arg1,int arg2,long arg3)
        {
            Intent intent;
            switch (arg2)
            {
            case 0:
                intent= new Intent (ClientStart.this,SystemSet.class);
                startActivity(intent);
                break;
            case 1:
                intent = new Intent (ClientStart.this,DeviceControl.class);
                startActivity(intent);
                break;
            case 2:
                intent = new Intent (ClientStart.this,Weather01.class);
                startActivity(intent);
                break;
            case 3:
                break;
                default:
                    break;
            }
        }
        }
    }
```

系统主界面的执行效果如图 29-2 所示。

▲图 29-2　系统主界面

29.4 系统设置

系统设置模块的功能设置当前使用者的系统信息，主要包括服务器电话、控制方式、短信服务、城市名称、更新频率、开启预案、服务器地址和端口信息等。本节将详细讲解实现本章系统设置模块的具体流程。

29.4.1 总体配置

编写文件 SystemConst.java，功能是实现系统的总体变量配置，这些变量包括服务器电话、控制方式、短信服务、城市名称、更新频率、开启预案、服务器地址和端口信息等。文件 SystemConst.java 的具体实现代码如下所示。

```
public class SystemConst {
    public static final String DB_NAME ="SmartHome_Client.db";
    public static final int DB_VERSION = 1;

    public static final String DB_TABLE_CONFIG = "system_config";
    public static final String KEY_ID="id";
    public static final String KEY_SERVER_TEL = "server_tel";
    public static final String KEY_CONTROL_METHOD = "control_method";
    public static final String KEY_CITY_NAME = "city_name";
    public static final String KEY_REFRESH_SPEED = "refresh_speed";
    public static final String KEY_WEATHER_SERVICE = "weather_service";
    public static final String KEY_LOCATION_SERVICE = "location_service";
    public static final String KEY_START_TIME = "start_time";
    public static final String KEY_SERVER_IP = "server_ip";
    public static final String KEY_SERVER_COM = "server_com";
}
```

29.4.2 系统总体配置

编写系统设置界面的布局文件 system_set.xml，功能是通过文本控件、文本框控件和单选按钮控件来显示当前系统的设置信息。文件 system_set.xml 的具体实现代码如下所示。

```
<AbsoluteLayout
android:id="@+id/widget0"
android:layout_width="fill_parent"
android:layout_height="fill_parent"
xmlns:android="http://schemas.android.com/apk/res/android"
>

<EditText
    android:id="@+id/server_tel"
    android:layout_width="200px"
    android:layout_height="27dp"
    android:layout_x="111px"
    android:layout_y="9px"
    android:textSize="12sp" >

</EditText>
<TextView
android:id="@+id/widget37"
android:layout_width="wrap_content"
android:layout_height="wrap_content"
android:text="&#26381;&#21153;&#22120;&#30005;&#35805;&#65306;"
android:layout_x="6px"
android:layout_y="21px"
>
</TextView>
<TextView
android:id="@+id/widget38"
```

```
android:layout_width="wrap_content"
android:layout_height="wrap_content"
android:text="&#30005;&#22120;&#25511;&#21046;&#26041;&#24335;&#65306;"
android:layout_x="5px"
android:layout_y="73px"
>
</TextView>
<RadioGroup
android:id="@+id/radiogroup"
android:layout_width="147px"
android:layout_height="14px"
android:orientation="horizontal"
android:layout_x="115px"
android:layout_y="64px"
>
</RadioGroup>
<TextView
android:id="@+id/location_text"
android:layout_width="wrap_content"
android:layout_height="wrap_content"
android:text="&#25552;&#20379;&#23450;&#20301;&#30701;&#20449;&#26381;&#21153;&#6530
6;"
android:layout_x="4px"
android:layout_y="133px"
>
</TextView>
<TextView
android:id="@+id/city_text"
android:layout_width="wrap_content"
android:layout_height="wrap_content"
android:text="&#22478;&#24066;&#21517;&#31216;&#65288;&#25340;&#38899;&#65289;&#6530
6;"
android:layout_x="4px"
android:layout_y="182px"
>
</TextView>

<EditText
    android:id="@+id/edit_city"
    android:layout_width="147dp"
    android:layout_height="30dp"
    android:layout_x="140dp"
    android:layout_y="171px"
    android:text="Beijing"
    android:textSize="12sp" >

</EditText>
<TextView
android:id="@+id/widget46"
android:layout_width="wrap_content"
android:layout_height="wrap_content"
android:text="&#26356;&#26032;&#39057;&#29575;&#65306;"
android:layout_x="6px"
android:layout_y="226px"
>
</TextView>
<TextView
android:id="@+id/widget49"
android:layout_width="wrap_content"
android:layout_height="wrap_content"
android:text="&#24320;&#21551;&#22825;&#27668;&#39044;&#26696;&#25552;&#31034;&#6530
6;"
android:layout_x="6px"
android:layout_y="265px"
>
</TextView>
<TextView
android:id="@+id/widget52"
android:layout_width="wrap_content"
```

```
android:layout_height="wrap_content"
android:text="&#39044;&#26696;&#24320;&#21551;&#26102;&#21051;&#65306;"
android:layout_x="6px"
android:layout_y="307px"
>
</TextView>
<TextView
android:id="@+id/widget56"
android:layout_width="wrap_content"
android:layout_height="wrap_content"
android:text="Server_IP:"
android:layout_x="6px"
android:layout_y="346px"
>
</TextView>

<TextView
    android:id="@+id/widget48"
    android:layout_width="wrap_content"
    android:layout_height="wrap_content"
    android:layout_x="78dp"
    android:layout_y="153dp"
    android:text="秒/次" />

<EditText
    android:id="@+id/frequence"
    android:layout_width="139px"
    android:layout_height="28dp"
    android:layout_x="118dp"
    android:layout_y="148dp"
    android:ems="10"
    android:text="60"
    android:textSize="12sp" />

<Button
    android:id="@+id/apply"
    android:layout_width="122dp"
    android:layout_height="wrap_content"
    android:layout_x="11dp"
    android:layout_y="340dp"
    android:text="应用系统设置"
    android:textStyle="bold" />

<TextView
    android:id="@+id/widget58"
    android:layout_width="wrap_content"
    android:layout_height="wrap_content"
    android:layout_x="153dp"
    android:layout_y="231dp"
    android:text="端口：" />

<EditText
    android:id="@+id/edit_ip"
    android:layout_width="83dp"
    android:layout_height="29dp"
    android:layout_x="70dp"
    android:layout_y="228dp"
    android:ems="10"
    android:text="127.0.0.1"
    android:textSize="12sp" />

<RadioButton
    android:id="@+id/radiobutton_blue"
    android:layout_width="wrap_content"
    android:layout_height="wrap_content"
    android:layout_x="119dp"
    android:layout_y="40dp"
    android:text="蓝牙" />
```

```
<RadioButton
    android:id="@+id/radiobutton_sms"
    android:layout_width="wrap_content"
    android:layout_height="wrap_content"
    android:layout_x="210dp"
    android:layout_y="40dp"
    android:checked="true"
    android:text="短信" />

<CheckBox
    android:id="@+id/checkbox01"
    android:layout_width="82dp"
    android:layout_height="28dp"
    android:layout_x="130dp"
    android:layout_y="86dp"
    android:checked="true"
    android:text="是" />

<EditText
    android:id="@+id/edit_com"
    android:layout_width="94dp"
    android:layout_height="30dp"
    android:layout_x="196dp"
    android:layout_y="232dp"
    android:ems="10"
    android:text="5432"
    android:textSize="12sp" />

<CheckBox
    android:id="@+id/checkbox_plan"
    android:layout_width="wrap_content"
    android:layout_height="30dp"
    android:layout_x="118dp"
    android:layout_y="170dp"
    android:text="是" />

<EditText
    android:id="@+id/planstart_time"
    android:layout_width="84dp"
    android:layout_height="34dp"
    android:layout_x="109dp"
    android:layout_y="196dp"
    android:ems="10"
    android:text="09:00"
    android:textSize="12sp" />

<Button
    android:id="@+id/reset"
    android:layout_width="124dp"
    android:layout_height="wrap_content"
    android:layout_x="148dp"
    android:layout_y="340dp"
    android:text="取消系统设置"
    android:textStyle="bold" />

</AbsoluteLayout>
```

编写文件 SystemSet.java，功能是获取设置界面文本框中的值和单选按钮的值，并将这些值保存在系统数据库中以实现系统设置功能。文件 SystemSet.java 的具体实现代码如下所示。

```
public class SystemSet extends Activity{

    private EditText serverTel;
    private EditText serverIp;
    private EditText serverCom;
    private RadioButton smsMethodView;
    private RadioButton blueToothMethodView;
    private EditText cityNameView;
    private EditText refreshSpeedView;
```

```
    private CheckBox weatherServiceView;
    private CheckBox locationServiceView;
    private EditText startTimeView;
    private Button applyBtn;
    private Button resetBtn;
    public static DBAdapter dbAdapter;


    public void onCreate(Bundle savedInstanceState) {
        super.onCreate(savedInstanceState);
        setContentView(R.layout.system_set);

        serverTel = (EditText)findViewById(R.id.server_tel);
        serverIp = (EditText)findViewById(R.id.edit_ip);
        serverCom = (EditText)findViewById(R.id.edit_com);
        smsMethodView=(RadioButton)findViewById(R.id.radiobutton_sms);
        blueToothMethodView=(RadioButton)findViewById(R.id.radiobutton_blue);
        cityNameView=(EditText)findViewById(R.id.edit_city);
        refreshSpeedView=(EditText)findViewById(R.id.frequence);
        weatherServiceView=(CheckBox)findViewById(R.id.checkbox_plan);
        locationServiceView=(CheckBox)findViewById(R.id.checkbox01);
        startTimeView=(EditText)findViewById(R.id.planstart_time);
        applyBtn =(Button)findViewById(R.id.apply);
        resetBtn = (Button)findViewById(R.id.reset);

        dbAdapter = new DBAdapter(this);
        dbAdapter.open();
        dbAdapter.LoadConfig();

        applyBtn.setOnClickListener(new View.OnClickListener(){
            public void onClick(View v) {
                SetNewData();
                UpdateUi();
            }
        });

        resetBtn.setOnClickListener(new View.OnClickListener(){
            public void onClick(View v){
                dbAdapter.LoadConfig();
                UpdateUi();
            }
        });

        UpdateUi();
    }

    private void SetNewData(){
        Config.ServerTel = serverTel.getText().toString();
        if(smsMethodView.isChecked()==true)
            Config.ControlMethod = "sms";
        else {
            Config.ControlMethod = "bluetooth";
        }
        if (blueToothMethodView.isChecked()==true)
            Config.ControlMethod = "bluetooth";
        else {
            Config.ControlMethod = "sms";
        }
        Config.CityName = cityNameView.getText().toString();
        Config.RefreshSpeed = refreshSpeedView.getText().toString();

        if (weatherServiceView.isChecked() == true)
            Config.WeatherService = "true";
        else {
            Config.WeatherService = "false";
        }
        if (locationServiceView.isChecked()==true)
            Config.LocationService = "true";
        else {
```

```
            Config.LocationService = "false";
        }
        Config.StartTime = startTimeView.getText().toString();
        Config.ServerIP = serverIp.getText().toString();
        Config.ServerCom = serverCom.getText().toString();

        dbAdapter.SaveConfig();
    }

    private void UpdateUi(){
        serverTel.setText(Config.ServerTel);

        if(Config.ControlMethod.equals("sms")==true){
            //smsMethodView.setChecked(false);
            //blueToothMethodView.setChecked(true);
        }
        locationServiceView.setChecked(Config.LocationService.equals
        ("true")?true:false);
        cityNameView.setText(Config.CityName);
        refreshSpeedView.setText(Config.RefreshSpeed);
        weatherServiceView.setChecked(Config.WeatherService.equals
        ("true")?true:false);
        startTimeView.setText(Config.StartTime);
        serverIp.setText(Config.ServerIP);
        serverCom.setText(Config.ServerCom);
    }
}
```

29.4.3 构建数据库

本系统中的设置信息是被保存在 SQLite 数据库中的，本实例通过文件 DBOpenHelper.java 来构建数据库的参数值，具体实现代码如下所示。

```
public class DBOpenHelper extends SQLiteOpenHelper{

    public DBOpenHelper(Context context,String name,
        CursorFactory factory,int version){
        super(context,name,factory,version);
    }
    //创建“系统设置”表结构
    private static final String DB_CREATE_CONFIG = "create table "
        + SystemConst.DB_TABLE_CONFIG + " (" + SystemConst.KEY_ID
        + " integer primary key autoincrement, " + SystemConst.KEY_SERVER_TEL
        + " text not null, " + SystemConst.KEY_CONTROL_METHOD + " text, "
        + SystemConst.KEY_CITY_NAME + " text, " + SystemConst.KEY_REFRESH_SPEED + " text,"
        + SystemConst.KEY_WEATHER_SERVICE + " text, " + SystemConst.KEY_LOCATION_SERVICE
+  " text, "
        + SystemConst.KEY_START_TIME + " text, "+ SystemConst.KEY_SERVER_IP +" text,
"+SystemConst.KEY_SERVER_COM+" text);";
    //创建“天气预案”表结构
    private final String DB_CREATE_RESERVEPLAN = "create table Reserve_Plan (_id integer
primary key autoincrement,"
        + "weather varchar(100), temperature varchar(100),"
        + "solution varchar(300));";
    @Override
    public void onCreate(SQLiteDatabase _db) {
        // TODO Auto-generated method stub
        //创建“系统设置”和“天气预案”表
        _db.execSQL(DB_CREATE_CONFIG);
        _db.execSQL(DB_CREATE_RESERVEPLAN);
        //系统设置默认数据加载
        Config.LoadDefaultConfig();
        //系统设置默认数据插入数据库表中
        ContentValues newValues = new ContentValues();
        newValues.put(SystemConst.KEY_SERVER_TEL, Config.ServerTel);
        newValues.put(SystemConst.KEY_CONTROL_METHOD, Config.ControlMethod);
        newValues.put(SystemConst.KEY_CITY_NAME,Config.CityName);
        newValues.put(SystemConst.KEY_REFRESH_SPEED, Config.RefreshSpeed);
```

```
        newValues.put(SystemConst.KEY_WEATHER_SERVICE, Config.WeatherService);
        newValues.put(SystemConst.KEY_LOCATION_SERVICE, Config.LocationService);
        newValues.put(SystemConst.KEY_START_TIME, Config.StartTime);
        newValues.put(SystemConst.KEY_SERVER_IP, Config.ServerIP);
        newValues.put(SystemConst.KEY_SERVER_COM, Config.ServerCom);
        _db.insert(DB_CREATE_CONFIG, null, newValues);
    }

    @Override
    public void onUpgrade(SQLiteDatabase _db, int _oldVersion, int _newVersion) {
        // TODO Auto-generated method stub
        _db.execSQL("DROP TABLE IF EXISTS"+SystemConst.DB_TABLE_CONFIG);
        _db.execSQL("DROP TABLE IF EXISTS Reserve_Plan");
        onCreate(_db);
    }

}
```

编写文件 DBAdapter.java，功能是根据用户的设置信息构建或更新数据库数据，具体实现代码如下所示。

```
public class DBAdapter {
    private SQLiteDatabase db;
    private final Context context;
    private DBOpenHelper dbOpenHelper;

    public DBAdapter (Context _context){
        context = _context;
    }

    public void open()throws SQLiteException{
        dbOpenHelper = new DBOpenHelper(context,SystemConst.DB_NAME,null,SystemConst.
DB_VERSION);
        try{
            db = dbOpenHelper.getWritableDatabase();
        }catch(SQLiteException ex){
            db = dbOpenHelper.getReadableDatabase();
        }
    }

    //从数据库查找系统设置表，加载数据到系统常量
    public void LoadConfig(){
        Cursor result =  db.query(SystemConst.DB_TABLE_CONFIG, new String[] { SystemConst.
KEY_ID, SystemConst.KEY_SERVER_TEL, SystemConst.KEY_CONTROL_METHOD,
                SystemConst.KEY_CITY_NAME, SystemConst.KEY_REFRESH_SPEED,
                SystemConst.KEY_WEATHER_SERVICE,    SystemConst.KEY_LOCATION_SERVICE,
SystemConst.KEY_START_TIME,SystemConst.KEY_SERVER_IP,SystemConst.KEY_SERVER_COM},
                SystemConst.KEY_ID + "=1", null, null, null, null);
          if (result.getCount() == 0 || !result.moveToFirst()){
             return;
           }
          Config.ServerTel = result.getString(result.getColumnIndex(SystemConst.KEY_
          SERVER_TEL));
          Config.ControlMethod = result.getString(result.getColumnIndex(SystemConst.
          KEY_CONTROL_METHOD));
          Config.CityName = result.getString(result.getColumnIndex(SystemConst.KEY_
          CITY_NAME));
          Config.RefreshSpeed =  result.getString(result.getColumnIndex(SystemConst.
          KEY_REFRESH_SPEED));
          Config.WeatherService = result.getString(result.getColumnIndex(SystemConst.
          KEY_WEATHER_SERVICE));
          Config.LocationService =  result.getString(result.getColumnIndex(SystemConst.
          KEY_LOCATION_SERVICE));
          Config.StartTime = result.getString(result.getColumnIndex(SystemConst.KEY_
          START_TIME));
          Config.ServerIP =result.getString(result.getColumnIndex(SystemConst.KEY_
          SERVER_IP));
          Config.ServerCom =result.getString(result.getColumnIndex(SystemConst.KEY_
          SERVER_COM));
```

```
        Toast.makeText(context, "系统设置读取成功", Toast.LENGTH_SHORT).show();

    }

    public void SaveConfig(){
        ContentValues updateValues =  new ContentValues();
        updateValues.put(SystemConst.KEY_CITY_NAME, Config.CityName);
        updateValues.put(SystemConst.KEY_CONTROL_METHOD, Config.ControlMethod);
        updateValues.put(SystemConst.KEY_LOCATION_SERVICE, Config.LocationService);
        updateValues.put(SystemConst.KEY_REFRESH_SPEED, Config.RefreshSpeed);
        updateValues.put(SystemConst.KEY_SERVER_COM, Config.ServerCom);
        updateValues.put(SystemConst.KEY_SERVER_IP, Config.ServerIP);
        updateValues.put(SystemConst.KEY_SERVER_TEL, Config.ServerTel);
        updateValues.put(SystemConst.KEY_START_TIME, Config.StartTime);
        updateValues.put(SystemConst.KEY_WEATHER_SERVICE, Config.WeatherService);
        db.update(SystemConst.DB_TABLE_CONFIG, updateValues,
            SystemConst.KEY_ID+"=1", null);
        Toast.makeText(context, "系统设置保存成功", Toast.LENGTH_SHORT).show();
    }

    public void close(){
        if (db!=null){
            db.close();
            db = null;
        }
    }
}
```

29.5 电器控制模块

在系统主界面中触摸按下“电器控制”图标后会来到如图 29-3 所示的界面。

由此可见，通过电器控制模块可以控制家居的温度、电灯和电扇。在本节的内容中，将详细讲解本章电器控制模块的具体实现流程。

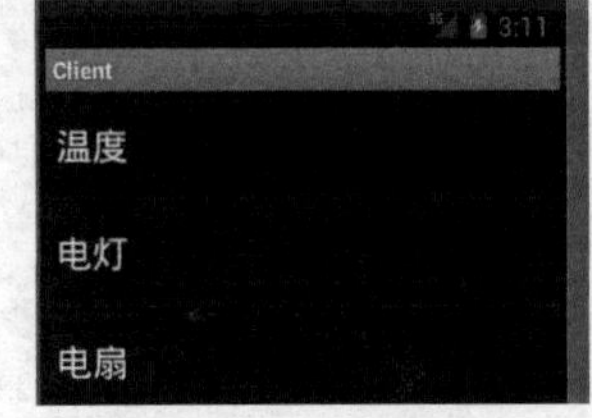

▲图 29-3 电器控制界面

29.5.1 电器控制主界面

编写布局文件 devicecontrol.xml 来实现电器控制模块的主界面，在主界面中通过 ListView 控件默认加载显示温度、电灯和电扇三个列表选项。文件 devicecontrol.xml 的具体实现代码如下所示。

```
<LinearLayout
 xmlns:android="http://schemas.android.com/apk/res/android"
 android:layout_width="fill_parent"
 android:layout_height="fill_parent">
 <ListView
 android:id="@+id/ListView01"
 android:layout_width="wrap_content"
 android:layout_height="wrap_content"
                            ></ListView>

</LinearLayout>
```

编写文件 DeviceControl.java 来监听用户触摸选择列表的选项，根据用户选择的选项来到指定的界面。文件 DeviceControl.java 的具体实现代码如下所示。

```
public class DeviceControl extends Activity {
    public int n;
    public void onCreate(Bundle savedInstanceState) {
        super.onCreate(savedInstanceState);
        setContentView(R.layout.devicecontrol);

/*       Button buttonSearch = (Button)findViewById(R.id.SearchState);
```

```
        Button buttonUpdate = (Button)findViewById(R.id.UpdateState);
        Button buttonBack = (Button)findViewById(R.id.back);

        final TextView textView = (TextView)findViewById(R.id.StateOutput);
        n=(int)Math.random()*2;

        buttonSearch.setOnClickListener(new View.OnClickListener() {
            public void onClick(View view) {
                if (n==1)
                {
                    textView.setText("灯是亮着的");
                }

                if (n==0)
                {
                    textView.setText("灯是关着的");
                }
            }
          });

        buttonUpdate.setOnClickListener(new View.OnClickListener() {
            public void onClick(View view) {

                if (n==1)
                {
                    n=0;
                }

                if (n==0)
                {
                    n=1;
                }
            }
          });
        buttonBack.setOnClickListener(new View.OnClickListener() {
            public void onClick(View view) {
                finish();
            }
          });*/
        ListView listView=(ListView) findViewById(R.id.ListView01);
        List<String>list=new ArrayList<String>();
        list.add("温度");
        list.add("电灯");
        list.add("电扇");
        ArrayAdapter<String> adapter=new ArrayAdapter<String>(this,android.R.layout.
        simple_list_item_1,list);

        listView.setAdapter(adapter);
        listView.setOnItemClickListener(new listener());


    }

    class listener implements OnItemClickListener{

        @Override
        public void onItemClick(AdapterView<?> arg0, View arg1, int arg2,
                long arg3) {
            // TODO Auto-generated method stub
            Intent intent;
            switch(arg2){
            case 0:
                intent= new Intent (DeviceControl.this,Temperature.class);
                startActivity(intent);
                break;
            case 1:
                intent = new Intent(DeviceControl.this,Lights.class);
                startActivity(intent);
```

```
                break;
            }
        }
    }
}
```

29.5.2 温度控制界面

当在电器控制主界面列表中选择“温度”选项后会来到温度控制界面，此界面的布局文件是temperature.xml，具体实现代码如下所示。

```
<AbsoluteLayout
android:id="@+id/widget0"
android:layout_width="fill_parent"
android:layout_height="fill_parent"
xmlns:android="http://schemas.android.com/apk/res/android"
>
<LinearLayout
android:id="@+id/widget36"
android:layout_width="wrap_content"
android:layout_height="wrap_content"
android:orientation="vertical"
android:layout_x="5px"
android:layout_y="3px"
>
</LinearLayout>
<Button
android:id="@+id/search"
android:layout_width="95px"
android:layout_height="wrap_content"
android:text="&#28201;&#24230;&#26597;&#35810;"
android:layout_x="11px"
android:layout_y="378px"
>
</Button>
<Button
android:id="@+id/update"
android:layout_width="102px"
android:layout_height="44px"
android:text="&#21047;&#26032;"
android:layout_x="113px"
android:layout_y="378px"
>
</Button>
<Button
android:id="@+id/back"
android:layout_width="99px"
android:layout_height="45px"
android:text="&#36820;&#22238;"
android:layout_x="221px"
android:layout_y="378px"
>
</Button>
<TextView
android:id="@+id/outcome"
android:layout_width="183px"
android:layout_height="44px"
android:text=""
android:layout_x="67px"
android:layout_y="160px"
>
</TextView>
</AbsoluteLayout>
```

编写文件Temperature.java，功能是监听用户的单击按钮在界面中显示当前温度，具体实现代码如下所示。

```
public class Temperature extends Activity {
```

```
    public void onCreate(Bundle savedInstanceState) {
        super.onCreate(savedInstanceState);
        setContentView(R.layout.temperature);

        Button buttonSearch = (Button)findViewById(R.id.search);
        Button buttonUpdate = (Button)findViewById(R.id.update);
        Button buttonBack = (Button)findViewById(R.id.back);

        final TextView textView = (TextView)findViewById(R.id.outcome);

        buttonSearch.setOnClickListener(new View.OnClickListener() {
            public void onClick(View view) {
            double n=Math.random()*100;
            textView.setText((int)n+"摄氏度");
            }
          });

        buttonUpdate.setOnClickListener(new View.OnClickListener() {
            public void onClick(View view) {
            double n=Math.random()*100;
            textView.setText((int)n+"摄氏度");
            }
          });
        buttonBack.setOnClickListener(new View.OnClickListener() {
            public void onClick(View view) {
                finish();
            }
          });
    }
```

温度控制界面的执行效果如图 29-4 所示。

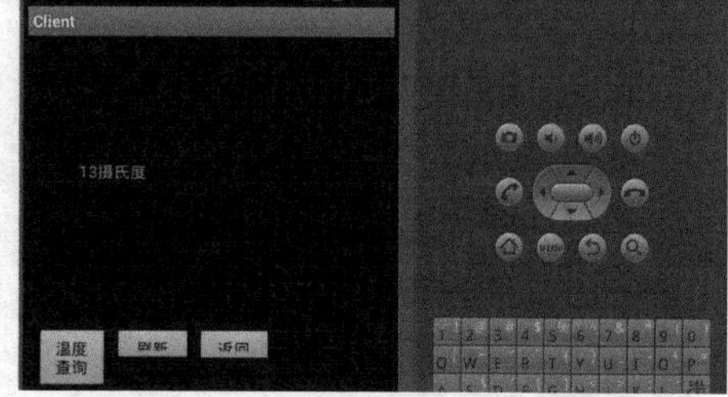

▲图 29-4　温度控制界面的执行效果

29.5.3　电灯控制界面

当在电器控制主界面列表中选择“电灯”选项后会来到电灯控制界面，此界面的布局文件是 light.xml，功能是在屏幕上方显示不同房间的电灯控制按钮，在下方显示操作按钮。文件 light.xml 的具体实现代码如下所示。

```
<AbsoluteLayout
android:id="@+id/widget0"
android:layout_width="fill_parent"
android:layout_height="fill_parent"
xmlns:android="http://schemas.android.com/apk/res/android"
>
<TextView
android:id="@+id/keting_text"
android:layout_width="wrap_content"
android:layout_height="wrap_content"
android:text="&#23458;&#21381;&#28783;"
android:layout_x="5px"
android:layout_y="5px"
>
</TextView>
<TextView
android:id="@+id/zhuwo_text"
android:layout_width="wrap_content"
android:layout_height="wrap_content"
android:text="&#20027;&#21351;&#28783;"
android:layout_x="5px"
android:layout_y="76px"
>
</TextView>
<RadioGroup
android:id="@+id/radiogroup2"
android:layout_width="165px"
android:layout_height="42px"
```

```
android:orientation="horizontal"
android:gravity="center"
android:layout_x="10px"
android:layout_y="112px"
>
<RadioButton
android:id="@+id/radiobutton2_on"
android:layout_width="wrap_content"
android:layout_height="wrap_content"
android:text="&#24320;"
android:textStyle="bold"
>
</RadioButton>
<RadioButton
android:id="@+id/radiobutton2_off"
android:layout_width="wrap_content"
android:layout_height="wrap_content"
android:text="&#20851;"
android:textStyle="bold"
android:checked="true"
>
</RadioButton>
</RadioGroup>
<RadioGroup
android:id="@+id/radiogroup1"
android:layout_width="163px"
android:layout_height="44px"
android:orientation="horizontal"
android:gravity="center"
android:layout_x="10px"
android:layout_y="33px"
>
<RadioButton
android:id="@+id/radiobutton1_on"
android:layout_width="wrap_content"
android:layout_height="wrap_content"
android:text="&#24320;"
android:textStyle="bold"
android:layout_gravity="center_vertical"
>
</RadioButton>
<RadioButton
android:id="@+id/radiobutton1_off"
android:layout_width="wrap_content"
android:layout_height="wrap_content"
android:text="&#20851;"
android:textStyle="bold"
android:checked="true"
android:layout_gravity="center_vertical"
>
</RadioButton>
</RadioGroup>
<Button
android:id="@+id/getstatus"
android:layout_width="70px"
android:layout_height="wrap_content"
android:padding="5px"
android:text="&#33719;&#21462;&#29366;&#24577;"
android:textStyle="bold"
android:layout_x="5px"
android:layout_y="300px"
>
</Button>
<Button
android:id="@+id/refresh"
android:layout_width="70px"
android:layout_height="wrap_content"
android:text="&#21047;&#26032;&#26174;&#31034;"
android:textStyle="bold"
```

```
android:layout_x="78px"
android:layout_y="300px"
>
</Button>
<Button
android:id="@+id/reset"
android:layout_width="70px"
android:layout_height="wrap_content"
android:text="&#37325;&#26032;&#35774;&#23450;"
android:textStyle="bold"
android:layout_x="150px"
android:layout_y="300px"
>
</Button>
<Button
android:id="@+id/back"
android:layout_width="70px"
android:layout_height="wrap_content"
android:text="&#36820;&#22238;"
android:textStyle="bold"
android:layout_x="220px"
android:layout_y="300px"
>
</Button>
</AbsoluteLayout>
```

编写文件 Lights.java 获取用户对不同房间电灯的控制按钮值，并监听用户触摸单击操作按钮值，通过这两个值实现对家居电灯的控制管理功能。文件 Lights.java 的具体实现代码如下所示。

```
public class Lights extends Activity implements OnClickListener{

    private static RadioButton radioButton01;
    private static RadioButton radioButton02;
    @Override
    public void onClick(View arg0) {
        // TODO Auto-generated method stub
        switch(arg0.getId()) {
        case R.id.back:
        finish();
        Intent intent = new Intent(Lights.this,DeviceControl.class );
        startActivity(intent);

        break;
        case R.id.reset:

        SendSMS("13889609674",newStatus());
        Toast.makeText(this,newStatus(),Toast.LENGTH_LONG).show();
        break;
        }

    }
    private String newStatus(){
        StringBuilder  ns = new StringBuilder();
        ns.append("10*keting*");
        if(radioButton01.isChecked())
                ns.append("1");
        if(radioButton02.isChecked())
                ns.append("0");
        return ns.toString();


    }
    private void SendSMS(String telnumer,String content){
        Intent intent = new Intent(
                "android.provider.Telephony.SMS_SEND");
        SmsManager smsManager = SmsManager.getDefault();
        PendingIntent  mpi = PendingIntent.getBroadcast(
                this,0,intent,PendingIntent.FLAG_ONE_SHOT);
        smsManager.sendTextMessage(telnumer,null,
```

```
            content,mpi,null);

    }
    @Override
    protected void onCreate(Bundle savedInstanceState) {
        // TODO Auto-generated method stub
        super.onCreate(savedInstanceState);
        setContentView(R.layout.light);
        Button backButton=(Button)findViewById(R.id.back);
        Button refreshButton=(Button)findViewById(R.id.refresh);
        Button getButton=(Button)findViewById(R.id.getstatus);
        Button setButton=(Button)findViewById(R.id.reset);
        radioButton01=(RadioButton)findViewById(R.id.radiobutton1_on);
        radioButton02=(RadioButton)findViewById(R.id.radiobutton1_off);
        backButton.setOnClickListener(this);
        refreshButton.setOnClickListener(this);
        getButton.setOnClickListener(this);
        setButton.setOnClickListener(this);
    }
}
```

电灯控制界面的执行效果如图 29-5 所示。

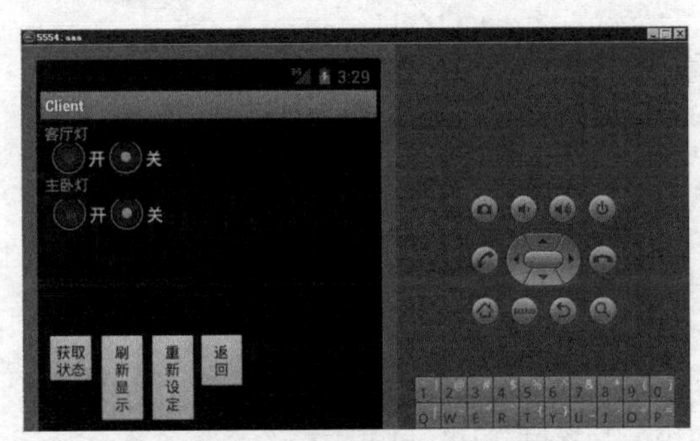

▲图 29-5　电灯控制界面的执行效果

29.6 预案管理模块

为了方便系统管理，在本智能家居系统中预先设置了预案管理模块。通过预案管理模块，可以快速地对天气情况、历史数据和系统设置信息进行浏览并管理。本节将详细讲解本系统预案管理模块的具体实现流程。

29.6.1 天气情况

编写文件 lx_weather.xml 在屏幕中构建一个天气预报界面，通过传感器获取当前的温度、湿度和 4 天的天气情况。文件 lx_weather.xml 的具体实现代码如下所示。

```
<LinearLayout xmlns:android="http://schemas.android.com/apk/res/android"
    android:orientation="vertical"
    android:layout_width="fill_parent"
    android:layout_height="fill_parent"
    >
    <TextView  android:id="@+id/tab_weather_current_condition"
        android:layout_width="fill_parent"
        android:layout_height="wrap_content"
        android:text="Temperature:60, Humidity:64%">
    </TextView>
        <TextView  android:id="@+id/tab_weather_current_wind"
        android:layout_width="fill_parent"
        android:layout_height="wrap_content"
        android:text="Wind:N at 2 mph, 2009-09-20 12:21:00 +0000">
```

```
    </TextView>

    <ImageView android:id="@+id/tab_weather_current_image"
       android:src="@drawable/sunny"
       android:layout_gravity="center_vertical|center_horizontal"
       android:layout_marginBottom="8dip" android:layout_marginTop="15dip"

android:layout_height="100dip" android:layout_width="100dip"/>

    <TextView  android:id="@+id/tab_weather_current_city"
       android:layout_width="fill_parent"
       android:layout_height="wrap_content"
       android:text="New York, NY"
       android:gravity="center_vertical|center_horizontal">
    </TextView>

    <LinearLayout android:layout_width="fill_parent"
       android:layout_height="fill_parent"
       android:orientation="horizontal" android:layout_marginTop="20dip">
      <LinearLayout  android:orientation="vertical"
          android:layout_width="fill_parent"
          android:layout_height="fill_parent"
          android:layout_weight="1">
          <TextView  android:id="@+id/tab_weather_d1_date"
             android:layout_width="fill_parent"
             android:layout_height="wrap_content"
             android:text="Mon"
             android:gravity="center_horizontal">
          </TextView>
          <ImageView android:id="@+id/tab_weather_d1_image"
                android:layout_marginTop="3dip"
                android:layout_marginBottom="3dip"
                android:src="@drawable/sunny"


android:layout_gravity="center_vertical|center_horizontal"
                android:layout_height="40dip"
                android:layout_width="40dip"/>
          <TextView  android:id="@+id/tab_weather_d1_temperature"
             android:layout_width="fill_parent"
             android:layout_height="wrap_content"
             android:text="15/29"
             android:gravity="center_horizontal">
          </TextView>
      </LinearLayout>
      <LinearLayout  android:orientation="vertical"
          android:layout_width="fill_parent"
          android:layout_height="fill_parent"
          android:layout_weight="1">
          <TextView  android:id="@+id/tab_weather_d2_date"
             android:layout_width="fill_parent"
             android:layout_height="wrap_content"
             android:text="Tue"
             android:gravity="center_horizontal">
          </TextView>
          <ImageView android:id="@+id/tab_weather_d2_image"
                android:layout_marginTop="3dip"
                android:layout_marginBottom="3dip"
                android:src="@drawable/sunny"


android:layout_gravity="center_vertical|center_horizontal"
                android:layout_height="40dip"
                android:layout_width="40dip"/>
          <TextView  android:id="@+id/tab_weather_d2_temperature"
             android:layout_width="fill_parent"
             android:layout_height="wrap_content"
             android:text="15/29"
             android:gravity="center_horizontal">
```

```
            </TextView>
        </LinearLayout>
        <LinearLayout  android:orientation="vertical"
            android:layout_width="fill_parent"
            android:layout_height="fill_parent"
            android:layout_weight="1">
            <TextView  android:id="@+id/tab_weather_d3_date"
                android:layout_width="fill_parent"
                android:layout_height="wrap_content"
                android:text="Wed"
                android:gravity="center_horizontal">
            </TextView>
            <ImageView android:id="@+id/tab_weather_d3_image"
                    android:layout_marginTop="3dip"
                    android:layout_marginBottom="3dip"
                    android:src="@drawable/sunny"

android:layout_gravity="center_vertical|center_horizontal"
                    android:layout_height="40dip"
                    android:layout_width="40dip"/>
            <TextView  android:id="@+id/tab_weather_d3_temperature"
                android:layout_width="fill_parent"
                android:layout_height="wrap_content"
                android:text="15/29"
                android:gravity="center_horizontal">
            </TextView>
        </LinearLayout>
        <LinearLayout  android:orientation="vertical"
            android:layout_width="fill_parent"
            android:layout_height="fill_parent"
            android:layout_weight="1">
            <TextView  android:id="@+id/tab_weather_d4_date"
                android:layout_width="fill_parent"
                android:layout_height="wrap_content"
                android:text="Thu"
                android:gravity="center_horizontal">
            </TextView>
            <ImageView android:id="@+id/tab_weather_d4_image"
                    android:layout_marginTop="3dip"
                    android:layout_marginBottom="3dip"
                    android:src="@drawable/sunny"

android:layout_gravity="center_vertical|center_horizontal"
                    android:layout_height="40dip"
                    android:layout_width="40dip"/>
            <TextView  android:id="@+id/tab_weather_d4_temperature"
                android:layout_width="fill_parent"
                android:layout_height="wrap_content"
                android:text="15/29"
                android:gravity="center_horizontal">
            </TextView>
        </LinearLayout>
    </LinearLayout>
</LinearLayout>
```

编写文件Lx_weather.java获取指定城市的天气信息，并监听用户触摸是否按下设备“MENU”按键。文件Lx_weather.java的具体实现代码如下所示。

```
public class Lx_weather extends Activity{
    final static int MENU_START_SERVICE= Menu.FIRST ;
    final static int MENU_STOP_SERVICE = Menu.FIRST + 1;
    final static int MENU_REFRESH = Menu.FIRST + 2;
    final static int MENU_QUIT = Menu.FIRST +3;

    private lx_DBAdapter dbAdapter ;

     @Override
```

```
    public void onCreate(Bundle savedInstanceState) {
        super.onCreate(savedInstanceState);
        setContentView(R.layout.lx_weather);

        dbAdapter = new lx_DBAdapter(this);
        dbAdapter.open();
        dbAdapter.LoadConfig();

}

 @Override
public boolean onCreateOptionsMenu(Menu menu){

    menu.add(0,MENU_START_SERVICE,0,"启动服务");
    menu.add(0,MENU_STOP_SERVICE,1,"停止服务");
    menu.add(0,MENU_REFRESH ,2,"刷新");
     menu.add(0,MENU_QUIT,3,"退出");
     return true;
}

@Override
public boolean onOptionsItemSelected(MenuItem item){
    final Intent serviceIntent = new Intent(this, lx_WeatherService.class);
    switch(item.getItemId()){
        case MENU_REFRESH:
            RefreshWeatherData();
           return true;
        case MENU_START_SERVICE:
            startService(serviceIntent);
           return true;
        case MENU_STOP_SERVICE:
            stopService(serviceIntent);
           return true;
       case MENU_QUIT:
           finish();
           break;
       }
       return false;
 }

private void RefreshWeatherData(){

    // 当前温度
       TextView currentCondition = (TextView)findViewById(R.id.tab_weather_current_
       condition);
       TextView currentWind = (TextView)findViewById(R.id.tab_weather_current_wind);
       ImageView currentImage = (ImageView)findViewById(R.id.tab_weather_current_image);
       TextView currentCity = (TextView)findViewById(R.id.tab_weather_current_city);

       String msgCondition = "";
       msgCondition += "Temperature: " + lx_Weather.current_temp + ", ";
       msgCondition += lx_Weather.current_humidity ;
       currentCondition.setText(msgCondition);

       currentWind.setText(lx_Weather.current_wind + ", " + lx_Weather.current_date_
       time);
       currentImage.setImageBitmap(lx_Weather.current_image);
       currentCity.setText(lx_Weather.city);

        // 预报：第 1 天
       TextView forcastD1Date = (TextView)findViewById(R.id.tab_weather_d1_date);
       ImageView forcastD1Image = (ImageView)findViewById(R.id.tab_weather_d1_image);
       TextView forcastD1Temperature = (TextView)findViewById(R.id.tab_weather_d1_
       temperature);

       forcastD1Date.setText(lx_Weather.day[0].day_of_week);
```

```
        forcastD1Image.setImageBitmap(lx_Weather.day[0].image);

        String msgD1Temperature = lx_Weather.day[0].high + "/" + lx_Weather.day[0]. low;
        forcastD1Temperature.setText(msgD1Temperature);

         // 预报：第 2 天
        TextView forcastD2Date = (TextView)findViewById(R.id.tab_weather_d2_date);
        ImageView forcastD2Image = (ImageView)findViewById(R.id.tab_weather_d2_image);
        TextView forcastD2Temperature = (TextView)findViewById(R.id.tab_weather_d2_
        temperature);

        forcastD2Date.setText(lx_Weather.day[1].day_of_week);
        forcastD2Image.setImageBitmap(lx_Weather.day[1].image);

        String msgD2Temperature = lx_Weather.day[1].high + "/" + lx_Weather.day[1].low;
        forcastD2Temperature.setText(msgD2Temperature);

         // 预报：第 3 天
        TextView forcastD3Date = (TextView)findViewById(R.id.tab_weather_d3_date);
        ImageView forcastD3Image = (ImageView)findViewById(R.id.tab_weather_d3_image);
        TextView forcastD3Temperature = (TextView)findViewById(R.id.tab_weather_d3_
        temperature);

        forcastD3Date.setText(lx_Weather.day[2].day_of_week);
        forcastD3Image.setImageBitmap(lx_Weather.day[2].image);

        String msgD3Temperature = lx_Weather.day[2].high + "/" + lx_Weather.day[2].low;
        forcastD3Temperature.setText(msgD3Temperature);

         // 预报：第 4 天
        TextView forcastD4Date = (TextView)findViewById(R.id.tab_weather_d4_date);
        ImageView forcastD4Image = (ImageView)findViewById(R.id.tab_weather_d4_image);
        TextView forcastD4Temperature = (TextView)findViewById(R.id.tab_weather_d4_
        temperature);

        forcastD4Date.setText(lx_Weather.day[3].day_of_week);
        forcastD4Image.setImageBitmap(lx_Weather.day[3].image);

        String msgD4Temperature = lx_Weather.day[3].high + "/" + lx_Weather.day[3].low;
        forcastD4Temperature.setText(msgD4Temperature);

    }

}
```

编写文件 Weather01.java 在屏幕顶端显示“天气预报”“历史数据”和“系统设置”三个选项卡，具体实现代码如下所示。

```
public class Weather01 extends TabActivity {
    /** Called when the activity is first created. */
    @Override
    public void onCreate(Bundle savedInstanceState) {
        super.onCreate(savedInstanceState);
        //setContentView(R.layout.main);

        TabHost tabHost=getTabHost();
        tabHost.addTab(tabHost.newTabSpec("TAB1").setIndicator("天气预报",getResources().
            getDrawable(R.drawable.tab_weather)).setContent(new Intent(this,Lx_
            weather.class)));
        tabHost.addTab(tabHost.newTabSpec("TAB2").setIndicator("历史数据",getResources().
            getDrawable(R.drawable.tab_history)).setContent(new Intent(this,Lx_
            history.class)));
        tabHost.addTab(tabHost.newTabSpec("TAB3").setIndicator("系统设置",getResources().
            getDrawable(R.drawable.tab_setup)).setContent(new
```

```
Intent(this,Lx_setup.class));
    }
}
```

编写文件 lx_WeatherAdapter.java，功能是在线获取当前指定城市的天气信息，具体实现代码如下所示。

```
public class lx_WeatherAdapter {
    public static void GetWeatherData() throws IOException, Throwable {
        String queryString = "http://www.google.com/ig/api?weather=" + lx_Config. CityName;
        URL aURL = new URL(queryString.replace(" ", "%20"));
         URLConnection conn = aURL.openConnection();
        conn.connect();
        InputStream is = conn.getInputStream();

          XmlPullParserFactory factory = XmlPullParserFactory.newInstance();
        factory.setNamespaceAware(true);
        XmlPullParser parser = factory.newPullParser();
        parser.setInput(is,"UTF-8");

        int dayCounter = 0;
        while(parser.next()!= XmlPullParser.END_DOCUMENT){
            String element = parser.getName();
            if (element != null && element.equals("forecast_information")){
                while(true){
                    int eventCode = parser.next();
                    element =  parser.getName();
                    if (eventCode == XmlPullParser.START_TAG){
                        if (element.equals("city")){
                            lx_Weather.city =  parser.getAttributeValue(0);
                        }else if (element.equals("current_date_time")){
                            lx_Weather.current_date_time = parser.getAttributeValue(0);
                        }
                    }

                    if (element.equals("forecast_information") &&
                            eventCode == XmlPullParser.END_TAG){
                        break;
                    }
                }
            }
            if (element != null && element.equals("current_conditions")){
                while(true){
                    int eventCode = parser.next();
                    element =  parser.getName();
                    if (eventCode == XmlPullParser.START_TAG){
                        if (element.equals("condition")){
                            lx_Weather.current_condition =  parser.getAttributeValue(0);
                        }else if (element.equals("temp_f")){
                            lx_Weather.current_temp = parser.getAttributeValue(0);
                        }else if (element.equals("humidity")){
                            lx_Weather.current_humidity = parser.getAttributeValue(0);
                        }else if (element.equals("wind_condition")){
                            lx_Weather.current_wind = parser.getAttributeValue(0);
                        }else if (element.equals("icon")){
                            lx_Weather.current_image_url = parser.getAttributeValue(0);
                            lx_Weather.current_image = GetURLBitmap(lx_Weather.current_
                            image_url);
                        }
                    }

                    if (element.equals("current_conditions") &&
                            eventCode == XmlPullParser.END_TAG){
                        break;
                    }
                }
            }
            if (element != null && element.equals("forecast_conditions")){
                while(true){
```

```
                int eventCode = parser.next();
                element =  parser.getName();
                if (eventCode == XmlPullParser.START_TAG){
                    if (element.equals("day_of_week")){
                        lx_Weather.day[dayCounter].day_of_week =  parser.getAttribute
                        Value(0);
                    }else if (element.equals("low")){
                        lx_Weather.day[dayCounter].low = parser.getAttributeValue(0);
                    }else if (element.equals("high")){
                        lx_Weather.day[dayCounter].high = parser.getAttributeValue(0);
                    }else if (element.equals("icon")){
                        lx_Weather.day[dayCounter].image_url = parser.getAttribute
                        Value(0);
                        lx_Weather.day[dayCounter].image = GetURLBitmap(lx_Weather.
                        day[dayCounter].image_url);
                    }else if (element.equals("condition")){
                        lx_Weather.day[dayCounter].condition = parser.getAttribute
                        Value(0);
                    }
                }

                if (element.equals("forecast_conditions") &&
                        eventCode == XmlPullParser.END_TAG){
                    dayCounter++;
                    break;
                }
            }
        }
    }
    is.close();
}
private static Bitmap GetURLBitmap(String urlString){

    URL url = null;
    Bitmap bitmap = null;
    try {
        url = new URL("http://www.google.com" + urlString);
    }
    catch (MalformedURLException e){
      e.printStackTrace();
    }
    try{
      HttpURLConnection conn = (HttpURLConnection) url.openConnection();
      conn.connect();
      InputStream is = conn.getInputStream();
      bitmap = BitmapFactory.decodeStream(is);
      is.close();
    }
    catch (IOException e){
      e.printStackTrace();
    }
    return bitmap;
  }
}
```

编写文件 lx_WeatherService.java，功能是根据用户选择的选项来启动或停止当前的天气服务活动，具体实现代码如下所示。

```
public class lx_WeatherService extends Service{
    private lx_DBAdapter dbAdapter ;
    private Thread workThread;
    private static ArrayList<lx_SimpleSms> smsList = new ArrayList<lx_SimpleSms>();
    private static int timeCounter = 1;

    public static void RequerSMSService(lx_SimpleSms sms){
        if (lx_Config.ProvideSmsService.equals("true")){
            smsList.add(sms);
        }
    }
```

```
private void SaveSmsData(lx_SimpleSms sms){
    if (lx_Config.SaveSmsInfo.equals("true")){
        dbAdapter.SaveOneSms(sms);
    }
}
@Override
public void onCreate() {
    super.onCreate();

    dbAdapter = new lx_DBAdapter(this);
    dbAdapter.open();

    Toast.makeText(this, "启动天气服务", Toast.LENGTH_LONG).show();
    workThread = new Thread(null,backgroudWork,"WorkThread");
}
@Override
public void onStart(Intent intent, int startId) {
     super.onStart(intent, startId);
     if (!workThread.isAlive()){
         workThread.start();
     }
}
@Override
public void onDestroy() {
     super.onDestroy();
     Toast.makeText(this, "天气服务启动停止", Toast.LENGTH_SHORT).show();
     workThread.interrupt();
}
@Override
public IBinder onBind(Intent intent) {
    return null;
}
private Runnable backgroudWork = new Runnable(){
    @Override
    public void run() {
        try {
            while(!Thread.interrupted()){
                ProcessSmsList();
                GetGoogleWeatherData();

                Thread.sleep(1000);
            }
        } catch (InterruptedException e) {
            e.printStackTrace();
        }
    }
};
private void ProcessSmsList(){
    if (smsList.size()==0){
        return;
    }
    SmsManager smsManager = SmsManager.getDefault();
    PendingIntent mPi = PendingIntent.getBroadcast(this, 0, new Intent(), 0);
    while(smsList.size()>0){
        lx_SimpleSms sms = smsList.get(0);
        smsList.remove(0);
        smsManager.sendTextMessage(sms.Sender,null,lx_Weather.GetSmsMsg(),mPi,null);
        sms.ReturnResult = lx_Weather.GetSmsMsg();
        SaveSmsData(sms);

    }
}
private void GetGoogleWeatherData(){
    Log.i("TIMER",String.valueOf(timeCounter));
    if (timeCounter-- < 0){
        timeCounter = Integer.parseInt(lx_Config.RefreshSpeed);
        Log.i("TIMER","NOW");
        try {
```

```
            lx_WeatherAdapter.GetWeatherData();
        } catch (IOException e) {
            // TODO Auto-generated catch block
            e.printStackTrace();
        } catch (Throwable e) {
            // TODO Auto-generated catch block
            e.printStackTrace();
        }
    }
  }
}
```

天气情况模块执行后面的效果如图 29-6 所示。

用户触摸按下设备“MENU”按键后的执行效果如图 29-7 所示。

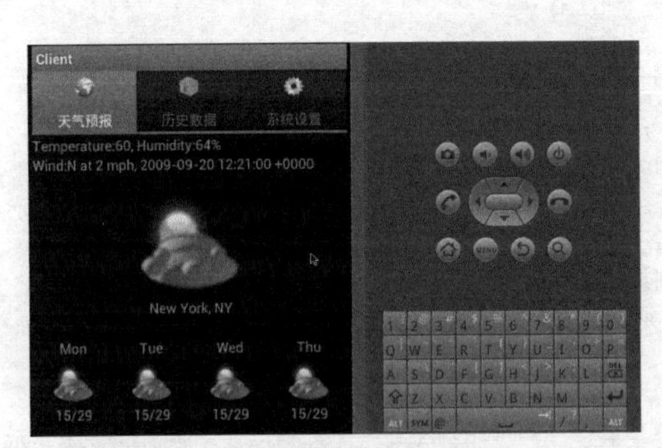

▲图 29-6　天气情况模块的执行效果

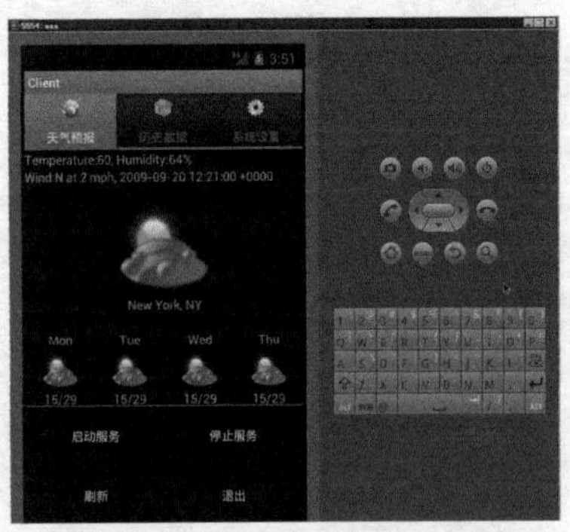

▲图 29-7　按下“MENU”按键后的执行效果

29.6.2　历史数据

当在屏幕顶部选择“历史数据”选项后会来到系统的历史数据界面，如图 29-8 所示。

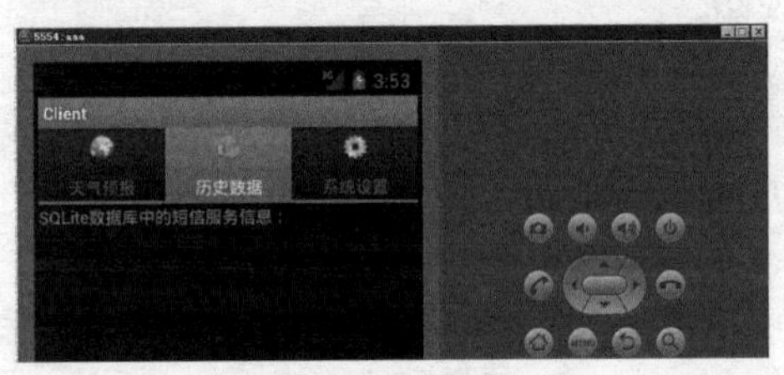

▲图 29-8　历史数据界面

编写文件 lx_history.xml，功能是通过 ListView 控件列表显示系统内的历史短信信息，具体实现代码如下所示。

```
<LinearLayout xmlns:android="http://schemas.android.com/apk/res/android"
    android:orientation="vertical"
    android:layout_width="fill_parent"
    android:layout_height="fill_parent"
    android:background="@drawable/black">

    <TextView  android:layout_width="wrap_content"
            android:layout_height="wrap_content"
            android:text="SQLite 数据库中的短信服务信息: ">
        </TextView>
    <ListView android:id="@android:id/list"
        android:layout_width="fill_parent"
        android:layout_height="wrap_content"
          android:layout_marginTop="2dip">
      </ListView>
</LinearLayout>
```

编写文件 Lx_history.java 获取系统数据库中的历史信息，并且根据监听用户触摸按下设备的“MENU”按键值来进行对应的操作。文件 Lx_history.java 的具体实现代码如下所示。

```
public class Lx_history extends ListActivity{
    final static int MENU_REFRESH = Menu.FIRST;
    final static int MENU_DELETE = Menu.FIRST+1;
    final static int MENU_QUIT = Menu.FIRST+2;

    private lx_DBAdapter dbAdapter ;
    private lx_SmsAdapter dataAdapter;
     @Override
        public void onCreate(Bundle savedInstanceState) {
            super.onCreate(savedInstanceState);
            setContentView(R.layout.lx_history);

            dbAdapter = new lx_DBAdapter(this);
            dbAdapter.open();

            dataAdapter = new lx_SmsAdapter(this);
            setListAdapter(dataAdapter);
     }

     @Override
     public boolean onCreateOptionsMenu(Menu menu){
         menu.add(0,MENU_REFRESH ,0,"刷新");
         menu.add(0,MENU_DELETE,1,"清空数据");
         menu.add(0,MENU_QUIT,1,"退出");
         return true;
     }

     @Override
     public boolean onOptionsItemSelected(MenuItem item){
         switch(item.getItemId()){
             case MENU_REFRESH:
                 lx_SmsAdapter.RefreshData();
                 setListAdapter(dataAdapter);
                return true;
            case MENU_DELETE:
                dbAdapter.DeleteAllSms();
                return true;
            case MENU_QUIT:
                finish();
                break;
            }
            return false;
     }
}
```

当触摸选择某一条历史数据时会调用文件 lx_SmsAdapter.java 显示这条数据的详细信息，具体实现代码如下所示。

```
public class lx_SmsAdapter extends BaseAdapter{
    private LayoutInflater mInflater;
    private static lx_DBAdapter dbAdapter ;
    private static lx_SimpleSms[] smsList ;

    public lx_SmsAdapter(Context context)
    {

      mInflater = LayoutInflater.from(context);
      dbAdapter = new lx_DBAdapter(context);
      dbAdapter.open();
      smsList = dbAdapter.GetAllSms();
    }

    public static void RefreshData(){
        smsList = dbAdapter.GetAllSms();
    }
```

```
    @Override
    public int getCount() {
        if (smsList == null)
            return 0;
        else
            return smsList.length;
    }

    @Override
    public Object getItem(int position) {
        if (smsList == null)
            return 0;
        else
            return smsList[position];
    }

    @Override
    public long getItemId(int position) {
        return position;
    }

    @Override
    public View getView(int position, View convertView, ViewGroup parent) {
        ViewHolder holder;

        if(convertView == null){
          convertView = mInflater.inflate(R.layout.lx_datarow, null);
          holder = new ViewHolder();
          holder.textRow01 = (TextView) convertView.findViewById(R.id.data_row_01);
          holder.textRow02 = (TextView) convertView.findViewById(R.id.data_row_02);

          convertView.setTag(holder);
        }
        else{
          holder = (ViewHolder) convertView.getTag();
        }


        if (smsList != null){
            String row01Msg ="（"+position+"）" +" 发送者: "+ smsList[position].Sender+",
            "+smsList[position].ReceiveTime;
            holder.textRow01.setText(row01Msg);
            holder.textRow02.setText(smsList[position].ReturnResult);
        }
        return convertView;
    }

    private class ViewHolder{
        TextView textRow01;
        TextView textRow02;
      }
}
```

29.6.3 系统设置

当在屏幕顶部选择“系统设置”选项后会来到系统设置界面，如图 29-9 所示。

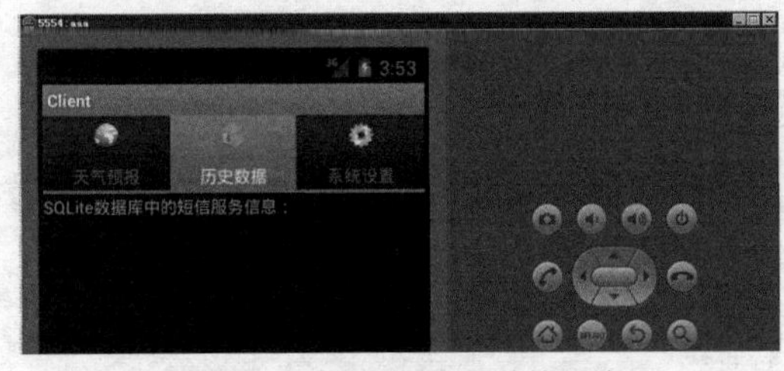

▲图 29-9　系统设置界面的执行效果

编写文件 lx_setup.xml，功能是通过文本框控件和复选框显示系统设置信息，具体实现代码如下所示。

```
<LinearLayout xmlns:android="http://schemas.android.com/apk/res/android"
    android:orientation="vertical"
    android:layout_width="fill_parent"
    android:layout_height="fill_parent"
    >
    <LinearLayout android:orientation="horizontal"
        android:layout_width="fill_parent"
        android:layout_height="wrap_content">
        <TextView  android:layout_width="wrap_content"
            android:layout_height="wrap_content"
            android:text="城市名称（拼音）："> 
        </TextView>
        <EditText android:id="@+id/tab_setup_city_name"
             android:layout_width="fill_parent"
             android:layout_height="wrap_content"
             android:text="Haerbin" >
        </EditText>
    </LinearLayout>

     <LinearLayout android:orientation="horizontal"
        android:layout_width="fill_parent"
        android:layout_height="wrap_content">
        <TextView
            android:layout_width="wrap_content"
            android:layout_height="wrap_content"
            android:text="更新频率：">
        </TextView>
        <EditText android:id="@+id/tab_setup_refresh_speed"
             android:layout_width="120dip"
             android:layout_height="wrap_content"
             android:numeric="integer"
             android:text="600" >
        </EditText>
        <TextView  android:layout_width="wrap_content"
            android:layout_height="wrap_content"
            android:text="秒/次">
        </TextView>
    </LinearLayout>

    <LinearLayout android:orientation="horizontal"
        android:layout_width="fill_parent"
        android:layout_height="wrap_content">
        <TextView  android:layout_width="wrap_content"
            android:layout_height="wrap_content"
            android:text="提供短信服务：">
        </TextView>

        <CheckBox android:id="@+id/tab_setup_sms_service"
            android:text="是"
            android:layout_width="wrap_content"
            android:layout_height="wrap_content"
            android:checked="true">
        </CheckBox>
    </LinearLayout>

    <LinearLayout android:orientation="horizontal"
        android:layout_width="fill_parent"
        android:layout_height="wrap_content">
        <TextView  android:layout_width="wrap_content"
            android:layout_height="wrap_content"
            android:text="记录短信服务数据信息：">
        </TextView>

        <CheckBox android:id="@+id/tab_setup_save_sms_info"
            android:text="是"
            android:layout_width="wrap_content"
```

```
        android:layout_height="wrap_content"
        android:checked="true">
    </CheckBox>
</LinearLayout>

 <LinearLayout android:orientation="horizontal"
    android:layout_width="fill_parent"
    android:layout_height="wrap_content">
    <TextView
        android:layout_width="wrap_content"
        android:layout_height="wrap_content"
        android:text="短信服务关键字：">
    </TextView>
    <EditText android:id="@+id/tab_setup_key_work"
        android:layout_width="120dip"
        android:layout_height="wrap_content"
        android:text="WR" >
    </EditText>
</LinearLayout>

 <LinearLayout android:orientation="horizontal"
    android:layout_width="fill_parent"
    android:layout_height="fill_parent"
    android:layout_marginTop="5dip">

    <Button android:id="@+id/tab_setup_apply"
        android:layout_width="wrap_content"
        android:layout_height="wrap_content"
        android:text="应用系统设置"
        android:layout_weight="1" android:layout_gravity="bottom">
    </Button>
    <Button android:id="@+id/tab_setup_cancel"
        android:layout_width="wrap_content"
        android:layout_height="wrap_content"
        android:text="取消系统设置"
        android:layout_weight="1" android:layout_gravity="bottom">
    </Button>
</LinearLayout>
</LinearLayout>
```

编写文件 lx_Config.java，功能是预先设置系统设置变量的初始值，具体实现代码如下所示。

```
public class lx_Config {
    public static String CityName;
    public static String RefreshSpeed;
    public static String ProvideSmsService;
    public static String SaveSmsInfo;
    public static String KeyWord;
    public static void LoadDefaultConfig(){
        CityName = "济南";
        RefreshSpeed = "60";
        ProvideSmsService = "true";
        SaveSmsInfo = "true";
        KeyWord = "无敌";
    }
}
```

编写文件 lx_DBAdapter.java，功能是根据用户的设置值对数据库进行更新，具体实现代码如下所示。

```
    public void SaveConfig(){
        ContentValues updateValues = new ContentValues();
        updateValues.put(KEY_CITY_NAME, lx_Config.CityName);
        updateValues.put(KEY_REFRESH_SPEED, lx_Config.RefreshSpeed);
        updateValues.put(KEY_SMS_SERVICE, lx_Config.ProvideSmsService);
        updateValues.put(KEY_SMS_INFO, lx_Config.SaveSmsInfo);
        updateValues.put(KEY_KEY_WORD, lx_Config.KeyWord);

        db.update(DB_TABLE_CONFIG, updateValues, KEY_ID + "=" + DB_CONFIG_ID, null);
```

```
        Toast.makeText(context, "系统设置保存成功", Toast.LENGTH_SHORT).show();

    }

    public void LoadConfig() {
    Cursor result =  db.query(DB_TABLE_CONFIG, new String[] { KEY_ID, KEY_CITY_NAME,
    KEY_REFRESH_SPEED,KEY_SMS_SERVICE, KEY_SMS_INFO, KEY_KEY_WORD},
           KEY_ID + "=" + DB_CONFIG_ID, null, null, null, null);
    if (result.getCount() == 0 || !result.moveToFirst()){
        return;
      }
    lx_Config.CityName = result.getString(result.getColumnIndex(KEY_CITY_NAME));
    lx_Config.RefreshSpeed = result.getString(result.getColumnIndex(KEY_REFRESH_SPEED));
    lx_Config.ProvideSmsService = result.getString(result.getColumnIndex(KEY_SMS_SERVICE));
    lx_Config.SaveSmsInfo =  result.getString(result.getColumnIndex(KEY_SMS_INFO));
    lx_Config.KeyWord = result.getString(result.getColumnIndex(KEY_KEY_WORD));

    Toast.makeText(context, "系统设置读取成功", Toast.LENGTH_SHORT).show();
}

    /** 静态 Helper 类，用于建立、更新和打开数据库*/
    private static class DBOpenHelper extends SQLiteOpenHelper {

        public DBOpenHelper(Context context, String name, CursorFactory factory, int
version) {
          super(context, name, factory, version);
        }

        private static final String DB_CREATE_CONFIG = "create table " +
          DB_TABLE_CONFIG + " (" + KEY_ID + " integer primary key autoincrement, " +
          KEY_CITY_NAME+ " text not null, " + KEY_REFRESH_SPEED+ " text," +
          KEY_SMS_SERVICE +" text, " + KEY_SMS_INFO +  " text, " +
          KEY_KEY_WORD + " text);";

        private static final String DB_CREATE_SMS = "create table " +
          DB_TABLE_SMS + " (" + KEY_ID + " integer primary key autoincrement, " +
          KEY_SENDER+ " text not null, " + KEY_BODY+ " text, " +
          KEY_RECEIVE_TIME +" text, " +   KEY_RETURN_RESULT + " text);";

        @Override
        public void onCreate(SQLiteDatabase _db) {
          _db.execSQL(DB_CREATE_CONFIG);
          _db.execSQL(DB_CREATE_SMS);

          //初始化系统配置的数据表
          lx_Config.LoadDefaultConfig();
          ContentValues newValues = new ContentValues();
          newValues.put(KEY_CITY_NAME, lx_Config.CityName);
          newValues.put(KEY_REFRESH_SPEED, lx_Config.RefreshSpeed);
          newValues.put(KEY_SMS_SERVICE, lx_Config.ProvideSmsService);
          newValues.put(KEY_SMS_INFO, lx_Config.SaveSmsInfo);
          newValues.put(KEY_KEY_WORD, lx_Config.KeyWord);
          _db.insert(DB_TABLE_CONFIG, null, newValues);

        }

        @Override
        public void onUpgrade(SQLiteDatabase _db, int _oldVersion, int _newVersion) {
          _db.execSQL("DROP TABLE IF EXISTS " + DB_TABLE_CONFIG);
          _db.execSQL("DROP TABLE IF EXISTS " + DB_CREATE_SMS);
          onCreate(_db);
        }
      }
```

系统设置界面的执行效果如图 29-10 所示。

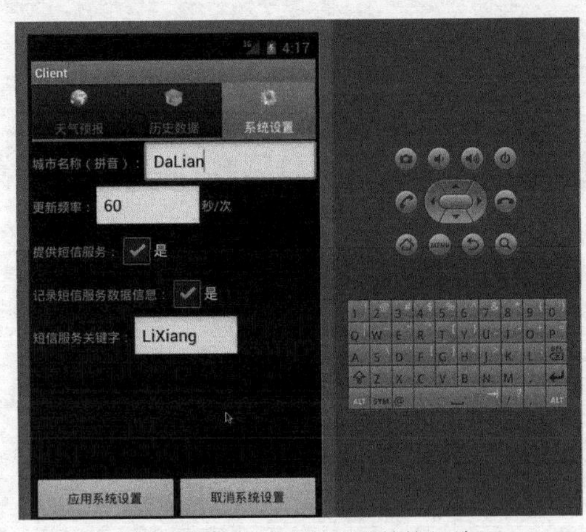

▲图 29-10　系统设置界面的执行效果

到此为止，本章智能家居系统全部介绍完毕。有关本实例的详细源码，请读者参阅本书附带程序中的源码。